Clinical Chemistry

Clinical Chemistry

Theory, Practice, and Interpretation

The late R. Richterich
and J. P. Colombo
*Department of Clinical Chemistry, University of Berne,
Inselspital, Berne, Switzerland*

With contributions from
C. Bachmann, J. Berüter, H. Ehrengruber, H. Keller, K. Lauber and
E. Peheim

JOHN WILEY & SONS
Chichester · New York · Brisbane · Toronto

Originally published under the title Klinische Chemie—Theorie, Praxis, Interpretation, 4. Auflage, by R. Richterich and J. P. Colombo
© Copyright S. Karger AG, Basel (Switzerland) 1978.

British Library Cataloguing in Publication Data:

Richterich, R.
  Clinical chemistry.
  1. Chemistry, Clinical
  I. Title   II. Colombo, J. P.
  616.07'56   RB40   80–40286

  ISBN 0 471 27809 2

Typeset by Preface Ltd, Salisbury, Wilts and printed in the United States of America by Vail-Ballou Press Inc., Binghampton, N.Y.

CONTRIBUTORS TO THIS VOLUME

CLAUDE BACHMANN, DR.MED.
Department of Clinical Chemistry, University of Berne, Inselspital, Berne

JOSEF BERÜTER, DR.SC.NAT.ETH
Department of Clinical Chemistry, University of Berne, Inselspital, Berne

JEAN-PIERRE COLOMBO, PROF. DR.MED.
Department of Clinical Chemistry, University of Berne, Inselspital, Berne

HANS EHRENGRUBER, DR.PHIL.NAT.
Data-processing Department, Inselspital, Berne

HERBERT KELLER, PROF.DR.MED. ET RER.NAT.
Institute for Clinical Chemistry and Haematology, Kantonsspital St. Gallen

KONRAD LAUBER, DR.CHEM.
Department of Medical Chemistry, University of Berne

EDGAR PEHEIM, DR.MED.VET.
Department of Clinical Chemistry, University of Berne, Inselspital, Berne

CONTENTS

viii

FOREWORD

It is a long time now since far-sighted and conscientious physicians were demanding, in the interest of the sick, an increased reliance on chemical methods as diagnostic aids. In Berne, for example, there was Hermann Sahli, Professor of Internal Medicine from 1882 to 1929. His *Textbook of Clinical Methods of Investigation* was numbered among the standard works on laboratory diagnostics for decades. His binding legacy has been faithfully honoured in the Berne Isolation Hospital. It was in continuance of Professor Sahli's principles that Roland Richterich founded an up-to-date Central Laboratory and opposed the previous collections of methods with a teaching aid, modern in its conception.

Fourteen years after the appearance of the 1st edition of the textbook *Clinical Chemistry; Theory and Practice*, this is now in its 4th edition. It is not just the external form which has changed; the textbook, which has now been translated into five languages, has been subjected to a thorough revision. Following the death of Professor Richterich, Privatdozent Dr.med. J. P. Colombo, Director of the Chemical Central Laboratory at the Berne Isolation Hospital since 1974, undertook to continue the work created by his predecessor in office. Together with a number of colleagues and co-workers he has assumed responsibility for the new revision. In this task they have been guided by the endeavour to retain as much of the original as was of proven value in respect of presentation and content but, at the same time, to give the best possible consideration to the rapid development in this technically sophisticated field.

'The present work does not seek to be regarded simply as a further collection of working instructions in the customary sense. It wants to go beyond this and act as a reliable, concise, and complete mentor for the physician, chemist, or pharmacist who wishes to work in this young and but recently independent field. To this end, it needed to incorporate not just a knowledge of methods ... but also biochemical and medical knowledge. The director of the clinical laboratory is a valuable and treasured specialist to his medical colleagues if he knows how to interpret the results of his investigations and can competently enter into stimulating discussion of the problems these have raised.' This statement from the Preface to the 1st edition is surely just as valid today. Textbooks which are designed to bring clinical and analytical aspects into closer association are now more than ever necessary in view of the role of clinical chemistry in linking different disciplines.

After years of what can only be described as an enormous development in breadth—fostered by the perfection of apparatus, automation, and electronic data processing—clinical chemistry is due for a phase of consolidation: quality must come before quantity, reliability and certainty are more important than mountains of stock data. Granted, clinical chemistry is an applied discipline, but for its further development as a science it is of crucial importance that it should not be pursued purely as a service, let alone as a lucrative business. Teaching and research must also contribute to the future 'image' of clinical chemistry and it is in the union of all these facets that clinical chemistry must find and maintain its rightful place between the clinical disciplines and the fundamental sciences. It is to be hoped that this work will make its due contribution towards the achievement of this goal.

Berne, March 1978   HUGO AEBI

PREFACE

Since the appearance of the last edition of this book in 1971 clinical chemistry has made considerable progress through the introduction of new methods and techniques. The development in the field of automation is reflected in the diversity of the instruments used. There have been manifold attempts, at both national and international level, to optimize methods, thus making a substantial contribution towards rendering the results of clinical–chemical analyses more comparable. Another consequence of this standardization of methods is that, in this book, 'cooking recipes' have been dispensed with wherever an appropriate optimized method was available. The introduction of SI units has been so successful in certain European countries that, for the principal metabolites, the concentration has been stated in both old and new units but, for the measurement of enzyme activity, the unit μmol min^{-1} has been provisionally retained.

As reliable information, the laboratory finding (the result of a clinical–chemical analysis) must be meaningfully keyed to the diagnosis. In order to clarify the clinical significance of the parameters investigated an attempt has been made, when describing a method, briefly to discuss the physiology, pathophysiology, and diagnostic relevance of the metabolite concerned. This means that even the reader who has not received a medical training can quickly apprise himself of the diagnostic relevance of a method, and it should also contribute to an intensification of the dialogue between laboratory and clinician. The book makes no claims to be comprehensive. The choice of methods was governed largely by practical considerations such as the frequency of the analyses and the practical feasibility of the methods in the author's laboratory. In many of these methods, detailed instructions have been dispensed with, just the principle is mentioned, and the reader is referred to the specialist literature.

In view of the wealth of material, the increasing specialization within individual fields, and the limitations of time, it was no longer possible for me, as it was for Roland Richterich, to write this book on my own. Establishing a unified presentation for all of the contributions was facilitated by the good agreement and the similarity of conception amongst the various authors.

It is both a pleasure and a duty for me to thank a large number of colleagues and co-workers for their valuable help. Prof. H. Bachofen of the

Pneumological Department of the Medical University Clinic, Berne, was kind enough to examine the chapter on oxygen. I am grateful to Prof. O. Oetliker of the Children's Clinic, University of Berne, for the method of determining the acid–base status in urine. Miss H. Dauwalder took a substantial share of the practical working-out of certain methods. Miss E. Lorenz, Miss H. Dauwalder, and Mr B. Käser helped me in the tedious and exhausting task of reading the proofs. My sincere thanks to them all. I owe a special debt of gratitude to Miss B. Schell and Miss P. Seiler who prepared the manuscript so reliably. The illustrations were prepared by Mr R. Stucki of the Children's Clinic, University of Berne. I am also indebted to Dr Th. Karger of Karger-Verlag and all of his colleagues who have contributed so circumspectly to the production and realization of this book.

Berne, March 1978   J. P. COLOMBO

1. GENERAL CLINICAL CHEMISTRY

1.1. GENERAL

1.1.1. CLINICAL CHEMISTRY

Clinical chemistry is that branch of medicine which is concerned with developing and carrying out chemical analyses of body fluids and other biological material for the diagnosis, therapy and prophylaxis of diseases. In terms of its content, clinical chemistry is not a systematic science. It has evolved from organic and physiological chemistry, the two disciplines used to investigate the material composition of the organism and its body fluids. These static approaches were replaced by the more dynamic biochemical approach, which led to the elucidation of the chemical and physiological metabolic processes, the basis of the *normal* life process. The scope of clinical chemistry is less restricted. Its most important role is the clarification of the *abnormal* metabolic processes which underlie the symptoms of an illness. The demands made upon it change from year to year, from laboratory to laboratory, from one doctor to another. Its frontiers with other sciences such as clinical pathology, haematology, immunology, and microbiology cannot be sharply drawn. A close collaboration with all these other fields, above all between the clinical chemist and the doctors, is a prerequisite for successful medical diagnostic work.

Clinical chemistry can scarcely be called a system; equally, it can scarcely be said to have a history. Since medicine began there have always been certain doctors who have attached special value to chemical analyses. For technical reasons the investigation of urine and excrement was initially the main interest.[1] But fruitful development was possible only when *detection* was replaced by *determination*, i.e. quantitative determinations replaced the qualitative.[2-4] The exact measurement—the basis of science—is also the basis of clinical chemistry.

It is odd that this field really developed only during the last 30 years, and the fact is unexplained. The analysis of electrolytes in plasma was possible 100 years ago, witness Schmidt's investigations during the last cholera epidemic in Dorpat;[5] his results for healthy and dehydrated patients show astonishing agreement with modern methods (Table 1).

However, these observations were not exploited, either pathogenetically or for diagnosis or therapy, and it was 50 years before the measurement of electrolyte concentrations was allotted its due significance; this was in the

3

4

Table 1. Results of electrolyte analyses in the plasma of a healthy person[5] in comparison with the mean of currently accepted reference values.

	Substance	Schmidt[5] 1850 mmol l^{-1}	Reference value ($\bar{x}$) 1962 mmol l^{-1}
Cations	Sodium	149.0	144.7
	Potassium	8.1	4.30
	Calcium	1.54	2.50
	Magnesium	1.49	1.58
Anions	Chloride	103.5	102.2
	Bicarbonate	28.6	26.5
	Sulphate	1.35	1.06
	Phosphate	1.52	2.24

field of paediatry.[6] Only a few clinicians, such as Sahli,[7] were quick to recognize the significance of the laboratory and made any valid contributions to investigating this field themselves.

The problems which have confronted clinical chemistry in recent years have been posed, for the most part, by the steady increase in the number of investigations (Fig. 1). Until recently it could be taken as a rule of thumb that, for a fixed number of beds, the number of chemical routine investigations doubles every 3–5 years. Today the growth rate seems to be

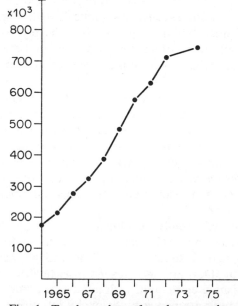

Fig. 1. Total number of analyses made per year at the Chemical Central Laboratory of the University of Berne Isolation Hospital.

slowing down; there is a distinct levelling off in the number of analyses. On the other hand, an ᵣexpansion of the investigation programme in the light of diagnostic refinements seems inevitable.

1.1.1.1. Increase in the number of analyses of a particular type

The majority of clinical–chemical methods are developed to be most reliable for an average number of analyses of about 5–10 determinations per day. If there should be 20 or even more than 50 such determinations a day, a remedy must be sought. The answer to this problem lies solely in automation. But even with few analyses the methods should be set up economically (less time spent, standardized equipment, fixed work place). This principle, which may be called 'mini-automation', deserves special emphasis.

1.1.1.2. Increase in the number of different methods of determination

A small or medium-sized laboratory cannot really handle more than 20–30 methods. Many clinics try to carry out a large number of determinations; this is misguided, uneconomic, and leads to unreliable results. A laboratory should take pride in its ability to perform a relatively small number of determinations exactly and reliably. However, in order to provide the doctor with a complete range of methods, a set number of smaller laboratories use the services of a central laboratory, manned with adequate and appropriately qualified personnel able to carry out the less frequently requested determinations.

1.1.1.3. Increase in especially complicated, time-consuming methods and methods requiring elaborate apparatus

The development of special laboratories is seen as the only answer to this problem. Since with these investigations the interpretation of the results is usually difficult too, an academic simultaneously engaged on research projects in this field must take on the responsibility of direction. One laboratory per country suffices for numerous special methods.

References

1. Colombo, J. P. and Richterich, R., *Die einfache Urinanalyse*, Huber, Bern, 1971.
2. Aebi, H., Die chemische Analyse im Dienste der Medizin einst und jetzt, *Schweiz. med. Wschr.*, **87**, 459 (1957).
3. Colombo, J. P., *Klinische Chemie. Ihre Entstehung und Entwicklung* Vorträge anlässlich der Einweihung des chemischen Zentrallabors Inselspital Bern, 1969.
4. Colombo, J. P., Richterich, R. and Bürgi, W., Zur Bedeutung der klinischen Chemie, *Praxis* **54**, 530 (1965).

5. Schmidt, C., *Charakteristik der epidemischen Cholera* Reyher, Leipzig-Mitau, 1850.
6. Gamble, J. L., Early history of fluid replacement therapy, *Pediatrics*, **11**, 554 (1953).
7. Sahli, H., *Lehrbuch der klinischen Untersuchungsmethoden für Studierende und praktische Ärzte*, Deuticke, Leipzig, 1894.

1.1.2. THE TASKS OF CLINICAL CHEMISTRY

Owing to the intensive development of clinical chemistry, the direction of a medium to large laboratory needs to be entrusted to a specially trained clinical chemist. In the German-speaking countries the German Society for Clinical Chemistry (Deutsche Gesellschaft für Klinische Chemie) and the Swiss Society for Clinical Chemistry (Schweizerische Gesellschaft für Klinische Chemie) have published guidelines to the appropriate training.[1,2] The work of the clinical chemist consists, in essence, of the following tasks which arise in connection with the development and control of a method:

1. It is the clinician's job to decide on the introduction of a new method of determination. First of all the medical specialists must be consulted to find out whether the method will lead to a substantial improvement in diagnosis. Methods which are not of general interest should not be implemented in the general laboratory but should be restricted to special projects.

2. After a thorough study of the literature on the methods currently in use, those methods are chosen which appear suited to the laboratory equipment and the available personnel. This often leaves several methods which have to be tried out before their relative merits can be assessed.

3. The chosen method is now set up and the individual steps are studied. A large number of analyses are carried out and finally the technique is scrutinized again, and, if necessary, modified. All of the chemicals, laboratory equipment and apparatus are precisely defined. Automation is used where possible. The method is now tested for reliability. If this proves satisfactory a permanent control system is set up.

4. The reference values of the particular laboratory must now be determined using voluntary patients; age, sex, etc., have to be taken into consideration. If there are sufficient data these are statistically evaluated and the reference range is established.

5. After the control system has been used (for example, in the determination of the reference value), the method is transferred to the routine laboratory and here great value should be attached to the daily control of the results.

6. The method is then passed on to the hospital. The doctors must be

made aware of the gist of the method, its limitations, the blood sample required, the technical difficulties, and the most important clinical literature. This is the joint task of the clinical chemist and the medical specialist.

7. In certain cases, e.g. with enzyme and hormone analyses, it will also be necessary for the clinical chemist to help with the interpretation of results. Often he is the only person who can give a correct interpretation in questions concerning the specificity of a finding. With most existing methods this task is no longer the sole responsibility of the clinician.

8. Practical experience shows that the life span of a method is about 5 years. After this period it is considered to be outmoded or at least due for revision. In view of the rapid technological development it is feared that in future a more frequent overhaul of methods will be necessary.

These considerations show that the development of a new method to the stage where it becomes a routine operation must be reckoned to take 6–12 months. Often a more rapid development only means that the method has to be modified later, entailing a consequent change in the reference value.

References

1. Spezialausbildung in klinischer Chemie, *Schweiz. Ärzteztg*, **52**, 1576 (1971).
2. Berufsbild des Klinischen Chemikers und Klinischen Biochemikers, *Z. klin. Chem. klin. Biochem.*, **7**, 644 (1969).

1.1.3. LABORATORY ORGANIZATION

1.1.3.1. Laboratory statistics

The foundation for all laboratory planning and laboratory organization is a comparable statistic on the number of tests performed. We count as tests all quantitative and qualitative investigations which lead to a result passed on from the laboratory to the doctor. Although such a statistic does not reflect the work expended on individual tests, nevertheless experience shows that the time spent on simple tests is roughly comparable to that needed for the more complicated and costly ones: also for similar laboratories the ratio between simple and more complicated tests is roughly the same. Finally, it should be borne in mind that with increasing automation the work involved in performing individual tests can change from year to year. So the use of such concepts as 'standard laboratory work hours', 'work units', 'work points', 'work load', and 'time–skill–frequency units' is, if anything, to be discouraged.[1–4] On the other hand, it is advisable to treat emergency tests as a separate class. In order to follow the degree of automation it is certainly also advantageous to classify the determinations into those performed manually and by automated means.

1.1.3.2. External communications

In the larger hospitals centralization has now prevailed. However, this presupposes that the laboratory is actually located centrally in the hospital so that paths of communication to and from the departments are as short as possible. For the transport of materials mechanical installations (lifts, paternoster, moving belt, or pneumatic tube) are used as far as possible. Investigations have shown that the use of pneumatic tube systems does not result in any significant changes in the blood.[5,6] With the majority of such systems the maximum acceleration—during braking—lies around $8000\,g$, i.e. comparable to that for a relatively slow centrifuge. All the same, with pneumatic tube systems the use of anticoagulated blood is recommended.

1.1.3.3. Internal organization

A skilful internal organization of the laboratory can make a substantial contribution to increased performance. Focal to the laboratory there should be a central receiving office for dealing with all clerical work and communications to the departments. Sample material is taken from this receiving office to the individual laboratory rooms by the shortest possible routes. A thorough standardization of forms and tube types greatly promotes a swift distribution.

1.1.3.4. Rationalization

From an organizational viewpoint the laboratory work may be resolved into the following three components:

1. *Material flow*, i.e. collecting samples in the departments, conveying them to the receiving office, and distributing them within the laboratory.
2. *Data flow*, i.e. the preparation and delivery of request cards, the labelling of the samples for patient identification, the paper-work at the receiving office and in the laboratory, the delivery of results to the departments, and storage of the data.
3. *Analytical work*, i.e. the preparation of samples and reagents, carrying out the determinations, ascertaining the results, and quality control.

Rationalizing measures can be applied at all three levels. The transport of materials may be speeded up by the above-mentioned aids, the introduction of modern means of data processing leads to considerable saving in respect of data flow, and the analytical work may be cut down by automation. Just where to begin the rationalization depends very much on local conditions.

More recent time studies have shown that the allocation of laboratory work is roughly as follows: transport 10%, preparation 30%, analysis 30%, ʼand paper-work 30%. Thus the clerical work takes up a considerable part of the time; in many cases the indications are that rationalization should begin here.

1.1.3.5. Rationalization of clerical work

In many places, even today, the name of the patient is hand written as many as five times for a single analysis, from the writing out of request cards and labelling of the tubes in the department to the issuing of the result forms. In many cases much time and labour may be saved by the introduction (in cooperation with the hospital administration) of an addressing system[7] or the use of gummed labels. A further aid, especially for the small laboratory, is the use of garnitures,[8] allowing the headings of the request card, result card, copy, and account, to be written simultaneously.

The use of an automatic copying machine can also relieve the work greatly. In this case the original of the case-history laboratory notes is kept in a laboratory card system. All results are entered cumulatively and the department receives a photocopy whenever new determinations have been made; the old case-history notes are thrown away. Another copy can be used for accounts.

1.1.3.6. Electronic data processing

Recent years have witnessed an ever increasing application of electronic data processing (EDP), even in the laboratory sector, resulting in about as many systems as laboratories. Which system is the most suitable depends on local circumstances, above all on the existing hardware (computers, storage banks, etc.) and software (programs); the choice also depends on whether an intermittent use of the computer will suffice ('off-line') or whether a permanent computer installation is necessary ('on-line'). Use of a computer considerably reduces the clerical work and greatly facilitates delivery of the results and the calculations for the individual determinations; there are fewer errors resulting from confusion of data.

Although the changeover to computers offers impressive possibilities, the problems of introducing such systems should not be underestimated. As illustrated in Fig. 2, several years preliminary work are usually necessary before the system is fully operational. The planning and installation of such systems entail close cooperation with other branches of the hospital—especially with administration—as well as the guidance of experts.

10

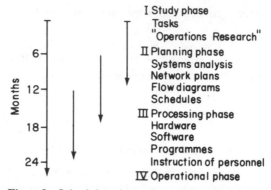

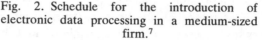

Fig. 2. Schedule for the introduction of electronic data processing in a medium-sized firm.[7]

References

1. Finch, E. P., Owen, S. E., and Byers, W. E., Clinical chemistry work-load control factors, *Clin. Chem.*, **2**, 427 (1956).
2. *Schedules of unit values for clinical laboratory procedures*, Dominion Bureau of Statistics, Ottawa, 1958.
3. Hainline, A., Time–skill–frequency (TSF) unit for reporting laboratory work-load, *Clin. Chem.*, **8**, 665 (1962).
4. College of American Pathologists, *Workload. Recording methods for clinical laboratory*, 1970.
5. Delbrück, A. and Poschmann, H., Über den Einfluss des Rohrposttransportes auf klinisches Untersuchungsmaterial unter verschiedenen Betriebsbedingungen, *Z. klin. Chem. klin. Biochem.*, **6**, 211 (1968).
6. Steige, H. and Jones, J. D., Evaluation of pneumatic tube system for delivery of blood specimen, *Clin. Chem.*, **17**, 1160 (1971).
7. Richterich, R. and Ehrengruber, H., Datenverabeiting im klinischen Laboratorium, *Naturwissenschaften*, **55**, 368 (1968).
8. Richterich, R. and Ehrengruber, H., Electronic data processing in the clinical laboratory, *Minn. Med.*, **52**, 69 (1969).

1.1.4. TERMINOLOGY IN THE CLINICAL CHEMISTRY LABORATORY

1.1.4.1. Introduction

Up to now the usage of particular terms, such as 'determination', 'test', 'assay', or 'analysis', has been arbitrary and non-standardized. The advent of EDP and automation in recent years has made the demands for an unequivocal, standardized terminology increasingly urgent. It is an essential precondition for working with the methods of systems analysis and operations research and for developing a language mutually comprehensible to the clinical chemist and the experts in these fields. An attempt to give more exact definition to a set of terms and processes relevant to the analytical laboratory ought therefore to be made.[1,2] Paramount is the key concept 'test'.

The word 'test' describes the whole set of operations necessary for obtaining a result, including both those applied to materials (determination) and information (data processing). If the tests are characterized by codings, then there will be as many codings as there are properties or components of the analysed system to be investigated.

These attempts at better definition do not always correspond with every-day usage. But this is not due to any real contradiction; rather, it stems from the indiscriminate and imprecise use of the expressions in the ordinary language of the laboratory.

1.1.4.2. Definitions and description of methods

In an area on the fringes of analytical science and information technology there has to be a clear differentiation between 'materials' and 'data'.

1.1.4.2.1. Materials

1.1.4.2.1.1. Specimen

A specimen is a liquid, solid, or gaseous material sent to the laboratory for closer characterization or investigation. The specimens dealt with in clinical chemistry are almost exclusively 'native material', i.e. biological material derived directly from the patient. If the specimen is representative, then the results obtained are also valid for the patient. The material is kept in a specimen cup or a specimen tube.

1.1.4.2.1.2. Sample

A sample is that part of the specimen which is used for the characterization or investigation. The vessel in which the sample is usually submitted for analysis is accordingly called the sample cup or sample tube. The differentiation between specimen and sample tube is important, since with certain analyses these are the same while with other equipment the sample must first be transferred from the specimen tube to a special sample cup.

The sample is representative of the specimen, and hence of the patient. A clear differentiation between specimen and sample is necessary because the responsibility for the specimen lies with the doctor, or with a hospital department, while responsibility for the sample lies exclusively with the laboratory. This is especially pertinent to the question of a flawless identification.

1.1.4.2.2. Data

1.1.4.2.2.1. Patient data

The patient data serve to identify the patient. In the ideal case, all information such as the name, forenames, sex, date of birth, and department is recorded in block capitals. Some laboratories work entirely with numbers but with such systems the risk of confusion is relatively high.

1.1.4.2.2.2. Request data

The request data are derived ultimately from the doctor and define the investigations or determinations to be made by the laboratory.

1.1.4.2.2.3. Specimen data

The specimen data characterize the material under investigation (e.g. blood, plasma, urine). Usually, additional information such as the time the blood sample was taken, the nature of the blood (e.g. venous, arterial), the condition of the patient (e.g. fasting), and type of collection (e.g. 24-hour collection), is necessary for a complete and unequivocal description.

1.1.4.2.2.4. Sample data

The sample data identify the individual samples. This is necessary when the investigation is performed not on the specimen direct but, e.g. following transference to a fresh vessel.

1.1.4.2.2.5. Result

A result is a qualitative or quantitative statement concerning a component or a property of a sample. If the sample is representative, then the result relates to the specimen or the patient.

1.1.4.2.3. Sample processing

The term 'sample processing' embraces all of the processes to which a sample is subjected from its arrival at the work-place until the conclusion of the desired investigations.

1.1.4.2.3.1. Assay

An assay is a process which leads to a physical quantity, a signal. With most of the older methods of investigation, such as gravimetry or titrimetry, a single assay frequently sufficed. With more recent methods of measurement several assays are usually necessary. Thus in photometry, for example, the calculation of a result often entails a total of four assays: a sample assay, a sample-blank assay, a reagent-blank assay, and a standard assay.

1.1.4.2.3.2. Determination

A determination is held to mean the full set of assays necessary for ascertaining a result. A determination yields a series of signals which have to be processed in further stages of the work.

1.1.4.2.4. Data processing

Data processing is the term used for the full set of all manipulations made in connection with data. The data processing may be done conventionally or electronically (EDP).

1.1.4.2.4.1. Data handling

Data handling covers the process of organizing the data adduced from data processing into other forms or transferring it to other data carriers. It includes, for example, the preparation of work lists, lists of results or result cards, and the labelling of sample tubes or cups.

1.1.4.2.4.2. Patient identification

Patient identification consists of all those processes which serve to link specific laboratory results with the patients from whom the samples are derived. Errors in patient identification can occur in the department, during transport, and in the laboratory (e.g. mistakes due to confusion, mistakes in writing); their frequency is difficult to assess.

1.1.4.2.4.3. Sample identification

The sample identification couples particular results with particular samples. Errors in sample identification occur exclusively in the laboratory; their frequency varies according to the laboratory organization. Since they are exclusive to the laboratory, such errors should be scrupulously avoided.

1.1.4.2.4.4. Calculations

As a rule the signals resulting from a determination require further mathematical processing. Such processing may include digital–analogue conversion (conversion of a signal on a continuous scale to a number),

14

linearization, logarithmization, and all calculation operations. These operations may be done *inter alia* with a calculator or with the aid of a computer.

1.1.4.2.5. General definitions

1.1.4.2.5.1. Methodology

The term 'methodology' or 'methodics' is used for indications of special principles selected in obtaining the results. 'Qualitative', 'with de-proteination', 'photometric', etc., are all examples of methodological statements.

1.1.4.2.5.2. Method

A method is an exact set of instructions for carrying out a determination. The directions must include a description of the treatment of the material (tests) as well as instructions for processing the data. Various types of method are provisionally defined.[4]

Definitive method, according to IFCC.[4] 'A method which after exhaustive investigation is found to have no known source of inaccuracy or ambiguity.' These methods are the most exact possible. They correspond in each case with the latest scientific findings. They yield the definitive, conclusive value, the best possible approximation to the true value of the constituent investigated. These are the province of the special laboratory.

Reference method according to IFCC.[4] 'A method which after exhaustive investigation has been shown to have negligible inaccuracy in comparison with its imprecision.' Its fitness should be confirmed by the analysis of the same constituent by the definitive method.

In addition to the above, yet other methods are also defined:[5]

Selected methods. These are methods of known dependability and practicability. They are methods which may be used by non-specialized laboratories and are recommended on the basis of a described chosen procedure. They yield the value of the selected method.

Acceptable methods. These are still frequently used in routine laboratories. Their precision and accuracy are known only to a limited degree. There may be great variation from one laboratory to the next.

Non-commendable methods. These are insufficiently reliable and/or practicable and on this account they should not be used.

In drawing up a method the use of the following guidelines is recommended:[5]

1	General
1.1	Title
1.2	Principle of the method
1.2.1	Reactions
1.2.3	Applications
1.3	Sampling
1.3.1	Sample collection
1.3.2	Sample preparation
2	Description of the method
2.1	Reagents
2.1.1	Detailing of the standards and reagents
2.1.2	Preparation of standard solutions or calibration
2.1.3	Stock solutions
2.1.4	Standard solutions
2.2	Equipment
2.3	Procedure
2.3.1	Safety measures
2.3.2	Practical procedure
2.3.3	Calibration
2.4	Calculation and statement of results
3	Criteria of reliability of the method
3.1	Determination of precision
3.1.1	Determination of reproducibility
3.1.2	Determination of precision for serial determination and from day to day
3.1.3	Investigation of precision under comparison conditions
3.2	Determination of accuracy
3.2.1	Limits of linearity of the calibration function
3.2.1.1	With the primary standards
3.2.1.2	By dilution of the sample
3.2.2	Contra-indications
3.2.3	Investigation of specificity (and selectivity)
3.2.3.1	Effect of various constituents present in the sample
3.2.3.2	Effect of other substances
3.2.4	Comparison of the selected method with the reference or definitive method
3.3	Determination of sensitivity
3.4	Determination of detectability
3.4.1	Blank trials
3.4.2	Calculation of the limit of detectability

1.1.4.2.5.3. Analysis

Analysis, in both practice and theory, signifies the splitting up of a whole into all its component parts (e.g. management analysis, business analysis, analysis of an instrument). In medicine, the expression 'analysis' is defined as follows: 'It particularizes the type and relationship of the symptons in conceptually the clearest possible form'.[2,3] In chemistry, the word has the same meaning and should only be used when a specimen is broken down ('separated') into all its possible parts (e.g. soil analysis, analysis of the electrolytes in plasma). The sum of all the individual components must yield the whole, i.e. 100%. In this sense, the expression 'analysis'—as correctly defined in most textbooks—is the logical opposite of synthesis. For these reasons, the term 'analysis' should not be used in analytical work as a synonym for 'determination' or 'test'.

1.1.4.2.5.4. Investigation

The meaning of the term 'investigation' is so general and non-specific that, if it is to be used at all in clinical chemistry, it should be confined to the description of a particular group of determinations (e.g. screening investigations).

References

1. Richterich, R. and Greiner, R., Analysatoren in der klinischen Chemie, *Z. klin. Chem. klin. Biochem.*, **9**, 187 (1971).
2. Hartmann, M., *Die philosophischen Grundlagen der Naturwissenschaften*, Fischer, Jena, 1948.
3. Gross, R., *Medizinische Diagnostik, Grundlagen und Praxis*, Springer, Berlin, 1969.
4. Provisional recommendation on quality control in clinical chemistry, IFCC, *Clin. chim. Acta*, **63**, F25 (1975).
5. Procedure of classification and choice of methods in clinical biology, *Annls Biol. clin.*, **34**, 236 (1976).

1.1.4.3. Statement of results

Modern clinical chemistry not only requires clear general definitions but also a unified set of concepts for the unequivocal description of a particular test or result. Dybkaer and Jørgensen[1] were the first to attempt to reduce the existing chaos to order. They insist that a result, i.e. a statement to the doctor, should consist not merely of an isolated number, but, besides identifying the patient in respect of person, place, and time, it should at least give information on the following seven groups of data:

1. On any special preparation of the patient, e.g. whether the patient was fasting or whether a special diet was administered.
2. On the specimen investigated, e.g. whether blood, plasma, or urine is concerned.
3. On the system or the components investigated. This could be a single component such as glucose or protein, but it might be an actual system such as the whole body, as in performing a creatinine clearance, for example.
4. On the method used; whether, for example, glucose was estimated with a glucose oxidase or a hexokinase/glucose-6-phosphate dehydrogenase.
5. On the units in which the result is expressed, e.g. whether mg per 100 mol or mmol l^{-1}.
6. On certain additional information, e.g. whether capillary or venous blood was used.
7. A numerical statement of the result obtained.

Unfortunately, these basic rules are neglected time and again, especially in clinical medicine, so that publications often fail to provide information on the condition of the patient, the methods used, etc.

If these seven points are considered more closely, it is seen that for most tests the first six requirements are constant and only the result varies. This observation may be applied, for example, to the coding of laboratory data by EDP. Fig. 3 shows an example of such an encoding for a particular test.[2]

18

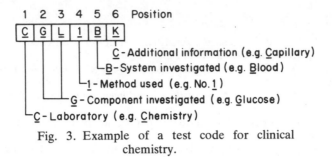

Fig. 3. Example of a test code for clinical
chemistry.

In such a scheme, not only would value be accorded to high redundancy, i.e. security of information, but at the same time the code would be designed on a mnemonic basis for easy learning and to this end it would not be too elaborate. The first letter signifies the laboratory, e.g. C for Chemistry or H for Haematology. The 2nd and 3rd letters define the system investigated, such as CC for Creatinine clearance, or the component, such as Gl for glucose, and here too, attempts to standardize abbreviations are being made.[3] In the 4th position a number shows the internal laboratory coding for the method used. Position 5 identifies the specimen, e.g. P for Plasma, U for Urine, and finally, the 6th position is available for additional information.

References

1. Dybkaer, R. and Jørgensen, K., *Quantities and units in clinical chemistry*, Munksgaard, Copenhagen, 1967.
2. Richterich, R. and Ehrengruber, H., A test code for electronic data processing in the clinical chemistry laboratory, *Clin. chim. Acta*, **22**, 417 (1968).
3. Baron, D. N., Moss, D. W., Walker, P. G., and Wilkinson, I. A., Revised list of abbreviations for names of enzymes of diagnostic importance, *J. clin. Path.*, **28**, 592 (1975).

1.2. UNITS OF MEASUREMENT

1.2.1. THE 'SYSTÈME INTERNATIONAL d'UNITÉS'

1.2.1.1. Introduction

Clinical chemistry is concerned with the measurement of chemical and physical conditions and changes in the living organism. Its methods are based on physical laws. For the description of measuring techniques and for stating the results of measurements, sensibly chosen, precisely defined units, so far as possible of world-wide acceptance, are used. Until recently there was no uniformity in the different fields . of science and technology—and this was true of clinical chemistry. For the expression of the *mass concentration*, one of the most important quantities in the clinical laboratory, there are more than 30 different units to choose from: $g\,l^{-1}$, $mg\,l^{-1}$, $\mu g\,l^{-1}$, $\gamma\,l^{-1}$, $ng\,l^{-1}$, $\mu\mu g\,l^{-1}$, g per 100 ml, mg per 100 ml, μg per 100 ml, γ per 100 ml, ng per 100 ml, $\mu\mu g$ per 100 ml, %, mg-%, μg-%, γ-%, % w/v, $g\,dl^{-1}$, $mg\,dl^{-1}$, $\mu g\,dl^{-1}$, $\gamma\,dl^{-1}$, $ng\,dl^{-1}$, $\mu\mu g\,dl^{-1}$, $g\,ml^{-1}$, $mg\,ml^{-1}$, $\mu g\,ml^{-1}$, $\gamma\,ml^{-1}$, $ng\,ml^{-1}$, $\mu\mu g\,ml^{-1}$, ppm, ppb.

A comprehensive standardization was needed. Together with the International Union of Pure and Applied Physics (IUPAP) and the International Union of Pure Applied Chemistry (IUPAC), the International Organization for Standardization (ISO) constructed a system of units which is gradually beginning to take over throughout the whole range of science and technology, the 'Système International d'Unités' (SI).

The pioneer work of introducing the SI units into clinical chemistry was accomplished by the Danes Dykbaer and Jørgensen.[1] In agreement with the International Federation of Clinical Chemistry (IFCC), a joint commission of the German, Austrian and Swiss Clinical Chemical Associations has published[2] recommendations valid for the German-speaking communities; as far as possible, their findings are followed throughout this book, but several of the common, older units are still used alongside the SI units.

1.2.1.2. Basic quantities

Seven basic quantities and the corresponding seven basic units form the framework of the 'Système International d'Unités'. These seven quantities are independent of one another (Table 2).

19

Table 2. SI basic quantities and their units.

Basic quantity			Basic unit	
Name	Symbol	Dimension	Name	Symbol
Length	l	L	Metre	m
Mass	m	M	Kilogram	kg
Time	t	T	Second	s
Electric current	I	I	Ampère	A
Thermodynamic temperature	T	Θ	Kelvin	K
Intensity of light	I	J	Candela	cd
Amount of substance	n	N	Mole	mol

1.2.1.3. Derived quantities

All the remaining physical and chemical quantities may be derived from the seven basic quantities by means of an equation.
Example.

$$\text{Force} = \text{mass} \times \text{acceleration} = \text{mass} \times \frac{\text{length}}{(\text{time})^2}$$

whence the SI unit of force is

$$1 \text{ kg } \frac{\text{m}}{\text{s}^2} = 1 \text{ kg ms}^{-2} = 1 \text{ N (Newton)}$$

Derived units in frequent use have their own names and symbols. Units which, as in the example of force, are derived from the seven basic quantities without recourse to a conversion factor are called coherent units. Units for which the derivation equation involves a numerical factor other than unity are non-coherent. The usual unit for density, $g \text{ ml}^{-1}$, for example, is non-coherent. The corresponding coherent unit would be $kg \text{ m}^{-3}$:$1 \text{ g ml}^{-1} = 1000 \text{ kg m}^3$.

1.2.1.4. Dimensions of a quantity

The dimension of each basic quantity is represented by a capital letter (Table 2). The dimensions of a derived quantity are given by the defining equation; they are always a product of powers of the dimenions of the basic quantities.
Example. The dimensions of force are $M \cdot L \cdot T^{-2}$, the dimensions of density are $M \cdot L^{-3}$.
The dimensions of a quantity are independent of any numerical factors. Ratios of quantities with the same dimensions have the dimension 1 (unity).

Table 3. Prefixes for units of measurement.

Exa-	E	10^{18}	Trillion	Milli-	m	10^{-3}	Thousandth
Peta-	P	10^{15}	Thousand billion	Micro-	μ	10^{-6}	Millionth
Tera-	T	10^{12}	Billion	Nano-	n	10^{-9}	Thousand millionth
Giga-	G	10^{9}	Thousand million (milliard $\doteq$ US billion)	Pico-	p	10^{-12}	Billionth
Mega-	M	10^{6}	Million	Femto-	f	10^{-15}	Thousand billionth
Kilo-	k	10^{3}	Thousand	Atto-	a	10^{-18}	Trillionth

Example. Relative density (e.g. the density of a gas in relation to the density of air):

$$\frac{M \cdot L^{-3}}{M \cdot L^{-3}} = 1.$$

Formerly such quantities were considered dimensionless.

1.2.1.5. Fractions and multiples of units; prefixes

The scale of the units may be adapted by means of a set of prefixes, thus catering for data involving very large or very small values. Each prefix corresponds to a factor of 1000^n or 10^{3n} (n is a whole number between -6 and $+6$).

Example. 1 megavolt (MV) $= 10^3$ kilovolts (kV) $= 10^6$ volts (V) $= 10^9$ millivolts (mV) $= 10^{-9}$ microvolts (μV).

The well known prefixes deci-, centi-, deca-, and hecto-, not listed in Table 3, do not conform with the basic concept of SI. They are admissible as before, but should be avoided as far as possible. Double prefixes such as millimicro- (mμ) and micromicro- ($\mu\mu$) are inadmissible.

Rule. In selecting the prefixes appropriate to the measurements, the prefix chosen is that which brings the numerical value between 1 and 1000; not 0.00035 g, but 350 μg; not 6350 μmol, but 6.35 mmol. Naturally, this rule does not apply to a closed set of data in which the individual values range over several orders of magnitude.

1.2.2. SI QUANTITIES AND UNITS IN CLINICAL CHEMISTRY

1.2.2.1. Length

The basic unit of length, the meter (m), is nowadays defined as 1 650 763.73 times the wavelength of the orange spectral line of the isotope ^{86}Kr (605.6 nm) *in vacuo*. Why has the krypton light ousted the time-honoured platinum rod? The reproducibility and precision of a basic unit should be equal to those attainable by the best available means.

Table 4. Precision of the different definitions of the metre.

1799	'Mètre des Archives'	±10 000 nm
1889	International prototype metre	±500 nm
1927	Red cadmium line	±20 nm
1960	Orange ^{86}Kr line (*in vacuo*)	±2 nm

Accordingly, the 'prototype metre' of 1799 has been redefined several times over the years—and with ever increased precision (Table 4).

In addition to the prefixes given in Table 3, the terms decimetre (dm) and centimetre (cm) are still admissible for expressing fractions of a metre. On the other hand, the terms micron (μ), millimicron (mμ) and micromicron ($\mu\mu$) are no longer to be used. They should be replaced by micrometre (μm), nanometre (nm) and picometre (pm), respectively. Also, the use of the Ångstrøm unit (1 Å = 100 pm = 0.1 nm) as formerly for the expression of atomic diametres and atomic distances is no longer recommended.

1.2.2.2. Mass

The basic unit of mass is the kilogram (kg), embodied in the standard kilogram kept at Paris (Pt/Ir cylinder; precision ±100 ng). The terms gamma (γ) for 10^{-6} g and millimicrogram (mμg) for 10^{-9} g should no longer be used and should be replaced by μg and ng, respectively. Mass is not to be confused with weight. A fixed quantity of matter has an invariant mass; it is dependent on geographical location and is therefore variable. According to the definition, the weight of a mass of 1 kilogram is equal to 1 kilogram, or better, 1 kilogram weight, only at sea level at latitude 45°. Consequently, whenever it is the amount of matter and not the action of the force on its support or suspension that is of interest, mass and not weight is the operative concept. The terms 'atomic weight' and 'molecular weight' ought to be replaced by 'relative atomic mass' and 'relative molecular mass', respectively. However, since these quantities are, in any case, ratios with the dimension 1, and the use of the old terms is strongly entrenched, the term 'molecular weight' (MW) is retained in this book.

1.2.2.3. Time

The basic unit of time is the second (s). Ultimately it derives from the rotation of the earth and until 1967 it was defined as the 86 400th part of the mean solar day (a solar day = the period between two consecutive solar transits of the meridian). Today the second is redefined as 9 192 631 times the period of oscillation between the two lines of the doublet hyperfine structure of the ground-state of atoms of the ^{133}Cs isotope.

The second differs from the other basic units in that use of the prefixes in Table 3 is confined almost exclusively to the expression of sub-divisions. For multiples of a second the non-coherent units minute (min), hour (h) and day (d) are still preferred to kilo- and megaseconds. The symbols ' for minute and " for second should no longer be used.

1.2.2.4. Temperature

The Centigrade (Celsius) scale in common usage is fixed by the freezing and boiling points of water at normal pressure. The degrees of the Kelvin scale (absolute temperature) are the same as those of the Centigrade scale (1 degK = 1 degC). Zero on the Kelvin scale (absolute zero) is at $-273.16 \,°C$; $K = °C + 273.16$. Clinical chemistry will continue to work with Centigrade temperatures as before.

1.2.2.5. Amount of substance

According to the early definition, 16.0000 g of oxygen were held to contain 6.023×10^{23} atoms (Loschmidt or Avogardo number, N); N molecules of any substance make 1 mole (gram molecule) of this substance. Today the concept 'mole' is of broader significance. As a first improvement the carbon isotope ^{12}C was chosen as reference because of the inhomogeneity of naturally occurring oxygen (isotopic mixture). The Loschmidt number (= number of atoms in 12 g of ^{12}C) was then $N = 6.02252 \times 10^{23}$. Depending on the requirements, mole may denote N molecules, N ions, N atoms, N electrons, N nucleons or even N formula units. In each case the mole represents a quantum of always the same number of identical units of matter.

Example. 1 mole of $H_2O \cong N$ molecules of $H_2O \cong 18.0253$ g of water; 1 mole of $PO_4^{3-} \cong N$ ions of $PO_4^{3-} \cong 94.9734$ g of phosphate ions; 1 mole of NaCl $\cong N$ formula units of NaCl $\cong 58.443$ g of salt (NaCl does not form a molecule in the customary sense); 1 mole of positive charge $\cong N$ units of positive charge, corresponding to 1 mole of Na^+ or 0.5 mole of Ca^{2+}.

It should be emphasized that 'Mole' is written with a capital M when it represents a magnitude and with a lower-case m (with or without a prefix) when it is used as a unit of measurement, following a number. 'Mole' may not be abbreviated.

The Mole acquired great significance in the earliest days of clinical chemistry. A large number of substances in our body fluids result from precursors via metabolic processes or are related to one another by chemical interaction. Metabolic routes are more readily surveyed, and anomalies easier to detect, if the metabolites involved in a system are expressed as molecular quantities (moles) and not as masses (grams).

Example. A person possesses 942 g of haemoglobin; this contains 32.2 g of porphyrin and 3.20 g of iron which binds 1.28 litres of oxygen gas. Translated into mole terminology: a person possesses 57.3 mmol of

Table 5. Haemoglobin and haemoglobin degradation in conventional and mole units.

	Conventional units	Mole units
Blood haemoglobin	15.8 g per 100 ml	9.80 mmol l^{-1}
Haemoglobin concentration	21.9 ml per 100 ml	9.80 mmol l^{-1}
O_2 capacity	21.9 ml per 100 ml	9.80 mmol l^{-1}
O_2 content (arterial)		
Whole body		
Haemoglobin	942 g	57.3 mmol
Haemoglobin-porphyrin	32.2 g	57.3 mmol
Haemoglobin-iron	3.20 g	57.3 mmol
Bound oxygen	1.28 litres	57.3 mmol
Daily haemoglobin degradation		
Haemoglobin	7.81 g	476 μmol
Bilirubin formed	278 mg	476 μmol
Bilirubin precipitated with gall	278 mg	476 μmol
Iron liberated	27 mg	476 μmol

haemoglobin (monomer); this contains 57.3 mmol of porphyrin and 57.3 mmol of iron which binds 57.3 mmol of oxygen (Table 5).

Use of the term 'equivalent' and the unit gram equivalent is no longer recommended. They are replaced by the Mole and a statement of the relevant elementary unit. For example, 1 gram equivalent of acid is 1 mole of dissociable hydrogen.

1.2.2.6. Volume

The unit of volume derived from the basic unit of length is the cubic metre (m^3). Clinical chemistry continues to work with the litre. The earlier definition of the litre, referred to 1 kg water at 4 °C, has been abandoned. A volume of 1 litre is identical with 0.001 m^3 or 1 dm^3. For fractions of a litre only the prefixes in Table 3 should be used; consequently, cm^3 or ccm should be replaced by ml and mm^3, and cmm or λ (lambda) by μl. For the most commonly used liquid the two sets of units most frequently used in the clinical laboratory are related as follows: 1 litre of $H_2O \cong 1$ kg; 1 ml of $H_2O \cong 1$ g; 1 μl of $H_2O \cong 1$ mg; and so on. For a given quantity of water the prefixes for mass and volume differ by one order of magnitude.

1.2.2.7. Pressure

Pressure is defined as force per unit area. The SI unit of force is the Newton (N) and 1 N is the force which accelerates a mass of 1 kilogram from rest to a speed of 1 metre per second in 1 second *in vacuo*. For comparison: 1 kilogram weight (the old unit of force) accelerates a mass of 1 kg in 1 s to 9.81 ms^{-1} (free fall at sea level and latitude 45°). Onc

Newton is also the force which a mass of 1/9.81 kg ≈ 100 g exerts upon its support.

The SI unit of pressure is 1 Newton per square metre or 1 Pascal ($1 \, Nm^{-2} = 1 \, Pa$). Because the Pascal is such an intractably small unit (1 Pa corresponds roughly with the pressure of a 0.1 mm column of water), the bar was introduced; 1 bar = 10^5 Pa. The bar is numerically almost equal to the previously used 'atmosphere'.

1 technical atmosphere (1 at = $1 \, kp \, cm^{-2}$) = 0.98066 bar

1 physical atmosphere (1 atm = 760 mmHg = 760 Torr) = 1.01324 bar

1 mmHg = 1 Torr = 1.33322 mbar = 0.133 kPa.

The kilopascal (kPa) is the relevant unit for expressing the partial pressures of gases in clinical chemistry.

1.2.2.8. Density

The density of a body is the quotient of its mass and its volume, in other words, its mass per unit volume. The SI unit of density, as directly derived, is kilograms per cubic metre. However, the non-coherent unit kilograms per litre ($kg \, l^{-1}$) (giving the same numerical result as grams per millilitre) may be used as before. Since the volume of a body is dependent on the temperature and pressure its density must also depend on these quantities.

The relative density of a body is the ratio of its density to that of a reference substance. Like other relative quantities it has the dimension of 1. If water at 4 °C is chosen as the reference substance then the relative density is numerically practically equal to the absolute density. The symbol d_4^{20} signifies 'relative density of a body at 20 °C referred to water at 4 °C'.

The specific gravity of a body is the quotient of its weight and its volume. Owing to the dependence of weight on locality, the specific weight is unsuited to the characterization of a body. Consequently, unless the actions of force are specifically concerned, which is hardly ever the case in chemistry, it is better to work with density rather than specific gravity.

1.2.2.9. Mass concentration

The mass concentration of a material component in a mixture is defined as the quotient of the mass of the component and the volume of the mixture. This magnitude is dependent upon temperature and pressure. The dimensions of mass concentration are $M \cdot L^{-3}$, the same as for density (total mass/total volume). The coherent SI unit of mass concentration is kilogram per cubic metre ($kg \, m^{-3}$). However, the litre is the preferred reference volume rather than the intractable cubic metre. Thus the recommended units are $kg \, l^{-1}$, $g \, l^{-1}$, $mg \, l^{-1}$, $\mu g \, l^{-1}$ and $ng \, l^{-1}$. All the other units listed in section 1.2.1.1 should disappear from the clinical laboratory as quickly as possible. All combinations with percent and per thousand are not only obsolete, they are also ambiguous.

For the characterization of mixtures of substances the mass concentration is more often replaced by the molecular concentration (mol l^{-1}). But in the following cases the mass concentration continues to fulfil its intended function: 1, a component of the mixture is inhomogeneous (molecules of different sizes); 2, the molecular weight of one of the components is unknown; 3, one material component is a reagent which is not in stoichiometric relationship with a reactant partner (e.g. trichloroacetic acid for protein precipitation). In recipes for the preparation of reagents the mass concentration should always be given in addition to the necessary molecular concentration, so that the laboratory personnel are spared the trouble of looking up the molecular weight. Balances are calibrated in grams, not in Moles.

1.2.2.10. Mass ratio

The proportions by mass of a mixture can also be referred to the total mass instead of to the total volume of the mixture (particularly meaningful in the case of solid mixtures). The obtained quotient of partial mass to total mass (kilogram/kilogram), the mass ratio, has a dimension of 1. It is independent of temperature and pressure. The expression of mass ratios as percent, per thousand, ppm, etc., is to be avoided, and so are the units $g\ kg^{-1}$, $mg\ kg^{-1}$, etc. If, for example, 1 kg of stool is found to contain 3.5 g of fat, the result should read: fat in stool, mass ratio 3.5×10^{-3}.

1.2.2.11. Volume ratio

If the mixture to be characterized consists of liquids or gases, then, in addition to the mass concentration or mass ratio, the volume ratio may be meaningful. It is the quotient of the volume of the component and the total volume of the mixture (litre/litre). The dimensions of volume ratio are 1. Arguments corresponding to those given in section 1.2.2.10 govern the use of the old units.

1.2.2.12. Molecular concentration

The molecular concentration is defined as the quotient of the molecular amount of a component (Mole) and the total volume of the mixture. The coherent SI unit is $mol\ m^{-3}$, but the unit $mol\ l^{-1}$ (with appropriate prefixes before mol) is customarily used. In cases of possible ambiguity the chemical formula of the relevant elementary unit should be given. The numerical value of the molecular concentration is dependent upon temperature and pressure.

The terms 'molar concentration', 'molarity' and 'normal concentration' and the symbols M and N for 'molar' and 'normal' should no longer be used.

For the characterization of a mixture of substances the molecular concentration is preferred to the other possibilities (sections 1.2.2.9–11) wherever it is meaningful. Above all, the Mole takes priority over the gram where components of a mixture occur as potential reaction partners or where components of the mixture are related to one another by metabolic processes (section 1.2.2.5). It is recommended that molecular concentrations (mmol l^{-1}, μmol l^{-1}) be used for all work with body fluids, insofar as this is possible. This recommendation will be observed to the utmost in this book. In the introduction to the determination of metabolite concentrations the calculation formulae will be given for both mg per 100 ml and mmol l^{-1} to facilitate the changeover. The following formulae are valid for the interconversion of mass and molecular concentrations:

Number of mol l^{-1} = Number of g l^{-1} divided by the 'molecular weight';
Number of g l^{-1} = Number of mol l^{-1} times the 'molecular weight'.

'Molecular weight' always denotes the mass of N elementary units. Conversion tables for new and old units and the corresponding *conversion factors* have been worked out[3-5] (Appendix 7).

For osmotic considerations the molecular concentration, Osmoles per litre, is used; 1 osmol (osm) is the Loschmidt number of osmotically active particles (free molecules or ions which cannot pass through a semi-permeable membrane).

1.2.2.13. Molecular concentration ratio

The molecular concentration ratio is the quotient of the molecular amount of a component in a mixture and the sum of the amounts of all the components. The dimensions of this quantity are 1 (mol/mol). Considerations analogous to those in section 1.2.2.12 apply to the use of older units. The molecular concentration ratio may also be expressed as the quotient of the number of elementary units (molecules, ions, etc.) of a material component and the number of the same kind of elementary units in the whole system.

1.2.2.14. Molality

The molality of a dissolved substance is the quotient of the molecular amount of this substance and the mass of the solvent. The unit of molality is mol kg^{-1}. In doubtful cases the chemical formula of the elementary unit should be given. The dimension osmol/kilogram solvent is termed the osmolality (cf. section 1.2.2.12).

1.2.2.15. Enzyme concentration

For work with catalysts, an eighth basic unit, the catal, is provided. The catal is defined as the quantity of catalyst which brings about the reaction

28

of 1 mole of substance in 1 second.[6,7] The catal has not yet caught on in clinical chemistry but attempts to introduce it are in progress. In this book enzyme concentrations will be stated as before in international units (U): 1 U is the amount of enzyme which produces reaction in 1 mole of substance in 1 minute under defined conditions. Enzyme concentrations are accordingly given in $U\,l^{-1}$.

References

1. Dybkaer, R. and Jørgensen, K., *Quantities and units in clinical chemistry*, Munksgaard, Copenhagen, 1967.
2. Gemeinsame Kommission der Schweizerischen, Österreichischen und Deutschen Gessellschaft für Klinische Chemie, Messgrössen und Einheiten in der Klinischen Chemie, *Z. klin. Chem. klin. Biochem.*, **12**, 180 (1974).
3. Stamm, D., Messgrössen und SI-Einheiten in der Klinischen Chemie, *Mitt. Dt. Ges. klin. Chem.*, **1**, 1 (1975).
4. Baron, D. N., Broughton, P. M. G., Cohen, M., Lansley, T. S., Lewis, S. M., and Shinton, N. K., The use of SI units in reporting results obtained in hospital laboratories, *J. clin. Path.*, **27**, 590 (1974).
5. Lehmann, H. P., Metrication of clinical laboratory data in SI units, *Am. J. clin. Path.*, **65**, 2 (1976).
6. IFCC methods for the measurement of catalytic concentration of enzymes, *Clin. chim. Acta*, **64**, F11 (1975).
7. Dybkaer, R., Quantities and units in enzymology, *Enzyme*, **22**, 91 (1977).

1.3 STATISTICS

1.3.1. INTRODUCTION

The assessment of laboratory methods, the evaluation of reference ranges and the interpretation of results—in the laboratory, just as at the bedside—all call for a knowledge of some of the basic concepts of statistics. According to Wald,[1] *statistics* consist of 'a set of methods allowing rational, optimal decisions to be made in cases of uncertainty'. Statistics find application whenever it is necessary to assess observations subject to so-called 'random' influences which, apart from their conformity to certain basic laws, are not more closely definable. This is the case in virtually all branches of research, especially in the empirical sciences. If, for example, we consider the results of a laboratory investigation, then the values may fluctuate for any of the following reasons:

—variation from person to person;
—temporal fluctuation: time of day, time of year;
—variation from one laboratory assistant to the next;
—differences between individual instruments;
—random errors of measurement.

Depending on the kind of question which is to be answered, one or other of these causes of variation will have to be eliminated as far as possible and this task is facilitated by mathematical statistics and the theory of probabilities.

In general, the tasks of statistics may be formulated as follows:

—the description of collective data;
—the estimation of unknown quantities;
—the assessment of differences;
—the testing of correlations.

In principle, statistics serve as a refinement of 'common sense', insofar as it places subjective findings and conclusions on an objective footing. This is done with the aid of probability theory. The *probability* of an event is a number between 0 (impossibility) and 1 (certainty); in practice, it is often given as a percentage. Statistics scarcely ever yield results which contradict common sense; results of that kind, 'proved' by statistics, should always be treated with suspicion and subjected to a very thorough testing!

29

Two fields of statistics may be distinguished, corresponding to the basic functions involved. *Descriptive* statistics are concerned with the description and characterization of data; examples occur daily in the newspapers and journals. The second branch, *inductive* (deductive, analytical) statistics, comes into its own when, on the basis of a small group of values conclusions are to be drawn concerning a larger system of comparable data; these conclusions are usually in the form of estimates or forecasts.

1.3.2. DESCRIPTION OF MEASUREMENT DATA

1.3.2.1. Graphical presentation of the values

In making measurements, the results are usually obtained in a temporal sequence. They are entered, as they occur in a so-called 'prime list'. This is generally non-ordered and is not easily scrutinized.

Example 1. Ten replicate determinations of the sodium concentration (mmol l^{-1}) in a pooled serum yielded the following results:

---

123   124   126   129   120   132   123   126   129   128   (prime list)

If the values are arranged according to their size, the data are already easier to inspect:

---

120   123   123   124   126   126   128   129   129   132

The largest and smallest values can easily be read off from this series. The difference of these two values is called the 'range':

$$R = x_{max} - x_{min} \tag{1}$$

There are as many values below the number 126 as there are above it. A simple graphical display shows the points on a diagram within the scale of possible values (Fig. 4).

If more data are available, arranging the values in order can be a very laborious task in itself.

Example 2. Determination of the glucose concentration (mg per 100 ml) in a control serum on 100 consecutive days (Table 6).

The values are tabulated into *classes* with virtually no loss of information. It is expedient to distribute the present data in about 10–20

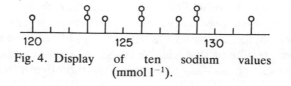

Fig. 4. Display of ten sodium values (mmol l^{-1}).

Table 6. Glucose concentration in a control serum (mg per 100 ml).

66	64	68	65	65	64	65	65	65	63	64	67	64	63	64
65	66	64	66	65	65	64	66	66	65	62	64	67	66	66
66	65	65	68	63	65	66	66	64	64	65	63	63	67	67
69	64	66	64	66	65	65	61	65	67	67	68	62	63	64
65	64	64	67	66	66	68	64	64	64	67	66	66	69	65
65	65	65	62	68	65	64	65	65	69	67	66	66	66	65
62	66	64	64	67	63	63	67	64	65					

classes. A working rule for the determination of the class width b derives the latter from the range R:[8]

$$b = \frac{R}{1 + 3.32 \log n} \tag{2}$$

where log signifies the logarithm to the base 10 and n signifies the number of observations.

Although it is always arbitrary, the formation of classes does provide a good survey of the distribution of the numbers over the total range of values. It is practicable to choose the classes so that their mid-values are whole numbers.

The class frequencies are most easily counted up in a *tally list*, as shown in Fig. 5 using the data of example 2.

For $n = 100$ and $R = 8$, equation 2 gives a value of $1.05 \approx 1$ for the class width, as clearly illustrated in this example.

The data may be illustrated further by plotting the class frequencies against the class mid-values on a coordinate system, as shown in Fig. 6.

A graphical presentation of this kind is called a *histogram*. The class frequencies are illustrated by vertical strips, each strip corresponding to a class. Histograms give a very clear picture and they are indispensable aids in the description of measurement data.

Class midvalue	Classes	Tally list	Absolute frequency	Relative frequency
61	60.5-61.5	I	1	0.01
62	61.5-62.5	IIII	4	0.04
63	62.5-63.5	ЖН III	8	0.08
64	63.5-64.5	ЖН ЖН ЖН ЖН II	22	0.22
65	64.5-65.5	ЖН ЖН ЖН ЖН ЖН I	26	0.26
66	65.5-66.5	ЖН ЖН ЖН ЖН	20	0.20
67	66.5-67.5	ЖН ЖН I	11	0.11
68	67.5-68.5	ЖН	5	0.05
69	68.5-69.5	III	3	0.03
Total			100	1.00

Fig. 5. Tally list for 100 glucose concentrations; tally scheme according to Riedwyl.[2]

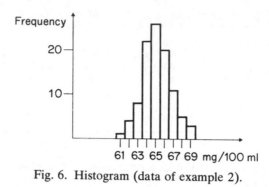

Fig. 6. Histogram (data of example 2).

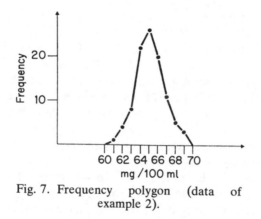

Fig. 7. Frequency polygon (data of example 2).

Figure 7 shows a further way of presenting the values. In this *frequency polygon* (empirical frequency function) the frequencies are plotted as points against the class mid-values and the points joined by straight lines.

Another way to present the values, and one which is more in line with the concepts of mathematical statistics, is to find the *cumulative frequency*

Table 7. Cumulative frequencies for example 2.

Upper class limit	Absolute frequency	Cumulative frequencies	
		Absolute	Relative
61.5	1	1	0.01
62.5	4	5	0.05
63.5	8	13	0.13
64.5	22	35	0.35
65.5	26	61	0.61
66.5	20	81	0.81
67.5	11	92	0.92
68.5	5	97	0.97
69.5	3	100	1.00

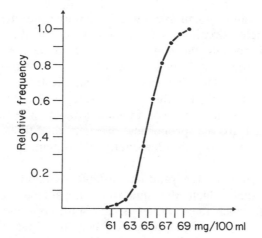

Fig. 8. Empirical distribution function (data of example 2). It is essential to plot the points against the upper class limits, not against the class mid-values.

(relative or absolute) of the classes. To do this, each class is coordinated not with the number of values lying within it but with the number of all the values which are smaller than or equal to the upper class limit. Table 7 shows the cumulative frequencies for example 2. The progressive summing is called cumulation.

If the relative cumulative frequencies are plotted against the upper class limits (not class mid-values!) the result is a characteristic curve known as the *empirical distribution function* (Fig. 8). It will be clear from the manner in which the empirical distribution function was obtained that its ordinate (variable on the vertical axis of the coordinate system) can only assume values between 0 and 1.

1.3.2.2. Random variable and distribution

Since measurement data are always subject to errors of measurement they must be considered as variables, their magnitudes governed by external influences (the errors of measurement, for instance); it is assumed that the fluctuations are 'purely random'. In such cases the statistician speaks of a *random variable* (chance variable, random quantity, stochastic variable, and similar terms). The scores from tossing a die, lottery numbers, even the physical measurements made in determining the speed of light, are all simple examples of random variables. Students' marks may also be considered as random quantities, and it is clear from this example that the fluctuations are not solely due to chance.

In general, it is possible to distinguish two kinds of random quantity according to the values it can assume. For example, the measured value for the sodium concentration in blood can be any conceivable decimal within a certain range (supposing the concentration could be measured accurately enough); on the other hand, the number of points shown on a die can only have the values 1–6. The first example is an illustration of a continuous variable and the second of a discrete variable. Since clinical chemistry is predominantly concerned with continuous variables, the following account will be restricted to the properties of such quantities. The reader is referred elsewhere[2,3] for an elementary treatment of discrete random quantities.

The characterization of a random variable demands a knowledge not only of the values which the quantity can assume, but also of the frequencies with which these values occur. With continuous variables, the relative frequency with which values occur in a particular class is an approximation to the probability of occurrence of the class.

Hence a continuous random variable is characterized by two pieces of information:

1. by the possible values it can assume (range of values);
2. by the probability that it lies within a given interval.

These two factors yield the *distribution function* of the random variable. Graphically, a distribution function resembles the cumulative frequency function of Fig. 8 but with the probability (i.e. an ideal parameter) replacing the relative frequency on the ordinate. The distribution function of a random variable assigns to each real number y the probability that the variable x assumes a value which is smaller than or equal to y. If the probability were to be plotted on the ordinate in a manner similar to that used for drawing the frequency polygon (Fig. 7), the result would be a graphical presentation of the density function (*distribution density*) of the random variable (for those mathematically inclined, the distribution function is the integral of the density function). The distribution or density function characterizes a random variable completely. Hence the significance of the distribution function: the variable is known (in terms of probability) if its distribution is known.

Strictly, when considering random quantities one ought to differentiate between the variables themselves and their realization, i.e. the value obtained each time. Thus, in example 1 the numbers 120, 123, 124, 126, 128, 129, and 132 are realizations of the random quantity 'sodium concentration in blood'.

1.3.2.3. Normal distribution

One special distribution is of fundamental significance: the *normal distribution* (more correctly 'Gaussian distribution' after the mathematician

C. F. Gauss, 1777–1855). This is given by the distribution density:

$$g(y) = \frac{1}{\sigma\sqrt{1\pi}} \cdot e^{-((y-\mu)/\sigma)^2/2} \tag{3}$$

In this equation, π and e are fundamental mathematical constants: $\pi = 3.14159\ldots$ (ratio of the circumference of a circle to its diameter), $e = 2.71828\ldots$ (basis of natural logarithms). The two parameters μ ('mean value') and σ ('standard deviation') characterize the function. Putting $\mu = 0$ and $\sigma = 1$ results in the *standard normal distribution* with the density given by

$$g(y) = \frac{1}{\sqrt{2\pi}} \cdot e^{-y^2/2} \tag{4}$$

The values of the standard normal distribution are listed in many tables (e.g. ref. 4). The graph of every normal distribution shows a familiar bell-like curve (Fig. 9).

One property of the normal distribution is that approximately 95% of all the values lie in the range $\mu - 2\sigma$ to $\mu + 2\sigma$. Its special significance stems from the circumstance that many actual phenomena can be described by the bell-like curve of the normal distribution and the fact that, for sufficiently large n, the mean values of any chosen measured quantities approximate to the normal distribution (the *central limit theorem* of probability theory).

A form of distribution which occurs more frequently is the *log-normal* distribution: a random variable has a log-normal distribution if the logarithms of the values are normally distributed. Following this excursion into the more theoretical aspects, the precise usage of the term 'random variable' will be abandoned for the remainder of the discussion; it will usually suffice to speak simply of 'variables'.

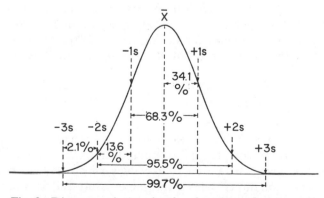

Fig. 9. Diagram of the density function of a normal distribution. $\bar{x}$ = mean value; s = standard deviation.

1.3.2.4. Statistical parameters

After the data have been listed in a table of values, arranged in order, and displayed in histograms, an attempt should be made to describe the numerical material by some characteristic numbers, i.e. descriptive parameters. A simple measure for identifying the approximate location of a variable is the *median* (middle value). This is the value in the middle when the data are ranged in order of their size. If the number of values is even, then the arithmetic means of the two middle values is taken. In example 1, the median is 126. The best known measure is the *mean*: the obtained values are summed and divided by the number of trials, n, i.e. it is the arithmetic mean of all the values. Expressed as an equation ($\bar{x}$ is pronounced 'x-bar'):

$$\bar{x} = \frac{1}{n}(x_1 + x_2 + \ldots + x_n) \quad \text{or} \quad \bar{x} = \frac{1}{n}\sum_{i=1}^{n} x_i \tag{5}$$

The symbol $\sum_{i=1}^{n}$ indicates the sum of all values x_i over the range $i = 1$ to $i = n$ (often written simply as Σ).

As example 3 shows, a distribution cannot be described simply by parameters for the location of the value.

Example 3. Realizations for two random variables, V_1 and V_2:

V_1	10	11	9	9	11
V_2	−990	1030	1000	−1000	10

In both cases the mean and the median are 10 but it can be seen at a glance that these two variables can scarcely have the same distribution. The frequency functions have different forms, V_2 showing a 'stronger dispersion' than V_1.

Parameters for describing the form of a distribution function are afforded by measures of dispersion. A first and very simple measure of dispersion has already been given: the range (equation 1). The best known and most important parameter for the dispersion is the *standard deviation*, s (SD). This is the square root of the *variance*, which is given by the equation:

$$s^2 = \frac{1}{n-1}\sum_{i=1}^{n}(x_i - \bar{x})^2 \tag{6}$$

According to equation 6 the mean is to be subtracted from each value measured, the differences are to be squared and the squared differences summed (so this gives extra weight to values which deviate considerably from the mean). Equation 7 yields the same result and is more suited to the numerical calculation (sometimes equations 6 and 7 are additionally modified by replacing $n - 1$ by n in the denominator; this makes little

Table 8. Statistical parameters (data of example 1).

Mean, x	126
Variance, s^2	12.89
Standard deviation, s	3.59
Coefficient of variation	0.028 = 2.8%
Standard error of the mean, SEM	1.14
Median	126
Range, R	12

difference if the number of trials is sufficiently large):

$$s^2 = \frac{1}{n-1}\left[\sum_{i=1}^{n} x_i^2 - \frac{1}{n}\left(\sum_{i=1}^{n} x_i\right)^2\right] \tag{7}$$

Apart from these, there are other equations for (approximate) calculation of the mean and the variance or standard deviation for classified data (cf. ref. 2). However, these will not be discussed because today computer programs which can calculate the parameters by the equations given, even when the data are very extensive, are available almost everywhere.

The *standard error of the mean* (SEM) is evaluated by dividing the standard deviation by $\sqrt{n}$. It verifies the intuitive expectation that an average of several measurements varies less than the individual values. It is sometimes used unjustifiably in publications because it makes the numerical value of the dispersion appear smaller.

The *coefficient of variation* is the ratio of the standard deviation to the mean:

$$CV = \frac{s}{x} \tag{8}$$

This is frequently expressed as a percentage, i.e. multiplied by 100. The coefficient of variation is well suited for comparing the dispersion where different units of measurement are involved. It is also significant because the standard deviation frequently increases in proportion to the mean. *Warning*: The coefficient of variation is only meaningful if all the observations are positive!

Table 8 contains the described parameters for the data of example 1.

Yet more parameters ('moments of higher order') are used for the more detailed characterization of the distribution function; the reader is referred to the specialist literature[5,6] for information on these.

1.3.2.5. Tests for distribution

In addition to the statistical parameters, the form of the distribution function for a variable will generally be of interest, above all, whether the

38

quantity is normally distributed. This may be checked arithmetically by applying the Kolmogorov–Smirnov significance test and, if necessary, the chi-squared (χ^2) test (cf. ref. 7).

In most cases it is simpler to test for agreement with the normal distribution graphically. This may be done by plotting the cumulative frequencies on a special paper, 'normal curve graph paper', or *probability graph paper* (obtainable in shops specializing in office equipment and in large stationers in two varieties: abscissae linear (semi-log) or logarithmic (log-log)). If the distribution is normal the lie of the points approximates to

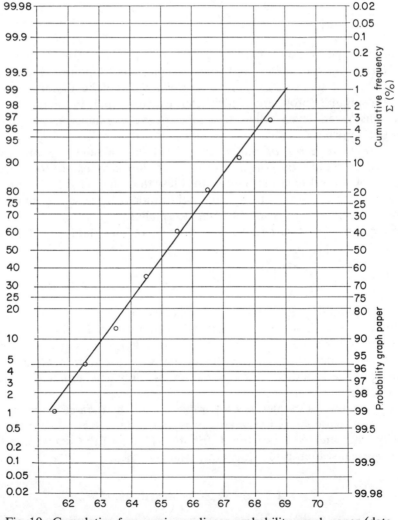

Fig. 10. Cumulative frequencies on linear probability graph paper (data of example 2).

a straight line. The smaller the standard deviation, the steeper is the line. Fig. 10 shows the plot for the data of example 2.

Approximate values of the mean and the standard deviation of a variable may be determined from the probability graph paper plot: the mean is the value of the abscissa for the cumulative frequency of 50%, and a rough estimate of the standard deviation is obtained by finding the 68% ordinate and taking the difference from the mean.

Testing for a log-normal distribution may be done in a similar fashion by forming the cumulative frequencies of the logarithms of the given values and plotting these on probability graph paper, or by plotting the cumulative frequencies of the given data directly onto log-log probability graph paper (logarithmic abscissa).

1.3.3. RELATIONSHIPS BETWEEN VARIABLES

1.3.3.1. Statement of the problem

When several variables are considered jointly the major point of interest, apart from the description of each individual variable, is whether the variables are related and what these relationships are. In such cases several measurements are considered simultaneously for each experimental subject or system: size and weight for different persons; the connection between the concentration of a substance and its photometric extinction (calibration curves); measurement of metabolite concentrations by different chemical methods.

Whereas in the preceding sections only one variable was observed at a time, here we are concerned with *connected* (correlated) (random) variables. Apart from the parameters for the location and dispersion of the individual variables, what is of interest here is information about the strength of the connection and the nature of the underlying relationship. Owing to their simplicity of form linear relationships are particularly important.

The most frequent, and most important, case is that for two variables and the following exposition will be confined to this case. The reader is referred to the literature[5,8] for a consideration of several variables.

1.3.3.2. Scatter plots

With two variables the measurements may be graphically represented as points in a plane by plotting one variable along the abscissa (x-axis) and the other as ordinate (y-axis). A diagram of this type is called a *scatter plot* (point diagram, correlation diagram). Examples are shown in Figs. 11 and 12. In one case the points fall fairly well on a straight line, in the other they form a disordered cluster. It is immediately apparent that in the first case there is a strong relationship (Fig. 11).

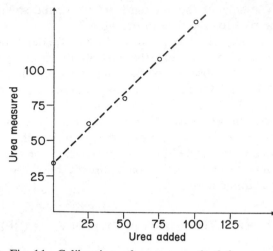

Fig. 11. Calibration of a new method for urea,
scatter plot.

Example 4. In setting up a new method for the determination of urea, known quantities of urea were added to serum with an unknown urea content and its concentration was measured. This experiment should show firstly whether the values obtained are 'correct', i.e. are not invalidated by other components of the serum. In addition, it should allow the range and validity of a Lambert–Beer law to be evaluated. The results in Table 9 were obtained.

The scatter plot for this example (Fig. 11) shows that the measured values lie in a close fit to a straight line, i.e. linearity is observed throughout the range of measurements.

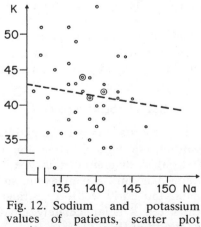

Fig. 12. Sodium and potassium
values of patients, scatter plot
(*K*-values multiplied by 10).

Table 9. Calibration points for a new method of urea.

	Urea added corresponding to . . . mg per 100 ml of serum				
	0	25	50	75	100
Urea found, mg per 100 ml	34	62	80	109	135

Example 5. The possibility of a relationship between the sodium and potassium values for a patient is to be checked by means of a random sample of 40 patients (the question has no theoretical basis; the example is only to illustrate the scatter plot). The values are presented in Table 10.

The scatter plot of the data in Table 10 is shown in Fig. 12. As was to be expected, this shows a disordered assemblage of points with no clear structure. There is no connection between sodium and potassium values.

To obtain an impression of the general nature of the connection there is no substitute for a clear, graphical display of the measured values in a scatter plot. The diagram shows whether the group of points may be fitted by a straight line or whether a curved line may have to be chosen. Sometimes it provides evidence of unforeseen systematic errors in the experimental procedure, as Natrella[6] demonstrates.

1.3.3.3. Correlation coefficient

To establish a measure for the strength of the correlation it is assumed that a point with a high x-value should also have a high y-value if there is a strong correlation between the two quantities. This can then be exploited by multiplying the deviations from the mean for x- and y-values and summing the products. The result is normalized by dividing by the standard deviations of x and of y and by the number of trials. The resulting

Table 10. Sodium and potassium values (mmol l^{-1}) of 40 patients (K-values multiplied by 10).

Na	K	Na	K	Na	K	Na	K
143	41	141	42	141	42	132	47
143	42	135	36	140	40	142	34
141	43	145	41	139	38	134	31
136	51	133	41	139	41	138	44
137	49	141	34	139	35	147	37
131	42	137	43	136	38	139	44
136	43	133	36	138	44	134	45
138	42	140	37	136	46	141	40
137	36	144	47	143	47	141	38
139	41	140	54	137	39	132	51

equation for the coefficient of correlation is:

$$r = \frac{1}{n-1} \cdot \frac{\Sigma (x_i - \bar{x})(y_i - \bar{y})}{s_x \cdot s_y} \tag{9}$$

The standard deviations s_x and s_y are calculated by using equation 6.

For making the actual calculations the following equation is more suitable:

$$r = \frac{\Sigma x_i y_i - \frac{1}{n}(\Sigma x_i)(\Sigma y_i)}{\sqrt{\left[\Sigma x_i^2 - \frac{1}{n}(\Sigma x_i)^2\right]\left[\Sigma y_i^2 - \frac{1}{n}(\Sigma y_i)^2\right]}} \tag{10}$$

The correlation coefficient may take values between -1 and $+1$. A value of 0 shows that there is no correlation between x and y; a value of $+1$ or -1 signifies 'perfect' correlation (positive or negative, respectively). In Fig. 11 $r = 0.998$ and in Fig. 12 $r = -0.114$, i.e. a very strong positive and a weak negative correlation, respectively.

r is a linear measure, so it can have a value close to 0 even when there is a strong correlation if this is non-linear (e.g. cyclic; for illustrations of this point, see ref. 2).

1.3.3.4. Regression

The pattern of points on the scatter diagram allows the nature of the correlation between the two variables to be characterized more closely. The simplest (and most important) case is that for which the points are scattered in more or less close fit with an imaginary straight line, the *regression line*. In this case we speak of a *linear regression*.

A straight line may be described mathematically as a linear function:

$$y = a + bx \tag{11}$$

Here, x is the independent and y the dependent variable. If the function $y = a + bx$ is drawn on a system of coordinates (x, y), then a is given by

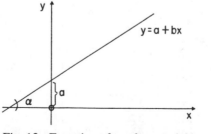

Fig. 13. Equation for the straight line.

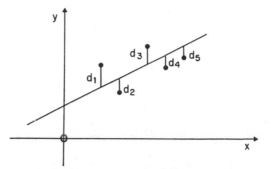

Fig. 14. Formation of the regression line: the
distances $d_1, \ldots, d_5$ are minimized.

the intercept of the straight line with the y-axis ($x = 0$) and b by the slope
of the line (Fig. 13); b characterizes the slope of the line relative to the
x-axis and is given by the tangent of the angle α: $b = \tan \alpha$.

The line of regression may be calculated by the method of least squares:
the line is so chosen that the distance of the points from the straight line is
minimized in the y-direction. This is shown in Fig. 14.

The method of least squares yields the following equations for b and a:

$$b = \frac{\Sigma(x_i - \bar{x})(y_i - \bar{y})}{\Sigma(x_i - \bar{x})^2} \tag{12}$$

$$a = \bar{y} - b\bar{x} \tag{13}$$

It is clear from equation 13 that the point with coordinates $(\bar{x}, \bar{y})$ must
always lie on the straight line since it satisfies the equation $\bar{y} = a + b\bar{x}$. For
calculating b, it is simpler to use another equation which yields the same
result:

$$b = \frac{\Sigma x_i y_i - \dfrac{1}{n}(\Sigma x_i)(\Sigma y_i)}{\Sigma x_i^2 - \dfrac{1}{n}(\Sigma x_i)^2} \tag{14}$$

Example. Calculation of a and b for Table 9 gives $a = 34.2$ and
$b = 0.996$, and the equation for the straight line is $y = 34.2 + 0.996x$. For
Table 10 the results are $a = 62.6$ and $b = -0.152$ and the equation for the
straight line is $y = 62.6 - 0.152x$. The lines are drawn in Figs. 11 and 12.

A regression line is frequently used to predict y on the basis of existing
values of x. It is intuitively clear that the less the points deviate from the
straight line, the more accurate this prediction will be. This is expressed
mathematically in the equation for the 'standard error of estimate of y on

Table 11. *Example 6*: summary of calculations for regression and correlation (data from Table 9; the equation numbers are given in parentheses).

Number of observations — $n = 5$

Sum of the x-values $0 + 25 + 50 + 75 + 100$ — $\Sigma x = 250$

Mean of the x-values (5) $250:5$ — $\bar{x} = 50$

Sum of the squared x-values $0^2 + 25^2 + 50^2 + 75^2 + 100^2$ — $\Sigma x^2 = 18\,750$

Variance of the x-values (7) $\frac{1}{4} \cdot (18\,750 - \frac{1}{5} \cdot 250^2)$ — $\text{var}(x) = 1562.50$

Standard deviation of the x-values $\sqrt{1562.50}$ — $s_x = 39.53$

Sum of the y-values $34 + 62 + 80 + 109 + 135$ — $\Sigma y = 420$

Mean of the y-values (5) $420:5$ — $\bar{y} = 84$

Sum of the squared y-values $34^2 + 62^2 + 80^2 + 109^2 + 135^2$ — $\Sigma y^2 = 41\,506$

Variance of the y-values (7) $\frac{1}{4} \cdot (41\,506 - \frac{1}{5} \cdot 420^2)$ — $\text{var}(y) = 1556.50$

Standard deviation of the y-values $\sqrt{1556.50}$ — $s_y = 39.45$

Sum of the products of x- and y-values $0 \cdot 34 + 25 \cdot 62 + 50 \cdot 80 + 75 \cdot 109 + 100 \cdot 135$ — $\Sigma xy = 27\,225$

'Covariance' of x and y $\frac{1}{4} \cdot (27\,225 - \frac{1}{5} \cdot 250 \cdot 420)$ — $\text{cov}(x, y) = 1556.25$

Correlation coefficient (9)
$$\frac{27225 - \frac{1}{5} \cdot 250 \cdot 420}{\sqrt{18750 - \frac{1}{5} \cdot 250^2} \cdot \sqrt{(41506 - \frac{1}{5} \cdot 420^2)}}$$ — $r = 0.998$

Coefficient of regression (14)
$$\frac{27225 - \frac{1}{5} \cdot 250 \cdot 420}{18750 - \frac{1}{5} \cdot 250^2}$$ — $b = 0.996$

Intercept on the y-axis (13) $84 - 0.996 \cdot 50$ — $a = 34.2$

Regression line (11) $y = 34.2 + 0.996x$

Standard error of estimate of y on x (15)
$$\sqrt{\frac{1}{3} \cdot [(34-34.2)^2 + (62-59.1)^2 + (80-84)^2 + (109-108.9)^2 + (135-133.8)^2]}$$ — $s_{y \cdot x} = 2.94$

x' ('scatter of points about the regression line'):

$$s_{y \cdot x} = \sqrt{\frac{\Sigma(y_i - y_i')^2}{n - 2}} \tag{15}$$

where y' symbolizes the value on the regression line corresponding to the given x-value; the division by $n - 2$ arises on theoretical grounds. $S_{y \cdot x}$ is a measure of the error in predicting y on the basis of x. This error of estimation is always smaller than, or at most equal to, the standard deviation of the y-values.

An estimate of y from x based on a regression line is the more accurate the closer x lies to the mean, $\bar{x}$. Hence it would be ill-advised, for example, to make any estimate of the size of y for $x = 100$ on the basis of a regression with x-values in the range 10–30. Outside the range considered, the accuracy of the prediction decreases rapidly.

As mentioned earlier in the discussion of correlation coefficients, not all correlations are linear. Sometimes the set of points is better fitted by a curve than a straight line. The simplest way of testing for this is visual inspection of the graph.

The straight line used for estimating y from x may not be used for the inverse process of estimating x from y. In this case the deviations of the points in the x-direction must be minimized. The two lines are only congruent for the case of an exactly linear relationship ($r = 1$).

Table 11 provides a comprehensive example for calculations concerning correlation and regression. Benninghaus[9] has provided a more detailed, yet readily understandable, treatment of these problems.

1.3.4. SIGNIFICANCE TESTS

1.3.4.1. Basis

Measurements in all fields of science and technology are subject to errors and uncertainties. The scientist by no means expects to obtain the same values each time he repeats his measurements. Nevertheless, the deviations should be within 'reasonable limits' if he is to be sure that the same variables are actually being measured, as far as possible, under the same conditions.

Recurrent questions include, 'what is the "true value" of a measured quantity?', or 'what is the difference between the true values of two or more quantities?'. For example, 'what percentage of the population has a particular disease?' or 'how do the results of the clinical–chemical tests depend upon the sex of the patient?'.

It is usually not possible to investigate all of the individuals in a population (statisticians also use the term *universe* or *entire group*, since the subject of their investigations need not be confined to persons). For example, it would hardly be possible or even sensible to determine the cholesterol concentration in blood for all men and women (even for a medium-sized town) in order to test for differences between the sexes. Statistics supply the requisite aid, the *representative sample*. A sample is a selection of individuals from a population. In order for it to be representative, two conditions must be fulfilled: (1) *random selection*—each individual in the population has an equal chance of being chosen (this may be assured, say, by using tables of random numbers); (2) *independence*—the choice of one individual must not affect either the choice of further individuals or their values.

One special type of representative sample is of particular significance in science: the independent repetition of an experiment.

Making the choice random allows the laws of probability theory to be applied to the representative sample and then it is possible to make assertions about the population which have a fixed probability of being valid. Since the conclusions drawn are not absolutely certain but are only probable, this also takes into account the inaccuracies of measurement expected at the outset.

If assertions are to be made about particular characteristics of the population (so-called 'population parameters') on the basis of a sample, then a *statistical test* is used. A statistical test, or simply 'test', is a way of testing *hypotheses* concerning the population. It allows the sample to be used as a basis for making decisions. Essentially, a test consists of a rule determining when, for particular samples, a hypothesis is to be accepted and when not. The decision is usually based on the value of a *test statistic* calculated from the sample data.

Tests of significance serve to establish meaningful ('significant') differences between a sample value and a hypothetically determined number (one sample case), or between values obtained from different samples (two- or multi-sample case). The test is conducted by first setting up the *null hypothesis*, H_0, that there is no difference between the values; this is then checked by means of the test statistics. The null hypothesis assumes that any differences between the sample values are attributable solely to random variations. It stands refuted if the test statistics fall in a range for which the values have a very low probability given that H_0 is valid ('region of rejection'). If the null hypothesis is rejected it may be concluded that its logical opposite, the *working hypothesis*, H_A (alternative hypothesis), is correct. Strictly, it is really H_A which is confirmed or not by the statistical test. It is a feature of the significance test that it provides no direct confirmation of the null hypothesis; all that can be said is that, in the absence of other findings, it should be accepted ('acquittal for lack of evidence').

If the test statistic lies in the region of rejection even though H_0 is correct, then rejection of the null hypothesis constitutes a 'type I error'. To avoid such an error, the region of rejection is so chosen that there is a very low probability (the 'probability of error', α) of its values being assumed if H_0 is valid. α is also termed the *level of significance* (threshold) of the test (often denoted by p in the medical literature).

The converse case, retention of H_0 although it is incorrect, constitutes a 'type II error'. The probability of making a type II error is denoted by β. α and β are mutually interfering: if α is made small, then β is large, and *vice versa*. α is usually fixed at the customary values of 0.05 (5%) or 0.01 (1%). The size of β will then depend on the nature of the test. Tests with small β (for small α) have a high 'confidence'. Statistical science endeavours to develop tests with high confidence under the most general conditions possible.

In general, two kinds of null hypothesis may be distinguished: in the *two-sided* (two-tailed) *case*, what is of interest is whether a measurement deviates from another value, regardless of the direction of the deviation. In the *one-sided case*, only differences in one direction are of interest, as, for example, in the question of whether the prospects of recovery are better with a new medicine than with one already in use. In the two-sided case, the region of rejection usually consists of two parts. For the hypothesis to be refuted, the test statistics must either be large enough to fall in the upper part or small enough to lie in the lower part of the region of rejection. With a one-sided test, the region of rejection has only one part. Consequently, for the same level of significance, this may be chosen to be larger than in the two-sided case.

1.3.4.2. Some methods of testing

Following the introductory discussion of the preceding section a few important methods of testing ought now to be presented. In this age of electronic data processing, computer programs for carrying out the necessary calculations for the statistical tests are available at all universities and the larger private computer centres. Hence it no longer seems to be so important to give a detailed account of how to perform the calculation. Instead, we shall give closer scrutiny to the assumptions necessary in order that the tests be applicable.

Fundamentally, two categories of data may be distinguished:

(a) *Quantitative data* have numerical values which can vary arbitrarily within certain limits. Example: sodium concentration in blood.
(b) *Qualitative* (nominal) *data* can only assume particular values. Example: sex (m/f); qualitative urine investigations.

Data of type (a) can be transformed into type (b) data by the formation of classes, this corresponding to a coarsening or loss of information. Different statistical tests are used for these two categories of data. Here we shall only consider quantitative data since in clinical chemistry these predominate. The specialist literature[7] should be consulted for tests applicable to qualitative data.

The following section considers a few tests for checking mean values. Additional tests, including tests for other population parameters, were simply and clearly expounded by Riedwyl.[2]

The procedure with each test is as follows. Values are selected from a population, the test statistics are calculated on the basis of this sample, and the statistical conclusions are drawn. Here it is absolutely essential that the individual measurements of the sample are independent, i.e. they do not influence one another. For example, one should not use several measurements of the same patients in a test to reveal differences between sick and healthy patients.

1.3.4.2.1. Testing the difference of two means

This test requires independent measurements from two mutually independent samples. The sizes of the sample populations are denoted by n_1 and n_2, respectively.

1.3.4.2.1.1. Wilcoxon test

Applicability. The Wilcoxon test is specially suitable for testing the difference of two means when the sample sizes are small ($n \leqslant 30$; with larger samples the procedure is laborious, although the test remains usable in principle). No assumptions are made concerning the distribution of the variables to be tested (the data do not have to be normally distributed).

Determination of the test statistic. The values for samples 1 and 2 are combined and collectively ranged in order of size. Each value is assigned a serial number: the smallest value is given the serial number 1, the second lowest 2, and so on. The highest value has the serial number $n_1 + n_2$. Where 'ties' occur (same value) the values are each assigned the mean value of the appropriate serial numbers. The test statistic T_1 is the sum of the serial numbers of the elements of sample 1.

Decision. Compare T_1 with the tabulated values appropriate to n_1 and n_2 (Geigy tables; ref. 4, pp. 124–127). Different tables are to be used according to the value of α chosen. For one-sided cases the probability of error may be halved. If T_1 lies outside the tabulated range it can be assumed that the means of the samples are different. The assumption is valid at level of significance α, e.g. $\alpha = 0.05$ (or 5% or 'with a certainty of 95%').

Example 7. The question to be tested is whether one medicine for the treatment of diabetes is better than another. The measured variable used is

Table 12. Percentage reduction of the blood-sugar content in diabetics.

Old medicine 12 cases	Serial number Wilcoxon test	New medicine 6 cases	Serial number Wilcoxon test
27	3	44	7.5
64	12	79	17
70	14	68	13
44	7.5	71	15
54	9	75	16
28	4	58	10
60	11		
24	2		
36	6		
83	18		
34	5		
16	1		

the percentage reduction in blood-sugar content achieved with the two medicines. Table 12 shows the percentages for each medicine.

The Wilcoxon test is applied (one-sided, since we require to know only whether the new medicine is better). The level of significance is chosen at $\alpha = 0.05$. After ordering the series, the serial sum $T_1 = 92.5$ is obtained for patients treated with the old medicine, and $T_2 = 78.5$ for the group treated with the new medicine. The tables of confidence limits (ref. 4, p. 124) give the critical value of the serial sum of the larger group 1 as 95 for $n_1 = 12$ and $n_2 = 6$; 92.5 is smaller than 95, and therefore the following conclusion can be drawn.

Result. The percentage decrease in the blood-sugar concentration is smaller with the old than with the new medicine.

1.3.4.2.1.2. z-test

Application. Testing the difference of two means when the samples are large (over 30). No particular assumption need be made concerning the distribution. According to the central limit theorem of probability theory, sample means are approximately normally distributed provided that the samples are large, regardless of the initial distribution.[2,10]

Test statistic. As test statistic we use the value

$$z = \frac{\bar{x}_1 - \bar{x}_2}{\sqrt{\dfrac{s_1^2}{n_1} + \dfrac{s_2^2}{n_2}}} \tag{16}$$

where $\bar{x}_1$, $\bar{x}_2$ are means and s_1, s_2 are standard deviations for the two samples.

Table 13. Frequently used values of the standard normal distribution.

α					
Two-sided	0.10	0.05	0.02	0.01	0.001
One-sided	0.05	0.025	0.01	0.005	0.0005
z_0-value	1.645	1.960	2.326	2.576	3.291

Decision. If the null hypothesis is valid, then z is normally distributed with mean 0 and variance 1. The values of this distribution may be found from tables (ref. 4, p. 31). Table 13 gives the relevant values.

If $|z|$ (the absolute value of z without regard to the sign) is equal to or exceeds the z_0 value corresponding to α, then it may be assumed with the corresponding degree of certainty that there is a difference between the means (two-sided case; the rarer one-sided case tests whether mean 1 is larger or smaller than mean 2; the assumption is confirmed if the test statistics obey $z \geqslant z_0$ or $z \leqslant -z_0$, respectively).

Example 8. Table 14 lists cholesterol values for patients over 17 years old; all of the values originate from the EDP laboratory system of the Bern Isolation Hospital. The question to be tested is whether there is a significant difference between the values for men and for women.

The calculation yields the following values:

	Men	Women	
n	43	34	
$\bar{x}$	215.7	215.8	
s_x^2	5287.7	3001.9	$z = 0.0069$
s_x	72.7	54.8	by equation 16

Hence there are no grounds for assuming a significant difference here between men and women as far as these patients are concerned. This is not

Table 14. Cholesterol values for patients.

Men									
227	199	219	230	252	295	133	209	225	285
216	170	314	181	124	341	286	194	197	216
231	205	254	308	218	307	265	236	174	237
210	215	304	129	261	221	164	159	203	258
165	350	204							

Women									
215	273	179	188	158	286	192	284	218	161
202	321	150	158	197	227	317	224	283	354
201	225	228	223	170	208	175	211	180	197
156	237	215	224						

to state unequivocally that there could not be a difference, and the result certainly cannot simply be transferred to healthy persons; published observations[11] have established a difference in such a case. Further examples of the use of the z-test were given by Riedwyl.[2]

1.3.4.2.1.3. t-test

Prefatory remark. The *t*-test is still used very frequently today, although the necessary assumptions are substantially more limiting (normal distribution, approximately equal variability of the two samples) and the tables more extensive than for the methods already mentioned. It is recommended that it be replaced by the z-test for $n \geq 30$ and the Wilcoxon test for $n < 30$.

Application. Testing of two means of samples from a normally distributed population. The variances of the populations must be about the same.

Test statistic.

$$t = \frac{\bar{x}_1 - x_2}{\sqrt{\dfrac{(n_1 - 1)s_1^2 + (n_2 - 1)s_2^2}{n_1 + n_2 - 2}}} \cdot \sqrt{\frac{n_1 n_2}{n_1 + n_2}} \tag{17}$$

For equal sample sizes ($n_1 = n_2 = n$):

$$t = \frac{x_1 - \bar{x}_2}{\sqrt{s_1^2 + s_2^2}} \cdot \sqrt{n} \tag{18}$$

Decision. The calculated statistic t is compared with t_0, the value found from the table for a level of significance α ($n_1 + n_2 - 2$ 'degrees of freedom'). If $|t| \geq t_0$, then it may be assumed with the corresponding certainty that the means are different (one-sided tests: $t \geq t_0$ or $t \leq -t_0$).

1.3.4.2.2. Other tests

Space does not permit the consideration of tests other than those for the testing of two means. For these the literature referred to in the following survey should be consulted.

1. Testing the standard deviations of two samples: *F*-test;[5] variance analysis.[12]
2. One-sample case: comparison of the mean of a sample with a theoretical value μ: procedures similar to those in Section 1.3.4.2.1.

   $n > 30$: z-test, $z = \dfrac{\bar{x} - \mu}{s_x} \cdot \sqrt{n}$.[2] $n \leq 30$: Wilcoxon sign serial sum test.[2,4]

3. One-sample case: tests on deviations from the random order, trend tests.[8]
4. One-sample case: testing for a particular distribution: Kolmogorov–Smirnov test[7] (however, it is easier to test for a normal distribution with probability graph paper).
5. Multi-sample case: testing of means from several samples: variance analysis.[3,5,12]
6. Qualitative data: checking frequencies with the χ^2 test (one-, two- and multi-sample case).[7]

1.3.4.3. Paired data

The considerations in Section 1.3.4.2 concerned independent samples where there was only one measurement per test subject. Another case of practical importance is that where two samples are linked as a pair, usually in such a way that there are two measurements deriving from each individual. Such samples have already been discussed generally in the consideration of regression and correlation (Section 1.3.3). The main interest here is the procedure of investigating individual test subjects *before and after* a specific treatment (example: measurement of the glucose concentration before and after a meal). When an individual acts as his own control, as it were, specially designed tests are possible. Their principle consists in taking the difference of each pair of values and testing this difference against 0 by the methods of the one-sample case.

Another application for paired data, which will not be given closer attention here, is the testing of correlation coefficients against 0 (cf. references, e.g. ref. 5).

1.3.4.3.1. Sign test

Application. Rapid checking of the differences derived from paired measurements. No assumptions regarding the distribution are necessary. It is not very sensitive. The test may also be used, and is particularly useful, when there are no quantitative data but it is possible to make assertions of the type 'improved' or 'worsened'.

Test statistics. The arithmetic signs of the differences, d_i, are counted (if $d_i = 0$ then the number of trials is reduced by 1) and the test statistic H_- put equal to the number of minus signs (H_+ could equally well be used).

Decision. H_- is compared with the appropriate tabulated values for the binomial distribution (ref. 4, pp. 105 and 106). If H_- is equal to the upper or lower limiting value, or lies outside the range, then it may be concluded that the means are different (certainty $1 - \alpha$).

Example 9. Inorganic phosphorus before and after breakfast (Table 15).

Table 15. Inorganic phosphorus before and after breakfast (personal measurements).

Patient number	Before	After	Difference
1	3.3	3.6	−0.3
2	4.0	4.2	−0.2
3	3.1	3.5	−0.4
4	3.0	3.7	−0.7
5	2.8	3.4	−0.6
6	2.5	2.8	−0.3

There are six minus signs, bringing H_- to the upper limit for rejection ($n = 6$, $\alpha = 0.05$, two-sided).

Result. After breakfast inorganic phosphorus is significantly higher than before breakfast ($\alpha = 5\%$).

1.3.4.3.2. Wilcoxon sign serial sum test

Application. Testing the differences of paired measurements to determine whether the means of the two samples are different. For $n \leqslant 50$ or thereabouts, otherwise the calculations become very lengthy.

Test statistic. The test statistic T_- is obtained by forming the pair differences and arranging them in order of their absolute size, then summing the serial numbers (cf. Section 1.3.4.2.1.1) associated with negative differences.

Decision. Consult the significance thresholds in the table (ref. 4, p. 128). Decision for a one- or two-sided case if T_- is equal to, or exceeds, the stated limits [e.g. the assumption that mean 1 is smaller than mean 2 (one-sided case) is feasible if T_- is equal to or exceeds the upper limit].

Example 10. If the differences in Table 15 are arranged according to their size, then value no. 2 is the smallest (serial number 1) and value no. 4 is the largest (serial number 6). The differences for numbers 1 and 6 are equally large; they are both assigned the serial number 2.5. Thus, the serial numbers are 2.5, 1, 4, 6, 5, 2.5. Since all the differences are negative, T_- is calculated as the sum of all the serial numbers: $T_- = 21$. The value is significant with $2\alpha = 5\%$.

1.3.4.3.3. z-test

Application. Testing of means for paired samples. Size n ($n > 50$).

Test statistic. Calculation of the pair-differences, d_i, their mean, $\bar{d}$, and variance, s_d^2:

$$z = \frac{\bar{d}}{s_d} \sqrt{n} \tag{19}$$

Decision. Checking z against the limit z_0 of the standard normal distribution appropriate to the level of significance α (Table 13; decision as given there).

1.3.5. CONFIDENCE AND TOLERANCE LIMITS

1.3.5.1. Confidence interval

If a series of measurements is carried out in order to determine the 'true value' of a quantity in the population, the measured data are not expected to reproduce the value exactly. In fact, the data will be expected to vary with each subsequent series of measurements. Statistically, this means that a sample of size n will not accurately reflect the circumstances obtaining in the population, not even if it is considered to be representative (cf. Section 1.3.4.1). Nevertheless it is natural to assume that the representative sample gives an approximation to the circumstances in the total population and the larger the sample size n, the better will that approximation be.

If, then, the true value cannot be determined exactly, what limits may be fixed to include the sought value with reasonable certainty? The answer is provided by the so-called 'confidence interval'. A 95% *confidence interval* for a true value (a 'parameter') of the population is a range within which the parameter is expected to lie with 95% certainty (for exact interpretation, cf. ref. 4). Naturally a 99% interval, or any other percentage interval, may be defined analogously.

In general, it is hardly possible to specify confidence intervals for parameters of the population unless the distribution of the measured variables is known. Fortunately, the situation with the means is more favourable. Once again the central limit theorem of probability theory is invoked, provided that the sample size is large enough (i.e. $n \geq 30$).[10] Small samples require a different treatment.[3,4]

For $n \geq 30$ the following confidence intervals are valid for the true mean, μ, of a variable x:

$$95\% \; interval: \; \bar{x} - 1.96 \frac{s_x}{\sqrt{n}} \leq \mu \leq \bar{x} + 1.96 \frac{s_x}{\sqrt{n}} \tag{20}$$

$$99\% \; interval: \; \bar{x} - 2.58 \frac{s_x}{\sqrt{n}} \leq \mu \leq \bar{x} + 2.58 \frac{s_x}{\sqrt{n}} \tag{21}$$

If, in the case of the 95% interval, the factor 1.96 is replaced by 2, the resulting formula is 'mean ± twice the standard error of the mean'. However, the intervals are *valid only for the mean*. A different procedure is necessary to establish limits for individual values of x (cf. Section 1.3.5.2).

The more general equation for a P% confidence interval for the mean, μ, of a population is

$$\bar{x} - z \cdot \frac{s_x}{\sqrt{n}} \leq \mu \leq \bar{x} + z \cdot \frac{s_x}{\sqrt{n}} \tag{22}$$

Here, z is that number for which the standard normal distribution assumes the value $(100 - P)$% (cf. Table 13). This equation shows that the interval becomes increasingly narrower with increasing n, which is intuitively clear.

Confidence intervals for other parameters will not be considered here. Equations for confidence intervals for the standard deviation are to be found in the Geigy tables,[4] for example (however, an approximately normal distribution has to be assumed in this case). In principle, every estimate of a parameter should be accompanied by a confidence interval, otherwise the result is not exactly reliable.

1.3.5.2. Tolerance limits

As mentioned above, where limits are to be defined for an individual value of x rather than the mean, a different procedure is necessary. The problem here is: what limits can reasonably be assumed to include 95% (or another percentage, say 90% or 99%) of the values of x? The question is answered by means of the so-called 'tolerance limits' and the word 'reasonably' is then replaced by a statement of probability.

There are two kinds of tolerance limits:[3,4]

1. 95% *tolerance limits without confidence probability* are limits which include *on average* 95% of the values of x (when the measurement, n values, is repeated very often);
2. 95% *tolerance limits with confidence probability* 0.99 are limits which, in 99 cases out of 100 (on average), include *at least* 95% of the values of x (i.e. there is a 0.99 probability that at least 95% of the cases are included).

For both kinds of tolerance limit it is assumed that 2.5% of the values lie beneath the lower and above the upper, limiting value. Analogous definitions hold for other percentages and confidence probabilities. Clearly, the second type of tolerance limit will yield broader intervals; it is the 'more conservative'. In practice, the tolerance limits may be calculated by the formula 'mean ± 2 standard deviations', but *only if two assumptions are made*: (a) the distribution of x must be approximately normal (the central limit theorem is not valid for percentiles!), and (b) the size of the sample must be sufficiently large ($n \geqslant 100$).

Provided that these conditions are fulfilled, the interval

$$\bar{x} \pm 1.96 s_x \tag{23}$$

may be selected as the 95% tolerance range (without confidence probability). For the 99% range, the z-factor for s_x has the value 2.58 (instead of 1.96); the x-factors corresponding to other percentages may be found from tables.

If the assumption of an approximately normal distribution is unwarranted, then attempts may be made to obtain normally distributed data by means of a suitable *transformation* of x; the best known transformation is $y = \log x$. The simplest way to test for normality is graphically (probability graph paper). The tolerance limits for x are obtained by back-calculation from the y values. (If the logarithms of a variable are normally distributed, then the variable itself is said to have a log-normal distribution.)

It is always safer to calculate tolerance limits without making assumptions about the distribution of x ('distribution-free' tolerance limits). The principle here is to take extreme values of the sample (e.g. the smallest and largest values but one, depending on the sample size) as tolerance limits. Reed *et al.*[13] showed that these limits are scarcely any broader than those obtained by assuming a normal distribution. The most important use of tolerance limits in clinical chemistry is for the determination of the reference ranges for particular analytical methods; in general, these should be evaluated as 95% tolerance limits without confidence probability.[4]

1.3.6. PROBLEMS PECULIAR TO CLINICAL CHEMISTRY

1.3.6.1. Comparison of two methods

A question which always arises in the introduction of a new analytical method is whether the new method is sufficiently reliable. The new procedure must be carefully tested. Here there are three questions to be answered:

Precision. The precision of a method manifests itself in the reproducibility of the results. One measure is the day-to-day reproducibility,[14] expressed as standard deviation or (better) coefficient of variation. Pooled serum is the suggested test material.[15]

Accuracy. To what extent does the method give the true value? This question can be resolved by means of a very reliable reference method or by weighing out known amounts of substance. The user can arbitrarily decide whether a particular deviation of the method from the true values is still acceptable (it is also possible to apply a significance test to the two means).

Equivalence of the new method with a reference method. This has to be decided by means of the analysis of patient samples. According to Barnett and Youden,[14] samples from at least 40 patients should be analysed and, of these, lest temporal variations should be neglected, not more than 5 on any one day. Each sample is to be investigated by both methods. The standard deviation of the differences gives the equivalence of the two methods. A graphical representation of the equivalence may be obtained by plotting the values from the two methods in a coordinate system (x = reference method; y = new method) and determining the regression line: ideally the line should have a slope of 45° and pass through the origin (bisecting the angle). Since this graphical method is a simple way of showing the relationship of the two methods to each other, it should never be neglected.

The user should decide for himself what deviations from 45° and what intercept he can tolerate. Of course, the coefficient of regression can also be tested statistically to find if it deviates from 1 ($\tan 45° = 1$), but this statistical statement is often meaningless for practical purposes.[16]

A special problem attaches to 'stray points', i.e. values that are very wide of the mark. There are statistical methods for testing whether a value of x_0 is a stray and can therefore be discarded (simplest: outside 4 standard deviations, calculated without x_0).[17] However, in discarding the stray it is important to check whether it is due to an error in the analysis, for if a method produces too many stray values this is a sure indication of its unreliability!

A more detailed introduction to the comparison of methods is given by Barnett and Youden.[14] Critical studies on the comparison of methods, based on simulated data, were presented by Westgard and Hunt.[16]

In the comparison of methods the graphical presentation of results and the determination of the line of regression (it may be sufficient to gauge the line by eye) is the most important step. In particular, the line of regression should be checked for linearity and tests should be made for the presence of 'suspected strays'. Statistics cannot decide whether the new method is acceptable or not. It only provides estimates of the type and size of the errors as most important aids to making the decision.

1.3.6.2. Estimating the sample size

In contrast to retrospective studies, which deal with the analysis of numerical data already collected, science frequently poses problems where the facts are to be clarified by an experiment. For this kind of prospective study the experiment can be carefully planned. Apart from establishing the sample size, the main object of *experiment planning*[18] is the whole organization of the experiment and the determination of the statistical methods of evaluation in advance. In this phase of the planning it is recommended that a statistician be consulted.

Here only the estimation of the necessary sample size will be treated in

58

detail. The question may only be answered provided supplementary information is available, namely:

1. The extent of the difference between two means (or, in the one-sample case, between mean and true values) which is to be detected by the experiment: $d = \mu_1 - \mu_2$. d must be determined by the user, and cannot be calculated statistically.
2. An estimate of the variability of the expected results, e.g. from preliminary experiments or from the literature (variance, s^2; or standard deviation, s).

As is always the case in statistics, the only statements that can be made are statements of probability. Two probabilities for errors must be considered.

Type I error. The probability of finding an apparent difference where in fact there is none (probability α).

Type II error. The probability of failing to discover a difference which is in fact present (probability β).

The estimate of the sample size n depends on the values selected for these two probabilities. Here it is necessary to distinguish whether deviations are to be detected on both sides (two-sided, commoner case) or only on one side (one-sided, rarer case). The equation is

$$n = k \cdot \frac{s^2}{d^2} \tag{24}$$

The factor k for the relevant α- and β-values can be found from Table 16.

For the two-sample case the sample sizes calculated according to equation 24 are to be doubled:

$$n = 2k \cdot \frac{s^2}{d^2} \tag{25}$$

This equation is applicable provided that the standard deviations of the two samples are approximately equal ($s_1 = s_2 = s$). For all other cases (α, β defined differently, unequal standard deviations) the reader is referred to the literature.[2]

Table 16. k-factor for estimating the sample size, calculated by an equation from Riedwyl[2].

Probability of detecting a real difference $1 - \beta$	Reliability (probability of not assuming a false difference) $1 - \alpha$	k-factor Two-sided	One-sided
0.95	0.95	12.995	10.823
0.95	0.99	17.814	15.770
0.99	0.95	18.372	15.770
0.99	0.99	24.031	21.647

It should again be emphasized that the derivation of equations 24 and 25 assumes normal distribution. Since in principle what is concerned here is the comparison of means, the assumption of normality can be dropped for $n \geqslant 30$.

Example 11. Barnett and Youden[14] compared two methods for the determination of glucose. Ten determinations with the two methods show a standard deviation of 9.46. What size sample must be chosen for the main study in order to detect differences of 5 mg per 100 ml with a probability of 95% at an accuracy of 5%?

Table 6 shows the relevant k-factor (two sided) to be 12.995. This gives

$$n = 12.995 \cdot \frac{9.46^2}{5^2} \approx 47$$

This value is in good agreement with the estimate of at least 40 measurements for comparison of methods, as suggested by Barnett and Youden[14] on other grounds (the fact that it is somewhat higher may be due to evaluating s from only 10 values). It should be noted that the comparison of methods is a one-sample case (testing the difference from 0).

A different procedure is adopted for estimating the sample size if tolerance limits and confidence probabilities are to be specified. In general, it will be found expedient to use the 'distribution-free' approach (i.e. without making assumptions about the distribution of the sample). For problems of this type, the reader is referred to the literature (ref. 4, p. 128), where the sample size can be taken straight from the table.

1.3.6.3. Determination of reference values

The determination of reference ranges is a recurrent problem in clinical chemistry. The nature of the problems involved is examined elsewhere (Section 1.4). Here we shall confine our attention to the statistical aspects.

There is no generally recognized method for ascertaining reference ranges. However, the procedure described here and in the following example ought to yield values adequate for practical purposes. Essentially it consists of three steps:

Step 1: Testing the collected data for homogeneity. In practice it is hardly possible to control all the causes of variation. Nevertheless, the graphical presentation of the values is indispensable (histogram); it can provide the first indications on how to group the data. The data should at least be tested for age and sex differences, if necessary by means of a significance test (Section 1.3.4). If no differences can be established the reference range can be determined for the set of data taken as a whole, otherwise it must be separated into its individual groups (in which case steps 2 and 3 must be carried out several times).

Step 2: Testing for normal distribution of the data. The histograms afford a first assessment of the general shape of the distribution. In order to test whether the values are normally (or log-normally) distributed, the percentage cumulative frequencies are calculated and plotted on probability graph paper (for a log-normal distribution logarithmic paper should be used).

Step 3: Determination of the reference range. In general, reference ranges should be determined as 95% *tolerance limits* without confidence probabilities (ref. 4, p. 156). If no excessive deviations from a normal distribution are revealed by the histograms or the probability graph paper plot, the 95% range can be calculated from the mean and the standard deviation ($\bar{x} \pm 1.96 s_x$. In the case of a log-normal distribution equation 23 is applied to the logarithms of the data and the limits of the range found by back-calculation. In other cases the 95% range can be determined without recourse to parameters by means of 'quantiles' (for this procedure, see ref. 4, pp. 128 and 161). Without going into the details of the method, this is an especially simple procedure for the practician: the size of the sample is fixed in advance at *200* values; after the data have been arranged in order of their size, the interval between the 6th and the 195th values is selected as the reference range. The range then contains just the middle 95% of the values. Fixing the sample size at 200 values does not impose too severe a limitation because the determination of the reference range is always a prospective procedure, hence the sample size can be chosen as desired. Of course, this applies only to homogeneous data; it would be going too far to require 200 values for each of several sub-groups. In this case the methods to use are those given in the Geigy tables (ref. 4, p. 128).

A few words more about collecting the data: for determining the reference ranges enough data should be collected to enable any statements which are to be made to fall within the region of statistical certainty. This is why 200 values are recommended, as mentioned above. Further, if statements are to be made about a total population, it is obvious that only one value per test person may be included in the step 3 calculations.

Example 12. In the context of a study for the determination of reference ranges at the Berne Isolation Hospital,[19] data were collected from a healthy Berne standard population and evaluated. The data for this study included a sample of 200 uric acid values for young men. The data are presented in Table 17, arranged in order of their size.

Limiting the study to young men excluded the possibility of age and sex differences, so it may be assumed that the data are homogeneous. This is confirmed by the histogram (computer output, Fig. 15).

Since Fig. 15 exhibits no marked deviations from the normal distribution, equation 23 was applied first; for $\bar{x} = 4.697$ and $s_x = 1.002$, this gives 95% range according to equation 23 = 2.73–6.66.

For the parameter-free determination according to the procedure described in step 3, the 6th value (2.79) and 195th value (6.74) were chosen, yielding 95% range, parameter-free = 2.79–6.74.

Table 17. Uric acid values from 200 young men (mg per 100 ml), chosen from the data of Käch et al.[19]

2.52	2.63	2.63	2.68	2.74	2.79	2.96	2.96	3.01	3.07	3.18	3.18	3.18	3.23	3.23	3.34
3.34	3.34	3.40	3.40	3.45	3.45	3.45	3.51	3.51	3.51	3.51	3.67	3.67	3.67	3.67	3.67
3.73	3.73	3.73	3.73	3.78	3.78	3.78	3.78	3.78	3.84	3.84	3.89	3.89	3.89	3.89	3.95
3.95	3.95	3.95	4.00	4.00	4.00	4.00	4.06	4.06	4.06	4.06	4.06	4.06	4.06	4.11	4.11
4.17	4.17	4.17	4.22	4.26	4.28	4.28	4.28	4.33	4.33	4.33	4.33	4.33	4.33	4.39	4.39
4.39	4.39	4.44	4.50	4.50	4.50	4.50	4.50	4.50	4.50	4.55	4.55	4.55	4.61	4.66	4.66
4.66	4.66	4.66	4.66	4.72	4.72	4.72	4.72	4.72	4.72	4.77	4.77	4.77	4.77	4.77	4.83
4.83	4.83	4.83	4.83	4.83	4.88	4.88	4.88	4.94	4.94	4.94	4.94	4.94	4.99	4.99	5.05
5.05	5.10	5.10	5.10	5.10	5.10	5.16	5.16	5.21	5.21	5.21	5.22	5.27	5.27	5.27	5.32
5.32	5.32	5.38	5.38	5.38	5.38	5.43	5.43	5.43	5.49	5.49	5.54	5.54	5.59	5.59	5.59
5.59	5.64	5.64	5.64	5.70	5.70	5.70	5.70	5.75	5.75	5.75	5.81	5.86	5.92	5.92	5.92
5.92	5.97	5.97	6.03	6.03	6.03	6.08	6.14	6.14	6.14	6.14	6.19	6.19	6.25	6.25	6.25
6.36	6.36	6.74	6.80	6.96	7.02	7.24	7.62								

HISTOGRAM OF VARIABLE   1 URIC ACID   SYMBOL X   COUNT 200   MEAN 4.697   ST. DEV. 1.002

INTERVAL NAME	Histogram	FREQUENCY INT.	CUM.	PERCENTAGE INT.	CUM.
2.50000		0	0	0.0	0.0
3.00000	xxxxxxxx	8	8	4.0	4.0
3.50000	xxxxxxxxxxxxxxx	15	23	7.5	11.5
4.00000	xxxxxxxxxxxxxxxxxxxxxxxxxxxxxxxx	32	55	16.0	27.5
4.50000	xxxxxxxxxxxxxxxxxxxxxxxxxxxxxxxxxxx	35	90	17.5	45.0
5.00000	xxxxxxxxxxxxxxxxxxxxxxxxxxxxxxxxxxxxx	37	127	18.5	63.5
5.50000	xxxxxxxxxxxxxxxxxxxxxxxxxxxxx	28	155	14.0	77.5
6.00000	xxxxxxxxxxxxxxxxxxxxxxxx	24	179	12.0	89.5
6.50000	xxxxxxxxxxxxxx	14	193	7.0	96.5
7.00000	xxxx	4	197	2.0	98.5
7.50000	xx	2	199	1.0	99.5
LAST	x	1	200	0.5	100.0

Scale: 5   10   15   20   25   30   35   40   45   50   55   60   65   70   75   80

Fig. 15. Histogram of uric acid values. Computer calculation using a program from Dixon.[20]

It is clear from this example that the reference ranges calculated by different methods are in good agreement, although the second method makes no kind of assumption about normal distribution. In particular, the width of the interval is almost equal, namely 3.93 in the first case and 3.95 in the second. The slight shift of the first range towards lower values is explicable in terms of the slight skewness of the histogram (Fig. 15): more values lie to the left of the mid-value than to the right. The range evaluated by equation 23 corresponds to the interval between the 5th and 194th values in Table 17. On theoretical grounds preference is to be accorded to the second range, since it is obtained without assumptions concerning normality, but for practical purposes it is reasonable to assume a reference range of 2.75–6.70.

References

1. Wald, A., *Statistical decision functions*, Wiley–Interscience, New York, 1950.
2. Riedwyl, H., *Angewandte mathematische Statistik in Wissenschaft, Administration und Technik*, Haupt, Berne, 1975.
3. Hays, W. L., *Statistics for the social sciences*, Holt International Edition, London, 1974.
4. Documenta Geigy, *Wissenschaftliche Tabellen*, 7th ed., Geigy, Basle, 1968.
5. Weber, E., *Grundriss der biologischen Statistik*, Fischer, Jena, 1961.
6. Natrella, M., *Experimental statistics*, National Bureau of Standards, Handbook 91, Washington, 1963.
7. Siegel, S., *Nonparametric statistics for the behavioral sciences*, McGraw-Hill/Kogakusha, Tokyo, 1956.
8. Sachs, L., *Statistische Auswertungsmethoden*, Springer, Berlin, 1969.
9. Benninghaus, H., *Deskriptive Statistik*, Teubner, Stuttgart, 1974.
10. Sahner, H., *Schliessende Statistik*, Teubner, Stuttgart, 1971.
11. Weisshaar, D., Enzymatische Cholesterinbestimmung, Norm-(Referenz)bereiche, *Medsche Welt, Stuttg.*, **26**, 940 (1975).
12. Linder, A., *Statistische Methoden*, Birkhäuser, Basle, 1964.
13. Reed, A. H., Henry, R. J., and Mason, W. B., Influence of statistical method used on the resulting estimate of normal range, *Clin. Chem.*, **17**, 275 (1971).
14. Barnett, R. and Youden, W. J., A revised scheme for the comparison of quantitative methods, *Am. J. clin. Path.*, **54**, 454 (1970).
15. Burnett, R. W., Accurate estimation of standard deviations for quantitative methods used in clinical chemistry, *Clin. Chem.*, **21**, 1935 (1975).
16. Westgard, J. O. and Hunt, M. R., Use and interpretation of common statistical tests in method-comparison studies, *Clin. Chem.*, **19**, 49 (1973).
17. Dixon, W. J. and Massey, F. J., *Introduction to statistical analysis*, McGraw-Hill, New York, 1969.
18. Fischer, R. A., *The design of experiments*, 7th ed., Oliver & Boyd, Edinburgh, 1960.
19. Käch, O., Sturzenegger, M. H., and Zbinden, M. A. E., *Hämatologische Richtwerte einer gesunden Berner Standardpopulation*, Inaug.-Diss., Universität Bern, 1976.
20. Dixon, W. J., *BMDP—biomedical computer programs*, University of California Press, Berkeley, 1975.

1.4. REFERENCE VALUES

1.4.1. INTRODUCTION

In biology, the concept of normality can only be defined arbitrarily. The definition is conditioned by the object under investigation on the one hand and the nature of the problem on the other. In clinical chemistry, as experience shows, the transition from health to sickness is a smooth one. Many illnesses are not recognizable in the asymptomatic, preclinical stage. There is no such thing as absolute health. For this reason it is no longer customary to speak of 'normal values' holding for an ideal, absolutely healthy population. The term *reference values* is proposed. These are the values for a particular quantity obtained in individuals with a defined state of health.[1]

As a rule, the 'reference range' comprises 95% of the found values, limited above and below. The value found for an individual is compared with these values in order to interpret it in this context.

1.4.2. FACTORS INFLUENCING REFERENCE VALUES

Every biological quantity is a variable. However, its variability is conditioned by certain laws and in fixing the reference range a knowledge of these is indispensable. Among the many factors which influence biological quantities, three groups may be broadly distinguished: genetic, extra-individual, and intra-individual factors.

1.4.2.1. Genetic factors

Distant relationship amongst human beings manifests itself in the existence of different races. In recent years biochemical anthropology has been able to demonstrate numerous metabolic defects which are restricted to particular ethnic groups.

Thus, sickle-cell anaemia, which is due to the presence of an abnormal haemoglobin (HbS) in the erythrocytes, occurs almost exclusively amongst black people. The frequency of the anomaly (in such populations) varies between 2 and 50%. The persistence of the gene can only be explained by assuming it to confer an improved life expectancy among heterozygotes (heterosis). In fact, it appears that the presence of the abnormal haemoglobin offers a certain protection against malarial infection. Also, the regions in which sickle-cell anaemia is common correspond closely to the regions where malaria is endemic.

As a consequence of increasing miscegenation, unequivocally racial differences are only observed in isolation. On the other hand, it must be remembered that interbred populations are always liable to sporadic occurrences of rare anomalies. Today there are about 2000 known metabolic anomalies, transmitted for the most part by recessive genes and hence only rarely erupting in a manifest illness, through the conjunction of two genes. Estimates permit the conjecture that every person is heterozygotic for about 10 such genes. If the heterozygotic condition itself results in slight metabolic disorders, as is known to be the case with numerous illnesses, then the reference values are subject to an error which is difficult to eliminate. An example of this is provided by uric acid values in families prone to gout.

Gout is undoubtedly a hereditary illness. In individual sufferers it leads to acute attacks with hyperuricaemia. If the members of the family are investigated, uric acid values above 6 mg per 100 ml are always found in a certain percentage of apparently healthy individuals. These patients with 'familial hyperuricaemia' appear in isolation as healthy individuals and so lead to distortion of the reference value.

Other common hereditary diseases where similar relationships ought to apply, are diabetes mellitus, muscular dystrophy, and cystic fibrosis (mucoviscidosis).

1.4.2.2. Endogenous (intra-individual) factors

Very many metabolites show a marked *age dependence*. Special interest centres on early childhood on the one hand and the alterations in advanced age on the other. The causes of this age dependence are only partially known; not only are truly endogenous factors involved but the differing diets and life-styles also play a part.

Next there are the *sex differences* to be considered. It is known that in animals, large sex differences are shown not only by systems directly controlled by hormones (sex characteristics) but also by enzymes which appear to be hormone-independent. Up to now, few comparable studies have been made with human beings. Here we must also mention the effects, almost always observable, of *gravidity* on serum components, as well as possible dependences on the menstrual cycle.

Finally, pathological conditions of an asymptomatic nature (e.g. anicteric hepatitis) should be counted as intra-individual factors. Up to now, variations showing a diurnal rhythm have been observed predominantly with hormones (e.g. cortisone), but other metabolites also appear to be subject to this kind of variation.[2-4]

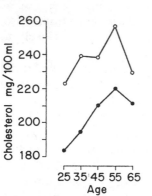

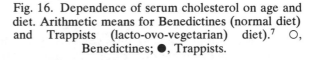

Fig. 16. Dependence of serum cholesterol on age and diet. Arithmetic means for Benedictines (normal diet) and Trappists (lacto-ovo-vegetarian) diet.[7] O, Benedictines; ●, Trappists.

1.4.2.3. Exogenous (extra-individual) factors

Foremost among the exogenous factors is the *diet*.[5,6] An illuminating example is provided by a comparison of serum cholesterol values among monks from two different orders, having similar life-styles but differing diets.[7] The Trappist diet is exclusively lacto-ovo-vegetarian, whereas the

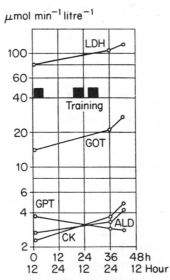

Fig. 17. Behaviour of some serum enzymes following physical training. LDH = lactate dehydrogenase; GOT = glutamate–oxaloacetate transaminase; GPT = glutamate–pyruvate transaminase; ALD = aldolase; CK = creatine kinase.[8]

Benedictines have an unrestricted diet. The results shown in Fig. 16 speak for themselves.

Bodily activity can likewise have a significant effect on serum components.[10] This is true, for example, of all serum enzymes deriving from the muscular system.[8,9] Fig. 17 shows the increase in creatinekinase, aldolase, and glutamate–oxaloacetate transaminase during bodily exertion. It is striking that glutamate–pyruvate transaminase, which originates predominantly in the liver, remains unchanged here. Even *psychological factors* can lead to great changes in serum concentrations. For example, we know that pepsinogen can increase to extremely high values in situations of stress and consequently, probably incorrectly, it was designated as aetiologically responsible for ulcers. This is also known of uric acid. The intake of *pharmaceuticals* also influences the concentration of various serum components.

References

1. Recommendations concerning the collection of reference values in clinical chemistry and activity reports, *Scand. J. clin. Lab. Invest.*, **35**, suppl., 144 (1975).
2. Zurbrügg, R. *Hypothalamic pituitary adrenocortical regulation*, Monogr. Paediat., Vol. 7, Karger, Basle, 1976.
3. Stamm, D., *Tagesschwankungen der Konzentration diagnostisch wichtiger Blutbestandteile*, Hab.-Schr., Universität Giessen, 1966.
4. Statland, B. E., Winkel, P., and Bokelund, H., Factors contributing to intra-individual variation of serum constituents. I. Within-day variation of serum constituents in healthy subjects, *Clin. Chem.*, **19**, 1374 (1973).
5. Statland, B. E., Winkel, P., and Bokelund, H., Factors contributing to intra-individual variation of serum constituents. II. Effects of exercise and diet on variation of serum constituents in healthy subjects, *Clin. Chem.*, **19**, 1380 (1973).
6. Zöllner, N., Auswirkungen des Ernährungszustandes, in Lang, Rick and Roka, *Optimierung der Diagnostik*, Springer, Berlin, 1973.
7. Barrow, J. G., Quinlan, C. B., Cooper, G. R., Whitner, V. S., and Goodloe, M. H. R., Studies in atherosclerosis. III. An epidemiologic study of atherosclerosis in Trappist and Benedictine monks; a preliminary report, *Ann. intern. Med.*, **52**, 368 (1960).
8. Colombo, J. P., Richterich, R., and Rossi, E., Serum-Kreatin-Phosphokinase: Bestimmung und diagnostische Bedeutung, *Klin. Wschr.*, **40**, 37 (1962).
9. Siest, G. and Galteau, M. M., Variations of plasmatic enzymes during exercise, *Enzyme*, **17**, 179 (1974).
10. Röcker, L., Schmidt, H.-M., and Motz, W., Der Einfluss körperlicher Leistungen auf Laboratoriumsbefunde im Blut, *Ärztl. Lab.*, **23**, 351 (1977).

1.4.3. DETERMINATION OF REFERENCE VALUES

As mentioned already, absolute health does not exist. There are dispositions for pathological developments in every individual. If reference values are to be ascertained from a so-called 'healthy' population, then it is necessary to define its state of health so as to know whether an individual

68

is to be included in, or excluded from, the population. Purely statistical methods are no longer adequate for this purpose. In consequence, various specifications have been drawn up which influence the reference values in both the broader and narrower sense and which should be borne in mind when the values are being determined.

1.4.3.1. Specifications

The Scandinavian Society for Clinical Chemistry and Clinical Physiology[1] has worked out a set of such specifications and they have been summarized by Sunderman.[2] The following points which influence reference values should be noted:

1. Characteristics of the reference population and its choice: sex, age, occupation, body-mass, height, habitus, genetic and ethnic background, geographical location, number of individuals questioned, methods of recruitment, basis for exclusion from or inclusion in the collective.
2. The environmental and physiological conditions under which the specimens were obtained: stress, physical activity, posture, diet (including consumption of alcohol and coffee), length of fasting, smoking habit, hospitalization vs. non-hospitalization, endocrine and reproductive status, including menses, pregnancy, intake of oral contraceptives, and medicines.
3. Collecting technique, type and storage of specimens: arterial, capillary or venous blood, use of tourniquet, anticoagulants, the time of collection, interval between collecting the blood and the separation of the plasma, transport of the specimen, temperature and duration of storage prior to analysis, thawing, freezing, haemolysis, urine fraction, or 24-hour urine.
4. Reliability of the method used for analysis: accuracy, precision, quality control, practicability.
5. The statistical methods employed in determining the reference range for application must be stated and so must the manner of fixing the reference range (e.g. 2.5–97.5 percentiles, $\bar{x} \pm 2s$ range).

These specifications ought to be observed as far as possible even though the requirements are not easily realized in practice. For this reason it is all the more worth striving to optimize and standardize the analytical methods so that the variability is not increased further.

Reference values do not represent any absolute values. The reference range for a particular clinical–chemical parameter can be very wide in the so-called 'healthy' population investigated, in comparison with serum measurements of this parameter in an individual considered in isolation. A particular value for an individual can lie outside his 'intra-individual' range, and consequently be of pathological significance, although it falls within the reference range of the so-called 'healthy' population. Although attempts have been made[3] to set up individual reference ranges, surely this will scarcely be possible in practice.

Consequently, in interpreting laboratory results in comparison with a particular reference range one should be aware of its relative significance.

1.4.3.2. Evaluation of the data obtained

As a first step it is wise to test the data for *sex* and *age differences*. Next there is the question of the *type of distribution* of the values. Contrary to the widely held view, the presence of a normal distribution is the exception rather than the rule. For realizing the distribution the graphic presentation of the values has proved useful (Section 1.3.2.5). If a normal distribution is present the evaluation of the data is simple. The arithmetic mean and the standard deviation are calculated (Section 1.3.2.3). In accordance with the definitions customary in biology, values within $\bar{x} \pm 2s$ are considered to lie in the reference range. If the distribution is complex, attempts may be made to evaluate the reference range by a parameter-free method. A simple, albeit crude, aid is to plot the percentage frequencies on linear or logarithmic probability graph paper and to evaluate the reference range graphically. An example of the determination of reference values is included in Section 1.3.6.3.

References

1. Recommendations concerning the collection of reference values in clinical chemistry and activity report, *Scand. J. clin. Lab. Invest.*, **35**, suppl., 144 (1975).
2. Sunderman, F. W., Jr., Current concepts of 'normal values', 'reference values', and 'discrimination values' in clinical chemistry, *Clin. Chem.*, **21**, 1873 (1975).
3. Harris, E. K., Some theory of reference values. I. Stratified (categorized) normal ranges and a method for following an individual's clinical laboratory values, *Clin. Chem.*, **21**, 1457 (1975).

1.5. RELIABILITY OF LABORATORY METHODS AND QUALITY CONTROL

1.5.1. INTRODUCTION

A knowledge of the reliability of the investigations carried out in the clinical chemical laboratory has a significance for practical medicine that ought not to be under-estimated. With increasingly refined diagnosis, the doctor relies more and more on the results of the laboratory investigations.[1] Consequently, a large part of the professional responsibility is shifted from the doctor to the laboratory personnel. Today, the decision as to whether a result is pathological or not can only be made in the laboratory. It is based on a knowledge of the reference value on the one hand, and on a knowledge of the reliability of the individual methods on the other. Knowledge of reliability implies periodic, e.g. yearly, testing of the precision of each individual technique, and continuous, daily monitoring of the method by means of a control system.

Several methods for testing reliability have been known to analytical chemistry for years,[2] but most of these suggestions fail to meet the particular requirements of the clinical laboratory. In particular, they pay too little attention to the following considerations:

1. Working conditions are generally much less favourable in the clinical laboratory than in the analytical laboratory (determinations at any hour of the day and night; the work is carried out by diversely trained assistants; laboratory apparatus and equipment is often inadequate and antiquated).

2. The described methods of testing and the statistical evaluation of the results are so complicated that the instructions are insufficient to enable an assistant to carry out a reliability test independently—and yet those who do the work are the very persons who should be capable of testing their methods and continually monitoring them, without any interference from the clinical chemist.

3. In recent years, centralization of laboratories has coincided with an unprecedented increase in analyses; as a result, it often happens that a very large number of analyses have to be carried out by the same person. This produces effects due to fatigue ('four-o-clock phenomenon') and hence leads to a substantial worsening of the results.

It is therefore important to recommend a simple method for continually testing the reliability of methods of determination.[3] This method should satisfy the theoretical and practical demands but at the same time it should be simple enough for a laboratory assistant to conduct the test and evaluate the results unaided.[4]

1.5.2. CONCEPTS AND DEFINITIONS

1.5.2.1. Types of error

Every laboratory result can be subject to error. A knowledge of the possible types of error is a precondition for successful quality control, for this is the only way to infer what corrective measures are appropriate. It has become customary to distinguish three types of error in medical–chemical laboratories.[5,6]

1.5.2.1.1. Gross errors

Carelessness in carrying out the analyses can lead to gross errors: confusion of specimens, samples, reagents, pipettes, photometer filters, technical defects (often hard to detect) in equipment, use of dirty apparatus. Gross errors should be avoided in advance by adopting suitable control measures. Errors due to confusion of sample material can be largely eliminated by appropriate laboratory organization.

1.5.2.1.2. Systematic errors

In the case of a systematic error, all of the results deviate from the correct value in the same direction (above or below). Systematic errors are not always easy to discover and their causes are many and various. Improperly prepared standard solutions (imprecise weighing, use of impure substances, etc.) lead to divergent results scattered about the correct value. Changes in the solutions, including standards, used for analysis, due to physical–chemical effects (contamination, oxidation, effect of light, etc.) and microbial processes, can lead to systematic errors, as can diverse behaviour of the biological material and standard in the chemical analysis. In running a standard, care should be taken to ensure an exact correspondence between the pipetted volumes (calibration of pipettes!). If, in photometric measurements, the extinction coefficient is used instead of a standard, systematic errors can occur, due to unsuitable equipment (photometer, pipettes, cuvettes), and these are often very difficult to recognize (control with primary standard).[7] The prescribed temperatures should be rigidly adhered to in all measurements, particularly with enzyme analyses.[7] As a rule systematic errors may be reduced, although this presumes some previous experience with the laboratory work.

Accuracy is the index for systematic errors.

1.5.2.1.3. Random error

Random errors must be taken into account in every analysis, even if the above-mentioned errors have been brought under control. The values obtained—and hence the results—are scattered about a mean. Theoretically, in this case, positive and negative deviations occur with equal frequency. The magnitude of the random errors is dependent on the number of single operations during an analytical run. The cause of random errors is the imperfect reproducibility of an analytical procedure. Various factors can influence the random error: temperature variation during incubation, number of pipettings, period elapsing between the final addition of reagent and reading the photometer, stability of the instrument. Mostly it is a matter of making small errors differently in each analysis. The random errors of an analytical run may be substantially reduced by automation of the individual steps of the analysis.

The index for random errors is *precision*. It is usually expressed by the standard deviation and the coefficient of variation. The precision (and the coefficient of variation) is worse the greater is the scatter of the individual values about a mean.

1.5.2.2. Reliability of the method

The concept of 'reliability' ('performance') is difficult to define since it embraces all of the following factors:

—specificity
—accuracy
—precision
—sensitivity
—practical testing over a long period of time.

A different weight has to be accorded to the individual factors depending on the practical, medical purpose, and the diagnostic significance of a particular method.

High-specificity methods. Already achieved in the determination of glucose (enzymatically with hexokinase), of urea (enzymatically with urease), and of uric acid (enzymatically with uricase).

Highly accurate methods. Accuracy is especially significant in hormone analysis and in determining metabolites such as lactate and pyruvate where their concentrations are easily changed by serum enzymes.

High-precision methods. The clinician requires highest precision in the determination of plasma electrolytes (sodium, potassium, chloride, bicarbonate, pH).

Highly sensitive methods. High sensitivity is particularly important in the determination of enzymes, trace elements (copper), and hormones.

This brief classification of the individual factors shows that the essential criteria for the choice of a particular method can only be discovered by the clinician and the clinical chemist working in cooperation.

1.5.2.2.1. Specificity

A specific method provides for the qualitative or quantitative determination of a single substance. Examples are the determination of glucose (with hexokinase), of urea (with urease), and of uric acid (with uricase). One of the goals of clinical chemistry is the replacement of non-specific group reactions ('blood-sugar determination' by the method of Hagedorn-Jensen) by absolutely specific methods. Testing for specificity ('spécificité', spezifität') is the task of the clinical chemist or the person who develops the method, not of the clinical routine laboratory. By rights the specificity of the individual methods should be stated in the operating instructions, thus, for instance, 'reducing substances by the Hagedorn-Jensen method', 'apparent creatinine by the Jaffé method' or 'apparent cerebrospinal protein by Kafka's method'.

1.5.2.2.2. Accuracy

The accuracy ('exactitude', 'richtigkeit') of a method states how much an individual value may deviate from the actual quantity present in the material under investigation. It is the index for systematic errors. Testing for accuracy is once again the task of the clinical chemist or the person who develops the method; it is not the concern of the routine laboratory. Three simple methods are available for the assessment of accuracy:

1. *Addition trial.* Known amounts of the substance to be analysed are added to the usual material for investigation, e.g. serum.
2. *Mixture trial.* A serum with a high concentration of the substance to be analysed is mixed in various proportions by volume with a serum having a low concentration of the same component.
3. *Comparison trial.* At least 50 sera are analysed by the method in question and simultaneous determinations are conducted using another method which is as reliable as possible.

The results for all three procedures may be statistically evaluated in the same way. In each case there is a series of expected results (incremental quantities of the analysed substance, or of the serum with known concentration, or the results of the comparison analyses, respectively). These expected results are contrasted with the observed results, conventionally by plotting the expected results on the abscissa and found values on the ordinate. The correlation coefficient is often used as a measure of the agreement or non-agreement between expectation and observation.

1.5.2.2.3. Precision

The precision ('précision', 'präzision') must be regarded as the most important criterion for the reliability of a method in the clinical chemical laboratory. What is meant here is the reliability of a particular method as carried out (1) by various assistants, (2) with varied apparatus, (3) on different days.

Precision is the index for random errors. Nevertheless, the precision should be found periodically for each method in every clinical laboratory. It ought to be done about once a year. It forms the basis for setting up the continual monitoring system.

Two concepts need to be distinguished:

1. *Serial precision* ('within-run' precision) (repeatability).[8] Measurement of the deviation of results within a series of analyses performed by one assistant with one set of equipment.
2. *Day-to-day precision* (reproducibility).[8] Measurement of the deviation of results for the same sample as obtained on different days by different assistants using different equipment.

As a rule the serial precision is the better.

Although analytical chemists often measure only the repeatability, in clinical chemistry it is essential to know the reproducibility. Since most quality-control systems are based on the precision, it would be unrealistic to use the repeatability. Practical experience shows that, as a rule, the reproducibility is only about half as great as the repeatability. As a measure of the precision, the standard deviation or the coefficient of variation may be used (for detailed treatments of the theory and practice of quality control, cf. refs. 9 and 10).

1.5.2.2.4. Sensitivity

The sensitivity ('sensibilité' 'empfindlichkeit') of a method corresponds to the smallest amount of the substance to be analysed that can be detected in a determination, i.e. can be differentiated from 0. For practical purposes it is equal to two standard deviations (testing the precision).

1.5.3. DETERMINATION OF PRECISION

1.5.3.1. Practical procedure

For the practician, testing the precision provides the most important method for testing the reliability of a method (it is presumed that the latter is highly accurate). The precision may be controlled by means of control samples of constant composition. Here an exact knowledge of the correct value for such a control sample is unnecessary. Consequently, it is possible to prepare the control sera in the laboratory but care should be taken to use only Australia-antigen-negative sera (risk of hepatitis).

Example. In a routine laboratory the precision is to be tested for 15 routine methods and a control system set up. On average 0.5 ml of serum is used per analysis, i.e. about 10 ml for one day's determinations or 4000 ml for the total of all the determinations over 400 days.

Laboratory preparation of the control serum. Instead of throwing away the daily (hepatitis-antigen tested) serum residues, these are pooled and stored frozen. It is also possible to use serum or plasma from blood or plasma preserves that are no longer utilizable. When about 4000 ml of serum, has been collected, this is thawed, mixed well and centrifuged. The sediment is thrown away and the clear centrifugate, kept thoroughly mixed, is poured into 10-ml test-tubes. These are numbered serially and stored frozen. The thawed serum must always be mixed well before analysis. Repeated freezing and thawing is inadmissible. If micro or ultra-micro methods are used the daily requirements of serum are much smaller.

Procedure. Every day, one of the tubes is thawed and the serum used for the routine analysis.

A control serum should be included in every series of analyses, e.g. after every 10 analyses. Where possible the identity of this control serum is concealed when it is sent for routine analysis (the best way is to invent a fictitious patient). This is the only way to ensure that the laboratory assistant does not treat it with special care. A control serum should always be run with single analyses in emergency work.

In recent years various commerical control sera have been developed.[11] These control sera have proved very valuable and make a substantial contribution to improving the reliability of methods in the clinical laboratory.[12] However, it seems important to clarify a few points which are often the subject of false assumptions:

1. With very few exceptions the value stated in the pack holds good only if the laboratory uses exactly the same method as that quoted in the pack.
2. Control sera should be used as standard solutions only in exceptional cases, otherwise there is no chance of eliciting discrepancies between the control serum and the standard prepared in the laboratory.
3. Above all, control sera are frequently used to provide a comparison value in the development and testing of new methods.

Control sera of known composition and known correct value are also used for the combined testing for precision and accuracy. These sera display a 'correct value range', quoted by the maker; the result obtained should vary only within this range. This only shows up gross inaccuracies. The criteria mentioned in Section 1.5.2.2 should be used to clarify these results; see also the use of standard solutions (p. 90).

1.5.4 STATISTICAL EVALUATION AND INTERPRETATION OF RESULTS

1.5.4.1. Precision sheet

For every method tested the first control serum result on each of 20 days is entered on a precision sheet. The data obtained are tested for distribution type by the procedure described on p. 38 and the arithmetic mean ($\bar{x}$), standard deviation (s) and coefficient of variation (CV) calculated as explained on p. 37.

If the values are 'normally' distributed, then it is known that 68.3% of all the analyses of the same serum lie between $\bar{x} \pm 1s$, 95.5% lie between $\bar{x} \pm 2s$ and 99.7% between $\bar{x} \pm 3s$. It is therefore to be expected that in consecutive analyses, out of 100 determinations, about 32 will fall outside $1s$, about 5 outside $2s$ and at most 1 outside $3s$. These rules form the basis for the control system outlined on p. 75.

For assessing the relative scatter the coefficient of variation may be used. It expresses how much the individual values vary about the mean: the smaller the scatter in comparison with the mean, the smaller is the coefficient of variation. With clinical–chemical analyses this should always be less than 10 and, if possible, less than 5. It allows a comparison of the precision of different methods of determining the same substances.

1.5.4.2. Desirable precision

It is difficult to say what degree of precision should be aimed for. Highly contradictory statements abound in the literature. For the present, a useful rule of thumb is that the coefficient of variation of the day-to-day precision should lie between 5 and 10% for enzyme analyses and between 3 and 5% for metabolite determinations. However, it is worth striving for values less than 2%, practically unattainable at the moment.

Since there is a connection between the scatter of the reference value and the scatter of the precision, various attempts have been made to express this mathematically. The best known equation is that of Tonks:[13]

$$\text{Totalled error (\%)} = \pm \frac{1/4 \text{ of the reference range}}{\bar{x} \text{ of the reference range}} \cdot 100$$

However, such equations ought to be used only as crude guide lines since there is a well founded suspicion that the present reference values are to some extent invalidated by the as yet insufficient precision of the methods.

1.5.4.3. Distribution-type of measurements of precision

Today it is usually assumed that errors of measurement are randomly distributed, i.e. a normal or Gaussian distribution is present. This may be taken as valid for most manual methods but it is still wise to test for the presence of a normal distribution. With analysers there is always the possibility of systematic deviations in a particular direction.

1.5.5. PERFORMING THE QUALITY CONTROL

1.5.5.1. Introduction

It is pointless to test the reliability of a method only to find after a few weeks that the results obviously no longer agree and that an error has occurred—perhaps in the preparation of a new reagent. The purpose of quality control is to allow such systematic deviations to be recognized at an early stage and to provide a running check on the reliability. Quality control has been used for decades in industry for checking manufactured goods[14] and has also penetrated analytical and clinical chemistry. The extent of current developments in this field may be seen from the number of recent publications as well as the constant battery of new suggestions for conducting and evaluating control analyses.[3-10] Briefly, the most important principles of making control analyses are as follows:

1. Daily use of a control serum in the reference range.[3]
2. Daily use of a control serum with abnormal concentration.[3]
3. Daily operation of a 'carry-over' analysis.[3] Every day a patient serum is frozen after carrying out a determination and analysed again on the following day.
4. Carrying out duplicate analyses for all determinations. Today this practice is discouraged in principle. Duplicate determinations should only be made in exceptional cases, e.g. emergency analyses. The contention that the result ought to be more reliable if two measurements are in agreement is entirely unfounded and gives the investigator a false sense of security. Duplicate determinations can only show up certain gross errors. They are unlikely to reveal an error due to confusion. Duplicate determinations do not provide an assessment of precision or accuracy. Consequently, in view of the extra time involved, the increased cost of reagents, and the larger quantities of sample material necessary, they are to be discouraged.
5. Daily evaluation of patient data. In this method all of the patient data are statistically analysed every day and the results compared with those of the previous day. This method allows particular results to be assessed only in consultation with the clinician. However, it does not comprise quality control in its narrower sense.

In practice, at least two of the five systems mentioned should always be used. The most suitable procedure is, on the one hand, to use a control serum in the reference range—this should never be dispensed with—and, on the other, to perform a carry-over analysis or to run a second control serum outside the normal range.

A special type of control system is provided by the *ring tests* (external quality control), in which a series of laboratories analyse the same sample. Such team experiments are very valuable and it is not rare for systematic deviations in a particular laboratory to be discovered in this way. Participation in such ring experiments—even if these are relatively expensive—can only be recommended.

It is clear that control analyses must be evaluated daily since otherwise their main function, the immediate, or at least early, recognition of errors, is lost. For this reason, complicated statistical methods only come into consideration if a computer is available for daily evaluation. The same applies to systems which use patient data as a control. Of the simple procedures, keeping a quality-control chart has proved particularly successful.

1.5.5.2. Quality-control chart of the Swiss Society for Clinical Chemistry (Schweizerischen Gesellschaft für klinische Chemie)

The Standardizing Commission of the Swiss Society for Clinical Chemistry has developed its own quality-control chart that can only be recommended (see Appendix 9). Leaflet No 2/70, *The quality-control chart*, is reproduced below by permission of the Commission.

---

1. Axioms
—The laboratory assistant himself performs a daily quality control on his routine analyses!
—No routine methods without quality control!
—No laboratory assistant without a quality-control chart!

2. Purpose
The analysis results are continually monitored *statistically* in respect of:
—*Precision.* For multiple analyses of the same sample the differences between individual values should be as small as possible.
—*Accuracy.* The found value should agree as closely as possible with the actual value.

3. Performing the quality control
3.1. Preliminary investigations
Throughout 20 series of investigations, the analysis to be controlled is performed in the routine manner by the chief assistant or another qualified laboratory assistant, working alone.
—A *standard solution* and a *commercial control serum* of known concentration are run with the series investigated.
—The following individual values are collected daily or from each series:
Measured values (extinction, emission, titres)

$\bar{x}$ (RB): blank (reagent blank value)
$\bar{x}$ (ST): standard less the blank
For exceptions, see 3.2.
Calculated analytical value (concentration, activity)
$\bar{x}$ (C): control serum
 The *means* of the three sets of numbers are calculated as follows:

$$\bar{x}(RB) = \frac{\Sigma\, x(RB)}{n}\ ; \qquad \bar{x}(ST) = \frac{\Sigma\, x(ST)}{n}\ ; \qquad \bar{x}(C) = \frac{\Sigma\, x(C)}{n}\ ,$$

i.e. the sum Σ of the individual values of a group is divided by the number of values.

 The *calculation of the standard deviation*, s, for blank $s(RB)$, standard $s(ST)$, and control serum $s(C)$ is accomplished using the following equations:

$$s(RB) = \sqrt{\frac{\Sigma\, x^2(RB) - n\bar{x}^2(RB)}{n-1}}\ ; \qquad s(ST) = \sqrt{\frac{\Sigma\, x^2(ST) - n\bar{x}^2(ST)}{n-1}}\ ;$$

$$s(C) = \sqrt{\frac{\Sigma\, x^2(C) - n\bar{x}^2(C)}{n-1}}\ .$$

i.e. the product of the square of the mean $\bar{x}^2$ and the sample number n is subtracted from the sum of the squares of the individual values x^2 and divided by the number of samples less 1 (number of degrees of freedom $= n - 1$). Then the square root of the result is taken.
 The *standard deviation, s*, is a measure of the precision of the method.

3.2. Use of the control chart in routine operation
—Each type of analysis and every control serum has its own control chart.
—Blank, standard, and control serum are run with every series of analyses (if necessary more than once a day).
—The measured values (extinction, etc.) are plotted as points in the 'blank' and 'standard minus blank' chart columns.
—The obtained analytical values (concentration, activity) are plotted as points on the 'control serum' chart column.
—The chart points are progressively joined up by lines.
—The horizontal scale of the individual columns is so chosen that the width of random scatter $\bar{x} \pm 2s$ occupies approximately half the column width.
—Before beginning the measurements the vertical lines $\bar{x} + 2s$ and $\bar{x} - 2s$ are drawn on the three chart columns.
—It is strongly recommended that control sera in the pathological range are also used. In this case the values are plotted on a second control sheet. The two columns to the left remain empty.
—If a method does not provide any blank or standard values, e.g. as in certain automatic methods, then the two left-hand columns remain empty.

3.3. Assessment
—*Blank value curve*. This provides information on the condition of the reagents, the reaction vessels and the measurement cuvettes.
—*Standard curve*. This supplies data on the usability of the reagents, the condition of the standard solution, and the measuring equipment and on the maintenance of the required reaction conditions.
—*Control serum curve*. This allows the quality of the analysis as a whole to be assessed.

3.4. Specifications for the blank, standard and control serum curves:
—they must lie within the $\bar{x} \pm 2s$ range;
—the calculated mean for the control serum, $\bar{x}(C)$, must lie as near as possible to the declared value or within the recommended range.

3.5. The specifications are not fulfilled:
—if one or more of the three curves should extend beyond the $\pm 2s$ lines;
—if 7 consecutive points on one or more curves chance to lie on the same side of the $\bar{x}$-line;
—if the $\bar{x}$-value of the control serum lies outside the recommended range.

If one of these three requirements is not satisfied the following procedure is necessary:
1. checking the *standard solution*;
2. checking the *procedure* (reagents, operating conditions);
3. *repetition of the series*.

The 'preliminary investigations' (para. 3.1) are to be performed with every *change of method*.

Control charts and leaflets can be obtained free of charge from the members of the Standardization Commission of the Swiss Society for Clinical Chemistry.

1.5.6. FREQUENT SOURCES OF ERROR

It has already been indicated that the reliability of the laboratory results leaves much to be desired. Results of ring experiments show that two thirds to three quarters of the analytical errors were conditioned by intra- and not inter-laboratory factors.[15] This emerges particularly clearly from international comparison studies. In the following paragraphs, therefore, attention will be drawn to the most important sources of error. It should be remembered that the errors accumulate with every procedure applied.

1.5.6.1. Incorrect treatment of the material under investigation

Errors may partly be traced to incorrect treatment of the material for analysis, before this even reaches the laboratory.

1. *Faulty blood-sampling* (p. 94).
2. *Mistaken identity.* It is difficult to determine the frequency of this source of error. However, it is probable that instances of confusion of blood and urine samples—during sampling during transport, or in the laboratory—are not rare.
3. *Too long a delay.* Numerous components undergo changes in concentration when blood is left standing.
4. *Dirty tubes.* Experience (particularly that based on enzyme determinations) shows that this represents a serious source of error.

1.5.6.2. Technical errors

The deviations of the results from the actual concentrations present (relative grossed error) amount to 4–6%. Estimates[16] show that, on average, an error of about 3% is attributable to faulty calibration, about 0.5–2% is due to actual errors of measurement, and about 1–5% to volumetric errors (Fig. 18). However, the proportions vary considerably from one method to the next. Whereas, for example, errors in flame photometry and bilirubin determinations are predominantly due to faulty calibration, volumetric errors are preponderant in methods involving many pipettings and transfers. The calibration has to be made with the greatest precision, the preparation of standard solutions with the greatest care (p. 90).

Errors of measurement are to a large extent determined by the quality of the apparatus used. With cheap photometers the measurements are subject to errors of 2–4% compared with 0.5% for high-grade apparatus. Criteria for testing photometers are to be found in Klein-Wisenberg's paper.[17] Large volumetric error is occasioned partly by incorrect calibration of equipment and partly by careless pipetting. The first source of error can be circumvented by using the same pipette for standard and sample. The individual error in pipetting depends on the laboratory assistant and thus it is hardly susceptible to control.

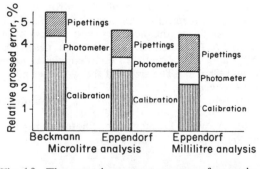

Fig. 18. The most important sources of error in laboratory methods.[16]

Different ways of performing individual manipulations (waiting time, pipette delivery speed, etc.) can also lead to slightly different results. Clear operating instructions help to eliminate this source of error.

Finally, errors of calculation must be mentioned. For methods used frequently it is advisable to have sheets on which the relationship between measured value (e.g. extinction) and result is tabulated or shown graphically.

However, the surest way to improve the reliability of the results and eliminate sources of error is the daily running of a control serum and keeping control charts.

References

1. Aebi, H., Die chemische Analyse im Dienste der Medizin einst und jetzt, *Schweiz. med. Wschr.*, **87**, 549 (1957).
2. Advisory Board, *Analyt. Chem.*, **34**, 364R (1962).
3. Richterich, R. and Colombo, J. P., Ultramikromethoden im klinischen Laboratorium. II. Die Bestimmung der Zuverlässigkeit von Laboratoriumsmethoden, *Klin. Wschr.*, **40**, 529 (1962).
4. Stamm, D., Qualitätskontrolle klinisch-chemischer Analysen, Thieme, Stuttgart, 1972.
5. Bürgi, W., Die Zuverlässigkeit klinisch-chemischer und hämatologischer Laboratoriumsanalysen, *Schweiz. med. Wscher.*, **102**, 367 (1972).
6. Büttner, H., Hansert, E., and Stamm, D., Auswertung, Kontrolle und Beurteilung von Messergebnissen, in Bergmeyer, *Methoden der enzymatischen Analyse,* Vol. 1, 3. Aufl. Verlag Chemie, Weinheim, 1974.
7. Eisenwiener, H.-G., Voraussetzungen zur Erzielung zuverlässiger klinisch-chemischer Analysenergebnisse. *Med. Lab., Stuttg.*, **29**, 171 (1976).
8. Whitby, L. G., Mitchell, F. L., and Moss, D. W., Quality control in routine clinical chemistry, *Adv. clin. Chem.*, **10**, 65 (1967).
9. Provisional recommendations on quality control in clinical chemistry. 1. General principles and terminology, *Clin. chim. Acta*, **63**, F25 (1975).
10. Provisional recommendations on quality control in clinical chemistry. 2. Assessment of analytical methods for routine use, *Clin. chim. Acta*, **69**, F1 (1976).
11. Stevens, J. F., Control materials for clinical biochemistry, *Ann. clin. Biochem.*, **10**, 133 (1973).
12. Provisional recommendations on quality control in clinical chemistry. 3. Calibration and control materials, *Clin. chim. Acta*, **75**, F11 (1977).
13. Tonks, D. B., A study of the accuracy and precision of clinical chemistry determinations in 170 Canadian laboratories, *Clin. Chem.* **9**, 217 (1963).
14. Moroney, J. M., *Facts from figures* 3rd ed., Penguin, London, 1958.
15. Gilbert, R. K. and Chir, B., The size and the source of analytic error in clinical chemistry, *Am. J. clin. Path.*, **61**, 904 (1974).
16. Büttner, H., Microliter analysis. Comparison of commercial apparatus in a routine laboratory, *Dt. med. Wschr.*, **88**, 910 (1963).
17. Klein-Wisenberg, A. von, Prüfung von Photometern für das medizinische Laboratorium, *Ärztl. Lab.*, **18**, 243 (1972).

1.6. LABORATORY ROOMS AND LABORATORY EQUIPMENT

1.6.1. LABORATORY ROOMS

The laboratory room should be well ventilated and brightly lit. Laboratories do not belong in the basement. White artificial light is preferred to yellow light. Furnishings and appointments should not be built into the laboratory as fixtures but should remain movable so that they can be regrouped as desired. One system which has proved extraordinarily successful is the standard construction principle whereby walls, installations, and appointments are all completely movable and can be assembled to another ground-plan at any time (Fig. 19). Many laboratory designs forget that today more and more of the laboratory work is being done at the office desk rather than the laboratory bench. Care should

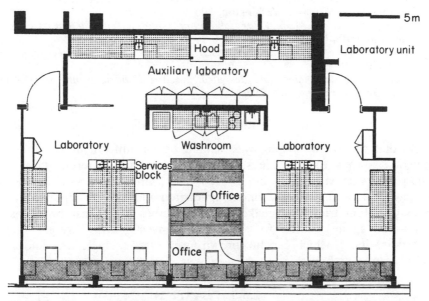

Fig. 19. Normal laboratory group consisting of two laboratories, two office booths, a wash-room, and an auxiliary laboratory (Architect: J. Iten, Chemical Central Laboratory, Berne, 1969).

therefore be taken to assure that a sufficient number of small offices are available.

Since in principle the clinical laboratory work is done sitting down rather than standing, the benches still used in some laboratories are no longer suited to a rational mode of working. Instead, it is best to use ordinary tables and chairs which allow a person to sit comfortably. There is a sink for every table: one basin is sufficient, connected to a central filter-pump. In addition, there needs to be an adequate number of electrical points nearby. Compressed air and gas are seldom necessary today so it suffices if just a few work-places are fitted with these services.

The spectrophotometer—shielded from direct incident light—belongs in the laboratory, not in an instrument room. The same applies to the usual laboratory balances. Each piece of equipment is permanently installed at a work-place. All the necessary reagents, glassware, tables, etc., stay here and the assistants move from place to place.

Each laboratory room needs to have a refrigerator, a water-bath (25–37 °C), adequate connections for filter-pumps, and a spectrophotometer. A unit for the preparation of demineralized (ion-exchange) water (DM-water) must be available.

References

1. MacFate, R. P., *Introduction to the clinical laboratory*, Year Book Publishers, Chicago, 1961.
2. *Planning the laboratory for the general hospital*, US Department for Health, Education and Welfare, Washington, 1963.
3. Rappaport, A. E., *Manual for laboratory planning and design*, College of American Pathologists, Chicago, 1960.
4. Schramm, W., *Chemische und biologische Laboratorien*, 3rd ed., Verlag Chemie, Weinheim, 1969.
5. Haeckel, R., *Rationalisierung des medizinischen Laboratoriums*, G-I-T-Verlag Ernst Giebeler, Darmstadt, 1976.

1.6.2. GLASSWARE

So that the stock of glassware should be as uniform as possible, it is practical always to buy the same model for each different type. Only equipment with standard joints is used. The following glassware is needed: several sizes of measuring cylinder, graduated and transfer pipettes, various sizes of volumetric flask, beakers and test-tubes in at least two sizes and sorts—a cheap sort for sensitive investigations (e.g. enzyme determinations), which can be thrown away after a single use, and also a robust type worth washing.

1.6.3. PLASTICS

Today much laboratory equipment is made out of plastic and it is therefore sensible to use plastic apparatus instead of glass. It has the

Table 18. Physical properties of plastics (from ref. 1).

| Plastic | Stability, °C | | Property |
	Heat	Cold	
Polypropylene	+100	−10	Unbreakable, resistant to boiling, sterilizable
Low-pressure polyethylene	+100	<20	Unbreakable, can be welded
High-pressure polyethylene	+80	<30	Unbreakable
Poly(vinyl chloride) (PVC)	+60	−25	Unbreakable
Polyamide (nylon)	+90	−30	Unbreakable, tough, sterilizable
Poly(methyl methacrylate) (acrylic resin)	+90	−30	Easily worked, clear as glass
Polytetrafluoroethylene	+260	−200	Unbreakable, non-flammable
Polymethane	+90	−30	Temperature stable
Polystyrene	+70	−35	Flexible, hard, clear as glass
Polycarbonate	+130	−100	Unbreakable, clear as glass

Table 19. Chemical stability of plastics (from ref. 1).

Plastic	Weak acids	Concentrated acids	Weak alkalis	Concentrated alkalis	Alcohols	Esters	Ketones	Petrol	Benzene	Petrol fuel	Mineral oils	Animal and vegetable oils	Chlorinated hydrocarbons	Ethers
Polypropylene	+	O	+	+	+	+	+	O	O	O	+	O	O	O
Low-pressure polyethylene	+	O	+	+	+	O	+	O	O	O	+	O	O	O
High-pressure polyethylene	+	O	+	+	+	O	+	O	O	O	+	O	O	O
Poly(vinyl chloride) (PVC)	+	+	+	+	+	O	O	+	O	O	+	+	O	O
Polyamide (nylon)	O	O	+	O	O	+	O	+	+	+	+	+	O	+
Poly(methyl methacrylate) (acrylic resin)	+	O	+	O	O	O	O	+	O	O	+	+	+	O
Polytetrafluoroethylene	+	+	+	+	+	+	+	+	+	+	+	+	+	+
Polymethane	O	O	O	O	+	+	+	+	+	+	+	+	O	+
Polystyrene	O	O	+	+	+	O	O	O	O	O	O	O	O	O
Polycarbonate	+	O	+	O	O	O	O	+	O	O	+	O	O	O

+ = Stable. O = Unstable to conditionally stable.

advantage of being unbreakable and does not act as an ion exchanger in the way that glass does. The physical properties of the most important products are listed in Table 18 and their chemical properties in Table 19.

Reference

1. Häufglockner, H., Physikalische und chemische Eigenschaften der Kunststoffe unter Berücksichtigung ihrer Anwendung, *G-I-T,* **10**, 156 (1966).

1.6.4. CLEANING LABORATORY EQUIPMENT

For ultramicro and enzyme determinations, absolutely clean glassware is indispensible. Chromic acid should not be used since traces of it inactivate enzymes. Soap, especially if the water is hard, leaves a film on the glass which is not removed even by repeated rinsing. Alkaline detergents (domestic washing-up liquids) etch the surface of the glass and are strong enzyme inhibitors. It is recommended that cleaning be done with neutral detergents followed by at least three rinsings with DM-water. If possible one of the commercially available laboratory washing-machines should be used. It should also be remembered that detergents contain large quantities of sodium and phosphate.

For many methods the assistant has to clean certain glassware himself at the work-place, since a high degree of cleanliness is never attained by washing up a diverse batch of apparatus. Particular recommendations:

Flame photometry. Use of disposable plastic tubes.

Ammonia determinations. The glassware is soaked in 0.1 N hydrochloric acid (about 4 h) and subsequently rinsed many times with distilled water.

Ultramicropipettes. After use, ultramicropipettes of any kind are filled with a mixture of hydrochloric acid and pepsin (1 spatula-tip of pepsin in about 10 ml of 0.5 N hydrochloric acid). This dissolves traces of protein. Subsequently rinse out with distilled water and allow to dry.

Cuvettes. Cleaning in a solution of *ca.* 50% sulphuric acid and 10% hydrogen peroxide, 1:1 by volume (**caution: highly corrosive**); leave to soak for about 2 h and finally rinse out well with distilled water.

The use of pipettes can be considerably reduced by employing pipetting and diluting apparatus, thus diminishing the work of washing-up.

1.6.5. TEMPERATURE CONTROL

Numerous chemical reactions are temperature dependent. For maintaining a steady temperature it is advantageous to use water-baths. Where possible the heater should be fitted with a rotary pump, not just with a rotor, and there should also be a device for reverse cooling. For routine working, each assistant requires at least one water-bath at 25–37 °C.

Labile reagents are stored in a refrigerator, either at 4 °C or frozen

$(-20\,^{\circ}\mathrm{C})$. Only plastic flasks should be used for freezing. In purchasing refrigerators it is advisable to look for as large a deep-freeze compartment as possible. For storing investigational material deep-freezers are more practical than chests.

Finally, drying cupboards (incubators) are needed, one set at $37\,^{\circ}\mathrm{C}$ (bacteriology) and one at $100\,^{\circ}\mathrm{C}$ (for drying glassware).

1.6.6. SAFETY IN THE LABORATORY

The work in the clinical chemistry laboratory involves certain dangers. The prevention of any kind of occupational injury presumes a knowledge of how to handle and use all the apparatus, equipment, chemicals, and radioactive substances used in the laboratory, and of the infectiousness of the investigational material. These preventative measures will not be discussed here. Information on this subject can be found in the literature.[1]

Reference

1. *Sicherheit im medizinischen Laboratorium. Ein Leitfaden*, Birkhäuser, Basle, 1976.

1.7. CHEMICALS AND SOLUTIONS

1.7.1. PURITY

The abbreviations 'puriss.' (*purissimum*, purest), 'pur.' (*purum*, pure), 'pract.' (*practicum*) and 'tech.' (*technicum*, technical) serve to characterize the purity of chemicals. The term 'pro anal.' (p.a., *pro analysi*) does not characterize the purity but merely states that the corresponding 'puriss.' preparation is especially suitable for analytical purposes. The specifications of purity have not been officially fixed but tend to vary from chemical to chemical and from one manufacturer to another. A substance of a particular degree of purity can be described in one firm's catalogue as 'puriss.' and in another as 'pur.'.

The highest specifications must be accorded to the chemicals in the clinical laboratory. In principle, only 'puriss.' preparations should be used, if possible, those giving the factory analysis. Since it is by no means rare for a colour reaction to be affected by the presence of a trace of impurity, it is advisable always to use chemicals of the same provenance for any particular reaction. The only chemicals to be bought are those with the original labels, factory packed, not in any bottle from the pharmacist or chemical store.

For a quick briefing on chemical composition, solubility, stability, and possible toxicity it is advisable to procure a chemical handbook, for example:

Stecher, P. G., *The Merck index of chemicals and drugs*, 8th ed. Merck, Rahway, 1968.
Rose, A. and Rose, E., *The condensed chemical dictionary*, 4th ed. Chapman and Hall, London, 1956.
Römpp, M. H., *Chemielexikon*, 7. Aufl., Franckh'sche Verlagshandlung, Stuttgart, 1972.

Since the same substance may often have several different names, when keeping a card index or arranging chemicals in cupboards it is wise to keep strictly to the nomenclature of whatever handbook is used.

1.7.2. TEST PACKS

In recent years, more and more 'ready-for-use' packs of reagents ('test packs') have come on the market. In principle there is nothing against using such packs, but it should be mentioned that the quality of the products is extremely varied. Hence it would be better to keep to the known brands and, in every case, to subject these packs to critical testing with control sera and by comparison with other methods before abandoning an established method in favour of a new one.

1.7.3. DISTILLED AND DEMINERALIZED WATER

The following grades of water are used in the clinical laboratory:[1,2]

---

1. Tap water (spring water, *Aqua fontana, Aqua communis*)
2. Distilled water (*Aqua destillata*)
3. Double distilled water (*Aqua bidestillata*)
4. Demineralized water (*Aqua demineralisata*, ion-exchange water, DM-water)

Since the conductivity of water is primarily dependent on the number of ions (electrolytes), its quality is most directly expressed by the specific conductivity in microsiemens (μS cm^{-1}). Alternatively, the reciprocal value, the specific electrical resistance, can also be used: the smaller the electrolyte concentration in the water, the smaller is the electrical conductivity and the greater is the specific electrical resistance. For comparison, the conductivity values for water of various grades of purity are collated in Table 20. This shows that DM-water takes precedence over distilled water in respect of its low concentration of electrolytes. Numerous units, some very cheap, are available for the preparation of DM-water and their capacity can be suited to the requirements. DM-water has one disadvantage: as a rule it is not sterile, and for this reason it should always be used immediately and should not be kept. However, bacterial impurities do not often interfere in analytical investigations, although the risk is greatest with enzyme analyses. For the preparation of standard solutions it is best to distil DM-water.

Table 20. Conductivity values for various grades of water.

	μS cm^{-1}
Tap water	290
Distilled water	5
Doubly distilled water (glass)	1
Doubly distilled water (quartz)	0.5
Demineralized water	0.2
Demineralized water, distilled	<0.2

References

1. Winstead, M., *Reagent grade water: how, when and why*, American Society of Medical Technology, Houston, 1967.
2. Appel, W., Wasser im klinisch-chemischen Laboratorium, *Mitt. Dt. Ges. klin. Chem.*, **8**, 105 (1977).

1.7.4. STANDARDS

The accuracy of clinical–chemical analyses ultimately depends on the quality of the standards or standard solutions (calibration standards) used. Every imperfection in a standard solution will affect all of the results calculated on the basis of this solution. So far this problem has received insufficient attention. By international agreement, the International Union for Pure and Applied Chemistry (IUPAC) distinguishes the following grades of purity:[1]

Grade A	Atomic weight standard
Grade B	Ultimate standard
Grade C	Primary standard: a commercially available substance of purity $100 \pm 0.02\%$
Grade D	Working standard: a commercially available substance of purity $100 \pm 0.05\%$
Grade E	Secondary standard: a substance of lower purity which can be standardized against primary grade C material

Obviously this terminology can only be applied conditionally to clinical chemistry, since in many cases it is not possible to standardize the relevant substances even at a grade D level. For this reason, the US National Bureau of Standards (NBS) has decided to create a new category, 'standard, clinical type'.[2] This usage is to be preferred to the expression 'clinical primary standard'.[3] The substances supplied by the NBS may be described as the purest materials so far and where possible they should be used as reference materials for standardizing other products. Up to now the following substances have been defined and may be bought from the NBS:[4] buffers for blood pH determinations, cholesterol, calcium, urea, glucose, sodium, potassium, bilirubin, uric acid, and creatinine. Independent suggestions for the standardization of haemoglobin cyanide and albumin are under consideration.

It is to be hoped that it will very soon be possible to establish purity criteria and to produce reference materials for other important standards. For the time being, for many substances all that can be done is to describe the purity as exactly as possible.

The Committee on Standards, Expert Panel on Nomenclature and Principles of Quality Control in Clinical Chemistry of the International Federation of Clinical Chemistry (IFCC) has suggested a more pragmatic definition for the primary and secondary standard, since the IUPAC definition is unsuitable for clinical–chemical purposes. In this definition the terms 'primary' and 'secondary' standard do not have the same meaning as in the IUPAC definition.[5]

A *primary standard solution* (calibration standard) is prepared by dissolving a weighed quantity of standard material in a certain volume of weight of a suitable solvent. The accuracy of the primary standard solution depends on the purity of the standard material and the other components as well as on the precision with which the solution is prepared.

A *secondary standard solution* is a solution for which the concentration is determined by a method of known reliability. The accuracy of the secondary solution therefore depends on the accuracy of the analysis. This, in turn, necessitates a standard, ultimately a primary standard.

References

1. Report prepared by the Analytical Standards Subcommittee, *Analyst*, **90**, 251 (1965).
2. Mears, T. W. and Young, D. S., Measurement and standard reference materials in clinical chemistry, *Am. J. clin. Path.,* **50**, 411 (1968).
3. Radin, N., What is a standard, *Clin. Chem.,* **13**, 55 (1967).
4. Meineke, W. W., Standard reference materials for clinical measurements, *Analyt. Chem.,* **43**, 28a (1971).
5. Provisional recommendation on quality control in clinical chemistry, IFCC, *Clin. chim. Acta,* **63**, F25 (1975).

1.7.5. BUFFER SOLUTIONS

1.7.5.1. General

In clinical chemistry various reactions, such as enzyme determinations, have to be carried out at constant pH. The acidic and basic products formed by the reaction must be buffered. The change in pH which they produce is diminished in the presence of a buffer. A buffer is a mixture of a weak acid and a salt having the same anion or a mixture of a weak base and a salt having the same cation.[1] Each buffer has an optimum range of action throughout which the buffer capacity of a solution depends on the total concentration of the buffer mixture as well as on the molar ratio of the two buffer components. The buffering of a solution is most effective when the molar concentration of the undissociated acid and of the free anion, or of the undissociated base and of the free cation, are equal, i.e. when there is e.g. 1 molecule of acetic acid to 1 molecule of sodium acetate. An acid–salt mixture shows optimum buffering when the pH is equal to the pK_a of the buffer acid (pK = negative logarithm of the dissociation constant). Consequently, each buffer has a specific optimum (Fig. 20).

Information on the preparation of buffer solutions is given in Appendix 1. For a closer study of buffer action, the literature should be consulted.[1,2]

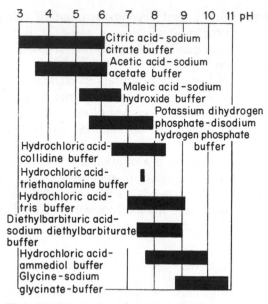

Fig. 20. Buffer ranges of individual buffer systems. Ammediol is 2-amino-2-methyl-propane-1,3-diol.

1.7.5.2. Ionic strength

In addition to the molarity and the pH, the ionic strength is often an important factor in determining whether a solution or a buffer is suitable. The ionic strength (I) is the unit of concentration for the electrolytic activity of a solution. This unit is of special significance if charged molecules are subjected to an electric field (electrophoresis). It takes into account both the ionic concentration and the charges on the ions.

Example. Let C be the molar concentration and Z the electric charge of the ions n. Then

$$I = \frac{\sum C_n \cdot Z_n^2}{2} = \frac{1}{2} \sum C_n \cdot Z_n^2$$

i.e. the molar concentration of each ion (C_n) is multiplied by the square of the electric charge on this ion (Z_n), the products are summed and the result is divided by 2. With mixtures of substances, e.g. buffers, allowance has also to be made for the degree of dissociation, α, of the individual components. The equation then reads

$$I = \frac{\sum (C\alpha)_n \cdot Z_n^2}{2}$$

The degree of dissociation, α, is the ratio of the ionized part to the total concentration of the acid or base. It is calculated as follows:

$$\alpha = \frac{K_{Acid}}{K_{Acid} + [H]} = \frac{10^{-pK_{Acid}}}{10^{-pK_{Acid}} + 10^{-pH}} \, ,$$

where K is the dissociation constant.

Examples.

1. Suppose the ionic strength of a 2 M (2 mol l^{-1}) ammonium sulphate solution is to be calculated:

$$I = \frac{\Sigma C_n \cdot Z_n^2}{2} = \frac{[(2 \cdot 2)(1^2)] + [(2)(2^2)]}{2} = \frac{12}{2} = 6$$

2. Barbiturate buffer for electrophoresis, 60 mmol l^{-1}, pH 8.6: 50 mmol l^{-1} diethylbarbituric acid, Na$^+$ salt; 10 mmol l^{-1} diethylbarbituric acid. In order to calculate the ionic strength, the degree of dissociation of barbituric acid has still to be considered. C for Na$^+$ = 0.05 mol l^{-1}, for barbiturate = 0.06 mol l^{-1} times α. The pK of diethylbarbituric acid is 7.98. Then,

$$\alpha = \frac{10^{-7.98}}{10^{-7.98} + 10^{-8.6}}$$

$$= \frac{1.047 \cdot 10^{-8}}{1.047 \cdot 10^{-8} + 2.512 \cdot 10^{-9}}$$

$$= \frac{1.047 \cdot 10^{-8}}{1.2982 \cdot 10^{-8}} = \underline{0.8065}$$

$$I = \frac{[(0.0484 \cdot 1)(1^2)] + [(0.05 \cdot 1) \cdot (1^2)]}{2}$$

$$= \underline{0.0492} \approx 0.05$$

References

1. Lauber, K., *Chemie im Laboratorium*, 3. Aufl. Karger, Basle, 1975.
2. Aeibi, H., *Einführung in die Biochemie*, 2. Aufl,, Karger, Basle, 1971.

1.8. COLLECTION OF BLOOD

1.8.1. GENERAL

1.8.1.1. Differences between venous and capillary blood

The blood used for routine clinical–chemical investigations is exclusively venous or capillary. Arterial blood is only required in special cases (e.g. analysis of blood gases). Hence first consideration must be given to the differences to be expected in the results according to whether venous or capillary blood is used in the investigations. Theoretically, with venous blood changes are particularly likely if the flow is restricted some time before collection. In taking capillary blood there is danger of haemolysis and of the influx of liquids from the interstitial into the intravascular space.

Kaplan et al.[1] compared the concentrations of cholesterol, chloride, glucose, sodium, potassium, urea, calcium, phosphorus, total protein, and albumin in venous and capillary blood. Only the glucose values were lower in venous than in capillary blood. No differences were detectable for any of the other substances.

Numerous authors have attempted to demonstrate differences in haemoglobin content between venous and capillary blood. The results are so contradictory that a substantial difference is improbable, even for this easily and precisely detectable substance.[2]

From these investigations it is to be inferred that the type of blood sampling is unimportant for the following substances: cholesterol, sodium, potassium, chloride, urea, total protein, calcium, albumin, and phosphorus. With glucose, lower values are to be expected for venous than for capillary blood (up to 22 mg per 100 ml),[3,4] while lactate and pyruvate concentrations are higher in venous than in capillary blood.[5]

References

1. Kaplan, S. A., Yuceoglu, A. M., and Strauss, J., Chemical microanalysis: analysis of capillary and venous blood. *Pediatrics,* **24**, 270 (1959).
2. Dacie, E. V., *Practical hematology*, 2nd ed., Churchill, London, 1956.
3. Bürgi, W., Oraler Glukosetoleranztest: interschiedlicher Verlauf der kapillären und venösen Belastungskurven, *Schweiz. med. Wschr.,* **104**, 1698 (1974).
4. Kreutz, F. H., Auswirkungen der Probenahme auf klinisch-chemische

Untersuchungsergebnisse, in Lang, Rick und Roka, *Optimierung der Diagnostik*, Springer, Berlin, 1973.
5. Richterich, R., Unpublished observations.

1.8.1.2. Use of fasting (post-absorptive) blood

A question which frequently arises is whether the determination of particular substances necessitates collection of fasting blood or whether the intake of nourishment is without a marked effect on the blood concentration.

Annino and Relman[1] investigated urea, glucose, bicarbonate, chloride, sodium, potassium, creatinine, total protein, cholesterol, calcium, phosphorus, and uric acid in 32 healthy subjects, fasting and 45 and 120 min after a breakfast. In the case of phosphorus only, a slight decrease was observed 45 min after the meal. Collection under fasting conditions is also correct for blood-glucose determination since the individual values show great deviations following a meal.

On the basis of these investigations, it may be assumed that fasting blood is not necessary in the determination of urea, bicarbonate, chloride, sodium, potassium, creatinine, total protein, cholesterol, calcium, and uric acid. In addition, personal observations show that no differences can be established between fasting blood and blood sampled after the subject has taken a light meal (breakfast) for amylase, ceruloplasmin, alkaline phosphatase, acid phosphatase, transaminases, leucine-amino peptidase, and cholinesterase. After substantial meals larger differences are likely, particularly with glucose, cholesterol, triglycerides and phosphorus (afternoon consultation hour!). The examination for lipids (risks involved!) should be investigated with fasting blood (10–12 h deprivation of food).

It should also be remembered that in many determinations lipaemia interferes with the chemical reactions and the photometer readings. For this reason alone it should be avoided. The daily intra-individual variations of single plasma components such as lipids, iron, urea, albumin, total protein, and chloride are another reason why one should aim to take the blood samples at a fixed time of day, best of all fasting.[2]

References

1. Annino, J. S. and Relman, A. S., The effect of eating on some of the clinically important chemical constituents of the blood, *Am. J. clin. Path.*, **31**, 155 (1959).
2. Winkel, P., Statland, B. E., and Bokelund, H., The effects of time of venipuncture on variation of serum constituents, *Am. J. clin. Path.*, **64**, 433 (1975).

1.8.1.3. Use of serum or plasma

The use of plasma instead of serum has the fundamental advantage that the haemolysis that occurs during blood collection coagulation, and transport is markedly smaller. Lum and Gambino[1] found no difference between serum and plasma for glutamate–oxaloacetate transaminase, glutamate–pyruvate transaminase, total bilirubin, urea, calcium, total CO_2, creatine kinase and glucose. They found statistically significant, but clinically unimportant, differences for the following parameters: alkaline phosphatase, albumin, phosphorus, sodium, triglycerides, and uric acid.[1] On the other hand, potassium was higher (about 0.38 mmol l^{-1}) in serum than in plasma, and total protein was higher in plasma by 0.24 g per 100 ml. Henny et al.[2] reached similar conclusions in their investigations. They found significant differences for total protein, phosphate, and glucose. No difference was established for potassium. Enzymes present in the thrombocytes, such as acid phosphatase, aldolase, and lactate dehydrogenase, can sometimes be liberated from the thrombocytes on coagulation.[3]

The reduced risk of haemolysis and the possibility of centrifuging the blood are important advantages of plasma preparation.[4]

References

1. Lum, G. and Gambino, S. R., A comparison of serum versus heparinized plasma for routine chemistry tests, Am. J. clin. Path., **61**, 108 (1974).
2. Henny, J., Houot, O., Steinmetz, J., and Siest, G., Comparaison des valeurs obtenues sur sérum ou plasma hépariné pour les examens biochimiques courants, Annls Biol. clin., **34**, 335 (1976).
3. Richterich, R., Colombo, J. P., and Weber, H., Ultramikromethoden im klinischen Laboratorium. VII. Bestimmung der sauren Prostata-Phosphatase, Schweiz. med. Wschr., **92**, 1496 (1962).
4. Solbach, H. G., Englhardt, A., and Merten, R., Unterschiede der Enzymaktivitäten in Serum und Plasma und ihre Bedeutung für die klinische Enzymdiagnostik., Klin. Wschr., **40**, 1136 (1962).

1.8.1.4. Differences between blood and plasma (serum)

Use of whole blood instead of plasma or serum is permissible only if the concentrations of the substance under investigation are approximately the same in the erythrocytes and in the plasma. This precondition is fulfilled, for example, by glucose and urea-nitrogen (Table 21). Nevertheless, the determination of these components in whole blood gives results differing from those obtained for plasma. This is due to the differing water contents of whole blood and plasma: plasma (serum) contains about 93% of water and whole blood about 81%. So for the same patients, the results obtained using plasma (serum) are about 12% higher than those determined with whole blood. Larger differences may be expected with marked increases in the haematocritical value.

It would be more correct to calculate all results with respect to the plasma water, but this would complicate the determination without significantly increasing the diagnostic value of the result. Still, it is always advisable to state whether a given result of reference range refers to whole blood, serum, or plasma.

Reference

1. Caraway, W. T., Chemical and diagnostic specificity of laboratory tests, *Am. J. clin. Path.*, **37**, 445 (1962).

1.8.1.5. Effect of bodily posture and venous stasis

The concentration of microscopic or macromolecular particles, and hence the erythrocytes, is dependent on the posture of the body and possible tourniquet application when the blood is taken. With low molecular weight particles these factors play a role only in exceptional cases. For blood taken from the cubital vein of sitting (15 min quiet sitting) and lying (10 min lying) test subjects, 10 macromolecular serum components (or components linked to macromolecular substances) and 4 serum enzymes (total protein, albumin, α_1-globulin, α_2-globulin, β-globulin, γ-globulin, cholesterol, bilirubin, calcium, inorganic phosphate, alkaline phosphatase, creatine kinase, lactate dehydrogenase, glutamate–oxaloacetate trans-aminase) show significant shifts in concentration of between 3 and 8% corresponding to a reduction. This is an order of magnitude which far exceeds the analytical tolerance achieved for these values and is therefore of clinical significance. Low molecular weight substances such as urea, glucose, uric acid, creatinine, sodium, and chloride show no changes on taking the blood from subjects in different postures.[1] Statland *et al.*[2] reported similar findings. Constriction of the vein for 3 min leads to an increase in total protein, iron, cholesterol, total lipids, glutamate–oxaloacetate transaminase, and bilirubin. Potassium decreases.

References

1. Röcker, L., Schmidt, H. M., Junge, B., and Hoffmeister, H., Orthostasebedingte Fehler bei Laboratoriumsbefunden, *Med. Lab., Stuttg.*, **28**, 267 (1975).
2. Statland, B. E., Winkel, P., and Bokelund, H., Factors contributing to intra-individual variation of serum constituents. IV. Effects of posture and prolonged tourniquet application on variation of serum constituents in healthy subjects, *Clin. Chem.*, **20**, 1513 (1974).

1.8.1.6. Haemolysis as a source of error

Haemolytic sera are a frequent source of error in clinical–chemical analyses. A haemolysis may be recognized by eye if the plasma haemoglobin concentration exceeds 20 mg per 100 ml.[1] With icteric sera the limiting concentration may be even higher. The causes of haemolysis are many and varied:[2] intravasal haemolysis (use of tourniquet!), too

Table 21. Differences between plasma and erythrocyte concentrations of a few components of clinical–chemical importance (from ref. 7).

Component, concentration	Erythrocytes	Plasma	Erythrocytes / Plasma
Lactate dehydrogenase, units	58 000.0	360.0	160.0
Acid phosphatase, units	200.0	3.0	67.0
Glutamate–oxaloacetate transaminase, units	31.5	0.8	40.0
Potassium, mmol l^{-1}	100.0	4.4	22.7
Glutamate–pyruvate transaminase, units	1.6	0.24	6.7
Residual reduction, mg per 100 ml	40.0	8.0	5.0
Magnesium, ion, mmol l^{-1}	5.5	2.2	2.4
Residual N, mg per 100 ml	44.0	25.0	1.8
Creatinine, mg per 100 ml	1.8	1.1	1.6
Glucose, mg per 100 ml	74.0	90.0	0.82
Urea-N, mg per 100 ml	14.0	17.0	0.82
Inorganic phosphorus, mg per 100 ml	2.5	3.2	0.78
Bicarbonate, mmol l^{-1}	19.0	26.0	0.73
Cholesterol, mg per 100 ml	139.0	194.0	0.72
Uric acid, mg per 100 ml	2.5	4.6	0.55
Chloride, mmol l^{-1}	52.0	104.0	0.50
Sodium, mmol l^{-1}	16.0	140.0	0.11
Calcium, ion, mmol l^{-1}	0.5	5.0	0.10
Cholesterol ester, mg per 100 ml	0.0	0.8	
Creatine kinase, μmol min^{-1} l^{-1}	0.0	0.8	

powerful an aspiration, mixing, blood spurting out, contamination (detergents, water), allowing whole blood to stand, intense cooling or heating of whole blood, powerful centrifuging. More frequent haemolyses are to be expected with capillary blood sampling in new-born babies owing to the high haematocrit and the greater fragility of the erythrocytes.[3]

The effects of haemolyses are consequently of a manifold nature. However, they may be broadly divided into three groups.

The *first source of error* which must be mentioned is the transference to the plasma of substances which are present in the erythrocytes at a higher concentration than in the plasma. The practical consequence is that the result is too high. Table 21 lists the concentrations in plasma and in the erythrocytes of a few important components. As this shows, the risk of an abnormally high result is greatest for the determination of lactate dehydrogenase, acid phosphatase,[4] glutamate–oxaloacetate transaminase, potassium, glutamate–pyruvate transaminase and creatinine.[5,6] The consequences of incorrect results in the determination of potassium are especially serious.

The error caused by the haemolysis can be calculated.[6,7] If:

[S] = concentration of the investigated substance in non-haemolytic plasma;

[E] = concentration of the investigated substance in the erythrocytes;

$[P_{Hb}]$ = haemoglobin concentration in the plasma in g per 100 ml;

p = the percentage deviation of the measured result from the true concentration (without haemolysis) in the plasma; then

$$p = 3 \cdot [P_{Hb}] \cdot \left(\frac{[E]}{[S]} - 1 \right)$$

Example. To illustrate the scale of such errors, suppose the haemolysis has resulted in a plasma haemoglobin concentration of 100 mg per 100 ml. The ratio of the intra-erythrocyte to plasma potassium concentration is 22.7 (Table 21). Then the deviation of the measured result from the actual concentration (in non-haemolytic plasma) for the patient is

$$p = 3 \cdot 0.1 \cdot (22.7 - 1.0) = +6.5\%$$

As a rule, the release of intra-erythrocytic substances is lower at 4 °C than at room temperature. Potassium appears to form an exception to this rule. In the case of inorganic phosphorus, in addition to the inorganic phosphorus released from the erythrocytes, allowance must be made for the phosphate esters which diffuse out at the same time; on standing, these are cleaved by the action of alkaline phosphatase present in the plasma.[6]

There is little fear of a haemolysis producing the opposite effect, i.e. yielding results which are too low, when the substance determined is present at a lower concentration in the erythrocytes than in the plasma; the extent of haemolysis would have to be exceptionally large. Consequently, this source of error is of importance only if whole blood is used for the determination instead of plasma (serum).

A *second group of sources of error* can come into play if photometric readings are taken in the short-wave region of the visible spectrum (300–500 nm). Since the molar extinction coefficient of haemoglobin in the Soret band (405 nm) is around 130 000 l mol^{-1} cm^{-1}, haemolysis results in a considerable extinction even at high dilution. Such an error can largely be avoided by including suitable sample blank values. A distortion of the result is unavoidable if the sample value and sample blank value are read at different pH values, for instance, in the determination of alkaline phosphatase and of bromosulphophthalein. Haemoglobin derivatives behave like indicators and have different extinction coefficients at different pH values, and this results in an error which is hard to eliminate.

Finally, a *third group of possible errors* in working with haemolytic sera is contingent upon the fact that haemoglobin can interfere with certain chemical reactions. Well known examples are its inhibition of diazotization in the determination of bilirubin and its interference in the enzymatic determination of cholesterol. Keeping all these potential sources of error in mind, it is easy to understand the present tendency to use plasma instead of serum in order to avoid the slight haemolysis which always occurs on coagulation.

100

References

1. Behrendt, H., *Chemistry of erythrocytes*,Thomas, Springfield, 1957.
2. Guder, W. G., Einfluss von Probennahme, Probentransport und Probenver-wahrung auf klinisch-chemische Untersuchungsergebnisse, *Ärztl. Lab.*, **22**, 69 (1976).
3. Michaelsson, M. and Sjölin, S., Haemolysis in blood samples from newborn infants, *Acta paediat. scand.*, **54**, 325 (1965).
4. Richterich, R., Colombo, J. P., and Weber, H., Ultramikromethoden im klinischen Laboratorium. VII. Bestimmung der sauren Prostata-Phosphatase, *Schweiz. med. Wschr.*, **92**, 1496 (1962).
5. Laessig, R. H., Hassemer, D. J., Paskey, T. A., and Schwartz, T. H., The effects of 0.1 and 1.0 per cent erythrocytes and hemolysis on serum chemistry values, *Am. J. clin. Path.*, **66**, 639 (1976).
6. Mather, A. and Mackie, N. R., Effects of hemolysis on serum electrolyte values, *Clin. Chem.*, **6**, 223 (1960).
7. Caraway, W. T., Chemical and diagnostic specificity of laboratory tests, *Am. J. clin. Path.*, **37**, 445 (1962).

1.8.1.7. Lipaemic plasma (serum)

If the neutral fat concentration in plasma is above 400 mg per 100 ml the plasma appears to be lipaemic. If the turbidity does not disappear during the chemical processing there is a risk of an abnormally high photometer reading. This applies particularly to readings taken in the short-wave region (300–500 nm). For such cases most authors recommend running a sample blank diluted with physiological saline solution to correspond with the sample and using this value to correct the extinction coefficient. However, such a procedure is incorrect since the degree of turbidity is altered by the colour reactions. The only course remaining is to attempt to eliminate the turbidity by deproteination, by extraction of the lipids, or by centrifugation (*ca.* 18 000 *g*).

1.8.1.8. Icteric plasma (serum)

With icteric sera the inherent colour has a pronounced effect on measurements in the short-wave visible region (400–500 nm). Nevertheless, these can usually be corrected by means of the appropriate sample blank. Certain chemical reactions, e.g. the Liebermann–Burchard reaction, produce changes in the colour of icteric sera (conversion of bilirubin to biliverdin). In such cases preliminary extraction of the components to be determined is the sole remaining recourse.

1.8.2. COLLECTION OF BLOOD

1.8.2.1. Collection of capillary blood

A technique for collecting capillary blood was described as early as 1921,[1] but such very small quantities of blood could not be used for routine analysis until the sensitivity of the methods of determination had been increased.

Capillary blood is taken from the tip of the finger, the heel (sucklings), or the lobe of the ear. First of all the site is rendered hyerpaemic by bathing with 70% alcohol, by warming in a bath at 45 °C, or by laying on warm compresses. After cleaning with 70% alcohol, it is advisable to coat the puncture sites with a hydrophobic cream (e.g. Hemo-lube). Enough cream to cover a pin-head is distributed over the skin with a sterilized lancet. The skin is then pierced deeply with the lancet through the cream. The blood is allowed to ooze out spontaneously until it has formed a large drop. After filling the collecting vessel the site of the puncture is covered with a sterile adhesive plaster.

1.8.2.2. Collecting vessels for capillary blood

Various principles and types of vessel have been described for collecting blood.

1.8.2.2.1. Collection of whole blood straight into the reaction vessel

The ideal methods are those in which whole blood can be used for the analysis. Thus, for the determination of the microhaematocrit and of haemoglobin, the blood is collected straight into the appropriate capillary. The same applies to blood-gas analyses. For the determination of glucose, 20 μl of blood are rinsed straight into 0.2 ml of perchloric acid. The determination of haemoglobin is also carried out by feeding the blood directly into the diluent solution.

1.8.2.2.2. Collection of blood in polyethylene vessels

Today glass vessels have been almost completely superseded by commercially available polyethylene vessels holding 0.4–0.5 ml of blood. The tubes are filled as full as possible and sealed air-tight with the accompanying lids.

1.8.2.2.3. Collection in heparinized glass capillaries

Technically the simplest method is to collect the blood in heparinized glass capillaries, 149 × 2 mm internal width (Natelson),[2] and short capillaries, 75 × 2 mm internal width (Caraway).[3] Both types are commercially available. After the blood has been collected they are closed with plastic caps. Collection is effected by gravity and capillary action: the capillaries are held almost horizontal to the blood drops. Mix well. With these capillaries it is easy to obtain ca. 100 μl of plasma. The collecting tubes are finally centrifuged in an ordinary centrifuge (by placing them in a centrifuge tube).

For pipetting out the serum or plasma, the polyethylene centrifuge cups can be cut with a razor-blade, the glass capillaries with an ampoule-saw, the cut being made at the plasma–cell boundary. With most types of micropipette the tip is fine enough for the serum or plasma to be drawn off directly.

References

1. Wright, A. E. and Colebrook, L., *Technique of teat and capillary glass tube, being a handbook for the medical research laboratory and the research world*, 2nd ed., Constable, London, 1921.
2. Natelson, S., *Microtechniques of clinical chemistry*, 2nd ed., Thomas, Springfield, 1961.
3. Caraway, W. T., *Microchemical methods for blood analyses*, Thomas, Springfield, 1960.

1.8.2.3. Collecting vessels for venous blood

One source of error of practical significance is the collection of blood in dirty tubes and closing them with even dirtier stoppers. If possible, the tubes should be used once only and then thrown away. Polycarbonate tubes (unbreakable in the post) are best suited for this purpose.

1.8.3. ANTICOAGULANTS

1.8.3.1. Introduction

A number of arguments favour the use of plasma instead of serum for chemical analyses. Anticoagulated blood can be centrifuged immediately without waiting until coagulation is complete. Haemolysis is always smaller for plasma than for serum; plasma haemoglobin amounts to around 3 mg per 100 ml and serum haemoglobin to around 30 mg per 100 ml. During coagulation there is an egress of substances from the erythrocytes, especially enzymes, e.g. acid phosphatase, and these result in a falsification of serum values. Intracellular substances are also released from the thrombocytes during the viscous metamorphosis and coagulation. Finally, plasma is the physiological body fluid and the absence of fibrinogen is not the only difference between serum and plasma. It may be mentioned that the oldest method of anticoagulation was the elimination of fibrinogen by defibrination. This method is no longer used today. The only modern methods of practical significance are the addition of oxalate, fluoride, citrate, EDTA, and heparin.

1.8.3.2. Oxalate

Lithium, potassium, and sodium oxalate prevent coagulation of the blood by binding the calcium. There ought to be little or no difference between the various salts. An oxalate concentration in blood of 2–3 mg ml^{-1} is recommended but in no circumstances may the concentration of 3 mg ml^{-1} be exceeded, otherwise haemolysis occurs and there is extensive shrinkage of the erythrocytes. It is convenient in practice to use 0.01 ml of a 20% solution per 1 ml of blood. Drying the solution is then unnecessary since the error of dilution amounts to less than 1%. Frequently the oxalate solution is evaporated, but a factor often overlooked in such cases is that oxalate is decomposed to carbonate on heating to 100 °C and the anticoagulant activity is consequently lost. Drying is best effected in a heating cupboard at 80 °C.

There are numerous disadvantages attendant upon the use of oxalate. It leads to a shrinkage of the erythrocytes, and consequently to a dilution of the plasma with erythrocyte water. The haematocrit of oxalate blood can be up to 8–13% lower than that of heparin blood, leading to errors of over 5% in the analysis of plasma components. Also, oxalate produces haemolysis, especially if present in high concentrations. Oxalate cannot be used in acid–base studies since it alters the pH considerably. Certain electrolyte analyses and certain methods of calcium determination are precluded, depending on the particular salt used. Oxalate also interferes with the determination of certain enzymes, e.g. lactate dehydrogenase. For these reasons the use of oxalate as anticoagulant in the clinical–chemical laboratory is not advisable.

1.8.3.3. Citrate

Citrate (5 mg ml^{-1} in blood) has occasionally been recommended. Its mode of action and its advantages and disadvantages are almost the same as for oxalate, but above all it has a very pronounced effect in shrinking the erythrocytes. For carrying out the sedimentation reaction, the adopted procedure is to mix 1 part of 3.8% sodium citrate (dihydrate) solution with 4 parts of blood.[1]

Reference

1. Jacobs, Ph., Letter to the editor: Concentration of citrate anticoagulant, *Am. J. clin. Path.*, **60**, 941 (1973).

1.8.3.4. ACD solutions

ACD solutions (*acidum citricum*/dextrose) are used for the preservation of blood samples. The concentration of the solution is chosen so as to maintain the vitality of the erythrocytes for as long as possible. Hence also the addition of glucose, which serves as a substrate for the energy-consuming processes of the erythrocytes. ADC solutions are ideal nutrient substrates for all microorganisms and are therefore unsuitable for work under non-sterilized conditions. However, they are excellent if erythrocytes are to be isolated for enzyme and metabolic investigations. Hence an ADC solution (4.7 g of citric acid, 16.0 g of trisodium citrate, 25.0 g of glucose, DM-water to 1000 ml), used in the ratio of 1 part of solution to 1 part of blood, is recommended for obtaining, preserving, and isolating erythrocytes. At 4 °C erythrocytes may be stored for a few days in such solutions without changes in the enzyme concentrations, and even at room temperature they may be relied upon not to change over 48 h (in the post).

1.8.3.5. Fluoride

Fluoride acts both as a coagulation inhibitor (by binding calcium) and as an inhibitor of glycolysis in the erythrocytes. A dosage of 2 mg ml^{-1} of sodium fluoride in blood is usually recommended. For the glucose determination whole blood treated with sodium fluoride can be kept for 60 min at 20 °C, or up to 6 h at 4 °C. Without the addition of fluoride the values can no longer be used after standing for only 30 min at 4 °C. Fluoride inhibits urease in urea analyses and, depending on the salt used, interferes with certain electrolyte determinations. The blocking of the metabolism results in certain intracellular components passing out of the erythrocytes. These considerations impose limits on the recommendations of fluoride as an anticoagulant for chemical analyses. Interferences have also been observed in the enzymatic cholesterol determination, leading to reduced values.[1]

Reference

1. Bachmann, C., Personal communication.

1.8.3.6. EDTA

Ethylenediaminetetraacetic acid (EDTA) forms a complex with calcium and in this way it acts as a coagulation inhibitor. Most frequently 1 mg of the disodium salt (dihydrate) is used per millilitre of blood. The disodium salt has the advantage of being more soluble. It is advisable to prepare a 1% solution, distribute this amongst the individual tubes and then to dry these at room temperature. EDTA has no effect on the volume of the erythrocytes. Interference is to be expected in the determination of certain electrolytes and of calcium, if this should be necessary. Little is known about the action of EDTA on enzymes. A strong inhibition of ceruloplasmin and an activation of acid phosphatase have been observed.

1.8.3.7. Heparin

Heparin is a 'physiological' anticoagulant already present in the blood in small concentrations as a normal constituent. Chemically it is a dextro-rotatory polysaccharide built from hexosamine and hexauronic acid units carrying sulphate-ester groups. It has an antilipaemic and an anticoagulant action but these effects do not always run parallel. The molecular weight of heparin lies between 10 000 and 20 000. The coagulation-inhibiting action of heparin is complex and multi-stage: it inhibits the activation of prothrombin to thrombin and the coagulation of fibrinogen to fibrin by thrombin, and, finally, it stabilizes the thrombocytes. Until the structure of heparin is better resolved, quantities should not be stated in units of weight but in biological units, i.e. in terms of the coagulation inhibition. Formerly the heparin was standardized so that 1 mg of the dry substance constituted 100 units. Today there are products with 120–150 units per mg dry weight. For the inhibition of coagulation a concentration of 75 units ml^{-1} of heparin in blood (ca. 0.75 mg ml^{-1}) is most suitable. To evaporate the heparin solutions it is permissible to heat them at 100 °C for 30 min. However, the higher the drying temperature, the less soluble is the residue. Heparin is commercially available as the sodium, potassium, lithium and ammonium salts. For obvious reasons we use the ammonium salt exclusively. The practical procedure is as follows: first, 300 mg of ammonium heparinate (corresponding to 30 000 units) is dissolved in 20 ml of DM-water. Each tube receives 50 μl (or 75 units) of solution per millilitre of blood. These solutions are evaporated at 90–100 °C. Prepared in this fashion, the tubes may be kept indefinitely.

The use of ammonium heparinate permits all electrolyte and acid–base analyses to be carried out, as well as the determination of all metabolites other than ammonia. As far as enzymes are concerned, ammonium heparinate only interferes with the analysis of acid phosphatase; this enzyme is rapidly inactivated in neutral to weakly alkaline media. As adumbrated in this account, ammonium heparinate is the anticoagulant of choice.

1.8.4. STORAGE OF SPECIMENS

1.8.4.1. Whole blood

The storage of investigational material is a difficult problem which has, as yet, received little attention. Keeping at room temperature (25 °C) for 24 h leads to the following changes in concentration: potassium, calcium, phosphate, and total protein increase; chloride, CO_2, bicarbonate and bilirubin decrease; sodium, cholesterol, and urea remain unchanged.[1] Whole blood can be kept for analysis for 4 h without additive. At room temperature (20 °C), just as at 4 °C, there are no significant changes in cholesterol, phosphate, urea, creatinine, triglycerides, GOT, GPT, alkaline phosphatase, and creatine kinase over this period. The LDH and HBDH activities increase. Glucose shows a reduction in concentration after only 30 min (personal observation). Whole blood may not be kept for longer periods, even at 4 °C. Keeping blood at room temperature (e.g. in the post) should be thoroughly discountenanced. There is no chemical additive which inhibits these diagnostically important changes. If samples are to be sent through the post or subjected to lengthy transportation at ambient temperatures it is imperative to centrifuge the blood first and to send only the plasma or serum.

1.8.4.2. Plasma and serum

If plasma, or serum that has been centrifuged immediately, is to be kept, the following rules apply:

1. At room temperature no changes of metabolites or enzymes are to be expected for 6 h.
2. At 4 °C serum or plasma may be kept sealed without change for up to 24 h.
3. If plasma or serum are to be stored for longer periods the samples should either be frozen or lyophilized.

Keller[1] has shown that, surprisingly, for plasma stored at room temperature (25 °C) the changes in concentration occurring in the first 24 h are relatively small: sodium, potassium, chloride, phosphate, cholesterol, and urea remain unchanged; protein, calcium, and creatinine increase; bilirubin, CO_2, and uric acid decrease. With creatinine, it is not

Table 22. Stability of enzymes in serum under different storage conditions; <10% reduction in activity during the time stated.

Enzyme	+25 °C (room temperature)	0 to +4 °C (cold storage)	−25 °C (deep-frozen)
Acid phosphatase	4 h[a]	3 days[b]	3 days[b]
Alkaline phosphatase	2–3 days[c]	2–3 days	1 month
Aldolase	2 days	2 days	Unstable[d]
α-Amylase	1 month	7 months	2 months
Cholinesterase	1 week	1 week	1 week
Creatine kinase, activated	2 days	1 week	1 month
γ-Glutamyl transpeptidase	2 days	1 week	1 month
Glutamate dehydrogenase	1 day	2 days	1 day
Aspartate aminotransferase	3 days	1 week	1 month
Alanine aminotransferase	2 days	1 week	Unstable[d]
Leucine arylamidase	1 week	1 week	1 week
Lactate dehydrogenase	1 week	1–3 days[e]	1–3 days[e]

[a]At pH 5–6.
[b]With addition of citrate or acetate.
[c]Increased activity possible.
[d]Enzyme unsuited for thawing.
[e]Depending on the relative proportions of isoenzymes in the serum.

the 'true creatinine' but the Jaffé-positive substances which appear to increase.[2]

On keeping for longer than 24 h, non-frozen samples are liable to bacterial growth. Previously, thymol fluoride, or an antibiotic, was generally used as bacteriostat. However, freezing or lyophilization is doubtless a better policy than the use of chemical additives. The following researches have been concerned with the preservability of frozen or freeze-dried serum or plasma: Walford et al.[3] have investigated the most important components of deep-frozen plasma after a period of 6 months. The only significant changes were with residual nitrogen, glucose, and alkaline phosphatase. The following components remained unchanged: albumin, globulin, total protein, urea, uric acid, creatinine, cholesterol, bilirubin, chloride, amylase, and acid phosphatase. Levey and Jennings[4] found no alteration in total protein, albumin, chloride, and urea for samples stored frozen for long periods, but the γ-globulin disappeared completely. Strumia et al.[5] monitored single components in dried and frozen plasma over a period of 10 years. Plasma albumin and plasma globulin did not change. The activity of prothrombin decreases slightly after 5 years. Electrophoretic investigations showed astonishingly small differences. Table 22 gives results on the stability of enzymes under different storage conditions.[6]

References

1. Keller, H., Lagerbedingte Fehler bei der Bestimmung von 11 Parametern in heparinisiertem Vollblut und Plasma, Z. klin. Chem. klin. Biochem., **13**, 217 (1975).
2. Kirberger, E. and Keller, H., Lagerungsbedingte Fehler bei Creatinin-Bestimmungen, Z. klin. Chem. klin. Biochem., **11**, 205 (1973).
3. Walford, R. L., Sowa, M., and Daley, D., Stability of protein, enzyme and nonprotein constituents of stored and frozen plasma, Am. J. clin. Path., **26**, 376 (1956).
4. Levey, S. and Jennings, E. R., The use of control charts in the clinical laboratory, Am. J. clin. Path., **20**, 1059 (1950).
5. Strumia, M. M., McGraw, J. J., and Heggestad, G. E., Preservation of dried and frozen plasma over a ten-year period, Am. J. clin. Path., **22**, 313 (1952).
6. Bergmeyer, H. U., Standardisation of enzyme assays, Clin. Chem., **18**, 1305 (1972).

1.8.5. INFECTIOUSNESS OF BLOOD

All investigational material arriving in the laboratory is potentially infected and represents a hazard to the laboratory personnel. But of all occupational illnesses in the clinical–chemical laboratory, the most serious and the only one of practical significance, is viral hepatitis B (serum hepatitis). All investigational material—blood, plasma, serum, urine, faeces—from hepatitis patients is highly infectious and even traces can result in infection by oral assimilation. It is probably that most of the people working in the laboratory contract a hepatitis infection even in their first year. Often this goes unnoticed since the majority of hepatitis diseases are anicteric and can only be confirmed by suitable laboratory tests, e.g. determination of glutamate–pyruvate transaminase and of hepatitis antigen. Scrupulous hygiene is the only prophylactic measure which can be recommended.[4,5] No coffee drinking or eating in the laboratory should be permitted.

Samples suspected of being hepatitic must be clearly distinguished, e.g. with a yellow sticker. The risk of sickness can be reduced by administering hepatitis B immunoglobulin.[6]

References

1. Serum hepatitis (hepatitis B): a major challenge to workers in infectious and iatrogenic disease, *Can. med. Ass. J.*, **110**, 974 (1974).
2. Grist, N,.R., Hepatitis in clinical laboratories: a three-year survey, *J. clin. Path.*, **28**, 255 (1975).
3. Krassnitzky, O., Pesendorfer, F., and Wewalka, F., Hepatitis und Laboratorium, *Med. Lab., Stuttg.*, **27**, 77 (1974).
4. Tesar, V., Laboruntersuchungen zur Hepatitis-Prophylaxe, *Z. med. Labortechn.*, **175**, 3 (1976).
5. *Sicherheit im medizinischen Laboratorium. Ein Leitfaden*, Birkhäuser, Basle, 1976.
6. Hepatitis B immune globulin—prevention of hepatitis from accidental exposure among medical personnel, *New Engl. J. Med.*, **293**, 1067 (1975).

1.8.6. DRUG INTERFERENCES

Misleading laboratory results are not always due to mistakes or to wrong analyses. Responsibility often lies with the doctor, whether it be a consequence of the faulty collection of material, inadequate provision for the limited durability of the test substance, or an ignorance of the interfering effects of drugs. In the following, an attempt will be made to particularize some of the well known interferences, using personal observations and information culled from the literature. These interferences are of three types:

1. Pharmacological action of a drug on a particular component, e.g. of ACTH (adrenocorticotropic hormone) on the blood-glucose concentration.
2. Pharmacological side-effects of a drug, such as the increase in plasma amylase after assimilation of morphine and its derivatives.
3. Chemical interference with the determination of a component, e.g. interference by phenothiazine metabolites in the determination of 5-hydroxyindoleacetic acid.

Often it is not clear which of these groups a particular effect should be assigned to. Also, up to now there have been very few systematic investigations on such interferences and, for the most part, we are left with chance observations, some of them poorly documented.

A tabular summary of some of the more important interferences may be found in Appendix 10. For extensive lists of interfering drugs see refs. 1 and 2.

References

1. Bibliography: drug interferences with clinical laboratory tests, *Clin. Chem.*, **18**, 1043–1304 (1972).
2. Young, D. S., Testaner, L. C., and Gibberman, V., Effects of drugs on clinical laboratory tests, *Clin. Chem.*, **21/5**, 1D–432D (1975).

1.9. URINE COLLECTION

1.9.1. GENERAL

1.9.1.1. Collection of urine

The type of collection and preservation of urine is designed to suit the needs of the morphological or chemical investigations planned.

Sediment. A reliable assessment of the components of urine sediment is only possible if the urine specimen fulfils the following conditions:

1. fresh (within 2 h of emptying bladder);
2. concentrated (specific gravity above 1010);
3. acidic (pH less than 6.0).

For the routine investigation of urine sediment these criteria are generally satisfied by the first morning urine. For the bacteriological investigations a mid-flow 'grab' (after cleaning the genitals) is required and can also be used directly for slide cultures.[4]

Chemical investigations. Apart from the electrolytes, all the products of excretion are subject to degradation by bacteria and fungi. It is therefore necessary to preserve urine if it is not going to be processed immediately. This can be done by prompt freezing (below $-20\ °C$) or by the addition of a 10% solution of thymol–isopropanol urine preservative;[1] in the light of recent investigations freezing is to be preferred.[6]

Special measures should be adopted with the following metabolites.

Sugar. For the identification of sugar, fresh urine, collected following a drug-free period of at least 48 h, is used (a sterile collection is best).

Porphyrin. Owing to its photosensitivity it should be protected against light and collected in the dark, adding $5\ g\,l^{-1}$ of sodium bicarbonate. It keeps for a few days.

Steroids. Urine collection with addition of *ca.* 5 g of boric acid per 24-h collecting period.

Amino acids. Only stable for a few days in urine; toluene or thymol is often used as an additive, leading to possible interference with the analyses. It is best to freeze individual portions of urine.

Catecholamines and serotonin. 24-h urine collection, preservation with concentrated perchloric acid (1 ml perchloric acid per 100 ml of urine). Stable for several days at pH 4. If the patient has eaten nuts prior to the collection, this results in interferences in the serotonin determination.

In addition, urine for chemical investigations should be collected sterile as far as possible. For many years catheter urine has been regarded as the specimen of choice, but catheterization is to be avoided where possible since there is a high risk of infection.[2,3]

1.9.1.2. Preservation of urine with thymol–isopropanol

The addition of toluene, thymol, and organic mercury salts during the collection of urine is little more than a ritual. The solubility of these substances is so low that bacteriostatic and fungistatic concentrations are never reached. The addition of mineral salts interferes with the determination of pH and titratable acids.

Nattalin and Mitchell[1] have pointed out that the low solubility of thymol can be increased by using a mixture of thymol and isopropanol. As a rule, 5 ml of a 10% solution of thymol in isopropanol suffices for one 24-h urine collection. The resulting concentration inhibits the growth of bacteria sufficiently to prevent any substantial changes in the metabolite concentrations.

Addition of thymol–isopropanol does *not interfere* with the following analyses: sodium, potassium, calcium (flame photometry), chloride (Schales and Schales), bicarbonate (van Slyke, Natelson), calcium (Shohl and Pedley), inorganic phosphorus (Fiske and Subbarow), ammonia (Nessler, Berthelot), amino acids (ninhydrin, chromatography), creatine and creatinine (Jaffé, van Pilsum), amylase (strong hydrolysis), urea (Berthelot), protein (biuret, heat coagulation test), sugar (reduction test, glucose oxidase), D-xylose (*p*-bromoaniline), urobilin, urobilinogen, porphobilinogen (Walenström), acetic acid (Gerhardt), acetone (Rothera), bilirubin (Fouchet), and indican (Obermayer and Jaffé).

The following determinations may *not be carried out* following the addition of thymol–isopropanol: 17-ketosteroids (Zimmermann), bile acids (Hay).

1.9.2. CHOICE OF REFERENCE PARAMETERS

The basis of every balance investigation is a reliable collection of urine over the whole 24-h period. It is advisable always to conclude the collection period with the first morning urine. For many chemical or

112

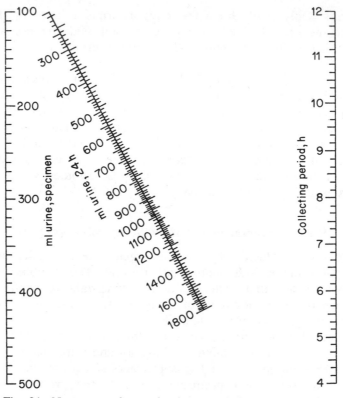

Fig. 21. Nomogram for evaluating the theoretical 24-h urine
volume from the collecting period (right-hand scale) and the
urine volume (left-hand scale) collected in this time. The value
of the right-hand scale (e.g. 6 h) is joined to the value on the
left-hand scale (e.g. 300 ml) by a straight line. The intersection
of the straight line with the central scale gives the theoretical
24-h excretion (1200 ml).

morphological investigations, urine is often collected for a few hours only
although this is not to be recommended. It is more practical and more
precise to note the time interval between two voidings rather than to
collect the urine over a fixed period of time. The theoretical 24-h volume
can easily be calculated from the quantity of urine and the time interval or
it may be read off from a nomogram (Fig. 21),[5] but it should be borne in
mind that individual excretion can show considerable variation in volume
over the 24-h period.

The 24-h collections give untrustworthy results, particularly if they are
carried out at home. The following methods should be used to control the
reliability.

1.9.2.1. Determination of the creatinine coefficient (cf. p. 691)

1.9.2.2. Determination of the specific gravity[4]

The commonest method for determining the specific gravity involves the use of an areometer (hydrometer, urinometer), but this yields results which are not very reliable. The refraction of light can be measured using a refractometer; the refraction is proportional to the number of dissolved particles. The Goldberg temperature-compensated refractometer (TS-meter, American Optical Company, Buffalo) has proved particularly good. A correction must be applied in cases of glucosuria or proteinuria.

References

1. Nattalin, L. and Mitchell, L. R., A new urine preservative, *Clin. chim. Acta,* **3**, 197 (1958).
2. Beeson, P. B., The case against the catheter, *Am. J. Med.,* **24**, 1 (1958).
3. Reber, H. and Massini, M. A., Verhütung von Katheterinfektionen, *Schweiz. med. Wschr.,* **95**, 551 (1965).
4. Colombo, J. P. and Richterich, R., *Die einfache Urinuntersuchung,* Huber, Berne, 1977.
5. Lippman, R. W., *Urine and the urinary sediment,* Thomas, Springfield, 1969.
6. Leach, C. S., Rambault, P. C., and Fischer, C. L., A comparative study of two methods of urine preservation, *Clin. Biochem.,* **8**, 108 (1975).

2. TECHNIQUE AND INSTRUMENTATION

2.1 SEPARATION AND DETERMINATION

The methods of analytical chemistry, and hence of clinical chemistry, may be divided—sampling apart—into the following categories:[1]

—methods for the *separation* of investigational material and/or for the isolation of a particular substance or group of substances;
—methods for *qualitative detection* and/or *quantitative determination* of a defined constituent or group of constituents.

In the field of clinical chemistry, direct methods, not necessitating a primary separation of the investigational material or removal of interfering substances (e.g. removal of albuminous body by deproteination), have been developed successfully for a large number of constituents of serum and urine.

These direct methods imply a reduction in the number of sample transfers and quantifying steps and, as a result, the precision of the method is increased. However, for constituents present in very low concentrations it is often necessary to perform an enrichment and/or an isolation. Chromatographic methods are frequently employed for this purpose, e.g. for steroid hormones, catecholamines, amino acids, pharmaceuticals, and many other substances.

2.1.1. GENERAL COURSE OF ANALYSES

In principle, chemical analytical methods consist of three steps:[2,11]

1. *Sampling.* This provides a random sample (portion) which may be assumed to be representative of the whole. The special difficulties encountered with biological material are dealt with elsewhere (p. 94).
2. *Analytical process.* This consists of various individual processes, the most important being:
   (a) Quantifying the investigational material and reagents. Since we are concerned for the most part with liquids, it is essential to work at constant temperature in order to avoid volumetric errors. In this step, care should be taken to ensure that there is no spillage of reagents or samples.
   (b) Mixing processes generally follow the quantifying stage. Here too, all spillages must be scrupulously avoided.

(c) Methods of separation serve to separate out interfering components or phases, or to isolate and/or enrich the desired constituent.

(d) Incubation at a prescribed temperature for a fixed period of time initiates the reactions and/or brings them to their conclusion.

3. *Obtaining the analytical results.* As a rule, one or more physical properties of the analytical material are measured by means of a suitable instrument. This may be a one-off, repeated, or continuous process. The instrument yields one or more signals which have to be converted to an analytical result. The calculations involved are of varied intricacy, ranging from a simple multiplication to more complex mathematical procedures.

2.1.2. WORKING VOLUMES

The quantities of sample material and reagent solutions required for an analysis in clinical chemistry have been systematically reduced. In 1850 Schmidt[3] required about 300 ml of blood for the determination of serum electrolytes. Today, one ten thousandth of this quantity (30 μl) is considered ample in every qualified clinical–chemical laboratory. Between 1950 and 1960, different research groups, working independently, developed clinical–chemical methods requiring very much smaller quantities of reagents and sample material than hitherto: the millilitre range gave place to the microlitre range.[4-9]

As regards nomenclature, there is no unity even today. It is not clear whether the quantity of investigational material used or the volume of the total assay is to be taken as a criterion. Table 23 shows a classification used by many authors. Most laboratories prefer the semimicro range for manual or part-automated methods.

These new techniques have resulted in a drop in sample consumption—which is more considerate for the patients, especially for sucklings and infants—and in the amount of reagent required, and hence, in the cost of expendable materials. The spatial requirements are also reduced since the apparatus can be made smaller. However, the most important advantage conferred by the microlitre technique is the increase

Table 23. Description of methods according to the volumes involved.

Method	Sample volume, ml	Total volume, ml
Macro	>0.5	>2
Semi-micro	0.5–0.1	2–1
Micro	<0.1–0.005	<1–0.1
Ultramicro	<0.005	<0.1

Table 24. Change in the surface area to volume ratio of a sphere with decreasing volume.

	Volume, μl				
	1000	500	100	50	10
Surface area, mm^2	484	305	104	66	22
Surface area to volume ratio	0.48	0.61	1.04	1.33	2.24

in the number of samples which can be processed: for a sphere, the ratio of surface to volume increases as the volume becomes smaller, i.e. the reacting surfaces are relatively greater at smaller volume. Table 24 shows this clearly. Because of this relative increase in surface with diminution in volume, temperature regulation, dosing, and mixing demand increasingly less time.

With manual methods it is not advisable to work with volumes less than 10 μl since the demands made on manual dexterity increase sharply in the lower ranges. With mechanized systems the limit of practicability may be reckoned to lie in the ultramicro (nanolitre) range.

2.1.3. MANUAL, MECHANIZED, AND AUTOMATED METHODS

With *manual methods*, each single analytical step of each single assay is conducted manually.

With *mechanized methods*, apart from the process control, all of the analytical steps are performed solely by means of mechanical (electromechanical, pneumatic, hydraulic, etc.) appliances. Semi-mechanical methods require manual (active) participation for individual steps, e.g. transference of the assay preparation to the instrument.

The term *automated method* should only be used if the method is monitored and guided by a (electronic) process control.[10] The process control can implement logical decisions, e.g. the analytical process can be repeated with a smaller sample volume if a signal exceeds a certain value, showing that the measurement lies outside the reliable range. Such automatic processors are not yet on the market although the technology for producing them exists and the possibilities have been realized in other fields.

References

1. Laitinen, H. A. and Harris, W. E., *Chemical analysis*, 2nd ed., McGraw-Hill, New York, 1975.
2. Richterich, R. and Greiner, R., Analysatoren in der klinischen Chemie. II. Terminologie und Klassifikation, *Z. klin. Chem. klin. Biochem.*, 9, 187 (1971).
3. Schmidt, C., *Charakteristik der epidemischen Cholera*, Mitau und Reyher, Leipzig, 1850.

4. Sanz, M. C., Ultra micromethods and standardization of equipment, *Clin. Chem.*, **3**, 406 (1975).
5. Schnitger, H., Vorrichtung zum schnellen und exakten Pipettieren kleiner Flüssigkeitsmengen, *Dt. Patentamt*, Auslegeschr. Nr. 1090449.
6. Mattenheimer, H., *Mikromethoden für das klinisch-chemische und biochemische Laboratorium*, de Gruyter, Berlin, 1961.
7. Natelson, S., *Microtechniques of clinical chemistry*, 2nd ed., Thomas, Springfield, 1961.
8. Seligson, D., An automatic pipetting device and its application in the clinical laboratory, *Am. J. clin. Path.*, **28**, 200 (1957).
9. Glick, D., *Quantitative techniques of histo- and cytochemistry*, Vols. 1 and 2, Interscience, New York, 1961, 1963.
10. Richterich, R. and Greiner, R., Analysatoren in der klinischen Chemie, *Z. klin. Chem. klin. Biochem.* **8**, 588 (1970).
11. Lauber, K., *Chemie im Laboratorium*, 3. Aufl., Karger, Basle, 1975.

2.2 GRAVIMETRY

Weighing, as an analytical method, plays only a subsidiary role in clinical chemistry. On the other hand, weighings for the preparation of reagent solutions are an important part of the laboratory work. Consequently, a knowledge of the fundamentals of the theory of weighing continues to be of considerable importance.[1-5]

2.2.1. FUNDAMENTALS OF THE THEORY OF WEIGHING

In general, the purpose of weighing is to determine the mass of a body. The mass—in the physical sense, the product of the volume and the density of a body—cannot be measured directly but must be determined indirectly, e.g. by weighing. The weight G is the force developed by a mass m under a gravitation g:

$$m = \frac{G}{g}$$

The earth's gravity is not distributed uniformly: at the equator the distance to the point of action of the force of gravity is greater than at the poles. At the same time, the rotation of the earth produces a centrifugal force which is greatest at the equator and zero at the poles. As a result of the combination of these two effects, 1 litre of water at 18 °C is about 5 g lighter at the equator than at the poles. In addition to these forces, every body is subject to the attractive forces of the sun and the moon.

2.2.2. BALANCES

The beam balance serves to compare the known mass of the weights with the unknown mass of the object weighed. This is accomplished by means of the balance of forces. Two types of balance are available to meet the requirements of the clinical–chemical laboratory, the *precision balance* and the *analytical balance*. As a rule, one of each is sufficient. A survey of operating range, limit of readability, and reproducibility is given in Table 25. A reproducibility of 0.005 g with the precision balance and 0.05 mg with the analytical balance ought to be sufficient. The precision balance can be located in the laboratory but the analytical balance should be in a

Table 25. Characterization of types of balance in relationship to their operating ranges.

Type of balance	Weighing range, g	Scale-reading limit, g	Reproducibility, g
Precision balances			
Electronic	30 000	1	±0.5
	15 000	0.1	±0.05
	3 200	0.1	±0.05
	1 200	0.01	±0.005
	320	0.001	±0.0005
Mechanical	20 000	2	±1
	10 000	1	±0.5
	5 000	0.1	±0.05
	2 200	0.01	±0.01
	160	0.001	±0.001
Analytical balances			
Electronic	160	1 mg	±1 mg
	160	0.1 mg	±0.1 mg
	160	0.01 mg	±0.02 mg
Mechanical	160	0.1 mg	±0.05 mg
	160	0.01 mg	±0.01 mg
Microbalances			
Electronic	3.3	10 μg	±10 μg
	3.12	1 μg	±1 μg
	3.10	0.1 μg	±0.3 μg
Mechanical	20	0.001 mg	±0.001 mg

special room and it is expedient to mount it on its own weighing-table, as shown in Fig. 22.

2.2.3. WEIGHING ERRORS

A knowledge of the possible sources of error is crucial to weighing technique.

(a) *Moisture film.* Every object carries a surface film of moisture corresponding with the humidity of the surrounding air. The lower the temperature of the object in comparison with its surroundings, the thicker is the film. The object to be weighed should therefore be placed upon the balance pan when its temperature is the same as that of its surroundings.

(b) Weighing-errors due to *dampness.* Many substances are hygroscopic and change their weight by absorbing water. It is therefore practical to dry every substance in a desiccator and to transfer it straight from the desiccator to the balance pan.

(c) Weighing errors due to *air buoyancy.* According to Archimedes'

123

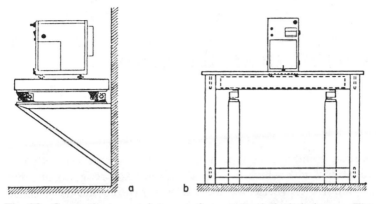

Fig. 22. Convenient appointment for an analytical balance. The table should not be larger than the balance, otherwise the extra surface at the operator's disposal can introduce further possible sources of error. (a) Wall console; (b) free-standing table.

Principle, the apparent loss in weight of a body is equal to the weight of the medium it displaces. Since, at 20° C, 1 ml of air weighs *ca.* 1.2 mg, inadmissibly large errors can occur with low-density bodies (e.g. plastics). In such cases, after weighing, the weight of the displaced air must be added to the apparent weight of the body.

(d) Weighing errors due to *electrostatic forces.* Electrostatically charged objects are subject to other forces in addition to their weight. Like charges on the object being weighed and on its surroundings lead to a repulsion, whereas unlike charges result in an attraction. A charged object (e.g. plastics) can therefore appear to be too heavy or too light. For electrically conducting bodies this can be remedied by discharging with earthed metal tongs. Non-conducting objects must be discharged by means of ionizing substances (e.g. a radioactive preparation).

(e) Weighing errors due to *foreign bodies.* The weight of foreign bodies and impurities can falsify the weight being measured. For this reason only absolutely clean weighing vessels should be used; the balance case and weighing pan must also be kept clean. The weighing vessels should not be placed on the pan by hand but should be lifted with tongs.

References

1. Plant, A. F., Microbalances, in Werner, *Microtechniques for the clinical laboratory*, Wiley, New York, 1976.
2. Balances and Weighing, in Foerst, *Ullmans Enzyklopädie der technischen Chemie*, Vol. II/1, 3. Aufl., Verlag Chemie, Weinheim, 1970.
3. Lauber, K., *Chemie im Laboratorium*, 3. Aufl., Karger, Basle, 1975.
4. Schraner, E., Wägetheorie, *Med. Lab., Stuttg.* **25**, 86, 111 (1972).
5. Bietry, L., Lexikon der wägetechnischen Begriffe, *Sonderdruck Chemische Rundschau*, 29, No. 25, 1976.

2.3. VOLUME QUANTIFICATION

The precision and accuracy of analytical methods are critically dependent on the reliability with which the liquid samples and reagents are quantified, and the smaller the precise volumes to be measured the more difficult it is to measure them out accurately.[1-6]

2.3.1. MACRO-PIPETTES

Volumetric apparatus is calibrated 'to deliver' (TD) or 'to contain' (TC). In addition to measuring cyclinders, volumetric flasks, etc., this includes pipettes. This *calibration* is calculated for water; departures from accuracy can be demonstrated even with dilute aqueous solutions.

A vessel calibrated for delivery (TD) holds a volume which exceeds the volume to be measured by as much as remains wetting the walls after emptying. The liquid remaining in a pipette after the bulk has run out is made up as follows:

1. drainage liquid, i.e. the part which flows out after the bulk;
2. the wetting liquid, i.e. the part which clings to the walls owing to the adhesion of the liquid.

This is why there is provision for a waiting period of 15–25 s after delivery of the bulk—the liquid collected in this period belongs to the measured volume. During delivery the pipette should be touched against the wall of the receiving vessel. The liquid finally remaining in the tip of the pipette does *not* belong to the measured volume but is part of the wetting residue and is taken into consideration in the calibration (*do not blow it out!*).

2.3.1.1. Transfer pipettes

Provided that the procedure is followed exactly as stipulated, transfer pipettes deliver the precise quantity of liquid specified on each pipette.

2.3.1.2. Graduated pipettes (measuring pipettes)

Provided that an exactly prescribed procedure is followed, graduated pipettes deliver a specific quantity of liquid, either the nominal volume of the pipette or a fraction of this volume, depending on the system of graduation.

Both transfer and graduated pipettes are calibrated to delivery (TD).

With graduated pipettes it is important to choose the pipette size in accordance with the quantity of liquid to be pipetted. If, for example, 1.5 ml is to be pipetted, a 2-ml pipette (on no account a 10-ml pipette) is the one to use. The widespread practice of running several part-volumes consecutively from a single graduated pipette is incorrect. This procedure is unreliable if exact results are to be achieved, since the pipettes are not calibrated for such usage.

Pipettes with damaged nozzles (tips) ought not to be kept in use but should be destroyed; this applies to both transfer and graduated pipettes.

2.3.1.3. Note on calibration

Both transfer and graduated pipettes are available in two types: with complete and with partial discharge (two-mark pipettes). With the latter type the calibration is based on the volume between the upper and lower marks. The same procedure must be observed as with the complete-discharge types: when the meniscus has reached the lower mark the pipette is stopped for 15–25 s and the meniscus again brought to the mark.

2.3.2. MICROLITRE PIPETTES

For volumes of less than 1 ml, but especially for volumes of less than 10 μl, microlitre pipettes should be used exclusively. The most important types are as follows:

1. *Capillary tubes* made of glass or plastic:

(a) Disposable end-to-end capillaries, also called 'length-calibrated capillaries', must be filled to both ends with the liquid to be measured. To do this they are held between thumb and forefinger, or with tongs, and filled by capillary action. To empty it, the filled capillary is transferred into a closed vessel containing a diluent and the contents are washed out into the diluent by vigorous shaking. Even if rubber teats are used for filling and/or emptying, the 'wash-out' principle should still be retained with capillaries of this type, especially where volumes of less than 100 μl are concerned.

(b) *'End-to-mark' capillaries* are filled only to a fixed mark. The Sahli pipette may be taken as representative (Fig. 23). Modifications include capillary pipettes with a bulge (for increasing the capillary diameter) and with finely tapered or bent tips. These are all 'wash-out' pipettes, generally re-usable. They are increasingly being replaced by disposable, glass, 'end-to-end' capillary tubes. These micro-pipettes are filled up to a ring mark by suction (with the aid of a suction tube) and emptied by blowing or washing out.

126

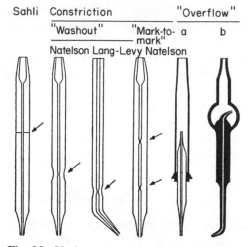

Fig. 23. Various types of microlitre pipette.

(c) *Disposable glass capillaries with dispenser* (Fig. 24) are filled with the aid of a piston, the latter being in direct contact with the sample or the reagent. For volumes of 1–5 μl the piston and piston-rod are identical; for volumes of 5–50 μl the piston is made of plastic and has a tight-fitting flange. The precision and accuracy attainable are around $\pm$ 0.5% for small volumes.

(d) In *constriction pipettes* (Fig. 23) the calibration mark is replaced by a narrowing of the capillary diameter. With slow suction the volume automatically adjusts itself to this point. These are also 'washout' pipettes.

2. With *overflow pipettes*, accurate filling is guaranteed by sucking up an excess volume: the superfluous liquid overflows. These pipettes afford very good reproducibility.

3. The *Sanz pipette* (Fig. 25a–c) is a modification of the overflow pipette; the receiving vessel is connnected to the supply bottle, which has flexible walls and can therefore be used as a pump. The Sanz pipette is available both as reagent and sample pipette.

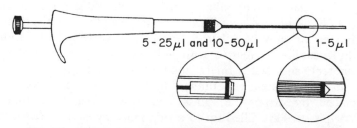

Fig. 24. Disposable glass capillary with piston dispenser.

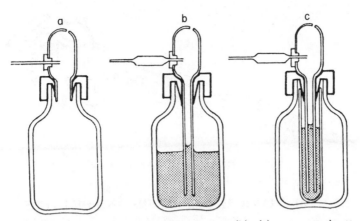

Fig. 25. Sanz pipettes. (a) Sample pipette; (b), (c), reagent pipettes.

4. *Travelling-plunger pipettes* are mechanized pipettes which measure out the required volume by means of a plunger or piston shifting between two stops. Fig. 26 shows the mode of action as exemplified by the Eppendorf pipette. The plunger is depressed and the liquid to be quantified is sucked (across an air cushion) into a disposable plastic tip. Residual liquid still held in the tip by surface tension on the reverse stroke is expelled by an additional stroke.

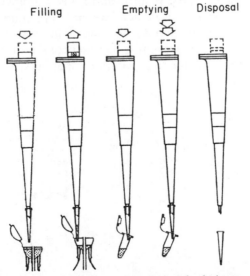

Fig. 26. Operation of an Eppendorf pipette.

128

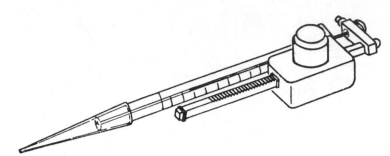

Fig. 27. Hamilton dispenser, manual.

2.3.3. MECHANICAL DISPENSERS AND DILUTERS

If the same reagent volume has to be measured out repeatedly, it is advisable to use a reagent dispenser. The following types may be distinguished:

1. *Valve-less dispensers* are available in different modifications.

(a) One large suction stroke of the plunger fills a reservoir and thereafter measured quantities can be delivered by stepwise feed of the plunger. A manually operated instrument is shown in Fig. 27; it may be used to suck up, e.g. 250 μl, and dispense this in 50 stages 5 μl a time. With a mechanized drive, the reagent is dispensed as illustrated in Fig. 28. The advantage of this method lies in the simplicity of the set-up and in the rapid working cycle. On the other hand, filling is time consuming, the stroke to diameter ratio is metrologically unfavourable, and the reservoir volume is limited.

(b) In principle, tube pumps can also be employed as dispensers, as realized in certain automated systems (Technicon AutoAnalyzer). This principle is applied to advantage in 'continuous-flow' systems whereby a stream of reagent is pumped continuously into the system.

(c) Finally, for feeding rotation analysers, valveless dispensers

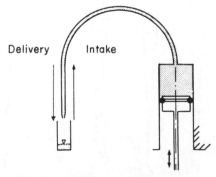

Delivery   Intake

Fig. 28. Valveless reagent dispenser.

which mechanically simulate the principle of the travelling-plunger pipettes are used. A (disposable) sample tip dips into a reservoir, the reagent is sucked up by one stroke of the piston, the pipette tip is swung over to the rotor, and the reagent carried in the tip is delivered by a second stroke of the piston.

2. *Valve dispensers* may be divided into systems in which the valve is controlled by the liquid itself and those in which the valve is controlled by other means. Manual dispensers are usually fitted with disk, ball or membrane valves which allow the liquid to pass in only one direction. Many different modifications are marketed. Naturally the valve is liable to give trouble: a hair or dust particle in the liquid being dispensed can temporarily lead to incorrect dosing and may pass unnoticed. The valve may jam through careless treatment, or become wholly or partially unusable due to the formation of a crust of reagent.

   Dispensers with externally controlled valves are generally motor-driven. The valve may be constructed as, e.g. a magnetic compression valve, a slide valve, or a rotary valve. With simple dispensers only a fixed, pre-set volume can be pipetted. Other dispensers permit dosing with several different volumes or allow for continuous (as opposed to stepwise) adjustment of the volume.

Sample/reagent dispensers ('diluters') suck up a specific quantity of sample on the first stroke and eject it together with a definite volume of reagent on the second stroke. These diluters are likewise available for manual or motorized operation. They are constructed on the same principle as the dispensers.

For further technical details, see the references to Section 2.1 on microlitre technique and the review treatments.[1-6]

2.3.4. CALIBRATION OF PIPETTES

Calibration of pipettes and dispensers is part of the regular control work of the clinical–chemical laboratory. The two most important methods are described in detail below.

2.3.4.1. Gravimetric calibration

A beaker is tared on the analytical balance. Then ten separate quantities of water are pipetted into the beaker and the weight is read to an accuracy of 0.1 mg each time. With pipettes or dispensers which are to be used mainly for plasma or serum, plasma or serum should be used instead of water, since the physical properties of these biological fluids differ appreciably from those of water. The results of the weighings are evaluated statistically.

Attention should be paid to the following: all pipettings and dispensings should be performed at a uniform temperature since the density of the liquid is temperature dependent; so that as little water as possible evaporates during weighing, the collecting vessel should be small, affording a small surface area; alternatively, the water is covered with a layer of liquid paraffin.

2.3.4.2. Photometric calibration

For photometric calibration a known volume of a coloured solution is pipetted into a known volume of solvent and the extinction is measured. Quinine sulphate, cobalt nitrate, nicotinamide and 4-nitrophenol have been recommended as colorants. A solution of 4-nitrophenol, which also serves as standard for the determination of alkaline and acid phosphatase, is particularly suitable. The concentration of this standard solution is 2 mmol l^{-1} (278.2 mg of p.a. 4-nitrophenol is dissolved in distilled water and made up to 1000 ml).

Example. Calibration of a 10-μl pipette (calibration factor).

To calculate the absolute volume of the pipette, 1 ml of standard solution is diluted with 100 ml of 0.1 mol l^{-1} sodium hydroxide solution, using a bureau-calibrated pipette and measuring flask. The extinction of this solution, measured against the sodium hydroxide solution used, serves as a reference value (E_{macro}). The experiment is repeated using the microlitre pipette to be calibrated, in this case 10 μl: a cuvette is filled with 1 ml of 0.1 mol l^{-1} sodium hydroxide solution and the extinction set at 0; then 10 μl of standard solution are added with the pipette to be calibrated, the solutions are mixed and the extinction (E_{micro}) is noted. This experiment is repeated ten times; the mean, standard deviation, and coefficient of variation of the extinction can be calculated from the separate results.

The *actual volume* of the pipette to be calibrated is calculated by multiplying the stated pipette volume by the following calibration factor:

$$\text{Calibration factor} = \frac{E_{micro}}{E_{macro}}$$

The difference between the actual volume of the calibrated pipette and the volume it ought to have corresponds to the *absolute error of the pipette*. The absolute error of the pipette has no significance for methods in which the standard and sample are dispensed with the same pipette. However, it is important in the case of all absolute measurements, e.g. if the molar extinction coefficient is to be used. It is clear that the macropipettes should be used for such calibrations, should all be calibrated absolutely, and the photometer should be a high-grade instrument.

For pipettes with other volumes the macro-experiment is varied accordingly. This procedure can be used for manually operated microlitre pipettes and, with appropriate modification, for reagent dispensers.

References

1. Kirk, P. L., *Quantitative ultramicroanalysis*, Wiley, New York, 1950.
2. Mattenheimer, H., *Micromethods for the clinical and biochemical laboratory*, Ann Arbor Science Publications, Ann Arbor, 1970.
3. Meites, S. and Faulkner, W. R., *Manual of practical micro- and general procedures in clinical chemistry*, Thomas, Springfield, 1962.
4. Natelson, S., *Techniques of clinical chemistry*, 3rd ed., Thomas, Springfield, 1971.
5. O'Brien, D., Ibbott, F. A., and Rodgerson, D. O., *Laboratory manual of pediatric microbiochemical techniques*, 4th ed., Harper & Row, New York, 1968.
6. Trautschold, J. und Löffler, G., Mikrotechniken, in Bergmeyer, *Methoden der enzymatischen Analyse*, 3, Aufl., Verlag Chemie, Weinheim, 1974.

2.4 METHODS OF SEPARATION

Physical methods of separation frequently used in the clinical–chemical laboratory are listed below.[1,27,28]

1. *Centrifugation (flotation)*
   Separation on the basis of differential particle density.
2. *Filtration, dialysis, gel filtration*
   Separation on the basis of differential particle size.
3. *Electrophoresis*
   Separation on the basis of differential electrical charge, size, and shape of the particles.
4. *Extraction, counter-current distribution*
   Separation on the basis of differential solubility.
5. *Chromatographic methods*
   Thin-layer, paper, gel, and ion-exchange methods, in addition to gas chromatography and high-performance liquid chromatography. Separation on the basis of several differential molecular properties, especially charge, solubility, affinity, and adsorption.

Many methods of separation are part of the actual methods of detection or determination. If, e.g., a mixture of dyes is separated into its individual components by thin-layer chromatography, visual inspection of the thin-layer plate allows a qualitative statement to be made as to which individual component dyes are present in the mixture. On the other hand, there are also numerous methods of separation which are only preliminary steps, having nothing to do with the analytical process in its narrower sense, e.g. separating off the blood cells to obtain the blood plasma. Systematization into methods of separation and methods of determination—expedient for teaching purposes—is often not possible in practice.

2.4.1. CENTRIFUGATION

Many clinical–chemical investigations begin with the separation of the cellular material from the liquid to be investigated. In principle this step may be accomplished by filtration or dialysis, but in practice centrifugation is used almost exclusively.

The centrifuge creates a force-field in which the denser particles of the material being centrifuged migrate more rapidly than those of lower density. A fractionation ensues, based on differential density. A unit of measurement for the characterization of the force-field is provided by terrestrial gravitation. Since this depends on the location of the mass concerned (cf. Section 2.1.1.) an average numerical value, the gravitational unit (acceleration due to gravity) has been adopted. It equals 981 ms^{-2} and is symbolized by g.

The performance of a centrifuge is essentially determined by the acceleration which can be produced with it. This is designated as relative centrifugal force, RCF, and is stated in $n \cdot g$. The RCF may be calculated from the following equation:

$$RCF = 1.118 \times 10^{-5} \cdot r \cdot (\text{rpm})^2 \cdot g$$

in which

RCF = the centrifugal acceleration in g, i.e. multiples of the terrestrial acceleration;

1.118×10^{-5} = a constant deriving from the angular velocity;

r = the radius in centimetres measured between the centrifuge axis and the centre of the inserted centrifuge tubes;

rpm = the number of revolutions per minute achieved.

By means of this equation the g-value achieved by a centrifuge may easily be calculated. However, it is also possible to use a nomogram (Fig. 29).

Microcentrifuges which are to be used (*inter alia*) for determining the haematocrit should reach 10 000 g. For separating off serum or plasma, strong fields of between 1000 and 2000 g are generally applied. For very fine sediments, fields of up to 10 000 g are necessary. For further preparative and analytical separations, ultracentrifuges are available, capable of attaining 100 000 g or more. The duration of centrifuging must be at least equal to the sedimentation period of the centrifuged particles. Under identical geometrical conditions, i.e. the same centrifuge, the same centrifuge-head, and the same or similar investigational material, the following, crudely approximate, relationship between the centrifugal acceleration and the duration of centrifuging is applicable:

$$t \approx \frac{k}{RCF}$$

where

t = the duration (minutes) of centrifuging required to cause sedimentation of the particles; and

k = a constant which takes into account *inter alia* the geometry of the centrifuge.

Thus the sedimentation period and the centrifugal acceleration stand in a reciprocal relationship to each other. This means that the centrifuging time

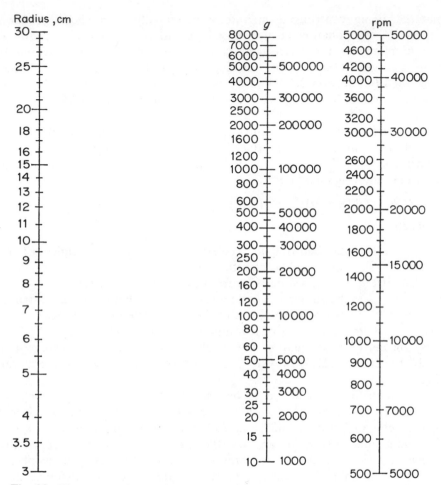

Fig. 29. Nomogram for evaluating the *g*-number of centrifuges. The radius (left-hand scale) is connected with the number of revolutions per minute (rpm) (right-hand scale). The intersection of the straight line with the central scale gives the *g*-number.

can be halved by doubling the acceleration, or *vice versa*. The sedimentation after 5 min at 10 000 g is practically the same as that for an identical system after 10 min at 5000 g. Hence the necessity for stating both the period of centrifuging and the centrifugal acceleration in g.

2.4.2. FILTRATION, DIALYSIS

Filtration through filter-paper is hardly ever used in the clinical laboratory nowadays and a brief look at the fundamentals is all that is required here. Filter paper has a pore width of about 5 μm. Consequently, the usual coarse-grained precipitates are retained whereas colloids and true solutions pass through unchanged. For corrosive materials, e.g. strong acids and alkalis, glass filter discs may be used to advantage; they are available in the following pore sizes:

Size	Average pore-diameter, μm
0	200–150
I	150–90
II	90–40
III	40–15
IV	15–5
V	1.5–1.0

The following information is useful for assessing the effect of the pore size:

pore diameter about 5 μm: all cells of body fluids are retained;
pore diameter about 0.4 μm: all polyform bacteria are retained;
pore diameter about 0.2 μm: all bacteria are retained.

The material remaining on the filter-paper is called the residue and the liquid which passes through is the filtrate.

In *dialysis*, the solution is put into a bag, tube, or sleeve made from a semipermeable membrane. This retains macromolecules but allows the solvent and small molecules to pass through. Since the process involved is essentially one of diffusion, the overall process is slow. Pressure can be applied to the solution under dialysis to speed up the dialysis process, and the method is then referred to as molecular filtration.[29] Membrane filters made of plastic foil with a uniform, continuous structure are available for this purpose. The pore width is guaranteed within narrow limits, and filters from 14 down to 0.025 μm are obtainable. The finest porosity filters even retain viruses.

An important field of application for molecular filtration in the domain of clinical chemistry is the enrichment of proteins in protein-poor biological fluids, e.g. urine or liquor. The water is expressed or sucked through a membrane, together with the small molecules and electrolytes, and thus the large molecules, especially the proteins, which cannot pass through the membrane are enriched.

2.4.3. ELECTROPHORETIC METHODS OF SEPARATION

2.4.3.1. Theoretical basis[2]

Electrophoretic separation relies on the differential speeds of migration of differently charged particles in an electric field. The migration speed is

136

primarily a function of the charge on the particle and the field strength applied. The most important application in clinical chemistry is the electrophoretic separation of proteins. Owing to their ampholytic nature their charge changes according to the pH of the surrounding medium. At the isoelectric point, when the positive and negative charges on a protein are equal, it does not migrate in an electric field. At pH values below that of the isoelectric point the positive charges on the protein predominate and it migrates to the cathode; at a pH above that of the isoelectric point it moves towards the anode. For the separation of serum proteins it is usual, in practice, to choose a pH above that of the isoelectric point (around 8.5).

The proteins to be separated are incorporated in a buffer so that the separation is carried out at a definitive pH value. Since the current is carried by the ions present, the current strength is greater the higher is the concentration of ions in the solution. However, the higher the concentration of buffer ions in comparison with the protein ions, the more current is carried by the buffer ions and the less by the protein ions, i.e. the protein ions migrate more slowly. Thus, in addition to the pH value of the buffer, its ionic strength (p. 92) is another factor determining the migration speed of the proteins.

The electrical resistance of the entire system is essentially dependent on the ionic strength of the buffer, i.e. at constant potential the current strength is smaller the less concentrated is the buffer solution. The current flowing through the system produces heat by the Joule effect; this is greater the greater is the flow of current per unit time. If the resistance of the medium is assumed to remain constant, then the potential must be raised if a higher current strength is desired. If, on the other hand, a constant voltage is maintained, then the intensity of the current increases during the course of the electrophoresis since the heat causes water to evaporate out of the system, resulting in a higher buffer concentration. However, this is also attended by a fall in the resistance of the system.

There are, therefore, a series of factors to be considered when optimizing the conditions for electrophoretic separations. A fundamental question is whether the current strength or the voltage is to be kept constant during the electrophoresis. As a guiding rule, with thin carriers a constant voltage is to be preferred, and with thick carriers (starch gel, agar

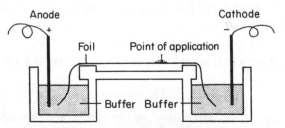

Fig. 30. Diagram of an electrophoresis apparatus.

gel) a constant current strength is preferable, in order to check the development of heat.

In *free (moving boundary) electrophoresis*, as developed by Tiselius,[2] the proteins migrate in an aqueous medium, moving from the anode to the cathode. The method does not give clear separations, i.e. the individual bands of proteins show extensive overlapping. For this reason, free electrophoresis has been abandoned in favour of *zone electrophoresis*[3-5] (Fig. 30).

However, this imposes fresh limitations on the system:

1. The water occupying the capillary pores of the carrier assumes a positive charge but the carrier surface is negatively charged (Fig. 31). If an electric field is applied, the water migrates to the cathode. The resulting flow of water and the flow of the positively charged buffer ions are thus opposed to the migration of the proteins. This phenomenon is called endosmosis. It depends on the nature of the carrier, the applied voltage, and the ionic strength and pH of the buffer. The endosmosis effect is decreased by addition of H^+ ions and increased by OH^- ions.
2. In addition, the flow of water produces a streaming potential, likewise opposed to the applied potential.
3. Depending on the nature of the carrier, but with paper especially, the path of the particle in the electric field is not rectilinear but is influenced according to the variable density of the carrier material. This phenomenon is called the 'barrier effect'.
4. There is a mutual interaction between the cellulose particles of the filter-paper and the protein molecules. Differential absorption effects act differently on different proteins. This further influences the free migration in the electric field. The chromatographic effects are substantially smaller with cellulose than with paper.
5. Finally, the viscosity of the buffer influences the migration of the

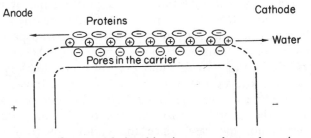

Fig. 31. Charge relationships in zone electrophoresis. In accordance with their charge the proteins migrate to the anode and water flows to the cathode by endosmosis. The carrier pores are negatively charged relative to the water.

proteins. The viscosity decreases with increasing temperature but it increases again if the concentration is raised by the evaporation of water.

2.4.3.2. Choice of support

The commonest supporting matrix in use today is *cellulose acetate film*. It has a highly homogeneous pore system and there is scarcely any adsorption of proteins. The separation distances are relatively short and the material can be made transparent for the purpose of evaluation (cf. Section 2.4.3.3). Another frequently used support medium is *agar gel*. Gels suppress convection especially well and, as a result, afford sharp bands. Many commercial gels are transparent and are consequently well suited to a photometric evaluation. Agar gel,[6] a multi-component substance, has numerous acid valencies, giving rise to considerable endosmosis. Agar gel binds cations to a certain extent. The principle field of application of agar gel is immunoelectrophoresis. *Agarose gel*,[7] obtained from agar, contains no agaropectin and the endosmosis is markedly less; it is the preferred support for the electrophoretic separation of lipoproteins.[8]

In contrast to the above-mentioned media, *starch gel*[9] is not a passive support. It acts as a molecular filter and fractionates the investigational material according to molecular size. Starch gel is used advantageously for the separation of haptoglobins and haemoglobins. *Polyacrylamide gel*[10] is increasingly used. Here, too, molecular-sieve effects are very much in evidence. Very efficient separations can be attained with these synthetic gels.

2.4.3.3. Detecting and quantifying materials separated by electrophoresis

To *reveal* the proteins separated be electrophoresis, the support is usually treated with certain dyestuffs, e.g. Ponceau S, amido black, bromophenol blue, and Coomassie blue. The protein bands adsorb these dyes considerably more strongly than the support, so that it is possible to elute the excess colour from the support completely, leaving the proteins to show as coloured fractions. By making the support transparent, either at the same time or later, transparent films may be produced, well suited for a photometric evaluation. To this end the foil is fed linearly through the light path and the extinction measured as a function of position. The areas beneath the resulting curve are integrated. The integrals ought to be proportional to the protein concentrations of the individual fractions. However, this is not the case, since the various blood-plasma proteins adsorb the dyes differentially and hence it is not possible to make an exact, quantitative statement.[11] It has even been recommended[12] that every kind of quantification on this basis should be rejected. Even so, a high reproducibility can be achieved by using a more refined technique. It must be made clear that the numerical results obtained are arbitrary, i.e. at best allowing a guess at the order of magnitude of the true concentrations.

In the electrophoretic separation of serum proteins the enzyme molecules contained in the serum are also fractionated.[13] Much of the serum enzyme activity is present in the form of isoenzymes which show differential mobility in the electric field. In order to show up the isoenzymes a 'sandwich technique' can be used. After the separation, a second support is placed over the first; this second medium, e.g. cellulose acetate or another gel, is impregnated with the appropriate reagents (for the isoenzymes of alkaline phosphatase, for example, it might be 4-nitrophenol in an alkaline buffer). The 'sandwich' is incubated for a certain time at a definite temperature and then taken apart when the isoenzymes show up on the upper side. Analogous techniques have been described for many isoenzymes.[14]

2.4.3.4. Special fields of application

1. *Immunoelectrophoresis*[6,15,16] is employed for the refined separation of proteins in serum, urine, or liquor. The principle is as follows: immediately after the electrophoretic separation a specific antiserum is introduced into a groove parallel to the migration track (Fig. 32a–c). The antiserum may be polyvalent, i.e. contain antibodies for all the proteins contained in human serum, or it may be monospecific and only respond to one type of protein or one protein fragment.

   The antiserum is either introduced directly into the antiserum groove or it is soaked onto a carrier, e.g. a narrow strip of filter-paper, which is then laid on the electrophoresis support. The (previously separated) proteins behaving as antigens diffuse towards the antibodies of the antiserum and precipitation occurs along the line of contact. Each precipitation arc corresponds to a particular antigen–antibody reaction.

2. In *rocket immunoelectrophoresis*[17] the support contains a monospecific antiserum. The (diluted) serum is applied at the cathode end and in the electric field it migrates towards the anode. That part of the serum which reacts with the antibody is progressively precipitated out, forming 'rockets' (Fig. 33). Supports containing antisera are commercially available. A series of standards are run alongside the sample and the heights of the standard rockets are used as a calibration graph, allowing the concentration of the sample to be read off. Related techniques are *transmigration electrophoresis* and two-dimensional electrophoresis.

3. In *isoelectric focusing*[18] the electric field is applied to a support (usually a polyacrylamide gel) furnished with a pH gradient. For example, the pH value may decrease continuously from the cathode

140

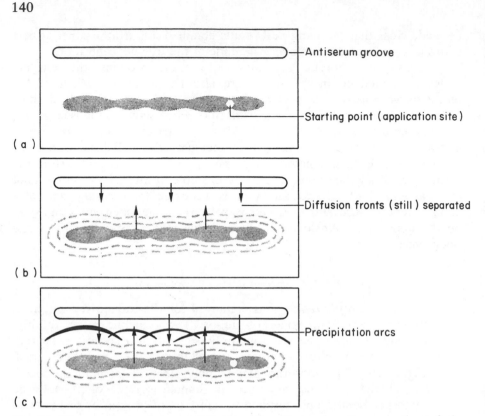

(a)

Antiserum groove

Starting point (application site)

(b)

Diffusion fronts (still) separated

(c)

Precipitation arcs

Fig. 32. Immunoelectrophoresis. (a) After the electrophoretic separation of the serum sample the antiserum is introduced into the antiserum groove; (b) antiserum and electrophoretically separated serum proteins diffuse towards each other; (c) formation of precipitation arcs.

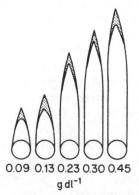

0.09 0.13 0.23 0.30 0.45
g dl⁻¹

Fig. 33. Rocket electrophoresis. The height of the 'rockets' is not linearly proportional to the concentration of the antigen applied (e.g. immunoglobulin G in human serum).

to the anode. The serum is applied to the alkaline region of the support; under the influence of the electric field an individual serum protein will migrate to that point on the pH gradient which corresponds to its isoelectric point—the proteins are focused at their isoelectric points. The sharp focusing results in a concentration of the protein fraction at this point.

Buffers which automatically produce a (natural) pH gradient in an electric field have been successfully developed;[19] they are made from a mixture of synthetic amino acids (Ampholine). Isoelectric focusing may be coupled with a subsequent immunodiffusion to afford separations of extraordinarily high quality.

2.4.4. EXTRACTION

In the clinical–chemical laboratory it is not uncommon to use extraction methods for the enrichment of compounds or for the removal of interfering substances. As a primary requirement, the substance to be extracted must be soluble not only in aqueous biological media but also in a phase which is immiscible (or only slightly miscible) with water, e.g. ether, chloroform, or carbon tetrachloride. If the aqueous phase is shaken with the organic solvent, the compound distributes itself between the two phases in accordance with its distribution coefficient. If the aqueous phase is repeatedly extracted with a fresh organic phase, a group of substances can be almost completely separated from the aqueous phase. By pooling all the organic fractions and concentrating the bulk, the substances—enriched in a smaller volume—are available for further analytical processing. The extraction may be discontinuous or continuous, given the appropriate set-up. Counter-current distribution represents a special form of extraction; details may be found in the literature.[20,27] Since liquid/liquid extraction suffers from several technical disadvantages, solid/liquid extraction is preferred. In this case the first step is the adsorption of the desired substance or group of substances (present, e.g., in serum or urine) on to a solid phase, e.g. an ion-exchanger, cellulose, or some other adsorbent material. In the second stage, the desired substance is dissolved from the solid phase with a liquid organic phase. Here again, the specialist literature should be consulted for more precise details.[30]

2.4.5. CHROMATOGRAPHIC METHODS

2.4.5.1. Introduction

In the field of clinical chemistry, chromatography is an important analytical method in those cases where the components to be determined have to be separated prior to quantitative determination.

The principle governing the chromatographic separation of mixtures is the distribution of the substances between two phases. One of these phases is fixed (stationary) and has a large surface area, the other is mobile and percolates through the stationary phase.

The *stationary phase* may be a solid (adsorbent) or a liquid. In the latter case, immobility is achieved by depositing the liquid as a thin film on a porous, fine-grained support. As the *mobile phase*, liquids or gases that are not miscible with the stationary phase may be used. The substances to be separated are brought on to the stationary phase by means of the mobile phase and are retained there. The individual components of the mixture have differing solubilities in the stationary phase (or are differentially adsorbed) and a primary distribution equilibrium (adsorption equilibrium) is established. The subsequent flow of mobile phase takes up fractions separating out from the stationary phase and carries them to the next section, where they are adsorbed again and a fresh phase distribution is established. This process repeats itself throughout the entire stationary phase (Fig. 34a).

Because the individual components of the mixture have different solubilities in the stationary phase (or are differentially adsorbed), they migrate through the stationary phase at different rates and the mixture is thus resolved into its components.

Chromatographic methods are subdivided into adsorption, ion-exchange, partition and gel chromatography, depending on the different types of

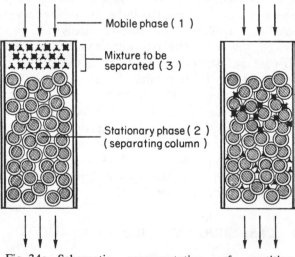

Fig. 34a. Schematic representation of partition chromatography. Left: at the beginning of the chromatographic separation. Right: after separation of the mixture. 1 = Mobile phase percolates through the stationary phase; 2 = stationary phase as a thin film deposited on grains of a support; 3 = substances to be separated.

interaction between the stationary phases and the substances being separated.

In *adsorption chromatography* the stationary phase (adsorbent) is a material such as active charcoal, silicate, alumina, calcium oxide, or calcium carbonate. The adsorption capacity is largely dependent on the surface properties of the adsorbent.

In *ion-exchange chromatography*, ion-exchange resins (synthetic organic resins with covalently bonded ionic groups) are used as the stationary phase. Ionic substances present in the mobile phase are bound to the resin by electrostatic forces.

In *partition chromatography*, the stationary phase is a liquid deposited on an inert carrier. The liquid stationary phase is usually water on a silica gel or cellulose carrier.

Separation by means of *gel chromatography* (molecular exclusion) depends on the different molecular sizes of the components being analysed. The stationary phase consists of small dextran spheres or deactivated glass particles with pores and channels of a size comparable to that of the molecules of the sample. Components with small molecules can penetrate the pores more easily and consequently spend longer in the stationary phase than large molecules which are excluded and are rapidly eluted through the relatively large spaces between the particles.

The techniques of importance for clinical chemistry are thin-layer chromatography, high-performance liquid chromatography and gas chromatography.

2.4.5.2. Thin-layer chromatography

In thin-layer chromatography[23,31] the stationary phase is a thin layer (0.1–0.5 mm) of fine-grained alumina or silica gel deposited on a carrier plate made of glass, plastic, or aluminium foil.

The solution of the mixture to be separated (1–5 µl) is applied, as a spot or band, at the starting line, about 10 mm from the edge of the plate. The prepared coated plate or foil is placed in a tightly closed tank, the bottom of which is filled (5–8 mm) with a suitable eluting solvent. The mobile phase diffuses up the plate by capillary action, effecting a separation. When the solvent front has travelled a sufficient distance the plate is removed from the tank and dried. If the zones are not immediately visible, the substances are detected and determined by the appropriate chemical and physical methods (observation under ultraviolet light) (Fig. 34b).

The basis of thin-layer chromatography is adsorption or partition, but ion-exchange processes are also involved.

A measure of the rate of migration of a substance is provided by the R_f *value*. This is evaluated from

$$R_f = \frac{\text{distance travelled by the component}}{\text{distance travelled by the solvent}}$$

144

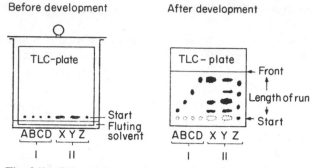

Fig. 34b. Schematic representation of thin-layer chromatography. A–D = reference substances in known concentrations; X–Z = mixtures assumed to contain substances A–D.

For a given mobile and stationary phase, the R_f value is constant for a particular substance.

In principle, when thin-layer chromatography is used for the qualitative and quantitative characterization of a mixture, a comparison substance or a series of standards should be chromatographed on the same plate as the substances in the mixture.

2.4.5.3. Gas chromatography

Gas chromatography[32–34] differs from other chromatographic methods in that a gas (N_2, He, H_2) is used as the mobile phase. For this reason the substances to be separated must be volatile or completely vaporized. The stationary phase consists of a high-boiling liquid carried on an inert, fine-grained support (silica gel, Celite) or on the walls of a capillary. Commonly used liquids of various polarity include paraffins, silicone oils, and polymers (e.g. Carbowax). Gas chromatography is a type of column chromatography since the stationary phase is in the form of a column packed in a glass or metal tube (length usually about 2 m, diameter 2–6 mm) (Fig. 34c).

The mixture of substances to be analysed is injected, and subsequently evaporated, at one end of the column via a sample port maintained at a high temperature (Fig. 34c). The gaseous mixture of substances is now flushed on to the column by the gaseous mobile phase and there it is resolved by partition. The substances to be separated must not be thermally unstable and they must have a sufficiently high vapour pressure. If necessary, more stable and more volatile derivatives must be made (e.g. by silylation) before carrying out the separation. Following resolution, the individual substances are detected at the end of the column, using physical methods (e.g. with a flame-ionization detector), giving rise to electrical signals which are finally amplified electronically.

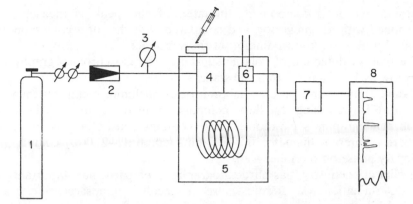

Fig. 34c. Schematic diagram of a gas chromatography train. 1 = Carrier-gas supply (gas cylinder); 2 = carrier gas flow regulator (needle valve); 3 = manometer; 4 = sample insertion port (injector); 5 = separating column filled with support on which the stationary phase (separating liquid) is deposited; 6 = detector, producing electrical signals whenever substances pass through; 7 = electronic amplifier; 8 = pen recorder. Parts 4, 5 and 6 are thermostatically controlled.

The signals produced in the detector by the eluted components are plotted against the retention time by means of a pen recorder (Fig. 34d). The retention time (t_R) is the period between the air or solvent peak up to the passage of the maximum concentration (peak) of the substance through the detector. It corresponds to the R_f value in thin-layer chromatography. Within a certain range, the quantity of the eluted substance is proportional to the detector signal (peak height), and this is exploited in the quantitative evaluation of the gas chromatogram. For the quantitative

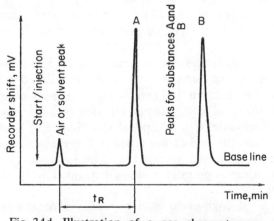

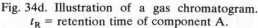

Fig. 34d. Illustration of a gas chromatogram. t_R = retention time of component A.

determination of a component, the area of the peak is measured or determined with an integrator, and compared with that of a corresponding standard peak for a known amount of the substance.

The limit of detection of volatile substances by gas chromatography can be as low as 10^{-12} g for certain substances.

Today, more and more laboratories, including clinical–chemical laboratories, are using capillary columns to increase the efficiency of gas-chromatographic separations. These columns consist of glass or metal capillaries (internal diameter 0.2–0.5 mm, length 15–200 m) coated with stationary phase on the inner walls.

In clinical chemistry, gas chromatography is of particular importance in the determination of steroids, organic acids, monosaccharides, and catecholamines.

It is possible to combine gas chromatography with another physical method, *mass spectrometry*. After removing the carrier gas, the components separated by gas chromatography may be fed to a mass spectrograph for characterization. The mass spectrogram then provides further information on the chemical structure of a component. In addition, a mass spectrometer can be used as a very sensitive and specific detector: the instrument is set to one or more mass numbers and the corresponding ion current is plotted as a mass fragmentogram. This method allows the determination of an individual component in the range 10^{-10} to 10^{-13} g.

The combined gas chromatography/mass spectrometry technique is also suited to metabolic investigations using stable isotopes. By this means it is possible to determine products of metabolism in blood and urine following the *in vivo* ingestion of compounds marked with stable isotopes.

2.4.5.4. High-performance liquid chromatography

High-performance liquid chromatography[21,22,35,36] is a column chromatographic technique in which the stationary phase occupies a pressure-resistant tantalum, steel, or glass tube (length 10–60 cm, diameter 2–6 mm).

The mixture of substances to be separated is fed to the end of the separating column via a sample port while the liquid mobile phase is flushed through the column by means of a high-pressure pump. After resolution the individual components are detected by physical methods, e.g. photometric measurement of optical absorbance at fixed wavelengths, as they emerge from the end of the column. Once again, the signal for the eluted components is plotted against the retention time by a pen recorder (Fig. 34e). The chromatogram is evaluated quantitatively exactly as in gas chromatography.

The advantage of high-performance liquid chromatography is that it allows one to choose the separative method—partition, adsorption, ion-exchange or gel chromatography—best fitted to a specific separation.

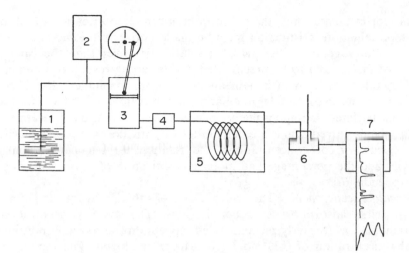

Fig. 34e. Schematic representation of a high-performance liquid chromatography train. 1 = Reservoir for mobile phase; 2 = gradient mixer; 3 = pump; 4 = sample port; 5 = separating column (under thermostatic control); 6 = detector; 7 = pen recorder.

Stationary phases consisting of particles with diameters ranging from 5 to 45 μm are available for all of these methods. In addition, the properties of the mobile phase can be varied by mixing a gradient.

Advantages of high-performance liquid chromatography are rapidity, good resolution, high sensitivity, and facility of column replacement. This method of separation is particularly suitable for the resolution of non-volatile components such as purine and pyrimidine nucleotides or nucleosides, amino acids, steroids, vitamins, lipids, and carbohydrates.

2.4.6. METHODS OF DEPROTEINATION

For certain clinical–chemical analyses it is often necessary to deproteinate the plasma or serum beforehand.

The commonest method of deproteination used in clinical chemistry is *precipitation of the protein as an insoluble salt.* In all precipitation methods the pH plays an important role. Proteins carry both positive and negative charges and the number of these charges is dependent on the pH. Thus, under strongly alkaline conditions (high pH), the proteins are present as anions with an excess of negative charges; under strongly acidic conditions (low pH), the positive charges are in excess and the proteins are present as cations. Between these two extremes there is a characteristic pH for every protein, the *isoelectric point* (IEP), at which there are equal numbers of positive and negative charges. The proteins are most labile and show the greatest tendency to precipitate at this isoelectric point. This is why acetic acid is added in the heat coagulation test for the detection of proteins in urine, for instance. Thus, the pH must be given close attention in all forms of deproteination. Distinction must be made between anionic and cationic precipitating agents. If the pH of a solution is lower than the isoelectric point of the protein to be precipitated, the latter is present as a cation and will therefore react with certain anions to form insoluble salts. The best known anions used for deproteination are picrate, molybdate, tungstate, sulphosalicylate, metaphosphate, perchlorate, and trichloroacetate. Of these, the preferred agents are perchlorate and trichloroacetate. To facilitate the calculation of concentrations, the relationship between the specific gravity and the percentage and molar concentration of these acids is given in Appendix 6.

Trichloroacetic acid. The advantage of this reagent is that the non-specific adsorption is generally low. The final concentration of $0.16 \text{ mol } l^{-1}$, as formerly recommended, appears to be too low; nowadays a final concentration of $0.31 \text{ mol } l^{-1}$ is usually employed. The mixture must not be centrifuged until at least 15 min after the addition of the trichloroacetic acid. The pH is around 1. Trichloroacetic acid does not precipitate proteins completely, e.g. it does not precipitate seromucoids. In cases of proteinuria, as little as 4–20% of the proteins are said to be precipitated. However, this is of little importance in practice. The chief disadvantage which should be mentioned is that trichloroacetic acid absorbs in ultraviolet light and must therefore be ruled out if ultraviolet spectrophotometric measurements are to be made.

Perchloric acid. The risk of explosions seems to be the main reason why many laboratories are reluctant to use this substance. However, explosions are hardly likely to occur with the 72% commercial preparation and the use of this acid in the cold may be recommended unreservedly. Under no circumstances should ashings be performed with perchloric acid in the clinical–chemical laboratory, however. A final concentration of $1.0 \text{ mol } l^{-1}$ is generally used, although smaller concentrations also yield satisfactory results. The use of ice-cold perchloric acid both accelerates the deproteination and makes it more complete. A waiting period of 10 min prior to centrifugation is sufficient. The pH of the supernatant liquid is around 0. The use of perchloric acid has the advantage that the excess acid may be quantitatively precipitated as potassium perchlorate ($KClO_4$) after the deproteination has been accomplished. This is effected by the addition of an equivalent quantity of potassium hydroxide, potassium carbonate (evolution of CO_2), or potassium phosphate. If the mixture is now strongly centrifuged a neutral supernatant is obtained which may be used for very sensitive (e.g. enzymatic) reactions. One disadvantage is that, just as with trichloroacetic acid, a series of glycoproteins (seromucoid, α_1-glycoprotein, and haptoglobins) fail to precipitate. In contrast to trichloroacetic acid, perchloric acid does not absorb in ultraviolet light.

If the pH of a solution is higher than the isoelectric point of the proteins, the latter are present as anions, i.e. they can form insoluble salts with certain cations. The *cations* used are ions of *heavy metals*, especially zinc, mercury, cadmium, uranium, iron, copper, and lead. The best known methods of this type are deproteination with zinc sulphate, and with zinc sulphate and barium hydroxide. The main disadvantage of all of these methods is that they almost always lead to the non-specific adsorption of substances on the precipitated proteins. Other methods of deproteination such as precipitation by dehydration or by means of specific antisera are less commonly used as routine clinical–chemical procedures. *Deproteination by dehydration* may be accomplished either by using solvents or by salting-out. In principle, both methods involve treatments with substances which form hydrates and thereby compete with the protein for the solution water. If a certain quantity of water is removed from the protein, it precipitates. The main solvents used are acetone, methanol, and ethanol. As Cohn has shown in his celebrated work on plasma fractionation, the deproteinating action of these solvents may be enhanced by working at lower temperatures. These methods are most particularly suited to preparative work and, as yet, they have been used only occasionally in clinical chemistry. In the salting-out procedure, water is removed from the protein by the addition of the salt at a pH as close as possible to the isoelectric point: the proteins precipitate out. These techniques are most commonly used in preparative protein chemistry. In clinical chemistry they are used only for the separation of fibrinogen, albumin, and globulin. In this context the precipitation of proteins by means of *antigen–antibody reactions* is of only theoretical interest. This method is particularly important in immunochemistry for the detection of certain proteins and in radioimmunoassays.

2.4.6.1. Volume displacement effect in deproteination

In 1927, Van Slyke et al.[24] reported that the determination of the total base in plasma following deproteination with trichloroacetic acid gave values 4% higher than by the direct determination without deproteination. They did not discover the reason for this. A few years later, Ball and Sadusk[25] obtained similar results in the determination of sodium. They explained the higher results (around 5% higher) following deproteination as being due to the volume displacement effects of the proteins. Bürgi et al.[26] carefully analysed the theory of this effect and showed that it also applies to the determination of glucose and inorganic phosphorus, and indeed, probably to all the low-molecular-weight substances in plasma.

Fig. 35 shows a diagram of a tube filled with plasma. In terms of physiological proportions, the total volume is mostly occupied by the volume of the liquid itself, but also to a small extent by the protein molecules present in the solution—these also occupy a certain volume. The space occupied by the proteins in solution corresponds to the partial specific volume $(\bar{V})$. $\bar{V}$ is a characteristic quantity for each protein and

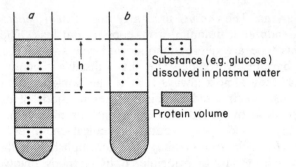

Fig. 35. Tube filled with blood plasma. The volume used for the determination (e.g. of glucose) corresponds to the height of the liquid columns. The concentration of glucose is lower or higher according to whether the proteins remain in the solution (a), or are centrifuged or precipitated out (b). For the sake of clarity, the actual volume occupied by the proteins (at a physiological concentration) is greatly exaggerated. h = volume used for the measurement.

corresponds to the reciprocal of the protein density:

$$\bar{V} = \frac{1}{d}\,\text{ml g}^{-1}$$

With plasma proteins, $\bar{V}$ lies between 0.700 and 0.750. The lowest values (down to 0.675) are observed with certain glycoproteins and the highest (up to 0.950) with lipoproteins. If a certain volume is pipetted out of a tube (Fig. 35a), the partial volume of the protein is included. After deproteination (Fig. 35b), the low-molecular-weight substances are more concentrated, and consequently the result must be higher.

The difference in the concentration may be calculated in the following manner:

If

C_1 = the concentration of a substance in plasma;

C_2 = the concentration of the substance in protein-free filtrate;

$dC = C_2 - C_1$

C_p = the concentration of the protein in g per 100 ml;

$\bar{V}$ = the average partial specific volume of the proteins;

then

$$dC = C_1\left(\frac{100}{100 - \bar{V}\cdot C_p} - 1\right)$$

The validity of this relationship has been confirmed by experiments using glucose and inorganic phosphorus. With glucose, the results of the direct determination were 5% and with inorganic phosphorus 6% lower than after a preliminary deproteination. Conversely, the results of a direct and indirect determination of a low-molecular-weight substance can be used to determine the average partial specific volume of the proteins.

These considerations are extremely important in the comparison of reference values obtained with and without deproteination. A difference of around 5% is to be expected and must be taken into account.

References

1. Morris, C. J. O. R. and Morris, P., *Separation methods in biochemistry*, 2nd ed., Pitman, London, 1976.
2. Tiselius, A., Electrophoresis, in Colowick and Kaplan, *Methods in enzymology*, Vol. 4, Academic Press, New York, 1957.
3. Laurell, A. H., Paper electrophoresis, in Colowick and Kaplan, *Methods in enzymology*, Vol. 4, Academic Press, New York, 1957.
4. Wunderly, C., *Die Papierelektrophorese*, Sauerländer, Aarau, 1954.
5. Kohn, J., Cellulose acetate electrophoresis and immunodiffusion techniques, in Smith, *Chromatographic and electrophoretic techniques*, Vol. 2, 2nd ed., pp. 84–146, Heinemann, London, 1968.
6. Grabar, P. and Burtin, P., *Immuno-elektrophoretische Analyse*, Elsevier, Amsterdam, 1964.
7. Barteling, S. I., A simple method for preparation of agarose. *Clin. chim. Acta*, **15**, 1002 (1969).
8. Noble, R. P., The electrophoretic separation of plasma lipoproteins in agarose gel, *J. Lipid Res.*, **9**, 963 (1968).
9. Smithies, O., Zone electrophoresis in starch gel, *Biochem J.*, **61**, 629 (1955).
10. Allen, R. C. and Maurer, H. R., *Electrophoresis and isoelectric focusing in polyacrylamide gel*, de Gruyter, Berlin, 1974.
11. Busse, V. and Dulce, H. J., Bestimmung von Korrekturfaktorn für die Farbstoff-Proteinbindung bei der Folienelektrophorese, *Z. klin. Chem.*, **7**, 486 (1969).
12. Laurell, C. B., Electrophoresis, specific protein assays, or both in measurement of plasma proteins, *Clin. Chem.*, **19**, 99 (1973).
13. Wieland, Th. and Pfleiderer, G., Multiple Formen von Enzymen, *Adv. Enzymol.*, **25**, 329 (1963).
14. Wieme, R. J., Messung von Enzymaktivitäten nach Elektrophorese, in Bergmeyer, *Methoden der enzymatischen Analyse*, 3. Aufl., Verlag Chemie, Weinheim, 1974.
15. Arquembourg, P. C., Salvaggio, J. E., and Bickers, J. N., *Primer of immunoelectrophoresis*, Karger, Basle, 1970.
16. Axelson, N. H., Kroll, J., and Weeke, B., *A manual of quantitative immuno-electrophoresis*, Universitätsforlaget, Oslo, 1973.
17. Laurell, C. B., Quantitative estimation of proteins by electrophoresis in agarose gel containing antibodies, *Analyt. Biochem.*, **15**, 45 (1966).
18. Latner, A. L., Isoelectric focusing in liquids and gels as applied to clinical chemistry, *Advances in clinical chemistry*, Vol. 17, Academic Press, New York, 1975.
19. Versterberg, O., Isoelectric focusing of proteins, in Colowick and Kaplan, *Methods in enzymology*, Vol. 22, Academic Press, New York, 1971.

20. Williams, B. L. and Wilson, K., *Principles and techniques of practical biochemistry*, Arnold Ltd., London, 1975.
21. Neher, R., *Steroid chromatography*, Elsevier, Amsterdam, 1964.
22. Randerath, K., *Dünnschichtchromatographie*, 2. Aufl., Verlag Chemie, Weinheim, 1965.
23. Stahl, E., *Dünnschichtchromatographie*, 2. Aufl., Springer, Berlin, 1967.
24. Slyke, D. D. van, Hiller, A., and Berthelsen, K. C., A gasometric micromethod for the determination of iodates and sulfates, and its application to the estimation of total base in blood serum, *J. biol. Chem.*, **74**, 659 (1927).
25. Ball, E. G. and Sadusk, J. F., A study of estimation of sodium in blood, *J. biol. Chem.*, **113**, 661 (1936).
26. Bürgi, W., Richterich, R., and Mittelholzer, M. L., Der Einfluss der Enteiwessung auf die Resultate von Serum- und Plasma-Analysen, *Klin. Wschr.*, **45**, 83 (1967).
27. Franke, R. and Thiele, K., *Physikalisch-chemische Methoden im klinischen Laboratorium*, Vol. 1, VEB Volk und Gesundheit, Berlin, 1969.
28. Rehfeld, N. and Reichelt, D., *Analytische und präparative Methoden der klinschen Biochemie*, Verlag Chemie, Weinheim, 1963.
29. Jacobs, St., Ultrafilter membranes in biochemistry, in Glick, *Methods of biochemical analysis*, Wiley, New York, 1974.
30. Sachs, A. and Kopfer, B., Neues Verfahren zur Extraktion lipophiler Stoffe aus wässerigen Lösungen, *GIT Fachzschr.*, **20**, 1209 (1976).
31. Kaiser, R. E., *Einführung in die Hochleistungs-Dünnschicht-Chromatographie*, Institut für Chromatographie, Bad Dürkheim, 1976.
32. Heftmann, E., *Chromatography: a laboratory handbook of chromatographic and electrophoretic methods*, 3rd ed., Van Nostrand-Reinhold, New York, 1975.
33. McFadden, W., *Techniques of combined gas chromatography/mass spectrometry*, Wiley-Interscience, New York, 1973.
34. Kaiser, R., *Chromatographie in der Gasphase*, Bibliographisches Institut, Mannheim, 1969.
35. Brown, Ph., *High pressure liquid chromatography. Biochemical and biomedical applications*, Academic Press, New York, 1973.
36. Deyl, Z., Macek, K., and Janak, J., *Liquid column chromatography. A survey of modern techniques and applications*, Elsevier, Amsterdam, 1975.

2.5. OPTICAL METHODS OF MEASUREMENT

2.5.1. PHYSICAL BASIS

Over the last 30 years clinical chemistry has been enriched by the development of an increasing number of photometric methods of quantitative analysis. Today, more than three quarters of all analyses performed in the routine clinical–chemical laboratory exploit photometric methods. Owing to the great significance of photometry its physical basis must be examined in some detail.

2.5.1.1. Electromagnetic radiation

Figure 36 shows the most important types of electromagnetic radiation. Each form of radiant energy is characterized by its wavelength, the range between 200 and 2000 nm being particularly important for absorption photometry. Radiation in the range 400–700 nm can be perceived by the human eye as colour. Light of shorter wavelength (200–400 nm), ultraviolet radiation, and light of longer wavelength (700–2000 nm), infrared radiation, cannot be detected by the human eye but are measurable with the appropriate photodetectors. The absorption process is

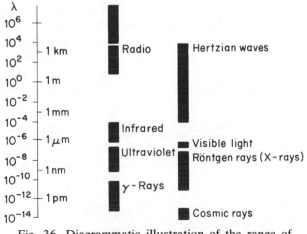

Fig. 36. Diagrammatic illustration of the range of electromagnetic vibrations.

154

basically as follows: a photon strikes a molecule, the absorber, and for an extremely short time this is raised to a higher energy level, the excited state; on returning to the initial state, heat is liberated:

$$A + h\nu \rightarrow A^* \rightarrow A + heat$$

where

A = absorber;
A^* = absorber in excited state;
h = Planck's constant;
ν = frequency.

Certain wavelengths are selectively absorbed, depending on the structure of the absorbing molecule. In each case, the wavelength of the emitted radiation (heat) is longer, i.e. of lower energy, than that of the exciting radiation.

In absorption photometry, what is measured is the difference between the intensity of the light beam incident on the sample and that emerging from the sample. In terms of semantic accuracy the expression 'absorption photometry' is therefore incorrect since it is the transmission and not the absorption which is measured.

The human eye is relatively ill-suited to judge differences in the intensities of light and colour. Instruments (colorimeters) in which the colour of the analysis solution is compared visually with that of a standard solution are hardly ever used nowadays, and the expression 'colorimetry' ought to be equally redundant.[4]

2.5.1.2. Dispersion of light

Most light sources emit light composed of a large number of different wavelengths. It is therefore called *polychromatic* light. This (white) light can be resolved, by various physical means (prism, diffraction grating), into the individual wavelengths of which it is composed. The projection of dispersed daylight on to a screen produces the well known spectrum from violet at the short-wave end to red at the long-wave end. The wavelengths of the individual colour ranges are listed in Table 26. If a solution appears

Table 26. Visual perception of wavelength ranges.

Wavelength range, nm	Colour perception
400–450	Violet
450–500	Blue
500–570	Green
570–590	Yellow
590–620	Orange
620–700	Red

coloured to the human eye, this is because it is able to absorb all the incident radiation apart from that (colour) which reaches the eye. Thus, a blue solution appears blue because the longer wave green, yellow, and red radiation is absorbed. Conversely, a red solution absorbs the shorter wave yellow, green, and blue radiation and allows only the red radiation to be transmitted. If this represents a single, discrete wavelength (or a narrow wave band) only, then this light is termed monochromatic. Only this *monochromatic* light is of interest in spectrophotometry, since the molecules absorb individual wavelengths to varying extents. The laws of photometry are valid only for the case of monchromatic light, or at least light composed of only a few, closely adjacent wavelengths.

2.5.2. ABSORPTION PHOTOMETRY

2.5.2.1. Bouguer–Lambert–Beer law

Let us assume a beam of monochromatic light of intensity I_0 passes into a homogeneous medium, e.g. a coloured solution or a coloured glass plate, of thickness d (e.g. 1 cm); it emerges with energy I_A. The ratio of emergent to incident light is termed the transmission, T:

$$\frac{I_A}{I_0} = T$$

T will assume its highest value, 1, if the incident light emerges completely unchanged. T will have the value 0 if no light passes through at all, and T will assume any value between 0 and 1 if part of the light fails to emerge because it has been absorbed within the medium.

The relationship between the decrease in the light intensity and the thickness (or length) of the absorbing medium (layer thickness) was investigated by Pierre Bouguer (1698–1758). He observed that layers of the same thickness and type, placed one behind the other, always absorbed a fixed proportion of the incident light. The principle is illustrated in Fig. 37: the light beam passes through three identical layers, all of which

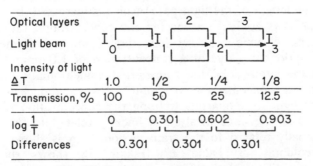

Fig. 37. Bouguer and Lambert's law.

Table 27. Laws of absorption photometry.

Bouguer and Lambert			Beer	
Number of layers	Transmission, T	$\dfrac{1}{T}$	Absorbance $= \log \dfrac{1}{T}$	Concentration
0	1.0	1	0.000	0
1	$\frac{1}{2}$	2	0.301	1
2	$\frac{1}{4}$	4	0.602	2
3	$\frac{1}{8}$	8	0.903	3
4	$\frac{1}{16}$	16	1.204	4
5	$\frac{1}{32}$	32	1.505	5

absorb the light in the same way. If the intensity of the incident light is given the value 1.0 and it is assumed that half of this light is absorbed (any other value could be chosen), then the emergent light has an intensity $I_1 = 0.5$. This passes on to the second layer, where a further half is absorbed to give a resultant emergent intensity $I_2 = 0.25$. After the third layer, $I_3 = 0.125$, and so on. This gives rise to an asymptotic curve defined by the following geometric series, which was mathematically formulated by Jean Henri Lambert (1728–1777),[2] on the basis of Bouguer's experiments: 1/2, 1/4, 1/8, 1/16, 1/32, etc. At the same time, the underlying law became apparent: the absorbing medium increases linearly $(1 + 1 + \ldots$ layers) while the intensity of the emergent light beam decreases (not linearly, but exponentially). Table 27 gives the number of layers and the decrease in T in the two left-hand columns, the central column gives the reciprocal of T as an aid to calculation, and the logarithms to the base 10 of these reciprocals are listed on the right. This series of numbers, the absorbance, optical density or extinction, shows a linear increase, which is to be expected since two layers absorb twice as much light (in terms of intensity) as one layer, and so on.

Whereas in the Bouguer–Lambert experiment it is the number of layers (and hence the length of the light path) which is increased, Beer[3] observed that the same laws applied on varying the concentration rather than the layer thickness. Assume a vessel (cuvette) of internal diameter 10 mm contains a solution of a dye at a concentration of 1.0 mmol l^{-1}. Suppose this reduces the intensity of the transmitted monochromatic light by half; then doubling the concentration to 2 mmol l^{-1} would reduce the energy of the emergent light to a quarter, trebling the concentration to an eighth, and so on. Thus, the concentration is linearly related to the absorbance and exponentially related to the transmission. This is shown in the two right-hand columns of Table 27. Figure 38 shows the graphical representation of these relationships.

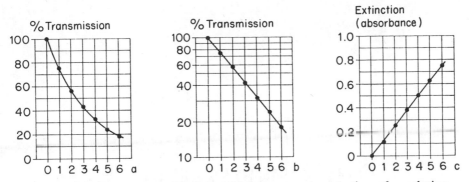

Fig. 38. Relationship between layer thickness or concentration of a solution (abscissa) and transmission or extinction of the solution (ordinate).

2.5.2.2. Absorbing substances

The determination of concentration by photometric methods depends on the availability of an absorber with known optical properties.[5,6] There are the following possibilities:

1. The substance to be determined has its own characteristic colour and absorbs in the visible or near-ultraviolet range, e.g. haemoglobin, bilirubin, carotinoids, proteins, and nucleic acids.
2. The substance to be determined forms a coloured compound with an added reagent in a one- or multi-stage process and there is a stoichiometric relationship between the original substance and the compound formed.
3. The substance to be determined changes the chemical structure of an added reagent in such a way that its optical properties are also changed in a characteristic manner.

2.5.2.3. Absorption spectrum

The absorption spectrum of a solution (or a transparent layer) can be measured with a spectrophotometer. This instrument may be set at any desired wavelength within a certain range, e.g. between 200 and 700 nm. With simple instruments a reference solution, containing the solvent but not the coloured compound, is first introduced at a certain wavelength, e.g. 700 nm, and the instrument adjusted to zero. The reference solution is then replaced with the sample solution and the absorbance measured at 700 nm. The wavelength is then changed to e.g. 690 nm and the sample again measured against the reference as blank, then at 680 nm and so on, down to 200 nm. With a recording double-beam spectrophotometer this time-consuming procedure is unnecessary: the blank reading is automatically set at zero, the wavelength range is scanned by the

158

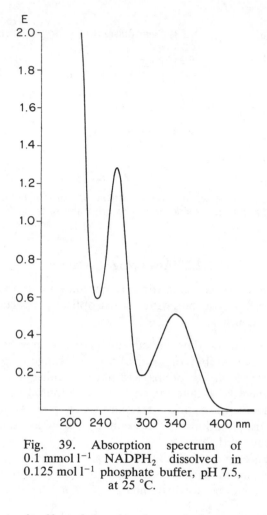

Fig. 39. Absorption spectrum of 0.1 mmol l^{-1} NADPH$_2$ dissolved in 0.125 mol l^{-1} phosphate buffer, pH 7.5, at 25 °C.

spectrophotometer itself, and the absorbance is drawn on a chart by means of an analogue recorder.

Such a spectrum is exemplified in Fig. 39. In this example, the compound (NADPH$_2$) gives a maximum at 340 nm and a minimum at 295 nm. Wavelengths suitable for the photometric determination of concentration are those at or close to the maximum; on the other hand, wavelengths in the neighbourhood of the minima, e.g. 295 or 400 nm, are unsuitable.

2.5.2.4. Molar extinction coefficient

For many compounds, the absorbance at a particular wavelength, measured under defined conditions, is a fixed, reproducible property. The spectrum in Fig. 39 was obtained for a 0.1 mmol l^{-1} solution of NADPH$_2$.

For any chosen wavelength it is possible to read off an absorbance which is specific for this compound and is valid under the conditions of measurement described (buffer, pH, solvent, temperature, etc.). The conversion of the extinction of the 0.1 mmol^{-1} solution to the (theoretical) extinction for a molar solution yields the molar extinction coefficient, ε, for a particular wavelength, λ. The symbol ε_λ always refers to a solution with a concentration of 1 mol in 1 litre. In general, the layer thickness, d, is 10 mm (1 cm) unless there is an express statement to the contrary.

All the attendant conditions such as the type of solvent, temperature, and pH, must be specified in reporting ε, since identical specific absorbances are to be expected only under identical conditions. If ε is known for a particular wavelength, then the absorbance may be calculated directly for any desired concentration of this compound, without the use of any comparison value.

Although 'absorbance' is the recommended term, the terms 'extinction' and 'optical density' are often employed. Other terms for the molar extinction coefficient are 'molar absorptivity' and 'molar absorbance index'.

2.5.2.5. Symbols and definitions used in photometry

T = transmission

E = extinction (absorbance)

I_0 = intensity of incident light

I_A = intensity of emergent light

FV = final volume of assay solution

SV = volume of sample used

MW = molecular weight (relative molecular mass)

1. Relationship between transmission and extinction:

   (a) $T = \dfrac{I_A}{I_0} = 10^{-E}$

   (b) $E = \log \dfrac{1}{T} = -\log T = \log \dfrac{1}{I_A/I_0} = \log \dfrac{I_0}{I_A}$

2. The molar extinction coefficient ε_λ = extinction of a 1 mol l^{-1} solution (c) at the wavelength λ and with a 1 cm layer thickness (d), under defined conditions (temperature, medium, etc.).

   $$\varepsilon_\lambda = \frac{E_\lambda}{cd}$$

3. (a) Calculation of the concentration in mol l^{-1} from E_λ by means of ε_λ:

   $$C = \frac{E_\lambda}{\varepsilon_\lambda d} \cdot \frac{FV}{SV} = \text{mol l}^{-1} \text{ sample}$$

   (b) To calculate the concentration in mg l^{-1}, the above must be multiplied by the relative molecular mass of the substance being determined:

$$C = \frac{E_\lambda \cdot MW \cdot 10^3}{\varepsilon_\lambda d} \cdot \frac{FV}{SV} = \text{mg } 1^{-1} \text{ sample}$$

(c) Measurement of the reaction rate, $\Delta E/\Delta t$, in μmol min^{-1}:
Reaction rate units, etc.

$$= \frac{10^6}{\varepsilon_\lambda d} \cdot \frac{FV}{SV} \cdot \frac{\Delta E}{\Delta t}$$

4. Another quantity used, albeit less and less frequently, is the *percentage extinction coefficient* (percentage absorptivity) (a), which is defined by

$$a = \frac{E}{C\% \cdot d}$$

where E = measured extinction, $C\%$ = concentration of the solution in g per 100 ml and d = layer thickness (with cuvettes, usually 1 cm).

Once the percentage absorptivity has been determined empirically, the concentration can be evaluated from a single measurement at the defined wavelength:

$$C\% = \frac{E}{ad}$$

Occasionally it is necessary to calculate the molar extinction coefficient from the percentage absorptivity and *vice versa*:

$$\varepsilon = \frac{aMW}{10} \quad \text{or} \quad a = \frac{\varepsilon \cdot 10}{MW}$$

where MW represents the relative molecular mass of the compound being determined.

2.5.2.6. Construction of the photometer

A spectrophotometer is used for measuring the absorption of light in the ultraviolet, visible, or infrared regions. It functions in the following way: by means of the appropriate components a particular wavelength or a narrow wave-band is selected from the light emitted by a light source within the spectrophotomer. This monochromatic light, usually in the form of a pencil of parallel rays, passes through a cuvette which holds the material under investigation. The intensity (energy) of the emergent light is measured.

Thus, the five basic components of a spectrophotometer (Fig. 40) are: 1, light source (emitter); 2, wavelength selector (monochromator); 3, cuvette; 4, detector, which converts the light energy to an electrical signal and feeds this via an amplifier to 5, a display system.

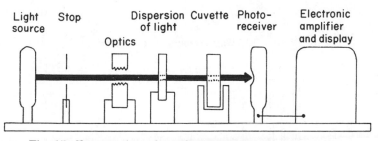

Fig. 40. Construction of an absorption spectrophotometer.

Stops and lenses are generally incorporated to provide the best possible collimation of the beam and to increase the light flux at the receiver.

2.5.2.6.1. Light sources

The function of a spectrophotometer and the quality of the measurements it affords are critically dependent on its light source (in combination with the rest of the components). A basic knowledge of the various types of lamp is therefore an important prerequisite for judging a spectrophotometer and using it purposefully. A first distinction may be made between those light sources which emit a continuous spectrum (continuum radiators) and those which emit at discrete wavelengths (line radiators).

2.5.2.6.1.1. Continuum radiators (Fig. 41)

The simplest light source of this type is a *tungsten-filament lamp*, which emits light with increasing intensity from about 300 nm, the maximum

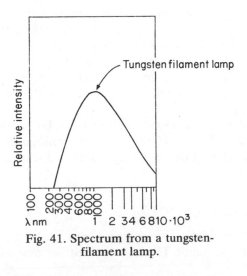

Fig. 41. Spectrum from a tungsten-
filament lamp.

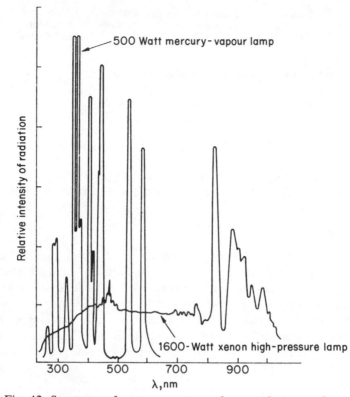

Fig. 42. Spectrum of a mercury-vapour lamp and a xenon lamp.

lying in the near infrared region. Owing to its very weak intensity at low wavelengths, this radiator is less suitable for the ultraviolet range.

The *halogen lamps* are also continuum radiators. Their spectrum is shifted slightly to the shorter wave region compared with that of the tungsten-filament lamp, but their intensity is substantially higher.

For the UV region, the *deuterium lamp* is a technically important source of radiation. This electric discharge lamp is filled with deuterium under low pressure, and emits a continuum from 165 to about 360 nm. Above this range the energy falls off rapidly so that the introduction of a second light source is necessary.

The *xenon high-pressure lamp* emits an irregular continuum (Fig. 42) of high energy from 200 to 10 000 nm. In contrast to the other types described so far, it requires a costly supply of current and has only a relatively short life.

2.5.2.6.1.2. Line radiators

Line radiators are electric discharge lamps which emit a line spectrum characteristic for the lamp. The *mercury-vapour lamp* is most frequently used. Its emission spectrum is shown in Fig. 42. The most important spectral lines of the mercury-vapour lamp are as follows:

nm	nm	nm
265	334	546
280	365	578
302	405	623
313	436	691

These very narrow, wavelength ranges can be isolated with relatively little difficulty and lead to a highly reproducible monochromacy.

The disadvantage of spectrophotometers which use line radiators as light sources is that only a limited number of wavelengths are available and consequently it is not possible to record the spectrum of a compound continuously.

2.5.2.6.2. Filters and monochromators

Filters and monochromators are used for isolating a desired wavelength from a continuum or a line spectrum.

A filter is characterized by its half-bandwidth (or half-width) (Fig. 43). This is the spectral range, in nanometres, within which the transmission is greater than half the value at the nominal wavelength. The smaller the half-width and the higher the transmittance at the maximum, the better is the filter. However, this is difficult to discern from the customary presentation of half-widths and does not fit the conditions obtaining in absorption spectrophotometry. For this reason, Netheler[18] suggested that

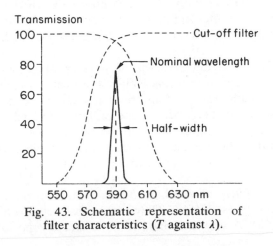

Fig. 43. Schematic representation of filter characteristics (T against λ).

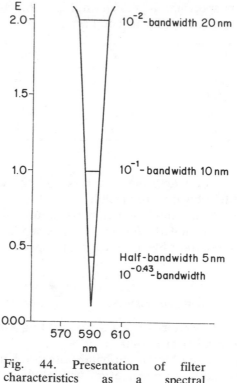

Fig. 44. Presentation of filter characteristics as a spectral transmission profile (E against λ).

filter characteristics should be reported as a spectral transmission profile (Fig. 44). The wavelength is plotted on the abscissa as before, but instead of the transmission, the extinction is plotted on the ordinate. This diagram shows how the quality of the filter falls as the extinction increases, since the half-width increases exponentially. A knowledge of this fact is crucially important for the interpretation of apparent deviations from the Lambert–Beer law (cf., Section 2.5.2.7.1.), caused by a high background extinction.

(a) *Glass filters* generally consist of a layer of coloured glass. They always allow transmission of a relatively wide spectral range and as a result they are not suitable for photometric purposes when combined with a continuum source. However, with a line radiator, monochromatic light can be produced even with a glass filter, if necessary, provided that there is a sufficient separation between the selected line and the neighbouring lines on either side and the filter can absorb the shorter and longer wavelength components.

(b) *Interference filters* have a higher maximum transmittance and smaller half-widths than glass filters. They are constructed on the following principle (Fig. 45): A semi-transparent metal film is vaporized on to a glass

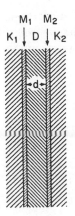

M₁ M₂
K₁ ↓ D ↓ K₂

Fig. 45. Diagram to show the construction of an interference filter. K_1 = Cut-off filter 1 acting as a support; M_1, M_2 = metal films; D = dielectric as separator; K_2 = cut-off filter 2 acting as a cover. $d = 2/\lambda$. (Modified from ref. 4).

or quartz plate. On top of this there lies a second layer composed of a dielectric, non-conducting material, e.g. magnesium fluoride, which in turn is coated with a thin semi-transparent metal film. As a protection against mechanical damage, the 'three-layer system' is backed with a glass or quartz plate. The dielectric layer has a thickness corresponding to half the wavelength of the light desired. If light falls on the surface of the filter, it is reflected to and fro between the two metal films. There is a reinforcement interference if the width of the separating layer is a whole-number multiple of $\lambda/2$. Only those wavelengths which fit this condition can pass through the filter, all others being reflected. Since, in addition to the desired wavelength, the multiples of this wavelength also pass through the filter, the undesired wavelengths must be absorbed by cut-off filters (Figs 44 and 45) opaque to light of higher and lower wavelengths.

(c) *Prisms.* The action of a prism consists in the refraction of light through the prism material, generally glass, quartz, lithium fluoride, or rock-salt. The dispersion of light by a prism is not linearly dependent on the wavelength but decreases with increasing wavelength. The spectral bandwidth is controlled by appropriate stops or diaphragms. The nominal wavelength is the average wavelength emerging from the exit slit. The nominal bandwidth is analogous to the half-width of a filter and includes 75% of the radiant energy in the centre of the band. A good quartz prism produces bandwidths between 0.5 and 1.5 nm over a wavelength range of 200–900 nm. The best prism photometers are equipped with double monochromators in order to reduce the scattered radiation to a minimum.

(d) *Gratings.* In stead of using filters or prisms, dispersion may be accomplished by means of a diffraction grating. This consists of a highly polished glass, metal, or plastic surface on which a large number of parallel lines have been ruled. High-quality diffraction gratings have around 2000 lines mm⁻¹. If white, polychromatic light falls on the grating it is analysed into its spectrum by diffraction in a way similar to that described

166

for interference filters. As with the latter, several different orders of spectra are dispersed by the grating. The appropriate optical arrangement ensures that for a given setting, only the lines of one order pass through the exit slit.

2.5.2.6.3. Cuvettes

For photometric measurement the sample must be transferred into a receptacle, the cuvette. The quality of the cuvette has a considerable effect on the measurements:

1. The length of the light path (layer thickness) is directly involved in the measurement. Cuvettes with variable-length light paths afford non-reproducible results.
2. The material of the cuvette (glass, quartz, plastic) should absorb as little as possible of the incident light (Fig. 46). For measurements below 320 nm, quartz glass is to be preferred.
3. A certain proportion of the incident beam is always reflected and scattered by the outer and inner walls of the cuvette, and this inevitably leads to deviations from the Lambert–Beer law. The proportion of this undesired reflection and scattering must be kept as low as possible. Impurities in the material of the cuvette, or damaged surfaces, increase the proportion of scattering and vitiate the results.

It has become customary to describe cuvettes in terms of their capacity as follows: *macrocuvettes* hold 2 ml or more, *semi-micro cuvettes* have capacities between 2 and 0.5 ml, and *microcuvettes* hold less than 0.5 ml. Figure 47 shows various types of cuvette in common usage. For most photometers, cuvettes with plane-parallel sides are provided. However, certain photometers or photometer systems use round (cylindrical) cuvettes. These act as cylindrical astigmatic lenses, making the calculation of the light path more difficult. However, if the light beam is optimally

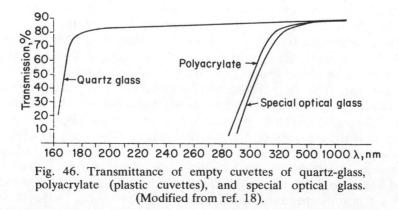

Fig. 46. Transmittance of empty cuvettes of quartz-glass, polyacrylate (plastic cuvettes), and special optical glass. (Modified from ref. 18).

Type		Volume needed for filling,ml
1. Normal cuvette (standard cuvette)		1.5-3.0
2. Semi-micro cuvette		0.5-1.0
3. Microcuvette		0.4-0.6
4. Suction semi-micro cuvette		0.5-1.0
5. Suction microcuvette		0.4-0.6
6. Flow-through microcuvette		Zirka 0.1
7. Round cuvette (for LKB 7600 analyser)		1.0-2.0
8. Eppendorf cuvette (for Enzymautomat 5030)		0.5-0.75

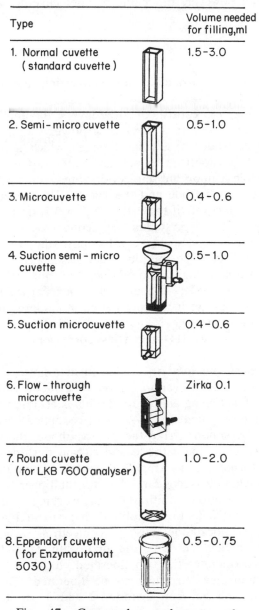

Fig. 47. Commonly used types of cuvette with a light path of 10 mm. 1–6 = Quartz or special optical glass; 1 and 2 = also plastic; 7 and 8 = only plastic. (Modified from ref. 19.)

aligned with the cuvette position, cylindrical cuvettes can also yield good results. As a rule, however, the plane-parallel cuvettes should be given preference.

2.5.2.6.4. Detectors

Various detectors are available for converting the light energy to electrical signals.

(a) the *barrier-layer cell* consists of three layers: the anode, a metal plate (usually iron), is covered with a layer of a polycrystalline semiconductor, e.g. silicon, copper oxide or selenium, on to which a thin, transparent layer, e.g. of gold, is vaporized to act as a cathode (collector electrode). When photons impinge upon the semiconducting layer a potential develops between the collector electrode and the anode. The current yield falls on prolonged usage but recovers if the cell is kept in the dark for some time.

(b) If the layer of polycrystalline semiconductor is replaced with a monocrystalline semiconductor (p-conducting), the resulting assembly constitutes a *photodiode*. The metallic anode can also be designed as a semiconductor (n-conducting). Photodiodes do not exhibit fatigue on prolonged illumination. On account of the particularly favourable signal-to-noise ratio they are increasingly used in modern photometers.

(c) If an amplification stage is integrated with the semiconductor, the result is a *phototransistor*. Its properties correspond with those of the photodiode.

(d) In the *photocell* (*phototube*) a light-sensitive layer, e.g. a mixture of silver and caesium oxide, is vaporized on to a metal cathode. Together with a metal wire functioning as the anode this is enclosed in an evacuated glass or quartz tube. A potential (50–200 V) is applied between the anode and the cathode. If photons strike the photocathode, these are converted to electrons, which leave the cathode and stream towards the anode. The electron current is electronically amplified, measured, and displayed.

(e) The *photomultiplier* (secondary-electron multiplier) is a combination of photocell and electron multiplier. Its working principle may be explained with reference to the simplified diagram in Fig. 48: a photon incident upon the photocathode causes two electrons to be liberated. The high-voltage electric field between the photocathode and a dynode (2) opposite strongly accelerates the two liberated electrons, which strike the dynode causing four new electrons to be liberated. These are further accelerated in the field between dynodes 2 and 3, strike dynode 3 and cause the emission of eight new electrons. These are accelerated to dynode 4, where sixteen electrons are produced, and so on until finally the geometrically augmented current reaches the anode. The sensitivity of the photomultiplier depends on the strength of the applied field. The higher the potential the greater is the electron current generated per photon. The photomultiplier is extremely sensitive and it responds to radiation ranging

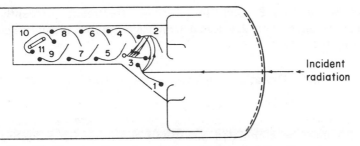

Fig. 48. Construction of a photomultiplier. 1–10 = Dynodes,
11 = anode.

from the far-ultraviolet to the infrared. Consequently, it also responds to
heat and (like all other photodetectors) produces a dark current. For this
reason the photomultiplier is thermostated in high-quality photometers. To
achieve the highest sensitivity the photomultiplier can also be cooled to
−80 °C so that the dark current is almost completely suppressed. At
present the photomultiplier is the best and most universally employable
photodetector.

2.5.2.6.5. Processing and issue of results

The detector produces an analogue signal, usually a change of voltage
over a specific period. This signal is electronically amplified and, in the
simplest case, registered directly by means of a galvanometer: the
galvanometer mirror, marked with longitudinal divisions, is projected on to
a matt screen. The latter bears a scale of extinction values, thus permitting
a parallax-free read-off. Instead of a *mirror galvanometer*, the simpler types
of photometers use a direct-reading instrument.

Continuous registration is usually accomplished by means of a
compensation recorder. So that this can also chart extinction values, most
photometers are equipped with a transmission/extinction converter which
computes the negative logarithms (base 10) of the signal values.

In the last few years an increasing number of digital read-out
photometers have come on the market,[19] inevitably at a higher price.
Digital read-out is more easily readable than analogue display but the main
advantage is the possibility of processing and storing the digital values.
Digitization converts the primary analogue values to a binary code, which
is processed by the appropriate electronic equipment to yield a multi-digit
numerical value. The digital read-out is effected with tubes or
semiconductors, and the results are recorded by a printer. With many
digital photometers it is possible to insert a factor that multiplies each
value before it is displayed. Again, several types make provision for storing
a null value and/or a reference value and make allowance for this in the
calculation prior to display. As a rule the digital values can also be fed

directly to an EDP unit for further treatment. Recently, digital photometers have been developed which incorporate a mini-calculator for processing the values.

2.5.2.7. Deviations from the Bouguer–Lambert–Beer law

In principle, this law is obeyed only under ideal conditions. In practice, small deviations ($< \pm 1\%$ of the measured value) are unavoidable. Larger deviations must be eliminated once their cause has been discovered. The fault may lie with the instrument or with the material being measured.

2.5.2.7.1. Errors attributable to the instrument

Imperfect monochromacy. The importance of monochromatic light may be discussed in terms of the example illustrated in Fig. 49.

The calibration series for a red solution is measured using monochromatic light of wavelength 492 nm, i.e. at the maximum, and of 546 nm, i.e. on the flank. The resulting calibration graphs are two straight lines with different slopes. A third series of measurements is made using a filter of large half-bandwidth (20 nm) with a maximum at 520 nm; the result is a curved line. The cause of this apparent deviation from the Lambert–Beer law may be investigated by means of a hypothetical experiment: let us assume that the filter used transmits two discrete wavelengths, one at the maximum for the absorber and one on the flank of the absorption curve. Further, let us assume that at the wavelength of maximum absorbance the transmittance is reduced by a factor of 0.5 for each change in concentration, i.e. we obtain the familar series 1/2, 1/4,

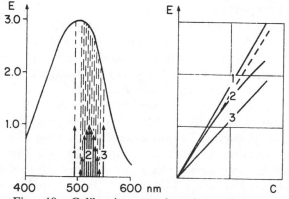

Fig. 49. Calibration graphs obtained using radiation of different bandwidths. 1 = Monochromatic light (492 nm); 2 = non-monochromatic light (filter with a maximum at 520 nm and a bandwidth of 20 nm); 3 = monochromatic light (546 nm).

Table 28. Importance of monochromacy, illustrated by a hypothetical two-wavelength filter.

C	T_{λ_1}	T_{λ_2}	$\dfrac{T_{\lambda_1} + T_{\lambda_2}}{2}$	E
1	1.0	1.0	1.0	0
2	0.500	0.750	0.625	0.204
3	0.250	0.563	0.406	0.391
4	0.125	0.422	0.273	0.564
5	0.063	0.317	0.190	0.721
6	0.031	0.273	0.134	0.873

1/8, etc. On the flank, where the absorbance is weaker, there is only a 75% diminution for each change in concentration, thus producing the series 3/4, 3/8, 3/16, etc. In Table 28, the first column gives the concentration, the next two columns the transmittances T_{λ_1} and T_{λ_2}. Adding the two transmittances and halving the sum gives the actual value of the transmittance, $(T_{\lambda_1} + T_{\lambda_2})/2$. The last column shows the calculated extinctions. The result of this hypothetical experiment is illustrated in Fig. 50.

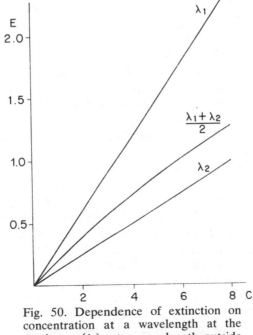

Fig. 50. Dependence of extinction on concentration at a wavelength at the maximum (λ_1), at a wavelength outside the maximum (λ_2), and with a (hypothetical) 'two-wavelength filter'. For details, see text.

As expected, E_{λ_1} and E_{λ_2} give straight lines with different slopes. Measurements using the two-wavelength filter result in a curve which deviates increasingly from the mean straight line and at higher concentrations runs parallel to E_{λ_2}. Analogous findings hold for polychromatic light. In this connection, it should be borne in mind that the half-bandwidth of filters, including interference filters, increases with decreasing extinction. Limited monochromacy may suffice for measurements of low absorbance but may be inadequate at higher absorbance.

Relative concentration error. The graph of transmission against concentration (Fig. 38) shows that at low concentrations a small error in the calibration solution (abscissa) has a large effect on the intensity of the transmitted light. Conversely, at high concentrations a small reading error (on the ordinate) has a large effect on the read-off value of the concentration. The optimal point, where the concentration error and the reading error are mutually minimized, lies at an extinction of 0.434.

2.5.2.7.2. Errors attributable to the sample

Temperature effects. Since aqueous (and other) solutions expand with increasing temperature, a change in the temperature of the absorbing solution always leads to a change in the extinction. In addition, a change in temperature results in changes in the degree of dissociation and/or the solubility, the extent of hydration, and many other factors which can influence the spectral properties. It follows that a description of a photometric measurement would be incomplete without a statement concerning the temperature of the sample. Conversely, all photometric measurements, regardless of whether they are static or kinetic, must be carried out at a defined temperature.

Sample instability. Many coloured compounds measured photometrically in the course of routine analysis undergo changes even in a relatively short period of time; the intensity of the colour may increase or decrease or there may be a shift in the absorption maximum. With samples showing this type of instability it is necessary to work to a fixed time schedule.

Turbidity. Sample turbidity is one of the most frequent causes of error, resulting in an incorrect extinction. Turbidity is usually caused by the use of an unsuitable source material, frequently by the use of milky sera. In such cases a photometric measurement can be performed only if it is possible to clarify the serum sample by centrifugation or other means.

Fluorescence. Many sera contain ingredients which fluoresce, especially at shorter wavelengths. Drugs and drug metabolites are a frequent cause of this phenomenon. Few photometers are equipped with a filter to exclude this fluorescent radiation from the detector and consequently the measured absorption may be incorrect.

2.5.3. FLUORIMETRY

As related above (p. 154), the absorption of light quanta causes the electrons of a molecule to be raised to a higher energy level for a brief period. When they fall back to the ground state, heat is liberated in association with a secondary radiation which is difficult to measure. However, many molecules emit a secondary radiation in the visible region, i.e. they fluoresce. This visually perceptible fluorescence radiation can be measured by photometric methods. An additional characteristic of *fluorescence* is that the exciting radiation is of higher energy (shorter wave) than the secondary radiation (longer wavelength).

Phosphorescence is a special type of fluorescence in which the return of the excited electrons to the ground state is delayed, resulting in an afterglow subsequent to illumination. This phenomenon should not be confused with chemiluminescence, which is produced by the conversion of chemical energy to light energy. Bioluminescence, a special case of chemiluminescence, depends on the ability of certain biological systems to convert chemical energy, usually in the form of adenosine triphosphate (ATP), to light energy. The best known examples are the 'luminous apparati' of glow-worms and fireflies—their photogenetic, enzymatic systems can also be used for analytical purposes.

Today phosphorescence, chemiluminescence, and bioluminescence play only a minor role in clinical chemistry. Also, as far as routine clinical analysis is concerned, fluorescence measurements are far less numerous than photometric measurements. On the other hand, fluorescence methods are indipensable in many research laboratories on account of their high sensitivity.[7–9]

2.5.3.1. Construction of the fluorimeter

The essential components of a fluorimeter are shown in Fig. 51. The light source for producing the primary beam is frequently a mercury-vapour lamp or a xenon high-pressure lamp. Filters, prisms, or gratings are used to select a monochromatic exciting radiation. This falls on the cuvette containing the sample. In many fluorimeters the secondary radiation, the fluorescent light, is measured at an angle of 90° to the primary beam, to prevent any primary radiation reaching the detector system. A particular range is selected from the secondary radiation by means of a filter, prism, or grating, and this is guided to the detector.

2.5.3.2. Measuring technique

Under constant conditions of measurement the intensity of the fluorescent radiation is determined by the concentration of the fluorescing

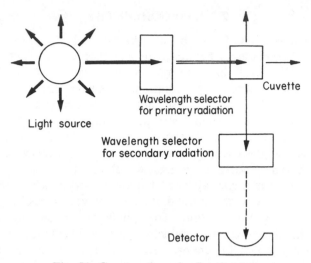

Fig. 51. Construction of a fluorimeter.

substance and the wavelength and intensity of the primary radiation. Fluorimetric determinations require that a set of standards are available so that a calibration graph may be drawn. The intensity of fluorescence in arbitrary units, e.g. in relative fluorescence units, is plotted on the ordinate. In principle, every fluorescent compound can also be determined photometrically since the exciting radiation is absorbed by the molecules. However, fluorescence determinations are often much (up to 1000 times) more sensitive than comparable photometric absorption methods. On the other hand, fluorimetry is highly sensitive to interferences and its application is consequently limited.

The main instrumental difficulties are due to variations in the intensity of the light source and to noise or drift in the detector. The integrity of the cuvette surface is critically important: it must not be damaged, however slightly, and must be extremely clean so as to obviate any non-specific fluorescence phenomena, e.g. from a fingerprint. Errors are also caused through insufficiently cleaned or dried cuvettes, the use of unsuitable cleansing agents, etc.

Compared with photometric methods, fluorescence phenomena are frequently much more strongly dependent on the conditions of the milieu, such as pH, temperature, solvent, buffer, and ionic strength. Exact standardization of conditions for fluorescence methods can be very difficult. Numerous substances have the property of suppressing the fluorescence of certain molecules. This property, known as 'quenching', can vitiate the analysis but in special cases it can be exploited as part of the method.

A linear relationship between concentration and intensity of fluorescence is usually observed over only a relatively narrow range of measurements.

In recent years, owing to these manifold difficulties, various fluorescence methods customarily used in routine analysis (e.g. for the determination of cortisone, catecholamines, oestrogens, and various drugs) have been replaced with other methods that are less liable to interference.

2.5.4. NEPHELOMETRY AND TURBIDIMETRY

If, in its passage through a medium, a light beam encounters particles with a refractive index differing from that of the medium, the light is scattered. The intensity of the scattered light is a function of the number of particles, their size and shape, the wavelength of the illumination, and the difference in the refractive indices of the particles and the medium. The scattered light is totally independent to the chemical nature of the particles.

Turbidimetry is analogous to photometry. It measures the portion of the light that is left to emerge from the cuvette after scattering. In *nephelometry*, on the other hand, the intensity of the scattered light is measured and it is thus analogous to fluorimetry, i.e. as a rule the measurements are made at right-angles to the incident light. In contrast to fluorimetry, however, in nephelometry the wavelength of the incident and emergent light is the same.

The choice of a particular wavelength in the visible spectrum is not critical provided that the solution is colourless. If, on the other hand, the matrix solution is coloured then the wavelength is chosen to correspond with an absorption minimum for the solution. If the particles themselves show a characteristic absorption, i.e. if they are coloured, the wavelength is chosen to correspond with their absorption maximum so as to combine the effects due to absorption and scattering.

In recent years nephelometric and turbidimetric methods have been restricted in nearly all clinical–chemical laboratories to the fully mechanized determination of immunoglobulins.[10] With these (continuous-flow) systems all the conditions, such as temperature, time interval, and surfaces of the reaction vessels, can be kept constant with minimal variation and standards are available for calibration. Consequently, it is possible to use nephelometric measurements in such determinations and achieve satisfactory reproducibility. More recently, interest in nephelometric methods has been revived by the development of a *laser nephelometer*[11] for the quantitative determination of proteins (Fig. 52). The light source is a helium–neon laser with a wavelength of 633 nm. The laser beam passes through the cuvette and is then totally absorbed in a light trap. Part of the laser light is scattered by the immunocomplexes in the cuvette. With a laser beam, the light scattered in the forward direction has a particularly high intensity. This effect is exploited in the instrument: in contrast to the conventional nephelometers (in which the scattered light is measured at an angle of 90°), the laser instrument only collects the light

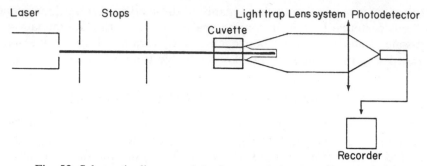

Fig. 52. Schematic diagram of the laser nephelometer (from ref. 11).

scattered in the forward direction; this is focused on to a photodetector by means of a lens system. The electrical signal from the photodetector is proportional to the intensity of the scattered light and can be registered by an analogue recorder or a digital voltmeter. The relatively high energy of the narrow-pencilled laser beam and the intense scattering in the forward direction produce a high intensity of scattered light. As a result, the method has a very high sensitivity, far surpassing that of conventional nephelometry.

2.5.5. FLAME-EMISSION PHOTOMETRY

If an aqueous solution of metal salts is sprayed into a flame as a fine mist, first the water evaporates, then the particles are vaporized to gaseous salt molecules, and finally these decompose to free atoms. If these atoms are heated above a critical temperature radiation is emitted as an *atomic band spectrum* (Fig. 53).

A brief explanation is that the electrons are raised to a higher energy level by absorbing thermal energy and on returning to the original state

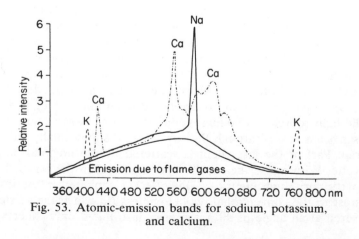

Fig. 53. Atomic-emission bands for sodium, potassium, and calcium.

this absorbed energy is liberated in the form of light quanta. Since alkali and alkaline-earth metals have only 1 or 2 electrons, respectively, in their outermost electron shells they are particularly easy to excite. However, the processes leading to excitation are complex and have not yet been fully explained.[12–14]

The atoms of several elements have a tendency to lose one or more electrons to produce ions. Ions can also be excited and the resulting emission is an *ionic spectrum*, but this is different from the atomic spectrum of the same element. For this reason the experimental conditions for taking the atomic-emission spectrum must be chosen so that the proportion of ions in the flame is constant and as small as possible.

Like atoms and ions, molecules and radicals can also be excited to emit light and the result is a broader emission band of low intensity. Over 50 metallic elements can be determined, qualitatively and quantitatively, by means of emission photometry but in clinical chemistry its use is restricted to the elements sodium, potassium, lithium, and, with limitations, calcium.

2.5.5.1. Construction of the flame photometer

The basic elements of a flame-emission photometer are the atomizer, burner, wavelength selector, and detector/amplifier/display instrument (Fig. 54). The atomizer–burner system serves to convert the aqueous solution of the sample into a mist, i.e. to create an aerosol, and to conduct this to a flame. So that the thermal changes may proceed as smoothly as possible the aerosol should consist of very small, uniform droplets and the flame should burn as quietly as possible. This is why most modern systems are

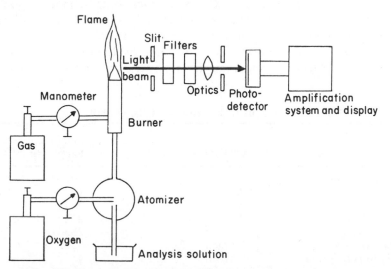

Fig. 54. Construction of an atomic-emission flame photometer.

equipped with an atomizer chamber where the larger droplets are precipitated and only a fine, even mist reaches the burner.

The optical system of the flame-emission photometer resembles that of the photometer. As a rule exchangable filters are used but more expensive instruments incorporate prisms or grating monochromators. The photodetectors and subsequent signal-amplification systems are also similar to those used in the absorption photometer.

2.5.5.2. Measuring technique

The operating conditions are optimal when the emission of the investigated element is directly related to its concentration in the sample solution. To achieve this, three factors demand careful control:

(a) The *flame temperature* should be chosen to give a high atomic emission at a low rate of ionization. The appropriate temperature differs from element to element. In the measurement of potassium atoms a high flame temperature is of no advantage since potassium atoms readily ionize. In this case it is advisable to use a propane/air mixture at about 1925 °C. At this temperature the sodium atoms are also sufficiently excited. On the other hand, it is expedient to use a hotter flame for the determination of calcium (Table 29). For routine clinical work compromise arrangements have been worked out that allow the measurement of sodium, potassium, and calcium in the same solution using a single flame temperature.

(b) The *concentration ratios* of the element to be measured and the other metals simultaneously present in the sample must be taken into consideration (Fig. 55): at low concentrations potassium is to a large extent ionized in the flame and this (ionized) portion is thus prevented from contributing to the atomic line. With increasing concentration the proportion of ionized potassium is smaller and as a result the calibration graph is bent upwards. However, the extent of this concave bending is also dependent on the concentrations of sodium or lithium atoms that are present. These elements suppress the ionization of potassium and thus

Table 29. Gaseous fuel mixtures for flame photometry.

Fuel	Oxidant	Maximum flame temperature, °C
Nitrous oxide	Air	1918
Propane	Air	1925
Hydrogen	Air	2045
Acetylene	Air	2325
Hydrogen	Oxygen	2660
Acetylene	Oxygen	3135
Hydrogen cyanide	Oxygen	4580
Hydrogen	Fluorine perchlorate	4000

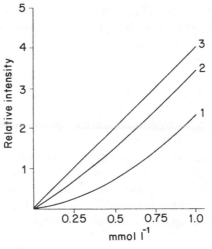

Fig. 55. Effect of different sodium concentrations on the potassium calibration graph: 1 = 0 mmol l^{-1}; 2 = 9 mmol l^{-1}; 3 = 18 mmol l^{-1}.

cause a straightening of the calibration graph for potassium emission. The addition of lithium to the sample has another advantage: the lithium emission may be used as a reference by a comparison detector in order to compensate for variations in the flame. Clearly, if this method is to be used there must be no lithium in the sample itself; in addition, the lithium determination is excluded.

(c) In choosing the *wavelength* for the measurements, care should be taken that the bands of the atoms present in the biological material do not overlap. The choice of optimal conditions for measurement is influenced by the type of photometer system. Since there are many empirical factors which affect the result it is imperative to keep strictly to the instruction manual pertaining to the instrument.

2.5.5.3. Interfering factors

Flame photometric measurements are more liable to interference than photometric measurements. Nevertheless, provided that the composition of the samples is relatively constant the method may be designed to avoid large deviations from the correct result. However, where the composition of the analytical material is very variable, e.g. urine, considerable deviations from the correct value may be encountered. Special attention should be given to three sources of interference:

(a) *Spectral interferences* result when several metals emit lines and/or bands which overlap or lie so close together that the optical system is unable to effect a separation. Either the measurement must be shifted to another band, i.e. another wavelength, or another optical system with higher resolution must be employed.

(b) With many elements *atomization in the flame* is delayed by various anions which form refractory compounds with the atoms to be measured. Thus, e.g., phosphate interferes with the atomization of calcium. Attempts are being made to overcome this interference by using hot flames. The alternative is to make the effect independent of the concentration ratios by the addition of excess phosphate ion.

(c) *Changes in the matrix*, i.e. changes in the physical properties of the carrier solution, can also be the cause of erroneous results. The most important properties concerned are the viscosity, density, vapour pressure, surface tension, temperature, and composition of the salt. To circumvent these effects as far as possible, a carrier solution of constant composition must be used.

2.5.6. ATOMIC-ABSORPTION PHOTOMETRY

The instrumental requirements for atomic-absorption photometry are similar to those of flame photometry.[15] Again, thermal energy is used to bring the material into the gaseous state prior to measurement. However, whereas in flame photometry the light emitted by excited atoms is measured, in atomic absorption it is the absorption of light by unexcited atoms, in other words, the number of atoms in the unexcited ground state, that is recorded. To this end, the atomic spectrum of the element to be measured is produced by means of a hollow-cathode lamp. The line spectrum emitted by this lamp penetrates the hot atomic cloud of one or more elements and is absorbed selectively by the unexcited atoms of the same element; the greater is the absorption, the more atoms of this element are present in the atomic cloud (Fig. 56). Hollow-cathode lamps are available for the analysis of more than 60 elements.[16,17]

It is possible to determine all elements which satisfy two requirements: (a) the element must possess an absorption line in the technically accessible region, i.e. between 200 and 800 nm, and (b) the element must be convertible into an atomic vapour. These conditions are not met in the case of noble gases, halogens, hydrogen, oxygen, nitrogen, carbon, phosphorus, or sulphur. Since the atomic lines are very narrow and the atomic spectrum is absolutely characteristic of a particular element, atomic absorption has an exceedingly high specificity.

2.5.6.1. Construction of the atomic-absorption spectrometer

The main structural elements of this instrument are the hollow-cathode lamp, an arrangement for producing an atomic cloud, a monochromator for

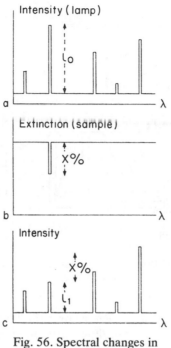

Fig. 56. Spectral changes in atomic absorption. (a) Atomic spectrum of a lamp; (b) extinction of a sample (%); (c) intensity (I_1) remaining after subtraction of (b).

selecting a particular resonance line and a detector/amplifier/display system (Fig. 57).

The *hollow-cathode lamp* is a gas-filled cylinder enclosing an anode and a cathode characteristic of the element. In principle, the determination of a particular element requires the use of the corresponding hollow-cathode lamp, but recently multi-element hollow-cathode lamps have been constructed, containing up to eight elements which may be determined with these lamps.

Formerly, the conversion of the element to the gaseous state was achieved by means of a burner, as used in atomic-emission spectrometry. Today, the flame is increasingly replaced by graphite cuvettes, affording flameless (electrothermal) atomic absorption. The liquid or solid sample is vaporized into a graphite tube which may be heated electrically; the axis of the tube forms the light path. On passing a current, the graphite cuvette is heated in three stages: first the solvent is evaporated off, then the sample is ashed at a higher temperature, and finally the material is converted to

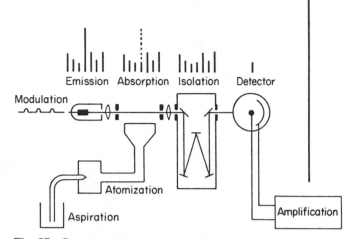

Fig. 57. Construction of an atomic-absorption spectrometer (from ref. 20).

atoms. This process lasts only a few seconds and results in the absorption of the line spectrum from the hollow-cathode lamp as recorded by the detector. Advantages of flameless atomic absorption are very small sample volumes, little sample preparation, lower noise level, and greater sensitivity.

2.5.6.2. Field of application

The application of atomic absorption in the clinical–chemical laboratory is confined to a few elements: calcium, magnesium, copper, and on occasion zinc, in serum and urine. Mercury and lead are occasionally determined in this way for the diagnosis of industrial poisoning. The use of atomic absorption for the determination of iron in biological fluids has been disappointing. Moreover, alkali metals are better determined by atomic emission. Matrix interference is not uncommon in atomic absorption.

References

1. Bouguer, P., *Essai d'optique sur la gradation de la lumière*, Paris, 1729; reprinted in *Les maîtres de la pensée scientifique*, Gauthier-Villars, Paris, 1921.
2. Lambert, H., *Photometrie, sive de mensura et gradibus luminis colorum et umbrae*, 1760; cited from Anding, E., *Lamberts Photometrie*, Engelmann, Leipzig, 1892.
3. Beer, A., Bestimmung der Absorption des rothen Lichts in farbigen Flüssigkeiten, *Annln Phys.*, **162**, 78 (1852).
4. Kortüm, G., *Kolorimetrie, Photometrie und Spektrometrie*, 4. Aufl., Springer, Berlin, 1962.

5. Netheler, H. G., Absorptionsphotometrie, in Bergmeyer, *Methoden der enzymatischen Analyse*, 3. Aufl., Verlag Chemie, Weinheim, 1974.
6. Villard, H. H., Merrit, L. L., and Dean, J. A., *Instrumental methods of analysis*, 5th ed., Van Nostrand, New York, 1974.
7. Elevitch, F. R., *Fluorometric techniques in clinical chemistry*, Little, Brown, Boston, 1973.
8. Udenfriend, S., *Fluorescence assay in biology and medicine*, Vols. 1 and 2, Academic Press, New York, 1962/1969.
9. Withe, C. E. and Argauer, R. J., *Fluorescence analysis, a practical approach*, Dekker, New York, 1970.
10. Eckmann, J., Robbins, J. B., Van den Hamer, C. J., Lentz, J., and Scheinberg, I. H., Automation of a quantitative immunochemical microanalysis of human serum transferrin: a model system, *Clin. Chem.*, **16**, 558 (1970).
11. Sieber, A. and Gross, J., Proteinbestimmung durch Laser-Nephelometrie, *Laboratoriumsblätter (Behring)*, **26**, 117 (1976).
12. Herrmann, R. and Alkemade, C. Th, *Flammenphotometrie*, 2. Aufl, Springer, Berlin, 1960.
13. Schuhknecht, W., *Die Flammenspektralanalyse*, Enke, Stuttgart, 1961.
14. Gaydon, A. G., *The spectroscopy of flames*, 2nd ed., Halsted, New York, 1974.
15. Massmann, H., Entwicklungsstand der Atomabsorptionsspektrometrie, *Angew. Chem.*, **86**, 542 (1974).
16. Robinson, J. W., *Atomic absorption spectroscopy*, 2nd ed., Dekker, New York, 1975.
17. Kirkbright, G. F. and Sargent, M., *Atomic absorption and fluorescence spectroscopy*, Academic Press, New York, 1974.
18. Netheler, H., Absorptionsphotometrie, in Bergmeyer, *Grundlagen der enzymatischen Analyse*, Verlag Chemie, Weinheim, 1977.
19. Haeckel, R., *Rationalisierung des medizinischen Laboratoriums*, GIT Verlag, Darmstadt, 1976.
20. Reynolds, R. J. and Aldous, K., *Atomic absorption spectroscopy*, Griffin, London, 1970.

2.6. ELECTROCHEMICAL METHODS

Of the numerous electrochemical methods used in chemical analysis, only potentiometric, amperometric, and coulometric methods of measurement are presently of practical importance in clinical chemistry. The physical basis of these methods can only be considered here in broad outline.

2.6.1. POTENTIOMETRIC METHODS OF MEASUREMENT

Analytic methods based on the null-current measurement of electrode potentials are called potentiometric methods.[1-4,8,9] The system is as follows: if a metal (e.g. a metal rod) is immersed in an aqueous solution of one of its salts it assumes a potential, the value of which is fundamentally related to the concentration (more correctly, the activity) of metal ions in the solution. Such potential-forming systems may be exemplified by silver/silver ions, copper/copper ions, etc., but non-metals such as halogens and oxygen can also form such systems with solutions of their ions. In the same way, reduction/oxidation systems and certain membrane systems are capable of creating a potential. All of these systems have one thing in common: the formation of the potential obeys the Nernst equation (for details, see recommended textbooks), bearing in mind that it is the ionic activity, not the ionic concentration, that is relevant to potential formation. The difference between the ionic concentration and ionic activity shows up, for example, in pH measurements of solutions theoretically expected to have very high ionic concentrations. The measured pH value is always higher than that calculated, i.e. the concentration of protons appears to be too low. This is also true for aqueous solutions of strong acids: they show increasing deviation from the 'theoretical' pH value with increasing concentration. This may be explained on the assumption that at higher concentrations ions of unlike charge exert an electrostatic attraction for one another. As a result, each cation is surrounded by a cloud of anions and *vice versa*, and this leads to a sort of electrostatic shielding so that the individual ions are no longer so 'active' as the free ions. In practice, ionic activities are usually measured with membrane electrodes. The membrane isolates the electrode space from the solution being measured. Its function is to allow selective diffusion of only those ions whose activity is to be determined. For H^+ ions a glass membrane is chosen (Fig. 58), but the

184

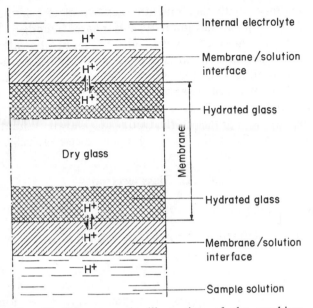

Fig. 58. Diagrammatic illustration of the working principle of a glass electrode (from ref. 8).

principle may be extended to numerous other ions by using specially formulated glass and/or various other materials as membrane. The following types of membrane are currently of importance in ion-selective electrodes:

1. Glass membranes.
2. Homogeneous and/or heterogeneous crystal membranes.
3. Membranes with neutral or electrically charged ligand groups.
4. Modified membrane electrodes.

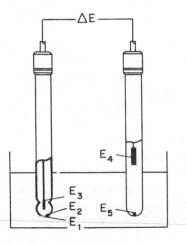

Fig. 59. Schematic illustration of a full cell for potentiometric measurement of ionic activity. E_1 = potential at the membrane exterior, dependent on the activity of the ions to be measured; E_2 = potential at the membrane interior, fixed by the solution within; E_3 = potential of the internal reference electrode (e.g. Ag/AgCl), likewise determined by the composition of the internal solution; E_4 = potential of the aqueous reference electrode, determined by the concentration of the reference electrolyte solution (saturated KCl solution, as a rule); E_5 = diffusion potential at the diaphragm of the outer reference electrode; ΔE = measured potential difference.

The electrodes made with these membranes are electrochemical half-cells characterized by a potential jump at the phase boundary between the electrolyte and the material of the electrode. In the ideal case, the magnitude of this potential jump is dependent solely on the activity of the ions to be determined in the sample solution. Since the individual potentials of these half-cells cannot be measured, they must be linked to form complete cells, one half-cell acting as the measuring electrode and the other serving as a reference electrode of known potential. This is illustrated schematically in Fig. 59 using a glass electrode cell as example.

2.6.1.1. Glass electrodes

2.6.1.1.1. pH electrodes

The pH value is defined as the negative logarithm to the base 10 of the hydrogen ion activity, a_{H^+}. Thus

$$pH = -\log a_{H^+}$$

An analogous relationship may also be formulated for other ions:

$$px = -\log a_x$$

where x signifies the activity of any chosen ions, e.g. Cl^-, K^+, S^{2-} Ca^{2+}.

The process resulting in the potential difference between the two sides of the membrane may be explained with reference to Fig. 58. A thin glass membrane separates two electrolyte solutions. After some time ionic interfaces are formed on both sides of the glass membrane—this is why new glass electrodes, fresh from the factory must be soaked (conditioned) for several hours. Next to each interface there is a layer of hydrated glass, followed by the central core of dry glass. In the glass electrode a conductor electrode dips into the internal electrolyte solution. To complete the electrical circuit, the outer electrolyte solution (the pH of which is to be measured) is connected with the inside of the glass electrode via the reference electrode.

The reference electrodes serve as substitutes for the classical (but impracticable) standard hydrogen electrode, which is conventionally assigned a potential of zero. The potential of the reference electrode is independent of pH and is known for each type from very precise measurements. Reference electrodes end in a diaphragm (porous pot, sintered-glass, etc.) which makes the connection with the sample solution. The most important reference electrodes are the following:

(a) *calomel electrode*. The potential is created by mercury and mercury(II) chloride (calomel). The metallic mercury is covered with a layer of calomel and the electrical connection is made by a platinum wire which dips into the mercury. Referred to the normal hydrogen electrode, the potential of a saturated calomel electrode is +244.4 mV at 25 °C.

(*b*) *silver chloride electrode*. Here the potential is formed by elemental silver and silver chloride. Referred to the normal hydrogen electrode, this electrode has a potential of $+198.9$ mV at 25 °C.

(*c*) *thalamide electrode*. In this case, the potential-forming pair is thallium amalgam and thallium chloride. Its potential is -576.6 mV at 25 °C, referred to the normal hydrogen electrode.

The pH meter is a sensitive voltmeter which measures the potential difference between the measuring electrode and the reference electrode. In order that (virtually) no current is consumed during the measurement, the voltmeter has a very high input resistance.

Alkali error. At lower hydrogen ion activity (high pH) and higher activity of small ions, e.g. sodium ions, it was formerly necessary to be aware of the so-called alkali error, the 'cross-sensitivity' of the electrode. The electrode response was not confined to H^+ ions; it also responded to, e.g. Na^+ ions, which made up part of the electrode potential. However, for the range in question (pH <12) the error is immeasurably small with modern glasses.

2.6.1.1.2. pNa electrodes

Types of glass showing a particularly large sodium error can be used to construct electrodes suitable for the measurement of sodium ion activity. However, these electrodes still respond to H^+ ions and this severely restricts their range of application.

2.6.1.2. Crystal membrane electrodes

In these sensors, also known as solid-state electrodes, the membrane consists of a single crystal or of a solid material possessing ionic conductivity (Fig. 60). A typical representative of this class is the silver iodide electrode, in which a crystal of silver iodide separates two electrolyte solutions. Exchange processes occur at the interfaces as described. This electrode may be used to measure silver and/or iodide ions. Silver iodide has a very low solubility in water. The relationship between the activity of the silver ions and the iodide ions is given by the solubility product of silver iodide and thus the lowest limit of detection is determined by the solubility of the membrane material. If the solution contains anions such as sulphite, which form silver salts with a solubility product lower than that of silver iodide, these cause interference and invalidate the measurement. Chloride ions may be measured with the silver chloride electrode, which has a polycrystalline moulding of silver chloride as the membrane. In the fluoride electrode a single crystal of lanthanum fluoride (LaF_3) is used. Similar electrodes are available for the determination of bromide, cyanide, thiocynate, sulphide, copper, cadmium, and lead ions.

188

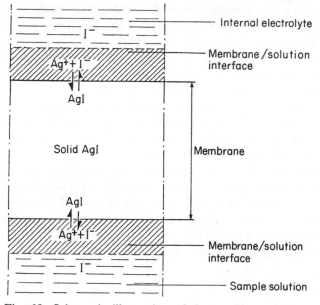

Fig. 60. Schematic illustration of the working principle of a crystal membrane electrode (here AgI) (from ref. 8).

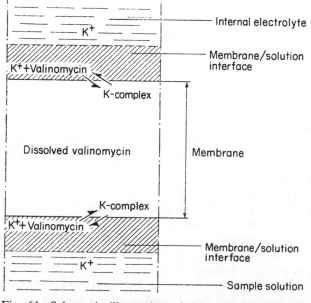

Fig. 61. Schematic illustration of the working principle of a valinomycin–potassium electrode (from ref. 8).

2.6.1.3. Ion-exchange membrane electrodes

With this type of electrode the membrane consists of an inert carrier which contains ion-exchange material. This may be a 'classical' ion exchanger or an electrically neutral ligand which binds the metal as a complex, as is the case with the valinomycin–potassium electrode (Fig. 61). The antibiotic valinomycin is incorporated in a plastic matrix. Since it is very poorly soluble in water it remains in the membrane. In the presence of potassium ions in aqueous solution a potassium–valinomycin complex forms at the interface between the membrane and the solution. Potassium penetrates the membrane, resulting in a potential difference between the two sides of the membrane. The measurement is made exactly as with the electrodes already described.

For the determination of calcium ions an electrode has been developed which contains the liquid ion exchanger dodecylphosphoric acid dissolved in dioctyl phenylphosphonate. This is incorporated in an inert plastic matrix. Inside the electrode the ion exchanger is in contact with a solution of calcium chloride. Electrodes of this type may be used to determine the activity of ionized calcium in blood, plasma, serum, and other biological fluids.

2.6.1.4. Modified membrane electrodes

The membrane electrodes described above may be modified by the introduction of a second membrane which separates the measuring zone from the rest of the sample solution. This second membrane may be selectively permeable by the substance to be measured, so that the latter diffuses into the actual measuring zone where it is determined. Another possibility is the production of a diffusible substance by reactions on the membrane; the reaction product penetrates to the measuring zone and is determined. The most important representatives of this type of sensor are the gas electrodes and the enzyme electrodes.

2.6.1.4.1. Gas electrodes

The gas electrode of greatest practical importance in medicine is that used for measurement of the carbon dioxide pressure in blood and other biological fluids.[5] If the Henderson–Haselbach equation (p. 281) is interpreted as a function of the pH value in dependence on the CO_2 pressure at constant bicarbonate concentration, we have the basis for a method of determining pCO_2 from the pH value. Special pCO_2 electrodes have been developed for this purpose; their principal features are illustrated in Figs 62 and 94. In principle they consist of a pH electrode clad with a plastic foil permeable only to CO_2. Between the glass membrane and the plastic foil there is a capillary fissure, the size of which

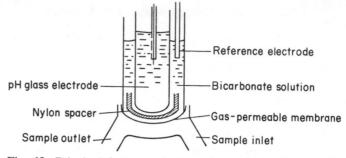

Fig. 62. Principal features of a gas electrode, in this case for the measurement of CO_2 pressure.

is fixed by a spacer, e.g. nylon webbing. The capillary fissure contains a very dilute solution of bicarbonate and a CO_2 equilibrium is established through the foil between this bicarbonate solution and the fluid under investigation. The pH of the internal solution is dependent only upon pCO_2 and is measured with the glass pH electrode. The more dilute the electrolyte solution, the more rapidly is equilibrium attained, although even today equilibration periods of up to 90 s must be tolerated.

The ammonia electrode is constructed similarly. In this case a fluorinated plastic is used as the gas-permeable membrane and the capillary fissure is filled with a dilute solution of an ammonium salt.

2.6.1.4.2. Enzyme electrodes

In recent years attempts have been made to combine the high specificity of enzyme-catalysed reactions with the advantages of electrode measurements.[6] This was feasible once it became possible to produce enzymes bound to carriers: an enzyme preparation of the highest possible purity (obtained from biological material) is bound to a synthetic carrier or support and thus rendered insoluble in water. In order to retain the activity, many kinds of support and various types of bonding are used. As in the gas electrodes, the actual electrode membrane is clad with a second membrane. For example, this may be urease polymerized in a polyacrylamide matrix, stabilized by a nylon net. If this electrode is immersed in a solution containing urea, the latter diffuses into the urease membrane. The NH_4^+ ions liberated by enzymatic hydrolysis are active at the electrode, which is selective for ammonium ions, and the resulting potential can be measured in the usual way (Fig. 63).

$$\underset{NH_2}{\overset{NH_2}{C}} = O + H_2O \xrightarrow{\text{Urease}} 2\,NH_3 + CO_2$$

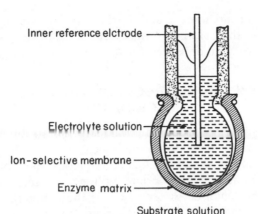

Inner reference elctrode

Electrolyte solution

Ion-selective membrane

Enzyme matrix

Substrate solution

Fig. 63. Diagram showing the construction
of an enzyme electrode.

Over 100 different enzyme electrodes have been described. There are still various technical difficultues which preclude their immediate application in clinical–chemical laboratories, but these are likely to be surmounted within the next few years.

2.6.2. AMPEROMETRIC METHODS FOR DETERMINING OXYGEN PRESSURE

These methods are based on the electrochemical reduction of oxygen.[7] Oxygen dissolved in water undergoes cathodic reduction at a specific potential:

$$O_2 + 2H_2O + 4e^- \rightarrow 4OH^-$$

The Clark electrode is based on this principle. A noble-metal electrode and a reference electrode are immersed in an electrolyte solution which is separated from the external solution by an oxygen-permeable membrane. Oxygen from the external solution diffuses into the electrolyte-filled capillary fissure of the pO_2 electrode. Since a potential difference of 0.7 V is maintained between the anode and the cathode, the oxygen is reduced at the noble-metal cathode and the resulting current is proportional to the partial pressure of the oxygen. Because the reduction potential is small there is no interference from other gases (Fig. 98).

The pO_2 electrode is used to measure the oxygen pressure of blood and other body fluids. Alternatively, the Clark electrode may be used to follow the course of chemical reactions which consume oxygen. Several years ago instruments were developed, e.g. for the determination of blood sugar, which measured the oxygen consumption in the glucose oxidase (GOD)-catalysed oxidation of glucose:

$$\beta\text{-D-Glucose} + H_2O + O_2 \xrightarrow{\text{GOD}} \text{Gluconolactone} + H_2O_2$$

Since this is a stoichiometric reaction, the oxygen consumption is directly proportional to the glucose concentration in the sample. Similar developments have been described for urea, ethanol, and galactose.

2.6.3. COULOMETRIC METHODS FOR THE DETERMINATION OF CHLORIDE IONS

The use of coulometry in clinical chemistry has so far been confined to the determination of chloride in biological fluids. The material under investigation is taken up in an acidic aqueous electrolyte solution of suitable composition and appropriate volume. In this assay, four silver wires dip into the 'chloride meter'; two of these function as a generator circuit, the other two as an indicator circuit. In the generator circuit the passage of current liberates silver ions from the silver wire anode. If the sample contains chloride ions, water-insoluble silver chloride is formed. Silver ions only remain in the solution when the chloride ions have been completely removed. Only a small current flows in the indicator circuit (also silver electrodes, as a rule) while there are no silver ions in the solution, but the instant free silver ions occur—because all of the chloride ions have been used up to form insoluble silver chloride—the amperometric current in this circuit shows a sharp increase. This activates a relay which cuts out the voltage of the generator circuit. The chloride meter measures the quantity of current consumed (product of the generator current strength and the period of current flow), allowing the concentration of chloride ions to be calculated directly.

References

1. Bücher, Th., Hofner, H., and Rouayrenc, J. F., Methoden mit Hilfe der Glaselektrode, in Bergmeyer, *Methoden der enzymatischen Analyse*, 3. Aufl., Verlag Chemie, Weinheim, 1974.
2. Camman, K., *Das Arbeiten mit ionenselektiven Elektroden*, 2. Aufl., Springer, Berlin, 1977.
3. Durst, R. A., *Ion-selective electrodes*, National Bureau of Standards Special Publication 314, Washington, 1969.
4. Koryta, J., *Ion-selective electrodes*, Cambridge University Press, Cambridge, 1975.
5. Koenig, W., *Klinish-physiologische Untersuchungs-Methoden*, Thieme, Stuttgart, 1972.
6. Guilbault, G. G., *Enzymatic methods of analysis*, Pergamon Press, Oxford, 1970.
7. Lessler, M. A. and Brierley, G. P., Oxygen electrode measurement in biochemical analysis, in Glick, *Methods of biochemical analysis*, Vol. 17, Wiley, New York, 1969.
8. Fiedler, F., Ionenselektive Elektroden, *Mitt. dt. Ges. klin. Chem.*, **2**, 109 (1971).
9. Meier, P. C., Ammann, D., Osswald, H. F., and Simon, H., Ion-selective electrodes in clinical chemistry, *Med. Progr. Technol.*, **5**, 1 (1977).

2.7. MECHANIZATION AND AUTOMATION

Apparatus which performs chemical analyses unaided is frequently described as automatic, although Richterich and Greiner[1] gave the following definition as early as 1960: 'The term automation should only be used when the quality of the products, whether their nature be substantial (in production, for instance) or informative (e.g. in sample identification), is continuously monitored by a feedback mechanism which makes appropriate adjustments to the process mechanism'. In the analysis machines used today, the analytical processes are performed with a fixed mechanism and consequently, following a suggestion from Richterich, they are better described as 'analysers'.

In principle, every analytical procedure can be mechanized by the development of suitable apparatus or the adaptation of combinations of existing commercial instruments. However, in many cases this would necessitate a high capital expenditure and a cost/benefit analysis would be negative.[2] Thus the development and installation of analysers is only sensible for those investigations where the number of assays regularly exceeds a critical value. The frequency distribution of the various clinical–chemical analyses shows that 10 methods suffice for about 50% of the required investigations and 25 methods cover more than 90%. Over 100 different methods are additionally required to deal with the remaining 5–8% of the tasks. The serial order of the frequency of individual methods varies only slightly from laboratory to laboratory or from clinic to clinic. The mechanization of short series thus leads to inordinate investment of capital with no saving in labour time.[2]

2.7.1. CLASSIFICATION OF ANALYSERS

In most present-day analysers, the different methods of investigation are carried out either consecutively (*single-channel analyser*) or simultaneously (*multi-channel analyser*). The latter may be broadly classified, further, as 'non-selective' and 'selective', according to their operating features. With *'non-selective'* analysers, the full programme of investigations is performed whether they are required or not, which is uneconomic in terms of consumption of sample and reagents and the total length of the analysis. However, owing to the fixed programme, the construction of these analysers is simpler and the increased cost of reagents can be partly offset by the lower price of the machine.

The *selective* machines may be programmed to perform only those tests required for each individual sample, and this demands appropriate means for telling the machine which tasks are to be performed. Naturally, the electronic and mechanical components are much more expensive for this type of analyser.

Various perspectives may be used to provide a more detailed classification of analysis machines. One classification frequently used is that of the American Society for Clinical Pathology (ASCP). This system, in virtual agreement with that of Richterich and Greiner,[3] distinguishes the following groups:

1. *Continuous-flow instruments.* The most important representative is the AutoAnalyzer (Technicon, Tarrytown, N.Y.) in its various modifications.
2. *Discrete-sample instruments.* The Auto-Chemist (AGA AB, Lund, Sweden) is reckoned to have been the first instrument of this type. Other representatives of this group are the GSA II (Greiner Electronics, Langenthal), the Vickers Multichannel 300 (Vickers, Malden), the AURA (Philips, Pye Unicam, Cambridge), the Acuchem-Microanalyzer (Ortho-Diagnostic Instruments, Raritan), the KDA Monitor (American Monitor Corporation, Dallas, Texas), and many others.
3. *Kinetic analysers.* This class includes the Analysenautomat 5020 (Eppendorf Gerätebau, Hamburg), the ABA 100 (Abbott Laboratories, Pasadena, California), the Enzyme Reaction Rate Analyzer 8600 (LKB, Stockholm, Sweden), the KA 150 (Perkin-Elmer Corp., Norwalk, Conn.), the Digecon System Model 10/11 (Sherwood Medical Industries, St. Louis, Mo.), the Kem-O-Mat (Coulter Electronics, Harpenden), and others.
4. *Analytical-pack analysers.* So far, the only instruments of this type to have come on to the market are the ACA (Du Pont, Wilmington, Del.) and the Stac (Technicon, Tarrytown, N.Y.).
5. *Centrifugal-parallel analysers.* These include the Gemsaec (Electronucleonics, Fairfield), Centrifi-Chem (Union Carbide Corp., New York), Roto-Chem (American Instruments, Silver Springs), and IL Multistat III (Instrumentation Laboratory, Lexington, Mass.).

2.7.2. WORKING PRINCIPLES

The working principles of these five types of analyser may be explained in terms of characteristic examples.

1. The *AutoAnalyzer*, conceived in 1957 by the physiologist Skeggs,[5] utilizes a principle fundamentally different from that of conventional

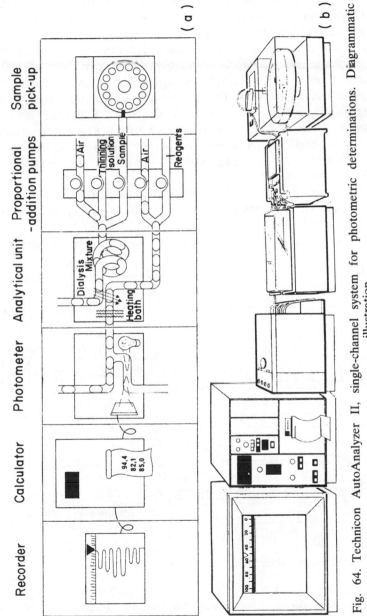

Recorder   Calculator   Photometer   Analytical unit   Proportional -addition pumps   Sample pick-up

Fig. 64. Technicon AutoAnalyzer II, single-channel system for photometric determinations. Diagrammatic illustration.

analytical methods, namely the 'automation of analytical methods by means of continuous-flow analysis' (Fig. 64).

In principle, the AutoAnalyzer represents a tube system through which there is a continuous flow of a stream of reagents and to which the investigational material is discontinuously introduced. The reagent/sample stream is subjected to different processes and is finally conducted to a measuring instrument, e.g. a photometer, fluorimeter, nephelometer, or flame photometer. However, this simple principle can only be realized if the stream of reagents and samples is segmented by the introduction of a tight sequence of air bubbles. This prevents the formation of an undesired streaming profile which would quickly deform the sample into a long thread as it flowed through the system of tubes. On the AutoAnalyzer principle, each segment represents a discrete analytical unit. Even so, a measurable spreading of the sample (overlap) is unavoidable in continuous-flow methods.[6,7]

Nowadays, the continuous-flow principle is used in many modifications for many kinds of application, including applications outside the field of clinical chemistry.[8]

2. The GSA III (Fig. 65) may be used as an example of a

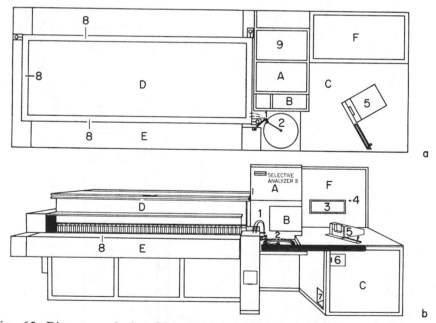

Fig. 65. Diagrams of the GSA Selective Analyzer II (Greiner Electronics). A = Compartment for process vessels; B = dosing compartment for sample processing; C = internal data processing; D = integrated refrigerator with dispensers and reagents; E = compartment for the process channel; F = methods programme unit. 1 = Sample doser/diluter for the two-assay method; 2 = sample carousel; 3 = reagent monitoring attachment; 4 = on/off switch; 5 = insertion

197

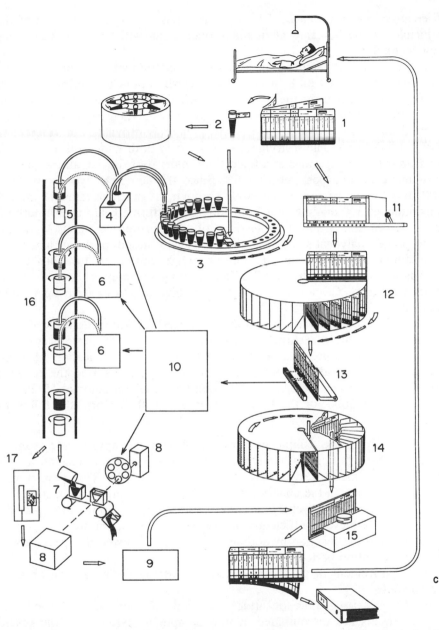

mechanism; 6 = emergency stop; 7 = issue of results; 8 = process channel with thermostat; 9 = measuring unit for photometer and flame-photometer. (c) Diagram to show functioning. 1 = Job card; 2 = sample cup; 3 = sample carousel; 4 = sample doser/diluter; 5 = process cup; 6 = reagent dispenser; 7 = photometer cuvette; 8 = photometer; 9 = process electronics; 10 = methods programme; 11 = test-card punch; 12 = data-input carousel; 13 = test-card reader; 14 = card-printer carousel; 15 = results printer; 16 = water-bath (37 °C); 17 = flame photometer.

discrete-sample instrument. The system[9] has been developed on the principle of mechanizing all manual operations without introducing any new technology.

The centrifuged sample is placed in the instrument without decanting. The request, marked on a special record card, is punch-coded manually at a punching machine on the instrument. All the remaining analytical steps are performed by the machine itself and the analytical results are finally printed on to an inserted request card. There is provision for connection with an EDP unit. The instrument can perform both end-point and kinetic measurements. It operates at a fixed time interval of 6 s, i.e. all the process cups move on one position after 6 s. Since the complete cycle consists of 100 stations, the passage of a sample through the machine takes 10 min. A maximum of 300 analyses per hour may be performed. Up to 30 methods may be chosen from a collection of 40 written methods and the appropriate analytical steps (quantity of reagent added, positions and times of addition, wavelength for photometric measurement, etc.) established on a programmer field. Up to 87 different reagents may be stored in the instrument at 4 °C and the sample can receive additions from up to four reagents.

3. Until recently, analysers specially designed for following the course of a reaction were used exclusively for the determination of enzyme activities, but there is an increasing tendency to extend the application of these instruments to the determination of substrates by measuring the reaction velocity. Kinetic analysers, e.g. the Eppendorf-Analysenautomat 50/20, the LKB-analyzer 8600, or the Perkin-Elmer KA 150, perform the following steps:[10]

i. Quantifying the sample and mixing it with the appropriate reagent. This may be a buffer, an auxiliary enzyme or some other reagent, and in the simplest case it serves only to dilute the sample. This first step is carried out either in the analyser itself or in a sample-preparation station.

ii. Incubation of the charge so that preliminary reactions may proceed and so as to achieve a sufficiently precise reaction temperature.

iii. Starting the reaction by adding an initiator, followed if necessary by a (short) incubation period.

iv. Measurement of ΔE over a fixed or variable period at one or more wavelengths.

v. After a fixed time or after a certain ΔE has been reached, the measurements are discontinued, a new sample is started, and the results are calculated for the first sample. The results are documented with an analogue recorder or with a calculator/print-out system. There is generally provision for connection to an EDP unit.

4. The *pack-analyser* ACA (Fig. 66) utilizes commercially prepared reagent envelopes which serve as process vessels.

'Packs' are available for over 30 different methods. The unquantified

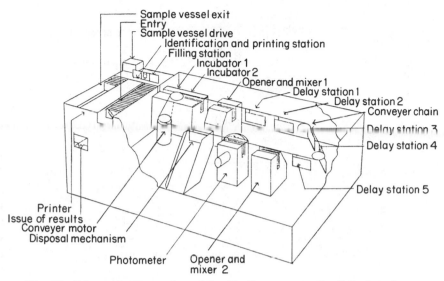

Fig. 66. Schematic illustration of the Du Pont automatic clinical analyser.

sample is contained in a sample carrier which bears the patient data written in capitals. This sample carrier is placed in the input tray of the machine and behind it are placed those reaction packs which correspond with the required tests. A new patient sample and new test packs may be placed behind these. On the instruction 'start', the machine takes an appropriate quantity of investigational material from the sample carrier, transfers it to the first pack, and simultaneously dilutes it with one of five different buffer solutions. The material and buffer now occupy the lower part of the envelope; the upper part contains dry reagents in separate sealed-in compartments. When a test pack has been filled in this way it is transferred from the filling station to the processing channel. Here its contents are warmed to 37 °C, then it passes to the first break-and-mix station where the first four reagent containers are opened and thoroughly mixed with the liquids in the envelope. After further incubation the three remaining reagent seals are broken at the second break-and-mix station and the contents of the packet thoroughly mixed. The 'pack' then reaches the photometer. Here the plastic material of the reaction envelope is deformed into a cuvette by hydraulic pressure between two quartz windows. Since the nature of the test is known to the control unit of the photometer, an appropriate filter is selected and a stat or kinetic measurement is made. With particular methods measurements are made at two different wavelengths. The photometer signals are fed to the calculator, converted to concentrations or activities, and the latter are printed on an ultraviolet photograph of the sample-carrier identification card.

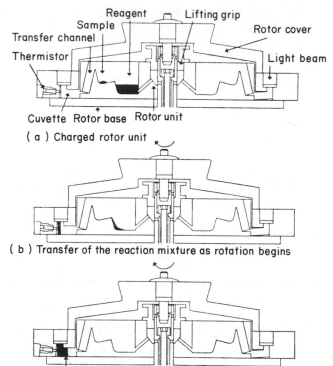

(a) Charged rotor unit

(b) Transfer of the reaction mixture as rotation begins

(c) Reaction mixture in the cuvettes during the analysis

Fig. 67. Centrifugal analyser.

5. The *centrifugal analysers* (Fig. 67) are based on a development made by Anderson.[11]

Each segment of a segmented centrifuge rotor contains one compartment each for the sample and reagent and a measuring cuvette. The sample and reagent are mixed by centrifugal force and transferred to the cuvette on the rim. In this way it is possible to carry out one type of test with 30 samples 'simultaneously'. The rotor can be charged manually or with a pipetting device; the filled rotor is then placed in the instrument. The process parameters appropriate for the chosen test are selected on the programmer: temperature, wavelength, measurement against a fixed or progressive blank, stat or kinetic measurement, time interval before the first measurement, time intervals for successive measurements, number of measurements desired, etc. When the rotor starts, the cuvettes fill with the reaction mixtures and the reactions are followed spectrophotometrically; from the numerous extinction values the results for each individual cuvette are calculated in a mini-computer.

Numerous reviews and monographs are available detailing the state of the art in the field of mechanization and automation in clinical chemistry.[12-16]

References

1. Richterich, R. and Greiner, R., Analysatoren in der klinischen Chemie. I. Maschinen, Automaten, Analysatoren, *Z. klin. Chem. klin. Biochem.*, **8**, 588 (1960).
2. Haeckel, R., *Rationalisierung des Medizinischen Laboratoriums*, GIT-Verlag, Darmstadt, 1976.
3. Richterich, R. and Greiner, R., Analysatoren in der klinischen Chemie. II. Terminologie und Klassifikation, *Z. klin. Chem. klin. Biochem.*, **9**, 187 (1971).
4. Couch, R. D., *Analyzers analyzed*, ASCP Commission on Continuing Education VII, no. 7 + 8, Chicago, 1970.
5. Skeggs, L. T., An automatic method for colorimetric analysis, *Am. J. clin. Path.*, **28**, 311 (1957).
6. Richterich, R., Greiner, R., and Küffer, H., Analysatoren in der klinischen Chemie. III. Beurteilungskriterien und Fehlerquellen, *Z. klin. Chem. klin. Biochem.*, **11**, 65 (1973).
7. Hjelm, M., Quality control of automated analytical systems in clinical chemistry, *Z. analyt. Chem.*, **243**, 781 (1968).
8. Snyder, L., Levine, J., Stoy, R., and Conetta, A., Automated chemical analysis: update of continuous flow approach, *Analyt. Chem.*, **48**, 942A (1976).
9. Greiner, R., Der Greiner Electronic Selective Analyzer GSA II. I. Beschreibung aus technischer Sicht, *Z. klin. Chem. klin. Biochem.*, **11**, 76 (1973).
10. Netheler, H. G., Die Automation der Analyse, in Bergmeyer, *Methoden der enzymatischen Analyse*, 3.Aufl., Verlag Chemie, Weinheim, 1974.
11. Anderson, M. G., Basic principles of fast analyzers, *Am. J. Clin. Path.*, **53**, 778 (1970).
12. Keller, H., *Automaten im klinischen Labor*, Franckhsche Verlagshandlung, Stuttgart, 1971.
13. Bergmeyer, H. U. and Klose, S., Analysenautomaten, in Bergmeyer, *Methoden der enzymatischen Analyse*, 3. Aufl., Verlag Chemie, Weinheim, 1974.
14. Mather, A., A critical approach to the evalution of automated systems in clinical chemistry, in Stefanini, *Progress in clinical pathology*, Vol. 6, Grune & Stratton, New York, 1975.
15. Westlake, G. E. and Bennington, J. L., *Automation and management in the clinical laboratory*, University Park Press, Baltimore, 1972.
16. Faust, U. and Keller, H., Automation der Analysentechnik im klinisch-chemischen Laboratorium, *Chem.-Ing.-Techn.*, **48**, 419 (1976).

2.8. ISOTOPE METHODS

2.8.1. GENERAL DISCUSSION OF ISOTOPES

Isotopes are nuclides of the same chemical element but with different mass numbers, i.e. atoms having the same number of protons but different numbers of neutrons (mass number = number of protons + neutrons per atom). The nucleus of an atom is stable only for precise ratios of neutrons to protons. If this ratio is out of balance, the nucleus can change, emitting particles to achieve a stable configuration. Consequently, isotopes are stable or unstable according to the ratio of protons to neutrons. The unstable isotopes change their configuration producing the so-called *radioactive radiation*. Such isotopes are also called radioisotopes. The natural abundances of the *stable* and *unstable isotopes* differ. Thus the relative frequencies of occurrence of the carbon isotopes are ^{12}C 98.9%, ^{13}C 1.11%, and radioactive ^{14}C traces; nevertheless, the last isotope is used, for example, in the archaeological dating of organic material. In addition, isotopes can also be produced artificially in reactors by neutron bombardment or nuclear fission.

At present the main isotopes of importance in clinical chemistry are the *β- and γ-emitting radioisotopes*. β-Radiation consists of negatively charged electrons, whereas γ-rays are not particles but electromagnetic waves.

The radiation emitted during the decay of an unstable isotope has an energy characteristic of the particular isotope. This is still often stated in electronvolts [1 eV = 1.602×10^{-19} J (joules)].

With *γ-emitters*, the electromagnetic waves emitted have characteristic 'energy lines' comparable to the emission lines of excited ions in flame photometry (discrete energy spectrum).

In contrast, the energies of the electrons produced by *β-emitters* are continuously distributed over a wide range up to a characteristic maximum energy. The energy-distribution pattern of a β-radiator is shown diagrammatically in Fig. 68a, from which it can be seen that the energy of the most frequently emitted electrons is about one third of the maximum energy. Although the energy ranges of different β-emitters may overlap, their different energy spectra allow their determination [using suitable 'windows' (see below)] even in a mixture of radioactive isotopes (double isotope methods).

Another property characteristic of the decay process of an unstable

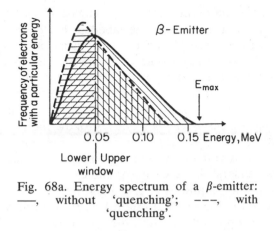

Fig. 68a. Energy spectrum of a β-emitter:
——, without 'quenching'; ---, with
'quenching'.

isotope is the half-life. This is the time in which the radioactive atoms decrease to half of their original number. Characteristic data for a few commonly used radioisotopes are given in Table 30. A valid measure of the activity of a radioisotope is provided by the number of disintegrations occurring in a fixed time, dimension T^{-1}, unit the curie (Ci): 1 Ci = 3.7×10^{10} disintegrations per second (dps); 1 Ci = 2.22×10^{12} disintegrations per minute (dpm).

2.8.1.1. Significance of isotopes

The significance of the isotope, whether it be a stable isotope of rare occurrence or an unstable, i.e. radioactive, isotope, stems from the fact that chemically it behaves to a large extent like the most frequently occurring natural isotope but is physically distinguishable from it, by reason of either

Table 30. Characteristic data for radioisotopes used in clinical chemistry.

β-Emitters	Maximum energy, MeV	Half-life
^{3}H	0.018	12.4 years
^{14}C	0.156	5568 years
^{35}S	0.167	87.5 years
^{32}P	1.71	14.4 days

γ-Radiators	Energy, keV (most important line)	Half-life days
^{125}I (^{125}Te)	35 (27)	60
^{131}I	364	8
^{57}Co	136	270
^{58}Co	810	71

its radioactivity or its different mass. By changing the isotopic ratio of an element in a molecule, the latter can be labelled. For example, if the ^{14}C in a molecule is enriched, possibly even in a particular position, then this labels the molecule and it is possible to follow its fate in the course of a process by observing the radioactivity emitted by the ^{14}C. In clinical chemistry isotopically labelled substances are used especially in radio-immunoassays and for very sensitive enzyme determinations *in vitro*. Their use in living organisms enables one to follow the morphological and metabolic behaviour of molecules by observing the enrichment of these traces of isotopes (tracers) or by seeking various labelled metabolites of administered precursors in the body fluids or in organ biopsies. Radioactive isotopes are injurious to the organism; consequently, they may only be administered in the smallest quantities. In working with radioactive isotopes the safety precautions for the protection of the laboratory personnel and the environment should be rigorously observed and their observance monitored. In many countries these precautions must comply with statutory regulations. Stable isotopes such as deuterium (^2H) or ^{13}C have no potentially harmful effects. They differ from the most abundant natural isotopes (^1H and ^{12}C) only in their mass. The development of mass spectrometry, allowing molecules or molecular fragments of differing mass to be quantified, has in recent years made the application of stable isotopes in clinical chemistry possible. Stable isotopes have been used, for example, as ideal internal standards of the quantification of international reference standards or for *in vivo* studies with larger quantities of tracer.[1,2] At present, however, these analytical methods are still the province of specially equipped laboratories.

2.8.2. MEASUREMENT OF RADIOACTIVE ISOTOPES

2.8.2.1. Principles

If β- or γ-rays collide with matter, their energy is taken up by the material. Depending on the substance concerned this may result in *ionization*, the triggering of a *chemical reaction*, or the *emission of photons*.

Ionization. The ionization of gas by radioactive radiation is detected in a *Geiger–Müller counter*.

Chemical reaction. The formation of metallic silver when radioactive radiation falls upon a photographic emulsion must be considered to be a photochemical effect. This method, *autoradiography*, serves mainly for locating the radioactive isotopes within a preparation (histological section, paper chromatogram, etc.). With this technique the effect of radiation on the emulsion may be cumulated over a long period (up to week-long exposures) and the background radiation may be effectively restricted to a minimum; consequently, even very small activities may be detected by autoradiography.

Emission of photons. The emission of photons due to the impact of corpuscular radiation (electrons) on matter is used for quantifying β-emission and γ-radiation in scintillation counters. Substances which exhibit this emission of flashes of light (ultraviolet region) are called *fluors*. Those used for the detection of *β-emission* are mainly heterocyclic hydrocarbons, oxazoles, oxadiazoles, and thiophenes (e.g. 2,5-diphenyloxazole, PPO). Since β-rays have only a very small penetrating power, it is advantageous to bring the fluors in close contact with the β-emitting molecules. This is done by adding the fluor to the radioactive sample in solution ('liquid scintillation counting'). Most of the primary fluors mentioned above exhibit an emission maximum below 400 nm. However, the optimum range of most photocathodes lies above 410 nm and, apart from this, it is not rare for the biological sample itself to absorb light below 400 nm. It is therefore an advantage to add a second fluor which emits at a wavelength above 410 nm when excited by the light flash from the primary fluor. A whole series of liquid fluor mixtures ('cocktails') are in use, differing from one another both in respect of the nature of the primary and secondary fluors and the solvent and solubilizing agent (to make them miscible with water). Detailed recipes may be found in the literature.[3]

Very short-wave electromagnetic radiation, such as *γ-rays*, penetrates matter more deeply than the corpuscular β-emission. Consequently, for γ-rays it is possible to use solid fluor crystals incorporated in the counter, held at a measurable distance from the sample. The crystals consist, for example, of thallium-activated sodium iodide. γ-Rays cause the release of secondary electrons which, in turn, act upon other electrons in the crystal resulting in scintillation.

2.8.2.2. Scintillation counters

The purpose of the scintillation counter is to count the number of light-flashes per unit time and thus measure the β- or γ-activity of a sample. The central part of β-counter is illustrated schematically in Fig. 68b.

The activity of the radiation is measured by converting the light flashes from the fluor to electrical impulses and counting these in conjunction with a timer. The conversion and amplification of the signal is effected by a *photomultiplier* (Fig. 48). The height of the electrical impulse is a measure of the radiation energy since the number of light flashes emitted simultaneously by the fluor is largely proportional to the radiation energy absorbed. The photons landing on the light-sensitive photomultiplier cathode liberate a corresponding number of electrons; these are multiplied and strike the anode, thus creating a current impulse (p. 168).

Other components of the scintillation counter are for improving the sensitivity, i.e. the signal-to-noise ratio, and for the measurement of selected energy ranges (window).

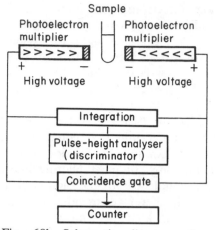

Fig. 68b. Schematic diagram of a scintillation counter.

A reduction in the noise is achieved by protecting the sample chamber from background radiation; cooling the system reduces the thermal motion of the electrons; by arranging two photomultipliers symmetrically about the sample and incorporating a *coincidence gate*, only those current impulses which derive from both photomultipliers simultaneously are fed to the counter. Impulses which are not due to the sample will only very rarely be produced in both photomultipliers simultaneously. The sensitivity of the measurement can be further increased by *summation* of the temporarily coincident current impulses from both sensors. The *pulse-height analyser* allows the measurement of current impulses within a prescribed range (window).

The measurement of impulse heights within certain limits and of different impulse-height ranges (and, consequently, of limited ranges of radiation energy) allows the selective determination of the activity of a particular isotope in the presence of another (double-isotope methods). This technique is also applied to the calculation of *quenching*. In counting the activity of samples containing β-emitters, the absorption of radiation energy or of photons is of importance ('quenching'): the components of the sample dissolved in the fluor liquid can absorb β-rays (electrons) or photons; thus, the solvent may not transfer the energy efficiently, or chemical reactions with the fluors may reduce (or induce: chemiluminscence) their fluorescence. As a result, fewer photons per impulse strike the photocathode and the energy spectrum appears to be shifted to a lower range (Fig. 68a). Thus, the determination of the degree of quenching or the count efficiency, for each sample, is a matter of considerable importance. The addition of an internal standard followed by a recount is troublesome and not very accurate. It is more usual to perform the count with an external standard and to calculate the ratio of the impulse rates for two energy ranges. The *external standard* is a γ-emitter which acts on the fluor causing the latter to emit Compton electrons, which are measured in a separate channel. These Compton electrons behave like high-energy γ-rays and they too are subject to quenching. After constructing an efficiency curve, using solutions with known quenching, the sample count efficiency can be determined by counting with and without the external γ-radiator. One requirement is that the sample volumes are the same as those used in making the calibration graph.

Another method suitable for determining the degree of quenching is calculation of the ratio of the count rates in different energy ranges. The measurement is made in an upper and a lower part of the energy range for the isotope (Fig. 68a). If the sample exhibits quenching there is an increase in the fraction corresponding to the lower window and a decrease in that of the upper window. The ratio of the impulse rates in the two ranges can thus be used as a measure for the energy absorption. Combination with the external standard method does not afford any significant improvement.

As a rule, modern counters offer various technical refinements. It is often possible to programme various count parameters such as the energy range and count period for discrete series. Additional computation programmes are included for special purposes. Combination of the scintillation counter with mini-computers allows greater flexibility than microprocessors with fixed programmes. Usually the instruments are equipped with automatic sample changers as an accessory. For an extended discussion of the principles and application of liquid scintillation counting, the reader is referred to the literature.[4,5]

2.8.2.3. Application

The principle application of radioisotopes in clinical chemistry is in the field of saturation analysis.

2.8.3. SATURATION ANALYSIS

2.8.3.1. Introduction

The term 'saturation analysis' was coined by Ekins.[6] In principle, the method is as follows. The amount of substance to be determined is made to react with a *limited* amount of a reagent which combines with it specifically and in known quantity. The binding reagent is saturated by the substance and a portion of the latter remains free, i.e. unreacted. Thus, the substance to be determined is present in both bound and free forms and the ratio of the quantities in these two forms is determined by the initial amount of the substance. By comparison with standards, the total amount of the substance may be concluded from this *ratio*. The latter is not measured directly. Instead, the following procedure is adopted: a small

208

Table 31. Commonly used tests and their names.

Labelling	Binding protein	Name
Radioisotope	Immuno-antibody	Radioimmunoassay
Enzyme	Immuno-antibody	Enzyme immunoassay
Radioisotope	Serum protein	Competitive protein binding assay
Radioisotope	Cytoplasmic receptor	Radioligand assay
Radioisotope	Cell-membrane receptor	Radioreceptor assay

quantity of labelled substance is added and the ratio is calculated from its distribution between the two separated phases (free and bound). The labelling must be done in such a way that the tracer substance can be determined even in the very smallest quantities. Suitable candidates are, e.g. radioisotopes of high specific activity, enzymes, or fluorescing groups. As binding reagents specific for the substances determined, proteins are used, e.g. immuno-antibodies or circulatory proteins. The multiple combinations of possible labelling and various binding proteins has resulted in a perplexing nomenclature (Table 31).

Another possibility is to label the binding protein and to add this in *excess*. After the reaction, that fraction of the binding protein which has not reacted is separated and the activity of the residual fraction is determined. If the marker is a radioisotope, the test is called an immunoradiometric assay. The subject has been reviewed by Woodhead *et al.*;[7] the following discussion is restricted to the radioimmunoassay.

2.8.3.2. Radioimmunoassays

The radioimmunoassay (RIA) was developed by two independent groups: Yalow and Berson[8] in the USA and Ekins[9] in the UK.

Principle. An unknown quantity of a ligand, i.e. in RIA an antigen (serum sample containing the substance, L, to be determined), and a known quantity of the radioisotope-labelled ligand (L^*) compete for a limited quantity of antibodies (AK) specific for the antigen (L and L^*). Following *incubation*, an equilibrium is established in accordance with the Law of Mass Action. The system then contains free, unlabelled ligand molecules (L), free labelled ligand molecules (L^*), unlabelled (LAB) and labelled (L^*AB) ligands bound to the antibody, as well as some free antibody (AB):

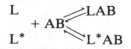

Following *separation* of the antibody-*bound* ligands from the *free* ligands, the ratio of labelled, bound to labelled, free ligands is determined. If the

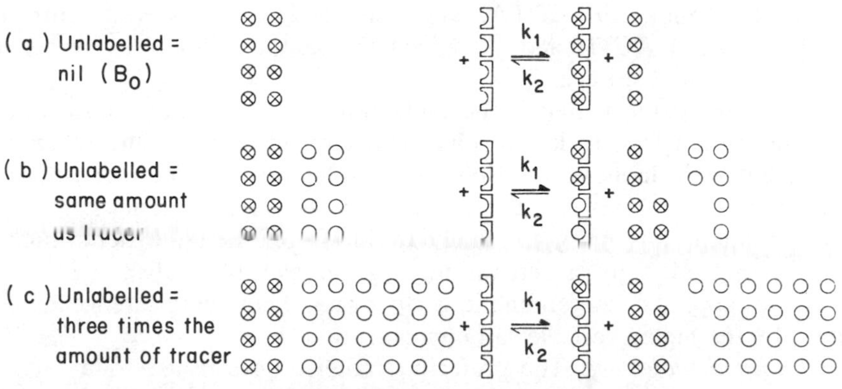

(a) Unlabelled = nil (B_0)

$+ \rightleftharpoons[k_2]{k_1} +$

(b) Unlabelled = same amount as tracer

$+ \rightleftharpoons[k_2]{k_1} +$

(c) Unlabelled = three times the amount of tracer

$+ \rightleftharpoons[k_2]{k_1} +$

Fig. 69. Principle of a radioimmunoassay. (a) Amount of unlabelled ligands = 0. Highest possible proportion of bound, labelled ligand (B_0). In the case where there is no reaction from right to left ($k_2 = 0$), 50% of the tracer is bound and 50% is free. (b) Concentration of the unlabelled ligand = concentration of the labelled ligand. If $k_2 = 0$, 25% of the labelled ligands are bound and 75% free. (c) The concentration of the ligands to be determined is three times as great as the tracer concentration. For $k_2 = 0$, 12.5% of the tracer is bound and 87.5% free.

amounts of labelled ligand and antibody are kept constant, then the fraction of labelled, bound ligands is smaller the higher is the concentration of the unlabelled ligands to be determined. Thus, an unknown amount of ligand can be determined by comparison with a set of standards (calibration graph). Figure 69 illustrates the process.

The use of radioactively labelled ligands with high specific activity and of specific antibodies allows the determination of substances in the femtomole range, often without any previous isolation from the physiological liquids. Detailed discussions including instruction in the types of calculation necessary can be found in the literature.[10-14]

Procedure. The procedure consists of the following steps:

1. Sample collection and preparation.
2. Addition of a precise quantity of labelled molecules to the substance to be measured.
3. Addition of a definite, limited quantity of antibody which reacts specifically with the substance.
4. Incubation.
5. Separation of the 'free' and antibody-bound molecules.
6. Measurement of the free or bound radioactive molecules.
7. Calculation by means of the calibration graph.
8. Checking (internal and external quality control).

1. *Sample collection and preparation*. In the determination of protein hormones, allowance must be made for proteases present in the serum and the samples should be collected in tubes containing inhibitors, e.g. trasylol,

dimercaptoethanol, or EDTA, and in certain cases (e.g. for the determination of ACTH and glucagon) the samples should be frozen for transportation and storage.

Depending on the material, the collection of blood from infusion tubes or catheters can lead to low results, owing to adsorption. The reaction is carried out with duplicate or triplicate samples (at the same concentration or as a dilution series). If the antibody is not specific and structurally closely related substances are likely to interfere, then the determination must be preceded by a separation, e.g. in steroid analyses. This also frequently applies to determinations on urine, where interference may be caused by the highly variable salt concentrations.

2. *Labelled molecules.* The preferred radioisotopes in use today are ^{125}I and, on occasion, tritium. Vitamin B_{12} is labelled with ^{57}Co. The γ-emitter, ^{125}I, is to be preferred to the ^{131}I isotope on account of its higher count efficiency and, especially, its longer half-life. Tritium is used to label ligands which cannot be labelled with iodine, either directly or indirectly (certain steroids).

The labelling is carried out by the chemical reaction of I^+ with a tyrosine or histidine residue of a protein or with a tyrosyl ester. The I^+ is produced by oxidation of sodium iodide with chloramine-T,[15] or, more elegantly, with lactoperoxidase and H_2O_2 *in situ*.[16] Since high activities of radioactive iodine are present, special protective measures must be taken when performing the iodination. The labelled ligands are expected to have a high activity and a high purity but, owing to the drastic reaction conditions, side-products are formed. It is therefore necessary to purify the labelled ligands from side-products and excess iodide by, e.g. column chromatography, prior to using them in the assay. Since radioisotopes cause ionization, this leads in time to a degradation of the tracer. For this reason it should be diluted and stored frozen (exception: tritium, stored at 4 °C). In addition, bovine serum albumin is frequently added to prevent adsorption on the walls of the vessel. The degradation of a labelled molecule may be recognized by an increase in *non-specific binding* in the assay. This is determined in each series by carrying out an analysis without addition of the antibody. It is that portion of the tracer which behaves as if it were bound to antibodies when there are no antibodies in the system.

3. *Antibodies.* An ideal antibody is expected to be highly specific yet not to differentiate between the labelled and unlabelled ligands. A high bonding affinity is an advantage since it allows a high dilution of the antibody in the assay and means that the same antiserum may be used to conduct a determination for a long time, in spite of the limited quantity of antisera available. Each new antiserum must be evaluated afresh, even if it is obtained from the same animal at a later date. Antisera are frequently ill-defined. In the commonest cases, the part of the protein which is susceptible to the antibody (amino acid groups) is now known. Often an antiserum will contain not only one antibody but several bonding sites for

the antigen, each site having a different affinity constant. Moreover, in the rarest cases for which this is known, the immunogenic part of the antigen is identical with the biologically active part. Consequently, if, e.g., fragments of the antigen hormone are still biologically active as hormones but do not react with the antibody, then the biological activity will not agree with the observed hormone concentration.[17]

Antibodies are prepared by injecting a suitable experimental animal (guinea pig, rabbit, goat, chicken) with a highly purified antigen (the substance to be determined) together with the complete Freund's adjuvant—a mixture of suspended mycobacteria, mineral oils, and an emulsifier. The antigens are proteins with molecular weights above 4000. Lower polypeptides or other molecules are first coupled to a larger molecule such as albumin. They then behave as haptens as well as antigens. After a few weeks the animal is given a booster injection of antigen and a few weeks later the animal's antiserum is tested for quality and collected for the assay.[18] An antiserum to be used in an assay should be tested for its antibody content (titre), its specificity, and its sensitivity in the assay, regardless of whether the antibody is a home preparation or has been acquired from another source.[19]

The antibody content of the antiserum is measured by its *titre*. An assay is carried out with a constant amount of tracer and various dilutions of the antiserum. The titre corresponds to that dilution of the antiserum which binds 50% of the tracer subjected to the assay. This is about the dilution to use in the RIA if the latter is conducted under equilibrium conditions and not sequentially. However, the titre is not a measure of the quality of the antibody.

In addition to the titre, it is therefore necessary to check the specificity of an antiserum. The aim is to establish whether, and to what extent, substances which are structurally closely related to the ligand to be determined react with the antibody or affect the binding of the ligand to the antibody and thus interfere with the assay. This can be discovered by mixture experiments using various sera and standards (dissolved in the ligand-free serum). It is important to test the antiserum for interference at different concentrations of the ligand in the definitively chosen assay system. With a heterogeneous antibody, it may be that at low concentrations of the ligand interfering molecules cannot offer measurable competition with the ligands for a high-affinity binding site, but on incubation with a higher ligand concentration, at which binding sites of lower affinity might also be occupied by ligands, the interference may be observable. The interference shows up on a linear calibration graph because the dilution series for the investigated serum does not then run parallel to the standard curve.

The sensitivity of an assay is partly dependent on the quality of the antiserum. In addition to the affinity constant of the antibody, which affects the slope of the calibration graph, the precision of the determination plays a vital role in defining the sensitivity. This can be determined graphically by making a Scatchard plot (abscissa = bound/free; ordinate = concentration of labelled plus unlabelled ligands). However, owing to the heterogeneity of most antisera, the graph is frequently not a straight line.[20,21]

4. *Incubation.* Different methods of adding the reagent and variations in the incubating conditions ramify the possibilities of radioimmunoassays. If the labelled and unlabelled ligands are added to the antibody at the same time, then one can either wait for equilibrium to be attained during incubation or separate the free and bound fractions beforehand. In this case a higher concentration of tracer is necessary[6] and the time factors must be carefully observed.

If the labelled and unlabelled ligand (sample) are *not* added to the antibody simultaneously we have what is known as a sequential assay. The advantages and problems associated with this procedure have been discussed by various authors.[11,22,23]

The incubation is very often conducted in the presence of serum albumin in order to reduce the non-specific adsorption. The albumin must not contain any ligand (T_4!) or carboxypeptidase impurities. The material of the tube also influences the non-specific binding.[24] The incubation temperature affects the rate of attainment of equilibrium: the binding of ligands proceeds more quickly at lower temperatures but the attainment of equilibrium is slower. Moreover, at lower temperatures and with a heterogeneous mixture of antibodies (the normal case), antibodies with a smaller binding affinity for the ligands can also participate in the reaction.[25] Mixing during incubation is usually carried out in rotors since powerful agitation leads to foaming. If the ligands in the sample are already bound to transport proteins, this bonding must be disrupted (cf. thyroxine RIA). The composition of the incubation solution can be responsible for non-specific effects and can also influence the method of separation, described in the next section. Ekins[6] provided an extended discussion of the optimization of incubation conditions.

5. *Separation of the free and bound fractions.* The separation of the free, labelled (and unlabelled) ligands from the antibody-bound, labelled (and unlabelled) ligands is an essential part of saturation analysis. If the separation is carried out before the equilibrium is attained then it is particularly important that the incubation period should be exactly the same for each sample. Care should also be taken to ensure that no changes in temperature disturb the ratio of bound to free fractions during the separation. The latter is effected by precipitation of either the free fraction or the ligand–antibody complex. There are a large number of methods available and the choice of method has a considerable bearing on the practical procedure entailed.

Precipitation of the free fraction is carried out by adsorption on silica, talc, active charcoal (if necessary, coated with dextran), or ion exchanger.

The alternative procedure, precipitation of the bound fraction, may be accomplished by well known methods for the precipitation of proteins, using such substances as ammonium or sodium sulphate, ethanol or poly(ethylene glycol) (if necessary, in combination with sodium sulphate).

The *double antibody* method of precipitation is elegant. If, for example, the first antibody (β- or γ-globulin) is derived from rabbits, then it is precipitated, together with the bound ligands, by means of a second antibody, deriving from guinea pigs (guinea pig anti-rabbit γ-globulin).

A further method, which simplifies the RIA procedure, is the *'solid-phase' method*. The antibodies are added to the assay solution as a solid—the antibodies are bound to particles, e.g. of cellulose or Sephadex. When the antibodies are in this form the bound fraction is much more easily centrifuged. The adherence of the antibodies to the walls of the test-tube makes it still easier to separate off the free fraction. However, the reproducibility of the binding to the walls of polystyrene tubes can vary. In this case the direct measurement of the non-specific binding is not possible. In spite of theoretical objections, such commercially prepared antibodies bound to tubes have assumed an important position. Yet another possibility is the bonding of antibodies to ferromagnetic particles or erythrocytes.[26,27]

Instead of centrifuging the precipitated antibodies, the separation may be accomplished with filters. This expensive approach is used *inter alia* in RIA automatic analysers.

6. *Measurement of the free or bound activity*. The following individual measurements have to be made: determination of the radioactivity of the duplicates or triplicates of the standard dilution series, of the total activity, of the bound activity (B_0) at a standard concentration of zero, of the non-specific bonding (bound without antibody), and of the free or bound activity of the sample duplicates. In the ideal case, the volumes of all the counted samples should be the same, since this, together with the detector geometry, the high-voltage stability of the photomultiplier, the amplifier, and the discriminator setting, influences the count-rate of the amount of radiation of specified activity. If the electronic dead-times are too long, the count-rate estimates are low.[28] The sensitivity of the method is ultimately limited by the precision of the counting in the scintillation counter. The procedure for counting γ-rays in β-counters has been described elsewhere.[29,30]

7. *Calculation*. The theory underlying the calculation of the quantity of unlabelled ligands is usually based on the assumption that the ligands react with only one bonding site. For the actual making of read-offs from a calibration graph, using one of the commonest graphical methods as outlined below, it is important that the variance of the measurements does not remain constant but changes over the range ('heteroscedasticity'; $\sigma\kappa\acute{\epsilon}\delta\alpha\sigma\iota\zeta$ = scatter). The scatter is least in the central region of the curve and greater in the marginal ranges. Hence is transpires, for example, that a

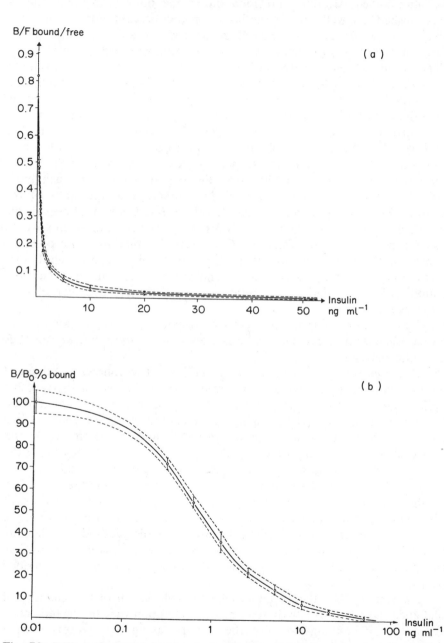

Fig. 70. Different possible ways of plotting the calibration graphs for saturation analyses, e.g. radioimmunoassays (for explanation, see text).

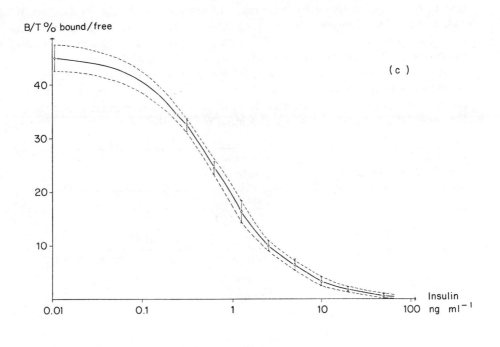

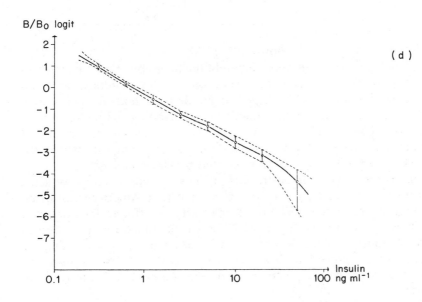

thryoxine determination in the reference interval has a lower scatter than one in the hypo or hyperthyroid range. Consequently, in the evaluation of patient data from various ranges, tests which assume a normal distribution are not generally applicable. If regression lines are drawn through linearized calibration graphs, the values must be weighted.

Before making a graph, the value of the non-specific bonding should be substracted from the count rates. The various standard concentrations may be plotted linearly or logarithmically against the free or bound count rates. However, the following plots are the most frequently used:

Ordinate Standard concentration	Abscissa Count-rate
Linear	$\dfrac{\text{Bound } (B)}{\text{Free } (F)}$
Logarithmic	$\dfrac{\text{Bound } (B)}{\text{Bound at standard concentration 0 } (B_0)}$
Logarithmic	$\dfrac{\text{Bound } (B)}{\text{Total } (T)}$
Logarithmic	$\text{Logit} \dfrac{B}{B_0} = \ln \dfrac{\dfrac{B}{B_0}}{1 - \dfrac{B}{B_0}}$

Examples of these plots are shown in Fig. 70.

The log/logit plot gives an approximately linear graph. Exact linearity is guaranteed only if the degree of saturation of the antibody does not change. Fernandex and Loeb[31] suggested another method of linearization and in recent years empirical functions, demanding larger computer programs, have been developed. Very useful corrections for the variance of B_0 and the non-specific bonding, two interfering factors to which the log/logit plot is sensitive, have been suggested by Hatch et al.[32]

8. *Quality control.* The dubious results of the ring experiments conducted by Marschner et al.[33] and Horn et al.[34] recall the first American results on haemoglobin determinations. The number of kits on offer to the consumer are too numerous to review. Even well known manufacturers provide kits which are obviously not the best that can be designed. Besides testing the practicability of the procedure, if kits are to be used, it is also necessary to check the quality of the assay personally before the method is brought into routine use.

A daily quality control, like that applied to routine determinations in clinical chemistry, should also be encouraged in respect of RIA.[6,35–38,47]

The internal quality check, using pooled sera or commercial

quality-control sera of declared concentration, should cover three ranges (low, average, and high). This is necessary because of the change in precision—heteroscedasticity—in the different concentration ranges. It is clinically sensible to make controls of precision and accuracy at those concentrations where a decision is attended by therapeutic consequences.

Because antibodies are ill-defined and are different proteins from one antiserum to another, not only must there by an external quality control through ring experiments and direct exchange between different laboratories, but the results should be discussed with the clinician in respect of their plausibility. Where bioassays can be carried out, comparison with these methods would also be desirable. Practical suggestions for the calculations and hints as to possible sources of error can be found in the literature.[36,39] Problems of standardization have been discussed by Bangham.[48]

9. *Application*. On account of their sensitivity, RIAs have led to an expansion of diagnostic possibilities in clinical endocrinology. A whole series of protein and non-protein hormones and of secondary effectors (AMP, prostaglandins) can be determined with this technique (as an example, see thyroid-gland diagnosis, p. 650). In addition, the monitoring of drug and vitamin concentrations in blood and the determination of narcotics in urine have been facilitated. RIAs have also been applied to the quantification of tumour-associated antigens, Australia antigen, immunoglobulin and coagulation factors. Monographs and reviews give a picture of the range and variety of the methods.[40–45,49]

In spite of the ever increasing popularity of RIA, the disadvantages of these assays should not be underestimated.[46] We would stress the brevity of the period for which the expensive tracer is usable (half-life!), the readily underestimated costs of the counter and semi-automation, as well as the spatial safety arrangements, the risk to health, and the danger of polluting the environment. For these reasons, close attention should be paid to the development of saturation analyses using non-radioactive labelled ligands, developments which are currently in progress.

References

1. Lawson, A. M., The scope of mass spectrometry in clinical chemistry, *Clin. Chem.*, **21**, 803 (1975).
2. Roboz, J., Mass spectrometry in clinical chemistry, *Adv. clin. Chem.*, **17**, 109 (1975).
3. Bakay, B., Methods of analysis of radioactive amino acids, in Nyhan, *Heritable disorders of amino acid metabolism*, Wiley, New York, 1974, p. 698.
4. Kobayashi, Y. and Maudsley, D. V., *Biological applications of liquid scintillation counting*, Academic Press, New York, 1974.
5. Parmentier, J. H. and Ten Haaf, F. E. L., Developments in liquid scintillation counting since 1963, *Int. J. appl. Radiat. Isotopes.*, **20**, 305 (1969).
6. Ekins, R. P., Radioimmunoassay and saturation analysis. Basic principles and theory, *Br. med. Bull.*, **30**, 3 (1974).

218

7. Woodhead, J. S., Addison, G. M., and Hales, C. N., The immunoradiometric assay and related techniques, *Br. med. Bull.*, **30**, 44 (1974).
8. Yalow, R. S. and Berson, S. A., Assay of plasma insulin in human subjects by immunological methods, *Nature, Lond.*, **184**, 1648 (1959).
9. Ekins, R. P., The estimation of thyroxine in human plasma by an electrophoretic technique, *Clin. chim. Acta*, **5**, 453 (1960).
10. Zettner, A., Principles of competitive binding assays (saturation analyses). I. Equilibrium techniques, *Clin. Chem.*, **19**, 699 (1973).
11. Rodbard, H. J., Ruder, H. J., Vaitukaitis, J., and Jacobs, H. S., Mathematical analysis of kinetics of radioligand assays: improved sensitivity obtained by delayed addition of labeled ligand, *J. clin. Endocr. Metab.*, **33**, 343 (1971).
12. Rodbard, D., Lenox, R. H., Wray, H. L., and Ramseth, D., Statistical characterization of the random errors in the radioimmunoassay dose–response variable. *Clin. Chem.*, **22**, 350 (1976).
13. Ekins, R. and Newman, B., Theoretical aspects of saturation analysis, *Acta endocr., Copenh.*, **147**, suppl., 11 (1970).
14. Ekins, R. P., Newman, G. B., and O'Riordan, J. L. H., Saturation assays, in McArthur and Colton, *Statistics in endocrinology*, Massachusetts Institute of Technology Press, Cambridge, Mass., 1970, p. 345.
15. Greenwood, F. C., Hunter, W. M., and Glover, J. S., The preparation of ^{131}I-labeled human growth hormone of high specific radioactivity, *Biochem. J.*, **89**, 114 (1963).
16. Thorell, J. I. and Johansson, B. G., Enzymatic iodination of polypeptides with ^{125}I to high specific activity, *Biochim. biophys. Acta*, **251**, 363 (1971).
17. Rosselin, G., Bataille, D., Laburthe, M., and Duran-Garcia, S., Hétérogénéité des hormones protéiques en radio-analyse, *Path. Biol., Paris*, **23**, 793 (1975).
18. Kruse, V., Production and evaluation of high-quality thyroxine antisera for use in radioimmunoassay, *Scand. J. clin. Lab. Invest.*, **36**, 95 (1976).
19. Rosa, U., Préparation, caractérisation et contrôle des antisérums, *Path. Biol., Paris*, **23**, 853 (1975).
20. Hollemans, H. J. G. and Bertina, R. M., Scatchard plot and heterogeneity in binding affinity of labeled and unlabeled ligand, *Clin. Chem.*, **21**, 1769 (1975).
21. Chamness, G. C. and McGuire, W. L., Scatchard plots: common errors in correction and interpretation, *Steroids*, **26**, 538 (1975).
22. Zettner, A. and Duly, P. E., Principles of competitive binding assays (saturation analyses). II. Sequential saturation, *Clin. Chem.*, **20**, 5 (1974).
23. Pratt, J. J. and Woldring, M. G., Radioimmunoassay specificity and the 'first-come, first-served effect', *Clin. chim. Acta*, **68**, 87 (1976).
24. Kubasik, N. B., Hall, J. L., and Sine, H. E., Selection of assay tubes for radioassay procedures, *Clin. Chem.*, **22**, 1745 (1976).
25. Malvano, R. and Rolleri, E., Antiserum characteristics and assay quality, *Path. Biol., Paris*, **23**, 863 (1975).
26. Hersh, L. S. and Yaverbaum, S., Magnetic solid-phase radioimmunoassay, *Clin. chim. Acta*, **63**, 69 (1975).
27. Luner, St. J., Continuous flow automated radioimmunoassay using antibodies attached to red blood cells, *Analyt. Biochem.*, **65**, 355 (1975).
28. Bell, T. K., High counting rates from ^{125}I in radioimmunoassays, *J. clin. Path.*, **27**, 860 (1974).
29. Zakhelman, E. and Ting, P., Counting of beta and gamma emitters used in clinical radioimmunoassays and competitive-binding assays by liquid scintillation counters, *Clin. Chem.*, **19**, 191 (1973).
30. Dixon, R. and Cohen, E., Simple modification of a liquid scintillation counter to decrease cost of and time for radioimmunoassay, *Clin. Chem.*, **22**, 1746 (1976).

31. Fernandez, A. A. and Loeb, H. G., Practical applications of radioimmunoassay theory. A simple procedure yielding linear calibration curves, *Clin. Chem., 21*, 1113 (1975).
32. Hatch, K. F., Coles, E., Busey, H., and Goldman, S., End-point parameter adjustment on a small desk-top programmable calculator for logit-log analysis of radioimmunoassay data, *Clin. Chem., 22*, 1383 (1976).
33. Marschner, I., Erhardt, F. W., and Scriba, P. C., Ringversuch zur radioimmunologischen Thyrotropinbestimmung (hTSH) im Serum., *J. clin. Chem. clin. Biochem., 14*, 345 (1976).
34. Horn, K., Marschner, I., and Scriba, P. C., Erster Ringversuch zur Bestimmung der Konzentrationen von L-Trijodthyronin (T_3) und L-Thyroxin (T_4) im Serum: Bedeutung für die Erkennung methodischer Fehlerquellen, *J. clin. Chem. clin. Biochem., 14*, 353 (1976).
35. Rodbard, D., Rayford, P. L., Cooper, J. A., and Ross, G. T., Statistical quality control of radioimmunoassays, *J. clin. Endocrin., 28*, 1412 (1968).
36. Rodbard, D., Statistical quality control and routine data processing for radioimmunoassays and immunoradiometric assays, *Clin. Chem., 20*, 1255 (1974).
37. Rodbard, D., Lenoz, R. H., Wray, H. L., and Ramseth, D., Statistical characterization of the random errors in the radioimmunoassay dose–response variable, *Clin. Chem., 22*, 350 (1976).
38. Midgley, A. R., Jr., Niswender, G. D., and Rebar, R. W., Principles for the assessment of the reability of radioimmunoassay methods (precision, accuracy, sensitivity, specificity), *Acta endocr., Copenh., 142*, suppl., 163 (1969).
39. Siddiqui, S. A. and Craig, A., Quality control of radioimmunoassay, *Med. Lab. Technol., 32*, 171 (1975).
40. Nieschlag, E. and Wickings, E. J., A review of radioimmunoassay for steroids, *Z. klin. Chem. klin. Biochem., 13*, 261 (1975).
41. Breuer, H., Hamel, D., and Krüskemper, H. L., *Methoden der Hormonbestimmung*, Thieme, Stuttgart, 1975.
42. Cleeland, R., Cristenson, J., Usategui-Gomez, M., Heveran, J., Davis, R., and Grunberg, E., Detection of drugs of abuse by radioimmunoassay: a summary of published data and some new information, *Clin. Chem., 22*, 712 (1976).
43. Broughton, A. and Strong, J. E., Radioimmunoassay of antibiotics and chemotherapeutic agents, *Clin. Chem., 22*, 726 (1976).
44. Skelley, D. S., Brown, L. P., and Besch, P. K., Radioimmunoassay, *Clin. Chem., 19*, 146 (1973).
45. Jaffe, B. M. and Behrman, H. R., *Methods of hormone radioimmunoassays*, Academic Press, New York, 1974.
46. Challand, G., Goldie, D., and Landon, J., Immunoassay in the diagnostic laboratory, *Br. med. Bull., 30*, 38 (1974).
47. Longley, W. J., Proficiency testing of radioimmunoassay, *Clin. Biochem., 9*, 109 (1976).
48. Bangham, D. R., Standardization in peptide hormone immunoassays: principle and practice, *Clin. Chem., 22*, 957 (1976).
49. Eckert, H. G., Die Technik des Radioimmunoassays, *Angew. Chem., 88*, 565 (1976).

3. METABOLIC INVESTIGATIONS

3.1. DETERMINATION OF ENZYMES

3.1.1. INTRODUCTION

3.1.1.1. General

The determination of enzymes has become an indispensible diagnostic aid. For numerous diseases, the enzyme determination cannot be replaced by any other diagnostic method. Nonetheless, enzyme diagnosis—as one of the youngest branches of clinical chemistry—presents many difficulties, both for the laboratory and for the doctor. The determination of enzymes demands a knowledge of the biochemistry and the kinetics of these catalysts,[1,2] and for the interpretation of the data the doctor must be familiar with the physiology and pathology of each individual enzyme.

3.1.1.2. Enzymes as proteins

All enzymes are proteins. They are labile and prone to inactivation and denaturation. Inactivation may be reversible but denaturation is always irreversible. It is therefore a matter of general principle that enzyme determinations should be done as quickly as possible. Enzymes are extremely sensitive towards heavy metals, soaps, detergents, and acids. Insufficiently clean glassware, particularly dispatch vessels, is one of the most frequent sources of error in enzyme determinations.

3.1.1.3. Enzymes as catalysts

Enzymes are distinguished from other proteins by their catalytic properties, i.e. their ability to accelerate spontaneous but extremely slow chemical reactions, producing a thousand to 10^3–10^6-fold increase in the rate. Grossly simplified, the enzyme reaction can be represented as follows:

$$S + E \xrightarrow{\ k\ } E + P$$

where

S = substrate;
E = enzyme;
P = product;
k = velocity constant.

223

As this representation shows, the enzyme does participate directly in the chemical reaction but it remains qualitatively and quantitatively unchanged. The constant k is characteristic for each enzyme. In principle, all enzyme reactions are reversible, i.e. the reaction may proceed from left to right or from right to left. For the determinations, however, the conditions (substrate concentration, pH, etc.) are chosen so that the equilibrium lies wholly to one side. For the following practical considerations, the reverse reaction may therefore be ignored.

Since, by definition, the enzyme-catalysed reaction can also occur spontaneously, albeit slowly:

$$S \rightarrow P$$

it is always necessary to include an appropriate blank value. This usually entails a parallel assay without enzyme (e.g. in the determination of alkaline phosphatase) or the addition of a reagent which destroys enzymes (e.g. in the determination of caeruloplasmin).

The amount of substrate conversion in unit time catalysed by an enzyme depends on various factors: substrate concentration, reaction temperature, hydrogen ion concentration, ionic strength, nature of the buffer, and, if applicable, effectors (cofactors, inhibitors, activators).

3.1.1.4. Enzyme activity and enzyme concentration

The difference between the often confused terms 'enzyme activity' and 'enzyme concentration' is illustrated in Fig. 71. Enzyme concentration

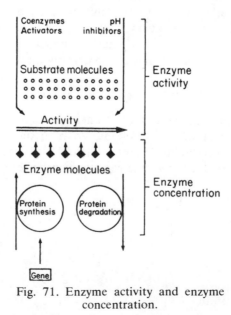

Fig. 71. Enzyme activity and enzyme concentration.

means the number of enzyme molecules (amount of enzyme) present in the reference unit (e.g. gram or litre). Since enzymes are proteins the result has to be expressed in mass units, i.e. grams or milligrams. However, the determination of enzyme concentration is difficult. The molecules must first be isolated and then quantified by the methods of protein chemistry. In special cases it is also possible to utilize specific antibodies (enzyme antisera) for the determination of enzyme concentration, but this technique is also very complicated. An exception is caeruloplasmin, which is relatively easy to determine by immunological and catalytic methods and the techniques of protein chemistry.

Biologically more important than the weight and technically much easier to determine is the catalytic action of the enzyme. It is therefore convenient to measure the enzyme activity rather than the amount of enzyme. A limited number of enzyme molecules are caused to act upon an excess of substrate molecules under defined conditions for a specific period (incubation period). The activity is measured by the amount of substrate converted per unit time, as indicated by either the decrease in the substrate ($-dS/dt$) or the increase in the products ($+dP/dt$):

$$\text{Enzyme activity} = -dS/dt = +dP/dt$$

In the determination of an enzyme activity it is assumed that enzyme activity is directly proportional to enzyme concentration:

$$-dS/dt = [E]k$$

This requirement is met in most methods, but one must always be prepared for the exception.

3.1.1.5. Kinetics

Figure 72 shows the course of an enzyme reaction as measured by the reaction velocity, v, in dependence on the substrate concentration (S). In region A of the curve, the enzyme is not yet saturated with the substrate and the reaction velocity is proportional to the substrate concentration. This is the only factor limiting the reaction and we are dealing with a *first-order reaction*. This may be represented by a straight line on a semi-logarithmic plot. Reactions of this type occur when insufficient substrate molecules pass into solution (e.g. forward reaction of creatine kinase with creatine as substrate),[4] and also with coupled enzyme reactions. In region B of the curve the enzyme is saturated with the substrate and the reaction velocity is dependent only on the enzyme concentration. This is a *zero-order reaction*. For most reactions of the diagnostically important enzymes in the plasma, the experimental conditions are designed to ensure a reaction of zero order. This means that the initial reaction velocity (v_0), i.e. the amount of substrate converted in unit time (dS/dt), remains

226

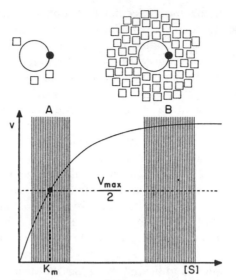

Fig. 72. Shape of the pS curve for an enzyme. Dependence of the reaction velocity on the substrate concentration. Case A: reaction is limited by the substrate. Case B: reaction is limited by the enzyme. $\bigcirc$ = Enzyme; $\square$ = substrate molecules. (From ref. 3.)

constant throughout the entire incubation period:

$$v_o = (dS/dt)k$$

This emerges from the plot of the conversion/time curve (Fig. 73). At a constant enzyme concentration the enzyme conversion is proportional to time, i.e. the reaction velocity remains constant. A closely related

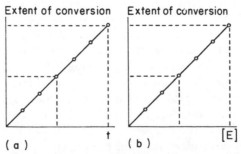

Fig. 73. (a) Relation between extent of conversion and time (conversion/time curve). (b) Relation between extent of conversion and amount of enzyme (conversion/enzyme curve).

statement is that of the enzyme–product rule: at constant time, the relationship between the amount of enzyme and the extent of conversion is linear. Under otherwise identical conditions, doubling the time or the amount of enzyme doubles the extent of conversion.

3.1.2. DETERMINATION OF ENZYMES

3.1.2.1. Two-point and kinetic determinations

In the older, so-called two-point methods (e.g. the determination of acid plasma phosphates, p. 664), the enzyme activity was determined by two assays, one corresponding to the blank and the second to the analysis. The difference in the amount of conversion in the presence and absence of the enzyme served as a measure of the enzyme activity. Such a procedure is dubious since it does not allow the kinetics of the experiment to be assessed. A few possible types of curves are shown in Fig. 74. Curve B corresponds to the hoped-for, ideal relationship, i.e. validity of the time–extent of conversion rule. Curves of type A are of rare occurrence—this is the so-called autocatlytic curve obtained with trypsin, for instance. Much more frequent—and much less to be desired—are curves of types C and D, however. At high enzyme activities a decrease in the amount of substrate or even the appearance of reaction products (so-called end-product inhibition) can lead to a decline in the enzyme activity. In such cases an apparently identical enzyme activity is obtained at time t despite a different initial reaction velocity (v_o). This danger is common to all two-point methods used in the clinical–chemical laboratory whenever the enzyme activity is high. With all pathological results it is therefore advisable to repeat the determination using half the quantity of enzyme or with half as long an incubation period. This frequently yields higher (and more correct) results than the first determination.

In modern kinetic analysis the progress of the reaction is continuously monitored, either manually or with the aid of a recording instrument. The type of curve is directly apparent from the measurements and aberrations

Extent of conversion

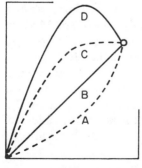

Fig. 74. Two-point determination. Possible conversion/time curves.

can be recognized immediately. These kinetic methods are therefore unquestionably preferable to the older two-point determinations. They will receive a thorough discussion in Section 3.1.4.

3.1.2.2. Effect of the amount of enzyme and the length of the incubation period

It is good policy to carry out enzyme determinations under optimum reaction conditions. Each component of the incubation mixture is chosen and standardized in such a way that the resulting activity cannot be increased by any further modifications.

A primary factor which frequently causes difficulty is the *incubation period*. With the older techniques it was common to incubate for 12 or even as long as 24 h. Today this would be bound to be an exception. As a general principle the incubation period should be chosen to be as short as possible, preferably not exceeding 1 h. If this rule is not observed there is a danger that the enzyme may suffer slow denaturation or the incubation mixture may become contaminated by bacteria. Since the measured results, as can readily be seen from the conversion/time equation, equal the product of the amount of enzyme and the period of incubation:

$$-dS = [E]dt \cdot k_1$$

there is usually no difficulty in reducing the time to a reasonable value by increasing the amount of enzyme.

3.1.2.3. Choice of substrate

The substrate specificity of the enzyme is generally not absolute. Thus, alkaline phosphatase cleaves β-glycerophosphate, phenylphosphate, glucose-6-phosphate, p-nitrophenylphosphate and α-naphthylphosphate. Since the 'natural', i.e. physiological, substrate for alkaline phosphatase is not known, there are no objections in principle to measuring the enzyme activity with any one of the above-mentioned substrates. The only decisive criteria here are that the results be reproducible and that the determination be technically as simple as possible.

A similar situation applies in the case of the peptidases. Here too the natural substrates are frequently unknown or are poorly defined. Modern protein research has shown that peptidases are often specific for the configuration of the amino acids in the neighbourhood of the cleavage site. As a result, attempts are being made to replace the substrate by definite peptides usually coupled with a dyestuff. At the same time there remains a series of enzymes with exceedingly high substrate specificity, so that the use of an atypical substrate is an irrelevant consideration anyway. Examples of such enzymes are those usually detectable in the optical test, such as the transaminases and creatine kinase.

3.1.2.4. Effect of the substrate concentration

The choice of a suitable substrate concentration is of great importance. If the substrate is in excess compared with the number of enzyme molecules, then the measured activity is a function of the enzyme concentration. However, if the substrate concentration is relatively small in comparison with the enzyme concentration, then, on thermodynamic grounds, there are fewer collisions between the enzyme and substrate molecules and the enzyme activity is lower. Finally, if the enzyme is in excess the activity is a function of the substrate concentration. This is the principle of 'enzymatic substrate analysis', as used in the clinical laboratory, e.g. for the determination of urea (with urease), glucose (with hexokinase), triglycerides, and cholesterol. If the activity of a constant amount of enzyme is measured at various substrate concentrations curves of the type shown in Fig. 72 are obtained. It is clear that the requirement of a constant reaction velocity is met only at high substrate concentrations. Where possible, therefore, the substrate concentration should be chosen so as to ensure that the maximum reaction velocity (V_{max}) is attained. As a preliminary to choosing this optimum substrate concentration the Michaelis constant, K_m, is calculated. This is a measure for the affinity between the substrate and the enzyme and it expresses the conditions applying in saturation kinetics. With increasing concentration of the substrate a point is reached at which the enzyme is saturated. The relationship between the substrate and the enzyme may be formulated as follows, applying the Law of Mass Action:

$$ V = \frac{V[S]}{K_m + [S]} = \frac{V}{1 + \frac{K_m}{[S]}} $$

where

V = maximum reaction velocity at substrate saturation (theoretical limit);
v = reaction velocity at a particular substrate concentration, [S].

If $v = V/2$, it follows that

$$ 0.5 = \frac{1.0[S]}{K_m + [S]} = 0.5K_m + 0.5[S]' = 1.0[S]K_m = [S] $$

K_m has the dimensions mol l^{-1}.
   That substrate concentration for which the reaction velocity is $V_{max}/2$ is known as the Michaelis constant (K_m). For the determination of K_m graphical methods are often used:
   Lineweaver–Burke plot (double reciprocal plot).[5-8] The Michaelis equation can be transformed so that there are reciprocals on both sides of

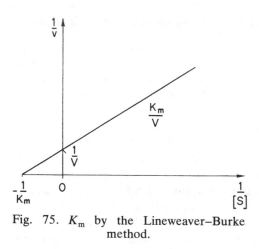

Fig. 75. K_m by the Lineweaver–Burke method.

the equation:

$$\frac{1}{v} = \frac{K_m + [S]}{V[S]} = \frac{K_m}{V[S]} + \frac{[S]}{V[S]}$$

$$\frac{1}{v} = \frac{K_m}{V}\left(\frac{1}{[S]}\right) + \frac{1}{V}$$

This equation corresponds to the expression $y = bx + a$, where a = intercept with the y-axis and b = slope of the curve. The intercept of the straight line with the abscissa ($y = 0$) corresponds to $-1/K_m$ (Fig. 75).

Eadie plot.[6] Here the slope is $-K_m$ (Fig. 76).

Example. Table 32 presents data as collected for the evaluation of K_m for the creatine kinase in plasma. These results are plotted by Lineweaver–Burk and Eadie methods (Fig. 77). The values of K_m obtained from the graphs are approximately 1 mmol l^{-1} (Lineweaver–Burk) and 1.07 mmol l^{-1} (Eadie).

A knowledge of the K_m value is of practical significance. If the substrate concentration is chosen to be 10 times as high as the K_m value, then the

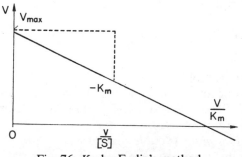

Fig. 76. K_m by Eadie's method.

Table 32. Final concentrations of creatine phosphate in the reaction mixture, as substrate for the evaluation of K_m for creatine kinase in plasma (method, p. 516).

S, mmol l^{-1} Final concentration	v, U l^{-1}	$\dfrac{1}{v}$	$\dfrac{1}{s}$, mmol l^{-1}	$\dfrac{v}{s}$
1	20.5	0.0488	1000	20.500
2.5	45.0	0.0222	400	18.000
5	73.5	0.0136	200	14,700
10	109	0.0092	100	10.900
20	128	0.0078	50	6.400
30	136	0.0074	30	4,533
40	142	0.0070	25	3.550
50	143	0.0069	20	2.860
100	141	0.0071	10	1.410

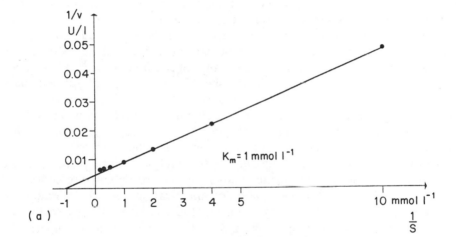

Fig. 77a Lineweaver–Burke plot for K_m for creatine kinase in plasma.

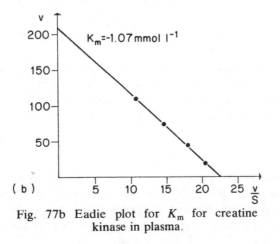

Fig. 77b Eadie plot for K_m for creatine kinase in plasma.

reaction velocity will be 90% of the maximum velocity. Such a substrate concentration often suffices for clinical purposes. The conditions of measurement must be so designed that the substrate remains in excess throughout the whole period of the observations. It may be taken as a general rule that not more than one fifth of the substrate introduced should react during the period of measurement. At the beginning of the reaction the substrate concentration should be high enough to guarantee a reaction of zero order so that both the conversion/time and conversion/enzyme curves are valid. This should be observed in optimizing enzyme methods.

3.1.2.5. Effect of temperature

Enzyme reactions are temperature sensitive, the enzyme activity increasing with increasing temperature. For a rise in temperature of 10 °C the enzyme activity is roughly doubled; otherwise expressed, a temperature rise of 1 °C produces about a 10% increase in the measured activity. The temperature of measurement must be strictly controlled and should not deviate by more than 0.05 °C from the prescribed value.[9] It has been suggested that enzyme reactions should be measured at standard temperatures such as 25, 30 and 37 °C and arguments and counter-arguments have been propounded for each such temperature. The advantage of carrying out an activity determination at 37 °C is the greater difference in the extinction over unit time and the consequent greater sensitivity. As a result, the frequency of the analyses can be higher and the amount of sample can be smaller than if the measurements are conducted at 25 °C. However, the possibility of inactivation and denaturation of certain enzymes[10] and coenzymes during the period of measurement has been cited as a disadvantage of using 37 °C as a measurement temperature. This has been challenged.[11] In 1971 the Deutschen Gesellschaft für Klinische Chemie (German Clinical Chemistry Society) fixed the measurement temperature at 25 °C[12] and in 1973 the Scandinavian Society for Clinical Chemistry and Clinical Physiology adopted a measurement temperature of 37 °C.[13]

The enzyme experts of the Standardization Commission of the International Federation on Clinical Chemistry (IFCC) have provisionally recommended a measurement temperature of 30 °C,[9] but the debate over the final, internationally acknowledged temperature is not yet over.

3.1.2.6. Effect of pH

There is also no firm rule for fixing the pH. The optimum hydrogen ion concentration must be determined experimentally in every case. Here it is important to remember that the temperature, the ionic strength, and the presence of activators and inhibitors can have a considerable effect on the optimum pH. The use of phosphate buffers is inadvisable since these frequently have a direct effect on the enzyme activity and they also form ideal culture media for bacteria and fungi. It is more advisable to use the amine buffers listed in Appendix 1, based on e.g. collidine, tris, ammediol, and triethanolamine.

3.1.2.7. Serum or plasma; haemolysis

With most enzymes it is irrelevant whether serum or plasma is used for the determination. On the other hand, this factor assumes great importance when the enzymes concerned are those liberated from the thrombocytes on coagulation of the blood. Thus, about one third of the acid serum phosphatase derives from the thrombocytes.[14] However, there are other enzymes present in the thrombocytes. Thus, Solbach et al.[15] found significantly higher values for aldolase and lactate dehydrogenase in serum than in plasma. On the other hand, no differences were detected for the transaminases. We prefer to conduct all enzyme determinations on plasma since the danger of haemolysis during the collection of plasma is less than that attending the collection of serum.

Erythrocytes are exceedingly rich in enzymes. An *intra-vitam* or *in vitro* haemolysis is therefore likely to be attended by liberation of enzymes. This is especially true for glutamate–oxaloacetate transaminase, lactate dehydrogenase, and acid phosphatase. Hence the current search for diagnostically interesting enzymes which do not occur in the erythrocytes (e.g. creatine kinase).

References

1. Richterich, R., *Enzympathologie*, Springer, Berlin, 1958.
2. Richterich, R., Enzymstoffwechsel, in Siegenthaler, *Klinische Pathophysiologie*, Thieme, Stuttgart, 1973, p. 169.
3. Aebi, H., *Einführung in die praktische Biochemie*, 2. Aufl., Karger, Basle, 1971.
4. Colombo, J. P., Richterich, R., and Rossi, E., Serum-Kreatin-Phosphokinase: Bestimmung und diagnostische Bedeutung, *Klin. Wschr.*, **40**, 37 (1962).
5. Bergmeyer, H. U., *Grundlagen der enzymatischen Analyse*, Verlag Chemie, Weinheim, 1977, p. 30.
6. Aebi, H., Kinetik, in Rauen, *Biochemisches Taschenbuch*, Vol. 2, Springer, Berlin, 1964, p. 159.
7. Dixon, M. and Webb, E. C., Enzyme kinetics, in *Enzymes*, 2nd ed., Longmans, London, 1967, p. 54.
8. Nylands, I. B. and Stumpf, P. K., *Outlines of enxyme chemistry*, Wiley, New York, 1958.
9. Provisional recommendation (1974) on IFCC methods for the measurement of catalytic concentration of enzymes, *Clin. chim. Acta*, **61**, 238 (1975).
10. Bergmeyer, H. U., Standardization of the reaction temperature for the determination of enzyme activity, *Z. klin. Chem. klin. Biochem.*, **11**, 39 (1973).
11. King, J., On a standard temperature for clinical enzymology, *Ann. clin. Biochem.*, **9**, 197 (1972).

12. Empfehlungen der Deutschen Gesellschaft für Klinische Chemie. Standardisierung von Methoden zur Bestimmung von Enzymaktivitäten in biologischen Flüssigkeiten, *Z. klin. Chem. klin. Biochem.*, **8**, 658 (1970).
13. Recommended methods for the determination of four enzymes in blood, *Scand. J. clin. Lab. Invest.*, **33**, 291 (1974).
14. Richterich, R., Colombo, J. P., and Weber, H., Ultramikromethoden im klinischen Laboratorium. VII. Bestimmung der sauren Prostata-Phosphatase, *Schweiz, med. Wschr.*, **92**, 1496 (1962).
15. Solbach, H. G., Englhardt, A., and Merten, R., Unterschiede der Enzymaktivitäten in Serum und Plasma und ihre Bedeutung für die klinische Enzymdiagnostik, *Klin. Wschr.*, **40**, 1136 (1962).

3.1.3. ENZYME UNITS

3.1.3.1. General

The unit of enzyme activity used in most laboratories was formerly μmol min^{-1} (U),[1,2] i.e. the amount of substrate converted per minute or the amount of products formed per minute. Recently a new unit has been recommended: the katal (kat), i.e. the amount of catalyst (e.g. an enzyme) which brings about the reaction of 1 mol s^{-1}.

$$1 \text{ kat} = 1 \text{ mol s}^{-1} = 60 \text{ mol min}^{-1} = 60 \times 10^6 \ \mu\text{mol min}^{-1} = 6 \times 10^7 \text{ U};$$

$$1 \text{ U} = 1 \ \mu\text{mol min}^{-1} = \frac{1}{60} \ \mu\text{mol s}^{-1} = \frac{1}{60} \ \mu\text{kat} = 16.67 \text{ nkat.}[3]$$

For the time being we shall use the unit U = μmol min^{-1}. The *volume activity* is the quantity corresponding to the enzyme activity per unit volume, e.g. U l^{-1}.[4] It is also termed the catalytic activity concentration.[11]

3.1.3.2. Activity and reference units in chemical investigations

For analytical and kinetic investigations, what is required is a way of expressing the 'chemical specific activity'. In the ideal case this corresponds to the number of substrate molecules reacting per minute per active centre of an enzyme molecule. However, since neither the molecular weight nor the number of active centres is usually known, use is made of one of the following approximations:

X (substrate) min^{-1} per g protein	Arbitrary unit; X corresponds to a measured quantity (extinction, drop in pH, etc.)
$-Q$ (substrate) h^{-1} per g protein	Manometric unit; $-Q$ in μl or μmol (1 μmol O_2 = 22.4 μl)
μmol (substrate) min^{-1} per g protein	Practical unit

mol (substrate) min^{-1} per 100 000 g protein	The molecular weight is arbitrarily taken to be 100 000
mol (substrate) min^{-1} per mol (enzyme)	Catalytic constant, molar activity
mol (substrate) min^{-1} per active centre	'Turnover number'

3.1.3.3. Activity and reference units in organ analyses

As a general principle, the unit

$$1 \text{ U} = 1 \text{ } \mu\text{mol (substrate) min}^{-1}$$

should be used as the unit of activity for biochemical and diagnostic investigations.

Greater difficulties attend the choice of a reference unit.[2,5] For the analysis of cellular material this ought to come as near as possible to the active cell mass. The following are the units most frequently used:

Fresh weight (f.w.)	Dependent on the degree of hydration of the tissue
Dry weight (d.w.)	Dependent on the lipid and carbohydrate content
Fat-free dry weight (f.f.d.w.)	Dependent on the percentage of carbohydrate and connective tissue
Nitrogen (N)	Dependent on non-protein N
Protein-N	Affected by metabolically inert protein (e.g. collagen)
Non-collagen protein	Affected by morphological heterogeneity (e.g. leucocyte infiltration)
Cells	Reliable method
DNS-cell equivalent	Reliable chemical method; cannot be used in cases of polyploidy

3.1.3.4. Activity and reference units in the analysis of body fluids

For the investigation of body fluids (serum, urine, etc.) the results should be expressed in

$$\text{U} = \mu\text{mol (substrate) min}^{-1}$$

If this is not possible the unit should be directly related to the physical quantity measured (e.g. Δ pH min^{-1}).

As a reference quantity we always use 1000 ml (1 l, litre). If 1.0 ml is used, the numerical values are often very small and the unit is consequently impractical. Since the temperature of measurement is such a critical quantity, this should always be stated in parentheses, thus e.g. μmol (substrate) min^{-1} l^{-1} (25 °C). Where necessary the direction of the measured reaction should be indicated (e.g. with creatine kinase).

3.1.3.5. Calculation of enzyme units: use of a standard

An obvious way of determining the activity is to run a parallel determination using an enzyme sample of known activity as a standard. Such enzyme standard or control sera are commercially obtainable so that it would be possible to operate this principle for some cases at least. However, experience has shown that for all these standard enzymes, the variation in activity is far too great to permit their use for calibration purposes. The use of enzymes with known activity for purposes of standardization must therefore be absolutely discountenanced.

However, instead of making the parallel determination with a known amount of enzyme, it may be performed with a known quantity of substrate (S) or products (P), depending on whether the decrease in the substrate or the increase in the products is measured. This is the customary procedure, e.g. in the determination of phosphatases. If equal volumes of sample and standard are used and the final volumes are the same for sample and standard assay, then the units are calculated by the following equation:

$$U \, l^{-1} = \frac{E(S)}{E(ST)} \cdot C_m \cdot 10^6 \cdot \frac{1}{t} \cdot \frac{FV}{SV} \, \mu\text{mol min}^{-1} \, l^{-1}$$

(for abbreviations, see Appendix 11, p. 745).

3.1.3.6. Calculation of enzyme units: use of the molar extinction coefficient

If the measurements are carried out photometrically, using monochromatic light (e.g. mercury line), absolutely calibrated pipettes, and plane-parallel cuvettes, then the extinction coefficient of either the substrate or the product may be used for the direct calculation of the enzyme activity. If dE min^{-1} is the change in extinction per minute for the sample, then the unit is calculated from

$$U \, l^{-1} = \frac{dE \text{ min}^{-1}}{\varepsilon d} \cdot 10^6 \cdot \frac{FV}{SV} \, \mu\text{mol min}^{-1} \, l^{-1}$$

3.1.3.7. Calculation of enzyme units: graphical method

The following rapid and simple graphical method may be recommended for all kinetic methods, particularly if the changes in extinction are traced directly by a recording instrument. The sole requirement is that there should be a linear relationship between the change in extinction and the time, i.e. the reaction should be of zero order. If the results are plotted directly on millimetre graph paper and the points joined by a straight line, then let v = length of abscissa (cm) corresponding to 1 min (2.0 cm is best), and b = length of ordinate (cm) corresponding to an extinction of 1.0 (40.0 cm is best). If, on the other hand, use is made of a direct recording instrument with linearization, then let v = speed of the paper feed in the recorder (cm min^{-1}) (best: 1200 mm h^{-1} = 2 cm min^{-1}), and b = width of paper (cm) corresponding to an extinction of 1.0 (best: 40.0 cm).

As illustrated in Fig. 78, the results are plotted and joined by a straight line. The angle between this line and the abscissa is measured with a protractor; the enzyme concentration can easily be calculated from the tangent of this angle (α):[6]

$$U\ l^{-1} = \tan\alpha \cdot \frac{10^6}{\varepsilon} \cdot \frac{FV}{SV} \cdot \frac{v}{b}\ \mu\text{mol min}^{-1}\ l^{-1}$$

Since ε, FV, SV, v and b are constant for a given system, the system constant is evaluated:

$$c = \frac{10^6}{\varepsilon} \cdot \frac{FV}{SV} \cdot \frac{v}{b}$$

Once this constant has been evaluated for a particular method, the calculation of the concentration is simplified:

$$U\ l^{-1} = \tan\alpha \cdot c\ \mu\text{mol min}^{-1}\ l^{-1}$$

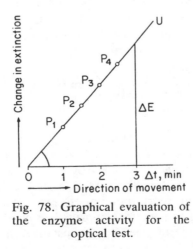

Fig. 78. Graphical evaluation of the enzyme activity for the optical test.

Example. In a determination of serum glutamate–oxaloacetate transaminase, let SV = sample volume = 0.1 ml and FV = final volume = 0.7 ml. The extinction coefficient for $NADH_2$ is 6.3×10^3 l mol^{-1} cm^{-1} at 340 nm and 3.40×10^3 l mol^{-1} cm^{-1} at 365 nm. Measurement is carried out at 365 nm using a path length of 1 cm.

1. *Reading manually.* The following extinctions were obtained at 1-min intervals:

t, min	E	ΔE min^{-1}
0	0.298	
1	0.279	0.019
2	0.257	0.022
3	0.240	0.017
4	0.215	0.025
Total		0.083
ΔE min^{-1}		0.0208

The differences between the extinctions are calculated and the arithmetic mean of the four resulting values is taken. Substituting all the data in the equation (p. 236) gives

$$U\,l^{-1} = \frac{0.0208}{3400 \cdot 1} \cdot \frac{0.7}{0.1} \cdot 10^6 \; \mu mol \; min^{-1} \, l^{-1}$$

$$U\,l^{-1} = 42.8 \; \mu mol \; min^{-1} \, l^{-1}$$

2. *Plotting on millimetre graph paper.* The extinctions are plotted straight on millimetre graph paper, using an abscissa scale of 2 cm for 1 min and an ordinate scale of 40.0 cm for an extinction of 1. The angle between the straight line and the abscissa is measured with a protractor and the tangent found from a set of trigonometric functions. In the present instance, the angle measures 22° 30′ and the corresponding tangent is 0.414. The data are now substituted in the equation:

$$U\,l^{-1} = 0.414 \cdot \frac{10^6}{3400 \cdot 1} \cdot \frac{0.7}{0.1} \cdot \frac{2}{40} \; \mu mol \; min^{-1} \, l^{-1}$$

$$U\,l^{-1} = 42.6 \; \mu mol \; min^{-1} \, l^{-1}$$

3. *Direct recording.* The determination is quickest if the extinctions are plotted by a linear recorder. The line of the measurements is extended, the angle measured, the tangent found from tables and multiplied by the system constant, not omitting the insertion of the feed velocity and the specific width of paper (corresponding to an extinction of 1.0) in the equation, or forgetting to make the appropriate changes to the equation if there are any changes, of speed, for instance, in the system. If the

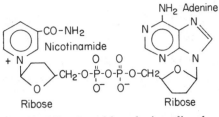

Fig. 79. Nicotinamide adenine dinucleotide (NAD).

measurements are made at another wavelength, then the appropriate extinction coefficient must be substituted.

In addition to the direct-recording instruments, some instruments are currently available which integrate $\Delta E/\Delta t$ automatically and express the enzyme units directly via a process calculator.

3.1.4. THE OPTICAL TEST

3.1.4.1. Principle

In 1936, Warburg and Christian[7] described a simple principle for the determination of hydrogen transfer enzymes. The basic measurement common to all these methods is that of the oxidation of $NADH_2$ (nicotinamide adenine dinucleotide, reduced form) or the reduction of NAD (nicotinamide adenine dinucleotide) (Fig. 79):

$$XH_2 + NAD \rightarrow X + NADH + H^+$$

Since $NADH_2$ absorbs strongly between 300 and 370 nm, in which region $NADH_2$ does not absorb at all (Fig. 80), the oxidation or reduction

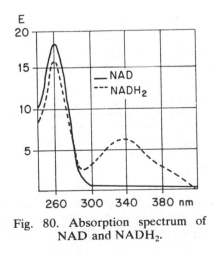

Fig. 80. Absorption spectrum of NAD and $NADH_2$.

240

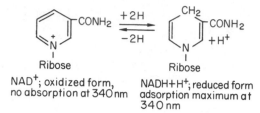

NAD⁺; oxidized form,
no absorption at 340 nm

NADH+H⁺; reduced form
adsorption maximum at
340 nm

Fig. 81. Structural changes at the nicotinamide residue.

may be followed by measurement of the extinction at a suitable wavelength
(334, 340, or 365 nm). The change from the aromatic to the quinoid form
of the pyridine ring results in a change in the ultraviolet absorption (Fig.
81).

The same principle can also be applied with another coenzyme,
nicotinamide adenine dinucleotide phosphate ($NADPH_2$ or NADP).

3.1.4.2. Molar extinction coefficients of $NADH_2$ and $NADPH_2$

Inaccuracies in the determination of standard solutions by means of the
$NADPH_2$-dependent enzyme reactions led Bergmeyer[8] to check the molar
extinction coefficients of the reduced pyridine nucleotides. The values are
given in Table 33.

However, from the point of view of the clinical–chemical laboratory, the
observed deviations from the earlier results ($<1\%$) are scarcely relevant.
Salient factors governing the ε value may be particularized as follows: .

ε is different for $NADH_2$ and $NADPH_2$.
ε is temperature dependent.
ε is dependent on the pH and ionic strength of the solution used for
measurement.
ε cannot be measured with sufficient accuracy at 37 °C owing to the
instability of the coenzyme.
The maximum of the $NADH_2$ or $NADPH_2$ absorption does not lie exactly
at 340 nm; 339 nm may be taken as a first approximation.

Table 33. Molar extinction coefficients of $NADH_2$ and $NADPH_2$ at
25, 30, and 37 °C; units: l mol⁻¹ cm⁻¹ (from ref. 8).

	Hg 334 nm[1] (334.15 nm)	340 nm (339 nm)	Hg 365 nm (365.3 nm)
$NADH_2$	6.18×10^3	6.3×10^3	3.4×10^3
$NADPH_2$	6.18×10^3	6.3×10^3	3.5×10^3

[1]Exact values at 25 °C:[9] 6.182×10^3($NADH_2$) or 6.178×10^3($NADPH_2$).
Error on 6.15×10^3 is $<0.1\%$.

The absorption maximum is temperature dependent.
The differences in the ε value conditioned by the above parameters are smallest by far at Hg = 334 nm.[8]

3.1.4.3. The simple optical test

If the enzyme to be measured requires the direct participation of NAD or $NADH_2$, then the substrate, enzyme, and coenzyme are mixed together in the cuvette and the change in extinction is followed directly. The determination of lactate dehydrogenase may serve as an example (Fig. 82a).

Pyruvate + NADH + H^+ → lactate + NAD^+

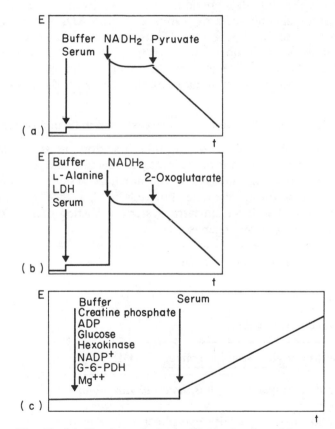

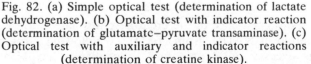

Fig. 82. (a) Simple optical test (determination of lactate dehydrogenase). (b) Optical test with indicator reaction (determination of glutamate–pyruvate transaminase). (c) Optical test with auxiliary and indicator reactions (determination of creatine kinase).

Serum, $NADH_2$, and a suitable buffer are pre-incubated at pH 7.5 until the extinction is steady (preliminary run). The reaction is started by the addition of pyruvate. The rate of the ensuing fall in the extinction is directly proportional to the enzyme activity.

3.1.4.4. The optical test with an indicator reaction

Even enzymes which do not react directly with the nicotinamide coenzymes may occasionally be estimated by combining the above test with an indicator reaction. Consider the determination of glutamate–pyruvate transaminase:

L-Alanine + 2-oxoglutarate → L-glutamate + pyruvate

The pyruvate formed can be determined by the addition of lactate dehydrogenase as in the previous reaction. Each mole of pyruvate produced from alanine oxidizes 1 mol 1^{-1} of $NADH_2$ to NAD and is reduced to lactate in the process. Thus the increase of extinction per minute is directly proportional to the rate of the transamination reaction (Fig. 82b). Similar systems are used for the determination of glutamate–oxaloacetate transaminase, of aldolase, and of pyruvate kinase.

3.1.4.5. The optical test with auxiliary and indicator reactions

Occasionally the product of a quantification reaction cannot be determined directly by means of an indicator reaction and it is necessary to interpolate a further enzyme system, the auxiliary system, as, for example, in the determination of creatine kinase. In this case the quantifying reaction is coupled with an indicator system consisting of hexokinase and glucose-6-phosphate dehydrogenase:

Quantifying reaction: *creatine kinase*

Creatine phosphate + ADP → creatine + ATP

Auxiliary reaction: hexokinase

ATP + glucose → glucose-6-phosphate + ADP

Indicator reaction: *glucose-6-phosphate dehydrogenase*

glucose-6-phosphate + $NADP^+$ → 6-phosphogluconate + . . .

If the rate of disappearance of the creatine phosphate is to be proportional to the formation of $NADH_2$ from NAD, the following substances must be present in excess: ATP, creatine phosphate, hexokinase, glucose,

glucose-6-phosphate dehydrogenase, and NADP. The course of the reaction is illustrated in Fig. 82c.

3.1.4.6. Kinetics of the optical test

For the direct or simple optical test, the relevant kinetics are as described at the beginning of this chapter, but coupling with an indicator reaction introduces complications, since here two enzymes are acting consecutively:

$$S \xrightarrow{k_1} P_1 \xrightarrow{k_2} P_2$$

where k_1 = rate constant for the quantifying reaction and k_2 = rate constant for the indicator reaction. The ratio k_1/k_2 determines whether the result actually corresponds to the activity of the system under measurement.

The conditions must be chosen so that the quantifying reaction is always rate determining. If all the components of the indicator system are present in a 1000-fold excess, i.e. $k_1/k_2 = 1/1000$, then the result deviates very little from the actual activity of the system being measured. Any such deviation is readily detected from the plot since it results in an S-shaped curve. It may be calculated[10] that for a 100-fold excess of the indicator system the result represents 96% of the effective activity of the system measured. For a 1000-fold excess, the corresponding value is 99.3%. This shows that the indicator system must have an activity at least 100 times greater than that of the enzyme to be quantified. If, in addition, an auxiliary system is used, then the relationships are even less predictable and it is advisable to ensure that the ultimate enzyme in the chain (the indicator enzyme) is present in at least a 1000-fold excess. Conditions for two-substrate enzyme reactions (e.g. aspartate aminotransferase) have recently been re-evaluated.[12]

3.1.4.7. Calculation of the activity of auxiliary and indicator enzymes

With all procedures based on the principle of the coupled optical test, it is advisable to calculate the prevailing activity of the auxiliary and indicator enzymes. For calculations involving stock and working solutions, let:

Spec. act. = specific activity of the enzyme in the stock solution in μmol min^{-1} mg^{-1};

(E) = concentration of the enzyme in the stock solution in mg protein l^{-1} (mg l^{-1});

ES = volume (in ml) of stock solution used in the preparation of the working solution;

WS = total volume (in ml) of the prepared working solution;

WV = volume (in ml) of working solution used per assay;

FV = final volume (in ml) of the assay solution.

Then the concentration of an auxiliary or indicator enzyme in the assay solution is given by

$$U\,l^{-1} = \text{spec. act. (E)} \cdot \frac{ES}{WS} \cdot \frac{WV}{FV} \, \mu\text{mol min}^{-1}\, l^{-1}$$

As a rule the results are sufficiently accurate for routine clinical–chemical work if this activity is up to 10 times the upper limit of the reference interval. If the latter is X (e.g. 20 U for glutamate–oxaloacetate transaminase) then the desired activity of the indicator enzyme (in this case, malate dehydrogenase) would be $10X$, i.e. 200 U. However, as explained above, in order to obtain an exact result at least a 1000-fold excess is necessary, i.e. the malate dehydrogenase must be present at an activity of 20 000 μmol min^{-1} l^{-1}. Calculation from the stock solution is as follows:

Simple test: indicator enzyme

$$1000X = \text{spec. act. (E)} \cdot \frac{ES}{WS} \cdot \frac{WV}{FV}, \quad \text{or} \quad ES = \frac{1000X \cdot WS \cdot FV}{\text{spec. act. (E)} \cdot WV}\,\text{ml}$$

Coupled test:
Auxiliary enzyme: calculation as for simple test.
Indicator enzyme:

$$10\,000X = \text{spec. act. (E)} \cdot \frac{ES}{WS} \cdot \frac{WV}{FV}, \quad \text{or} \quad ES = \frac{10\,000X \cdot WS \cdot FV}{\text{spec. act. (E)} \cdot WV}\,\text{ml}$$

3.1.4.8. Practical procedure

Today there are over 100 enzymes which can be rapidly and conveniently determined by means of the optical test. Since, in all these determinations, the procedure is basically the same, in the following we shall merely indicate a few points concerning the determination of the enzymes of greatest clinical–chemical importance.

For carrying out the optical test, the first requirement is a photometer providing monochromatic light at wavelengths of 334, 340, or 365 nm. For enzyme determinations, the temperature of the cuvette holder must be thermostatically controlled at 25, 30, or 37 °C. To maintain the temperature precisely, a reverse-cooling mechanism is obligatory. We find the semi-micro cuvette (diameter 1 cm) most suitable for enzyme determinations since this makes for a considerable saving of the relatively expensive substrates, enzymes, and coenzymes. A recording instrument with linearization or a process calculator is a great advantage in carrying out the optical test: the determinations can be performed more quickly and the results are more accurate. Thus, with a recording instrument a determination of glutamate–pyruvate transaminase takes 1–2 min, compared with 3–6 min for reading the extinction manually.

Buffer. Various buffers are recommended for the measurement of enzyme activity (phosphate, tris, triethanolamine buffers, etc.), but it is most important to check that the buffers are chemically inert and do not exert an enzyme-inhibiting action.

Coenzymes. All of these are very expensive and consequently they should never be dissolved in other than small quantities. They are best kept frozen. $NADH_2$ is most stable in slightly alkaline solution, while NADP can be protected against decomposition by the addition of nicotinamide. Recent investigations have shown that $NADH_2$ undergoes a very rapid, spontaneous decomposition producing coenzyme analogues which exert an inhibiting action on certain lactate–dehydrogenase isoenzymes.

Substrates. Most substrate solutions may be kept in a frozen condition for anything from weeks to months. However, turbid solutions should be thrown away immediately.

Enzymes. These are very expensive and therefore they should only ever be dissolved in small quantities. The ammonium sulphate suspensions should never be diluted with anything but distilled water—not with buffer. In this form they may be kept for weeks at $4\,°C$. It is advisable to measure the activity of the stock solutions from time to time. This helps to avoid unpleasant surprises (insufficient enzyme activity in the auxiliary and indicator systems).

Before embarking on the enzyme determinations, the required quantities of working solutions are prepared. The appropriate working solution is then incubated with the sample (e.g. serum) as a preliminary to the determination. During this time, any contaminating coenzymes and substrates react with the appropiate enzymes. When this pre-incubation is over, the mixtures are transferred to the photometer cuvettes and the extinction observed for 1–2 min. If this is absolutely steady the reaction is started by adding the specific substrate and the changes in extinction over a few time intervals are read off or recorded. If the extinction is not quite constant during the preliminary run but undergoes a steady slight decrease (or increase), the determination can be performed anyway, provided that the changes in extinction (dE/dt) in the preliminary run are subtracted from the changes in extinction during the test.

With most optical tests a blank determination (without substrate) is superfluous since side-reactions do not occur, provided that the substrates and auxiliary enzymes are pure. However, they do occur on occasion, as indicated by the inconstancy of extinction values during the preliminary run. In this case the procedure may be repeated using new reagents or the determination is conducted using the changes of extinction during the preliminary run as a blank, as mentioned above. In doubtful cases it is possible to run a blank without substrate. The optical test is extremely versatile. The methods described are designed so that the period of measurements is as short as possible but they do require the use of relatively large amounts of serum (0.1–0.3 ml). If only small amounts are available it is possible to carry out the determinations with 0.02–0.1 ml but the period of observation must then be extended in proportion and the calculation factors must be modified accordingly.

3.1.4.9. Calculation of results

With the optical test, the results are calculated on the basis of the molar extinction, never by running standards. The basic equation is as follows:

$$U\,l^{-1} = \frac{[\Delta E(P)\,\mathrm{min}^{-1} - \Delta E(L)\,\mathrm{min}^{-1}]}{\varepsilon d} \cdot \frac{FV}{SV} \cdot 10^6\,\mu\mathrm{mol\ min}^{-1}\,l^{-1}$$

If there is no blank, the term $\Delta E(L)\,\mathrm{min}^{-1}$ vanishes and for cuvettes with a path-width of 1.0 cm, the d in the denominator vanishes too. The value of the extinction coefficient (e.g. for $NADH_2$) is that appropriate for the wavelength employed:

334 nm, $6.18 \times 10^3\,l\,\mathrm{mol}^{-1}\,\mathrm{cm}^{-1}$
340 nm, $6.3 \times 10^3\,l\,\mathrm{mol}^{-1}\,\mathrm{cm}^{-1}$
365 nm, $3.4 \times 10^3\,l\,\mathrm{mol}^{-1}\,\mathrm{cm}^{-1}$

The final volume, FV, and the sample volume, SV, depend on the design of the experiment. The constant

$$c = \frac{FV \cdot 10^6}{SV}$$

is first calculated for each method used. Then, for calculating the units it is only necessary to multiply this constant by the change of extinction per minute ($dE\,\mathrm{min}^{-1}$):

$$U\,l^{-1} = [\Delta E(P)\,\mathrm{min}^{-1} - \Delta E(L)\,\mathrm{min}^{-1}]c\ \mu\mathrm{mol\ min}^{-1}\,l^{-1}$$

Table 34. Conversion factors for enzyme units[2].

(a) *Definitions*
Old μmol unit: substrate conversion in μmol l^{-1} ml^{-1}
Bücher unit: an extinction change of 0.100 ($\lambda = 366$ nm) ml^{-1} per 100 s
Wroblewski unit: an extinction change of 0.001 ($\lambda = 340$ nm) min^{-1} per 3 ml

(b) *Factors*

Given	Sought (multiply by)			
	μmol h^{-1} ml l^{-1}	Bücher unit	Wroblewski unit
μmol h^{-1} ml^{-1}	—	0.915	34.6	16.7
Bücher unit	1.097	—	37.7	18.2
Wroblewski unit	0.0289	0.0265	—	0.483
μmol min^{-1} l^{-1}	0.060	0.055	2.07	—

An example of the calculation is given on p. 238. Obviously, the change in extinction must always be measured over several time intervals.

Instead of noting the changes in extinction by hand, the results may either be plotted as a graph or registered with an automatic recorder (the calculation of the units is then performed according to the instructions given on p. 238), or directly calculated by electronic instruments.

A few conversion factors for changing from the older to international units are collected in Table 34.

References

1. *Enzyme nomenclature*, Elsevier, Amsterdam, 1973.
2. Richterich, R., Schafroth, P., Colombo, J. P., and Temperli, F., Die Wahl von Enzymeinheiten bei diagnostischen Untersuchungen, *Klin. Wschr.*, **39**, 987 (1961).
3. Provisional recommendations (1974) on IFCC methods for the measurement of catalytic concentration of enzymes, *Clin. chim. Acta*, **61**, F11 (1975).
4. Bergmeyer, H. U., Bernt, E., Krassel, M., and Michal, G., Ermittlung von Messergebnissen, in Bergmeyer, *Methoden der enzymatischen Analyse*, Vol. 1, Verlag Chemie, Weinheim, 1974, p. 329.
5. Uniform expression of tissue analysis, in Knox, *Enzyme patterns in fetal adult and neoplastic rat tissues*, Karger, Basle, 1972, p. 58.
6. Weber, H. and Richterich, R., Eine einfache und rasche graphische Methode zur Berechnung von Enzymaktivitäten, *Klin. Wschr.*, **41**, 665 (1963).
7. Warburg, O. and Christian, W., Pyridin, der Wasserstoff-übertragende Bestandteil von Gärungsfermenten (Pyridin-Nukleotiden), *Biochem. Z.*, **287**, 291 (1936).
8. Bergmeyer, H. U., Neue Werte für die molaren Extinktions-Koeffizienten von NADH und NADPH zum Gebrauch im Routine-Laboratorium (see corrected Supplement), *Z. klin. Chem. klin. Biochem.*, **13**, 507 (1975).
9. Ziegenhorn, J., Senn, M., and Bücher, Th., Molar absorptivities of β-NADH and β-NADPH, *Clin. Chem.*, **22**, 151 (1976).
10. Bergmeyer, H. U., Die chemischen Grundlagen der enzymatischen Analyse, *Ärztl. Lab.*, **7**, 261 (1961).
11. Bergmeyer, H. U., Grundlagen der enzymatischen Analyse, Verlag Chemie, Weinheim, 1977, p. 11.
12. Bergmeyer, H. U., Evaluation of optimum conditions of two-substrate enzyme reactions, *J. clin. Chem. clin. Biochem.*, **15**, 405 (1977).

3.1.5. ENZYMATIC SUBSTRATE ANALYSIS

In the measurement of enzyme activity the enzyme is mixed with a specific substrate and the conversion of the substrate is followed. This is a measure for the enzyme activity, i.e. the amount of enzymically active molecules. As a rule, the substrate is present in excess and the enzymic activity is unknown. Conversely, it is also possible to measure the concentration of a substrate by means of enzymatic conversion. In this case, however, it is the enzyme which must be in excess so that sufficient substrate is converted per unit time. The advantage of enzymatic substrate determination is the specificity of the determination for the particular metabolite concerned. For example, the following metabolites can be determined in plasma using enzymatic analysis: glucose (glucose oxidase; hexokinase, glucose dehydrogenase), uric acid (uricase), lactate (LDH), pyruvate (LDH), urea (urease), ammonia (GLDH), acetoacetate (3-hydroxybutyrate dehydrogenase, *q.v.*).

The indicator reactions can be measured according to the following principles: decrease in the extinction due to the conversion of a light-absorbing substrate into a non-absorbing product, e.g. observation of the degradation of uric acid at 293 nm by uricase to allantoin, CO_2 and H_2O_2 which do not absorb at this wavelength; increase in the extinction due to an increased absorption of light by the products or by a coloured compound from the latter in a second reaction, e.g. cleavage of urea by urease and measurement of the liberated ammonia by means of the Berthelot reaction; determination of glucose with glucose oxidase/peroxidase, with conversion of a colourless, reduced substance to an oxidized dye; consumption or formation of $NADH_2$ or $NADPH_2$, directly or indirectly coupled with the principal reaction as indicator reaction; e.g. measurement of glucose with glucose dehydrogenase (direct coupling) or with hexokinase/glucose-6-phosphate dehydrogenase (indirect coupling).

The conditions of the reaction are arranged so that the reaction goes to completion in an acceptable time: *end-point method*.

If the molar extinction coefficient of the reacted substrate (e.g. uric acid at 293 nm) or of the formed products or of the coupled products of the indicator reaction (e.g. $NADH_2/NADPH_2$, reduced dye/oxidized dye) is known, then the concentration of the substrate can be evaluated directly. If this is not the case, then it is necessary to run a standard solution at the same time as the sample. In enzymatic substrate analysis, depending on the reaction kinetics and the amount of enzyme, the substrate is completely converted only up to a certain concentration. This limiting concentration must be precisely determined and samples with higher concentrations must be diluted accordingly.

Provided that certain reaction conditions are fulfilled, the enzymatic substrate analysis may also be measured kinetically:

Enzyme–kinetic substrate determination. The concentration of a substrate can also be determined by measuring the rate of the enzyme-catalysed reaction if the rate of reaction is a linear function of the substrate concentration (this condition is fulfilled at the beginning of a first-order reaction). This means that for a one-substrate reaction, the substrate concentration is very much smaller than the Michaelis constant, $[S] \gg K_m$. In this case, the rate of the enzymatic conversion, the enzyme activity, is determined. This is proportional to the unknown concentration of substrate. Since these measurements are extremely sensitive, the experimental conditions must be rigidly adhered to.[1] Corresponding methods have been described for the determination of glucose.[2,3]

References

1. Bergmeyer, H. U., Determination of concentrations by kinetic methods, in Bergmeyer, *Methods of enzymatic analysis*, Vol. 1, Academic Press, New York, 1974, p. 131.
2. Faust, U., Keller, H., and Becker, J., Kurzzeit-Substratbestimmung mit enzymkinetischer Methode. Optimierung der Reaktionsparameter, *Chem. Rdsch.*, **26**, No. 47 (1973).
3. Keller, H., Faust, U., and Becker, J., Enzymatische Glukose-Bestimmung durch Kurzzeitmessung der Reaktionskinetik, *Chem. Rdsch.*, **23**, No. 47 (1973).

3.1.6. ENZYMES AS MARKERS; ENZYME IMMUNOASSAYS

In histochemistry and immunochemistry, enzymes have long been used as markers for the detection of antibodies or antigens. The combination of enzyme marking with the methods of saturation analysis is embraced by the term 'enzyme immunoassay' (EIA).[1,2,11] The methods are analogous to the radioimmunoassays (RIA) (p. 208) but instead of using a radioisotope-labelled tracer molecule, the antigen (AG) or the antibody (AB) is covalently bound to an enzyme and thus used as a tracer. Instead of measuring the radioactivity, the enzyme activity is measured.

The chemical *coupling* of the enzyme to the antigen or antibody is effected, e.g., with dialdehydes (glutaraldehyde), sodium periodate, dimaleimide, or anhydrides.[3] The chain length of the coupling agent influences the behaviour of the enzyme-labelled molecule in the assay. As *enzymes*, vegetable or bacterial enzymes are used; these are pure, stable, of high activity, and are easily determined. Also, and most particularly, they are not (or scarcely) present to any measurable extent in physiological fluids.[4,5] They include, e.g., the peroxidases, lysozyme (EMIT—urine assays) and the glucose oxidases. The use of enzymes which occur in body fluids can lead to interferences.

With EIA there are fewer alternatives for the *separation* [necessary in saturation analysis (p. 207)] of the free from the bound portion than with RIA. In EIA the marker molecules are larger and there must be no change in the enzyme protein as a result of the separation. Consequently, precipitation with ammonium sulphate, for example, is out of the question, and the separation is generally effected by solid phase methods, i.e. the antigen or the primary or secondary antibody are adsorbed on particles (cellulose, Sephadex), test-tube walls, or test plates. Such EIA methods are accordingly named ELISA ('enzyme-linked immunosorbent assays'). The measurement of the enzyme in the free or bound phase can be done visually by means of a colour reaction (yes–no answer), photometrically, or fluorimetrically.

The sensitivity of EIA is comparable to that of RIA (down to 1 amol) (e.g. ref. 6), but there are considerable advantages: the labelled reagents have a better stability, radiation protection is unnecessary, and there is no need to purchase a counter. However, the possibility of inhibition or activation of the enzyme reactions by substances in the physiological liquids—particularly with the so-called *competitive methods*—constitutes a disadvantage.

The principles of the *EIA methods* currently of importance are illustrated schematically in Table 35. These methods are used for the determination of haptens or proteins, including immunoglobulins pertaining to the body.[7] In the assay, these substances function as antigens. The determination of immunoglobulins (antibodies present in the body) in this way is often denoted by the term ELISA, used in its narrower sense. In addition, EIA may also be used for the determination of antibodies as such in the system (Table 35: Section 2.2.1).

Among the competitive EIA methods, the *EMIT methods* (enzyme-multiplied immunoassay technique)[8] are especially noteworthy (Table 35: Section 1.2). The method is used for the determination of small molecules (until now only haptens) such as, e.g. antiepileptics, opiates (urine), thyroxine, and digoxin. A separation of the free from bound enzyme activities is not necessary. In this technique only certain enzymes are used for coupling to antigens. Their activity in the antigen–enzyme complex is modified by the bonding of the antigen to its antibody so that the resulting enzyme–antigen–antibody complex is enzymically inactive. This is thought to be due to steric hindrance of the substrate or changes in the conformation of the enzyme. On the other hand, it is also possible for the bonding to the antibody to cause an activation of the enzyme (example: malate dehydrogenase). If an unknown quantity of free antigen is mixed with a fixed quantity of enzyme antigen and a limited quantity of

---

Table 35. Principles of the different enzyme immunoassays.

Symbols

AG = antigen, AB = antibody, E = enzyme

$\boxed{\text{AG}}$, $\boxed{\text{AB}}$ = insoluble antigen or antibody, bound to the solid phase

ag = amount (unknown) of antigen to be determined
ab = amount (unknown) of antibody to be determined

---

1. *Antigen marking with competition*

1.1.   Separation of soluble and solid phases.

$$\begin{array}{c} \text{ag} \\ \text{E–AG} \end{array} + \boxed{\text{AB}} \leftrightarrow \begin{array}{c} \text{ag–}\boxed{\text{AB}} \\ \text{E–AG–}\boxed{\text{AB}} \end{array}$$

The greater the amount of ag, the lower is the enzyme activity in the solid phase.

1.2. With modification of the enzyme activity by bonding of E–AG to AB (enzyme-multiplied immunoassay technique, EMIT)

$$\begin{array}{c} \text{ag} \\ \text{E–AG} \end{array} + \boxed{\text{AB}} \leftrightarrow \begin{array}{c} \text{ag–}\boxed{\text{AB}} \\ \text{E–AG–}\boxed{\text{AB}} \end{array}$$

E–AG enzymatically active,
E–AG–AB enzymatically inactive.
Enzyme activity is paralleled by the amount of ag, no separation necessary.

Table 35. cont.

2. *Marking the antibody*

2.1. With competition

$$\frac{ag}{\boxed{AG}} + AB\text{--}E \leftrightarrow \frac{ag\text{--}AB\text{--}E}{\boxed{AG}\text{--}AK\text{--}E}$$

The greater the amount of ag, the lower is the enzyme activity in the solid phase.

2.2. Without competiton

2.2.1. Sandwich method for determining an antigen (large protein)

$$\boxed{AB} + ag \rightarrow \frac{\boxed{AB}}{\boxed{AB}\text{--}ag} + AB\text{--}E \rightarrow$$

Enzyme activity in line with amount of ag

washing away the soluble AB–E excess

$$\rightarrow \frac{\boxed{AB}}{\boxed{AB}\text{--}ag\text{--}AB\text{--}E}$$

2.2.2. Determination of immunoglobulins

$$\boxed{AG} + ab_1 \rightarrow \frac{\boxed{AG}}{\boxed{AG}\text{--}ab_1} + AB_2E$$

ab_1 = immunoglobulin
AG = antigen for ab_1
AB_2 = anti-ab

washing away the soluble AB_2–E excess

$$\rightarrow \frac{\boxed{AG}}{\boxed{AG}\text{--}ab_1\text{--}AB_2\text{--}E}$$

Enzyme activity in line with amount of ab

2.3. CELIA (competitive enzyme-linked immunoassays)

$$\frac{ag}{\boxed{AG}} + AB_1 \leftrightarrow \frac{ag\text{--}AB}{\boxed{AG}\text{--}AB_1}$$

AB_1 = AB against AG or ag; from species 1

wash

$$\frac{\boxed{AG},}{\boxed{AG}\text{--}AB_1} + AB_2$$

AB_2 = anti-AB_1, from species 2, in excess 2 bonding sites

wash

$$\frac{\boxed{AG},}{\boxed{AG}\text{--}AB_1\text{--}AB_3} + AB_2\text{--}E$$

AB_3 = soluble cyclic peroxidase antibody from species 1

wash

$$\frac{\boxed{AG},}{\boxed{AG}\text{--}AB_1\text{--}AB_2\text{--}AB_3\text{--}} + urea$$

E = peroxidase, urea for dissociation of AB_3–E in line with amount of ag

separation, enzyme activity in the supernatant

antibody, the antigens compete for bonding to the antibody. A separation is not necessary since the labelled antigens bound to the antibody are enzymically inactive, i.e. they are no longer tracers. Up to now, EMIT appears to be less sensitive than the other EIA or RIA methods. The method has been automated and this has improved its precision. It is particularly recommended for serial investigations (e.g. screening of drugs). EMIT is also known as homogeneous enzyme immunoassay, in contrast to the inhomogeneous EIA which require a separation.

All the methods mentioned so far have been competitive methods. The sandwich method described below is an example of a *non-competitive method* (Table 35: Section 2.2.1). The antigen to be determined (quantity unknown) is bound to an undissolved antibody which is present in excess. The enzyme-tagged antibody is added and this couples to another site on the bound antigen. The excess is washed away and the enzyme activity of the undissolved AB–AG–AB–E complex is determined. This activity increases in line with the AG concentration. The method is only suitable for an AG with two bonding sites, usually larger proteins (e.g. human chorionic gonadatrophin, HCG). Unspecific compounds of AB and AB–E can lead to erroneous results (at high concentrations of rheumatic factor).[7]

The CELIA method (competitive enzyme-linked immunoassay), which is not yet widely disseminated, has the advantage that the enzyme-tagged AB (3.AB) reacts with a non-specific 2.AB, so that potentially it may be used for various AG determinations.[9] The antiperoxidase–peroxidase system used (cyclic AG–AB complex with two antibody and three peroxidase molecules)[10] helps to bring about a multiplication effect by means of its AB/enzyme ratio.

EIA methods have found *application* in the determination of plasma proteins, especially immuno-antibodies, hormones, pharmaceuticals, and bacterial toxins.[3,5,6]

References

1. Engvall, E. and Perlmann, P., Enzyme-linked immunosorbent assay (ELISA). Quantitative assay of immunoglobulin G, *Immunochemistry*, **8**, 871 (1971).
2. Weemen, B. K. van and Schuurs, A. H. W. M., Immunoassay using antigen–enzyme conjugates, *FEBS Lett.*, **15**, 232 (1971).
3. Van der Waart, M., Bosch, A., Wolters, G., Kuijpers, L., Weemen, B. van, and Schuurs, A., Enzyme-immunoassay—a new diagnostic tool, *Chem. Rdsch.*, **28**, 1 (1977).
4. Scharpé, S. L., Cooreman, W. N., Blomme, W. J., and Laekeman, G. M., Quantitative enzyme immunoassay: current status, *Clin. Chem.*, **22**, 733 (1976).
5. Wisdom, G. B., Enzyme-immunoassay, *Clin. Chem.*, **22**, 1243 (1976).
6. Kato, K., Hamaguchi, Y., Okawa, S., Ishikawa, E., Kobayashi, K., and Katunuma, N., Enzyme immunoassay in rapid progress, *Lancet*, **i**, 40 (1977).
7. Weemen, B. K. van and Schuurs, A. H. W. M., Prinzip des Enzym-Immunoassay, in Bergmeyer, *Grundlagen der enzymatischen Analyse*, Verlag Chemie, Weinheim, 1977, p. 96.

8. Rubenstein, K. F., Schneider, R. S., and Ullmann, E. F., 'Homogeneous' enzyme immunoassay. A new immunochemical technique, *Biochem. biophys. Res. Commun*, **47**, 846 (1972).

9. Yorde, D. E., Sasse, E. A., Wang, T. Y., Hussa, R. O., and Garancis, J. C., Competitive enzyme-linked immunoassay with use of soluble enzyme/antibody immune complexes for labeling. I. Measurement of human choriogonadotropin, *Clin. Chem*., **22**, 1372 (1976).

10. Sternberger, L. A. and Petrali, J. P., The unlabeled antibody enzyme method: immunocytochemistry of hormone receptors at target cells, in Feldmann, *Immunoenzymatic techniques*, North-Holland, Amsterdam, 1976, p. 43.

11. Schuurs, A. H. W. and Weemen, B. K. van, Enzyme-immunoassay, *Clin. chim. Acta*., **81**, 1 (1977).

3.2. WATER

3.2.1. HOMEOSTASIS OF BODY FLUIDS

In water and electrolyte metabolism, homeostasis, or the law of preservation of equilibrium, is maintained in three linked mechanisms:[1,2]

1. preservation of pH or of the acid–base ratio;
2. preservation of the ionic composition;
3. preservation of the osmolality.

In every assessment of water and electrolyte metabolism, these three aspects must first be analysed separately and then considered in terms of their mutual interdependence. Often, the primary disturbance is confined to a single system but, owing to secondary effects and feedback, all three systems are always affected in the end.

3.2.2. WATER COMPARTMENTS OF THE BODY[1,2]

The organism is by no means an osmotically open system; it is much more correctly represented as a series of compartments, separated by interfacial membranes which resist the free diffusion of particles. Only a very few substances such as urea and ammonia are distributed throughout the total body water. The most important water compartments are shown in Fig. 83. The total body water amounts to between 50 and 73% of the body weight, the remaining 27–50% corresponding to dry matter. This body water is primarily divided between the intracellular (36–50%) and extracellular space (14–23%), separated by the cell membrane as interface. The extracellular space may be further sub-divided into an interstitial space (10–18%) and an intravascular space (4–5%), the compartments being bounded in this case by the walls of the blood vessels.

The capacities of the individual compartments are measured by determining the distribution volume: the subject is fed or injected with a substance which is distributed to a specific compartment. The capacity of the compartment may then be calculated from the subsequent dilution of the substance. For measuring the total body water, heavy water (D_2O), antipyrine, or urea may be used. The extracellular space roughly corresponds to the distribution volume of inulin, saccharose, chloride, or ^{24}Na, while the intravascular space may be measured by labelling plasma

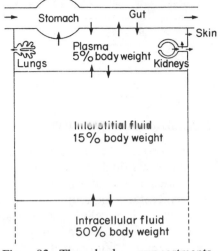

Fig. 83. The body compartments
according to Gamble.[1]

protein with Evans Blue or ^{131}I, or by labelling the erythrocytes with ^{51}Cr, ^{32}P, or ^{55}Fe.

Since the different interfaces (cell membrane, vascular wall) have different permeabilities, the composition of the solutions in the individual compartments is also different. The distribution of the most important components is shown in Table 36. Since the vascular walls are permeable to electrolytes there is scarcely any difference in the electrolyte content of the

Table 36a. Electrolyte composition of the body compartments.

		mmol l^{-1} water		
	Electrolyte	Intravascular space	Interstitial space	Intercellular space
Cations	Sodium	154	147	35
	Potassium	5	4	115
	Calcium (ion)	5	3	5
	Magnesium (ion)	2	2	27
	Total	166	156	182
Anions	Bicarbonate	29	30	10
	Chloride	112	114	25
	Phosphate (ion)	2	2	80
	Sulphate (ion)	1	1	20
	Organic acids (ion)	5	8	—
	Proteinate (ion)	17	1	47
	Total	166	156	182

Table 36b. Average daily water balance for a healthy adult.

Intake	ml	Excretion	ml
Direct water intake	1500	Urine volume	1600
Water in diet	750	Stool	50
Water from metabolism	250	Breath, perspiration	850
Total	2500	Total	2500

intravascular and interstitial space. On the other hand, the protein content of the intravascular space is much higher than that of the interstitial space (preservation of the colloidosomotic or oncotic pressure in the vascular system to balance the hydrostatic pressure). However, the electrolytic composition differs greatly in the intracellular and extracellular space. In the former, the preponderant cation is potassium, and phosphate and proteinate are the major anions. In contrast, sodium and chloride are the preponderant cation and anion in the extracellular space. Table 36b shows the water balance for an adult.

References

1. Gamble, J. L., *Chemical anatomy, physiology and pathology of extracellular fluid*, 6th ed., Harvard University Press, Cambridge, 1960.
2. Pitts, R. F., *Physiology of the kidney and body fluids*, Year Book Medical Publishers, Chicago, 1963.

3.3. ELECTROLYTES

3.3.1. INTRODUCTION

The results of electrolyte determinations were previously expressed in mg per 100 ml and similar units. However, as electrically charged particles, electrolytes are subject to the principle of electroneutrality and consequently these ways of expressing the results lacked significance and were misleading. According to the principle of electroneutrality, the number of positively charged particles (cations) in a solution must always be equal to the number of negatively charged particles (anions). The unit of measurement for electrically charged particles used to be the gram-equivalent; hence a solution had always to contain as many gram-equivalents of cations as of anions. In Gamble's well known ionogram[1] the cation column must therefore be just as large as the anion column. The concept of the gram-equivalent is no longer used in physics, chemistry, or clinical chemistry, and has been replaced by the mole, based on a specific formula weight. This makes no fundamental difference either to the numerical values or the graphic principle of the ionogram; it is only necessary to make the appropriate choice of formula unit for the individual components.

Monovalent ions. For monovalent ions such as sodium, potassium, and chloride, the gram-equivalent is identical with the mole. In clinical chemistry, blood bicarbonate must also be included among the monovalent ions since under physiological conditions only a single hydrogen ion is dissociated: $H_2CO_3 \rightarrow HCO_3^- + H^+$.

Divalent cations. For divalent cations, 1 mole is numerically twice as great as 1 gram-equivalent. For the purposes of electrolyte metabolism studies, this difference may be circumvented by the appropriate choice of formula unit. Thus, for calcium, the appropriate formula unit reads 'Calcium (Ca^{2+}) 0.5', giving the same result whether the answer is expressed in moles or gram-equivalents. The same applies to magnesium, although for both calcium and magnesium it must be remembered that the degree of ionization depends on the protein concentration and the pH and therefore it is fundamentally more correct to give the results in mmol l^{-1} or mg per 100 ml. To be correct, then, the statement of results in milligram-equivalents l^{-1} or 0.5 mmol l^{-1} is only permissible if what is measured is actually the ionized calcium or magnesium. With SI units

257

258

(p. 19), quantity of substance is stated independent of the degree of dissociation or ionization as mmol l^{-1}.

Polyvalent anions. The situation is especially complicated with the polyvalent anions phosphate, sulphate, and proteinate since the degree of ionization is very strongly dependent on pH. It is therefore better in this case to state the results in mmol l^{-1} or mg per 100 ml. Nomograms are a help insofar as they use empirical conversion factors which presuppose a physiological pH. All former values may be retained without further modification provided that the corresponding formula units are declared, e.g. phosphorus, inorganic (ion), protein (ion). If the formula units are defined in this way, the results in mmol l^{-1} are identical with those in milligram-equivalents l^{-1}. With SI units, the quantity of substance is stated independent of the degree of dissociation or ionization as mmol l^{-1}.

In the ionic medium of the intracellular space, the preponderant cations are potassium and magnesium, and proteinate and phosphate are virtually the only anions. In the extracellular space sodium and calcium are the preponderant cations and chloride and bicarbonate the major anions. Owing to the different composition of the ionic milieus noticeable differences in concentration are produced at the cell membranes. These gradients have to be maintained by the expenditure of energy. Consequently, in the stationary state there is an equilibrium between passive diffusion (in the direction of the gradient) and active transport (against the gradient).

The ionic composition of the plasma is often expressed in the form of a two-column diagram, the ionogram, as used by Gamble[1] (Fig. 84). This takes electroneutrality into account and gives an idea of the quantitative distribution of the ions.

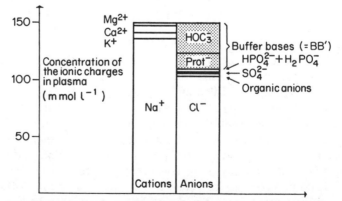

Fig. 84. Ionogram. Electrolytic composition of the plasma according to Gamble[1] as modified by Siggaard-Andersen.[2]

3.3.2. SODIUM

3.3.2.1. Introduction

In contrast to potassium, sodium is at its highest concentration in the extracellular space. Extracellular sodium amounts to 142 mmol l^{-1} and constitutes about 98% of the sodium content of the body; of this, 55% occurs in the bones. The total sodium content of an adult weighing 70 kg amounts to 4200 mmol, i.e. 60 mmol kg^{-1}. The daily sodium intake is 90–250 mmol. About 95% of the sodium is excreted through the kidneys and the remainder through stool and sweat. With a normal diet the excretion is generally balanced by the intake. Measurement of sodium in serum is made in connection with disturbances of the water balance.[7]

3.3.2.2. Choice of method

Despite intensive research, attempts to develop a specific and sensitive method for the photometric determination of sodium in body fluids have so far proved unsuccessful. Chemical methods must therefore be regarded as obsolete and should, where possible, be replaced by physical methods. Flame photometry is the method currently preferred for routine measurements but the manufacturer's instructions should be rigidly adhered to. The determination with the sodium electrode is just as quick and specific and is more sensitive but this method is only likely to offer serious competition with flame photometry if an equally specific potassium electrode is developed. Since it is possible that part of the sodium ions are present in a bound form, further investigations with the sodium electrode are awaited with interest, for the amount of ionized sodium is likely to be of much more biological significance than the total sodium. In recent years there have been occasional references to the use of atomic-absorption photometry for sodium determination; however, this technique has scarcely any advantages over flame photometry. Neutron-activation analysis serves as a reference method for very exact determinations.

3.3.2.3. Reference values

The sodium values in serum are normally distributed.[3] Up to 15 years of age, differences between the age groups are not significant, nor are there

Table 37. Standard values for the concentration of sodium in various body fluids (mmol l^{-1}).

Plasma	Gastric juices	Bile and pancreatic secretion	Small intestine secretion	Ileostomy	Diarrhoea	Sweat
140	60	140	110	130	60	60

260

Table 38. Behaviour of sodium concentration in serum from sucklings, children of various ages, and adults (mmol l^{-1}).[3]

Age group	Number	Scatter	$\bar{x} \pm s$
Up to the 1st month	40	134–143	139.0 ± 2.27
1st–3rd month	21	135–143	138.7 ± 2.03
3rd–6th month	21	135–142	138.7 ± 2.22
6th month–1st year	35	135–142	138.5 ± 1.94
1–3 years	72	135–143.5	139.1 ± 1.99
3–6 years	102	132–144	139.5 ± 1.97
6–10 years	118	136–146	139.7 ± 1.79
10–15 years	109	134–143	139.4 ± 1.98
Adults	65	132.5–145.5	140.0 ± 2.38
Adult men	31	136.5–145.5	140.6 ± 2.27
Adult women	34	136–144	139.7 ± 2.09
Adults (personal results)	314	133–142	138 ± 2.32

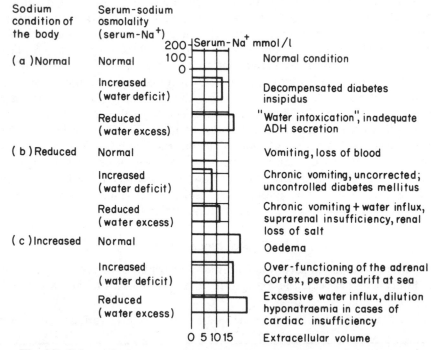

Fig. 85. Schematic representation of the disturbances of the salt and water balance according to Truniger.[4] Conditions 4–9 may be named as follows according to the classical EDH nomenclature (Eu-, De-, Hyperhydration): 4 = isotonic dehydration; 5 = hypertonic dehydration; 6 = hypotonic dehydration; 7 = isotonic hyperhydration; 8 = hypertonic hyperhydration; 9 = hypotonic hyperhydration.

any sex differences for sodium concentrations. It is not known whether the reference intervals change during the course of life. The values remain normal during pregnancy. Table 37 gives standard values for the occurrence of sodium in various body fluids, and Table 38 shows variations with age.

3.3.2.4. Comments on the method

If sodium and potassium are to be determined in plasma, ammonium heparinate is the only anticoagulant which may be used.

Table 39. Changes in the sodium concentration in plasma (according to ref. 5).

Observed changes and causes	Examples
Reduced concentration Water retention, or water and Na^+ retention with a disproportionally larger retention of water, leading to expansion of the extracellular fluid	1. Reduced ability to excrete water, as in kidney failure or in oedema 2. Inadequate secretion of ADH by the pituitary body (e.g. metabolic reaction to trauma, head injury), or of substances similar to ADH in cases of non-endocrinal tumours 3. Excessive water intake, as with psychogenic diabetes insipidus
Abnormal water and Na^+ losses with insufficient or untimely replacement	1. Gastro-intestinal losses such as occur in vomiting, or in aspiration of the contents of the gut, diarrhoea, drainage of fistulae 2. Renal, salt-losing condition as in suprarenal insufficiency, diuretic phase in acute kidney failure 3. Immoderate sweating with replacement of lost water
Disturbances leading to a migration of the Na^+ into the cells; possibly conditioned by disturbance of the exchange mechanism of Na^+ and K^+	1. Chronic debilitating illnesses leading to 'sick cell syndrome' 2. Potassium loss due to sundry causes (Table 41)
Elevated concentration Loss of water leads to elevated plasma Na^+ with shrinkage of the volume of the extracellular fluid	1. Inadequate intake during unconsciousness, in elderly people, paralysis 2. Excessive loss with insufficient replacement, as in e.g. diabetes insipidus (sundry causes) or with diseases of the kidney (e.g. advanced kidney damage)
Na^+ overloading or immoderate retention, leading to increased plasma sodium with increase in the volume of the extracellular fluid	1. Overloading occasioned by excessive infusion of hypertonic solutions of Na^+ 2. Excessive retention, occasioned by overdoses of Na^+ containing steroids or hyperaldosteronism

3.3.2.5. Diagnostic significance

Pathological changes in sodium concentration generally accompany changes in the water balance.[8] Consequently, this gives rise to combined disturbances for which the measurement of sodium, and of the osmolality, in the serum is of diagnostic significance (Fig. 85). A few diagnostically important changes in the sodium concentration in plasma are listed in Table 39.

3.3.3. POTASSIUM

3.3.3.1. Introduction

The potassium content of the body amounts to about 50 mmol kg^{-1}, of which the potassium in plasma accounts for 0.4%. The concentration of potassium in the extracellular space (including the intravascular space) lies between 3.3 and 5.3 mmol^{-1}. About 50–100 mmol of potassium are ingested every day; 5–10 mmol are excreted through the gut and 30–90 mmol through the kidneys. Potassium participates in exchange processes between the extracellular and intracellular spaces, e.g. if glucose is ingested, potassium accompanies the glucose into the cell. Furthermore, there can be intracellular replacement of H^+ ions by K^+ ions and *vice versa*. In an acidosis (H^+ ion excess), H^+ ions pass into the cell and K^+ ions pass from the intracellular to the extracellular space. The measurement of serum potassium affords only limited information on the changes of the potassium content.[4,7]

3.3.3.2. Choice of method

Today potassium is mainly determined by flame photometry. The procedure in the instruction manual should be followed closely. Under no circumstances should glass vessels be used for dilution (glass acts as an ion exchanger). Investigations on ion-specific electrodes are already under way.[6]

3.3.3.3. Reference values

Spot tests which we carried out on healthy children of different ages gave results all of which lay within the range of values determined by Liappis.[3] The determination in plasma gives lower values (smaller haemolysis) than those in serum. In contrast to sodium, the potassium concentration shows statistically significant differences for different age groups, both between sucklings and children and between children and adults (Table 40).

Table 40. Ranges for the potassium concentration in serum for sucklings and children of various ages, and of adults (mmol l^{-1}).[3]

Age group	Number	Scatter	$\bar{x} \pm s$
Up to 1st month	40	3.7–5.9	4.9 ± 0.40
1st–3rd month	21	4.3–5.4	4.8 ± 0.32
3rd–6th month	21	3.9–5.7	4.8 ± 0.47
6th month–1 year	35	3.8–5.4	4.7 ± 0.39
1–3 years	72	3.8–5.1	4.4 ± 0.36
3–6 years	102	3.6–5.5	4.2 ± 0.31
6–10 years	118	3.6–5.4	4.3 ± 0.34
10–15 years	109	3.6–5.2	4.3 ± 0.31
Adults	65	3.6–5.0	4.1 ± 0.29
Adult men	31	3.6–5.0	4.18 ± 0.34
Adult women	34	3.6–5.0	4.07 ± 0.31
Adults (personal results)	312	3.4–4.5	95% interval

3.3.3.4. Comments on the method

1. If the potassium is determined in plasma, then ammonium heparinate is the only anticoagulant which may be used.

2. Erythrocytes contain about 25 times as much potassium as the plasma. Consequently, haemolytic serum or plasma may not be used for potassium determinations. The serum or plasma must be separated from the erythrocytes within an hour, otherwise incorrect, high values are obtained.

3.3.3.5. Diagnostic significance

Table 41 provides information on the interpretation of the most common pathological potassium values.

References

1. Gamble, J. L., *Chemical anatomy, physiology and pathology of extracellular fluid*, 6th ed., Cambridge University Press, Cambridge, 1960.
2. Siggaard-Andersen, O., *The acid–base status of the blood*, 4th ed., Munksgaard, Copenhagen, 1974.
3. Liappis, N., Die Natrium-, Kalium- und Chlorid-Konzentration im Serum von Säuglingen, Kindern und Erwachsenen, *Mschr. Kinderheilk.*, **120**, 138 (1972).
4. Truniger, B., *Wasser- und Elektrolythaushalt*, 4. Aufl., Thieme, Stuttgart, 1974.
5. Withby, L. G., Percy-Robb, I. W., and Smith, A. F., *Lecture notes on clinical chemistry*, Blackwells, Oxford, 1975, p. 38.
6. Marsoner, H. J. and Harnoncourt, K., Potentiometrische Bestimmung der Kaliumkonzentration im Plasma, *Ärztl. Lab.*, **23**, 327 (1977).
7. Grobecker, H., Klinisch-pharmakologische Aspekte der Therapie von Störungen des Elektrolythaushaltes, *Infusionstherapie*, **1**, 11 (1973/74).

Table 41. Changes in the potassium concentration in plasma (from ref. 5).

Observed changes and causes	Examples
Diminished concentration Inadequate intake or immoderate loss or a combination of both these factors	1. Gastro-intestinal tract: insufficient intake (e.g. following an operation), immoderate loss (vomiting, diarrhoea, loss through fistulae) and pharmaceutical treatment (misuse of purgatives, ion-exchange resins) 2. Kidneys: diuretic therapy (e.g. thia acids), renal tubular acidosis, diuretic phase in acute kidney failure, Bartter syndrome **Warning**: Owing to the exchange of K^+ between the intracellular fluid, a low plasma K^+ is not necessarily a characteristic symptom of the conditions described above.
Hormonal disturbances leading to increased loss or to a redistribution of K^+ in the body, or both	1. Insulin. Poorly supervised diabetes mellitus, especially during the recovery phase following diabetic coma. Insulin overdose and islet cell tumours 2. Corticosteroids. Cushing syndrome and steroid therapy 3. Aldosterone. Hyperaldosteronism
Other causes leading to exchange of K^+ from the extracellular to the intracellular fluid	1. Metabolic or respiratory alkalosis 2. Hereditary periodic paralysis
Elevated concentration Reduced ability to excrete the daily intake	1. Kidney insufficiency (e.g. acute kidney failure)
Hormonal disturbances	1. Insulin. Poorly supervised diabetes mellitus, particularly during the development of a diabetic coma 2. Corticosteroids. Insufficiency of the adrenal cortex
Increased fragility of the leucocytes	1. Leukaemia. Increased plasma $[K^+]$, vanishes if the separation of the plasma from the cells is conducted with sufficient care
Artefacts	1. If the plasma or serum remains in contact with the blood cells for long 2. Haemolysis $[K^+]$

8. Truniger, B., Störungen des Elektrolythaushaltes—Messgrössen, Nomenklatur und Störfaktoren, in F. W. Ahnefeld, H. Bergmann, C. Burri, W. Dick, M. Halmagyi, and E. Rügheimer, *Wasser-, Elektrolyt- und Säuren-Basen-Haushalt*, Springer, Berlin, 1977, p. 38.

3.3.4. CHLORIDE

3.3.4.1. Introduction

Chloride is the anion of highest concentration in the plasma and in the extracellular fluid. The chloride content of an adult (70 kg) amounts to about 2000 mmol. About 12% of this corresponds to the intracellular portion and 88% to the extracellular portion, of which about 30% is deposited in the bones. The chloride intake is dependent on the diet and is 100–260 mmol per 24 h. Of this, about 80–95% (120–240 mmol per 24 h) is excreted through the kidneys. Together with sodium, chloride is responsible for maintaining the normal osmolality in the extracellular space. In addition, it participates in the regulation of the acid–base balance and the blood pH. Chloride and bicarbonate often act in a complementary manner in establishing electroneutrality with the cations. There may be an exchange of chloride for bicarbonate ions between the erythrocytes and the plasma, and this is known as 'chloride shift'.

3.3.4.2. Choice of method

The most reliable results are obtained with the physical methods of determination using the silver electrode (coulometry)[1] or the silver chloride electrode (potentiometry).[2] However, these require the purchase of special instruments. Colorimetric methods[3] are also employed; these are based on the following principle. On adding chloride-containing analytical material to an aqueous solution of mercury(II) thiocyanate, mercury(II) chloride is formed and thiocyanate ions are liberated. These form a complex with iron(III) ions, and this may be quantified spectrophotometrically:[4]

$$2Cl^- + Hg(SCN)_2 \rightleftharpoons HgCl_2 + 2SCN^-$$
$$SCN^- + Fe^{3+} \rightleftharpoons Fe(SCN)^{2+}$$
$$\text{red colour}$$

This method may be used in automated analysers under precise standard conditions,[5] but in our experience it is not suitable for manual operation. Mercurimetric titration[6,7] has proved particularly simple and dependable.

3.3.4.2.1. Mercurimetric titration

3.3.4.2.1.1. Principle[6]

The chloride ions are titrated with mercury(II) nitrate in acidic solution, which produces undissociated mercury(II) chloride. Diphenylcarbazone may be used as indicator; this gives a violet colour in the presence of excess mercury(II) ions. No deprotcination is necessary.

3.3.4.2.1.2. Reagents

Mercury(II) nitrate solution, 5 mmol l⁻¹: 1.083 g of red mercury(II) oxide (HgO) is dissolved in 11.0 ml of concentrated nitric acid and diluted to 1000 ml with DM water. The solution may be kept indefinitely.

Diphenylcarbazone solution, 0.02 mol l⁻¹: 500 mg of diphenylcarbazone are dissolved in 96% ethanol and diluted to 100 ml with further 96% ethanol. The solution should be kept in a refrigerator.

Chloride standard, 0.1 mol l⁻¹: 7.455 g of potassium chloride are dissolved in DM water and made up to 1000 ml with further DM water. Chloroform (1 ml) is added. This solution keeps indefinitely if frozen.

3.3.4.2.1.3. Procedure

	Sample (S)	Standard (ST)
DM water, ml	1.0	1.0
Serum, ml	0.1	—
Standard, ml	—	0.1
Diphenylcarbazone solution	5 drops	5 drops

Titrate the sample and standard to the same end-point (change to violet).

3.3.4.2.1.4. Calculation

$$\text{Concentration} = \frac{\text{ml (S)} \cdot 100}{\text{ml (ST)}} \text{ mmol } l^{-1}$$

3.3.4.3. Reference values

These are given in Table 42.

3.3.4.4. Comments on the method

1. The tendency to overtitrate in the mercurimetric analysis may be overcome by titrating both sample and standard to the same end-point.

2. The determination may be carried out with or without a preliminary deproteination. Following deproteination, the end-point colour is blue; in the presence of proteins it is violet. The results are identical.

3. If a microtitrator is available, all the volumes may be reduced to a fifth; 10–20 µl of serum or plasma are then sufficient for the determination.

4. The determination should be made within 2 h of collecting the blood. If this is not possible, then the plasma should be separated from the erythrocytes as quickly as possible.

Table 42. Behaviour of chloride concentration in the serum of sucklings and children of various ages, and of adults (mmol l^{-1}).[8]

Age group	Number	Scatter	
Up to 1st month	40	94–109	103.0 ± 3.10
1st–3rd month	20	97–109	103.7 ± 3.13
3rd–6th month	21	97–108	104.1 ± 2.40
6th month–1 year	35	100–108	103.7 ± 2.41
1–3 years	71	97–109	103.4 ± 2.37
3–6 years	102	97–110	103.4 ± 2.54
6–10 years	117	96–109	102.6 ± 2.49
10–15 years	109	92–107	101.8 ± 2.67
Adults	65	97–107	103.7 ± 1.90
Adult men	31	97–107	103.5 ± 1.67
Adult women	34	100–107	104.0 ± 1.75
Adults (personal results)	314	97–108	95% interval

3.3.4.5. Diagnostic significance

Changes in the chloride concentration in the plasma often run parallel to changes in the sodium concentration since they are subject to the same regulation (p. 262). In a few cases this is not so, however. Owing to the different electrolytic composition of the gastric juices, chronic loss of gastric fluid (e.g. in pyloric stenosis) can lead to hypochloraemia which is much larger in relation to the corresponding sodium loss. Such patients may also exhibit dehydration and metabolic hypochloraemic alkalosis. Increased chloride concentrations not paralleled by increased sodium concentrations arc mainly found in the following changes: disturbances of the acid–base balance, such as metabolic acidosis produced by chronic kidney insufficiency (Lightwood–Albright syndrome with loss of bicarbonate), treatment with carboanhydride inhibitor, uretersigmoidostomy (re-absorption of chloride from the gut) or in a respiratory alkalosis occasioned by immoderate, assisted breathing. Also, one must always remember the possibility of an excessive dosing with physiological saline solution (contains 154 mmol l^{-1}).

References

1. Cotlove, E., Trantham, H. V., and Bowman, R. L., An instrument and method for rapid, accurate and sensitive titration of chloride in biologic samples, *J. Lab. clin. Med.*, **51**, 461 (1958).
2. Seligson, D., McCormick, G. J., and Sleeman, A., Electrometric method for the determination of chloride in serum and other biologic fluids, *Clin. Chem.*, **4**, 159 (1958).
3. Kim, E. K., Waddell, L. D., and Logan, J. E., Observation on diagnostic kits for the determination of chloride, *Clin. Biochem.*, **5**, 214 (1972).
4. Schoenfeld, R. G. and Lewellen, C. J., A colorimetric method for determination of serum chloride, *Clin. Chem.*, **10**, 533 (1964).

5. Küffer, H., Richterich, R., Kraft, R., Peheim, E., and Colombo, J. P., Die Bestimmung des Chlorids im Plasma und Serum (Quecksilber-2-Thiocyanatmethode mit dem Greiner Electronic Selective Analyzer GSA II), *Z. klin. Chem. klin. Biochem.*, **13**, 203 (1975).
6. Schales, O., Chloride, *Stand. Meth. clin. Chem.*, **1**, 37 (1953).
7. Schales, O. and Schales, S. S., A simple and accurate method for the determination of chloride in biological fluids, *J. biol. Chem.*, **140**, 879 (1941).
8. Liappis, N., Die Natrium-Kalium- und Chlorid-Konzentration im Serum von Säuglingen, Kindern und Erwachsenen, *Mschr. Kinderheilk.*, **120**, 138 (1972).

3.4. OSMOLALITY

3.4.1. INTRODUCTION

Besides the concentrations of constituent ions and substances, the general, so-called 'colligative', or concentrative properties of the body fluids are also frequently of interest. These properties include, e.g. the osmolality, the specific weight, the electrical conductivity, the refractivity, the viscosity, and the surface tension. These properties result from the combined action of the many different kinds of particles present and their concentrations. The osmolality has proved to be a particularly valuable parameter because it is almost wholly dependent on the number of the dissolved particles and is independent of their nature and their degree of dissociation. If two aqueous solutions are separated by a membrane permeable to water, then, provided the chemical activity of the water is the same on both sides of the membrane, an osmotic equilibrium is established. For an 'ideal' solution, the chemical activity of the dissolved particles is a simple function of their concentration and depends only upon the relative proportions of the dissolved particles and the solvent. The chemical activity of the water can be measured by the depression of the freezing point. For dilute physiological fluids, for which the properties are close to 'ideal', the depression of the freezing point beneath that of water is directly proportional to the number of moles of the dissolved particles ('osmole') per unit mass of water or volume of solvent. However, at higher concentrations the osmotic activity falls off, so that the osmolality can only be calculated with the aid of so-called osmotic coefficients (ϕ) which are dependent on the concentration; thus, for univalent ions at their customary plasma concentrations, ϕ lies between 0.91 and 0.93. In contrast with the electrolyte balance, conservation of osmolality involves not just the electrically charged particles but all the particles.

The biological significance of the osmolality stems from the fact that a large part of the work performed by the cells and organs is 'osmotic' work, i.e. the transport of osmotically active particles serving to maintain a constant osmolality. This osmotic work is primarily dependent on the ratio of the osmolalities of the two compartments concerned in a transport of particles. By analogy with the concept of molarity, the *osmolarity* corresponds to the concentration of osmotically active particles in 1 litre of body fluid, the *osmolality* to the concentration of such particles in 1 kg of solvent or water. The conversion from osmolarity to osmolality is analogous to that from molarity to molality. With today's modern osmometers the osmolality is measured directly.

Molarity. The concentration of a substance in moles per litre of the investigational material. All determinations performed directly on native material, i.e. without prior deproteination, yield results corresponding in principle to a molarity.

Molality. The concentration of a substance in moles per kilogram of solvent, e.g. water. If a sample is first deproteinated and the determination is carried out on the aqueous supernatant liquid then the result obtained is in the sense of a molality, i.e. in moles per kilogram of solvent or supernatant.

A series of approximations are available for converting molarities to molalities and *vice versa*. For example, if the protein concentration is known, then the concentration of plasma water may be estimated by the following equation:

$$\text{Plasma water (g l}^{-1}) = 984 - (0.718 \times \text{protein in g l}^{-1})$$

The difference between molarity and molality in plasma is largely a consequence of the volume-displacement effect of proteins and lipids. Since the small, osmotically active molecules are only dissolved in the plasma

Table 43. Molarity, molality, and absolute and relative osmolality of a normal plasma (specific gravity 1.027) (from ref. 1).

Component	Molarity, mmol l^{-1}	Molality, mmol kg^{-1} water	Osmolality, mosm kg^{-1} water	Osmolality, %
Sodium	142.0	151.6	139.0	48.2
Potassium	5.0	5.3	4.9	1.7
Calcium (ionic)	5.0	5.3	2.4	0.7
Magnesium (ionic)	2.0	2.1	1.0	0.3
Total cations	154.0	164.3	147.3	50.9
Chloride	103.0	109.9	101.8	34.6
Bicarbonate	28.0	29.8	27.4	9.2
Phosphate (ionic)	2.0	2.1	1.1	0.4
Sulphate (ionic)	1.0	1.1	0.5	0.2
Proteinate (ionic)	16.0	17.1	1.0	0.3
Acid residue (ionic)	4.0	4.3	3.9	1.2
Total anions	154.0	164.3	135.7	45.9
Urea-N	5.35[a]	5.71[a]	5.3	1.8
Glucose	3.89[b]	4.15[b]	4.1	1.4
Total			292.4	100.0

[a]Corresponding to a urea-N of 15.0 or 16.0 mg per 100 ml.
[b]Corresponding to a glucose concentration of 70.0 or 74.7 mg per 100 ml.

water, the water volume for a protein concentration of 70 g l^{-1} is about 5–6% smaller than the total volume, i.e. the molality is about 5–6% larger than the molarity. The extent of this difference for the plasma electrolytes, according to whether the result is referred to 1 litre of plasma or 1 kg of plasma water, can be seen from Table 43.

The osmometers used in clinical–chemical laboratories today, operating on the principle of depression of the freezing point, measure the osmolality directly. Consequently, in the following, the reference values are stated in mosm kg^{-1} water. In cases of hypernatraemia and hyperlipidaemia the plasma osmolality is preferred to the osmolarity; and in urine especially, which contains neither proteins nor lipids, it is in any case the osmolality which is measured. Thus, for the direct comparison of urine and plasma there is no choice but osmolality. The normal serum osmolality amounts to about 285 ± 8 mosm kg^{-1} water and is dependent on the number of dissolved particles. There is no significant difference between plasma and serum osmolality since fibrinogen is not osmotically active. Sodium and sodium salts, with about 275 mosm kg^{-1} water, make up the larger part of the total osmolality. Glucose, potassium, and urea contribute only about 10 mosm kg^{-1} water (Table 43). As a general rule it may be assumed that if sodium, chloride, bicarbonate, glucose, and urea lie within the reference interval, then the osmolality per kg water is normal. Since proteins have a high molecular weight, their contribution—2–4 mosm kg^{-1} water—is even smaller, i.e. the serum osmolality would still remain in the normal range if all the proteins were removed from the serum. For the same reason, lipids, at their normal concentration, contribute almost nothing to the total osmolality. On the other hand, hyperlipidaemia can interfere in the measurement of the depression of the freezing point owing to the high viscosity and the results may be too high.[2]

Of the exogenous substances, alcohol is by far the most important in contributing to an elevation of the osmolality. There is a linear relationship between the blood-alcohol concentration and osmolality.[3] For 100 mg of alcohol to 100 ml of blood, the osmolality is raised by about 22 mosm kg^{-1} water.[4]

Apart from measuring the depression of the freezing point, the osmolality may be calculated. Of 13 different methods of calculation the following equation has given the best agreement with measurements of the freezing-point depression:[5]

$$\text{Osmolality} = 1.86 \text{ Na (mmol l}^{-1}) + \text{glucose (mmol l}^{-1})$$
$$+ \text{ urea-N (mmol l}^{-1}) + 9$$

Boyd et al.[6] were able to show that the difference between the calculated and measured osmolality can be greater under various conditions of stress. The increase is obviously created by unknown, abnormal metabolites, and this is prognostically unfavourable for the patient. As yet it has not been possible to identify these unknown substances.[7]

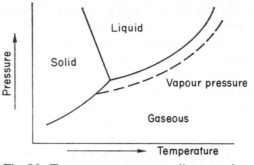

Fig. 86. Temperature–pressure diagram of a
solution of definite concentration.

3.4.2. CHOICE OF METHOD

Figure 86 is a schematic temperature–pressure diagram for different
states of matter of a solution. An increase in the osmolality, i.e. an increase
in the concentration of osmotically active particles, gives rise to the
following changes: 1, the vapour pressure falls; 2, the freezing point is
lowered; and 3, the boiling point rises. In principle, any of these properties
may be used for measuring the osmolality, but in practice the measurement
of the depression of the freezing point (cryoscopy) has gained almost
exclusive acceptance (the osmolality is occasionally determined by
measuring the fall in the vapour pressure). In each case, the instrument used
(osmometer) is calibrated with solutions of known osmolality.

3.4.2.1. Principle of cryoscopy

Formerly, the depression of the freezing point was measured according to
Beckmann's principle, by slowly cooling a solution, the addition of seed
crystals, and measurement of the freezing point with a thermometer. Today

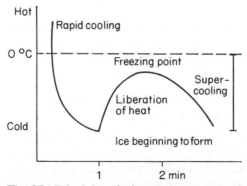

Fig. 87. Principle of the measurement of
the osmolality of a solution by supercooling.

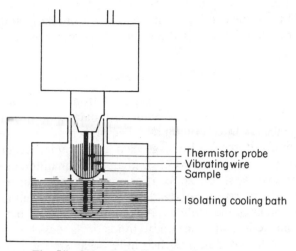

Fig. 88. Schematic diagram of a cryoscope.

the measurements are usually carried out by the method of initial supercooling.[8] The principle is illustrated in Fig. 87, and a cryoscope in Fig. 88. The sample is subjected to rapid, thermoelectric cooling (Peltier effect) in a cooling bath at about −6 °C. Crystal formation is prevented by the vibration of a metal wire. On the formation of ice, heat is slowly liberated, leading to an increase in temperature, resulting finally in a temperature plateau. The temperature is measured electronically by means of a thermistor and the result is obtained directly in mosm kg^{-1} water. The difference between 0 °C and the temperature of the plateau is used as a measure for the depression of the freezing point. An ideal, aqueous solution with a concentration of 1 mol kg^{-1} or 1 osmol kg^{-1} solvent, produces a freezing-point depression of 1.858 °C. For plasma, the freezing-point depression lies around 0.56 °C and hence corresponds to an osmolality of about 300 mosm kg^{-1} water.

3.4.2.2. Reagents

Standard solutions. As a general principle, measurement of the osmolality is carried out following preliminary calibration of the instrument using standard solutions with concentrations lying roughly within the expected range.[8] The precise composition of calibration solutions can be found from tables.[8] For practical purposes, the following four solutions, prepared by weighing out recrystallized and dried sodium chloride, should suffice:

g per 100 ml	mosm kg^{-1} H$_2$O	g per 100 ml	mosm kg^{-1} H$_2$O
0.870	277	0.921	292
0.898	286	0.960	305

There should be a linear relationship between osmolality and depression of the freezing point, at least over the range of these concentrations.

3.4.2.3. Reference values

The osmolality is fundamentally different from all the other parameters concerned with the electrolyte and acid–base balance in that it is identical in all the body cells and true body fluids.[8] The preservation of constant osmotic pressure throughout the entire organism is more important (and this is phylogenetically readily explicable) than the maintenance of specific electrolyte or hydrogen ion concentrations. Actually, the preservation of constant osmotic pressure can lead to passive (Donnan equilibrium) or active displacement of electrolytes. As a consequence of the equal osmotic pressure of all cells and true body fluids it is possible to draw direct conclusions as to the condition of the entire organism from investigations on the plasma. The molarity, molality, and osmolality of a typical, normal plasma are listed in Table 43. This clearly shows that the sodium and chloride concentration together make up the greater part of the effective osmolality. The fraction corresponding to urea-N and glucose is small in healthy persons but these two components may rise to considerable values in cases of uraemia or hyperosmolar coma.

Table 44. Serum and urine osmolalities (mosm kg^{-1} water) for new-born babies and adults.

Age group	Sex	Number	$\bar{x} \pm s$	Reference interval	References
Plasma					
New-born[a]					
1st day	m, f	23	280 ± 8.4	266–295	10
2nd day	m, f	19	277 ± 8.8	265–293	10
3rd day	m, f	19	273 ± 9.3	258–289	10
4th day	m, f	17	277 ± 9.3	262–292	10
5th day	m, f	18	280 ± 8.5	262–297	10
Adults	m, f	50	289 ± 4	281–297	8
Urine					
1st day	m, f	18		9–418[b]	10
2nd day	m, f	18		100–352	10
3rd day	m, f	25		58–393	10
4th day	m, f	27		49–363	10
5th day	m, f	28		37–216	10
6th day	m, f	11		39–264	10
Adults	m, f			50–1200[b]	
	m			767–1628[b]	11
	f			433–1146[b]	11

[a]Diet of cow's milk.
[b]24-hour urine.

For healthy subjects, the scatter of the normal values is very slight. The lowest value observed so far was for a patient suffering from an insufficiency of the adrenal cortex and measured 260 mosm kg^{-1}.[9] The highest published value, 457 mosm kg^{-1},[9] was observed with a patient suffering from very severe dehydration. The values for erythrocytes and whole blood are identical with those for plasma. The 'true' body fluids, for which the universal principle of equal osmotic pressure holds, are ascites, cerebrospinal fluid, oedoema, and pericardial and pleural effusions.[9] These fluids are isotonic with plasma, regardless of their electrolyte compositions. There is a series of secretions which do not belong to the group of 'true' body fluids; their osmolality may be hypotonic (sweat, saliva, amniotic fluid), isotonic (hepatic bile), or hypertonic (cystic bile) with respect to the plasma. Urine may be hypotonic, isotonic, or hypertonic.

Since urine samples are frequently investigated for osmolality in the clinical–chemical laboratory, a few standard values for the osmolality in urine are given in Table 44. The excretion of osmotically active particles is strongly influenced by the diet and the functioning of the kidneys.

3.4.2.4. Comments on the method

1. Earlier publications reported a higher osmolality within the cells in comparison to plasma but these findings are attributable to technical artefacts.[7] The major cause is autolysis prior to measurement; this can result in apparently higher concentrations due to the liberation of osmotically active particles.

2. The samples should not be diluted for osmolality measurement, since the osmotic activity (ϕ) increases with increasing dilution and the results appear to be higher.

3.4.2.5. Diagnostic significance

High values of the osmolality:[4,12,13] 1, hypernatraemia (*q.v.*); 2, hyperglycaemic (e.g. diabetes; hyperosmolar, hyperglycaemic non-ketoacidotic coma); 3, uraemia; 4, presence of unknown, osmotically active particles; 5, combinations of points 1–4.

Low values. A hypo-osmolality is usually an expression of a hypo-natraemia since sodium makes up the largest part of the osmotically active substances (cf. 'Changes of the sodium concentration', p. 262).

References

1. Chughtai, M. A. and Hendry, E. B., Serum electrolyte, urea and osmolality in cases of chloride depletion, *Clin. Biochem.*, **1**, 91 (1967).
2. Barlow, W. K., Volatiles and osmometry, *Clin. Chem.*, **22**, 1231 (1976).
3. Champion, H. R., Baker, S. P., Benner, C., Fisher, R., Chaplan, Y. H., Long,

W. B., Cowley, R. A., and Gill, W., Alcohol intoxication and serum osmolality, *Lancet,* **i**, 1402 (1975).

4. Loeb, J. N., The hyperosmolar state, *New Engl. J. Med.,* **290**, 1184 (1974).
5. Dorwart, W. V. and Chalmers, L., Comparison of methods for calculating serum osmolality from chemical concentrations and the prognostic value of such calculations, *Clin. Chem.,* **21**, 190 (1975).
6. Boyd, D., Addis, H., Chilimindris, C., Lowe, R., Folk, F., and Baker,, R., Utilization of osmometry in critically ill surgical patients, *Archs Surg., Chicago,* **102**, 363 (1971).
7. Burg, B. von and Lemann, D., Die Serumosmolalität bei Intensivpflegepatienten in Abhängigkeit von osmotischer Belastung und Nierenfunktion. Eine prospektive Studie, *Diss. Universität Bern,* 1975.
8. Hendry, E. B., Osmolarity of human serum and of chemical solutions of biological importance, *Clin. Chem.,* **7**, 156 (1961).
9. Hendry, E. B., The osmotic pressure and chemical composition of human body fluid, *Clin. Chem.,* **8**, 246 (1962).
10. Feldman, W. and Drummond, K. N., Serum and urine osmolality in normal full-term infants, *Can. med. Ass. J.,* **101**, 595 (1969).
11. Weissman, N. and Pileggi, V. J., Inorganic ions, in Henry, Cannon and Winkelman, *Clinical chemistry,* Harper & Row, New York, 1974, p. 739.
12. Warhol, R. M., Eichenholz, A., and Mulhausen, R. O., Osmolality, *Archs intern. Med.,* **116**, 743 (1965).
13. Eklund, J., Granberg, P. O., and Hallberg, D., Clinical aspects of body fluid osmolality, *Nutr. Metab.,* **14**, suppl., 74 (1972).

3.5. REGULATION OF THE ACID–BASE STATUS

3.5.1. BASIC CONCEPTS

3.5.1.1. Acids and bases

Arrhenius' 'neutralization theory' of 1887[1] was the first attempt to define acids and bases in terms of their chemical composition.

Acids are hydrogen-containing compounds which, on dissolution in water, ionize and afford hydrogen ions.

Bases are hydroxyl-containing compounds which furnish hydroxyl ions (OH^-) in aqueous solution.

According to these definitions, acids and bases react in aqueous solution with the formation of water and a salt (e.g. $NaOH + HCl \rightleftharpoons Na^+ + OH^- + Cl^- \rightleftharpoons H_2O + NaCl$), hence the term 'neutralization theory'. The weaknesses of this theory as applied to biological systems is not so much that it is only valid for aqueous solutions as in the unsatisfactory definition of bases. Numerous compounds, e.g. ammonia, behave as bases but do not contain hydroxyl ions.

Expressions such as 'fixed acids' and 'fixed bases', deriving from this period and still used in many clinics today, are relatively meaningless and make no small contribution to the continuing confusion of terminology in the literature dealing with acids and bases. The term 'fixed bases' should be replaced by 'cations' and the term 'fixed acids' should be wholly avoided on account of its ambiguity.

In 1923 Bronsted[2] developed his 'proton theory' of acids.

Acids are compounds which furnish hydrogen ions or protons (H^+) (proton donors).

Bases are compounds which accept hydrogen ions or protons (proton acceptors).

According to this definition, acids consist of a hydrogen ion or proton (H^+) and an acid radical (A^-). It should be observed that according to this currently held theory, metal hydroxides, such as sodium and potassium hydroxide, are not bases. However, in aqueous solution they behave as bases since hydroxyl ions are formed on ionization (e.g. $NaOH + H_2O \rightleftharpoons Na^+ + OH^- + H_2O$); these hydroxyl ions are proton acceptors and hence bases ($OH^- + H^+ \rightleftharpoons H_2O$). Both acids and bases may be classified into molecular, anionic, and cationic types. A few examples

Table 45. Different types of acids and bases.

Type	Acid	Base
Molecular	$H_2SO_4 \rightarrow HSO_4^- + H^+$ $R—COOH \rightarrow R—COO' + H^+$	$HN_3 + H^+ \rightarrow NH_4^+$ $R—NH_2 + H^+ \rightarrow R—NH_3$
Anionic	$HSO_4^- \rightarrow SO_4^{2-} + H^+$ $HCO_3^- \rightarrow CO_3^{2-} + H^+$	$R—COO^- + H^+ \rightarrow R—COOH$ $OH^- + H^+ \rightarrow H_2O$
Cationic	$H_3O^+ \rightarrow H_2O + H^+$ $NH_4^+ \rightarrow NH_3 + H^+$	$^+H_3N.CH_2.CH_2NH_2 + H^+$ $\rightarrow {}^+H_3N.CH_2.CH_2.NH_3^+$

are given in Table 45, which also shows that there are compounds, e.g. water and proteins ($RNH_2^+ + H^+ \leftrightarrows RNH_3^+$ or $RCOO^- + H^+ \leftrightarrows RCOOH$), which can function both as proton donors and proton acceptors. Such compounds are said to be amphoteric and in some cases are called 'zwitterions'. According to the Bronsted concept, sodium and potassium are not bases but cations, and chloride and sulphate are not acids but bases.

From the physiological point of view it is necessary to distinguish two categories of acids and bases.

Volatile acids and bases. These are acids and bases which may occur in the gaseous state at room temperature. The concentration of these acids and bases is regulated almost exclusively by diffusion processes controlled by the lungs. Pulmonary diseases lead to disturbances in the metabolism of these volatile acids and bases. Examples are carbonic acid and ammonia.

Non-volatile acids and bases are those which do not occur in the gaseous state at room temperature. Their concentration cannot therefore be regulated by the lungs and control must be effected exclusively by non-respiratory organs such as the kidneys, the gut, and the skin. Diseases of the kidneys can lead to disturbances in the metabolism of these acids and bases (e.g. in renal acidosis). Sulphuric and phosphoric acids are examples of such acids. In the more recent literature, non-volatile acids and bases are occasionally referred to as 'fixed acids and bases'. However, this terminology ought no longer to be used since it is misleading.

Those disturbances in the acid–base system which involve volatile acids and bases are appropriately termed *respiratory disturbances*. If the organism suffers an accumulation or abnormal loss of non-respiratory acids and bases, the disturbance is said to be a *non-respiratory* or *metabolic disorder*.

3.5.1.2. Acidosis and alkalosis

The preservation of a near-constant hydrogen ion concentration is of greatest importance for the vitality of the cells. In the course of evolution the range of optimal hydrogen ion concentrations for biochemical processes has been progressively narrowed. For human beings the intracellular optimum lies around 80 nmol l^{-1}. Since in the higher organisms there is a much greater danger of intoxication by H^+ than by OH^-, it would appear significant that the reference interval for the extracellular space and the plasma lies lower than that for the intracellular space. In blood, values between 35 and 42 nmol l^{-1} pass as normal.

In inanimate nature, the hydrogen ion concentration varies between 0 and 10^{-14} mol l^{-1}. Since such figures are inconvenient to work with, and numerous physical and chemical processes are directly related not to the absolute value of H^+ but to its logarithm, Sørensen[3] introduced the concept of pH. In the following discussion, it should be remembered that the blood pH varies within very narrow limits, namely between 7.38 and 7.45, or 35 and 42 nmol l^{-1}. Although there is no strict correspondence between the pH in the extracellular and intracellular compartments, nevertheless the determination of the blood pH provides some useful information on the intracellular situation. For an exact analysis of the acid–base conditions in sick persons, the pH determination is therefore indispensable. The clinician characterizes the pH condition by means of the terms acidosis and alkalosis.

Acidosis describes the condition in which the blood pH is lower than the normal range (an increase in the H^+ concentration); *alkalosis* is the opposite condition in which the blood pH is higher than the normal range (a decrease in the H^+ and a consequent increase in the OH^- concentration).

The expressions ought only to be used if the pH has actually been measured. Respiratory and metabolic alkaloses and acidoses are distinguished according to the origin of the disorder (p. 310).

3.5.1.3. Dissociation of acids and bases

According to the Arrhenius theory,[1] electrolytes dissociate in aqueous solution to produce ions (e.g. $NaCl \leftrightharpoons Na^+ + Cl^-$). However, in 1918 Bjerrum[4] showed that a fundamental distinction must be made between strongly and weakly ionizing or dissociating electrolytes.

Strongly dissociating electrolytes are molecules which ionize almost completely in aqueous solution (e.g. $H_2SO_4 \rightleftharpoons HSO_4^- + H^+ \rightleftharpoons SO_4^{2-} + 2H^+$). Weakly dissociating electrolytes do not dissociate completely into their ions in aqueous solution (e.g. $H_2CO_3 \leftrightharpoons HCO_3^- + H^+ \leftrightharpoons CO_3^{2-} + 2H^+$). Clearly, the difference between the two groups is of a quantitative not qualitative, nature, and it must be susceptible to numerical characterization. In aqueous solution an acid dissociates according to the following reaction:

$$HA \rightarrow H^+ + A^-$$

Then, according to the Law of Mass Action, the dissociation constant is given by

$$K = \frac{[H^+][A^-]}{[HA]}$$

i.e. there is an equilibrium between the concentration of ionized and non-ionized molecules. For highly ionized compounds K is large, for weakly dissociated compounds it is small. However, the numerical value of K may be very large or very small and it is therefore an impracticable expression. Since $[H^+]$ occurs in the dissociation constant equation and, nowadays, this quantity is almost always given as a pH value, an analogous expression for K, the 'pK' has been derived. By taking logarithms, the previous equation transforms to

$$\log K = \log \frac{[H^+][A^-]}{[HA]}$$

or

$$\log K = \log[H^+] + \log \frac{[A^-]}{[HA]}$$

This may be written as

$$-\log[H^+] = -\log K + \log \frac{[A^-]}{[HA]}$$

As we know, $-\log[H^+]$ is identical with the pH. By making an analogous substitution of pK for $-\log K$ (pK is the negative logarithm to the base 10 of the dissociation constant), we obtain the usual form of the dissociation equation:

$$pH = pK + \log \frac{[A^-]}{[HA]}$$

In practical terms, this means that the dissociation constant pK is equal to the pH if half of the molecules are dissociated, for the ratio term, $\log[A^-]/[HA]$ is then log 1/1, i.e. 0, since log 1 = 0. Hence the pK is equal to the pH in that case. The higher the pK of an acid, the smaller is its degree of dissociation; the smaller the pK, the greater is the dissociation of a base. Thus, e.g. the pK of phosphoric acid is 2 ($H_3PO_4 \rightleftharpoons H_2PO_4^- + H^+$) or 7 ($H_2PO_4^- \rightleftharpoons HPO_4^{2-} + H^+$), that of acetic acid is 4.75 ($CH_3COOH \rightleftharpoons CH_3COO^- + H^+$), of ammonia 9.3 ($NH_4^+ \rightleftharpoons NH_3 + H^+$), and of water 15.7 ($H_2O \rightleftharpoons OH^- + H^+$).

3.5.1.4. Physiological buffer systems

Solutions which contain a weak acid or base in addition to its salt are called buffers. They are able to accommodate the addition of H^+ or OH^- without any marked change in the pH.

Theoretically, water has a pH of 7.0. If HCl is added to produce a solution containing 0.001 mol l^{-1} of HCl, the pH changes to 3.0. A mixture containing 0.1 mol l^{-1} of acetic acid and 0.1 mol l^{-1} of sodium acetate has a pH of 4.73. If hydrochloric acid is added to produce a final concentration of 0.001 mol l^{-1}, the pH only falls to 4.72. The mixture of acetic acid and sodium acetate is a buffer.

Both the intra- and extra-cellular compartments are equipped with buffer systems for preserving the pH. While the intracellular buffers have been little investigated, those in the intravascular compartment may be determined by titration. As shown in Table 46, about 80% of the buffer systems of the blood are located in the erythrocytes (predominantly haemoglobin buffer). The remainder are distributed among the plasma proteins (14%), the bicarbonate–CO_2 system (6%), and the phosphate system (1.5%).[5]

On consideration of Table 46, it seems difficult at first to understand why physiologists and clinicians should be concerned almost exclusively with the bicarbonate system. The answer is that the bicarbonate system is the only physiological buffer which contains a component, carbon dioxide, which is gaseous at room temperature and which can therefore be breathed out through the lungs $(H_2CO_3 \rightleftharpoons CO_2 + H_2O)$. Consequently, an analysis of this system allows a direct statement to be made on the respiratory condition of the acid–base balance. An increase or decrease in the CO_2 or of its measured value, the pCO_2, can occur only as a result of changes in the pulmonary function. The HCO_3^- concentration of the plasma can be regulated via the kidneys (p. 310).

The Henderson–Hasselbalch equation. As early as 1909, Henderson[6] recognized that the equation

$$K = \frac{[H^+][A^-]}{[HA]}$$

was also valid for physiological relationships and he used it to describe the

Table 46. Contribution of the individual buffer systems to the buffer capacity of whole blood (from ref. 5).

	Total buffer capacity, %
Blood	100
Cells	79
Plasma	21
Proteinate	13.6
Bicarbonate	6.1
Phosphate	1.5

behaviour of the bicarbonate buffer system:

$$K = \frac{[H^+][HCO_3^-]}{[H_2CO3]}$$

Shortly afterwards, Sørensen[3] introduced the concept of pH and it fell to Hasselbalch[7] to unite the two concepts and propose an equation which is known today as the Henderson–Hasselbalch equation:

$$pH = pK + \log \frac{[HCO_3^-]}{[H_2CO_3]}$$

where

pH = pH measured in arterial blood;
pK = dissociation constant of the bicarbonate system;
$[HCO_3^-]$ = 'true' bicarbonate concentration in mmol l^{-1}; and
$[H_2CO_3]$ = carbonic acid concentration in mmol l^{-1}.

Since carbonic acid (H_2CO_3) decomposes into H_2O and CO_2, and CO_2 is a gas, its concentration is given by the partial pressure of carbon dioxide

Table 47. Solution of the Henderson–Hasselbalch equation (two measured values are given, the third parameter is required to be found; either the bicarbonate concentration $[HCO_3^-]$ or the total CO_2, $\Sigma[CO_2]$, may be used as a measure of the bicarbonate concentration).

Sought	Known	Formula	Equation number
1. Measured value of bicarbonate: $[HCO_3^-]$ in mmol l^{-1} = bicarbonate			
Formula: $pH = pK + \log \dfrac{[HCO_3^-]}{\alpha\, pCO_2}$			
pH	pCO_2 $[HCO_3^-]$	$pH = pK + \log \dfrac{[HCO_3^-]}{\alpha\, pCO_2}$	1
pCO_2	pH $[HCO_3^-]$	$pCO_2 = \dfrac{[HCO_3^-]}{\alpha\, \text{antilog}\,(pH - pK)}$	2
$[HCO_3^-]$	pH pCO_2	$[HCO_3^-] = \alpha\, pCO_2[\text{antilog}(pH - pK)]$	3
2. Measured value of bicarbonate: $\Sigma[CO_2]$ in mmol l^{-1} = total carbon dioxide			
Formula: $pH = pK + \log \dfrac{\Sigma[CO_2] - (\alpha\, pCO_2)}{\alpha\, pCO_2}$			
pH	pCO_2 $\Sigma[CO_2]$	$pH = pK + \log \dfrac{\Sigma[CO_2] - (\alpha\, pCO_2)}{\alpha\, pCO_2}$	4
pCO_2	pH $\Sigma[CO_2]$	$pCO_2 = \dfrac{\Sigma[CO_2]}{[\text{antilog}\,(pH - pK + 1]\alpha}$	5
$\Sigma[CO_2]$	pH pCO_2	$\Sigma[CO_2] = \alpha\, pCO_2[\text{antilog}(pH - pK) + 1]$	6

(pCO₂) in mmHg multiplied by a proportionality constant (solubility coefficient). The equation can also be written as follows:

$$pH = pK + \log \frac{[HCO_3^-]}{pCO_2 \cdot \alpha}$$

Since, in addition to two constants (pK = 6.100; α = 0.0306 mmol l⁻¹mmHg⁻¹), this equation contains three variables (pH, [HCO₃⁻] and pCO₂), at least two measured values are required for the complete characterization of the acid–base status. The third value can then be calculated arithmetically. It is immaterial which two of the three unknowns are determined in the laboratory. Table 47 shows the nature of the calculation. In this connection it should always be borne in mind that neither α nor pK is a constant in the literal sense, but each is dependent on various factors such as temperature, pH, and lipid content of the plasma.[8] In plasma, under normal conditions and at 37 °C, α = 0.0306 and pK = 6.10.[9]

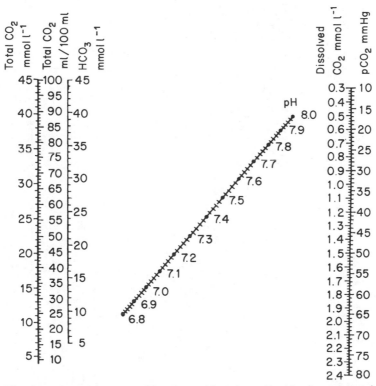

Fig. 89. Nomogram coupling the total carbon dioxide (in mmol l⁻¹ or ml per 100 ml), bicarbonate (in mmol l⁻¹), pH, the dissolved carbon dioxide (in mmol l⁻¹) and the partial pressure of carbon dioxide (in mmHg) (from ref. 10).

284

Since the calculation procedure is fairly time consuming, the unknown value is usually determined from tables or nomograms.

Line nomograms for expressing the Henderson–Hasselbalch equation. For calculating the unknowns in the Henderson–Hasselbalch equation or for performing conversions, line nomograms are particularly convenient. Figure 89 shows McLean's nomogram,[10] which is, in fact, a graphical expression of equation 1 in Table 47. One of the two outside lines expresses total CO_2 in mmol l^{-1} and the other pCO_2 in mmHg. A glance at the equation shows that these two parameters are in inverse proportion and the relationship

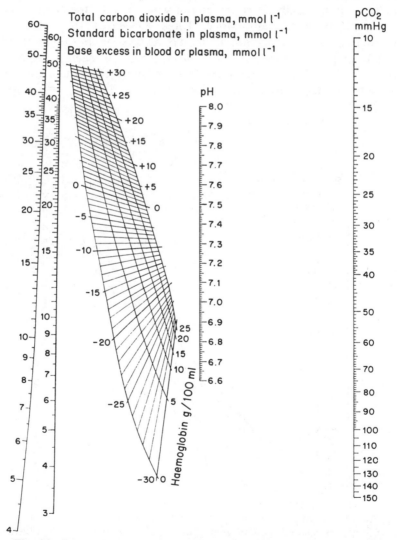

Fig. 90. Blood acid–base line nomogram, blood temperature 37 °C
(from ref. 9).

between them is linear. Substituting values for pCO_2 and total CO_2 permits the construction of a pH scale. The scale of the total CO_2 in ml per 100 ml is calculated directly from the values in mmol l^{-1} by means of the equation

$$\text{ml per 100 ml } CO_2 = \frac{\text{mmol } l^{-1}\ CO_2 \times 22.26}{10}$$

Similarly, the dissolved $[CO_2]$ is calculated from the pCO_2:

$$[CO_2]p \text{ (mmol } l^{-1}) = 0.0306\ pCO_2 \text{ (mmHg)}$$

If two of the values expressed on the nomogram are known, then the other value can be read off. The two known values are joined by a straight line and the intercept of this line with the third scale gives the desired value. For example, if $pCO_2 = 65$ mmHg and total $CO_2 = 27$ mmol l^{-1}, then the total CO_2 comes to 60 ml per 100 ml, the bicarbonate to 24.5 mmol l^{-1}, the pH to 7.21, and the dissolved CO_2 to 1.94 mmol l^{-1}.

More accurate results are obtained with the Siggaard–Andersen line nomogram[9] shown in Fig. 90. This is based on equation 4 of Table 47, and uses a linear pH scale and a logarithmic pCO_2 scale. These are the only two scales which are in an exact mathematical relationship to each other. The other three scales shown here have been corrected on the basis of empirical data. The base excess is discussed later (p. 308). The procedure for using this nomogram is analogous to that for the McClean nomogram.[10]

Cartesian nomogram for expressing the Henderson–Hasselbalch equation. Although line nomograms are convenient for making calculations, they give a poor spatial representation of the actual relationships which show up better on Cartesian nomograms. Of the many that have been suggested, two will be discussed in more detail.

Perhaps the best known is the simple plot of pH against bicarbonate, introduced by Van Slyke[11] and publicized mainly in the monograph by Davenport.[12] This is shown in Fig. 91. If the pH is plotted on the abscissa and the bicarbonate concentration on the ordinate, using linear scales, then

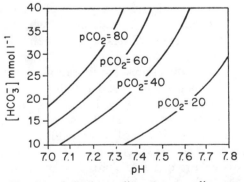

Fig. 91. Acid–base diagram according to
van Slyke.[11]

286

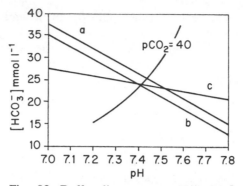

Fig. 92. Buffer lines on a pH/(HCO₃⁻]
plot. (a) 'True' plasma from reduced
blood; (b) 'true' plasma from oxygenated
blood; (c) separated plasma from
oxygenated blood.

the different pCO_2 values form isobars. If the pH and bicarbonate value of
a particular serum or blood sample are determined, one obtains the 'buffer
curve' for the sample; this usually approximates to a straight line (Fig. 92).
We shall discuss the significance of this buffer curve later.

Astrup[13] used a pH/log pCO_2 nomogram to represent the acid–base
relationships. As shown in Appendix 5, the abscissa is a linear pH scale
and the ordinate a logarithmic pCO_2 scale. The bicarbonate values lie on
straight lines falling from upper left to lower right. This diagram is
particularly useful in cases where the pCO_2 is determined by means of a
pH measurement following equilibration with known partial pressures of
CO_2.

Other nomograms are based on the $pCO_2/\Sigma[CO_2]$ system, the
pCO_2/bicarbonate system, the $[HCO_3^-]/pCO_2$ system, the log
$pCO_2/\log[HCO_3^-]$ system, the log $pCO_2/\log\Sigma[CO_2]$ system, the
pCO_2/buffer base system and the pCO_2/log standard bicarbonate system.

Buffer curves and the type of blood collection. The buffer curve is
obtained by measuring the pH and pCO_2 of a blood sample. The slope of
this buffer curve depends on the manner in which the investigational
material was collected. Figure 92 shows the behaviour of three blood
samples obtained in different ways. Curve a is obtained from the analysis
of reduced blood (blood collection—equilibration of the whole blood with
the desired pCO_2—anaerobic separation of erythrocytes and
plasma—analysis of the plasma). Curve b results from treating oxygenated
blood in exactly the same way. That the slope of the curves is largely
conditioned by the buffer action of haemoglobin may be seen from a
comparison of curves a and b with curve c. Curve c shows the behaviour of
plasma which has been brought into contact with various pCO_2 after prior
separation from the cells. From these investigations it may be concluded
that the bicarbonate values of the plasma agree with the buffer capacity of

the blood only if whole blood is equilibrated with oxygen prior to analysis and the separation of the plasma is carried out subsequently.

pCO₂ as a measure of respiratory disturbances. Carbon dioxide is produced as the final product of oxidative processes in the cells. It is bound peripherally by various buffers in the plasma and, above all, in the erythrocytes, and transported to the lungs by the venous blood. In the lungs, carbon dioxide (in part spontaneously, in part by the action of the enzyme carbonate anhydratase) passes out of the erythrocytes and into the plasma of the pulmonary capillaries. From thence it diffuses into the alveoli and the expired air. The partial pressure of CO_2 in the blood is the sole index for the diffusion pressure. When the diffusion process is disturbed (e.g. in chronic emphysema), not all the CO_2 passes to the alveolar air and there is an increase in the pCO_2 in the blood. Conversely, hyperventilation results in the expiration of abnormally large amounts of CO_2 and the pCO_2 in the blood falls. On the basis of these considerations the determination of the pCO_2 allows one to conclude directly whether or not there is a respiratory disturbance of the acid–base balance.

The converse conclusion is probably of greater practical importance: in the absence of pulmonary disease, the pCO_2 can be taken as normal and if this is the case, then the Henderson–Hasselbalch equation reads

$$pH = pK + \log \frac{[HCO_3^-]}{k\alpha}$$

or

$$pH = \log[HCO_3^-]k'$$

On this assumption, then, there is a direct proportionality between pH and bicarbonate concentration, i.e. the equation with three unknowns reduces to one of two unknowns. But only one measured value is necessary for the solution of such an equation—the pH or the bicarbonate concentration. This is the reason why formerly it was common practice only to carry out a bicarbonate concentration determination (as total CO_2) in order to characterize the acid–base status. It was assumed that the pulmonary function, and hence the pCO_2, was normal and thus the presence of an elevated or lowered pH could be concluded directly from the total pCO_2 result.

'True bicarbonate'. There is confusion over the meaning of the term 'true bicarbonate'. In the ideal case, it should represent a value which is independent of respiratory changes and reflects metabolic alterations only. The following quantities are used for this purpose:

(a) *Total CO_2 in mmol l^{-1}*: the CO_2 liberated from carbonic acid and bicarbonate in plasma or serum, using blood collected anaerobically.

(b) *Bicarbonate concentration in mmol l^{-1}*: the bicarbonate present in the plasma from blood collected anaerobically. This corresponds to the difference between the total CO_2 and the physically dissolved carbon

288

Table 48. Effect of oxygenation and pCO_2 on various methods for measuring the 'true bicarbonate' (from ref. 13).

	pCO_2		
	20 mmHg	40 mmHg	80 mmHg
Oxygenated blood			
Total CO_2	16.8	22.2	30.0
CO_2-binding capacity	18.0	22.2	27.5
Standard bicarbonate	21.0	21.0	21.0
Reduced blood			
Total CO_2	19.6	25.7	34.8
CO_2-binding capacity	21.0	25.7	31.9
Standard bicarbonate	21.0	21.0	21.0

dioxide. The value is always calculated, the basis being measurements with the van Slyke and Neill manometric methods.

(*c*) *CO_2 binding capacity in mmol l^{-1} (alkali reserve)*: the total CO_2 of plasma which has been separated from the cells at a natural pH and equilibrated with a pCO_2 of 40 mmHg.

(*d*) *Standard bicarbonate in mmol l^{-1}*:[9] the bicarbonate concentration in plasma from blood which has been equilibrated with a pCO_2 of 40 mmHg and with oxygen until the haemoglobin is fully saturated at 37 °C.

A simple but impressive experiment by Astrup[13] (Table 48) shows that numerically different values may be obtained as a measure of the 'true bicarbonate', according to the degree of oxygen saturation and the pCO_2 present. All of the methods give results in fair mutual agreement following oxygenation of the blood and using normal pCO_2, but they deviate considerably from one another if these preconditions are not fulfilled. Only the technique of Astrup gives values which are independent of pCO_2 and oxygenation. This carries substantial advantages, one being that it is also possible to use venous blood for the analysis.

The transatlantic acid–base debate. At one time a lively discussion of the acid–base status was provoked by an article in *The Lancet*[16] which attempted to bring order into the chaos of acid–base terminology. The resulting discussion led to a certain clarification of concepts and more and more laboratories adopted the concepts of Astrup.[13] Schwartz and Relman[14] did not hold with this opinion, however; they published a series of experiments in which they contested that any conclusions could be drawn about the behaviour of blood *in vivo* from experiments involving *in vitro* equilibration. Their opinion was supported by other workers.[8,15]

References

1. Arrhenius, S., Über die Dissociation der in Wasser gelösten Stoffe, *Z. phys. Chem.*, **1**, 631 (1887).

2. Bronsted. J. N., Bemerkungen über den Begriff der Säuren und Basen, *Recl Trav. chim. Pays-Bas Belg.*, **42**, 718 (1923).
3. Sørensen, S. P. L., Über die Messung und Bedeutung der Wasserstoffionen-Konzentration bei biologischen Prozessen, *Ergebn. Physiol.*, **8**, 254 (1909).
4. Bjerrum, N., Die Dissoziation der starken Elektrolyte, *Z. Electrochem.*, **24**, 321 (1918).
5. Ellison, G., Straumfjord, J. V., and Hummel, J. P., Buffer capacities of human blood and plasma, *Clin. Chem*, **4**, 452 (1958).
6. Henderson, L. J., Das Gleichgewicht zwischen Basen und Säuren im tierischen Organismus, *Ergebn. Physiol.*, **8**, 254 (1909).
7. Hasselbalch, K. A., Die Berechnung der Wasserstoffzahl des Blutes aus freier und gebundener Kohlensäure desselben, und die Sauerstoffbindung des Blutes als Funktion der Wasserstoffzahl, *Biochem. Z.*, **78**, 112 (1916).
8. Howorth, P. J. N. RIpH revisited, *Lancet*, **i**, 253 (1974).
9. Siggaard-Andersen, O., *The acid–base status of the blood*, 4th ed., Munksgaard, Copenhagen, 1974.
10. McClean, F. C., Application of the law of chemical equilibrium (law of mass action) to biological problems, *Physiol. Rev.*, **18**, 495 (1938).
11. Slyke, D. D. van, Studies of acidosis. XVII. The normal and abnormal variations in the acid–base balance of the blood, *J. biol. Chem.*, **48**, 153 (1921).
12. Davenport, H. W., *The ABC of acid–base chemistry*, 5th ed., University of Chicago Press, Chicago, 1969.
13. Astrup, P., A new approach to acid–base metabolism, *Clin. Chem.*, **7**, 1 (1961).
14. Schwartz, B. and Relman, A. S., A critique of the parameters used in the evaluation of acid–base disorders, *New Engl. J. Med.*, **268**, 1382 (1963).
15. Acids, bases, and nomograms, *Lancet*, **ii**, 814 (1974).
16. Creese, R., Neil, M. W., Ledingham, J. M., and Vere, D. W., The terminology of acid–base regulation, *Lancet*, **i**, 419 (1962).

3.5.2. COLLECTION OF BLOOD SPECIMENS FOR INVESTIGATING THE ACID–BASE STATUS AND THE BLOOD GASES

The collection of blood for acid–base investigations is the subject of much confusion and many misunderstandings. We shall, therefore, deal with it here, although many of the points have already been made elsewhere (p. 100). The important factors are the actual collection of the blood, preservation of the sample, centrifugation, the collecting vessels, and the choice of anticoagulants.

3.5.2.1. Collection of the blood

Arterial blood. The most accurate results are doubtless obtained by using arterial blood collected from the arteria brachialis, radialis, or femoralis. Punctures of the arteria jugularis yield deviant results. The blood should be collected anaerobically. Arterial puncture should only be performed by a doctor.

Arterialized capillary blood. The blood may be taken from the finger-tip, the ear-lobes, or the heel. The expression 'arterialized' is not well chosen, but what is meant is an increased flow of blood, produced by warming or vasodilatory drugs, so that the chemical composition of the blood is as near as possible to that of arterial blood. The blood is arterialized by heating the collection site for at least 5 min either with a heat-lamp or with water at 45 °C, until the skin shows a deep flush. The same effect may be produced by using vasodilators such as Trafuril or Priscol. After the hyperaemia has been induced, the skin is cleaned with 70% alcohol. To obtain a better drop, the skin may be lightly covered with sterile Hemolube cream prior to puncture. These preliminaries are time consuming. It is most important that the puncture should be deep enough and that the blood should flow quickly and freely. The first drops of blood are discarded. Squeezing or 'milking' the puncture is liable to give unreliable values for the pH, pCO_2, and pO_2.[1-3]

Venous blood. If neither arterial nor capillary blood can be collected, then it is possible to use venous blood for the analysis of acid–base status, provided that special care is taken. The patient must lie quietly for at least 5 min with head and torso upright and must be kept warm (skin temperature around 35 °C). It is best to collect the blood without using a tourniquet, but if one is used it should not be removed before the blood is collected.[4] The first 10 ml are used for acid–base status measurements. 'Pumping' with the hand is absolutely out of the question.

On the basis of our own experience, we unequivocally advise against the use of venous blood for the determination of the pH and pCO_2; such samples may only be used for the determination of standard bicarbonate or total CO_2. Assistant personnel are rarely able to accomplish the collection in a reliable manner, observing all the necessary conditions. Arterial blood collection is doubtless the method of choice, but in practice, its application is strictly limited; it is replaced by collection of capillary blood from the ear-lobes. Comparative measurements on capillary and arterial blood of healthy patients show that, in respect of the blood gases, capillary blood collected from hyperaemic ear-lobes behaves like arterial blood and that for 80% of all determinations, the difference between capillary and arterial blood for the carbonic acid and oxygen partial pressures was less than 3 mmHg.[5] No systematic deviation has ever been observed.

The technique of collecting the capillary blood is critical to the accuracy of measurements made with such blood. Unless hyperaemia is induced, the capillary blood for the ear-lobes contains a venous blood component of 2–7% and consequently the pO_2 can be up to 10 mmHg lower when there is no hyperaemia.[5] The change in pCO_2 is only very slight.

3.5.2.2. Collecting vessels

Blood for the analysis of acid–base status ought to be collected anaerobically, but this expression is misleading for, as biochemists understand it, it implies that the procedure should be accomplished under complete exclusion of air, a condition impossible in practice and one which is also superfluous since brief contact with the air does not lead to any significant changes in the blood. As a rule, the risk of a loss of CO_2 arises not in the actual collection of the blood but in the preservation of the samples.

Glass syringes. The syringes must have an air-tight fit and, if necessary, they are 'oiled' with paraffin oil. The dead space between the needle and the syringe is filled with heparin. Any air in the syringe after collecting the blood is expelled immediately and, provided that this is done thoroughly, there is no need to close the aperture with a stopper since the surface area is much too small to lead to CO_2 losses by diffusion.

Plastic syringes. These are preferable to glass syringes since they generally have a better fit and do not require lubricating, and also they can be discarded after use. The collecting procedure is the same as for glass syringes.

Vacuum syringes. Arterial or venous blood can also be collected in heparinized vacuum syringes, but they have to be filled completely otherwise the collection is likely to be attended by a decrease in pCO_2 and an increase in the pH. Complete filling can be a difficult business with these syringes.

Heparinized capillaries. The size of the capillary is selected in accordance with the sample volume required for the analytical instrument.

Of these techniques, the only ones which may be recommended are collection in plastic syringes and in heparinized capillaries. In every case, it is important that the collecting vessel be completely filled and should not contain any dead space. The transference of the blood drop, e.g. on the ear-lobe, into the heparinized capillary constitutes no problem as far as pO_2 is concerned provided that the blood is taken from the centre of the blood drop since atmospheric oxygen can diffuse into the drop only very slowly. This is not true for the carbonic acid, however, which is readily lost. In order to ensure a relatively air-tight seal around the ear-lobe, some authors recommend the use of funnel capillaries.[5,6]

A few remarks on the problem of using paraffin oil seem appropriate here. Paraffin oil is supposed to prevent the diffusion of CO_2 from the blood into the surrounding air. This is true provided that the surface of the blood is small and it is covered with a thin layer of oil, but if a lot of paraffin oil is used, e.g. 1 ml of paraffin oil to 5 ml blood, this is so only if the sample remains absolutely still and undisturbed—a condition never fulfilled in practice. If the samples are transported, shaken, or even centrifuged, then there is liable to be considerable loss of CO_2 by diffusion into the paraffin oil.[1] Since blood is never left undisturbed in routine operations, it is safer to dispense with any addition of paraffin oil and to fill the vessels as completely as possible.

3.5.2.3. Preservation of the samples

In routine work it is often not possible to carry out the analyses immediately after the blood has been collected although one should certainly strive to do so under all circumstances. Ways must therefore be found to prevent changes in the chemical composition of the samples. After collection of the blood, the metabolism of the erythrocytes, and, in particular of the leucocytes, continues, producing mainly lactate, pyruvate, and other acidic end-products. Consequently, the pH drops and CO_2 is driven off. Attempts to block the metabolism by addition of glycolysis inhibitors have led to unsatisfactory results. Either the inhibition is inadequate or else there is complete inhibition leading to a metabolic block with loss of cellular potassium so that the pH of the samples rises.[1] The addition of fluoride, monoiodoacetate, etc., should therefore be abandoned.

It is generally agreed that there is at present only one sure method of reducing the glycolysis to a minimum: keeping the blood at 4 °C or on ice. This is also suitable for transportation. Our own experiments have shown that on keeping in a refrigerator (4 °C) or on ice, the pO_2 falls by about 2% during the first 30 min, and by about 8% after 60 min, while pCO_2 falls by about 3% and 8%, respectively. After 60 min the pH falls by about 0.008 unit. In contrast, on keeping at 38 °C, there is liable to be a fall of 0.04–0.08 unit in the pH, an increase in pCO_2 of 5 mmHg and a fall in bicarbonate of 2 mmol 1^{-1}.[1] The discrepancies are even greater for samples with abnormally high leucocyte counts (leukaemia).

Professionally acceptable results are guaranteed by transportation on ice, preservation in a refrigerator, and performing the analyses within 30 min of collecting the blood. All other results should be ignored.

3.5.2.4. Influence of the temperature of centrifugation

The following considerations are more theoretical than practical, but nonetheless they ought to be mentioned. The coagulation of the blood has no effect on the pH, and the pH values in plasma and in serum are identical. Identical values are also obtained for whole blood on the one hand, and for plasma or serum on the other, provided that the fresh blood is centrifuged at body temperature. Even so, the measured pH value is 0.01 unit lower for whole blood than for plasma but this is due to a technical artefact (the effect of the cells on the salt-bridge of the reference electrode). However, if the blood is centrifuged at a temperature lower than body temperature, the pH values in plasma are too high. This is probably due to the degree of ionization of the blood proteins, which is strongly pH dependent. Since cells contain more protein than plasma, a decrease in temperature produces a greater change in the blood pH than in the plasma pH. The pH of plasma falls by 0.01–0.012 unit per degree centigrade, and that of blood by 0.015 unit. If, e.g. whole blood has a pH of 7.40 at 38 °C, on cooling to 4 °C this rises to 7.90. If this blood is centrifuged at 4 °C, then the plasma pH will also be 7.90, falling to 7.56 on reheating to 38 °C, not to the expected value of 7.40 since the plasma is less sensitive to temperature changes than the blood.

The practical conclusion to be drawn from this is that the plasma pH should only be measured if the blood has been centrifuged at body temperature. Ice-cold blood ought never to be centrifuged. Samples which have been kept at 4 °C must be warmed at least to room temperature before centrifugation.

3.5.2.5. Choice of anticoagulants

Heparin is the only anticoagulant suitable for investigations of the acid–base status since all others produce changes in the acid–base parameters.

References

1. Siggaard-Anderson, O., The acid–base status of the blood, 4th ed., Munksgaard, Copenhagen, 1974.
2. Gambino, S. R., Oxygen, partial pressure, *Stand. Meth. clin. Chem.*, **6**, 171 (1970).
3. Müller-Plathe, O., *Säure-Basen-Haushalt und Blutgase*, Thieme, Stuttgart, 1973.
4. Gambino, S. R., Comparisons of pH in human arterial, venous and capillary blood, *Am. J. clin. Path.*, **32**, 298 (1959).
5. Reichel, G., Blutgase, *Diagnostik*, **6**, 371 (1973).
6. Scherrer, M. and Valenzuela, A., Kritische Überprüfung der Methode von Astrup zur Blut-pCO$_2$-Bestimmung, *Praxis*, **55**, 917 (1966).

3.5.3. BLOOD pH

3.5.3.1. Choice of method

The pH of the blood is usually measured potentiometrically with the glass electrode (p. 186). The development of temperature-controlled capillary glass electrodes has made the measurement of blood pH a routine operation in the clinical laboratory.

3.5.3.2. Principle

If a glass membrane is placed between two solutions with differing hydrogen ion concentrations, an electrical potential is established. The experimental system used for measuring this potential by means of the glass electrode is shown diagrammatically in Fig. 93. On one side of the membrane there is a solution of known H$^+$ concentration (usually 0.1 N HCl). This is coupled to the potentiometer via a silver–silver chloride electrode. The solution of unknown pH is placed on the other side of the

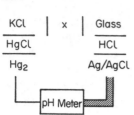

Fig. 93. Diagram of the double cell system for measuring the blood pH x = blood of unknown pH.

membrane and is connected with the reference electrode (here a calomel electrode) via the so-called bridge solution (usually a saturated KCl solution). The circuit is completed by the connection of the calomel electrode to the potentiometer. In principle, there is a potential at every phase boundary, but by appropriate design of the electrical components it is generally possible to eliminate all potentials except that between the sample and the glass. In most apparatus for pH measurement, the critical point is the junction between the sample solution and the bridge solution. The best results are obtained if there is open communication between the bridge solution and the sample, i.e. there is so-called liquid–liquid junction.[1]

The precise control of the temperature is important in all pH measurements. There are three places at which a change in temperature can effect the measured pH value: in the potentiometer, at the salt bridge, and in the solution to be measured. It must be possible to control the temperature at all three places. Measurement of the pH is not an absolute measurement but is relative, and thus a further requirement is the use of very accurate standard solutions for calibration.

3.5.3.3. Standard pH solutions

The most important precondition for the accurate measurement of pH is the use of accurately prepared standard solutions. For obvious reasons, buffer solutions are particularly suitable for this purpose. After thorough research, the US National Bureau of Standards has defined a practical pH scale based on a series of buffer mixtures; this scale has now been accepted almost everywhere. These standards may be prepared in the laboratory, if necessary, with chemicals bought from the National Bureau of Standards but also obtainable from commercial firms. Since, for the most part, these solutions do not keep well, special attention should be paid to manufacturer's guarantees and dates of expiry. For physiological investigations the acid 'equimolar phosphate buffer' and the basic 'phosphate buffer for blood pH' have been developed. Both buffers have an ionic strength of 0.100. They are prepared as follows.

Basic buffer (phosphate buffer for blood). 1.179 g of potassium dihydrogen phosphate (KH_2PO_4) and 4.302 g of disodium hydrogen phosphate (Na_2HPO_4) are dissolved in ammonia and carbon dioxide-free DM-water to make 1000 ml at 25 °C.

Table 49. pH values for the buffer standards.

°C	Alkaline buffer	Acidic buffer
20	7.429	6.881
25	7.413	6.865
30	7.400	6.853
35	7.389	6.844
38	7.384	6.840
40	7.380	6.838
45	7.373	6.834

Acidic buffer (equimolar phosphate buffer). 3.39 g of potassium dihydrogen phosphate (KH_2PO_4) and 3.53 g of disodium hydrogen phosphate (Na_2HPO_4) are dissolved in ammonia and carbon dioxide-free DM-water to make 1000 ml at 25 °C.

The salts used are first dried for 2–6 h at 120 °C and should then be kept in a desiccator. The pH values obtained with these buffers at various temperatures are listed in Table 49. Unfortunately, the stability of these buffers leaves much to be desired. Some authors recommend the addition of thymol or keeping the buffer at 4 °C (do not freeze!). Nevertheless, an opened bottle of buffer should not be used for longer than 1 week, since otherwise there is a risk that the pH will change as a result of bacterial or fungal infection. If a clean procedure is adopted (never pipette directly from the bottle) the solutions may be used for 1 week even if stored at room temperature.

3.5.3.4. Apparatus

pH meter. The instrument used for measuring the blood pH should be a pH meter expressly manufactured for this purpose. The sensitivity should be at least 0.002 pH unit at 37 °C, and the accuracy ±0.01 pH unit. The temperature of the thermometer must be controlled to within ±1 °C.

Cell circuit. The electrodes and the salt-bridge must be thermostatically controlled to within ±0.2 °C. The volume of blood sample required should not exceed 100 µl. In practice, the liquid–liquid junction is the only suitable bridge (Fig. 93).

3.5.3.5. Procedure

For technical details on how to operate the instrument, the manufacturer's instructions should be followed closely. A few frequent sources of error are indicated below:

1. Extreme cleanliness is absolutely essential to making accurate pH measurements. This applies to the whole apparatus but most particularly to the electrodes and the salt-bridge.

2. For each series of measurements the pH meter must first be calibrated with the basic buffer (pH 7.384). After rinsing, the pH of the acidic buffer (pH 6.840) is measured. This measurement should not deviate from the expected value by more than 0.005 pH unit. If the difference is greater than this and the buffers are faultless, then the potentiometer must be corrected until the values are satisfactory.
3. After about 10 determinations the calibration should be checked with the basic buffer.
4. The electrode must be absolutely clean. Cleaning is best done with a solution of pepsin in 0.1 N hydrochloric acid or with a neutral detergent. Overnight, or during interruptions, the electrode is immersed in the basic buffer.
5. The electrode must be rinsed with water or, if necessary, detergent solution between consecutive determinations. Blood should never be used for rinsing, not even that of the next sample (erroneous result!). After rinsing, the electrode should be dried before the next measurement.
6. Care should be taken that there are no air bubbles between the sample and the bridge solution.
7. The bridge solution should be kept clean.
8. Fluctuations of the potentiometer needle are sometimes due to electrically charged plastic clothing (underwear, for example).

3.5.3.6. Reference values

The reference values of the pH in blood have been the subject of a great number of investigations, whence it emerges that the values of the individual authors are in remarkably good agreement. However, this is partly to be explained by the fact that the pH value only shows up large deviations; if the results were expressed as H^+ concentration in nmol l^{-1}, they would be seen to be less satisfactory. With a careful technique, the value for adults lies between 7.34 and 7.47 pH units. A few representative investigations are collated in Table 50, which shows that the results in

Table 50. Reference values for blood pH.

Sample, sex	Number	$\bar{x} \pm 2s$	Reference interval	References
Arterial blood, full-term new-born	45	7.35 ± 0.26		6
Arterial blood, adults (m, f)	51	7.426 ± 0.026	7.400–7.452	4
Arterial blood, adults (m, f)	18	7.41 ± 0.04	7.37–7.45	5
Capillary blood, f.	20	7.405 ± 0.030	7.375–7.435	2
Capillary blood, m	20	7.413 ± 0.022	7.391–7.435	2
Venous blood (m, f)	55	7.398 ± 0.020	7.378–7.418	7

arterial blood and in capillary blood are in relatively good agreement. However, this will only apply if the blood is collected with careful precautions. Gambino[3] found an average difference of 0.0076 pH unit; in contrast, the average difference between venous and arterial pH in plasma was 0.0128 pH unit.

3.5.3.7. Physiological variability

There is no statistically detectable sex difference. During pregnancy the scatter of the values is increased but they still remain approximately in the range of the norm.[8] New-born babies show a slight physiological acidosis which is more marked amongst those born prematurely.[9] The relatively wide scatter of the reference values for new-born babies slowly narrows and after 6 years of age the range is the same as for adult values.[10]

3.5.3.8. Comments on the method

1. The importance of precise temperature control becomes apparent once it is known that a temperature fluctuation of 2 °C at the electrode results in a pH fluctuation of 0.02 unit for plasma and 0.03 unit for whole blood.

2. If the pH is measured at a temperature different from the body temperature, the result must be corrected. This applies particularly to investigations in cases of hypothermia. The value may be corrected by means of the following empirical factor:[11]

$$\Delta pH = -0.0146\Delta \ ^\circ C$$

However, this Rosenthal factor, $\Delta pH/\Delta \ ^\circ C$, is not absolutely constant but is a function of pH and bicarbonate concentration; for children its value is not 0.0146 but 0.0128. The temperature correction of the pH may be carried out using a nomogram (Fig. 99, p. 322).

References

1. Bühler, H., Grundlagen und Probleme der pH-Messung, *G-I-T Fachz. Labor*, **19**, 641 (1975).
2. Siggaard-Anderson, O., *The acid–base status of the blood*, 4th ed., Munksgaard, Copenhagen, 1974.
3. Gambino, S. R., Comparisons of pH in human arterial, venous and capillary blood, *Techn. Bull. Registry med. Technol.*, **29**, 136 (1959).
4. Millar, R. A. and Marshall, B. E., Acid–base changes in arterial blood associated with spontaneous and controlled ventiliation during anaesthesia, *Br. J. Anaesth.*, **37**, 492 (1965).
5. Fillmore, S. J., Shapiro, M., and Killip, T., Arterial oxygen tension in acute myocardial infarction. Serial analysis of clinical state and blood gas changes, *Am. Heart J.*, **79**, 620 (1970).
6. Fenner, A., Busse, H.-G., Junge, M., and Müller, R., Acid–base parameters

298

and alveolar–arterial oxygen tension gradients in healthy neonatal and postneonatal infants, *Eur. J. Pediat.*, **122**, 69 (1976).

7. Gambino, S. R., Normal values for adult human venous plasma pH and CO_2 content, *Am. J. clin. Path.*, **32**, 294 (1959).
8. Stenger, V., Andersen, T., Eitzman, D., Gessner, I., Padua, C. de, and Prystowsky, H., The oxygen, carbon dioxide and hydrogen-ion concentrations in the arterial and uterine venous blood of pregnant women, *Am. J. Obstet. Gynec.*, **87**, 1037 (1963).
9. Prod'hom, L. S., Levison, H., Cherry, R. B., Drorbaugh, J. E., Hubbell, J. P., and Smith, C. A., Adjustment of ventilation, intrapulmonary gas exchange, and acid–base balance during the first day of life. Normal values in well infants of diabetic mothers, *Pediatrics*, **33**, 682 (1964).
10. Gambino, R. S., PH and pCO_2, *Stand. Meth. clin. Chem.*, **5**, 169 (1965).
11. Rosenthal, T. B., The effect of temperature on the pH of blood and plasma *in vitro*, *J. biol. Chem.*, **173**, 25 (1948).

3.5.4. pCO_2 IN THE BLOOD

3.5.4.1. Introduction

The disturbances of the acid–base status show (p. 309) that the measurement of the carbon dioxide partial pressure (pCO_2) is of important diagnostic significance.

3.5.4.2. Choice of method

For the determination of the pCO_2, the following methods are the most frequently used:

1. Measurement of the pH and $[HCO_3^-]$ or $\Sigma[CO_2]$, and calculation of pCO_2 by means of the Henderson–Hasselbalch equation (cf. Table 47 and the nomograms in Figs. 89 and 90).
2. Direct measurement with the pCO_2 electrode.
3. pH measurement after equilibration with two gas mixtures of known pCO_2.

3.5.4.3. Blood CO_2: measurement with the pCO_2 electrode

Principle. The pCO_2 electrode is a modified glass electrode (Fig. 94) which measures the pH of a bicarbonate solution. The latter is separated from the blood sample by a plastic foil (PTFE, nylon). This membrane is permeable to low-molecular-weight neutral particles such as CO_2 but impermeable to charged ions such as HCO_3^-. CO_2 diffuses through the membrane in both directions, depending on the difference in partial pressures on either side. The electrolyte solution (usually $NaHCO_3$ and NaCl) on the inside is equilibrated with CO_2 diffusing from outside and this changes the pH of the bicarbonate–NaCl solution:

$$CO_2 + H_2O \rightleftharpoons H_2CO_3 \rightleftharpoons H^+ + HCO_3^-$$

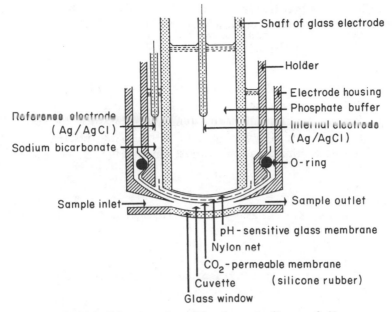

Fig. 94. Diagram of a pCO_2 electrode (from ref. 1).

Thus the pCO_2 of the blood sample can be determined directly by means of a pH measurement. A 10-fold increase in the pCO_2 corresponds roughly to a decrease of 1 pH unit.

Measurement. Various instruments are available, some with combined cuvettes for measuring pH, pCO_2, and pO_2. Calibration and measurements should be made exactly as described in the manufacturer's instructions.

3.5.4.4. Comments on the method

1. An electrolyte solution of proven utility is a solution of 5 mmol l^{-1} $NaHCO_3$ and 20 mmol l^{-1} NaCl saturated with AgCl.[1]
2. The pCO_2 meter is a sensitive galvanometer. The electrode should exhibit an equal sensitivity over the range 5–200 mmHg pCO_2.[1]
3. The instrument is calibrated with a CO_2 gas concentration of 4–5 and 9–10%. As a rule the instrument should be calibrated with both gases several times a day, preferably before every measurement. There are some modern instruments which are self-calibrating before each measurement. The partial pressure of the calibration gases is calculated by the methods given on p. 000. For the calibration with CO_2 it is not necessary to take the gas/blood factor into account (p. 321).
4. The electrode should be thermostatically controlled to ±0.1 °C. The pCO_2 increases by 4–5% per degree centigrade.[1]
5. If the patient's body temperature differs from the temperature at

which the blood is measured, the *in vivo* pCO_2 can be calculated by means of the following empirical equation:

$$\Delta \log pCO_2 = +0.021\Delta \,°C \qquad (1)$$

or be read off from a nomogram (Fig. 99).

6. From repeated serial measurements, the standard deviation of the blood pCO_2 measurement is 0.5–1.0 mmHg[1] and the coefficient of variation is 1–2.5%. The corresponding values for our own serial measurements are 0.9 mmHg and 1.9%. The difference between duplicate determinations was at most 1.3 mmHg.

Reference

1. Siggaard-Andersen, O., The acid–base status of the blood, 4th ed., Munksgaard, Copenhagen, 1974).

3.5.4.5. Blood pCO_2: pH measurement after equilibration with two gases of known pCO_2

3.5.4.5.1. Principle

The principle may be best explained with reference to the pH/log pCO_2 nomogram shown in Appendix 5. If the pH of a blood sample is measured following equilibration with various partial pressures of CO_2, all the measured values fall on a straight line running from top left to bottom right (curve II → III → I). To establish the gradient of this line, at least two results are necessary, one at high pCO_2 (II) and one at a low pCO_2 (I). If, in addition, the original pH of the blood sample is known, then the original pCO_2 may be determined by taking a horizontal through the point of intersection (C) of the vertical above the original pH (I) with the line (III). Siggaard-Andersen[1] and Astrup have perfected this method, both theoretically and technically, to such a degree that it is possible to make routine pCO_2 measurements with relatively little trouble.

3.5.4.5.2. Apparatus

pH meter; apparatus for blood measurements; unit for equilibrating the blood samples with gases of different pCO_2 at constant temperature and water-vapour pressure;[1] barometer; gas cylinder with a high pCO_2, containing about 9% CO_2 and 91% O_2; gas cylinder with a low pCO_2, containing about 4% CO_2 and 96% O_2

3.5.4.5.3. Reagents

Standard buffer solutions (p. 294).

3.5.4.5.4. Preparation

Analysis of the gas mixture. The gaseous contents of both the gas mixtures must be accurately known and, if necessary, they must be analysed in the laboratory. If this is not possible, one of the approximation methods must be used (see 'Comments on the method').

Correction for barometric pressure. The pCO_2 of a gaseous mixture is dependent both on the percentage composition of the gases and on the barometric pressure:

$$pCO_2 = \frac{B - 47}{100} \cdot \%CO_2 \qquad \text{or} \qquad \%CO_2 = \frac{pCO_2 \cdot 10}{B - 47}$$

where

B = barometric pressure in mmHg (measured);
47 = water vapour pressure at $37\,°C$ in mmHg (47.08 to be exact);
$\%CO_2$ = percentage of CO_2 in the gaseous mixture.

The actual pCO_2 present must be determined before each series of analyses, either by means of the equation or the nomogram in Appendix 4.

3.5.4.5.5. Procedure

1. The pH of capillary blood is measured at $37\,°C$ using a capillary glass electrode. This value corresponds to the actual pH.
2. A sample is equilibrated in a thermostatically controlled equilibration chamber, with a gas mixture with low pCO_2 for 3 min at $37\,°C$.
3. In another thermostatically controlled equilibration chamber, a second sample is equilibrated for 3 min with a gas at high pCO_2.
4. The pH of the sample equilibrated with low pCO_2 is measured.
5. The pH of the sample equilibrated with high pCO_2 is measured.

3.5.4.5.6. Calculation

It is advantageous to use printed pH/log pCO_2 nomogram sheets for calculating the results. The following *example* is illustrated graphically in Appendix 5.

1. Barometric pressure = 766 mmHg.
2. CO_2 content in cylinder I = 4.4%. Calculation of the pCO_2 by means of the above equation gives a partial pressure of 31.6 mmHg. A horizontal line is drawn from the ordinate corresponding to 31.6 mmHg pCO_2 (arrow A).
3. CO_2 content in cylinder II = 9.2%. Calculation with the same equation gives a partial pressure of 66.2 mmHg. A horizontal line is drawn from the point corresponding to 66.2 mmHg on the ordinate (arrow B).

4. A vertical (arrow 2) is drawn from the pH value (7.07) measured after equilibration with 31.6 mmHg CO_2.
5. A vertical line (arrow 3) is drawn from the pH value (7.29) measured after equilibrating with 66.2 mmHg.
6. A straight line is drawn through the intersections (points I and II) of the pH lines for low and high pCO_2 (arrows 2 and 3) with the lines (arrows A and B) for the corresponding pCO_2 values.
7. A vertical line is drawn from the measured, actual pH (7.18) of the blood (arrow 1).
8. A horizontal line is drawn from the intersection (point III) of this actual-pH line with the sloping line. The pCO_2 value is given by the intersection of this horizontal (arrow C) with the ordinate. This value corresponds to the actual pCO_2 in mmHg (46).
9. The intersection of the straight line II → III → I with the base-excess curve gives the excess in mmol l^{-1} (−11).
10. The intersection of the straight line II → III → I with the buffer-base curve gives the buffer base in mmol l^{-1} (33.5).
11. The intersection of the straight line II → III → I with the horizontal logarithmic standard bicarbonate line (40 mmHg pCO_2) gives the standard bicarbonate in mmol l^{-1} (16).
12. The base-excess is subtracted from the buffer base [33.5 − (−11) = 33.5 + 11 = 44.5]. Using the obtained value the haemoglobin value (8 g per 100 ml) can now be read off from a section of the buffer-base curve. However, this result is very imprecise (±3 g per 100 ml).
13. The actual bicarbonate concentration in mmol l^{-1} is obtained by drawing a line (arrow D) from point III at an angle of 45° to the horizontal standard bicarbonate line (pCO_2 40 mmHg). The intersection of this line with the standard bicarbonate line corresponds to the actual bicarbonate concentration (16.5 mmol l^{-1}).
14. For calculating the total CO_2, the following equation is used: $\Sigma[CO_2]$ = actual bicarbonate + (0.03 times the actual pCO_2). Thus in the present example:

$$\Sigma[CO_2] = 16.5 + (0.03 \cdot 46) = 17.88$$

3.5.4.5.7. Collection of the blood (p. 289)

3.5.4.5.8. Comments on the method

1. The concentration of the CO_2 used for equilibration should always be checked in the laboratory, since it is not uncommon for this to deviate considerably from the declared value. The concentration of the calibration gases must be checked with Scholander's apparatus or by gas chromatography.
2. If the body temperature of the patient differs from the temperature of

Table 51. Reference values for the pCO_2 in blood.

Sample	Sex	Number	$\bar{x} \pm s$	Reference interval	References
Adults					
Arterial blood	m	50	41.5 ± 4.4	37.0–46.0	2
	m	13	38 ± 4.8	32.6–43.4	3
	f	50	39.8 ± 5.2	34.3–45.3	2
Capillary blood	m	20	39.3 ± 5.2	33.5–45.1	1
	f	20	36.4 ± 5.6	30.4–42.4	1
Venous blood	m	7	48.0	44–53	4
	f	8	44.5	38–48	4
Children					
Capillary blood					
10th–90th day		24	36.0 ± 8.0	28.0–44.0	5
90th–360th day		47	34.9 ± 6.4	28.5–41.3	5
1st–6th year		26	37.3 ± 7.6	29.7–44.9	6
7th–12th year		33	38.0 ± 5.2	32.8–43.2	6
13th–17th year		57	41.3 ± 3.1	38.3–44.4	6
Arterial blood					
Full-term new-born		45	31.4 ± 10.0		7

the measurements, the pCO_2 present *in vivo* may be calculated by means of the following empirical equation:

$$\Delta \log pCO_2 = +0.021\Delta \ °C$$

or by reading off the value from a nomogram (Fig. 99).

3.5.4.5.9. Reference values

The reference values for the pCO_2 are listed in Table 51. It may be noted that there is good agreement between the values for arterial and capillary blood and that the norms for venous blood are significantly higher.

3.5.4.5.10. Physiological variability

The pCO_2 is always higher in venous blood than in arterial blood. For women, pCO_2 is about 4 mmHg lower than for men. Similarly, for children, the values lie somewhat lower than for adults. In new-born babies the pCO_2 is also lower owing to the prevalent mixed acidosis; it is interesting that the values for premature babies are a little closer to the norm.[8]

References

1. Siggaard-Andersen, O., The acid–base status of the blood, 4th ed., Munksgaard, Copenhagen, 1974.

2. Møller, B., The hydrogen ion concentration in arterial blood. A clinical study of patients with diabetes mellitus and diseases of the kidneys, lungs, and heart, *Acta med. scand.*, **165**, suppl., 348, 1 (1959).
3. Goldring, R. M., Cannon, P. J., Heinemann, H. O., and Fishman, A. P., Respiratory adjustment to chronic metabolic alkalosis in man, *J. clin. Invest.*, **47**, 188 (1968).
4. Gambino, S. R., Normal values for adult human venous plasma pH and CO_2 content, *Am. J. clin. Path.*, **32**, 294 (1959).
5. Bartels, O. and Wenner, J., Standard-Bicarbonat, pH, CO_2-Druck im 'arterialisierten' Blut gesunder Säuglinge nach der Neugeborenenperiode bis zum Ende des ersten Lebensjahres, *klin. Wschr.*, **43**, 437 (1965).
6. Cassels, D. E. and Morse, M., Arterial blood gases and acid–base balance in normal children, *J. clin. Invest.*, **32**, 824 (1953).
7. Fenner, A., Busse, H.-G., Junge, M., and Müller, R., Acid–base parameters and alveolar–arterial oxygen tension gradients in healthy neonatal and postneonatal infants, *Eur. J. Pediat.*, **122**, 69 (1976).
8. Prod'hom, L. S., Levison, H., Cherry, R. B., Drorbaugh, J. E., Hubbell, J. P., and Smith, C. A., Adjustment of ventilation, intrapulmonary gas exchange, and acid–base balance during the first day of life. Normal values in well infants of diabetic mothers, *Pediatrics*, **33**, 682 (1964).

3.5.5. METABOLIC COMPONENTS

3.5.5.1. Total CO_2 and bicarbonate

3.5.5.1.1. Choice of method

An assessment of metabolic disorders of the acid–base status is provided by the determination of either the total CO_2 or of the bicarbonate. The relationship between these two quantities is characterized by the following equation:

$$\Sigma[CO_2] = [HCO_3^-] + [\alpha pCO_2]$$

For the accurate conversion of total CO_2 to bicarbonate or *vice versa*, it would really be necessary to know either the pH or the pCO_2, but since the dissolved CO_2 only fluctuates between 0.6 and 1.8 mmol l^{-1} for a pCO_2 between 20 and 60 mmHg this error can be neglected. For practical purposes, mmol l^{-1} bicarbonate = mmol l^{-1} total CO_2. This assumption is also made in the relationships expressed in the nomograms (pp. 283 and 284). Numerous methods for the determination of the metabolic components have been described and the most important of these will be briefly mentioned.[1]

Gasometry. Gasometry was applied to the analysis of blood gases in 1912 by Barcroft and Haldane. In 1917 Van Slyke developed the 'volumetric' apparatus and, together with Neill in 1924, the 'manometric' apparatus. The CO_2 is driven off with acid, collected *in vacuo*, and the quantity of CO_2 determined by volumetric or manometric methods. For the clinical–chemical laboratory the Kopp–Natelson micromanometer[2] is particularly convenient; this apparatus permits a determination of the total CO_2 to be carried out on a 30-μl sample.

Microdiffusion. The gas liberated by the acid can also be isolated by the principle of microdiffusion and precipitated as barium carbonate. The measurement is performed by titration.[3]

Photometry. In a technique described for use in an AutoAnalyzer the CO_2 liberated by the acid is adsorbed in an alkaline buffer containing phenolphthalein. The change in colour serves as a measure for the CO_2 content.[4]

Gas chromatography. With a gas chromatograph, all the blood gases may be determined in a single operation, but the method is very expensive.

Titrimetry. The first titrimetric method for the determination of bicarbonate was introduced by Van Slyke *et al.*[5] in 1919. A known quantity of acid is added to the serum and a proportionate amount of acid is consumed in expelling the CO_2. The remaining acid is determined by back-titration to the original pH of the plasma, using an indicator such as neutral red of phenol red. Today, the end-point should be determined with a pH meter if possible.

Indirectly from pH using CO_2 at 40 mmHg. If the blood is equilibrated with CO_2 at 40 mmHg, then the resultant pH is directly proportional to the standard bicarbonate concentration.[6] This method is technically simple, rapid, and accurate.

Indirectly from pH and pCO_2. The bicarbonate can easily be calculated if the pH and the pCO_2 are known (cf. Table 47). In the blood-gas analyzers available today, this calculation is performed automatically by electronic means.

Enzymatic measurement. An enzymatic method of measuring the CO_2 has been suggested, utilizing phosphoenolpyruvate carboxylase according to the following reaction:

$$\text{Phosphoenolpyruvate} + HCO_3^- \xrightarrow[Mg^{2+}]{\text{PEP carboxylase}} \text{oxaloacetate} + \text{inorganic phosphate}$$

$$\text{Oxaloacetate} + NADH + H^+ \xrightarrow{\text{malate dehydrogenase}} \text{malate} + NAD^{+}\text{[7]}$$

The change in extinction at 340 nm (or at 334 or 365 nm) is proportional to the CO_2 concentration. Instead of $NADH_2$, diazonium salts are also used. Experience of this method is as yet inadequate.

References

1. Lustgarten, J. A., Creno, R. J., Byrd, C. G., and Wenk, R., Evaluation of contemporary methods for serum CO_2, *Clin. Chem.*, **22**, 374 (1976).
2. Natelson, S., *Microtechniques of clinical chemistry*, 2nd ed., Thomas, Springfield, 1961.
3. Conway, E. J., *Microdiffusion analysis and volumetric error*, Lockwood, London, 1962.

306

4. Skeggs, L. T., An automatic method for the determination of carbon dioxide in blood plasma, *Am. J. clin. Path.*, **33**, 181 (1960).
5. Slyke, D. D. van, Stillman, E., and Cullen, G. E., Studies of acidosis. XIII. A method for titrating the bicarbonate content of the plasma, *J. biol. Chem.*, **38**, 167 (1919).
6. Siggaard-Andersen, O., *The acid–base status of the blood*, 4th ed., Munksgaard, Copenhagen, 1974.
7. Forrester, R. L., Vatai, L. J., Silverman, D. A., and Pierre, K. J., Enzymatic method for determination of CO_2 in serum, *Clin. Chem.*, **22**, 243 (1976).

3.5.5.2. Standard bicarbonate

3.5.5.2.1. Principle

As early as 1916, Haselbalch suggested determining the non-respiratory component of the bicarbonate system by measuring the pH in blood which had previously been equilibrated with CO_2 at a partial pressure of 40 mmHg. This proposal was obviously forgotten and it was not until 1956 that it was taken up again, by Jørgensen and Astrup, and developed as a simple and rapid method for the determination of the standard bicarbonate. The theory of the method derives from the graph of pH (abscissa) against pCO_2 (ordinate), shown in Appendix 5. At a pCO_2 of 40 mmHg all the bicarbonate values lie on a line parallel with the abscissa. The relationship is only valid for complete oxygenation of the blood and the imposition of a definite pCO_2, if possible 40 mmHg. The respiratory component is thereby eliminated so that the measured pH reflects only the metabolic component.[1]

The determination of the standard bicarbonate from the pH and pCO_2 by equilibrating with two gases of known pCO_2 has already been considered (p. 301). The standard bicarbonate can also be determined if the following quantities are known:[1] the actual pH and pCO_2, the haemoglobin concentration, and the O_2 saturation. Instead of measuring the last quantity, it can also be evaluated from the pH and the pCO_2 using the Severinghaus blood-gas calculator.[2] Certain commercially available instruments which measure the pH, pCO_2, and pO_2 simultaneously automatically calculate the standard bicarbonate for a given haemoglobin concentration.

Table 52. Reference values for standard bicarbonate ($mmol\,l^{-1}$).

Sample, sex	Number	$\bar{x} \pm 2s$	Reference interval	References
Arterial blood	51	23.3 ± 1.80	21.5–25.3	3
Capillary blood, m	20	24.7 ± 2.2	22.5–26.9	1
Capillary blood, f	20	24.0 ± 2.0	21.8–26.2	1

3.5.5.2.2. Reference values

These are given in Table 52.

References

1. Siggaard-Andersen, O., *The acid–base status of the blood*, 4th ed., Munksgaard, Copenhagen, 1974.
2. Severinghaus, J. W., Blood gas calculator, *J. appl. Physiol.*, **21**, 1108 (1966).
3. Millar, R. A. and Marshall, B. E., Acid–base changes in arterial blood associated with spontaneous and controlled ventilation during anaesthesia, *Br. J. Anaesth.*, **37**, 492 (1965).

3.5.5.3. Buffer base and base excess

3.5.5.3.1. Definitions

In contrast to all other bicarbonate values, the standard bicarbonate has the advantage of being independent of the pCO_2 and the degree of oxygenation of the haemoglobin. However, in common with all other bicarbonate values, it suffers from the disadvantage that it gives no direct information (in mmol l^{-1}) as to how much acid or base is responsible for the departure from the norm, the reason being that the bicarbonate buffer system is only responsible for about 75% of the buffering of the blood. For the clinician, however, it would be useful to have a direct measure of the metabolic component so that he might apply the appropriate quantitative therapy. The concept 'buffer base', introduced by Singer and Hastings,[1] is an attempt to provide such a measure.

The buffer base is the sum of all buffer anions (including haemoglobinate) present in 1 litre of blood (in mmol l^{-1}).

Table 53. Buffer base, base excess and standard bicarbonate at normal and low haemoglobin concentration after the addition of acid and base (from ref. 2).

Quantity measured	Addition		
	None	10 mmol acid	10 mmol base
Haemoglobin, 15 g per 100 ml			
Buffer base, mmol l^{-1}	46.2	36.2	56.2
Base excess, mmol l^{-1}	0	−10.0	+10.0
Standard bicarbonate, mmol l^{-1}	22.9	15.9	30.4
Haemoglobin, 7.5 g per 100 ml			
Buffer base, mmol l^{-1}	43.5	33.5	53.5
Base excess, mmol l^{-1}	0	−10	+10
Standard bicarbonate, mmol l^{-1}	22.9	15.4	31.0

308

One disadvantage of this quantity is that its absolute value is strongly dependent on the haemoglobin concentration. This shows up very clearly in the simple experiment presented in Table 53. Accordingly, Astrup et al.[2] sought to find another concept which shows directly how much acid or base must be added to the blood in order to correct a metabolic deviation: the base excess.

The base excess includes the bases which can be determined by titrating the blood to a normal pH, expressed as the difference from a normal standard bicarbonate of 24.2 mmol l^{-1} at normal pCO_2 of 40 mmHg and complete oxygenation of the blood. The advantage of this quantity is that it is not dependent on the haemoglobin content.

3.5.5.3.2. Principle

Buffer base and base excess are obtained as secondary results in the determination of pH and pCO_2 by the methods of Astrup et al.[2] and Siggaard-Andersen.[3] Calculation is made exclusively with the nomogram of Appendix 5, the curves of which were determined empirically from titration measurements.

3.5.5.3.3. Reference values

These are given in Table 54.

3.5.5.3.4. Physiological variability

The base excess in venous blood is about 2 mmol l^{-1} higher than in arterial blood.[5] This difference disappears on arterialization of the blood *in vitro*[3] or arterialization of the vein *in vivo*. The marked negative values for new-born babies is symptomatic of the prevalent physiological acidosis. Even at the age of 2–3 years the values are about 2 mmol l^{-1} lower[6] and this difference only disappears at puberty. The sex difference is scarcely significant. As a consequence of the acid lost in the stomach, an 'alkaline tide' may be observed after a meal, the base excess increasing by 3–4 mmol l^{-1}.[3] In cases of organic acidaemia the base excess is not always negative.[7]

Table 54. Base excess (mmol l^{-1}).

Sample	Sex	Reference interval	References
Capillary blood	m	−2.7 to +2.5	3
	f	−3.4 to +1.4	3
Arterial blood		−2.9 to +1.2	4

References

1. Singer, R. B. and Hastings, A. B., An improved clinical method for the estimation of disturbances of the acid–base balance of human blood, *Medicine, Baltimore*, **27**, 223 (1948).
2. Astrup, P., Siggaard-Andersen, O., Jørgensen, K., and Engel, K., The acid–base metabolism. A new approach, *Lancet*, **i**, 1035 (1960).
3. Siggaard-Andersen, O., *The acid–base status of the blood*, 4th ed., Munksgaard, Copenhagen, 1974.
4. Millar, R. A. and Marshall, B. E., Acid–base changes in arterial blood associated with spontaneous and controlled ventilation during anaesthesia, *Br. J. Anaesth.*, **37**, 492 (1965).
5. Gambino, S. R., Normal values for adult human venous plasma pH and CO_2 content, *Am. J. clin. Path.*, **32**, 294 (1959).
6. Cassels, D. E. and Morse, M., Arterial blood gases and acid–base balance in normal children, *J. clin. Invest.*, **32**, 824 (1953).
7. Biervliet, J. P. van, Stekelenburg, G. J. van, Duran, M., and Wadman, S. K., Base excess and organic acidaemia, *Lancet*, **ii**, 1518 (1974).

3.5.6. DISTURBANCES OF THE ACID–BASE BALANCE

If laboratory values for the pH and the pCO_2 are available, then the most frequent disturbances in the acid–base balance may be readily classified on the basis of the considerations made (Table 55).[1] The only important points are a precise definition of the terms acidosis and alkalosis, and the assumption that there can only be a primary abnormality of the pCO_2 in cases of respiratory disturbances.

However, it is rare for the disturbances to be so clear cut and combined disturbances are more frequent since the shift in the acid–base balance immediately triggers renal and pulmonary compensation mechanisms. With this classification it turns out that the pCO_2 assumes a governing role, while the definition of the metabolic changes results *per exclusionem*.

Pathological changes in the acid–base status occur if there is imbalance between the production, or exogenous supply, and elimination of H^+ ions. The accumulation or loss of H^+ in the body fluids does not lead unequivocally to a change in the pH since this may be compensated for, to a certain degree, by the buffer systems present.

Table 55. Classification of the acid–base disturbances.

I. pH acids	Acidosis
1. pCO_2 increased	Respiratory acidosis
2. pCO_2 normal	Metabolic acidosis
II. pH alkaline	Alkalosis
1. pCO_2 lowered	Respiratory alkalosis
2. pCO_2 normal	Metabolic alkalosis

3.5.6.1. Acidosis, alkalosis

3.5.6.1.1. Respiratory acidosis

Causes. Respiratory acidosis is due to a reduction in CO_2 excretion by the lungs. This may be occasioned by mechanical hindrance to ventilation (obstruction, emphysema, pulmonary fibrosis, pneumothorax, central or peripheral paralysis of the breathing musculature), disturbance of the blood supply in cases of pulmonary stasis and embolism, and disturbances of the diffusion (pulmonary oedema), this being rare since, intrinisically, CO_2 diffuses well.

Secondary disturbances. The pCO_2 rises (hypercapnia) and the HCO_3^-/H_2CO_3 ratio falls leading to a reduction in pH. H^+ ions are taken up by the cells, whereupon potassium passes into the extracellular space with a possibility of hyperkalaemia. The kidneys also strive to compensate for the pH change. The pCO_2 in the tubular cells rises and the reaction

$$CO_2 + H_2O \rightleftharpoons H_2CO_3 \rightleftharpoons HCO_3^- + H^+$$

proceeds from left to right.

The tubular H^+ secretion and bicarbonate resorption are elevated. Since the pH in the tubule cells is low, owing to acidosis, there is an increased production of ammonia. This regulation mechanism, which functions relatively slowly, is mainly active in chronic respiratory acidosis. Acute respiratory acidoses are compensated for not by the kidneys, but by the other buffer systems.

Respiratory adjustment. Initially, the breathing centre is stimulated by the high pCO_2 but it becomes accustomed to it and hypoxia is then the chief stimulus to ventilation.

3.5.6.1.2. Metabolic acidosis

Causes.
1. Increase in H^+ ion production. If the cells are inadequately supplied with glucose, there is an increased conversion of fat with an accumulation of keto acids such as 3-hydroxybutyric acid and acetoacetic acid. Such a situation may occur, e.g. in chronic starvation and with diabetic ketoacidosis. The impairment of the aerobic metabolism of the cells can lead to an increase in lactic acid. This may occur, e.g. in cases of hypoxia, conditions of shock, and with severe anaemia.
2. Decrease in the excretion of H^+ ions by the kidneys. This can be due to an inadequate production of NH_3, e.g. in cases of chronic kidney insufficiency. But it may also be caused by an inadequate secretion of H^+, as found in renal tubular acidosis (distal type).
3. Loss of bicarbonate from the extracellular space. This is relatively rare and is found in disturbances of the proximal bicarbonate

reabsorption in the kidneys (renal tubular acidosis, proximal type), also in cases of loss of bicarbonate through the gut in chronic diarrhoea, and with biliary or pancreatic fistulae.

Secondary disturbances. There is a fall of pH and bicarbonate concentration in the plasma. The exodus of potassium from the cells in exchange for H^+ ions produces hyperkalaemia. The decrease in pH stimulates breathing and the pCO_2 falls below the normal value, leading to respiratory compensation. Renal compensation is possible only if the kidneys do not participate in the genesis of the acidosis. The result is an increase in the reabsorption of bicarbonate, of titratable acid, and an increased secretion of NH_4^+ owing to an increased production of ammonia. The urinary pH falls and the excretion of H^+ ions increases (p. 699).

3.5.6.1.3. Anion gap

The anion gap is a coarse parameter often used in the clinic for the evaluation of the metabolic acid–base relationships. It is based on the numerical equivalence of the anions and cations. The anion gap is calculated as

$$[Na^+] + [K^+] - ([Cl^-] + [HCO_3^-])$$

and amounts to about 6–20 mmol l^{-1}.[1,2,3]

An anion gap occurs in ketoacidosis, in cases of increased excretion of organic acids (organic aciduria), in congenital or inherited metabolic disorders, in kidney insufficiency, treatment with diuretics, and in other conditions.

3.5.6.1.4. Respiratory alkalosis

Causes. Respiratory alkalosis results from an increased CO_2 excretion by the lungs. It may be occasioned by the following: psychogenic hyperventilation, e.g. in psychovegetative lability; hypoxia; pulmonary diseases involving blocking of the alveolar capillaries are accompanied by a diminution in the diffusion of oxygen. The pO_2 in the blood sinks, stimulating the O_2 chemoreceptors in the carotid and the aorta, leading to stimulation of the breathing centre with increased exhalation of CO_2.

Secondary disturbances. The pCO_2 falls (hypocapnia) and the pH rises. Hydrogen ions pass out of the cells to be replaced by potassium ions, resulting in hypokalaemia. The kidneys attempt to correct the alkalosis. Because the pCO_2 in the renal tubule cells is low, there is a decrease in secretion of hydrogen ions and a lessening of the bicarbonate reabsorption. In urine, the titratable acid and production of ammonia are decreased and the pH is raised (p. 699). The increase in the pH in the blood can lead to a fall in the ionized fraction of the calcium in the plasma and hyperventilatory tetanus may ensue.

3.5.6.1.5. Metabolic alkalosis

Causes. It may result from a loss of hydrogen ions or an accumulation of bases, particularly of bicarbonate. This is particularly the case with chronic vomiting of acid gastric juices, and also in hypokalaemic alkalosis. With hypokalaemia, produced e.g. as a result of an enteric loss of potassium, potassium passes out of the intercellular into the extracellular space, while hydrogen ions invade the cellular space. This results in a rise in pH in the extracellular space. In spite of this H^+ ions are lost in the urine, leading to a paradoxical aciduria and an intensification of the metabolic alkalosis. The addition of too much alkali (sodium bicarbonate and other bases), e.g. in the treatment of stomach acidity, and the over-correction of metabolic acidoses can likewise occasion a metabolic alkalosis.

Secondary disturbances. The pH and bicarbonate concentration in the plasma increase. Intracellular hydrogen ions are exchanged by extracellular potassium ions and a hypokalaemia results. Respiratory compensation: the increased pH causes a slackening of the breathing with a concomitant increase of the pCO_2, but as a rule this respiratory compensation is limited by lack of O_2. For this reason, pulmonary compensation is usually inadequate. Provided that the metabolic alkalosis is not caused by a primary loss of potassium through the kidneys, then the latter may participate in compensating for the metabolic alkalosis. There is a reduction of H^+ ion secretion and of NH_3 production. In urine the titratable acid and the excretion of NH_4^+ are decreased and the pH is alkaline. If the bicarbonate concentration in the plasma exceeds 28 mmol l^{-1}, there may be excessive excretion of HCO_3^- (as sodium salt) in the urine. This may result in a loss of sodium with the risk of extracellular dehydration.

References

1. Truniger, B., *Wasser-und Elektrolyt-Haushalt*, 3. Aufl., Thieme, Stuttgart, 1971.
2. Witte, D. L., Rodgers, J. L., and Barrett, D. A., The anion gap: its use in quality control, *Clin. Chem.*, **22**, 643 (1976).
3. Zilva, J. F., The anion gap, *Lancet,* **ii**, 34 (1977).

3.6. OXYGEN IN THE BLOOD

3.6.1. BASIC CONCEPTS

3.6.1.1. Oxygen transport

A very small part of the oxygen (O_2) in the blood is physically dissolved. In the main it is transported bound to haemoglobin (Hb). The Hb molecule has a molecular weight of 64 458 and an Fe(II) moiety is enclosed in each haem residue. Each iron atom binds one molecule of O_2 without any change in valency. This process is known as oxygenation.

O_2 transport consists of those processes whereby O_2 is carried from the inhaled air to the places where it is consumed inside the cells. The amount of O_2 available per minute is determined by the product of the arterial O_2 content and the heart-beat volume per minute. The exchange of gases in the lungs, the heart-beat volume, the Hb concentration of the blood, the Hb–O_2 affinity and the O_2 consumption of the body all affect the oxygen transport. To make this intelligible, we ought first to define a few terms.

3.6.1.2. Partial pressure of the oxygen; oxygen tension

In a mixture of gases, each individual gas exerts a pressure corresponding with its percentage in the gas mixture and this is called the *partial pressure* (p_{gas}) for the gas concerned.

An *example* will show how the partial pressure is calculated: the composition of air is as follows: 20.98% O_2, 0.04% CO_2, 78.06% N_2, and 0.92% inert gases. The barometric height at sea level is 760 mmHg. Accordingly, the partial pressures of the individual gases would be:

$$pO_2 = \frac{760 \cdot 20.98}{100} = 160 \text{ mmHg, } 0.3 \text{ mmHg for } pCO_2, 600 \text{ mmHg for}$$

pN_2 and inert gases.

A gas diffuses from a place of higher partial pressure to one of lower partial pressure.

In calculating the partial pressures in the lungs and in the gas bottles used for calibrating measuring instruments, attention has also to be paid to the water-vapour pressure, since this may be a component of the gas mixture.

314

Example. A gas mixture in the lungs is under a pressure of 708 mmHg (Berne) and has the following composition: 12.3 vol.-% O_2, 5.2 vol.-% CO_2, 6.6 vol.-% water vapour, and 76 vol.-% N_2. Calculation gives the following partial pressures:

$$
\begin{array}{ll}
pO_2 \dfrac{708 \cdot 12.3}{100} = & 87.0 \text{ mmHg} \\[2mm]
pCO_2 \quad = & 36.8 \text{ mmHg} \\[1mm]
pH_2O \quad = & 46.7 \text{ mmHg} \\[1mm]
pN_2 \quad \underline{= 538.1 \text{ mmHg}} \\[1mm]
\quad\quad\quad 708.6 \text{ mmHg}
\end{array}
$$

Figure 95 shows the average partial pressure in the lungs and circulation. The pO_2 is the 'driving force' for the oxygenation of the blood. Normally, in the lungs the partial pressure of the O_2 is equalized between the alveoli and the pulmonary capillaries. The venous mixed blood flows into the lungs with a pO_2 of about 40 mmHg and leaves with a pO_2 of about 95 mmHg. The contact time in the lungs is extremely short (less than 1 s).

3.6.1.3. Oxygen capacity, content, affinity, and saturation

If a gas such as air is brought into contact with a liquid which contains no oxygen, then O_2 molecules diffuse into the liquid until its pO_2 is identical with the pO_2 of air. The amount of 'dissolved' O_2 in the liquid is directly proportional to the pO_2. Thus, when involved in the distribution equilibrium in the liquid, the gas has the same partial pressure as in the gas

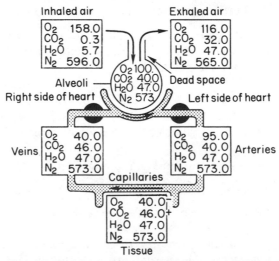

Fig. 95. Partial pressure of various gases in the body at a barometric pressure of 760 mmHg (from ref. 1).

phase. The amount of O_2 by volume which is dissolved in the liquid, e.g. plasma, is additionally dependent on the *absorption* or *solubility coefficient*, α. This is dependent on the nature of the gas and the liquid, and the temperature. α expresses how many millilitres of a gas (under normal conditions) are taken up by 1 ml of a liquid at a particular temperature when the gas concerned is a component of a gaseous mixture exerting a pressure of 760 mmHg.[2] Accordingly

$$V(ml) = \frac{\alpha}{760} \cdot p \cdot 100$$

are dissolved in 100 ml of liquid.

For O_2 in plasma at 37 °C α is 0.0214. Therefore, at a pO_2 of 100 mmHg, the volume of the dissolved O_2 would be

$$O_2(ml) = \frac{0.0214 \cdot 100 \cdot 100}{760} = 0.28 \text{ ml per 100 ml of plasma.}$$

For whole blood, α_{O_2} is 0.02356 at 37 °C. Siggaard-Andersen[6] gives the solubility coefficients in mole units instead of volume units and defines them as the number of millimoles of a gas which dissolves in 1 litre of liquid at a partial pressure of 1 mmHg: thus, for O_2 in plasma the value is 0.00126 mmol l^{-1} mmHg^{-1}.

The higher the partial pressure, the more gas dissolves. This relationship is linear (Fig. 96, curve B), but if the solution, e.g. plasma, contains Hb, the curve is sigmoid (Fig. 96, curve A). At any pO_2 far more O_2 can be bound by Hb than by plasma. At a pO_2 of 250 mmHg the Hb curve (A) is horizontal, i.e. no more O_2 molecules can be bound by the Hb. This maximum quantity is the O_2 *capacity*; obviously it varies with the Hb content. A maximum of 1.39 ml of O_2 is bound per gram of Hb A_1, i.e. 20.8 ml of O_2 (identical with 20.8 vol.-%) at 15 g of Hb per 100 ml. This

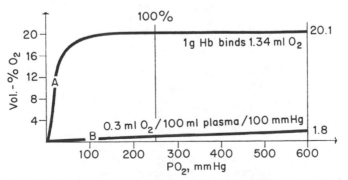

Fig. 96. Comparison between the dissociation curve of haemoglobin (15 g per 100 ml) at pH 7.40 and 38 °C (curve A) and the amount of oxygen dissolved in the plasma (curve B) for O_2 partial pressures from 0 to 600 mmHg (from ref. 3).

does not take into account CO-Hb (smokers!), Met-Hb, and Sulph-Hb, since these cannot be used for O_2 transport.

The O_2 capacity must be distinguished from the *O_2 content* of whole blood. This is the sum of the physically dissolved O_2 and that bound by Hb, expressed in volume percent. If the Hb concentration, O_2 saturation, and pO_2 are known, the O_2 content may be calculated as follows:

$$O_2 \text{ content(vol.-\%)} = \frac{O_2\text{(vol.-\%)} \cdot \text{Hb(g/100 ml)} \cdot 1.39}{100} + 0.0031 \cdot pO_2$$

$$\left(0.0031 = \text{ml } O_2/100 \text{ ml blood/mmHg} = \frac{0.02356}{760} \cdot 100\right)$$

The fraction of Hb in whole blood which is charged with O_2 corresponds to the *percentage oxygen saturation* (sO_2 %).

$$O_2 \text{ saturation(\%)} = \frac{O_2 \text{ content(vol.-\%)} - \text{physically dissolved } O_2\text{(vol.-\%)}}{O_2 \text{ capacity(vol.-\%)}} \cdot 100$$

The following *example* will show the practical operation of these concepts.[3]

An arterial blood sample might have a pO_2 of 50 mmHg, an sO_2 of 90% and 10 g of Hb per 100 ml; 90% saturation means that at a pO_2 of 50 mmHg, 9 g of Hb are charged with O_2 (90% of 10 g). In other words, 90% of the capacity of the Hb present in the blood is being used for O_2 transport, i.e. this blood sample contains 12.51 ml of O_2 bound to Hb (1.39 ml $\times$ 9 g). In the same blood sample, the O_2 content amounts to

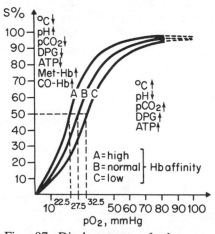

Fig. 97. Displacement of the oxy-haemoglobin dissociation curve by temperature, pH, pCO_2, 2,3-diphospho-glycerate (DPG), ATP, Met-Hb, and CO-Hb. B at 38 °C, pH 7.4, pCO_2 35.0 mmHg (from ref. 3).

12.21 ml, i.e. 12.66 ml bound to Hb and 0.15 (0.003 × 50 mmHg) dissolved in the plasma. Naturally, the physiological significance of 90% saturation is quite different if the blood sample contains 20 g of Hb instead of 10 g of Hb per 100 ml. For the same saturation, the amount of O_2 transported by Hb is twice as great (1.39 × 18 = 25.02 ml of O_2). The O_2 content would then be 25.17 ml in 100 ml of whole blood (25.02 ml of O_2 bound to Hb + 0.15 ml of O_2 dissolved in plasma).

The ability of the blood to transport oxygen is also characterized by another property, the O_2 *affinity*. This is a measure of the O_2 uptake of the Hb at a particular pO_2. The oxyhaemoglobin dissociation curve (Fig. 97) expresses this relationship. The position of the dissociation curve is shifted by various factors.[4,5] In order to fix the position of a particular dissociation curve, the so-called P_{50} is determined; this is the pO_2 at which the Hb is 50% saturated. A low P_{50} signifies a high affinity of the Hb for O_2 and *vice versa*. Displacement of the dissociation curve to the left means the Hb has a high affinity for O_2, the release of O_2 is made more difficult, and the O_2 uptake is facilitated. A displacement of the dissociation curve to the right signifies a smaller affinity, the release of O_2 is easier, and the O_2 uptake is more difficult. Table 56 lists the factors which can shift the dissociation curve. The change in the O_2 affinity of the Hb occasioned by an increase in hydrogen ion concentration is known as the Bohr effect. The bonding of H^+ ions in the deoxygenation of the Hb is called the Haldane effect.[6,7]

The following practical points derive from the shape of the dissociation curve.

Since the curve is flat in its upper ranges, the statement that, e.g. the O_2 saturation for a patient receiving supplementary oxygen is 95–98%, tells one nothing about his O_2 reserve. To evaluate the minimal O_2 supplement required, it is necessary to reduce the dosage stepwise until there is a marked decline in the saturation, e.g. down to 90% or even lower. In contrast, if the pO_2 (also called oxygen tension) is determined, this provides an immediate insight into the patient's reserve. A pO_2 of

Table 56. Factors affecting the Hb–O_2 affinity (from ref. 4).

Increased affinity	Reduced affinity
Increased temperature	Decreased temperature
Increased H^+ ion concentration (acidosis)	Reduced H^+ ion concentration (alkalosis)
Increased DPG and ATP	Reduced pCO_2
Increased Hb concentration	Reduced DPG and ATP
Increased ionic strength	Reduced Hb concentration
Abnormal Hb	Reduced ionic strength
Cortisone, aldosterone	Abnormal Hb
Age of the erythrocytes?	Carboxy-Hb
	Met-Hb
	Age of the erythrocytes?

318

Table 57. Standard values for certain O_2 parameters in whole blood at 37 °C.

	Arterial	Venous
pO_2, adults	80–90 mmHg	20–40 mmHg
pO_2 for persons over 65 years	75–85 mmHg	
pO_2, new-born	60–70 mmHg	
O_2 saturation	95–98%	40–70%
O_2 content, men	17.5–23 vol.%	
O_2 content, women	16–21.5 vol.%	
P_{50}	27 mmHg	

150 mmHg would indicate that the O_2 supplement might be diminished, while a value of 80 mmHg would show that the supplement was in the correct range. If the O_2 supplement is diminished further, then the saturation also falls. Thus, the pO_2 determination allows an adequate setting for the patient's O_2 supplement without first having to go to the trouble of an undersaturation. The area beneath curve B in Fig. 96 represents the O_2 dissolved in the plasma. This increases linearly with increasing partial pressure. At a pO_2 of 100 mmHg it amounts to 0.3 vol.-%. In hyperoxia experiments the O_2 dissolved in the plasma is increased to about 2 vol.-%. In order to assess such experiments, the pO_2 is necessary since the O_2 saturation is 100% in any case and this tells one nothing. Table 57 gives a few standard values in whole blood for the parameters discussed in this section.

References

1. Ganong, W. F., *Review of medical physiology*, Lange Medical Publications, Los Altos, 1965, p. 515.
2. Müller-Plathe, O., *Säure–Basen-Haushalt und Blutgase. Pathophysiologie, Klinik, Methodik*, Thieme, Stuttgart, 1973.
3. Duc, G., Assessment of hypoxia in the newborn. Suggestions for a practical approach, *Pediatrics*, **48**, 469 (1971).
4. Shappell, S. D. and Lenfant, C. J. M., Adaptive, genetic, and iatrogenic alterations of the oxyhemoglobin dissociation curve, *Anesthesiology*, **37**, 127 (1972).
5. Arturson, G., Garby, L., Robert, M., and Zaar, B., The oxygen dissociation curve of normal human blood with special reference to the influence of physiological effector ligands, *Scand. J. clin. Lab. Invest.*, **34**, 9 (1974).
6. Siggaard-Andersen, O., *The acid–base status of the blood*, 4th ed., Munksgaard, Copenhagen, 1974.
7. Roughton, F. J. W., Transport of oxygen and carbondioxide, in Fenn and Rahn, *Handbook of physiology. Respiration*, Vol. 1, American Physiological Society, Washington, 1964, p. 767.

3.6.2. MEASUREMENT OF THE OXYGEN PARTIAL PRESSURE

3.6.2.1. Principle

Today, polarography is the predominant method for measuring the physically dissolved oxygen in whole blood. This is based on amperometric measurement of the electric current produced by the reduction of molecular O_2. The electrode, developed by Clark *et al.*,[1] consists of a platinum cathode and a silver/silver chloride (Ag/AgCl) anode (Fig. 98) The surface of the electrode is covered with a semipermeable membrane which permits the passage of non-ionizing gases such as O_2, but is impervious to proteins, blood cells, and ions. This membrane consists of polypropylene, polethylene, or PTFE.

If a negative potential (referred to the non-polar Ag/AgCl anode) of -0.6 to -0.7 V is applied to the cathode, then the O_2 is reduced according to the following reaction:

$$O_2 + 2H^+ + 4e^- \rightarrow H_2O_2 + 2e^- \rightarrow 2OH^-$$

The flow of electrons produces a current proportional to the O_2 diffusing into the system. OH^- combines with H^+ to form water, so the pH in the electrolyte solution increases. This increase is compensated for by the addition of buffer. Silver ions are produced at the anode:

$$Ag \rightarrow Ag^+ + e^-$$

and are precipitated as AgCl. The O_2 concentration at the cathode is zero

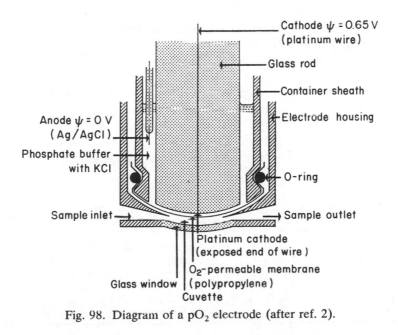

Fig. 98. Diagram of a pO_2 electrode (after ref. 2).

since the O_2 is constantly reduced. The amount of O_2 diffusing into the system is dependent on the pO_2 gradients, i.e. the pO_2 of the sample and the diffusion coefficient of the O_2 in the sample, as well as the thickness of the membrane.

3.6.2.2. Measurement

Various instruments are available for the measurement of pO_2, some with combined cuvettes for the measurement of pH and pCO_2. The calibration and measurements should be carried out in exact accordance with the manufacturer's instructions. In the main, the measurements are made with freshly collected arterial blood, but arterialized capillary blood may also be used. Measurements on venous blood are exceptional.

3.6.2.3. Reference values

These are given in Table 58. Further reference values may be found in the book by Siggaard-Andersen.[2]

3.6.2.4. Comments on the method

1. The electrolyte solution is the conducting element between the platinum electrode and the reference electrode. Gambino[4] used a mixture of equal parts of 0.1 mol l^{-1} KCl and 0.5 mol l^{-1} $NaHCO_3$. A mixture of phosphate buffer (0.5 mol l^{-1}, pH 7.0) with KCl (0.13 mol l^{-1}) saturated with AgCl is also used.[2]

2. The pO_2 meter is a sensitive ammeter yielding a polarization potential of -6.5 V independent of the current strength.

3. The cuvette should be suitable for small volumes of blood (100–200 μl) and the electrode should be thermostatically controlled to ± 0.1 °C. If the blood has been kept cold it is essential to wait until it has reached the measuring temperature of 37 °C. As a rule this takes only a few seconds in the cuvette. The more modern instruments have an electrical temperature indicator which shows when the sample has reached 37 °C and is ready for measurement. The error of measurement can

Table 58. Reference values for arterial O_2 at rest for healthy, male non-smokers, valid for Berne, at an average barometric height of 712 mmHg.

Age, years	Number	Reference values, mmHg	Reference
20–30	10	86 ± 5.0	3
40–50	8	81.0 ± 7.5	
>60	7	74.5 ± 9.5	

amount to 10% $°C^{-1}$.[2] Air bubbles in the vessels used for collecting the blood lead to gross errors.[5]

4. If the body temperature of the patient differed from 37 °C when the blood was collected, and the pO_2 is measured at 37 °C, a correction must be applied. Apart from the temperature difference, the correction factor is dependent on the O_2 saturation, which must also be measured beforehand and taken into account. Kelman and Nunn[6] constructed a nomogram for the temperature correction of pH, pCO_2, and pO_2 for a measurement temperature of 37 °C (Fig. 99).

5. As a rule, the instrument is calibrated with a gas mixture of known O_2 concentration. The pO_2 can be calculated from this as follows: barometric pressure minus water-vapour pressure at the measurement temperature (37 °C) multiplied by the gas content in volume-percent:

$$pO_2 = \frac{(pAtmos - pH_2O) \cdot \text{vol.-}\%}{100}$$

Example. pAtmos = 712 mmHg (Berne elevation above sea-level); pH_2O (37 °C) = 47 mmHg; gas mixture contains 12% of O_2:

$$\frac{(712 - 47) \cdot 12}{100} = 80 \text{ mmHg}$$

When calibrating with gases, it should be remembered that the diffusion conditions are not the same as those present with the O_2-containing liquid. Consequently, the results given by the pO_2 electrode will not be exactly identical for blood or gas at the same O_2 pressure, and a correction factor must be introduced for the measurement of blood. This can be evaluated as follows: a blood sample is equilibrated with a calibration gas mixture at 37 °C and water-vapour saturation. The sample must not be allowed to come into contact with air during the measurement. It is transferred to the cuvette and the instrument is set to the calculated value of the pO_2 of the calibration gas (= value a). The cuvette is then rinsed and subsequently filled with the calibration gas without changing the setting. A higher value (= value b) is measured. The measurement is repeated several times and the results are averaged. The gas/blood factor is then a/b.

Another way of calculating the gas/blood factor is to derive it from the gas/water factor, which is determined as described above. According to Severinghaus,[7] the gas/blood factor is given by 1.7 × gas/water factor −0.7.

The calculated pO_2 for the calibration gas is multiplied by the gas/blood factor to give the corrected pO_2. The factor is supplied by some instrument manufacturers. It has also been suggested that a solution of glycerine in water, equilibrated with a known pO_2, may be used to determine the gas/blood factor.

Certain authors found no difference whether the calibration was performed with moist, slowly flowing gas or with tonometricated blood.[9] Isherwood et al.[10] advised against standardizing with a gas/water mixture.

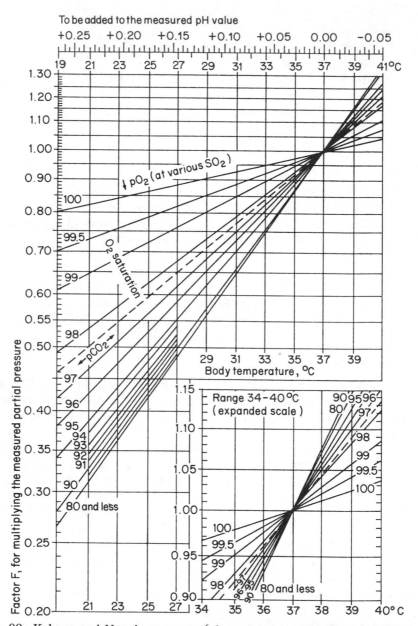

Fig. 99. Kelman and Nunn's nomogram[6] for correcting pH, pCO_2, and pO_2 for a measuring temperature of 37 °C. Abscissa = body temperature of sample donor. Ordinate = factor F by which the measured partial pressure is to be multiplied. Uppermost scale: pH to be added to the measured pH value. At the lower right, the range 34–40 °C has been drawn on an expanded scale. The factor for pCO_2 is given by the dashed line; F for pO_2 is found by means of the O_2 saturation lines. Example of the use of the nomogram: the following measurements are made at 37 °C: pCO_2 = 40 mmHg, pO_2 = 80 mmHg, pH = 7.40, O_2 saturation = 96%. What would be the correct values for a body temperature of 30 °C? The multiplication factor F for pCO_2 (dashed line) is 0.74; pCO_2 (30 °C) = 29.6 mmHg; F for pO_2 (line for 96% saturation) is 0.68; pO_2 (30 °C) = 54.4 mmHg; pH (top abscissa) + 0.103; ph 30 °C = 7.503.

With certain electrodes, the gas/blood factor lies within the error of measurement of the electrodes and may therefore be neglected.[11]

6. Repeated measurements on oxygen/nitrogen mixtures over a period of 4 months gave a coefficient of variation for the pO_2 values of 1.34% (1.55% for high pO_2).[10] For serial measurements on arterial blood, we obtained a coefficient of variation of 2.0–2.3%.

7. The precision and accuracy of pO_2 measurements depend on various factors: collection of the blood, preservation, transport and mixing of the blood, temperature of the patient, diffusion properties of the pO_2 standards, the membrane, and electrical and physical properties of the electrodes.[10] Many of these factors are outside laboratory control. The laboratory quality control of the pO_2 determinations is consequently essentially a matter of calibrating the instruments and working according to instructions. Although artificial gas–liquid mixtures have been recommended,[12] their suitability for calibration and quality control purposes is limited. In our opinion, blood is the only material worthy of consideration for the quality control of pO_2 measurements. The accuracy of the pO_2 values is best tested[8] with fresh blood which has been equilibrated with a gas content pO_2 in a tonometer at 37 °C.[13]

8. Weisbrot et al.[14] found divergences for pCO_2 of up to 11 mmHg and for pO_2 of up to 12 mmHg for two laboratories using different instruments. Our own investigations with two different instruments showed average differences of 1.1 mmHg (maximum 4.7 mmHg) using arterial blood. The mean difference between duplicated determinations on the two instruments was 1.05 or 0.95 mmHg (maximum 3.9 mmHg).

9. Recently, attempts have been made, especially with new-born infants, to measure the pO_2 transcutaneously with electrodes.[15]

10. A nomogram for evaluating the pO_2 and the O_2 saturation at various heights above sea-level has been published by Schnabel et al.[16]

References

1. Clark, L. C., Wolf, R., Granger, D., and Taylor, Z., Continuous recording of blood oxygen tensions by polarography, *J. appl. Physiol.*, **6**, 189 (1953).
2. Siggaard-Andersen, O., *The acid–base status of the blood*, 4th ed., Munksgaard, Copenhagen, 1974.
3. Bachofen, H., Hobi, H. J., and Scherrer, M., Alveolar–arterial N_2 gradients at rest and during exercise in healthy men of different ages, *J. appl. Physiol.*, **34**, 137 (1973).
4. Gambino, R., Oxygen, partial pressure electrode method, *Stand. Meth. clin. Chem.*, **6**, 171 (1970).
5. Müller, R. G., Lange, G. E., and Beam, J. M., Bubbles in samples for blood gas determinations, *Am. J. clin. Path.*, **65**, 242 (1976).
6. Kelman, G. R. and Nunn, J. F., Nomograms for correction of blood pO_2, pCO_2, pH, and base excess for time and temperature, *J. appl. Physiol.*, **21**, 1484 (1966).
7. Severinghaus, J. W., Measurements of blood gases: pO_2 and pCO_2, *Ann. N.Y. Acad. Sci.*, **148**, 115 (1968).

324

8. Gibson, P. F., Quality control in blood gas analysis, *Lab. Equip. Dig.*, (January 1974).
9. Holmes, P. L., Green, H. E., and Lopez-Majano, V., Evaluation of methods for calibration of O_2 and CO_2 electrodes, *Am. J. clin. Path.*, **54**, 566 (1970).
10. Isherwood, D. M., Isherwood, D. R., and Annan, W., Factors affecting the precision and accuracy of pO_2 measurements using the Clark electrode, *Clin. chim. Acta*, **42**, 295 (1972).
11. Bird, B. D., Williams, J., and Whitwam, J. G., The blood gas factor: a comparison of three different oxygen electrodes, *Br. J. Anaesth.*, **46**, 249 (1974).
12. Veefkind, A. H., Camp, R. A. M. van den, and Maas, A. H. J., Use of carbon dioxide- and oxygen-tonometered phosphate–bicarbonate–chloride–glycerol– water mixtures for calibration and control of pH, pCO_2, and pO_2 electrode systems, *Clin. Chem.*, **21**, 685 (1975).
13. Chalmers, C., Bird, B. D., and Whitwam, J. G., Evaluation of a new thin film tonometer, *Br. J. Anaesth.*, **46**, 253 (1974).
14. Weisbrot, I. M., Kambli, V. B., and Gorton, L., An evaluation of clinical laboratory performance of pH–blood gas analyses using whole-blood tonometer specimens, *Am. J. clin. Path.*, **61**, 923 (1974).
15. Huch, R., Lübbers, D. W., and Huch, A., Reliability of transcutaneous monitoring of arterial pO_2 in newborn infants, *Archs Dis. Childh.*, **49**, 213 (1974).
16. Schnabel, K. H., Schmidt, W., Schultz, V., and Gladisch, W., Nomographische Darstellung der Beziehung zwischen der Höhe über N.N. und dem arteriellen Sauerstoffpartialdruck des Menschen in Abhängigkeit vom Lebensalter, *Respiration*, **32**, 277 (1975).

3.6.3. MEASUREMENT OF THE OXYGEN SATURATION

3.6.3.1. Principle

In a blood sample, that portion of the Hb which is charged with O_2 is expressed by the percentage saturation (sO_2)

Various methods are available for determining this quantity. It may be calculated approximately from the O_2 dissociation curve (p. 316) if the pO_2, pH, and the temperature are known. This is usually done with the aid of the calculator devised by Severinghaus.[1] Other nomograms have been considered on the basis of more recent data. Figure 100 shows that of Rahn, corrected according to Severinghaus, as employed in the Berne isolation hospital.[1,11,12] All those factors which affect the O_2 dissociation curve also affect the calculation of the O_2 saturation and make the construction of such nomograms all the more difficult.[2]

The O_2 saturation can be determined instrumentally in the classical manner: gasometrically using the method of Peters and van Slyke.[3] This is precise but inconvenient and time consuming; it is still recommended as a reference method, however.[4,5] Recently, other methods based on reflection photometry[6] or transmission spectrophotometry[7] have been used. The former has the advantage that the O_2 saturation in whole blood can be measured directly, whereas with transmission spectrophotometry, the whole

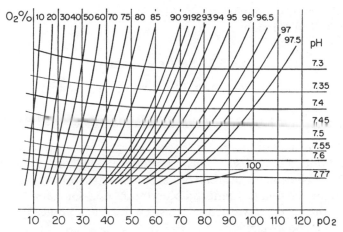

Fig. 100. Relationship between pO_2, sO_2 and blood pH. The O_2–CO_2 diagram of Rahn and Fenn[11] has been modified by Severinghaus[1] in accordance with more recent data for the O_2 dissociation curve. The dissociation curve is used for checking evaluated parameters of the acid–base status.

blood or packed erythrocytes must be haemolysed first. Nevertheless, transmission spectrophotometry is to be preferred since it is based on the Lambert–Beer law. The principle of the method consists in measuring the difference in transmission of the haemolysate at different wavelengths. The choice of wavelengths differs from one instrument to the next but it is based on the different absorption spectra of the individual haemoglobin derivatives (Fig. 128). If the measurement is carried out at two wavelengths, as a rule these are 505 nm (isosbestic point for Hb and oxi-Hb) and 598 nm (Hb absorption higher than that of HbO_2), or 650 and 805 nm (infrared). Instruments which also take HbCO into account make the measurements at three wavelengths [548 nm (isosbestic point for Hb, HbO_2, and HbCO), 568 nm, and 578 nm].

Example 1. Found values: pH = 7.45, pO_2 = 94 mmHg; consequently the O_2 saturation must lie between 97 and 97.5%.

Example 2. Found values: pH = 7.33, O_2 saturation = 35%; consequently the pO_2 must be about 22 mmHg.

A change in the absorption, e.g. at 578 nm, in comparison with the absorption at 548 nm, is due to the oxy-Hb concentration. This is compounded as follows:

$$[Hb_{total}] = [Hb] \times [HbO_2] \times [HbCO]$$

whence

$$\%HbO_2 = \frac{[HbO_2]}{[Hb] + [HbO_2] + [HbCO]} \cdot 100$$

The measurements should be made in exact accordance with the manufacturer's instructions.

3.6.3.2. Reference values

The oxygen saturation in arterial blood normally amounts to 95–98%.

3.6.3.3. Comments on the method

1. Since the measurement is based on the spectral curves of Hb derivatives, the filters used must reproduce the desired wavelengths exactly. As a control, measurements should be performed on completely oxygenated and completely reduced blood from non-smokers.

2. Filling the cuvettes with blood must be done anaerobically and bubble formation must be avoided.

3. Methylene blue and eosin interfere with the O_2 measurement but cardiogreen does not.[4]

4. Arterial capillary blood can be used with certain instruments.[8,9] Theil et al.[9] found good agreement between arterial and capillary O_2 saturation in children, over the saturation range 21–100%. The value for capillary blood lay on average 1.49%, and in the extreme cases up to 6%, lower than that for arterial blood.

5. The error of measurement with transmission spectrophotometry is stated as ±0.2%.[4]

6. For the arterial O_2 determination, the coefficient of variation for duplicate determinations was 1.17%, compared with 1.58% for capillary blood.[9]

7. Agreement between reflection and transmission measurements is good over the entire range of saturation.[5]

8. If the O_2 saturation is measured at two wavelengths only, then incorrect values may be obtained at very high concentrations of HbCO (intoxication above 20% HbCO). The interference is negligible with values of up to 7%, as present in smokers.[10]

References

1. Severinghaus, J. W., Blood gas calculator, *J. appl. Physiol.*, **21**, 1108 (1966).
2. Müller-Plathe, O., *Säure–Basen-Haushalt und Blutgase*, Thieme, Stuttgart, 1973.
3. Peters, J. P. and Slyke, D. D. van, *Quantitative clinical chemistry. I. Interpretations*, 2nd ed., Williams & Wilkins, Baltimore, 1946.
4. Scherrer, M., Küng, J., and Mösli, P., Vergleich der mit dem IL-CO-Oximeter-Modell 182 und der nach van Slyke ermittelten O_2-Sättigung des Blutes. Einfluss von Methämoglobin und anderen-Farbstoffen, *Schweiz. med. Wschr.*, **101**, 1399 (1971).
5. Thomas, E. V., Anema, R. J., and McNamara, J. J., Comparison of three methods for determining oxygen saturation in arterial blood, *Am. J. med. Techn.*, **41**, 360 (1975).

6. Zijlstra, W. G., *A manual of reflection oximetry*, Van Gorcum, Assen, 1958.
7. Drabkin, D. L., Spectroscopy, photometry, spectrophotometry, *Med. Phys.*, **2**, 1039 (1950).
8. Siggaard-Andersen, O., *The acid–base status of the blood*, 4th ed., Munksgaard, Copenhagen, 1974.
9. Theil, E., Stocker, F., and Weber, J. W., Sauerstoffsättigung im Kapillarblut, *Helv. paediat. Acta*, **27**, 321 (1972).
10. Maas, A. H. J., Hamelink, M. L., and Leeuw, R. J. M. de, An evaluation of the spectrophotometric determination of HbO_2, HbCO and Hb in blood with the CO-Oximeter IL 182, *Clin. chim. Acta*, **29**, 303 (1970).
11. Rahn, H. and Fenn, W. O., *A graphed analysis of the respiratory gas exchange. The O_2-CO_2 diagram*, American Physiological Society, Washington, 1955.
12. Bachofen, H., Pneumologische Abteilung, Med. Universitätsklinik, Berne, personal communication.

3.7. TRACE ELEMENTS

3.7.1. COPPER

3.7.1.1. Introduction

The human body contains 100–150 mg of copper, the highest concentration being in the liver (6.9 ± 1.7 μg g^{-1} moist weight). The daily requirement is about 1–2 mg. In the blood, copper is transported by means of caeruloplasmin, an α_2-globulin which binds 8 mol of Cu(II) per mole; 90% of the plasma copper is bound to caeruloplasmin and about 7% to albumin, while a further small fraction is bound to sulphur-containing amino acids. The latter two contribute to the direct-acting copper. Copper is a component of a number of enzymes (cytochrome a, catalase, tyrosinase, monoamino-oxidase, ascorbic acid oxidase, and uricase). Apart from these, a number of organ-specific proteins containing copper are known (hepatocuprein of the liver, erythrocuprein of the erythrocytes and the bone marrow, cerebrocuprein of the central nervous system). The metabolism of copper is closely connected with that of iron; thus, copper promotes the reabsorption of iron from the digestive tract and is also necessary for the synthesis of haemoglobin. Copper is mainly eliminated into the gut through the bile. On average, the renal copper clearance amounts to 0.123 ± 0.117 ml min^{-1} per 1.73 m^2.[7] The concentration of copper in the hair of the head is 11.7 ± 2.6 μg g^{-1} in men and 15.4 ± 4.0 μg g^{-1} in women.[7]

3.7.1.2. Choice of method

There are several photometric methods for the determination of copper in biological media and they are all based upon the chelation of copper to form highly coloured complexes. However, since the introduction of atomic-absorption photometry into the clinical–chemical laboratory, these methods have passed into the background. We shall therefore confine ourselves to guidance on the preparation of serum (plasma) and urine for atomic-absorption measurements. The method for serum has been used successfully in the author's laboratory for several years. The method for urine derives from Pirke and Stamm[1] and this—adapted to a smaller sample volume—has also been in routine service in the author's laboratory for some years. Methods suitable for those laboratories which do not have an atomic-absorption photometer may be found in the literature.[2] Serum and plasma must be deproteinated prior to atomization in the flame. Owing to rapid contamination of the burner by protein residues, the methods which dispense with deproteination can hardly be recommended. The normal concentration of copper in urine is too low for direct atomization, so the copper is chelated amd then extracted into an organic solvent. Measurement in an organic solvent is many times more sensitive than in water (more effective atomization). It is not necessary to ash the urine before complexing the copper.

3.7.1.3. Principle

Serum (plasma) is deproteinated with trichloroacetic acid and the protein-free supernatant is aspirated into the flame. A light beam emitted by a copper-cathode lamp passes through the flame and its intensity is reduced by the copper atoms; the extinction is measured. Urine is treated with sodium diethyldithiocarbamate and the copper complex so formed is extracted with isobutyl methyl ketone. The extract is sprayed into the flame and the extinction measured. The instrument is calibrated with a dilution series of aqueous standard solutions of a copper salt which have been treated in the same way as serum or urine.

3.7.1.4. Reagents

Deproteinating agent. Dissolve 60 g of trichloroacetic acid in DM-water and make up to 1 litre; the solution keeps indefinitely.

Sodium diethyldithiocarbamate, 222 mmol l^{-1}. Dissolve 5 g in DM-water and make up to 100 ml; this solution may be kept for several months at 4 °C in a polythene bottle.

Isobutyl methyl ketone, p.a. grade.

*Copper stock solution,*100 mg l^{-1} (a) or 2 mmol l^{-1} (b). Dissolve 50 mg (a) or 63.54 mg (b) of copper turnings in 1–2 ml of concentrated nitric acid and make up to 500 ml with DM-water; the solution keeps indefinitely.

Copper working standard solution; 100 μg per 100 ml (a) or 20 μmol l^{-1} (b). Dilute 1 ml of the stock solution (a) or (b) to 100 ml with doubly distilled water; prepare freshly each week.

3.7.1.5. Procedure

1. *Serum, plasma.* Serum or plasma (1 ml) is diluted with 2 ml of water in a centrifuge tube and deproteinated with 2 ml of trichloroacetic acid.

	Sample, ml	Blank, ml	Standard, ml
Plasma, serum	1	—	—
Working standard	—	—	1
Doubly distilled water	2	3	2
Trichloroacetic acid	2	2	2

Centrifuge after 15 min; spray the supernatant into the flame and record the extinctions.

2. *Urine.* The urine samples are adjusted to pH 6–5.5, treated with sodium diethyldithiocarbamate in centrifuge tubes and extracted with isobutyl methyl ketone.

	Sample, ml	Blank, ml	Standard, ml
Urine, pH 6	10[1]	—	—
Doubly distilled water	10	10	9
Working standard	—	—	1
Sodium-diethyl dithiocarbamate solution	0.2	0.2	0.2
Isobutylmethyl ketone	5	5	5

[1]For Morbus-Wilson patients receiving therapy, use urine diluted 10-fold.

Stopper the vessels and shake vigorously for 2 min; centrifuge; break up any emulsion of the urine sample in the organic phase with a glass rod and centrifuge again; spray the supernatant into the flame and record the extinction.

3.7.1.6. Calculation

The following formulae are valid for a linear calibration graph (extinction proportional to concentration):
Plasma copper:

(a) $\dfrac{E_S - E_B}{E_{ST} - E_L} \cdot 100 \ \mu g$ per 100 ml

(b) $\dfrac{E_S - E_B}{E_{ST} - E_L} \cdot 20 \ \mu mol \ l^{-1}$

Urine copper:

(a) $\dfrac{E_U - E_B}{E_{ST} - E_B} \cdot 100 \ \mu g \ l^{-1}$

(b) $\dfrac{E_U - E_B}{E_{ST} - E_B} \cdot 2 \ \mu mol \ l^{-1}$

3.7.1.7. Reference values

These are given in Table 59.

3.7.1.8. Comments on the method

1. The volume ratio of serum to trichloroacetic acid and the volume of serum used can be varied according to the sensitivity of the

Table 59. Reference values.[3]

	Serum		Urine	
	μg per 100 ml	μmol ml^{-1}	μg per 24 h	μmol per 24 h
New-born infants	30–54	4.7–8.5		
Children	77–185	12.1–29.0	6–54	0.09–0.85
Men	78–142	12.3–22.4	28–70	0.44–1.10
Women	84–156	13.2–24.6		

atomic-absorption spectrophotometer. When undiluted serum is mixed with the precipitating agent, care must be taken to ensure a fine distribution by intense agitation during the precipitation.[3]

2. Adding caesium chloride as ionization inhibitor[3] scarcely increases the sensitivity.

3.7.1.9. Diagnostic significance[4]

1. An elevated serum copper level is observed in infections, leukaemia, lymphomas, lymphogranulomatosis, hypo- and hyperthyrosis, collagenosis (lupus erythematodes, acute rheumatic fever), lipidoses, with diphenylhydantoin theraphy, and on ingestion of oestrogen and contraceptives.

2. A low serum copper is found in cases of Wilson's disease. The caeruloplasmin content of the serum is greatly reduced. The serum copper which is not bound to caeruloplasmin is increased, and so is the excretion of copper in the urine.

The hepatolenticular degeneration is now considered to be a consequence of a chronic copper intoxication which has its cause in a congenital autosomal–recessive, inherited metabolic disorder. The disease is accompanied by an accumulation of copper in various organs, above all in the liver, the basal ganglia of the brain, the nerves, and the muscles. The pathogenesis is not clear. Patients with Wilson's disease show a retarded uptake and an increased retention of copper in the liver, the incorporation of the copper into caeruloplasmin is defective, and there is a reduced transference of the copper from the liver into the circulation. At the same time there is a diminished excretion of copper through the bile and an increased excretion in urine. Wilson's disease often manifests itself even in youth. Clinically there is a cirrhosis of the liver, a usually bilateral degeneration of the basal ganglia and other parts of the central nervous system, and, in advanced cases, of the pathognomic Kaiser–Fleischer–Corneal ring; the latter may be absent in purely hepatic forms.[6] If copper-chelating drugs (e.g. D-penicillamine) are administered, the accumulated organ copper can be eliminated via the kidneys.[5,6]

A lowering of the serum copper is also occasioned by an acute leukaemia in remission following treatment with ACTH or prednisone, iron-deficiency anaemia in children, nephrosis (loss of caeruloplasmin into the urine), kwashiorkor, and with 'kinky-hair syndrome'.[8]

3. An increased excretion of copper in the urine is met with in Wilson's disease and with proteinuria. Monitoring the copper excretion is only indicated for the diagnosis and control of therapy in Wilson's disease.

References

1. Pirke, K. M. and Stamm, D., Die Messung der Kupferausscheidung im Harn mit der Atomabsorptionsspektrophotometrie, *Z. klin. Chem. klin. Biochem.*, **8**, 449 (1970).
2. *Standard methods of clinical chemistry*, Vol. 4, Academic Press, New York, 1963, p. 57.
3. Campenhausen, H. von and Müller-Plathe, O., Bestimmung des Serum-Kupfers mit der Atomabsorptionsphotometrie, *Z. klin. Chem. klin. Biochem.*, **13**, 489 (1975).
4. Eastham, R. D., *Interpretation klinisch–chemischer Laborresultate*, translated and edited by R. Richterich and J. P. Colombo, Karger, Basle, 1970.
5. Sass-Kortsak, A., Wilson's disease, *Pediat. Clins N. Am.*, **22**, 963 (1975).
6. Tschumi, A., Colombo, J. P. and Moser, H. Die Wilsonsche Krankheit in der Schweiz, Klinische, genetische und biochemische Untersuchungen, *Schweiz. med. Wschr.*, **103**, 89 (1973).
7. Mertz, D. P., Wilk, G., and Koschnick, R., Renale Ausscheidungsbedingungen von Kupfer beim Menschen, *Z. klin. Chem. klin. Biochem.*, **11**, 15 (1973).
8. Mollekaer, A. M., Kinky hair syndrome, *Acta paediat. scand.*, **63**, 289 (1974).

3.7.2. CAERULOPLASMIN: ENZYMATIC DETERMINATION WITH *p*-PHENYLENEDIAMINE

3.7.2.1. Choice of method

A large number of methods for the determination of caeruloplasmin have been described: immunochemical, spectrophotometric and enzymatic. Ravin[1,2] devised a simple photometric method based on the oxidation of *p*-phenylenediamine by oxygen with caeruloplasmin as enzyme. The technique described here is an adaptation of Ravin's method to micro-quantities and for use with a linear-recording photometer.[3] The determination may also be carried out by the two-point method.

3.7.2.2 Principle

Caeruloplasmin catalyses the oxidation of the colourless *p*-phenylenediamine to a blue–violet dye. The process is followed photometrically. A blank value is obtained by inhibiting the enzyme with azide.

3.7.2.3. Reagents

Acetate buffer, 212 *mmol* l^{-1}, pH 5.6. Dissolve 1.34 ml of glacial acetic acid + 26.44 g of sodium acetate trihydrate in DM-water and make up to 1000 ml; the solution keeps indefinitely if frozen.

Substrate solution, 8 *mmol* l^{-1}. Dissolve 36 mg *p*-phenylenediamine dihydrochloride in 25 ml of acetate buffer; the pH should be 5.6, otherwise it should be adjusted with NaOH. Prepare freshly, shortly before use; the solution is very sensitive to light.

Sodium azide solution, 460 *mmol* l^{-1}. Dissolve 3 g of sodium azide in DM-water and make up to 100 ml; the solution keeps well.

3.7.2.4. Procedure

3.7.2.4.1. Kinetic determination

Mixture. Pre-incubate 0.7 ml of substrate for 3 min at 37 °C; add 50 µl of serum.

(a) The increase in extinction at 546 nm is recorded for about 3 min with a linear recorder.

$$\text{Caeruloplasmin concentration} = \frac{v}{b} \cdot 1354 \cdot \tan\alpha \text{ mg per 100 ml}$$

where v = speed of paper and b = width of paper; cf. p. 237.

(b) The increase in extinction may be recorded manually for 3–4 min.

$$\text{Caeruloplasmin concentration} = \frac{\Delta E}{\min} \cdot 1354 \text{ mg per 100 ml}$$

3.7.2.4.2. Two-point method

	Sample, ml	Blank, ml
Substrate solution	1.0	1.0
Sodium azide solution	—	0.2
Serum	0.02	0.02

Incubate for 15 min at 37 °C.

Sodium azide solution	0.2	

Read off the extinction with respect to water at 546 nm after 15 min.

$$\text{Caeruloplasmin concentration} = 237(E_S - E_B) \text{ mg per 100 ml}$$

Table 60. Reference values.[4]

Age	Serum caeruloplasmin, mg per 100 ml
New-born infants	3–17
Up to 14 days	1–19
Up to 30 days	8–28
Up to 3 months	6–54
Up to 6 months	21–69
Up to 12 months	27–67
Over 1 year	30–58
Pregnant women	40–90

3.7.2.5. Specificity

The good agreement between the results obtained from the enzymatic and the non-enzymatic chemical methods and the immunological determination indicate that the method is highly specific.

3.7.2.6. Reference values

These are given in Table 60.

3.7.2.7. Comments on the method

1. The method has been calibrated with purified human caeruloplasmin. If the measurements cannot be made at 546 nm then it is necessary to run a parallel assay on a standard serum and to base the calculation on the caeruloplasmin content of the standard.

2. Various authors recommend the addition of EDTA, which stabilizes the substrate solution and gives a smaller blank. However, EDTA inhibits caeruloplasmin and the results tend to be low, so we have dispensed with this addition.

3. Haemolytic serum should be discarded since there is no guarantee of a linear relationship with these sera.

4. The caeruloplasmin in serum keeps fairly well. At room temperature the activity falls by 10–20% in 24 h. If frozen, it remains stable for at least 1 month (the pure enzyme loses activity even when frozen).

3.7.2.8. Diagnostic significance[4,5]

1. An elevated serum caeruloplasmin is found in infections, neoplasms, in post-operative conditions, on medication with oestrogens and contraceptives, and in schizophrenia.

2. A lowered serum caeruloplasmin level is found in Wilson's disease, kwashiorkor, and, sporadically, with chronic hepatitis or cirrhosis of the liver and similar diseases.

References

1. Ravin, H. A., Rapid test for hepatolenticular degeneration, *Lancet*, **i**, 726 (1956).
2. Ravin, H. A., An improved colorimetric enzymatic assay of caeruloplasmin, *J. Lab. clin. Med.*, **58**, 161 (1961).
3. Colombo, J. P. and Richterich, R., Zur Bestimmung des Caeruloplasmins im Plasma, *Schweiz. med. Wschr.*, **94**, 715 (1964).
4. Eastham, R. D., *Interpretation klinisch–chemischer Laborresultate*, translated and edited by R. Richterich and J. P. Colombo, Karger, Basle, 1970.
5. Pfeiffer, H. and Colombo, J. P., Diagnostische Bedeutung des Caeruloplasmin bei Leberkrankheiten, *Epatologia*, **17**, 383 (1971).

3.7.3. IRON

3.7.3.1. Introduction

Dietary iron must be reduced to the divalent form in the intestinal lumen. The Fe(II) is absorbed by a 'trapping' mechanism in the intestinal wall and released into the plasma after being oxidized to Fe(III). In the plasma it is bound to a β-globulin, transferrin; 1 molecule of transferrin (siderophilin) binds 2 atoms of Fe(III). It is in this form that the iron is transported between the organs. About 50 μg of iron is excreted daily in the urine. A small part of this acts as a coenzyme in the cells, especially in conjuction with electron-transfer enzymes. A further part is stored in an inactive form. A specific protein, apoferritin, bonds with iron to form ferritin, a metal proteid consisting of 23% of iron by weight. This reserve of iron is predominantly in the reticulo-endothelium. In bone marrow during erythropoiesis, the Fe(III) is reduced to Fe(II), the latter being used for the synthesis of haemoglobin. Myoglobin, another O_2-binding protein, is produced in the muscle by a similar process. Iron metabolism is illustrated in Fig. 101.

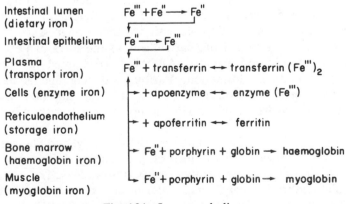

Fig. 101. Iron metabolism.

Iron is present in the blood plasma in at least the following four forms:

Haemoglobin iron. 1 mg of haemoglobin contains 3.4 μg of iron which cannot be sundered by dilute acid. Even if the blood is collected with the utmost care, 100 ml of plasma still contains about 5 mg of haemoglobin or 17 μg of iron. On average, serum contains 30 mg of haemoglobin or 102 μg of haemoglobin iron per 100 ml. The haemoglobin passes into the serum during the coagulation of the blood.

Transferrin iron. 100 mg of transferrin binds about 120 μg of iron. Transferrin is not a homogeneous protein but is present in various genetically determined forms. Its function is to render the toxic iron ions inactive by binding them. As a result, there are usually many more iron-bonding sites than iron atoms in the plasma (the so-called latent iron-binding capacity).

Enzyme iron. Iron also occurs as a component of cell enzymes in the plasma, for example in plasma catalase. Since the enzymes which contain iron are predominantly concerned with the respiratory chain, i.e. they are integral components of cell organelles, they only pass into the plasma in small quantities.

Free iron ions. In iron poisoning there is not enough iron-binding protein to fix all the iron and the excess iron ions are toxic.

For a crude assessment of the iron metabolism, the determination of the total iron in the serum is of paramount importance. A knowledge of the so-called iron-binding capacity is also of interest as this is a measure for the transferrin content of the serum (p. 341).

3.7.3.2. Choice of method

The spectrophotometric methods for the determination of serum iron can be analysed into the following three steps:

(*i*) *Cleavage of the iron from the transferrin.* The iron bound to protein is not very reactive and must first be liberated. This may be done either with acids or with a detergent. Cleavage of the iron with acids is often combined with deproteination. We prefer to use the detergent 'Teepol', as recommended by Sanford;[1] a deproteination is then unnecessary.

(*ii*) *Reduction of the iron.* Iron, particularly in the form of Fe(II), forms various intensely coloured chelates. All the Fe(III) must be reduced to Fe(II) prior to complex formation. Various reducing agents have been recommended, such as hydroquinone, hydrazine, sulphite, ascorbate, thioglycolate, and hydroxylamine. The method to be described below uses dithionite.[2]

(*iii*) *Formation of the chelate complexes.* The dipyridyl and phenanthroline derivatives, well known since the turn of the century, have relatively low extinction coefficients and are hardly ever used nowadays. The introduction of bathophenanthroline and its water-insoluble sulphonate was a considerable advance. The iron chelate with tripyridyltriazine has an absorption maximum about 10% higher than that of the bathophenanthroline complex, but its position is less favourable for photometers employing a mercury lamp (593 mm). Also, chelate formation with this ligand is more sensitive to pH than for bathophenanthroline sulphonate.

The fewer the manipulations and transferences, the more accurate is the iron determination. The method recommended here, first described by Sanford[1] and modified by Lauber,[2] has the advantage that the determination may be carried out directly in the photometer cuvette.

3.7.3.3. Principle

(a) $Transferrin(Fe^{III})_2 \xrightarrow{\text{Teepol}} Transferrin + 2Fe^{3+}$.

(b) $Fe^{3+} + dithionite \longrightarrow Fe^{2+} + sulphite$.

(c) $Fe^{2+} + 3(bathophenanthroline\ sulphonate)$
$$\longrightarrow Fe^{II}(bathophenanthroline)_3;$$

3.7.3.4. Reagents

Teepol reagent (iron-free). Dissolve 50 g of sodium dithionite ($Na_2S_2O_4$) + 10 g of magnesium sulphate ($MgSO_4 \cdot 7H_2O$) in 800 ml of DM-water and immediately mix with 1200 ml of Teepol-610. Then, at once mix in 500 ml of sodium hydroxide solution (15 g of NaOH per 500 ml of DM-water). After 10–15 min, centrifuge the turbid solution. All of the iron impurities are thereby converted into $Fe(OH)_2$, which is carried down as sediment with the $Mg(OH)_2$ precipitate. Separate off the supernatant and adjust its pH to 5.4–6.2 with glacial acetic acid. Portions adequate for one day's work are preserved at 4 °C in completely filled, tightly stoppered vessels. These may be kept for at least 2 months. The reagent loses its reducing action on exposure to air.

Chelating reagent. Dissolve 1.7 g of bathophenanthrolinedisulphonic acid, disodium salt, in DM-water and make up to 100 ml. The sulphonate can also be prepared as follows: dissolve 100 mg of bathophenanthroline in 0.5 ml of chlorosulphonic acid and heat over a low flame. Cool the brown solution with water and then mix with 10 ml of DM-water. Redissolve the precipitate which forms by gentle heating. The solution keeps indefinitely.

Iron standard. (a) 100 μg of iron per 100 ml: dissolve 50 mg of pure iron in a mixture of 1 ml each of concentrated HCl and concentrated HNO_3, and dilute to 500 ml with DM-water. Dilute 1 ml of this stock solution to 100 ml with doubly distilled water. If frozen in glass vessels, this solution keeps indefinitely.

(b) 20 μmol l^{-1} of iron: 55.85 mg of iron are treated as described for (a).

3.7.3.5. Procedure

The determination is carried out in semi-micro cuvettes with a 1-cm path length and 0.5 ml capacity at 500–550 (546) nm (serial order I, II, III).

	Assay, cuvette		
	I: blank, ml	II: standard, ml	III: sample, ml
DM-water	0.2	—	—
Teepol reagent	0.5	—	—
Mix well and set the extinction at 0			
Chelating reagent	0.01	—	—
Mix well and read off the extinction $E(B)$			
Standard	—	0.2	—
Teepol reagent	—	0.5	—
Mix well and set the extinction at 0			
Chelating agent	—	0.01	—
Mix well and read off the extinction $E(ST)$			
Serum	—	—	0.2
Teepol reagent	—	—	0.5
Mix well and set the extinction at 0			
Chelating reagent	—	—	0.01
Mix well and read off the extinction $E(S)$			

3.7.3.6. Calculation

(a) With standard 100 μg per 100 ml:

$$\text{Iron concentration} = 100 \cdot \frac{E(S) - E(B)}{E(ST) - E(B)} \, \mu g \text{ per } 100 \text{ ml}$$

(b) With standard 20 μmol l^{-1}:

$$\text{Iron concentration} = 20 \cdot \frac{E(S) - E(B)}{E(ST) - E(B)} \, \mu mol \ l^{-1}.$$

(c) The calculation may also be made using the extinction coefficient.
$a_{546\,nm} = 3850$

$$\text{Concentration} = \frac{E(S) - E(B)}{ad} \cdot \frac{FV}{SV} \cdot 10^6 \ \mu g \text{ per 100 ml}$$

where FV = final volume and SV = sample volume

$$= \frac{E(S) - E(B)}{3850.1} \cdot \frac{0.71}{0.20} \cdot 10^6 \ \mu g \text{ per 100 ml}$$

$$= E(S) - E(B) \cdot 922 \ \mu g \text{ per 100 ml}$$

$$\varepsilon_{546\,nm} = 21502 \ 1 \ \text{mol}^{-1} \ \text{cm}^{-1}$$

$$\text{Concentration} = E(S) - E(B) \cdot 165 \ \mu\text{mol} \ 1^{-1}.$$

In this case the standard is used as a control.

3.7.3.7. Stablity

Serum and plasma may be kept for at least 4 days at room temperature and for at least 1 week at 4 °C.[3]

3.7.3.8. Specificity

Bathophenanthroline sulphonate is specific for Fe(II). The determination does not include any haemoglobin iron which may be present (iron is not liberated from haemoglobin by Teepol).

3.7.3.9. Reference values

These are given in Table 61.

3.7.3.10. Comments on the method

1. The determination may be performed on serum or plasma as desired.
2. Instead of making up the mixtures in the cuvettes they may be prepared beforehand in tubes (greater risk of contamination).
3. If sufficient serum is available it is advisable to scale up all of the given volumes three-fold and to work with macrocuvettes (easier mixing and less risk of contamination).
4. With lipaemic sera it is necessary to wait 15 min after adding the Teepol reagent; this gives the sample time to clear. If this is not done there is a risk of downwards 'trailing' indications.
5. If the reagents have been prepared correctly, the blank values for the extinction should be less than 0.01.
6. Haemolysis does not interfere provided the haemoglobin concentration is not above 500 mg per 100 ml (too highly coloured).

Table 61. Reference values for serum iron.

Age	µg per 100 ml	µmol l^{-1}	Reference
Umbilical cord blood	17–237	3.0–42.4	4
New-born infants	28–108	5.0–19.4	5
Up to the 2nd month	84–180	15.0–28.7	5
Up to the 3rd month	48–136	8.6–24.4	5
Up to the 1st year	22–118	3.9–21.1	5
Up to the 2nd year	21–119	3.8–21.3	5
Up to the 14th year	46–230	8.2–41.2	5
After the 14th year, women	45–165	8.0–29.6	5
After the 14th year, men	55–175	9.8–31.3	5
First 0.5 month	63–201	11–36	10
1st month	58–172	10–31	10
2nd month	15–159	3–29	10
4th month	18–164	3–29	10
6th month	28–135	5–24	10
9th month	34–135	6–24	10
12th month	35–155	6–28	10
Men	36–158	6.4–28.3	6
Adults, men + women (author's laboratory)	52–179	9.3–32.2	

7. Bilirubin (icteric sera) does not interfere.

8. With normal sera, the extinction reaches its final value a few seconds after the addition of the chelating reagent, but if the patient is being treated for iron the extinction may creep upwards over a longer period (competition between the chelating ligands of the drugs and the reagent). In this case it is necessary to wait until the extinction has stabilized if the determination is to provide a value for the total iron including that from the drugs.

9. Both EDTA and dextran interfere in the determination (do not use any plasma from EDTA – blood!); no interference is caused by ammonium heparinate, however.

3.7.3.11. Diagnostic significance[5]

1. The following points may be noted concerning the physiological scatter of serum iron values: pregnancy leads to lowered serum iron. Shortly after birth the iron in the serum is relatively high. In elderly persons the values are lowered. The concentration of the serum iron is strongly dependent on the time of day. The values are low at night, rise towards morning and reach a maximum at around 14.00 hours.[7,8] For a single individual, the within-day variation of the concentration may be 13% and the day-to-day variation up to 27%.[3,8,9]

2. Elevated serum iron is found in intensive parental donation of iron, repeated blood transfusion, haemochromatosis, haemolytic anaemia, hepatopathia, nephritis, and anaemias refractory to treatment.

3. The iron is lowered in iron-deficient anaemia, megaloblastic anaemia in remission, infections, neoplasms, nephrosis, treatment with ACTH or corticoids, kwashiorkor, and post-operative conditions.

References

1. Sanford, R., A new method for determination of the serum iron, *J. clin. Path.*, **16**, 174 (1963).
2. Lauber, K., Bestimmung von Serumeisen und Eisenbindungskapazität ohne Enteiweissung, *Z. klin. Chem. klin. Biochem.*, **3**, 96 (1965)..
3. Ramsay, W. N. M., Plasma iron, in *Advances in clinical chemistry*, Vol. 1, Academic Press, New York, 1958, p. 1.
4. Weippl, G., Pantlitschko, M., Bauer, P., and Lund, S., Normal values and distribution of single values of serum iron in cord blood, *Clini. chim. Acta*, **44**, 147 (1973).
5. Eastham, R. D., *Interpretation klinisch–chemischer Laborresultate*, Karger, Basle, 1970.
6. Megraw, R. E., Hritz, A. M., Babson, A. L., and Carroll, J. J., A single-tube technique for serum total iron and total iron binding capacity, *Clin. Biochem.*, **6**, 266 (1973).
7. Wiltin, K. W. F., Kruithof, J., Mol, C., Missgré, B. O. S., and Eigk, H. G. van, Diurnal and nocturnal variations of the serum iron in normal subjects, *Clini. chim. Acta*, **49**, 99 (1973).
8. Statland, B. E., Winkel, P., and Bokelund, H., Variation of serum iron concentration in young healthy men: within-day and day-to-day changes, *Clin. Biochem.*, **9**, 26 (1976).
9. Fiet, J., Passa, Ph., Dubos, G., Tabuteau, F. and Dreux, C., Etude des valeurs de référence et du cycle nyothéméral du fer sérique, *Annls Biol. clin.*, **35**, 305 (1977).
10 Saarinen, U. M. and Siimes, M. A., Developmental changes in serum iron, total iron-binding capacity, and transferrin saturation in infancy, *J. Pediatr.*, **91**, 875 (1977).

3.7.4. IRON-BINDING CAPACITY

3.7.4.1. Introduction

The *total iron-binding capacity*, TIBC, signifies the total quantity of iron in μg (or μmol) which can be bound by the transferrin in 100 ml (or 1 litre) of serum.

The *latent* or *unsaturated iron-binding capacity* (UIBC) is the additional quantity of iron in μg (or μmol) which can be bound by the transferrin in 100 ml (or 1 litre) of serum, over and above the serum iron already present (Fig. 102).

Serum iron + unsaturated iron-binding capacity = total iron-binding capacity

The ratio

$$\frac{\text{serum iron} \cdot 100}{\text{total iron-binding capacity}}$$

342

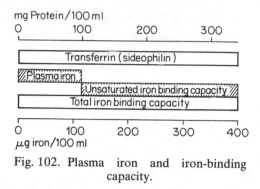

Fig. 102. Plasma iron and iron-binding capacity.

is termed the percentage transferrin saturation. Since 1 mol of transferrin binds 2 mol of iron, then

$$X \; \mu\text{mol } l^{-1} \text{ total iron-binding capacity} = 0.5 \; X \; \mu\text{mol } l^{-1} \text{ of transferrin}$$

3.7.4.2. Choice of method

Several methods of determination are available, both for total and for unsaturated iron-binding capacity.[1] They may be classified into three groups:

1. Photometric determination of the free-iron excess or of the iron bound to protein following addition of iron.
2. Measurement of the radioactivity in the serum protein after addition of ^{59}Fe.
3. Immunological methods. With the immunological and radioimmunological methods the transferrin is determined directly. With the other methods, depending on the conditions, the other proteins participating in binding the iron may be included in the determination. Since, with patients' sera, such foreign proteins are scarcely of any consequence, the non-immunological methods may also be used for diagnostic purposes. Because the immunological techniques are relatively expensive, we have chosen two methods from the first group which are suitable even for laboratories with less-sophisticated equipment: Ramsay's method[2] for the total iron-binding capacity and, for the unsaturated iron-binding capacity, a slightly modified procedure by Gemba et al.[3] The two methods correlate well (personal investigations). On the other hand, the method of Gemba et al.[3] correlates well with a radioisotope method.

3.7.4.3. Total iron-binding capacity

3.7.4.3.1. Principle

The serum sample is treated with an excess of Fe^{3+} ions, resulting in saturation of the transferrin. The iron which is not bound to protein is adsorbed on a magnesium carbonate suspension and centrifuged off. The iron bound to protein is determined on the supernatant.

3.7.4.3.2. Reagents

Iron(III) chloride solution, 25 mg l⁻¹ (92 μmol l⁻¹ of iron). Dissolve 25 mg of $FeCl_3.6H_2O$ in 0.005 mol l^{-1} HCl and make up to 1 litre with further acid; the solution keeps indefinitely.

Magnesium carbonate, 'basic'. E.g. magnesium hydroxide carbonate, Merck)

Reagents and standard for iron determination (p. 337).

3.7.4.3.3. Procedure

Mix 0.1 ml of serum (plasma) with 0.2 ml of FeCl₃ solution and allow to stand for 5 min, then add 10 mg of magnesium carbonate. Shake the samples thoroughly every 5 min over a period of 30 min. Centrifuge strongly. Determine the iron in 0.1 ml of supernatant (as described on p. 338).

3.7.4.3.4. Calculation

(a) Total iron-binding capacity using as standard 100 μg of iron per 100 ml:

$$\frac{E(S) - E(B)}{E(ST) - E(B)} \cdot \frac{0.3}{0.1} \cdot 100 \ \mu g \text{ per 100 ml}$$

(b) Total iron-binding capacity using as standard 20 μmol l⁻¹ iron:

$$\frac{E(S) - E(B)}{E(ST) - E(B)} \cdot \frac{0.3}{0.1} \cdot 20 \ \mu mol \ l^{-1}$$

3.7.4.3.5. Stability

Serum (plasma) may be kept for at least 4 days at room temperature, for 1 week at 4 °C, and indefinitely if frozen.

3.7.4.3.6. Reference values

These are given in Table 61a.

3.7.4.3.7. Comments on the method

1. It is advisable to test the adsorptivity of the magnesium carbonate used (no iron should be detectable in the supernatant from a test performed with iron standard and magnesium carbonate but no serum).

Table 61a. Reference values for the total iron-binding capacity.

Age	μg per 100 ml	μmol l$^{-1}(\bar{x} \pm s)$	Reference
First 0.5 month	191 ± 43	34 ± 8	5
1st month	199 ± 43	36 ± 8	5
2nd month	246 ± 55	44 ± 10	5
4th month	300 ± 39	54 ± 7	5
6th month	321 ± 51	58 ± 9	5
9th month	341 ± 42	61 ± 7	5
12th month	358 ± 38	64 ± 7	5
Adults	280 + 400	50 − 72	4

2. Certain lyophilized control sera yield values by this method which are too high.[1] The results can be strongly dependent on the amount of iron added. In redissolved lyophilizates more iron appears to be bound to proteins other than transferrin.

3. Neither bilirubin nor haemoglobin interferes in the determination.

3.7.4.3.8. Diagnostic significance[4]

1. An elevated total iron-binding capacity in association with a simultaneously elevated serum iron is found with hepatopathia, active cirrhosis of the liver, increased degradation of the blood (e.g. haemolytic anaemias), haemochromatosis, excessive intake of iron, iron-rich diet with a low phosphate intake, and taking of contraceptives.

2. An elevated total iron-binding capacity in association with a lowered serum iron occurs in pregnancy, loss of blood, and occasionally, with diminished glomerular filtration.

3. The total iron-binding capacity is reduced in infections (the fall in serum iron is proportionately greater than the fall in total iron-binding capacity), perniciosa in remission, uraemia, carcinomatosis, nephrotic syndrome, scurvy, haemolytic anaemia, and rheumatoid arthritis.

3.7.4.4. Unsaturated iron-binding capacity

3.7.4.4.1. Principle

A known quantity of iron is added to the serum sample so that the transferrin is saturated. The excess iron not bound to the transferrin is determined at pH 8.6. At this pH the iron bound to the protein does not react with the chelating agent.

3.7.4.4.2. Reagents

Nitrilotriacetate/tris buffer. Dissolve 120 g of tris(hydroxymethyl)amino-methane + 80 mg of nitrilotriacetic acid [tris(carboxymethyl)amine], Complexone I, in about 800 ml of DM-water; adjust the pH to 8.6 with HCl ($6 \text{ mol } l^{-1}$) and make up to 1 litre with DM-water.

Nitrilotriacetate–iron complex solution. Iron stock solution: dissolve 50 mg of pure iron in a mixture of 1 ml each of concentrated HCl and concentrated HNO_3 and dilute to 500 ml with DM-water. Take 1.67 ml of iron stock solution [(a) 100 mg l^{-1} of iron or (b) 2 mmol l^{-1} of iron] and make up to 100 ml with nitrilotriacetate/tris buffer.

Sodium ascorbate.

Colour reagent. Dissolve 0.5 g of bathophenanthroline disulphonate in 100 ml of DM-water.

3.7.4.4.3. Procedure

	Reagent blank, ml	Sample, ml	Standard, ml
Nitrilotriacetate	1.5	—	—
Iron nitrilotriacetate	—	1.5	1.5
Sodium ascorbate	10–30 mg	10–30 mg	10–30 mg
Serum	—	0.5	—
Doubly distilled water	0.5	—	0.5

Allow to stand at room temperature for 10 min, then read off the extinction for the serum sample $E(S)_1$ at 546 (500–550)nm.

Colour reagent	0.05	0.05	0.05

Allow to stand for 20 min, then read off the extinction

	$E(RB)$	$E(S)_2$	$E(ST)$

3.7.4.4.4. Calculation

(a) Unsaturated iron-binding capacity using the iron stock solution of 100 mg l^{-1}:

$$\left(1 - \frac{E(S)_2 - E(S)_1 - E(RB)}{E(ST) - E(RB)}\right) \cdot 500 \ \mu\text{g per 100 ml}$$

(b) Unsaturated iron binding capacity, using the iron stock solution of 2 mmol l^{-1}:

$$\left(1 - \frac{E(S)_2 - E(S)_1 - E(RB)}{E(ST) - E(RB)}\right) \cdot 100 \ \mu\text{mol } l^{-1}$$

3.7.4.4.5. Stability

See under 'Total iron-binding capacity' (p. 342).

3.7.4.4.6. Reference values

These are 180–260 μg per 100 ml or 32–56 μmol l^{-1} (65% of the total iron-binding capacity).

The values for the percentage transferrin saturation appear to be lower, particularly in childhood.[5,6]

References

1. Haeckel, R., Haindl, H., Hultsch, E., Mariss, P., and Oellerich, M., Comparison of 8 different colorimetric, radiochemical and immunological procedures for the determination of iron binding capacity, *Z. klin. Chem. klin. Biochem.*, **11**, 529 (1973).
2. Ramsay, W. N. M., Plasma iron, in *Advances in clinical chemistry*, Vol. 1, Academic Press, New York, 1958, p. 17.
3. Gemba, A., Shimoyama, C., and Hara, F., Simplified spectrophotometry for the determination of serum unsaturated iron binding capacity, *Clin. chim. Acta*, **48**, 85 (1973).
4. Eastham, R. D., *Interpretation klinisch–chemischer Laborresultate*, translated and edited by R. Richterich and J. P. Colombo, Karger, Basle, 1970.
5. Saarinen, U. M. and Siimes, M. A., Developmental changes in serum iron, total iron-binding capacity, and transferrin saturation in infancy, *J. Pediatr.*, **91**, 875 (1977).
6. Koerper, M. A. and Dallman, P. R., Serum iron concentration and transferrin saturation in the diagnosis of iron deficiency in children: normal developmental changes, *J. Pediatr.*, **91**, 870 (1977).

3.8. ENERGY METABOLISM

3.8.1. PYRUVATE AND LACTATE

3.8.1.1. Pyruvate: enzymatic determination

3.8.1.1.1. Choice of method

The method of choice is the enzymatic determination using lactate dehydrogenase as indicator enzyme. This technique is specific for pyruvate and yields very reliable results. The older colorimetric methods are less specific.

3.8.1.1.2. Principle

The determination is effected on the principle of enzymatic analysis utilizing the lactate dehydrogenase system:

$$\text{Pyruvate} + \text{NADH} + \text{H}^+ \xrightarrow[\text{pH 7.5}]{\text{LDH}} \text{L}(+)\text{-Lactate} + \text{NAD}^+$$

At a physiological pH, the equilibrium is shifted strongly towards the right and thus the above system can be applied directly to the determination of pyruvate.

3.8.1.1.3. Reagents

Perchloric acid solution, 6.6 g per 100 ml (0.66 mol l^{-1}). Dilute 5.7 ml of 70% perchloric acid to 100 ml with DM-water. The solution keeps indefinitely.

Potassium carbonate solution, 5 mol l^{-1}. Dissolve 69 g of anhydrous potassium carbonate in DM-water and make up to 100 ml. The solution keeps indefinitely.

Triethanolamine–EDTA buffer, 0.5 mol l^{-1}, pH 7.5. Dissolve 46.4 g of triethanolamine hydrochloride in *ca.* 400 ml of doubly distilled water, add 2.0 g of EDTA-Na$_2$H$_2$.2H$_2$O, adjust the pH to 7.5 with 5 mol l^{-1} NaOH solution and make up to 500 ml with doubly distilled water.

NADH$_2$ solution, 3 mmol l^{-1}. Dissolve 2.1 mg of nicotinamide adenine dinucleotide, reduced form (disodium salt), in 1 ml of 0.01 N NaOH solution (labile in acidic solution).

Lactate dehydrogenase solution, 1.5 *mg ml⁻¹* (*minimum activity* 300 *μmol min⁻¹ mg⁻¹*). Dissolve 15 mg of pure enzyme in 10 ml of DM-water.

Working solutions. Pyruvate I: mix 1 part of $NADH_2$ solution with 2 parts of buffer solution; prepare freshly each day. Pyruvate II: mix 2 parts of lactate dehydrogenase solution with 18 parts of buffer solution; prepare freshly each day.

3.8.1.1.4. Collection of the blood

The pyruvate/lactate ratio changes within seconds of collecting the blood. The most reliable results are therefore obtained if the blood is collected from the unconstricted vein directly into ice-cold perchloric acid. To this end, prepare test-tubes containing 3.0 ml of perchloric acid; these are tared (W_1). Allow about 1 ml of whole blood to flow from the canulus direct into the perchloric acid, mix well, and re-weigh the tubes (W_2).

3.8.1.1.5. Deproteination

Centrifuge strongly, draw off 2.0 ml of supernatant and add 0.1 ml of potassium carbonate solution. Mix well. The pH should lie between 7 and 8. Centrifuge strongly. The supernatant is used for the determination of pyruvate and lactate.

3.8.1.1.6. Stability of the sample

Pyruvate is not stable in either acidic or neutral extracts. The smallest losses have been observed in a neutral extract at 4 °C. Freezing produces considerable losses.[1]

3.8.1.1.7. Conditions for measurement

Triethanolamine, 11.14 mmol l⁻¹; EDTA, 1.11 mmol l⁻¹; $NADH_2$, 0.14 mmol l⁻¹; LDH, 21 mg l⁻¹ (21 μg ml⁻¹); final volume, 0.7 ml.

3.8.1.1.8. Procedure

In photometer cuvettes	Supernatant	0.5 ml
	Pyruvate I	0.1 ml

Read off the extinction (E_0) at 365 or 340 nm

Initiation	Pyruvate II	0.1 ml

Read off the extinction at 2-min intervals until constant (E_1)

3.8.1.1.9. Calculation

Correction factor for deproteination.

$$k = \frac{3.0 + (W_2 - W_1)}{W_2 - W_1} \cdot \frac{2.1}{2.0} = \frac{3.0 + (W_2 - W_1)}{W_2 - W_1} \cdot 1.05$$

Calculation of the change in extinction. The change in extinction (ΔE) amounts to

$$\left[(E_0 - E_1) - \frac{E_0}{7} \right],$$

where $E_0/7$ has to be subtracted to correct for the dilution after the addition of the lactate dehydrogenase buffer solution (pyruvate II).

Calculation of the concentration, $\lambda = 365$ nm ($\varepsilon = 3400$).

$$\text{Concentration} = \frac{\Delta E}{\varepsilon d} \cdot \frac{FV}{SV} \cdot 10^6 \cdot k \ \mu\text{mol l}^{-1}$$

$$= \frac{E}{3400 \cdot 1} \cdot \frac{0.7}{0.5} \cdot 10^6 \cdot k \ \mu\text{mol l}^{-1}$$

$$= \Delta E \cdot 412 \cdot k \ \mu\text{mol ml}^{-1}$$

$$= \left[(E_0 - E_1) - \frac{E_0}{7} \right] \cdot 0.432 \cdot \frac{3.0 + (W_2 - W_1)}{W_2 - W_1} \ \text{mmol l}^{-1}$$

3.8.1.1.10. Specificity

The method is specific for pyruvate.

3.8.1.1.11. Reference values

In order to obtain reliable results for lactate and pyruvate determinations, it is essential the two following preconditions be fulfilled:

1. The patient must rest for 1–2 h. Even the slightest exertion leads to a rise in the lactate and pyruvate values in arterial and venous blood.
2. Collection of the blood direct into perchloric acid. If the blood is allowed to stand, the lactate concentration increases within seconds, while the pyruvate concentration falls.

Depending on the clinical requirements, the determination may have to be conducted on arterial blood since the values for venous blood can reflect local metabolic changes in the peripheral musculature and may deviate from the ratios in the entire organism.

Reference values for venous blood. Pyruvate, 0.049 ± 0.009 mmol l^{-1};[2] 0.048 ± 0.028 mmol l^{-1} (personal observations).

3.8.1.1.12. Comments on the method

1. If the test does not proceed to a conclusion, i.e. if the final plot is not horizontal, then the result is evaluated graphically by extrapolation.

2. Freshly prepared standard solutions with pyruvate concentrations between 0.2 and 2 mmol l^{-1} are suitable for monitoring the method.

3.8.1.1.13. Diagnostic significance

See Lactate, p. 352.

References

1. Von Korff, R. W., Purity and stability of pyruvate and α-ketoglutarate, in *Methods in enzymology*, Vol. 13, Academic Press, New York, 1969, p. 519.
2. Landon, J., Fawcett, J. K., and Wynn, V., Blood pyruvate concentration measured by a specific method in control subjects, *J. clin. Path.*, **15**, 579 (1962).

3.8.1.2. Lactate: enzymatic determination

3.8.1.2.1. Choice of method

For the determination of L(+)-lactate, enzymatic methods have superseded non-specific reactions. In addition to the photometric measurement of the increase in $NADH_2$, described in detail below, a method employing membrane-bound enzymes and automated electrochemical measurement has been reported. Cytochrome b_2 serves as indicator enzyme as shown in the following scheme:[1]

$$\text{Lactate} + 2[Fe(CN)_6]^{3-} \xrightarrow{\text{Cytochrome } b_2} \text{Pyruvate} + 2[Fe(CN_6]^{4-} + 2H^+$$

$$2[Fe(CN_6)]^{4-} \xrightarrow{\text{Platinum}} 2Fe(CN_6)^{3-} + 2e^-$$

3.8.1.2.2. Principle

The determination of lactate is accomplished with lactate dehydrogenase, the same enzyme system as used in the determination of pyruvate. However, the position of the equilibrium is so unfavourable for determining the lactate that it is necessary to remove the reaction products, pyruvate and H^+, from the system. This is done by converting the pyruvate into its hydrazone and absorbing the protons in a strongly alkaline buffer:

L(+)-Lactate
NAD$^+$ $\xrightarrow[\text{pH 9.5}]{\text{LDH}}$ Pyruvate hydrazone
Hydrazine   NADH + H$^+$
Pyruvate   H$_3$O$^+$

The increase in the extinction at 365, 334, or 340 nm is then directly proportional to the lactate concentration present.

3.8.1.2.3. Reagents

Perchloric acid solution, 0.66 *mol l⁻¹* (p. 347).
Potassium carbonate solution, 5 *mol l⁻¹* (p. 347).
Glycine–hydrazine buffer, pH 9.0. Suspend 7.5 g of glycine, 5.2 g of hydrazine sulphate and 0.2 g of ethylenediaminetetraacetate (disodium salt, dihydrate) in DM-water. Add 51 ml of sodium hydroxide solution (2 mol l⁻¹) and make up to 100.0 ml with DM-water.
NAD solution, 30 *mmol l⁻¹*. Dissolve 20 mg of nicotinamide adenine dinucleotide in 1 ml of DM-water.
Lactate dehydrogenase, 1.5 *mg ml⁻¹* (p.348).
NaOH, 2 *mol l⁻¹*.

3.8.1.2.4. Blood collection, deproteination

See p. 348.

3.8.1.2.5. Conditions for measurement

Glycine	667 mmol l⁻¹
Hydrazine	267 mmol l⁻¹
EDTA	3.6 mmol l⁻¹
NAD	4.0 mmol l⁻¹
LDH	0.2 mg ml⁻¹
Final volume	1.5 ml

3.8.1.2.6. Procedure

In test-tubes	Sample, ml	Reagent blank, ml
Buffer	1.0	1.0
NAD solution	0.2	0.2
DM-water	—	0.1
Supernatant	0.1	—
Lactate dehydrogenase solution	0.2	0.2

Place in a water-bath at 25 °C for 30 min. Read off the extinction of the sample and the reagent blank against water at 365 nm.

3.8.1.2.7. Calculation

Correction for deproteination. (See p. 349.)
Calculation of the change in extinction.

$$\Delta E = E(\text{S}) - E(\text{B})$$

Calculation of the concentration, $\lambda = 365\ nm$ $(\varepsilon = 3400)$.

$$\text{Concentration} = \frac{\Delta E}{\varepsilon d} \cdot \frac{FV}{SV} \cdot 10^6 \cdot k\ \mu\text{mol}\ l^{-1}$$

$$= \frac{E(S) - E(RB)}{3400 \cdot 1} \cdot \frac{1.50}{0.1} \cdot 10^6\ k\ \mu\text{mol}\ l^{-1}$$

$$= E(S) - E(RB) \cdot 4.41 \cdot 10^3 \cdot k\ \mu\text{mol}\ l^{-1}$$

$$= E(S) - E(RB) \cdot 4.63 \cdot \frac{3.0 + (W_2 - W_1)}{W_2 - W_1}\ \text{mmol}\ l^{-1}$$

Calculation of the concentration, $\lambda = 340$ nm $(\varepsilon = 6300)$.

$$\text{Concentration} = E(S) - E(RB) \cdot 2.50 \cdot \frac{3.0 + (W_2 - W_1)}{W_2 - W_1}\ \text{mmol}\ l^{-1}$$

3.8.1.2.8. Specificity

The enzymatic method is highly specific for L(+)-lactate. L(−)-Lactate is not reduced by means of lactate dehydrogenase. In contrast to most of the chemical methods, the enzymatic method is not responsive to any other metabolites, such as methylglyoxal.

3.8.1.2.9. Reference values

Venous blood <1.78 mmol l^{-1}.[2]
Arterial blood 0.66–0.84 mmol l^{-1}.[9]
The ratio of plasma to erythrocytes amounts to 1.37.

3.8.1.2.10. Comments on the method

1. If no constant end-value is reached, the activity of the lactate dehydrogenase is too small.
2. For monitoring the method, standard solutions with lactate concentrations between 1 and 10 mmol l^{-1} are suitable.
3. The concentration of L-lactate is directly proportional to the change in extinction for concentrations up to 10 mmol of lactate per litre of blood. For higher concentrations the quantity of the supernatant must be reduced.
4. Gutmann and Wahlefeld[3] recommend lower concentrations of buffer and co-substrate than indicated here.

3.8.1.2.11. Diagnostic significance

1. Elevations of L-lactate in the plasma result from an increased formation and diminished removal and degradation (liver) or excretion (kidneys) of lactate.[7]

The increased formation of lactate from pyruvate is a consequence of increase in pyruvate and/or an increase of the $NADH_2/NAD$ concentration. The relative increase of lactate compared with pyruvate (lactate/pyruvate > 10) points to an increased $NADH_2/NAD$ ratio in the cytoplasm (excess lactate) and the clinical prognosis is bad.

An increase in lactate and pyruvate most frequently results from tissue hypoxia through anaerobic glycolysis. If this is accompanied by a condition of shock, the latter diminishes the removal to the liver and the excretion through the kidneys. Ethanol leads to an increase in lactate due to the increased formation of $NADH_2$.

Apart from this, there are congenital enzyme defects of the intermediary metabolism which lead to pyruvate/lactate acidosis (glycogenose type I, pyruvate carboxylase deficiency, phosphoenol pyruvate carboxykinase defect, fructose-1,6-diphosphatase deficiency, defects of the pyruvate–hydrogenase complexes and of the thiamine triphosphate synthesis). The extremely rare D-lactate acidosis cannot be detected by the enzymatic lactate determination.

Probably the most important indication for a lactate determination in adults is with acidotic diabetics. In such cases it is necessary to ascertain the lactate acidosis and to distinguish it from the ketoacidosis which requires a different treatment. In particular, in recent years lactate acidosis has been observed as a side-effect of biguanides (frequency: 10% of diabetic ketoacidoses). Detection of a ketosis does not exclude the presence of a lactate acidosis (3-hydroxybutyrate, acetoacetate).[4,5,8]

A useful clue to the presence of a lactate acidosis (or increase in other anions) is provided by evaluating the anion gap (p. 311). For a lactate acidosis, the difference between cations and anions lies above 15 or 20 mmol l^{-1}:

Normal: $Na^+ - (Cl^- + HCO_3^-) < 15$ mmol l^{-1};
$(Na^+ + K^+) - (Cl^- + HCO_3^-) < 20$ mmol l^{-1}.

The difference depends on anionic proteins, sulphates, phosphates, and the dissociated moiety of organic acids.

2. Raisis et al.[6] established an elevation of lactate and of the lactate/pyruvate ratios in cerebrospinal fluid in cases of increased cerebral pressure.

References

1. Ditesheim, P.-J. and Bossart, H., Premiers essais de mesures du L-lactate plasmatique au moyen du 'Lactate Analyzer 5400', Schweiz. med. Wschr., 106, 1598 (1976).
2. Laudahn, G., Fermentaktivitäten und Konzentration von Stoffwechselzwischenprodukten im Blut bei Leber- und Herzkrankheiten, Klin. Wschr., 37, 850 (1959).
3. Gutmann, I. and Wahlefeld, A. W., L-(+)-Lactate. Determination with lactate

dehydrogenase and NAD, in Bergmeyer, *Methods of enzymatic analysis*, Vol. 3, Verlag Chemie, Weinheim, 1974, p. 1464.
4. Conlay, L. A. and Loewenstein, J. E., Phenformin and lactic acidosis, *J. Am. med. Ass.*, **235**, 1575 (1976).
5. Searle, G. L. and Siperstein, M. D., Lactic acidosis associated with phenformin therapy. Evidence that inhibited lactate oxidation is the causative factor, *Diabetes*, **24**, 741 (1975).
6. Raisis, J. E., Kindt, G. W., McGillicuddy, J. E., and Miller, C. A., Cerebrospinal fluid lactate and lactate/pyruvate ratios in hydrocephalus, *J. Neurosurg.*, **44**, 337 (1976).
7. Alberti, K. G. M. M. and Nattrass, M., Lactic acidosis, *Lancet*, **ii**, 25 (1977).
8. Wittmann, P., Haslbeck, M., Bachmann, W. and Mehnert, H., Lactacidosen bei Diabetikern unter Biguanidbehandlung, *Dt. med. Wschr.*, **102**, 5 (1977).
9. Hohorst, H. J., Enzymatische Bestimmung von L(+)-Milchsäure, *Biochem. Z.*, **328**, 509 (1957).

3.8.2. KETONE BODIES

3.8.2.1. Introduction

The ketone bodies are acetoacetic acid (3-oxobutyric acid) and its metabolites, 3-hydroxybutyric acid and acetone.

Acetoacetic acid is produced from 2-acetyl-coenzyme A (acetyl-CoA) by way of the following reactions:

2-Acetyl-CoA $\rightleftharpoons$ CoA + Acetoacetyl-CoA

Acetoacetyl-CoA + Acetyl-CoA $\rightarrow$ CoA + Hydroxymethyl-glutaryl-CoA

Hydroxymethylglutaryl-CoA $\rightarrow$ Acetoacetate + Acetyl-CoA

Hydroxymethylglutaryl-CoA $\rightarrow$ Acetoacetate + Acetyl-CoA

Acetoacetic acid is reduced to 3-hydroxybutyrate (hydroxybutyrate hydrogenase) or decarboxylated to acetone. An accumulation of the three ketone bodies acetoacetic acid, β-hydroxybutyric acid, and acetone is called *ketosis*. With high acetyl-CoA and/or diminished CoA in the mitochondria, larger amounts of acetoacetic acid are formed (example: an increased degradation of fatty acids or other consumption of CoA, e.g. formation of propionyl-CoA). A diminution of the breakdown of acetyl-CoA by way of the citric acid cycle occurs when there is an increased build-up of glucose (gluconeogenesis) from oxaloacetate, which is then less available for the formation of citric acid.

The investigation of the ketone bodies in urine is predominantly concerned with the determination of the acetoacetic acid.[1] Quantitative determinations in urine after conversion of the acetoacetic acid and 3-hydroxybutyric acid into acetone are difficult to interpret. The ketone acetone, which is soluble in fat, is excreted by a mechanism different from that of the acids.[2]

A suitable method for the quantitative assessment of ketosis is the enzymatic determination of the acetoacetic acid and 3-hydroxybutyric acid in blood.

3.8.2.2. Acetoacetic acid in blood: enzymatic determination

3.8.2.2.1. Principle

At pH 6.9, D-3-hydroxybutyrate dehydrogenase (3-HBDH) catalyses the reduction of acetoacetic acid to 3-hydroxybutyrate in the presence of excess $NADH_2$. The decrease in the extinction of $NADH_2$ is measured.

$$\text{Acetoacetate} + NADH + H^+ \underset{}{\overset{\text{3-HBDH}}{\rightleftharpoons}} \text{3-Hydroxybutyrate} + NAD^+$$

Since certain 3-HBDH preparations show lactate and malate dehydrogenase activity, these enzymes arc added. In this way the pyruvate and oxaloacetate present in the blood are eliminated beforehand.

3.8.2.2.2. Reagents

Phosphate buffer, 0.5 *mol* l^{-1}, *pH* 6.9. (See Appendix 1.)

$NADH_2$ solution, 17 *mmol* l^{-1}. Dissolve 12.1 mg of $NADH_2$ sodium salt in 1.0 ml of DM-water.

Lactate dehydrogenase (LDH). 100 mg of crystalline suspension from rabbit muscle in 10 ml of ammonium sulphate solution (3.2 mol l^{-1}). Specific activity *ca*. 500 U mg^{-1}. Working solution, 2500 U ml^{-1}: dilute 0.5 ml of stock solution with 0.5 ml of 0.5 mol l^{-1} phosphate buffer, pH 6.9.

Malate dehydrogenase (MDH). 100 mg of crystalline suspension from pig heart in 10 ml of 3.2 mol l^{-1} ammonium sulphate solution. Specific activity *ca*. 1100 U mg^{-1}. Working solution, 5500 U ml^{-1}: dilute 0.5 ml of stock solution with 0.5 ml of 0.5 mol l^{-1} phosphate buffer.

3-Hydroxybutyrate dehydrogenase (3-HBDH). 5 mg in 1 ml of 3.2 mol l^{-1} ammonium sulphate. Specific activity *ca*. 3 U mg^{-1}. Activities of impurities: LDH < 0.1%; MDH < 5%. Use the undiluted solution.

Perchloric acid, 0.66 *mol* l^{-1}. See p. 347.

Tripotassium phosphate, 0.8 *mol* l^{-1}. Dissolve 27 g of $K_3PO_4.7H_2O$ in 100 ml of DM-water.

3.8.2.2.3. Blood collection and deproteination

The blood is collected from the unconstricted vein direct into ice-cold perchloric acid and immediately deproteinated. Each centrifuge tube receives 2 ml of perchloric acid and is then weighted (W_1). Allow about 2 ml of whole blood to flow from the unconstricted vein directly into the ice-cold perchloric acid. Mix the contents well and weigh again (W_2). Centrifuge briskly and withdraw the clear supernatant (0.8 ml of this supernatant I is used for each determination of 3-hydroxybutyric acid and of acetoacetic acid).

3.8.2.2.4. Adjustment of the pH

A 0.8-ml volume of the clear supernatant I is neutralized with K_3PO_4 until a pH of 6.8–7.2 is achieved (supernatant II). Note the amount of K_3PO_4 required. Allow supernatant II to stand in the ice-bath for 20 min and then centrifuge off the precipitated potassium perchlorate. Different amounts of this supernatant II are used for the measurement, depending on the expected acetoacetate concentration: 400 μl for normal and 200 μl for higher than normal concentrations.

3.8.2.2.5. Measurement

Conditions. Final concentrations in the reaction mixture for normal acetoacetate concentrations: phosphate, 89 mmol l^{-1}; NADH$_2$, 0.30 mmol l^{-1}; LDH, 89 U ml^{-1}; MDH, 196 U ml^{-1}; 3-HBDH, 268 mU ml^{-1}; final volume, 0.56 ml. For increased acetoacetate concentrations: phosphate, 303 mmol l^{-1}; NADH$_2$, 0.258 mmol l^{-1}; LDH, 76 U ml^{-1}; MDH, 167 U ml^{-1}; 3-HBDH, 227 mU ml^{-1}; final volume, 0.66 ml; 37 $^\circ$C; 334 nm.

3.8.2.2.6. Procedure

The following amounts of each ingredient are mixed in a thermostated cuvette (1 cm diameter) to be used with an Hg-line photometer coupled to a recording instrument:

	Determination in normal range, ml	Determination for elevated concentration, ml
Supernatant II	0.4	0.2
Phosphate buffer, pH 6.9	0.1	0.4
NADH solution	0.01	0.01
LDH	0.02	0.02
MDH	0.02	0.02

Allow to stand at room temperature for 4 min. Then place the cuvette in the photometer and record the extinction for 5 min at 37 $^\circ$C and wavelength 334 nm.

Start the reaction by adding 0.01 ml of HBDH solution

Record until the decrease in extinction becomes linear and parallel with the decrease before the reaction was started (5–10 min).

3.8.2.2.7. Calculation

Calculation of the correction factor.

$$K_1 \text{ for the deproteination} = \frac{2.0 + (W_2 - W_1)}{(W_2 - W_1)}$$

$$K_2 \text{ for the adjustment of pH} = \frac{0.80 \text{ ml} + \text{ml K}_3\text{PO}_4}{0.80 \text{ ml}}$$

Calculation of the concentration, $\lambda = 334\ nm$ ($\varepsilon = 6180$ mol l^{-1} cm^{-1}).

$$\text{mmol } l^{-1} = \frac{\Delta E}{\varepsilon d} \cdot \frac{FV}{SV} \cdot 10^3 \cdot K_1 \cdot K_2$$

$$= \Delta E K_1 K_2 \cdot 0.226$$

Assay for normal concentrations.

$$\text{mmol } l^{-1} = \frac{\Delta E}{6180 \cdot 1} \cdot \frac{0.56}{0.40} \cdot 10^3 K_1 K_2$$

$$= \Delta E K_1 K_2 \cdot 0.226$$

Assay for elevated concentrations.

$$\text{mmol } l^{-1} = \frac{\Delta E}{6180 \cdot 1} \cdot \frac{0.66}{0.20} \cdot 10^3 K_1 K_2$$

$$- \Delta E K_1 K_2 \cdot 0.534$$

3.8.2.2.8. Reference values for acetoacetic acid in whole blood

Age	mmol l^{-1}	Reference
Adults	0.017 ± 0.098	3
New-born infants	0.041 ± 0.007	4
1st–2nd day	0.012 ± 0.019	5
3rd day	0.249 ± 0.106	5
4th–10th day	0.279 ± 0.116	5
	0.161 ± 0.131	5

358

3.8.2.2.9. Comments on the method

1. The pH of the mixture in the cuvette is critical and should lie between 6.8 and 7.2.
2. The amount of acetoacetic acid present can be estimated semi-quantitatively beforehand. Persson[5] recommends that this be done with Ketostix in the deproteinated, neutralized supernatant (a value up to + positive on the colour scale corresponds to an acetoacetate concentration of about 0.8 mmol l^{-1}).
3. The enzymatic determination of acetoacetic acid has also been described for determinations in urine.[8]

3.8.2.2.10. Diagnostic significance

A ketosis is found in hunger and other catabolic conditions following an unbalanced, fat-rich nutrition and diabetes mellitus. In childhood, apart from these, acetonaemic vomiting leads to ketosis.[6] Functional ketotic hypoglycaemia and metabolic disturbances (enzyme defects of gluconeogenesis, ketotic hyperglycinaemia syndrome) are rare causes of ketosis. Acetoacetic acid is increased with a deficiency of succinyl-CoA-3-keto acid transaminase.[7]

3.8.2.3. 3-Hydroxybutyric acid in blood: enzymatic determination

3.8.2.3.1. Principle

At alkaline pH (8.9–9.5), D-3-hydroxybutyrate dehydrogenase (3-HBDH) catalyses the oxidation by NAD of 3-hydroxbutyric acid to acetoacetic acid. The formation of $NADH^+$ is measured. The lactate and malate in the blood are eliminated by a preliminary reaction with NAD^+ in the presence of added lactate dehydrogenase (LDH) and malate dehydrogenase (MDH).

$$\text{3-Hydroxybutyrate} + NAD^+ \underset{\text{pH 7}}{\overset{\substack{\text{3-HBDH} \\ \text{pH 9}}}{\rightleftharpoons}} \text{Acetoacetate} + NADH + H^+$$

3.8.2.3.2. Reagents

Glycine buffer, 0.5 *mol* l^{-1}, *pH* 10.0. Dissolve 3.75 g of glycine in DM-water. Adjust the pH to 10.0 with NaOH (2 mol l^{-1}). Make up to 100.0 ml with DM-water. Monitor the pH.
NAD-solution, 70 *mmol* l^{-1}. Dissolve 46.5 mg of NAD in 1.0 ml of DM-water.
LDH. See acetoacetic acid determination.
MDH. See acetoacetic acid determination.

3-HBDH. See acetoacetic acid determination.
Perchloric acid. See acetoacetic acid determination.
K_3PO_4 *solution*. See acetoacetic acid determination.

3.8.2.3.3. Blood collection

See acetoacetic acid determination.

3.8.2.3.4. Adjustment of the pH

Add K_3PO_4 solution to 0.8 ml of supernatant I until the pH lies between
9.0 and 9.5. Make a note of the amount of K_3PO_4 required. Allow to stand
in the ice-bath for about 20 min, then centrifuge off the crystals of
potassium perchlorate. The amount of this supernatant II used for
the determination depends on the expected concentration of
3-hydroxybutyrate: 300 μl for measurements in the normal range, 200 μl
for higher concentrations.

3.8.2.3.5. Measurement

Conditions. Final concentrations in the reaction mixture for normal
ranges are glycine buffer 167 mmol l^{-1}, NAD 4.66 mmol l^{-1}, LDH
83 U ml^{-1}, MDH 183 U ml^{-1}, 3-HBDH 0.5 U ml^{-1}, and final volume
0.58 ml; for higher concentrations, glycine buffer 286 mmol l^{-1}, NAD
4 mmol l^{-1}, LDH 71 U ml^{-1}, MDH 157 U ml^{-1}, 3-HBDH 0.43 U ml^{-1}, and
final volume 0.68 ml; 37 °C, 334 nm.

3.8.2.3.6. Procedure

The following quantities of reactants are placed in a thermostated
cuvette (1 cm diameter) for measurement with an Hg-line photometer with
a coupled recorder:

	Measurement in normal range, ml	Measurement for elevated concentration, ml
Supernatant II	0.3	0.2
Glycine buffer	0.2	0.4
NAD	0.04	0.04
LDH	0.02	0.02
MDH	0.02	0.02

Allow to stand at room temperature for 4 min. Place the cuvette in the
photometer and record the extinction for 5 min at 37 °C and a wavelength
of 334 nm.

Start the reaction by adding 0.02 ml of 3-HBDH.

Record the increase in extinction until it is linear and parallel to that before the reaction. The time taken is 20–30 min or longer.

3.8.3.2.7. Calculation

Calculation of the correction factors. For deproteination:

$$K_1 = \frac{2.0 + (W_2 - W_1)}{W_2 - W_1}$$

For the adjustment of the pH to 9–9.5:

$$K_2 = \frac{0.80 + \text{ml } K_3PO_4}{0.80}$$

Calculation of the concentration, $\lambda = 334\ nm$ $(\varepsilon = 6180\ mol\ l^{-1}\ cm^{-1})$.

$$\text{mmol } l^{-1} = \frac{\Delta E}{\varepsilon d} \cdot \frac{FV}{SV} \cdot 10^3 K_1 K_2$$

Assay for normal concentration.

$$\frac{\Delta E}{6180.1} \cdot \frac{0.600}{0.300} \cdot 10^3 K_1 K_2$$

$$= \Delta E K_1 K_2 \cdot 0.324$$

Assay for elevated concentration.

$$\frac{\Delta E}{6180.1} \cdot \frac{0.700}{0.200} \cdot 10^3 K_1 K_2$$

$$= \Delta E K_1 K_2 \cdot 0.566$$

3.8.2.3.8. Reference values

These are given in Table 62.

Table 62. Reference values for 3-hydroxybutyric acid in whole blood.

Age	mmol l^{-1}	Reference
Adults	0.058 ± 0.17	3
New-born infants	0.106 ± 0.014	4
1st–2nd day	0.082 ± 0.054	5
3rd day	0.662 ± 0.402	5
4th–10th day	0.565 ± 0.255	5
	0.470 ± 0.411	5

3.8.2.3.9. Comments on the method

1. The pH of the reaction is critical and should lie between 8.9 and 9.5. It is advisable to check the pH after completion of the reaction.
2. At high values, the addition of hydrazine to remove the acetoacetate produced leads to higher results.[4]

3.8.2.3.10. Diagnostic significance

1. The causes of ketosis have been mentioned in connection with the determination of acetoacetic acid.
2. Physiologically, the ratio of 3-hydroxybutyrate to acetoacetate lies between 2 and 4 and reflects the intramitochondrial redox condition, while e.g. the lactate to pyruvate ratio is also influenced by the redox condition in the cytosol of the cell.

References

1. Colombo, J. P. and Richterich, R., *Die einfache Urinuntersuchung*, (Huber, Berne, 1977, p. 123.
2. Bachmann, C., Baumgartner, R., Wick, H., and Colombo, J. P., Quantitative gaschromatographic determination of short chain aldehydes and ketones in the urine of infants, *Clin. chim. Acta*, **66**, 287 (1976).
3. Bergmeyer, H. U. and Bernt, E., Enzymatische Bestimmung von Keton-Körpern im Blut, *Enzymol. biol. clin.*, **5**, 65 (1965).
4. Bach, A., Métais, P., and Jaeger, M. A., Etude critique de la détermination enzymatique de l'acétoacétate et du D-(−)-béta-hydroxybutyrate par la méthode de Williamson, *Annls. Biol. clin.*, **29**, 39 (1971).
5. Persson, B., Determination of plasma acetoacetate and D-β-hydroxybutyrate in newborn infants by an enzymatic fluorometric micromethod, *Scand. J. clin. Lab. Invest.*, **25**, 9 (1969).
6. McGarry, J. D. and Foster, D. W., Regulation of ketogenesis and clinical aspects of the ketotic state, *Metabolism*, **21**, 471 (1972).
7. Tildon, J. T. and Cornblath, M., Succinyl-CoA: 3-ketoacid CoA-transferase deficiency, *J. clin. Invest.*, **51**, 493 (1972).
8. Oudheusden, A. P. M. van, Fischer, M., Hilvers, A. G., Triet, A. J. van, Vals, G. H. van, Wimmer, P., and Koller, P. U., Teststreifen zum Nachweis von Ketonkörpern im Harn, *Diagnostik*, **9**, 14 (1976).

3.9. CARBOHYDRATES

3.9.1. GENERAL

Of all nutritional substances, carbohydrates together with fats and proteins, are the most important source of energy. The basic unit of the carbohydrates is the monosaccharide, and of the many monosaccharides the groups of greatest physiological importance are the hexoses (6 carbon atoms) and the pentoses (5 carbon atoms).

The hexoses of physiological importance are glucose, fructose, and galactose. These sugars act as reducing agents and can be detected with Fehling's or Benedict's solution.[1,2] Ribose (in RNA) and deoxyribose (in DNA) are among the important pentoses.

The disaccharides are molecules built up of two monosaccharides. The physiologically important disaccharides are saccharose (fructose and glucose), lactose (galactose and glucose), and maltose (glucose and glucose), of which only the last two have reducing properties.

Polysaccharides are long-chain carbohydrates, represented in the vegetable kingdom by starch, a mixture of amylose (straight chain) and amylopectin (branched chain). Glycogen, a component of animal tissue, has a very highly branched structure. In both polysaccharides the basic structural unit is glucose.

3.9.2. Glucose

3.9.2.1. Introduction

Most cells are dependent on the utilization of glucose, particularly the cells of the brain and the muscles. In the metabolism of glucose, however, it is the liver which assumes a central role.

After hydrolysis by the digestive enzymes, the carbohydrates are mainly present in the form of monosaccharides. After reabsorption, the glucose in the cells is converted to glucose-6-phosphate and is then either stored as glycogen (glycogenesis) or used to provide energy via glycolytic degradation (glycolysis) and the citric acid cycle (Fig. 103). An important alternative metabolic path for glucose is the oxidation of hexose monophosphate in the pentose phosphate cycle.

Numerous hormones participate in the regulation of the blood sugar.

362

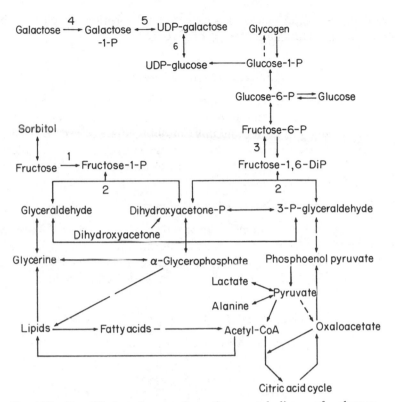

Fig. 103. Simplified scheme for the metabolism of glucose, galactose, and fructose. The numbers indicate enzymes known to be subject to genetically conditioned defects. 1, Fructokinase; 2, fructose-1-phosphate aldolase; 3, fructose-1,6-diphosphatase; 4, galactokinase; 5, galactose-1-phosphate uridyl transferase; 6, uridine diphosphate galactose 4-epimerase.

Besides the essential insulin, glucagon, the growth hormone, the glucocorticoids, and adrenalin all take part in the mechanism of regulation.

In a condition of fasting, the glucose content of the blood is maintained by the breakdown of glycogen (glycogenolysis) and the formation of glucose from amino acids (gluconeogenesis). Free fatty acids and ketone bodies act as alternative sources of energy during fasting.

The commonest disturbance of the glucose metabolism is diabetes mellitus. This is a metabolic disease caused by a partial or complete deficiency in insulin and which is consequently accompanied by hyperglycaemia. This can be evident in the fasting condition and/or following a glucose feed. The deficient intracellular glucose metabolism results in a ketosis which, in severe cases, degenerates into an acidosis (p. 358).

Increased concentrations of insulin in the blood lead to hypoglycaemia (lowered blood glucose), but this condition may also be observed in

364

endocrinological disturbances, inadequate reabsorption of glucose in the intestine, and in inadequate storage of glycogen.

3.9.2.2. Choice of method

Central to all clinical–chemical investigations on carbohydrate metabolism is the determination of the 'blood sugar'. Three types of method are available for determining this immediate source of energy for the organism: determination of the reducing substances, determination of the total carbohydrate, and enzymatic determination of the blood glucose.

3.9.2.2.1. Determination of reducing substances

A knowledge of the chemical properties of glucose is important for understanding the different methods of determination.

In solution, glucose (an aldose) is present in various forms. Thus, in neutral or weakly acidic media it forms a half-acetal which has a cyclic structure:

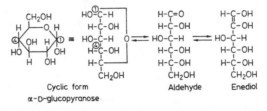

Cyclic form
α-D-glucopyranose

The aldehyde form of glucose is very unstable and only very small amounts of it exist in aqueous solution. In alkaline solution the equilibrium shifts to the enediol form, which is very strongly reducing. The letter D (L) is a conventional description of the position of the hydroxyl group on carbon atom 4, while the designation α(β) refers to the position of the hydroxyl group on the asymmetric carbon atom 1.

In connection with the detection of glucose, pride of place must be given to *polarimetry*, since this was the first method to be adopted in routine clinical work. Although it does not constitute a reduction test, the method deserves to be mentioned here since it is still used occasionally for the crude appraisal of glucosuria in diabetics. The optical activity of the sugar was observed by Biot in 1817 and it was he who constructed the first polarimeter. This instrument can be used to measure the magnitude of the rotation of linearly polarized light by an optically active solution of glucose: the degree of rotation is proportional to the concentration of the sugar. Today, the application of polarimetry is severely limited. Apart from substances such as amino acids, deriving from the body itself, other substances, above all antibiotics, can lead to completely wrong results.[3,4]

The first quantitative chemical methods for the determination of glucose were based on the ease of oxidation of the enol form of glucose. If glucose is heated in alkaline solution with the yellow hexacyanoferrate(III) (ferricyanide)ion, the colourless hexacyanoferrate(II) (ferrocyanide)ion is formed by reduction:

$$Fe^{III}(CN)_6^{3-} \xrightarrow[\text{(Glucose)}]{+e^-} Fe^{II}(CN)_6^{4-}$$

$$Cu^{2+} \xrightarrow[\text{(Glucose)}]{+e^-} Cu^{I}$$

Divalent copper is also reduced by glucose under similar conditions to give monovalent copper, as shown above, but the reactions are non-stoichiometric. Nevertheless, under rigorously standardized conditions the amount of glucose present is proportional to the reduction products formed. Older methods for determining glucose were based on the photometric measurement of the hexacyanoferrate(III) consumed[5] or determination of the monovalent copper ions with molybdophosphoric acid.[6]

The *reduction methods* are not specific for glucose but estimate it together with the remaining reducing sugars and other substances such as creatine, creatinine, uric acid, glutathione, and ascorbic acid.[7] For this reason the reduction test should no longer be used for the quantitative determination of blood sugar.

The qualitative reduction tests for the detection of urine sugar with Fehling's or Benedict's solution[1,2] and with the commercial Clinitest tablets[8] are still of significance, however. These tests are based on the reduction of divalent copper ions by glucose in hot, alkaline solution. The reaction mixtures contain polyvalent hydroxycarboxylic acids (e.g. sodium potassium tartrate) as complexing agents to prevent the precipitation of $Cu(OH)_2$. In hot, alkaline solution, the reduced monovalent copper subsequently converts to the orange–red copper(I) oxide (Cu_2O), which is precipitated.

Again, the main disadvantage of all the qualitative reduction tests lies, as with the methods for quantitative determination, in their low specificity. False-positive results can be produced by endogenous products of metabolism, especially by medicaments such as ascorbic acid, thiamine, penicillin, p-aminobenzoic acid, caronamide, PAS, streptomycin, isoniazid, and tetracycline.[9] Reduction tests should only be used if the presence of a non-glucosuric mellituria is suspected (e.g. in new-born infants) (7).

3.9.2.2.2. Determination of the total carbohydrate

This type of method is concerned with the formation of typical coloured derivatives of the monosaccharides which are susceptible to photometric measurement.

If glucose is heated in the presence of strong acids, water is split off and

366

a furan derivative is produced:

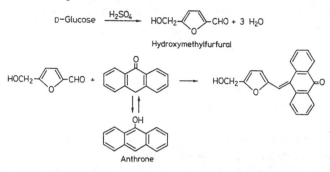

Hydroxymethylfurfural

Anthrone

The hydroxymethylfurfural so formed may be condensed with phenols such as anthrone[10] on heating. This produces intensely coloured dyes which can be estimated photometrically. This method is not specific for glucose since other hexoses, e.g. fructose and galactose, can also form furfurals. Since polysaccharides (inulin, glycogen, starch) are initially hydrolysed to monosaccharides on heating with acid, the method is suitable for the determination of the total carbohydrate.

The anthrone method is of importance in the determination of the polyfructosan inulin (see p. 385). If deproteination is followed by incubation at 37 °C with a small volume of concentrated anthrone solution, the colour reaction is specific for fructose.[11]

The determination of the blood sugar with aromatic amines, encompassing the aldoses, is another chromogenic method which is still used relatively frequently even today, since it is very simple to carry out. With this technique, glucose is condensed to glycosylamine by heating with aniline[12] or o-toluidine[13,14] in the presence of glacial acetic acid:

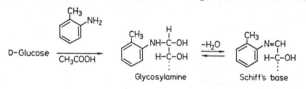

Glycosylamine   Schiff's base

The reaction product probably consists of a mixture of glycosylamine and the corresponding Schiff's base in equilibrium with one another.

The toluidine method[13] can be performed without deproteinating the plasma but the drawback is that in addition to glucose, other sugars such as mannose, galactose, and fructose yield differential amounts of chromogens.[15] We ought also to mention that by slightly modifying aniline to p-bromaniline, a fairly specific method for pentoses can be developed.[16]

3.9.2.2.3. Enzymatic determination of the blood glucose

The technology of enzyme purification has made possible the development of new methods in recent years. These techniques permit the exclusive determination of glucose and therefore, for precise determinations, they are to be preferred to the other methods. Seniority goes to the determination of glucose with *glucose oxidase/peroxidase (GOD/POD)*:[17]

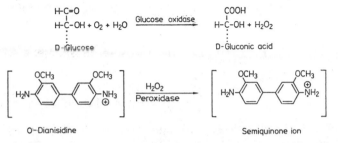

In a first stage, the enzyme GOD catalyses the oxidation of D-glucose to D-gluconic acid and hydrogen peroxide. The addition of POD and the electron-acceptor *o*-dianisidine produces a coloured compound in an indicator reaction; the colour of the product derives from a semiquinone ion. Other commonly used electron acceptors are *o*-tolidine,[18] the ammonium salt of 2,2'-azinodi-[3-ethylbenzthiazoline sulphonic acid-(6)],[19] 3-methyl-2-benzothiazolinone hydrazone/*NN*-dimethylaniline,[20] and, in the Trinder modification,[43,44] phenol/4-aminophenazone (PAP). Although GOD is virtually absolutely specific for glucose, one disadvantage of this method is the relatively non-specific nature of the indicator reaction with POD. For the determination of glucose in urine particularly, numerous substances interfere and lead in some cases to high and in other cases to low, incorrect results.[4] The interference is particularly bad with uric acid, ascorbic acid, homogentisic acid, adrenaline, hydroquinone, and bilirubin glucuronide.

The qualitative testing for urine glucose with commerical test papers is based on the same principle as the GOD/POD methods. Although numerous substances can interfere here also,[9] the test is suitable for a semi-quantitative assessment of the course of glucosuria in known diabetics.

Finally, as an interesting principle which may be applied to the GOD method, we may mention the determination of the rate of oxygen consumption during the glucose oxidation, using an *oxygen-specific electrode*.[21] The hydrogen peroxide formed in the reaction is removed by two subsequent reactions by adding catalase/ethanol and iodide/molybdate.

$$H_2O_2 + CH_3CH_2OH \xrightarrow{\text{Catalase}} CH_3CHO + H_2O$$

$$H_2O_2 + 2H^+ + 2I^- \xrightarrow{\text{Molybdate}} I_2 + 2H_2O$$

This method permits the determination of glucose in whole blood. In the GOD/POD method, the whole blood should first be deproteinated. Deproteination is unnecessary for the determination in plasma, serum, or cerebrospinal fluid.

Nowadays the *hexokinase method* is regarded as reference method for the determination of glucose.[15,22] Under optimal conditions of measurement this technique is distinguished by the following advantages: high specificity, great sensitivity, rapidity of operation and exceptional accuracy, and the measurements may be made over a wide range. In this method glucose is first phosphorylated with ATP, catalysed by hexokinase:

$$\text{D-Glucose} + \text{ATP} \xrightarrow{\text{Hexokinase}} \text{D-Glucose-6-phosphate} + \text{ADP}$$

$$\text{Glucose-6-phosphate} + \text{NADP}^+$$

$$\xrightarrow{\text{G-6-PDH}} \text{6-Phosphogluconate} + \text{NADPH} + \text{H}^+$$

The glucose-6-phosphate so formed reacts with NADP in the presence of glucose-6-phosphate dehydrogenase (G-6-PDH) to form 6-phosphogluconate and $NADPH_2$. Each molecule of phosphorylated glucose produces one molecule of $NADPH_2$ and the latter is measured photometrically at a wavelength of 334, 340, or 365 nm. Although fructose and mannose are also phosphorylated in the initial reaction, the method is nevertheless absolutely specific for glucose since only the glucose-6-phosphate reacts under catalysis by G-6-PDH in the indicator reaction. Thanks to its freedom from interference, the hexokinase method is the preferred method today. It yields excellent results with plasma, serum, cerebrospinal fluid, and urine, and deproteination is unnecessary.

The hexokinase method may even be used for the determination of glucose in whole-blood haemolysates.[23] First a haemolysate is prepared by adding digitonin and maleic acid; this may be kept for 1 week at 2–8 °C. For this determination, the measurement of the $NADPH_2$ must be carried out at 365 nm.

Finally, we may mention the determination of blood sugar with *glucose dehydrogenase* (G-DH).[24,45] This enzyme derives from *Bacillus megaterium* and, as a catalyst, it is specific for the oxidation of β-D-glucose to D-gluconic acid-δ-lactone in the presence of NAD:

$$\alpha\text{-D-Glucose} \xrightarrow{\text{Mutarotase}} \beta\text{-D-Glucose}$$

$$\beta\text{-D-Glucose} + \text{NAD}^+$$

$$\xrightarrow{\text{G-DH}} \text{D-Gluconic acid-}\delta\text{-lactone} + \text{NADH} + \text{H}^+$$

Here, the increase in $NADH_2$ is proportional to the concentration of β-D-glucose. Since in aqueous solution 36% of the glucose is in the form of the α-isomer, the enzyme mutarotase is added to the reaction mixture to catalyse the conversion to the β-isomer. Like the hexokinase method, this method is highly specific but it has one advantage over the former method: the reduced coenzyme is produced in the main reaction itself and an auxiliary reaction is therefore unnecessary.

3.9.2.3. Total carbohydrate: aniline method

3.9.2.3.1. Principle

If aldoses and ketoses are heated with aminobenzenes in glacial acetic acid they react to produce dyes (p. 366). If aniline is used, the maximum intensity of colour lies at 340 nm.[12] The method is suitable for the determination of total carbohydrate in blood, plasma, serum, cerebrospinal fluid, and effusions. The samples must be deproteinated before the determination.

3.9.2.3.2. Reagents

Trichloroacetic acid, 0.61 *mmol l^{-1}* (10%). Dissolve 10 g of trichloroacetic acid in DM-water and make up to 100 ml. The solution keeps indefinitely at room temperature.

Aniline reagent. Dissolve 6.0 g of aniline (p.a. grade) in glacial acetic acid and make up to 100 ml with further acetic acid. The solution keeps for several months provided that air is excluded. A slight yellowing causes no inteference.

Glucose standard, 100 *mg* per 100 *ml* (5.5 *mmol l^{-1}*). Dissolve 110 mg of glucose monohydrate in a cold, saturated solution of benzoic acid in DM-water and make up 100 ml with the same solution. The solution keeps indefinitely.

3.9.2.3.3. Procedure

	Sample, ml	Reagent blank, ml	Standard, ml
Trichloroacetic acid	0.2	0.2	0.2
Blood, plasma, serum	0.02	—	—
DM-water	—	0.02	—
Glucose standard	—	—	0.02

Allow to stand for a short time, then centrifuge strongly

Additional solutions:			
Aniline reagent	1.0	1.0	1.0
Supernatant	0.1	0.1	0.1

Heat for exactly 10 min in a boiling water-bath, cool with running water and record the extinction, compared with water, between 340 and 440 nm. The colour is stable for about 3 h.

3.9.2.3.4. Calculation

$$\text{Concentration} = \frac{E(S) - E(RB)}{E(ST) - E(RB)} \cdot 100 \text{ mg per 100 ml}$$

$$\text{Concentration} = \frac{E(S) - E(RB)}{E(ST) - E(RB)} \cdot 5.5 \text{ mmol } l^{-1}$$

3.9.2.3.5. Specificity

The method is not specific for glucose but responds to all the ketoses and hexoses present.[15] Hence the results should not be used if the patient is under fructose infusion.

3.9.2.3.6. Reference values

80–120 mg per 100 ml (plasma).
4.44–6.66 mmol l^{-1} (plasma).

3.9.2.3.7. Comments on the method

1. With the volumes given, it is advisable to take the reading at about 400 nm (e.g. 405 nm). The sensitivity of the method can be almost doubled by reading the extinction at 365 nm, i.e. only 0.01 ml of blood is required for each analysis.
2. For concentrations greater than 500 mg per 100 ml, the volume of supernatant should be halved and the result multiplied by 2.
3. The method can also be used for the determination of total carbohydrate in the supernatant from a perchloric acid deproteination.
4. If the method gives any difficulties, the acetic acid should be changed.
5. For the technique of collecting blood for blood-sugar determinations, see p. 374.
6. The enzymatic glucose determinations and determinations of carbohydrate by the aniline method can be performed with a single blood sample of 20'μl of blood. This is a great advantage, particularly in disaccharide tolerance tests in children.

3.9.2.4. Glucose: enzymatic determination with hexokinase and glucose-6-phosphate dehydrogenase

3.9.2.4.1. Principle

The determination of glucose with hexokinase and glucose-6-phosphate dehydrogenase is currently the preferred method. It is suitable for the specific determination of glucose in plasma, serum, cerebrospinal fluid, urine, haemolysates, and exudates;[4,22,23] it is quick to perform and no deproteination is necessary. The calculation is made on the principle of the absolute value by means of the extinction coefficients. The following processes are involved:

Primary system.

$$\text{Glucose} + \text{ATP} \xrightarrow{\text{Hexokinase}} \text{Glucose-6-phosphate} + \text{ADP}$$

Indicator system.

$$\text{Glucose-6-phosphate} + \text{NADP}^+$$

$$\xrightarrow{\text{Glucose-6-phosphate dehydrogenase}} \text{6-Phosphogluconate} + \text{NADPH} + \text{H}^+$$

Provided that ATP, hexokinase, NADP, and glucose-6-phosphate dehydrogenase are in excess, one molecule of $NADPH_2$ formed [in contrast to NADP, this has a high extinction at 340 nm (334–366 nm)] corresponds to one molecule of glucose. The analyses may be carried out at a rate of over 3 per minute if an automatic cuvette and a recorder are used.

3.9.2.4.2. Reagents

Glucose standard solution, 250 mg per 100 mol (13.9 mmol l^{-1}). Dissolve 275 mg of glucose monohydrate in a cold, saturated solution of benzoic acid and make up to 100 ml with further benzoic acid solution. The solution keeps indefinitely at room temperature or 4 °C.

Triethanolamine hydrochloride buffer, 300 mmol l^{-1}, *EDTA 5 mmol l^{-1}*, *pH 7.5*. Dissolve 55.70 g of triethanolamine hydrochloride and 1.86 g of EDTA (disodium salt, dihydrate) in DM-water and make up to about 800 ml. Adjust the pH to 7.5 and then make up to 1000 ml with DM-water. The solution keeps indefinitely at 4 °C.

Magnesium sulphate solution, 200 mmol l^{-1}. Dissolve 4.94 g of magnesium sulphate heptahydrate in DM-water and make up to 100 ml.

ATP solution, 160 mmol l^{-1}. Dissolve 9.68 g of adenosine triphosphate (trisodium salt, trihydrate) in DM-water and make up to 100 ml.

NADP solution, 12 mmol l^{-1}. Dissolve 920 mg of nicotinamide adenine dinucleotide phosphate (sodium salt) in 100 ml of nicotinamide solution (50 mmol l^{-1}, 611 mg per 100 ml). The solution keeps indefinitely at 4 °C.

Enzyme solution: hexokinase 50 μmol min^{-1} l^{-1}, glucose-6-phosphate dehydrogenase 10 μmol min^{-1} l^{-1}. Dissolve 5 mg of hexokinase (minimal activity > 100 μmol min^{-1} l^{-1}, free from phosphohexose isomerase) and 1 mg of glucose-6-phosphate dehydrogenase (minimal activity > 100 μmol min^{-1} l^{-1}, free from phosphogluconate dehydrogenase) in a solution of ammonium sulphate (1.0 mol l^{-1}, 13.22 g per 100 ml), to make 10 ml. The solution may be kept for a few weeks at 4 °C.

Working solution. A solution sufficient for 1000 analyses is obtained by using the volumes stated in parentheses: magnesium sulphate solution, 2 parts (100 ml); ATP solution, 1 part (50 ml); NADP solution, 2 parts (100 ml); and triethanolamine buffer, 15 parts (750 ml). Check the pH and adjust to 7.5. The solution keeps for at least 1 week at 4 °C.

3.9.2.4.3. Measurement

Conditions. The final concentrations in the reaction mixture are as follows: triethanolamine buffer, 200 mmol l^{-1}, pH 7.5; EDTA, 3.34 mmol l^{-1}; MgSO$_4$, 17.8 mmol l^{-1}; ATP, 7.12 mmol l^{-1}; NADP, 1.07 mmol l^{-1}; hexokinase, 44.5 mg l^- (890 μmol min^{-1} l^{-1}); glucose-6-phosphate dehydrogenase, 8.92 mg l^{-1} (178 μmol min^{-1} l^{-1}). Other conditions are: temperature of measurement, 37 °C; wavelengths, 334, 340, or 365 nm; $d = 1$ cm.

3.9.2.4.4. Preparation of the samples

Plasma, serum, cerebrospinal fluid. With this method, the optimal range for the glucose concentration is between 5 and 700 mg per 100 ml. For glucose concentrations above 700 mg per 100 ml, 1 part of plasma should be diluted with 9 parts of physiological saline solution. Alternatively, 10 μl of plasma may be used (multiply the result by 2). For glucose concentrations less than 20 mg per 100 ml, use a 40 μl sample and divide the result by 2.

Urine. The urinary glucose concentration is first estimated by means of a rapid strip test. If the concentration is greater than 500 mg per 100 ml, 1 part of urine is diluted with 9 parts of water and the result must be multiplied by 10. If the measured result exceeds 5 g per 100 ml, then 1 part of urine should be diluted with 99 parts of water and the result multiplied by 100.

3.9.2.4.5. Procedure

Solutions (in photometer cuvettes)	ml
Working solution	1.00
Sample	0.02

Read off the extinction: E(B).

Enzyme solution	0.10

After 1–3 min at 37 °C, read off the extinction at 334, 340, or 365 nm: E(S).

3.9.2.4.6. Calculation of the glucose concentration

General equations.

$$\text{Concentration} = \frac{E(S) - E(B)}{\varepsilon d} \cdot \frac{FV}{SV} \cdot 1000 \text{ mmol l}^{-1}$$

$$\text{Concentration} = \frac{E(S) - E(B)}{\varepsilon d} \cdot \frac{FV}{SV} \cdot \frac{MW}{10} \cdot 1000 \text{ mg per 100 ml}$$

For measurements at 365 nm ($\varepsilon = 3500 \ l \ mol^{-1} \ cm^{-1}$).

$$\text{Concentration} = \frac{E(S) - E(B)}{3500 \cdot 1} \cdot \frac{1.12}{0.02} \cdot \frac{180.16}{10} \cdot 1000 \text{ mg per 100 ml}$$

$$\text{Concentration} = [E(S) - (B)] \cdot 288.3 \text{ mg per 100 ml}$$
$$= [E(S) - E(B)] \cdot 16.0 \text{ mmol l}^{-1}$$

For measurements at 334 nm ($\varepsilon = 6180 \ l \ mol^{-1} \ cm^{-1}$).

$$\text{Concentration} = [E(S) - E(B)] \cdot 163.2 \text{ mg per 100 ml}$$
$$= [E(S) - E(B)] \cdot 9.06 \text{ mmol l}^{-1}$$

For measurements at 340 nm ($\varepsilon = 6300 \ l \ mol^{-1} \ cm^{-1}$).

$$\text{Concentration} = [E(S) - E(B)] \cdot 160.14 \text{ mg per 100 ml}$$
$$= [E(S) - E(B)] \cdot 8.888 \text{ mmol l}^{-1}$$

3.9.2.4.7. Specificity

The specificity of coupled enzymatic methods is determined by the specificity of the individual enzymes. The first enzyme, hexokinase, is relatively non-specific; besides glucose, fructose, mannose, and glucosamine are all converted to their respective phosphate esters. In contrast, glucose-6-phosphate dehydrogenase is highly specific for glucose-6-phosphate.[4] We should expect the combination of the two enzymes to be very highly specific for glucose. Thus, the phosphohexose isomerase contamination ought not to exceed 0.05% since otherwise the phosphorylated fructose (fructose-6-phosphate) produced by the hexokinase will be converted to glucose-6-phosphate. The glucose-6-phosphate dehydrogenase used must be free from 6-phosphogluconate dehydrogenase, otherwise the reaction with NADP will proceed further and the results will appear to indicate abnormally high glucose values. Provided that these conditions are met, the method may be considered highly specific.

3.9.2.4.8. Reference values

For values in plasma see p. 376 and in urine see p. 709.

3.9.2.4.9. Comments on the method

1. Contrary to earlier findings, deproteination of the samples is not necessary.[22]

2. The calculation of the results should always be done by using the extinction coefficients. The occasional analysis of standard solutions only serves to monitor the activity of the enzymes in the primary and indicator systems.

3. The method is excellently suited to the precise determination of the urinary glucose concentration.[4]

4. With plasma or serum, the results are, on average, 3–5% higher than those obtained by the glucose oxidase/peroxidase method.[4]

5. With certain commercial preparations for the glucose determination there are no optimal concentrations for enzyme and coenzyme. In these cases, the reaction may take longer to run its course.

6. None of the following substances interfere with the determination:[22,25] sulphonamides, antibiotics, tuberculostatics, cytostatics, digitalis derivatives, morphine derivatives, vitamins, insulin, sulphonylurea, glucuronic acid, creatinine, p-aminohippuric acid, heparin, oxalate, EDTA, fluoride, sodium iodacetate, uric acid, proteins, or phosphate.

7. The determination may be carried out with the Greiner Electronic Selective Analyzer GSA II.[25]

3.9.2.5. Collection and preservation of blood specimens for glucose determinations

3.9.2.5.1. Collection of the blood

The following three techniques are all in current use for collecting blood for glucose determinations:

Capillary whole blood. See p. 100.

Capillary plasma. See p. 100.

Venous blood. Apply a gentle venous (not arterial!) tourniquet. Use a sterile needle and a dry, sterile syringe. Remove the tourniquet as soon as the blood flows into the syringe. Let the blood flow slowly into a tube, with no additive if the determination is to be carried out straight away, otherwise into a special tube containing additive. Mix the contents of the special tube well by repeated inversion (not shaking). Preparation of the blood-sugar tubes in the laboratory is scarcely worth the trouble and it is advisable to use commercial products (tubes treated with ammonium heparinate and sodium fluoride).

3.9.2.5.2. Preservation of the blood

When whole blood is kept, there is always the risk of glycolytic degradation of the glucose by the erythrocytes and leucocytes (a high increase in the *in vitro* degradation of glucose is observed particularly with patients suffering from myelogenous leukaemia; in extreme cases, the degradation can amount to 10–20 mg per 100 ml per hour).[26–28] This process can be prevented by any of the following three methods:

1. Collection into the deproteination solution. If, e.g. capillary blood is collected directly into the deproteination solution, the glucose concentration remains stable for several days even at room temperature.
2. Immediate separation of the plasma or serum. Here, 'immediate' means that the blood (capillary or venous) is centrifuged within 30 min of collection. After centrifugation, the plasma or serum should immediately be transferred to a fresh tube. Using this procedure, the glucose concentration in the plasma or serum remains steady for at least 24 h if the samples are kept at 4 °C (refrigerator). Investigations[29] substantiated that the glucose concentration in plasma or serum, preserved under sterile conditions, remains constant for some time.
3. Addition of glycolysis inhibitors. If there is no possibility of carrying out the determination directly or separating the plasma or serum immediately, *anticoagulants* and *glycolysis* inhibitors must be added.

For this purpose, the addition of 2 mg of sodium fluoride + 1 mg of potassium EDTA (Complexone) or 75 IU of ammonium heparinate per 1 ml of blood has proved best. In order to avoid a volume effect, the glycolysis inhibitor and the anticoagulant must be in the form of a powder or present as a residue from the evaporation of a solution. As anticoagulants, chemists use predominantly ammonium heparinate and haematologists EDTA (ethylenediaminetetraacetate, Complexone). First a stock solution is prepared, containing 400 mg of potassium EDTA in 20 ml of water and 50 μl of this solution are added to the tube for each 1 ml of blood. The solution is then evaporated at room temperature (do not heat!). To prepare the ammonium heparinate solution, about 300 mg (*ca.* 30 000 IU) of the substance are dissolved in 20 ml of water. The volume of this solution pipetted into each tube is likewise 50 μl per 1 ml of whole blood. This solution may be evaporated at 90 °C.

As the glycolysis inhibitor for blood-glucose analyses, the only substance which may still be recommended is sodium fluoride. A mixture is prepared by adding 800 mg of sodium fluoride to each of the stock solutions of anticoagulant described above. We advise against the use of other substances, e.g. oxalate, since these lead to shrinking or swelling of the erythrocytes, resulting in a displacement of water. When used in the suggested concentration, sodium fluoride produces no interference in the enzymatic analysis by the hexokinase method.[4,22] Our own experiments have shown that whole blood for glucose determinations, when treated with sodium fluoride, can be kept for up to 60 min at 20 °C and up to 6 h at 4 °C. Without additives, the analysis results are worthless after a delay of only 30 min.

Special difficulties arise with samples which have to be sent through the post when the time between collecting the blood and performing the analysis is liable to be relatively long. The following procedure is recommended for *samples to be dispatched*.

Collection of venous blood in special tubes containing anticoagulant and glycolysis inhibitor. Centrifugation of the plasma and transference to a fresh tube. Sealing the tubes carefully with plastic stoppers (not corks). If whole blood is dispatched there is a risk of haemolysis. This interferes in numerous analyses and can even lead to incorrect results (potassium, inorganic phosphorus, enzymes). It is a wise precaution to use only unbreakable plastic tubes for sending through the post; this makes a special packaging superfluous.

3.9.2.6. Reference values for the glucose concentration in blood

The results of blood-glucose determinations depend on four groups of factors:

1. the use of whole blood, plasma, or serum;
2. the use of capillary or venous blood;
3. a possible deproteination; and
4. the method used.

Note to 1. The following simple considerations will suffice to show that results obtained with whole blood must deviate from those in plasma or serum: 1, glucose is only soluble in the aqueous phase of the blood or plasma; 2, the water content of plasma is about 91% but whole blood contains 79% water. This alone will account for a difference of about 10% between the results of whole blood and plasma analyses—even with deproteination. Thus, it transpires that the glucose concentration in plasma is about 10–20 mg per 100 ml higher than in the erythrocytes.[30,31] Even greater differences are to be expected if the blood is loaded with glucose.

Various authors[32,33] have given equations for the interconversion of results for whole blood and plasma (serum). One such equation is

Plasma glucose = 6 + (1.15 × blood glucose)

We advise against using such equations since they are valid only for a particular method and have no absolute significance.

Note to 2. There have been conflicting results concerning the difference between the capillary and venous concentrations. Some authors observed no significant differences[34] while others found the values for capillary blood[35] and haemolysates[23] to be markedly higher. According to a more recent investigation,[36] the normal range of concentrations in venous plasma were found to be about 17% lower than in capillary plasma.

Note to 3. The difference between results obtained with and without preliminary deproteination is attributable to the so-called volume displacement effect of the plasma proteins (p. 149).[37] If the plasma is deproteinated, then the determination is performed on the plasma water, i.e. the result corresponds to a molality, but if the plasma is used directly, the concentration of the glucose must be lower since about 5% of the pipetted volume belongs to the plasma protein and the amount of plasma water, containing the glucose, is correspondingly smaller. In principle, the result corresponds to a molarity.

Note to 4. The following recommendations may be made with regard to the choice of method. Today, from the standpoint of the biochemist, whole blood should be replaced with plasma where possible. Whole blood yields results which differ appreciably from those for plasma or serum (see Note to 1). The erythrocytes can contain substances which interfere with most methods of determination; these are irrelevant if plasma or serum is used. If whole blood is used, it must always be deproteinated, except for the hexokinase method, when it is unnecessary.[38] If serum or plasma is used, a deproteination is superfluous with most methods. This cuts out an additional source of error. In view of the numerous factors which influence the results, it is not surprising that they should differ according to the procedure and method employed.[15]

For the above reasons, no absolutely valid norms can be given at present for the glucose concentration of fasting blood. In consequence, the following numbers should be construed as *standard values* combined from all methods:

Normal range	50–100 mg per 100 ml	2.78–5.55 mmol l^{-1}
Limiting range	100–130 mg per 100 ml	5.55–7.22 mmol l^{-1}
Pathological range	>130 mg per 100 ml	>7.22 mmol l^{-1}

Representative reference values derived from an investigation by the glucose oxidase method without deproteination in venous and capillary blood are collected in Table 63.[36] They ought to differ little from the values obtained with the hexokinase/glucose-6-phosphate dehydrogenase method. The low values for new-born infants are worthy of attention. For persons between the ages of 5 and 80, capillary blood concentrations are roughly between 60 and 100 mg per 100 ml and the venous plasma concentrations between 60 and 80 mg per 100 ml. The blood-glucose concentration does not increase with increasing age but the scatter becomes greater. No sex differences are observable.

3.9.2.7. Diagnostic significance

3.9.2.7.1. Blood-glucose concentration[39]

For a healthy adult, the body glucose content lies between 10 and 20 g. Most of this is in the extracellular fluid. In the intracellular compartment,

Table 63. Fasting concentrations of glucose in capillary and venous plasma; determination with glucose oxidase/peroxidase without deproteination.[36]

		Capillary blood plasma			Venous blood plasma		
		$n = 20$		Reference interval $\bar{x} \pm 2s$	$n = 20$		Reference interval $\bar{x} \pm 2s$
	Age	$\bar{x}$	s		$\bar{x}$	s	
New-born infants	6 h	32.2	13.13	5.94–58.46			
	5 days	44.1	15.63	12.84–75.36			
Children	1–2 years	72.0	19.75	32.5–111.5	63.3	13.02	37.3–89.4
	3–4 years	75.1	11.53	52.0–98.2	65.4	7.60	50.2–80.6
	5–6 years	78.9	4.96	69.0–88.8	67.9	3.30	61.3–74.5
	7–9 years	81.8	7.48	66.8–96.8	69.7	4.92	60.0–79.7
	10–12 years	80.2	9.22	61.8–98.6	68.7	6.22	56.0–80.9
	13–15 years	78.2	7.68	62.8–93.6	72.4	5.10	57.2–77.6
	16–20 years	82.4	4.97	72.5–92.4	70.1	3.30	63.6–76.7
Adult men	21–30 years	79.5	5.06	69.4–89.6	68.3	3.32	61.6–74.9
women	21–30 years	83.4	5.20	73.0–93.8	70.7	3.40	64.0–77.6
men	31–50 years	83.6	7.55	68.5–98.7	71.0	4.97	61.0–80.9
women	31–50 years	83.8	7.68	68.4–99.2	71.1	5.07	60.9–81.2
men	51–80 years	84.3	8.58	67.1–101.5	71.5	5.62	60.2–82.7
women	51–80 years	83.5	8.19	67.1–99.9	70.9	5.40	60.1–81.7
Adults	20–80 years	83.0	7.04	68.9–97.1	69.8	4.58	60.6–78.9
		$n = 120$			$n = 120$		

glucose is present either as phosphate ester or as glycogen. Glucose in the circulation has a half-life of 40 min (= 'turnover' of 200 mg min^{-1}).

3.9.2.7.2. Physiology

High value is attributable to an elevated concentration of circulatory adrenalin under conditions of stress such as severe bodily exertion and excitement or fear.

Low value. 1, Normal pregnancy may be attended by slight hypoglycaemia. 2, New-born infants of diabetic mothers can display a severe hypoglycaemia. 3, The blood-glucose concentration falls to its lowest value 2–4 h after birth and re-attains the birth value on the 3rd–5th day.

3.9.2.7.3. Pathology

High values: *hyperglycaemia*
 Elevated concentration of P-adrenalin
  following injection of adrenalin
  shock
  phaeochromocytoma, during a seizure

possibly in severe thyrotoxicosis

burns—the hyperglycaemia may last from hours to days

Diabetes mellitus

  insulin-deficient type

  insulin-resistant type

Diseases of the pituitary gland and adrenal cortex

  Cushing's syndrome, frequent complication: insulin-resistant diabetes

  gigantism and acromegaly; an insulin-resistant diabetes may be present

  in the early stages; hypopituitarism develops later

  ACTH injections result in the liberation of 11-oxysteroids, which lead

  to an increased gluconeogenesis and inhibition of carbohydrate

  utilization

Diseases of the pancreas

  acute (and in individual patients, chronic) pancreatitis rarely, with

  extensive carcinomas of the pancreas

  vitamin B_1 deficiency; Wernicke encephalopathy

In hyperglycaemias, the elevated glucose concentration can lead to an increased osmotic pressure which results in an expansion of the extracellular space at the expense of the intracellular space.

Low values: *hypoglycaemia in adults*

  Increased production of insulin

    islet cell tumour of the pancreas

      hyperplasia

      adenoma

      carcinoma

    insulin overdose, e.g. if a diabetic misses a meal

    dumping syndrome following gastrectomy; a rapid reabsorption of

    carbohydrates in the small intestine results in liberation of insulin

    which persists even when the greater part of the carbohydrates is

    reabsorbed and already stored

    functional hypoglycaemia; possibly there is an abnormally rapid

    reabsorption

    lesions in the hypothalamus; this can lead to vagal stimulation of the

    islet cells in the pancreas; this is rare, however

  Deficiency of insulin antagonists

    insufficiency of the adrenal cortex (Addison's disease)

    hypopituitarism

  Insufficient storage or mobilization of glycogen

    hepatopathias

      acute infections

        viral hepatitis (rare)

        yellow fever

      poisons

        organic arsenic compounds

carbon tetrachloride
chloroform
cinchophen
phosphorus
cirrhosis of the liver, terminal stage (individual patients)
metastasic liver (rare)
glycogenosis, especially with a deficiency of glucose-6-phosphatase
renal glycosuria (very rare)

Low values: *hypoglycaemia in children*
  Normoinsulinaemic forms
    diminished reabsorption of glucose:
      fasting
      vomiting
      diarrhoea
      malabsorption in cases of coeliac syndrome
      hypothyrosis
      enzyme defects
        starch intolerance
        lactose intolerance
        saccharose–isomaltose intolerance
        glucose and galactose reabsorption distrubance
        large abdominal tumours
    diminished glycolysis, diminished glycogen reserve, diminished gluconeogenesis
      pituitary insufficiency, adrenal cortex insufficiency
      diminished elimination of cetechol amines
      idiopathic infantile hypoglycaemia (McQuarrie)
      new-born hypoglycaemia
      alcohol hypoglycaemia
      galactosaemia
      glycogenosis
      fructose intolerance
      ketotic hypoglycaemia
      large abdominal tumours
    diminished consumption.
      lactation (only with animals)
      renal glycosura
      fever, exhaustion
      large abdominal tumours
      hyperinsulinaemic forms
        prediabetes
        leucine-sensitive hypoglycaemia
        children of diabetic mothers
        islet cell adenoma and β-cell hyperplasia

α-cell deficiency syndrome (relative hyperinsulinism)
uncertain forms
salicylate intoxication
disturbances of the central nervous system

Neonatal hypoglycaemia occupies a special position since it must be clearly differentiated from the physiologically low glucose values at this stage of life;[40-42] hypoglycaemia of new-born infants is defined in Table 64.

References

1. Fehling, H., Quantitative Bestimmung des Zuckers im Harn, *Arch. Physiol. Heilk.*, **7**, 64 (1848).
2. Benedict, S., The detection and estimation of glucose in the urine, *J. Am. med. Ass.*, **57**, 1193 (1911).
3. Schmidt, F. H., Fehlmessungen der Harnglucose durch Polarisation, *Dt. med. Wschr.*, **92**, 2025 (1967).
4. König, R., Dauwalder, H., and Richterich, R., Vergleichende Bestimmungen der Uringlucosekonzentration mit der Polarimetrie und einer enzymatischen Methode (Hexokinase/Glucose-6-Phosphatdehydrogenase), *Schweiz. med. Wschr.*, **101**, 860 (1971).
5. Hoffman, W. S., Rapid photometric method for determination of glucose in blood and urine, *J. biol. Chem.*, **120**, 51 (1937).
6. Benedict, S. R., The analysis of whole blood. II. The determination of sugar and of saccharoids (non-fermentable copper-reducing substances), *J. biol. Chem.*, **92**, 141 (1931).
7. Colombo, J. P. and Richterich, R., *Die einfache Urinuntersuchung*, Huber, Berne, 1977.
8. Cook, M. H., Free, A. H., and Giordano, A. S., The accuracy of urine sugar tests, *Am. J. med. Technol.* **19**, 283 (1953).
9. Teuscher, A., Richterich, R., Bürgi, W., Dettwiler, W., and Zuppinger, K., Neue schweizerische Richtlinien zur Diagnose des Diabetes mellitus, *Schweiz. med. Wschr.*, **101**, 345 (1971).
10. Morris, D. L., Quantitative determination of carbohydrates with Dreywoods anthrone reagent, *Science*, **107**, 254 (1948).

Table 64. Definition of hypoglycaemia in new-born infants.

	Blood glucose, determined enzymatically, mg per 100 ml
Normal weight full term:	
First 72 h	<30
Thereafter	<40
Premature births, defective births:[a]	
1st week	<20
Thereafter	<40

[a]For this purpose the Swiss neonatology group includes underweight new-born (=defective birth = light for dates) in the group of prematurely born infants.

382

11. Davidson, W. D. and Sackner, M. A., Simplification of the anthrone method for the determination of inulin in clearance studies, *J. Lab. clin. Med.*, **62**, 351 (1963).
12. Lorenz, K., Blutzucker-Schnellbestimmung mit Anilin-Eisessig, *Z. klin. Chem.*, **1**, 127 (1963).
13. Hultmann, E., Rapid specific method for determination of aldosaccharides in body, *Nature, Lond.*, **183**, 108 (1959).
14. Abraham, C. V., A modified *o*-toluidine reagent for glucose analysis. *Clin. chim. Acta*, **70**, 209 (1976).
15. Passey, R. B., Gillum, R. L., Fuller, J. B., Urry, F. M., and Giles, M. L., Evaluation and comparison of 10 glucose methods and the reference method recommended in the proposed product class standard 1974, *Clin. Chem.* **23**, 131 (1977).
16. Roe, J. H. and Rice, E. W., A photometric method for the determination of free pentoses in animal tissues, *J. biol. Chem.*, **173**, 507 (1948).
17. Richterich, R. and Colombo, J. P., Vereinfachte enzymatische Bestimmung der Blut-Glucose mit 20 Mikroliter Blut, *Klin. Wschr.*, **40**, 1208 (1962).
18. Middleton, J. E., Preparation and investigation of a stabilized glucose oxidase, peroxidase reagent for estimating glucose, using *o*-tolidine with an alkylaryl sulphonate and polyethylene glycol, *Clin. chim. Acta*, **22**, 433 (1968).
19. Werner, W., Rey, H. G., and Wielinger, H., Über die Eigenschaften eines neuen Chromogens für die Blutzuckerbestimmung nach der GOD/POD-Methode, *Z. analyt. Chem.*, **252**, 224 (1970).
20. Gochman, N. and Schmitz, J. M., Application of a new peroxide, indicator reaction to the specific, automated determination of glucose with glucose oxidase, *Clin. Chem.*, **18**, 943 (1972).
21. Kadish, A. H., Litle, R. L., and Sternberg, J. C., A new and rapid method for determination of glucose by measurement of rate of oxygen consumption, *Clin. Chem.*, **14**, 116 (1968).
22. Richterich, R. and Dauwalder, H., Zur Bestimmung der Plasmaglucose-konzentration mit der Hexokinase/Glucose-6-Phosphatdehydrogenase-Methode, *Schweiz. med. Wschr.*, **101**, 615 (1971).
23. Schmidt, F. H., *Enzymatische Teste zur Schnelldiagnose. 3. Int. Donau-Symp. Diabetes mellitus 1973*, Mandrich, Vienna, 1973, p. 567.
24. Leybold, L. and Rick, W., Eine Glucosedehydrogenase für die Glucose-Bestimmung in Körperflüssigkeiten, *Z. klin. Chem. klin. Biochem.*, **13**, 101 (1975).
25. Richterich, R., Küffer, H., Lorenz, E., and Colombo, J. P., Die Bestimmung der Glucose in Plasma und Serum (Hexokinase/Glucose-6-Phosphatdehydro-genase-Methode) mit dem Greiner Electronic Selective Analyzer GSA II, *Z. klin. Chem. klin. Biochem.*, **12**, 5 (1974).
26. Falcon-Lesses, M., Glycolysis in normal and in leukemia blood, *Archs intern. Med.*, **39**, 412 (1927).
27. Field, I. B. and Williams, H. E., Artifactual hypoglycemia associated with leukemia, *New Engl. J. Med.*, **265**, 946 (1961).
28. Messeloff, C. R., Stolz, C., and Schoenfeld, M. R., Factitious hypoglycemia in chronic myelogenous leukemia, *N.Y. St. J. Med.*, **64**, 551 (1964).
29. Ruiter, J., Weinberg, F., and Morrison, A., The stability of glucose in serum, *Clin. Chem.*, **9**, 356 (1963).
30. Caraway, W. T., Chemical and diagnostic specificity of laboratory tests: effect of hemolysis, lipemia, anticoagulants, medications, contaminants and other variables, *Am. J. clin. Path.*, **37**, 445 (1962).
31. Clauvel, M., Schwarz, K., and Terrier, E., Comparaisons des taux de glucose

sanguine, plasmatique et érythrocytaire, *Revue fr. Etud. clin. biol.*, **10**, 753 (1965).

32. Standardization of the oral glucose tolerance test. Report of the Committee on Statistics of the American Diabetes Association, *Diabetes*, **18**, 299 (1969).
33. McDonald, C. W., Fisher, G. F., and Burnham, C., Differences in glucose determination obtained from plasma or whole blood, *Publ. Hlth Rep., Wash.*, **79**, 515 (1964)
34. Zender, R. and Nolen, S., L'analyse de glucose sanguine. Comparaison de quatre méthodes, Valeurs normales dans le sang capillaire et veineux et dans le plasma veineux, *Méd. Lab.*, **19**, 214 (1966).
35. Norval, M. A., Kennedy, R. I., and Berkson, J., Blood sugar in newborn infants, *J. Pediat.*, **34**, 342 (1949).
36. Bürgi, W., Richterich, R., Mittelholzer, M. L., and Monstein, S., Die Glucosekonzentration im kapillären und venösen Plasma bei direkter enzymatischer Bestimmung, *Schweiz. med. Wschr.*, **97**, 1721 (1967).
37. Bürgi, W., Richterich, R., and Mittelholzer, M. L., Der Einfluss der Enteiweissung auf die Resultate von Serum- und Plasma-Analysen, *Klin. Wschr.*, **45**, 83 (1967).
38. Stork, H. and Schmidt, F. H., Mitteilung über eine enzymatische Schnellmethode zur Bestimmung des Blutzuckers in 5 μl Kapillarblut ohne Enteiweissung und ohne Zentrifugation, *Klin. Wschr.*, **46**, 789 (1968).
39. Eastham, R. D., *Interpretation klinisch–chemischer Laborresultate*, translated and edited by R. Richterich and J. P. Colombo, Karger, Basle, 1970.
40. Milner, R. D. G., Neonatal hypoglycemia. A critical reappraisal, *Archs. Dis. Childh.*, **47**, 679 (1972).
41. Bossi, E., Neonatale Hypoglykämie, *Schweiz. Rdsch. Med.* **64**, 1214 (1975).
42. Zuppinger, K. A., *Hypoglycemia in childhood*, Karger, Basle, 1975.
43. Trinder, P. A., Determination of glucose in blood using glucose oxidase with an alternative oxygen acceptor, *Ann. clin. Biochem.*, **6**, 24 (1969).
44. Ziegenhorn, J., Neumann, U., Hagen, A., Bablok, W., and Stinshoff, K., Kinetic enzymatic method for automated determination of glucose in blood and serum, *J. clin. Chem. clin. Biochem.*, **15**, 13 (1977).
45. Gerbig, K., Eine neue hochspezifische und praktikable Methode zur Glucose-Bestimmung mit Glucose-Dehydrogenase, *Med. Lab., Stuttg.*, **29**, 1 (1976).

3.9.3. FRUCTOSE

3.9.3.1. Introduction

Fructose makes up about one sixth to one third of the total carbohydrate intake. This ketohexose is a component of the disaccharide cane sugar which is cleaved into glucose and fructose in the intestines. Fructose is incorporated in the degradation of carbohydrates (Fig. 103) by way of the enzymes fructokinase, fructose-1-phosphate aldolase, and fructose-1, 6-diphosphatase. Under normal conditions small concentrations are detectable in the blood and urine. Higher values may be found with congenital enzyme-defect diseases where the defective enzyme is one of the above mentioned, participating in fructose metabolism;[1] such diseases are usually accompanied by a fructosuria.

3.9.3.2. Choice of method

Chromogenic methods, such as those described for glucose, are unsuitable for the determination of fructose since glucose and related compounds interfere. One exception is the determination of inulin, which is useful for measuring the extracellular space and for evaluating the extent of glomerular filtration. Inulin is a polyfructosan which is determined as fructose by the anthrone method after deproteination of the plasma and subsequent hydrolysis.

In the older methods,[2] glucose and other alkali-labile chromogens were destroyed by first boiling in alkali; the resulting solution was then used for the colour reaction. However, as Davidson and Sackner[3] showed, boiling in alkali is superfluous since the specificity of the method for fructose can be significantly increased by using a small volume of concentrated anthrone solution and incubating at 37 °C. Fructose can also be determined as the fructose–zirconyl chloride complex after heating with a solution of zirconium oxychloride ($ZrOCl_2$).[4] Measurement is made either of the absorption of the complex at 334 nm or of its fluorescence at 410 nm (excitation at 335 nm). The reaction is non-specific but the absorption of the glucose complex is only one eighth as strong. Nevertheless, the best specificity for fructose· is provided by an enzymatic method.[5] The determination in whole blood necessitates a preliminary deproteination.

$$\text{D-Fructose} + \text{ATP} \xrightarrow{\text{Hexokinase}} \text{Fructose-6-phosphate} + \text{ADP}$$

$$\text{Fructose-6-phosphate} \xrightarrow{\text{PGI}} \text{Glucose-6-phosphate}$$

$$\text{Glucose-6-phosphate} + \text{NADP}^+$$

$$\xrightarrow{\text{G-6-PDH}} \text{6-Phosphogluconate} + \text{NADPH} + \text{H}^+$$

The first stage of this method consists in the conversion of D-fructose to D-fructose-6-phosphate by means of the non-specific hexokinase which likewise catalyses the conversion of glucose. However, the phosphohexose isomerase (PGI) from yeast, added in the secondary reaction, specifically catalyses the formation of glucose-6-phosphate from fructose-6-phosphate. In the indicator reaction the glucose-6-phosphate so formed is reacted with NADP and glucose-6-phosphate dehydrogenase, being converted to 6-phosphogluconate and $NADPH_2$; the increase in $NADPH_2$ is measured photometrically.

Yet another method for the determination of fructose should be mentioned;[6] this is based on a highly specific fructokinase from *Streptomyces*. Here too, the phosphorylation is coupled with an NADP reduction via secondary reactions.

Finally, fructose can also be determined enzymatically using sorbitol dehydrogenase (SDH) at pH 12 (triethanolamine buffer):[7]

$$\text{D-Fructose} + \text{NADH} + \text{H}^+ \xrightarrow{\text{SDH}} \text{D-Sorbitol} + \text{NAD}^+$$

The reaction is slow (*ca.* 60–70 min), which is a disadvantage, but on the other hand it appears to give more precise values than the hexokinase method in the range below 4 mg per 100 ml.[7]

3.9.3.3. Diagnostic significance: fructosuria

1. Alimentary (particularly with hepatopathia), after taking honey and grapes.
2. Essential fructosuria, a congenital metabolism anomaly with no clinical significance.
3. Fructose intolerance: after oral administration of fructose (beware!) the blood-fructose concentration increases and simultaneously there is a steep fall in the blood-glucose concentration.

3.9.3.4. Inulin (fructose): anthrone method

3.9.3.4.1. Principle

The anthrone method is suitable for the determination of the glomerular filtration. For measuring the plasma and urine content, the samples are first deproteinated with trichloroacetic acid. This is followed by the addition of modified anthrone reagent and incubation at 37 °C for 50 min before making the photometric measurement.

3.9.3.4.2. Reagents

Trichloroacetic acid, 0.61 *mol* l^{-1} (10%). Dissolve 10 g in DM-water and make up to 100 ml. The solution keeps indefinitely.

Anthrone reagent. Add 500 ml of concentrated sulphuric acid slowly to 130 ml of DM-water and dissolve 500 mg of anthrone (recrystallized twice from hot, glacial acetic acid) in the solution. Shake to dissolve. The solution may be kept for at least 2 days in a dark bottle at 4 °C. A slight yellowing does not interfere.

Inulin standard, 5 *mg per* 100 *ml.* Dilute the commercially available inulin solution with trichloroacetic acid (0.61 mol l^{-1}) until the desired concentration is obtained. The solution keeps indefinitely at room temperature.

3.9.3.4.3. Deproteination

(a) *Serum, plasma.* Add 0.1 ml of plasma or serum to 1.0 ml of trichloroacetic acid. Allow to stand for 5 min and centrifuge strongly.

(b) *Urine.* Add 0.02 ml of urine to 2.0 ml of trichloroacetic acid.

3.9.3.4.4. Procedure

	Sample, ml	Reagent blank, ml	Standard, ml
Anthrone reagent	1.0	1.0	1.0
Supernatant	0.1	—	—
Trichloroacetic acid	—	0.1	—
Standard, 5 mg per 100 ml	—	—	0.1

Cover the tube with a glass bead and incubate for 50 min, at 37 °C in a water-bath. The mixture is then allowed to stand 15 min, at room temperature and the extinction measured against DM-water at 630 nm. The colour is stable for about 3 h.

3.9.3.4.5. Calculation

Plasma, serum.

$$\text{Concentration} = \frac{E(S) - E(RB)}{E(ST) - E(RB)} \cdot C(ST) \cdot \frac{FV(\text{depr.})}{SV(\text{depr.})} \text{ mg per 100 ml}$$

$$= \frac{E(S) - E(RB)}{E(ST) - E(RB)} \cdot 5 \cdot \frac{1.1}{0.1} \text{ mg per 100 ml}$$

$$\text{Concentration} = \frac{E(S) - E(RB)}{E(ST) - E(RB)} \cdot 55 \text{ mg per 100 ml}$$

Urine.

$$\text{Concentration} = \frac{E(S) - E(RB)}{E(ST) - E(RB)} \cdot C(ST) \cdot \frac{FV(\text{depr.})}{SV(\text{depr.})} \cdot \text{ mg per 100 ml}$$

$$= \frac{E(S) - E(RB)}{E(ST) - E(RB)} \cdot 5 \cdot \frac{2.02}{0.02} \text{ mg per 100 ml}$$

$$\text{Concentration} = \frac{E(S) - E(RB)}{E(ST) - E(RB)} \cdot 505 \text{ mg per 100 ml}$$

3.9.3.4.6. Specificity

Glucose scarcely reacts under the above conditions and the reaction is therefore fairly specific for fructose or inulin.[3]

3.9.3.4.7. Comments on the method

In clearance investigations, a null value is subtracted from the result to correct for the possible presence of positive substances in the plasma or urine.

1. After the addition of the trichloroacetic acid (deproteination), the mixture should be thoroughly mixed by shaking, since there is a risk of small quantities of inulin being carried down with the protein (due to the relatively high molecular weight of inulin).

2. The method should not be used if the blood-glucose concentration exceeds 200 mg per 100 ml.

3. The Lambert–Beer Law is obeyed for inulin concentrations in plasma of up to 100 mg per 100 ml and in urine of up to 5 g per 100 ml. In clearance investigations, values of 30–60 mg per 100 ml are obtained in plasma at an average urine concentration of 400 mg per 100 ml.

4. The method is also suitable for the determination of fructose in plasma and urine (fructosuria, fructose intolerance). In this case, fructose is used as a standard.

References

1. Gitzelmann, R., Bärlocher, K., and Prader, A., Hereditäre Störungen im Fructose- und Galaktosestoffwechsel, *Mschr. Kinderheilk*, **121**, 174 (1973).
2. Young, M. K. and Raisz, L. G., An anthrone procedure for the determination of inulin in biological fluids, *Proc. Soc. exp. Biol. Med.*, **80**, 771 (1952).
3. Davidson, W. D. and Sackner, M. A., Simplification of the anthrone method for the determination of inulin in clearance studies, *J. Lab. clin. Med.*, **62**, 351 (1963).
4. Schlegelova, J. and Hruska, K. J., Determination of fructose by the zirconyl chloride reaction, *Analyt. Biochem.*, **79**, 583 (1977).
5. Bernt, E. and Bergmeyer, H. U., D-Fructose, in Bergmeyer, *Methods of enzymatic analysis*, Vol. 3, Academic Press, New York, 1974, p. 1304.
6. Sabater, B. and Ascnsio, C., Fructose determination with a specific fructokinase, *Analyt. Biochem.*, **54**, 205 (1973).
7. Schaub, J., Universitätskinderklinik, Munich, personal communication.

3.9.4. GALACTOSE

D-Galactose is a component of milk sugar and other polysaccharides as well as of glycoproteins and glycolipids. In the human organism the free hexose is keyed into the glucose-degradation train (Fig. 103) via the enzymes galactokinase, galactose-1-phosphate uridyl transferase, and uridine diphosphate galactose 4-epimerase. Small quantities are detectable in the blood and urine after the intake of galactose or lactose. Since the sugar undergoes conversion in the liver, the galactose determination may be used as a test of liver function in various illnesses.[1] High galactose concentrations in the blood and urine of new-born infants following normal feeding indicate congenital enzyme-deficiency diseases[2] affecting galactose metabolism.

For the determination of galactose the enzymatic method using galactose dehydrogenase has yielded good results.[3]

$$\beta\text{-D-Galactose} + NAD^+ \xrightarrow[\text{dehydrogenase}]{\text{Galactose}} \text{D-Galactono-}\delta\text{-lactone} + NADH + H^+$$

In the presence of NAD, galactose dehydrogenase from *Pseudomonas fluorescens* specifically catalyses the oxidation of β-D-galactose to D-galactono-δ-lactone. The quantity of $NADH_2$ formed during the reaction is proportional to the amount of galactose present and can be determined by photometry. Galactose dehydrogenase is specific for β-D-galactose. Since, in solution, D-galactose is present as a mixture of the α- and β-isomers the incubation must be long enough for all the galactose to be reacted. The formation of the β-form can be accelerated by the addition of mutarotase, allowing the use of shorter incubation periods.

3.9.4.1. Diagnostic significance

Elevated values in plasma:

1. In healthy subjects following a galactose-rich feed.
2. In hepatopathia (individual patients).
3. In galactosaemia, galactokinase deficiency.

References

1. Rommel, K., Böhmer, R., and Adam, W. E., Der intravenöse Galaktosetoleranztest als Leberfunktionsprobe, *Schweiz. med. Wschr.*, **97**, 484 (1967).
2. Gitzelmann, R., Bärlocher, K., and Prader, A., Hereditäre Störungen im Fructose- und Galaktosestoffwechsel, *Mschr. Kinderheilk.*, **121**, 174 (1973).
3. Rommel, K., Bernt, E., Schmitz, F., and Grimmel, K., Enzymatische Galaktosebestimmung in Blut und oraler Galaktose-Toleranztest, *Klin. Wschr.*, **46**, 936 (1968).

3.10. NITROGEN METABOLISM

3.10.1. SURVEY

Tables 65 and 66 provide a survey of nitrogen-containing metabolites relevant to the analysis of body fluids.

3.10.2. TOTAL NITROGEN

Table 65. Nitrogen-containing substances in human plasma.

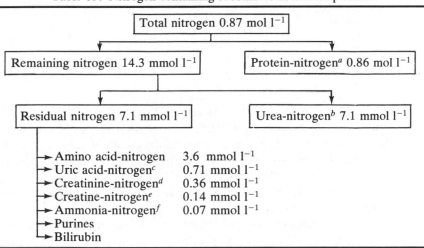

Total nitrogen 0.87 mol l^{-1}

Remaining nitrogen 14.3 mmol l^{-1}

Protein-nitrogen[a] 0.86 mol l^{-1}

Residual nitrogen 7.1 mmol l^{-1}

Urea-nitrogen[b] 7.1 mmol l^{-1}

Amino acid-nitrogen	3.6 mmol l^{-1}
Uric acid-nitrogen[c]	0.71 mmol l^{-1}
Creatinine-nitrogen[d]	0.36 mmol l^{-1}
Creatine-nitrogen[e]	0.14 mmol l^{-1}
Ammonia-nitrogen[f]	0.07 mmol l^{-1}
Purines	
Bilirubin	

[a]Protein = protein-nitrogen · 6.54. (on average, proteins contain 15% of nitrogen).
[b]Urea = urea-nitrogen · 2.14.
[c]Uric acid = uric acid-nitrogen · 3.0.
[d]Creatinine = creatinine-nitrogen · 2.69.
[e]Creatine = creatine-nitrogen · 3.12.
[f]Ammonia = ammonia-nitrogen · 1.22.

3.10.3. UREA AND AMMONIA

The determination of ammonium ions, whether in the free form (e.g. urinary and blood ammonium) or as the end-product of chemical manipulations (e.g. ashing, hydrolysis) or of enzymatic reactions (e.g. urease treatment), is one of the most important and most frequently used

390

Table 66. Nitrogen-containing substances in human urine.

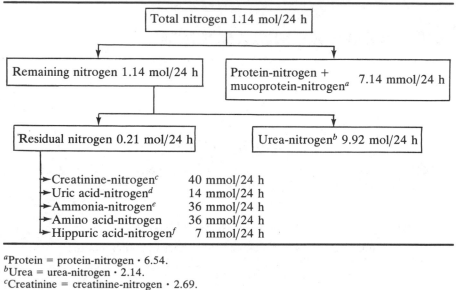

[a]Protein = protein-nitrogen · 6.54.
[b]Urea = urea-nitrogen · 2.14.
[c]Creatinine = creatinine-nitrogen · 2.69.
[d]Uric acid = uric acid-nitrogen · 3.0.
[e]Ammonia = ammonia-nitrogen · 1.22.
[f]Hippuric acid = hippuric acid-nitrogen · 12.79.

methods in the clinical chemistry laboratory. Whereas formerly titrimetric or gasometric methods were in relatively common use, the scene is now dominated by photometric and enzymatic methods. It is scarcely possible to predict whether these methods will be partly replaced by the use of ammonium-specific electrodes.

The following descriptions are confined, in the main, to the determination of ammonium ion by means of photometry.

1. *Reaction with Nessler's reagent.* Ammonium ions react with Nessler's reagent in alkaline solution to produce a colloidal, orange–yellow substance: $2(HgI_2 + 2KI) + NH_4OH + 3NaOH = OHgNH_2I + 3H_2O + 4KI + 3NaI$. This reaction was used for the routine determination of ammonia for many years and a large number of modifications have been described. However, the reaction is very susceptible to interferences causing turbidity (even on the addition of so-called stabilizers) and non-specific reactions. Also, the Lambert–Beer law is rarely obeyed. The sensitivity of the method is lower ($\varepsilon = 3000$) than that of the Berthelot reaction. For these reasons most laboratories have now abandoned the use of Nessler's reagent.

2. *Photometric determination by the Berthelot method.* Over 100 years ago Berthelot observed that a coloured compound was formed when ammonium ions were treated with phenol and hypochlorite. The reaction is extremely sensitive and the extinction coefficient (ε) is about 20 000. A

further advantage is its high specificity for ammonium ions. The reaction is slow but the rate can be increased considerably by the addition of coupling agents, e.g. sodium nitroprusside. Numerous different procedures have been described.

Patton and Crouch[1] studied the reaction kinetics of the Berthelot reaction and indicated the essential factors to be optimized. The reaction steps are:

I. (a) $NH_3 + HOCl \longrightarrow H_2NCl + H_2O (pH > 7.5)$

    Hypochlorous   Chloramine
    acid

(b) $[Fe(CN)_5NO]^{2-} + 2OH \rightleftharpoons [Fe(CN)_5NO_2]^{4-} + H_2O \rightleftharpoons$

    Nitroprusside   Nitritopentacyanoferrate

$$[Fe(CN)_5H_2O]^{3-} + NO_2^-$$

                                 Aquopentacyanoferrate

II. $[Fe(CN)_5H_2O]^{3-} + H_2NCl \longrightarrow$ Complex + Phenol $\longrightarrow$

(in alkaline medium)

The formation of monochloramine is most rapid at pH 10.5, very slow at pH > 11.5, and below pH 10.5 monochloramine decomposes quickly. Thus the reaction should be carried out at a pH between 10.5 and 11.5. Since the formation of aquopentacyanoferrate has an equilibrium constant of $3 \cdot 10^{-4}$, either a large excess of nitroprusside must be added or, better, aquopentacyanoferrate at roughly the same number of moles as the highest expected ammonia value. The sensitivity of the reaction may be increased by substituting the phenol ring with electron donors in position 2. Patton and Crouch[1] recommend 2-chlorophenol. Salicylate (R = COOH) has proved sufficiently sensitive for clinical purposes.[2]

3. *Enzymatic determination with glutamate dehydrogenase (GLDH)*. At present, the only method permitting an absolute measurement of ammonium ions is the determination which employs the optical test:

NH_4 + NADH + H^+ + 2-oxoglutarate

$$\xrightarrow{\text{GLDH, ADP}} NAD^+ + \text{glutamate} + H_2O$$

This method must now be regarded as the reference method for the determination of ammonia.

4. *Other methods*. There are few reports on the effectiveness of the so-called rubazonic acid method. Another interesting but likewise

little-studied method is the measurement of ammonia by means of its catalytic effect on the peroxidase reaction of hydrogen peroxide with o-dianisidine.

5. *Ion-selective electrodes.* See p. 189.

References

1. Patton, C. J. and Crouch, S. R., Spectrophotometric and kinetics investigation of the Berthelot reaction for the determination of ammonia, *Analyt. Chem.*, **49**, 464 (1977).
2. Lorentz, K., Mechanismus und Spezifität der Indophenolreaktion zur Ammoniakbestimmung. I. pH-Abhängigkeit, Kinetik, Inhibitoren, anorganische Verbindungen, *Z. klin. Chem. klin. Biochem.*, **5**, 291 (1967).

3.10.3.1. Urea nitrogen: urease cleavage and determination by the Berthelot method

3.10.3.1.1. Choice of method

Today, the determination of urea is based almost exclusively on one of the following four principles:

1. Reaction of urea with α-diketones and their oximes.
2. Determination of the ammonia liberated on treatment with urease.
3. Determination with ion-selective electrodes.
4. Measurement of urea in the optical test, using urease and glutamate dehydrogenase.

1. The reaction of urea with α-diketones has been known for many years. It is not specific for urea and can also be used for the detection of citrullin, allantoin, creatinine, arginine, proteins, etc. The following substances have been suggested as reagents: diacetyl, diacetyl monoxime, diacetyl dioxime, phenylpropanedione, phenylpropanedione monoxime, α-isonitrosopropiophenone, heptoxime, nioxime, and diacetyl monoxime glucuronolactone. These methods have numerous disadvantages: inadequate specificity, inadequate stability of the coloured product, invalidity of the Lambert–Beer law, and the necessity for deproteination and heating to 100 °C. An extensive discussion concerning the optimization of the individual reagents was given by Siest.[1]

2. Urea is cleaved by the enzyme *urease*, which occurs almost exclusively in plants (leguminosae) and bacteria, to give carbon dioxide and ammonia. Apart from urea, no other substances native to the body are known to be cleaved in this way. Hence the use of urease confers a very high specificity on this method. The liberated ammonia is determined with the Berthelot reagent.

3. Methods employing urease fixed in a membrane and determination of the ammonia with ion-selective electrodes are still at the development stage.

4. The combination of urease with glutamate dehydrogenase is most elegant.

5. Other photometric methods, e.g. with p-dimethylaminobenzaldehyde or xanthydrol, have not found widespread acceptance and are not specific. The measurement of urea by microdiffusion, gasometry, or electrochemical methods is generally too troublesome for routine tasks.

3.10.3.1.2. Principle

The urea is cleaved to ammonia and carbon dioxide by pre-treatment with urease:

$$\begin{array}{c} NH_2 \\ \diagdown \\ \diagup \\ NH_2 \end{array} C{=}O + H_2O \xrightarrow{\text{Urease}} 2\,NH_3 + CO_2$$

The liberated ammonia is determined by the Berthelot reaction.

3.10.3.1.3. Reagents

Salicylate–nitroprusside solution, sodium salicylate 187 $mmol\ l^{-1}$, *sodium nitroprusside* 2 $mmol\ l^{-1}$. Dissolve 3 g of sodium salicylate and 60 mg of sodium nitroprusside in DM-water and make up to 100 ml. The solution keeps for about 6 months in the dark at 4 °C or for at least 1 month at room temperature.

Hypochlorite stock solution, NaOCl 1.1 $mol\ l^{-1}$. This solution can be obtained commercially, or prepared according to the directions of Weller.[2] It keeps for at least 1 month at room temperature.

Sodium hydroxide solution, NaOH 12.5 $mol\ l^{-1}$. This keeps indefinitely at room temperature.

Hypochlorite–sodium hydroxide solution, NaOCl 550 $mmol\ l^{-1}$, *NaOH* 6.25 $mol\ l^{-1}$. Equal portions of the hypochlorite stock solution and the sodium hydroxide solution (12.5 $mol\ l^{-1}$) are mixed together. The solution keeps for about 6 months in the dark at 4 °C or at least 1 month at room temperature.

Urea-nitrogen standard solution, 20 mg per 100 ml, 7.13 $mmol\ l^{-1}$. Dissolve 428 mg of urea in cold, saturated benzoic acid solution (in DM-water) and make up to 1000 ml with further benzoic acid solution. The solution keeps indefinitely provided it is sterile.

EDTA buffer, 27 $mmol\ l^{-1}$, pH 6.5. Dissolve 1 g of ethylenediaminetetraacetate (disodium salt, dihydrate) in 99 ml of DM-water, adjust the pH to exactly 6.5 with sodium hydroxide solution, and make up to 100 ml with DM-water. The solution keeps indefinitely.

Urease solution, 57–114 $U\ mg^{-1}$. Dissolve 25 mg of urease in EDTA buffer and make up to 50 ml with further buffer. The solution keeps for a few weeks at 4 °C.

394

3.10.3.1.4. Procedure

	Sample, ml	Sample blank, ml	Standard, ml	Standard blank, ml
Urease solution	0.1	—	0.1	—
Serum, plasma	0.02	0.02	—	—
Standard solution	—	—	0.02	0.02
Incubate for 20 min at 37 °C or 30 min at room temperature				
Salicylate solution	2.5	2.5	2.5	2.5
Urease solution	—	0.1	—	0.1
Hypochlorite solution	2.5	2.5	2.5	2.5

After 10 min at room temperature read the extinction against DM-water between 540 and 590 nm (546 nm).

3.10.3.1.5. Calculation

$$\text{Concentration} = \frac{E(S) - E(SB)}{E(ST) - E(STB)} \cdot C(ST) \text{ mg per 100 ml}$$

$$\text{Concentration} = \frac{E(S) - E(SB)}{E(ST) - E(STB)} \cdot \begin{array}{l} 20 \text{ mg per 100 ml (urea nitrogen)} \\ 7.13 \text{ mmol } l^{-1} \text{ (urea nitrogen)}, \end{array}$$

$$\text{Concentration} = \frac{E(S) - E(SB)}{E(ST) - E(STB)} \cdot \begin{array}{l} 42.8 \text{ mg per 100 ml urea} \\ 15.25 \text{ mmol } l^{-1} \text{ urea}. \end{array}$$

3.10.3.1.6. Reference values

Adults, $n = 347$, 2.5- to 97.5-percentiles
Urea nitrogen 7.8–21.4 mg per 100 ml, 5.57–15.3 mmol l^{-1}.
Urea 16.7–45.9 mg per 100 ml, 2.78–7.64 mmol l^{-1}.

3.10.3.1.7. Specificity

The determination using urease is highly specific. No physiological compounds other than urea are known to be cleaved by this enzyme. The method can be used for the direct (i.e. without preliminary deproteination) determination of urea in body fluids.

3.10.3.1.8. Comments on the method

1. The Lambert–Beer law holds for concentrations (urea nitrogen) of up to about 140 mmol. For higher concentrations, the sample volume should

be reduced to a half (10 μl) and the result multiplied by 2. At low concentrations it is advisable to make the reading at 578 nm, or even at 645 nm for very low concentrations.

2. The urea nitrogen in serum and plasma is stable for at least 24 h at room temperature if the samples are stored under sterile conditions.

3. If the result is to be expressed as urea (in mg) rather than as urea nitrogen, the result should be multiplied by the factor 2.14 (60.06/28.02). To convert from mmol urea-nitrogen to mmol urea, the nitrogen value should be halved. If SI units are used, the result is no longer expressed as urea-nitrogen but as mmol l^{-1} urea.

4. Urease is an SH enzyme and is therefore very sensitive to traces of heavy metals. The life of the solution is considerably improved by the use of an EDTA buffer.

5. Measurements on the standard, reagent blank, and standard blank must be made once for each series of determinations.

6. *Urine*. So as to avoid bacterial cleavage of the urea, the urine specimens are either treated with 5 ml of thymol–isopropanol or individual protons are deep-frozen. Turbid urea should first be filtered. The concentration of urea nitrogen in urine is about 100 times greater than that in plasma, so urine should be diluted 1:100 before the determination and the result multiplied by 100.

3.10.3.1.9. Diagnostic significance

1. The plasma urea-nitrogen is above all a function of the protein balance. This shows up very clearly in Table 67.

The values are significantly higher with a protein-rich than with a protein-poor diet.

The normal range may be regarded as 4.7–23.0 mg of urea nitrogen per 100 ml. The concentration of urea-nitrogen in the urine is dependent on the protein feed, the functioning of the kidney, and the urine time-volumes. On average, 9–17 g of urea-nitrogen are excreted per day. The dependence on protein feed is even more marked with sucklings (Table 68).

2. In all studies to date, a higher plasma urea-nitrogen concentration has

Table 67. Plasma urea-nitrogen and protein feed.

Protein feed	Urea-nitrogen, mg per 100 ml	
	$\bar{x}$	$\bar{x} \pm 2s$
0.5 g/day/kg body weight	8.95	6.25–11.6
1.5 g/day/kg body weight	17.9	11.3–24.4
2.5 g/day/kg body weight	21.1	14.4–27.7

Table 68. Healthy sucklings, 1–3 months,[3] urea nitrogen in capillary blood.

Diet	Urea-nitrogen, mg per 100 ml $\bar{x} \pm s$	n
Human milk	10.6 ± 2.6	12
Modified cow's milk preparation	22.1 ± 3.7	16
Additional solid food	24.2 ± 5.6	33

been found for men than for women; this is probably because, in most populations, the men's diet is richer in proteins than the women's. With increasing age there is a slight increase in the plasma urea-nitrogen concentration; it is not known whether this is connected with a slight reduction in kidney function.

3. An elevation of the urea is found in cases of kidney insufficiency when the quantity of the glomerular filtrate is reduced to about a fifth of the norm.

4. A diminution of the urea in plasma is found in hereditary disturbances of the urea synthesis. This leads to hyperammonaemia.[4,5]

References

1. Siest, G., Etude des réactions colorées de l'urée et des dérivés carbamides avec des alphadicétones, applications biologiques, *Thèses, Faculté de Pharmacie, Université de Nancy*, 1966.
2. Weller, N., The use of Berthelot's reaction to ammonia in clinical chemistry. I. Study of the reaction and its sensitivity, *Röntg.-LabPrax.*, **15**, L77 (1962).
3. Davies, D. P. and Saunders, R., Blood urea. Normal values in early infancy related to feeding practices, *Archs Dis. Childh.*, **48**, 563 (1973).
4. Colombo, J. P., Congenital disorders of the urea cycle and ammonia detoxication, *Monogr. Paediatr.*, Vol. 1, Karger, Basle, 1971.
5. Bachmann, C., Urea cycle, in Nyhan, *Heritable disorders of amino acid metabolism*, Wiley, New York, 1974, p. 361.

3.10.3.2. Ammonia

The only methods of ammonia determination which can be recommended at present are the enzymatic technique using glutamate dehydrogenase and the determination by means of the Berthelot reaction.

With solutions low in proteins, e.g. urine or cerebrospinal fluid, the ammonia determination may be performed directly, i.e. it is unnecessary to isolate the ammonia beforehand. With solutions which contain proteins at a significant level, e.g. blood, plasma, or serum, there is a risk that 'labile' ammonia may be cleaved, leading to incorrect, high results. The risk is particularly great when, as in blood, the concentration of ammonia is very small relative to other nitrogen-containing compounds. Regardless of the method used, it transpires that if blood is analysed shortly after collection the ammonia concentration is small but the value increases continually as

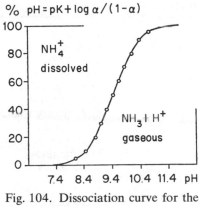

Fig. 104. Dissociation curve for the
NH_4^+/NH_3 system.

the blood is allowed to stand. It is assumed that the ammonia is formed as a result of enzymatic and hydrolytic cleavage of adenine nucleotides and glutamine. Experience over the last 70 years has shown that with increasingly improved methodology, the reference value for the blood ammonia concentration has decreased lower and lower. For routine clinical–chemical studies this does not matter provided that precise reference values are known for the method used; in practice it is found that the results obtained by a particular method have very good reproducibility even if the reference values are relatively high.

The NH_4^+/NH_3 system has a pK of around 9.3 (Fig. 104). Consequently, a high pH is necessary for methods based on the determination of ammonia as NH_3, and this can lead to the cleavage of labile ammonia. In addition to the microdiffusion method,[1] which is still used in many places, the important methods are isolation by ion exchange[2] and especially the more recent enzymatic determination of NH_4 using glutamate dehydrogenase as indicator enzyme. At present, the determinations with ion-selective electrodes still require large amounts of blood. In this case, ammonia is determined as NH_3, necessitating a high pH, achieved by mixing with a buffer. Variations in the pH attained affect the dissociation of the ammonia and hence the results.[3,4] Table 69 provides a survey of the methods of determination.

References

1. Richterich, R. and Colombo, J. P., Ultramikromethoden im Klinischen Laboratorium. 3. Prinzip der Ammoniak-Bestimmung durch Mikrodiffusion und Nesslerisierung, *Ärztl. Lab.*, **8**, 129 (1962).
2. Fenton, J. C. B. and Williams, A. H., Improved method for the estimation of plasma ammonia by ion exchange, *J. clin. Path.*, **21**, 14 (1968).
3. Guilbault, G. G. and Tarp, M., A specific enzyme electrode for urea, *Analytica chim. Acta*, **73**, 355 (1974).

398

Table 69. Methods for the determination of ammonia in blood.[5]

Principle	Reaction
1. Driving off the NH_3 by	Distillation
	Aeration
	Diffusion/microdiffusion
Determination after trapping, by	Back-titration
	Colorimetry with Nessler's reagent, ninhydrin
	Berthelot reaction, hypobromite/phenosoframine
	Coulometry
2. Adsorption of NH_4^+ on ion-exchange resins followed by determination on the eluate by	Colorimetry (see above)
3. Direct determination, after deproteination, by	Berthelot reaction
	Enzymatic reaction
	Fluorimetric determination
4. Direct determination in EDTA–plasma	Enzymatic reaction

4. Attili, A. F., Autizi, D., and Capocaccia, L., Rapid determination of plasma ammonia using an ion specific electrode, *Biochem. Med.*, **14**, 109 (1975).
5. Colombo, J. P., Congenital disorders of the urea cycle and ammonia detoxication, *Monogr. Paediat.*, Vol. 1, Karger, Basle, 1971.

3.10.3.3. Enzymatic determination of ammonia

3.10.3.3.1. Principle[1,2]

2-Oxoglutarate $+ NH_4^+ + NADPH + H^+$

$$\xrightarrow{\text{GLDH}} \text{L-Glutamate} + NADP^+ + H_2O$$

The indicator enzyme is glutamate dehydrogenase (GLDH), and the decrease in $NADPH_2$ is measured. ADP is added to prevent inactivation of GLDH, allowing the reaction to be carried out at the optimum pH of 8.6. The addition of ADP increases the rate of conversion of GLDH.

3.10.3.3.2. Collection of blood

Venous blood is collected without stasis in a plastic one-way syringe and transferred to a plastic tube containing Na_2EDTA (1 mg per 1 ml of blood). After centrifugation the plasma (not serum!) may be kept on ice for up to 2 h.[3,4]

3.10.3.3.3. Reagents

Solution 1. Triethanolamine buffer, 130 mmol l^{-1}, pH 8.6, containing 2-oxoglutarate (13 mmol l^{-1}) and ADP (1.3 mmol l^{-1}).

Reaction solution. $NADPH_2$ is dissolved in solution 1 immediately before the determination to give a concentration of 0.1 mmol l^{-1} (0.2 μmol per 2.0 ml of solution).

Glutamate dehydrogenase in 50% (w/w) glycerol $\geqslant$ 5 U ml^{-1}.

3.10.3.3.4. Measurement

Conditions. Final concentrations in the reaction mixture after the first addition of GLDH: $NADPH_2$ 66 μmol l^{-1}, 2-oxoglutarate 8.6 mmol l^{-1}, ADP 0.86 mmol l^{-1}, GLDH $\geqslant$ 33 U l^{-1}, triethanolamine buffer 86 mmol l^{-1}, pH 8.6.

3.10.3.3.5. Procedure

Temperature of measurement, 25 °C. Wavelength, 340 or 334 nm. One-way cuvettes, 1 cm in diameter.

	Sample, ml	Reagent blank, ml
Reaction solution	1.0	1.0
EDTA–plasma	0.5	—

Mix, and after exactly 10 min read off the extinction, E_1, against air, of the sample and reagent blank

GLDH solution	0.01	0.01

Mix, read off E_2 after exactly 10 min more

GLDH solution	0.010	0.010

Mix, measure the extinction E_3 after a further 10 min

3.10.3.3.6. Calculation

$$(E_1 - E_2) - (E_2 - E_3) = \Delta E_{\text{sample}} \text{ or } \Delta E_{\text{RB}}$$

Since the final volume for the reagent blank is smaller than that for the sample, ΔE_{RB} must be corrected to the final volume of the sample:

$$\Delta E_{\text{RB}} \cdot \frac{\text{Final volume RB}}{\text{Final volume S}} = \Delta E_{\text{RB}} \cdot 0.67 = \Delta E_{\text{RB corr.}}$$

$$\text{Concentration} = \frac{\Delta E_{\text{sample}} - \Delta E_{\text{RB corr.}}}{\varepsilon d} \cdot \frac{FV}{SV} \cdot 10^6 \ \mu\text{mol ml}^{-1}$$

Measurement at 340 nm, $\varepsilon = 6300$: $\Delta E_{\text{sample}} - \Delta E_{\text{RB corr.}}$ 479.4 μmol l^{-1}
at 334 nm, $\varepsilon = 6180$: $\Delta E_{\text{sample}} - \Delta E_{\text{RB corr.}}$ 488.7 μmol l^{-1}
at 365 nm, $\varepsilon = 3500$: $\Delta E_{\text{sample}} - \Delta E_{\text{RB corr.}}$ 862.9 μmol l^{-1}

Conversion factors.
μmol l^{-1} to μg NH$_3$ per 100 ml = ×1.703
to μg NH$_3$ per 100 ml = ×1.804
to μg NH$_3$-N per 100 ml = ×1.401

3.10.3.3.7. Reference values:[2] ammonia (NH$_3$)

Plasma (men) 15–60 μmol l^{-1}; 25–102 μg per 100 ml
Plasma (women) 11–51 μmol l^{-1}; 19–87 μg per 100 ml

3.10.3.3.8. Comments on the method

1. Since making up the reagent blank with distilled water can give rise to problems, the blank is measured without the addition of water and the result corrected for the difference in final volume. If a blank is to be performed with the addition of distilled water, the latter should be boiled again immediately before use.

2. The glycerine with its dissolved GLDH clarifies lipaemic plasma. Consequently, more glycerine-containing enzyme-solution must be added after the reaction and a correction must be applied for this decrease in the extinction, and this should take into account the dilution occasioned by the addition of GLDH.

3. If the initial extinction is too high or the results are higher than 350 μg per 100 ml, a smaller sample should be used. Whatever dilution (with ammonia-free water) is made, the final results (not the difference in the sample extinction) should be multiplied by the dilution factor.

4. Muscular activity should be prevented prior to sample collection since ammonia is liberated by the action of adenosine deaminase.

3.10.3.3.9. Diagnostic significance

Physiologically, the ammonia concentration is higher for new-born infants than for adults or children.

The ammonia concentration is raised in cases of liver insufficiency (hepatic coma). Hyperammonaemia occurs following portacaval shunt operations so that blood from the intestines is not poisoned as it circulates the liver, but it can also result from intestinal bleeding. Extensive dermatoses can lead to slight increases in ammonia. Apart from these, hyperammonaemia may result from certain inherited metabolic disturbances in children. This is primarily due to defects in the enzymes of the urea synthesis and secondary hyperammonaemia with organic acidaemias (propionic acid acidaemia, etc.).[5,6]

References

1. Da Fonseca-Wollheim, F., Bedeutung von Wasserstoffionenkonzentration und ADP-Zusatz bei der Ammoniakbestimmung mit Glutamatdehydrogenase. Verbesserter enzymatischer Ammoniaktest, 1, *Mitt. Z. klin. Chem. klin. Biochem.*, **11**, 421 (1973).
2. Da Fonseca-Wollheim, F., Direkte Plasmaammoniakbestimmung ohne Enteiweissung. Verbesserter enzymatischer Ammoniaktest, 2, *Mitt. Z. klin. Chem. klin. Biochem.*, **11**, 426 (1973).
3. Kornmüller, K. J. and Müller-Plathe, O., Einflüsse der Materialentnahme und -verarbeitung auf die Ammoniakkonzentration in Blut, Plasma und Serum. Vergleich zwischen enzymatischer Bestimmung und Mikrodiffusionsmethode, *Ärztl. Lab.*, **22**, 257 (1976).
4. Colombo, J. P. and Peheim, E., personal communication.
5. Bachmann, C., Urea cycle, in Nyhan, *Heritable disorders of amino acid metabolism*, Wiley, New York, 1974, p. 361.
6. Colombo, J. P., Congenital disorders of the urea cycle and ammonia detoxication, *Monogr. Paediat.*, Vol. 1, Karger, Basle, 1971.

3.10.4. PROTEIN METABOLISM

3.10.4.1. Plasma proteins

Human plasma contains hundreds of different proteins, the structures of most of which have not yet been fully resolved. At present, the following groups may be differentiated in terms of their function:

Proteins with a carrier or transport function
1. Binding of water (colloid–osmotic function): albumin
2. Non-specific binding: albumin (fatty acids, bilirubin, pigments, drugs)
3. Iron-binding proteins: transferrin
4. Haem-binding proteins: albumin (methaemalbumin), haemopexin (haemopexin haem)
5. Haemoglobin-binding proteins: haptoglobins
6. Lipid-binding proteins: lipoproteins (α_1-pre-β-, β-lipoprotein, chylomicrons)
7. Copper-containing proteins: caeruloplasmin
8. Zinc-containing proteins: Zn-α_2-glycoprotein
9. Vitamin-binding proteins: B_{12}-binding proteins (transcobalamin), retinol-binding proteins

Enzymes and enzyme-inhibiting proteins
1. Enzymes: cholinesterase, transaminases, phosphatases
2. Enzyme inhibitors: antitrypsin

3. Coagulation factors: fibrinogen, factors II–XIII
4. Fibrinolysis system: plasminogen
5. Complementary system: C_1–C_8, properdin

Hormones and hormone-binding proteins
1. Proteohormones: LH (the hormone stimulating the corpus luteum). FSH (follicle-stimulating hormone), TCSH (interstitial-cell stimulating hormone), gonadotrophin
2. Hormone-binding proteins: thyroxine-binding protein (binds thyroxine, triiodothyronine), transcortin (binds corticoids)

Proteins with an antibody function
1. Antibodies: IgG, IgM, IgA, IgE
2. Blood-group antibodies: isohaemaglutinine

These proteins are distinguished on the basis of their chemical properties, catalytic abilities, and immunological specificities. The most important plasma proteins are listed in Table 70, from which are omitted about a dozen coagulation factors, 100 enzymes, and a few proteohormones and hormone-binding proteins.

In all studies on plasma proteins the determination of *total proteins* is a primary consideration. Obviously, in view of the heterogeneity of the individual proteins, such a study is of limited clinical significance. Historically it was achieved by the separation of total proteins by *salting-out*. Two fractions were obtained, the albumins and the globulins, chemically distinguished as follows:

albumins: water-soluble, precipitation with ammonium sulphate at full saturation;

globulins: water-insoluble, precipitation with ammonium sulphate at half-saturation.

The results were used to calculate the albumin/globulin ratio; a diminution in albumin is usually accompanied by an increase in the globulin, and these changes are intensified in the ratio. It is no longer of importance, however.

The introduction of *paper electrophoresis* was of decisive significance for the further separation of proteins. Other support materials, such as cellulose acetate film, agarose, polyacrylamide, and agarose-containing antibodies (rocket electrophoresis), have supplanted paper. With these means it has been possible to separate the globulin fraction into about 20 fractions. Immunological methods have provided a sensitive and accurate way of selectively determining the individual proteins in a mixture (immunoprecipitation, fluorescence immunoassay, radioimmunoassay).

The function of proteins is measured by means of enzyme determinations. The enzymic activity often, but not always, runs parallel with the amount of protein.

Table 70. Quantitative data for the most important human plasma proteins.[1-3]

	Molecular weight	Electrophoretic mobility, pH 8.6, I = 0.1	pI	Peptide moiety, %	Nitrogen content, %	Biuret (albumin = 100%)	a(280 nm)	Average plasma concentration, g per 100 ml
Pre-albumin	61 000	7.6	4.7	99	16.7	97	13.2	0.03
Albumin	67 000	5.92	4.9	100	16.6	100	5.8	3.5
Orosomucoid	44 000	5.2	2.7	62	15.2	112	8.9	0.06
High-density lipoproteins (HDL)	180–435 000			42–55				0.36
α_1-Antitrypsin	45 000	5.42	4.0	86	16.3	116	5.3	0.29
Transcortin	55 700			86			7.4	0.007
Retinol-binding protein	21 000							
Transcobalamin I	56 000							
Gc-globulin	50 800			96	16.2	105		0.04
Caeruloplasmin	130–160 000	4.6	4.4	89	16.9	104	14.9	0.04
α_2-HS-glycoprotein	49 000							
Haptoglobin	85 000	4.5	4.1	81	15.4	98	12.0	0.20
α_2-Macroglobulin	820 000	4.2	5.4	92	16.7	107	8.1	0.25
Thyroxine-binding globulin (TBG)	36 500–45 000	4.0		85			4.0	0.002
Serum cholinesterase	348 000	3.1		76				0.001
Low-density lipoproteins (LDL)	$5–20 \cdot 10^6$			22				0.6
Haemopexin	57–80 000	3.1		77	17.6	112	16.5	0.10
Transferrin	80–90 000	3.1	5.9	95	17.1	99	11.2	0.30
Fibrinogen	341 000	2.1	5.8	97			13.8	0.30
IgG (γG-globulin)	150 000	(1.2)	5.8–7.2	97	16.5	100	13.8	1.25
IgA (γA-globulin)	150 000	2.1		92	17.4		13.4	0.21
IgM (γM-globulin)	900 000	2.1		89	16.7		13.3	0.125
IgD (γD-globulin)	170 000	2.1		88				0.02
IgE (γE-globulin)	190 000	2.1		89				

Table 71. Survey of the commonest methods.

	Extinction (210 nm)	Folin	Kjeldahl	Extinction (280 nm)	Biuret
Principle	Determination of the number of peptide links	Biuret + determination of tyrosine and tryptophan	Precipitation, ashing, determination of N as ammonia-N	Determination of the protein-bound tyrosine and tryptophan	Determination of the number of peptide links
Sensitivity, g	0.5	1	1	10	100
Specificity (protein)	Good	Fair	Good	Fair	Good
False-positive results	None	Tyrosine, tryptophan, phenols, uric acid, p-aminosalicylic acid	None	Tyrosine, tryptophan, uric acid, xanthine	Amine buffer
Effect of the nature of the proteins	Small	Large	Small	Large	Small
Technique	Complicated	Demanding	Complicated	Simple	Simple

3.10.4.2. Determination of proteins

A large number of methods are available for the determination of proteins in biological material and they differ in their sensitivity, specificity, and complexity. Table 71 lists the commonest methods used today. Since the specificity of the biuret method surpasses that of all other methods, we have accorded it preference for use in the clinical–chemical laboratory. However, we shall give brief attention to the other methods for the sake of completeness.

3.10.4.2.1. Specificity and accuracy

The specificity and accuracy of the individual methods emerge from a closer consideration of Table 71. The characteristic feature of all proteins is the peptide linkage ($-CO-NH-$) and the greatest sensitivity is therefore to be expected with those methods which measure the number of peptide links: the biuret method and measurement of the extinction at 210 nm. Differences between the individual proteins are smallest with these methods. Methods which measure predominantly the fraction of aromatic amino acids (tyrosine, tryptophan), such as the technique of Folin–Ciocalteu and measurement of the extinction at 280 nm, show a great variation in values from one protein to the next, varying by as much as a factor of three. The Kjeldahl technique measures the percentage nitrogen; this also varies considerably from protein to protein.

3.10.4.3. Standardization of methods of protein determination

To obtain accurate results suitable reference and standard substances must be available for calibrating the method. In this respect the field of protein determinations is in a sorry shape. Difficulties arise even with regard to the definition of the term 'protein'. As shown in Table 70, it is rare for proteins to be built exclusively of peptides. It is much more usual for them to consist of a peptide part and a non-peptide component which contains carbohydrates (glycoproteins), lipids (lipoproteins), heavy metals (metalloproteins), etc. We are therefore faced with the question of whether the results of protein analyses should be referred to the peptide moiety or to the whole molecule. This is no mere academic discussion, for the peptide moiety can vary from 7.3% (α_2-lipoproteins) to 99.0% (albumin). Up to a few years ago the calculation factors were chosen so that the result corresponded to the whole molecule. However, according to more recent and well founded suggestions this is incorrect, at least for the measurement of mixtures of proteins. Here the result should be confined exclusively to the peptide moiety. The reasons for this are predominantly of a technical nature as follows.

In the determination of the protein concentration by physical methods, e.g. by measurement of the specific gravity or by means of refractometry, in principle it is the dry substance, i.e. the total molecular weight, which is measured. However, because of their inadequate specificity and accuracy and the numerous sources of error (turbidities, bilirubin, high concentrations of uric acid and glucose), these methods are no longer used today.

If the percentage of aromatic amino acids is used as a basis for measurement, as, for example, in the measurement of the extinction at 280 nm (Table 70), or in the method of Folin–Ciocalteu, then, depending on the protein, very different results are obtained (Table. 71). These methods should therefore be used only for particular purposes (e.g. elution in column chromatography).

Immunological methods (whether these be photometric or utilize linear immunodiffusion, radial immunodiffusion, or electroimmunoassay) are always based on comparison measurements on proteins with known and standardized antigen properties. They are suitable for the determination of individual proteins under precisely standardized conditions; consequently, they are of little relevance to the present problem.

The classical Kjeldahl method measures the proportion of nitrogen. However, the existing confusion derives from the numerous different Kjeldahl factors (in some cases the differences are considerable). These contradictions may be attributed to the above-mentioned discrepancy in the definition of proteins. The Kjeldahl method should be calibrated with urea or with tris(hydroxymethyl)aminomethane (tris).

If the measurement of proteins is based on the number of peptide linkages—and at present this is the only correct procedure—whether by the biuret method or measurement of the extinction at 210 nm, then it is exclusively the peptide moiety which is measured and this should be reflected in the result. These methods are therefore currently the methods of choice and should be used for calibration purposes.

Besides these methodological considerations, the availability of suitable reference substances is an important factor in the standardization of a method. For this purpose, bovine serum albumin is recommended since its structure and properties are accurately known and it can be prepared in a stable form. The most important preconditions which must be met by such a product are as follows:[5,6] weighing out on the basis of the dry weight (drying at $110\,°C$ for $24\,h$ or to constant weight *in vacuo*); immunoelectrophoresis or electrophoresis is on cellulose acetate film—less than 1–2% of other proteins; ultracentrifuge or acrylamide electrophoresis—proportion of monomer at least 95%; impurities—carbohydrates less than 0.05%, lipids less than 0.5%, ash less than 0.1%, heavy metals less than 0.02%; optical properties of a 7% solution—$E_{405\,nm}$ less than 0.2 (haematin), $E_{500\,nm}$ and $E_{600\,nm}$ smaller than 0.05, no peak between 450 and 650 nm, $E_{280\,nm}/E_{250\,nm}$ greater than 2.0; molecular weight (including 1 molecule of oleic acid) 64 694; nitrogen concentration 16% (Kjeldahl factor 6.25).

In addition, bovine serum albumin has the advantage of cheapness and it may be prepared Australia-antigen free. In contrast to serum, a stable solution may be prepared by heating at 70 °C for 30 min.

These proposals signify substantial progress in the analytical chemistry of the proteins. An important disadvantage of bovine serum albumin is that it cannot be used as a calibration substance for the determination of human albumin with dyes. According to the sources cited, albumin preparations of declared concentration—if these should ever be necessary—need not be used for calibration every day. They are much better used to calibrate control sera or peptide solutions for routine usage.

3.10.4.4. Total protein: spectrophotometry at 280 and 260 nm

3.10.4.4.1. Suitability

In work with tissue extracts the nucleic acids interfere in the determination of proteins at 280 nm since they show a characteristic absorption at this wavelength. However, a typical feature of nucleic acids is that the extinction at 260 nm is higher than that at 280. Thus it is possible to estimate both the proteins and the nucleic acid concentration by simultaneous measurement at 260 and 280 nm.

3.10.4.4.2. Principle

Warburg and Christian[7] were the first to show that by using a simple equation or nomogram it is possible to evaluate the concentrations of protein and nucleic acids from the extinction of a solution at 280 and 260 nm. The determination utilizes the principle of the absolute measurement.

3.10.4.4.3. Apparatus

Spectrophotometer providing wavelengths of 260 and 280 nm, and quartz cuvettes.

3.10.4.4.4. Reagent

Sodium chloride solution, 154 $mmol\ l^{-1}$ (0.9%).

3.10.4.4.5. Procedure

The extinctions of the material under investigation are measured at 260 and 280 nm against sodium chloride solution.

3.10.4.4.6. Calculation

According to Layne.[8]

Concentration $= [1.55 E(280)] - [0.76 E(260)]$ mg ml^{-1}

Reading the results straight from the nomogram (Fig. 105).

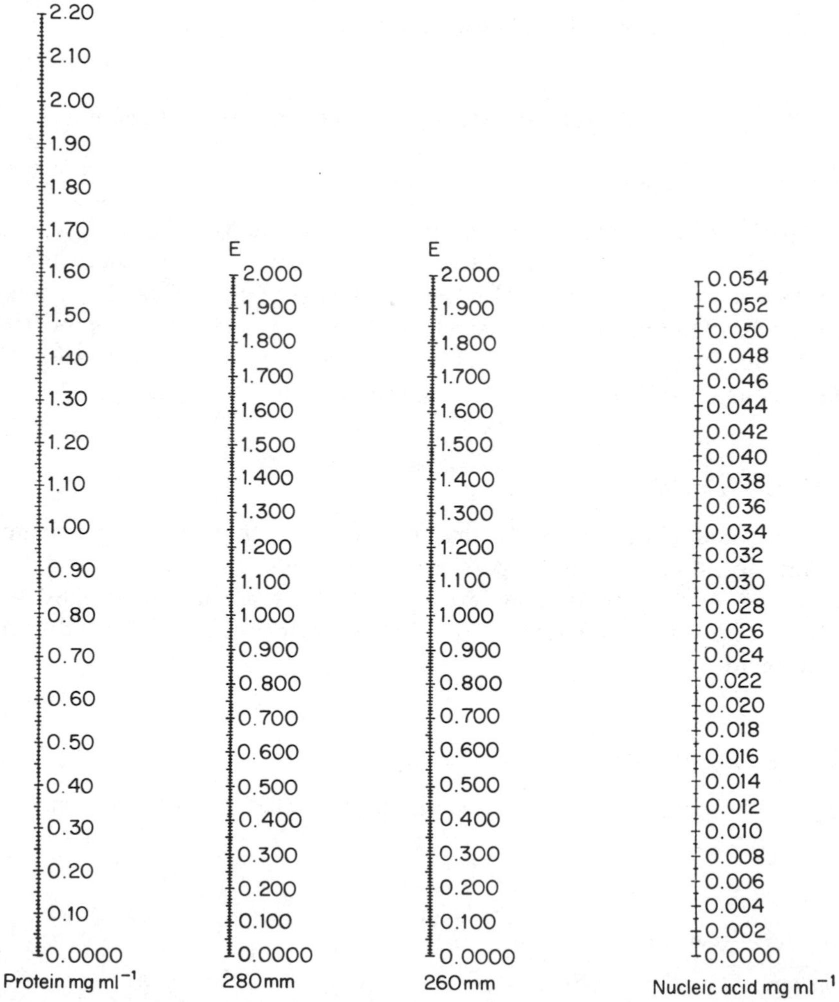

Protein mg ml^{-1}	E 280 mm	E 260 mm	Nucleic acid mg ml^{-1}
2.20			
2.10			
2.00			
1.90			
1.80			
1.70			
1.60	2.000	2.000	0.054
1.50	1.900	1.900	0.052
1.40	1.800	1.800	0.050 / 0.048
1.30	1.700	1.700	0.046
1.20	1.600	1.600	0.044
1.10	1.500	1.500	0.042 / 0.040
1.00	1.400	1.400	0.038 / 0.036
0.90	1.300	1.300	0.034 / 0.032
0.80	1.200	1.200	0.030
0.70	1.100	1.100	0.028 / 0.026
0.60	1.000	1.000	0.024
0.50	0.900	0.900	0.022
0.40	0.800	0.800	0.020 / 0.018
0.30	0.700	0.700	0.016
0.20	0.600	0.600	0.014
0.10	0.500	0.500	0.012 / 0.010
0.0000	0.400	0.400	0.008
	0.300	0.300	0.006
	0.200	0.200	0.004
	0.100	0.100	0.002
	0.0000	0.0000	0.0000

Fig. 105. Nomogram for the evaluation of the concentrations of protein and nucleic acid from the extinction of a solution at 280 and 260 nm. The readings at 280 and 260 nm are joined with a straight line. The point of intersection of this line with the left-hand scale gives the protein concentration in mg ml^{-1} and the intersection with the right-hand scale gives the concentration of nucleic acids in mg ml^{-1}.

3.10.4.4.7. Specificity

The method is not specific for proteins and nucleic acids but also responds to aromatic amino acids, purine bases, peptides, and nucleotides.

3.10.4.4.8. Comments on the method

For nucleic acid concentrations greater than 20% the deviation of the result from the actual protein concentration is very large.

3.10.4.5. Total protein: biuret method

3.10.4.5.1. Choice of method

A large number of physical methods (determination of the specific weight, refractometry, ultraviolet spectrophotometry) and chemical methods (Kjeldahl, Ninhydrin, Folin–Lowry) are available for the determination of total protein in plasma. The biuret method surpasses all these techniques since no special apparatus is necessary, only one reagent is required and that is stable, the specificity is high, and the labour involved is slight. For these reasons it has also been included in the Standard Methods of Clinical Chemistry of the American Association of Clinical Chemists.[5]

3.10.4.5.2. Principle

Substances with at least two peptide linkages:

$$R-\overset{\overset{\displaystyle O}{\|}}{C}-NH-R'-\overset{\overset{\displaystyle O}{\|}}{C}-NH-R''$$

give a violet coloration with copper salts in alkaline solution. The simplest compound to give a positive reaction is the substance biuret, which may be prepared by heating urea:

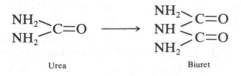

Urea   Biuret

The biuret reaction depends on the formation of a complex between a copper(II) ion and four peptide-N atoms under alkaline conditions. It is specific for peptides, polypeptides, and proteins and gives negative results with ammonia, urea, amino acids, and other simple nitrogen-containing compounds. For proteins, the maximum extinction lies at 456 nm but for peptides this is shifted towards shorter wavelengths. Many proteins show scarcely any difference in the position of the maximum or the extinction coefficients. Thus the values obtained from the biuret reaction are not falsified, even with highly pathological plasmas. Over a wide range of measurements, the intensity of the colour is proportional to the number of peptide linkages and thus the Lambert–Beer law is obeyed. The present method is based on Weichselbaum's modification[9] of the method developed by Kingsley.

The main disadvantage of the conventional biuret methods is the problem of the sample blank. Because of the relatively low sensitivity of the method, slight turbidities (Tyndall effect) are almost always encountered. This source of error may be largely eliminated by the addition of urea (final concentration 6 mol l^{-1}) to the biuret reagent. If the appropriate photometers, providing absolute measurements, are used, the results are best calculated by means of the extinction coefficient. Having regard to the statements made on p. 405 concerning the standardization of the method, the percentage extinction coefficient, a, for bovine serum albumin is 3.12 at 546 nm. In this case it is necessary to use a suitable standard solution only occasionally. With photometers which do not yield absolute measurements it is necessary to run a standard with each series of samples.

3.10.4.5.3. Reagents

Base reagent. Dissolve 36 g of urea in 2 mol l^{-1} NaOH and make up to 100 ml with further NaOH solution.

Biuret reagent. Copper sulphate 34 mmol l^{-1}, potassium iodide 21.7 mmol l^{-1}, sodium carbonate 472 mmol l^{-1}, sodium citrate 349 mmol l^{-1}, urea 5.99 mol l^{-1}. Dissolve 8.5 g of $CuSO_4$ and 3.6 g of potassium iodide separately in about 50 ml of DM-water. Dissolve 87 g of sodium citrate and 50 g of sodium carbonate in about 500 ml of DM-water and, while stirring, add the $CuSO_4$ solution and finally the potassium iodide solution. Dissolve 360 g of urea in this solution, with constant stirring, and make up to 1000 ml with DM-water.

Biuret solution. According to the number of analyses that are to be performed, mix 1 part of base reagent with 1 part of biuret reagent.

Standard albumin solution, 7 g per 100 ml. A suitable starting material is a bovine serum albumin conforming with the criteria listed on p. 406. Dissolve 7 g of albumin (dry weight) in 0.02 mol l^{-1} potassium chloride solution, the pH of which has been previously adjusted to 6.5–6.8 with sodium bicarbonate, and make up to 100 ml with further potassium chloride solution. Under sterile conditions this solution may be kept for months at 4 °C. The standard solution is best kept in ampoules which are used only on the day they are opened. The percentage extinction coefficient, a, is 3.12, measured with the biuret reagent described.

3.10.4.5.4. Measurement

Conditions. The final concentrations in the reaction mixture are $CuSO_4$ 16.6 mmol l^{-1}, KI 10.6 mmol l^{-1}, Na_2CO_3 231.3 mmol l^{-1}, Na citrate 171 mmol l^{-1}, urea 2.94 mol l^{-1}, and NaOH 0.98 mol l^{-1}.

3.10.4.5.5. Procedure

	Sample, ml	Reagent blank, ml
Plasma, serum	0.02	—
DM-water	—	0.02
Biuret solution	1.00	1.00

Mix well, allow to stand, read off the extinction at 546 nm against DM-water between 30 and 40 min after mixing

3.10.4.5.6. Calculation

Using a protein standard solution.

$$\text{Concentration} = \frac{E(S) - E(RB)}{E(ST) - E(RB)} \cdot C(ST) \text{ g per 100 ml}$$

By means of the extinction coefficient ($a = 3.12$).

$$\text{Concentration} = \frac{E}{ad} \cdot \frac{FV}{SV} = \frac{E}{3.12 \cdot 1} \cdot \frac{1.02}{0.02} \text{ g per 100 ml}$$

$$\text{Concentration} = E(S) - E(RB) \cdot 16.3 \text{ g per 100 ml}$$

3.10.4.5.7. Specificity

The biuret method is largely specific for peptide linkages and thus not only proteins but peptides also give a positive reaction. The maximum extinction is shifted towards the ultraviolet region for lower molecular weight peptides. The only non-specific positive reaction of practical importance is that with amine buffers, particularly tris and glycine. Such buffers should be avoided in studies which are to include a protein determination.

3.10.4.5.8. Investigational material

When heparin is used as anticoagulant no difference is observed between plasma and serum, but with oxalate, citrate, or fluoride slightly lowered protein concentrations are found owing to loss of water from the

Table 72. Reference values for children.[10]

Age	n	$\bar{x}$	s	$\bar{x} \pm 2s$
New-born	92	7.16	0.96	5.24–9.08
Up to 3 months	21	6.86	0.74	5.38–8.34
4–6 months	51	7.08	0.74	5.60–8.56
7–9 months	33	7.08	0.66	5.76–8.40
10–12 months	28	7.22	0.75	5.72–8.72
13–19 months	16	7.16	0.54	6.08–8.24
1½–3 years	11	7.20	0.70	5.80–8.60
Adults	28	7.69	0.49	6.71–8.67

erythrocytes. In serum and plasma the protein concentration remains constant for weeks at 4 °C or for years if frozen.

3.10.4.5.9. Reference values

Adults: 6.82–8.87 g per 100 ml.
Our values, using a human protein standard: 2.5- to 97.5-percentiles, $n = 315$.
Reference values for children are given in Table 72.

3.10.4.5.10. Physiological variability

As a result of the increased haem concentration, the plasma protein concentration is slightly higher for blood collected from patients lying down as opposed to standing, the maximum being observed 2–4 h after lying down. Venous and capillary blood give almost identical results. The values for premature births are lower than for full-term births; the slightly lower values for new-born infants rise to attain adult concentrations in the 3rd year of life. There is no, or at most a negligible, decrease in old age. No sex difference is detectable. The plasma protein concentration show no postprandial change, nor are any variations detectable for the time of year or daily rhythm. In contrast, there is a rise in the protein concentration of 6–12% following brief, intensive physical exericse—probably as a result of the increased haem concentration.

3.10.4.5.11. Comments on the method

1. The biuret reaction is a stoichiometric reaction with colour constancy, i.e. the values of the extinction for a particular protein are constant under identical optical conditions. It is therefore necessary to run a standard only if no absolute-measuring photometer is available. For the evaluation of the results from the extinction at 546 nm it is advisable to prepare a table so that the results may be read off directly.

2. The use of a sample blank, e.g. by diluting the sample with a biuret reagent not containing copper sulphate, leads to an over-correction, i.e. the results are too low. Probably the biuret reaction is accompanied by a clarification of the turbidity almost always present and visible owing to the Tyndall Effect. Deproteination with trichloroacetic acid cannot be recommended either, since it is often non-quantitative. The addition of urea almost always clears the turbidity and renders the sample-blank assay superfluous. If the solution remains turbid despite these measures, attempts should be made to eliminate this by brief shaking (30 sec) with 1 ml of diethyl ether followed by powerful centrifugation. If the turbidity still persists, then 20 mg of potassium cyanide (highly poisonous) may be added to decolorize the biuret reaction but this method may also lead to over-correction.

3. If dextrans (Macrodex, Rheomacrodex) are present in the plasma samples, turbidities ensue which cannot be clarified with urea or extracted with ether. In this case it is advisable to precipitate the proteins with trichloroacetic or perchloric acid and subsequently dissolve the proteins in the biuret reagent. It is still simpler to eliminate the dextran by centrifugation.

4. Neither bilirubin in concentrations up to 30 mg per 100 ml, nor haemoglobin in concentrations up to 100 mg per 100 ml, interfere in the determination.

5. The extinction coefficient is not identical for all biuret methods but is partly dependent on the chemical composition of the reagent.

6. The colour of the copper–protein complex is not completely stable but there is a very slow increase in the extinction even after 30 min; however, it amounts to less than 5% per hour and is therefore insignificant. Use may be made of this property to determine poorly soluble proteins (e.g. organ extracts, microsome fractions). The analytical material is triturated with the biuret reagent and the solution allowed to stand at room temperature for about 24 h. In this case it is necessary to run a standard.

7. The biuret–peptide complex has an additional absorption maximum in the ultraviolet region and this may also be used for measurements (especially of peptides), although glucose and individual amino acids and buffers (tris) interfere.

8. The biuret method can also be used for the determination of peptides following a preliminary precipitation of the proteins. A method for the determination of urine peptides was described by Balikov et al.[11]

9. The IFCC experts recommend the Doumas method[5] for the determination of proteins in *serum*. Since serum does not suffer from interference by turbidity as plasma does, it is not necessary to add urea in the Doumas method. Further differences in the measurement conditions are a higher concentration of sodium hydroxide ($0.6 \ mol \ l^{-1}$), a temperature of $25 \pm 1 \ °C$ and measurement of the extinction at 540 nm.

Calculation after correction of the reagent blank by the analysis blank ($a = 2.918$). The lyophilized standards are dissolved in NaCl (9 g l^{-1}), pH 6.8, containing 0.5 g l^{-1} of NaN_3.

10. The method of Lowry et al.[12] represents a further modification for the determination of minute amounts of dissolved or precipitated proteins: the proteins are reacted with copper in alkaline medium and the protein–copper complex so formed is determined by reduction of molybdophosphate phosphotungstate reagent.

3.10.4.5.12. Diagnostic significance

The detection of specific plasma proteins serves for the diagnosis of disturbances in the formation of proteins or of loss of proteins.[2,4] Deficiencies in the supply of precursors (protein under-nourishment, malabsorption) lead only to a slight diminution of the plasma proteins. With acute infections or trauma with breakdown of the tissue, particular proteins (e.g. fibrinogen → elevated sedimentation rate, C-reactive protein) are elevated while others (albumin, transferrin) are diminished. The decrease is especially marked with chronic infections. In this case immunoglobulins increase. However, apart from the detection of antigen-specific immunoglobulins, these changes are not meaningful for differential diagnosis.

On the other hand, the formation of abnormal proteins (paraproteins), e.g. in cases of plasmacytoma, is of diagnostic importance. Over 100 hereditary disorders leading to a deficiency in specific proteins are known (inborn errors of metabolism). This applies to proteins with a carrier and transport function (analbuminaemia, bisalbuminaemia, caeruloplasmin deficiency = Wilson's disease, haemoglobinopathia TBG deficiency), to enzymes (carbohydrate, amino acid, and fat metabolism, hormone metabolism, and coagulation disturbances) and to antibodies (antibody-deficiency syndrome).

The investigation of selective proteins in kidney affections allows an insight into the disturbance concerned. The loss of low-molecular-weight proteins (albumin) with disturbances in the glomerulum and the extent of reabsorption in the tubulus are identified by the appropriate investigations of plasma and urine.

References

1. Weast, R. C., Selby, S. M., and Hodgman, C. D., *Handbook of chemistry and physics*, 46th ed., Chemical Rubber Co., Cleveland, 1965–1966.
2. Kawai, T., *Clinical aspects of the plasma proteins*, Springer, Berlin, 1973.
3. Hitzig, W. H. and Jako, J., Fractions of the plasma proteins, in Curtius and Roth, *Clinical biochemistry, principles and methods*, Vol. 2, de Gruyter, Berlin, 1974, p. 1479.
4. Alper, C. A., Plasma protein measurements as a diagnostic aid, *New Engl. J. Med.*, **291**, 287 (1974).

5. Doumas, B. T., Standards for total serum protein assays—a collaborative study, *Clin. Chem.*, **21**, 1159 (1975).
6. Peters, T., Jr., Proposals for standardization of total protein assays, *Clin. Chem.*, **14**, 1147 (1968).
7. Warburg, O. and Christian, W., Isolierung und Kristallisation des Gärungsfermentes Enolase, *Biochem. Z.*, **310**, 384 (1941).
8. Layne, E., Spectrophotometric and turbidimetric methods for measuring proteins, *Meth. Enzym.*, **3**, 447 (1957).
9. Weichselbaum, C. T., An accurate and rapid method for the determination of proteins in small amounts of blood serum and plasma, *Am. J. clin. Path.*, **16**, 40 (1946).
10. Josephson, B. and Gyllenswärd, C., The development of the protein fractions and of cholesterol concentration in the serum of normal infants and children, *Scand. J. clin. Lab. Invest.*, **9**, 29 (1957).
11. Balikov, B., Lozano, E. R., and Castello, R. A., The assay of urinary peptides using a biuret reagent, *Clin. Chem.*, **4**, 409 (1958).
12. Lowry, O. H., Rosebrough, N. J., Farr, A. L., and Randall, R. J., Protein measurement with the Folin phenol reagent, *J. biol. Chem.*, **193**, 265 (1951).

3.10.4.6. Albumin; spectrophotometry with bromocresol green

3.10.4.6.1. Introduction

The isolated determination of albumin is still used for diagnostic purposes in many places, although simple methods of determination seldom give accurate results even when the only question is whether the albumin is low; moreover, quantification of this protein does not take the diagnosis very far (reduction in the acute phase of various diseases, infections, and traumas). Nevertheless, owing to its relatively high molar concentration in the plasma ($0.68 \, mmol \, l^{-1}$) it plays an important pathophysiological role in regulating the colloidosmotic pressure (*ca.* 14 mmHg). Thus, a reduction in the albumin concentration leads to oedaemas. Since the isoelectric point is 4.9 it occurs in blood predominantly as the anion and can bind cations. On the other hand, it is also able to bind anions and in this way it manifests a type of detergent function. By this means, fat-soluble substances are kept in solution in the blood. Consequently it plays an important part as a transport protein (Ca^{2+}, uric acid, thyroid hormone, free fatty acids, bilirubin, and various pharmaceuticals).

Albumin is synthesized in the liver. The rate of degradation is proportional to the plasma concentration.

3.10.4.6.2. Choice of method

The reference method for the determination of albumin in plasma is the rocket electroimmunoassay method designed by Laurell.[1,2] Salting-out methods and precipitation by organic solvents are too costly. Electrophoresis using support materials which do not exert as strong an adsorption as paper is frequently used (cellulose acetate film). The differential staining after separation (denaturation!) can lead to poor reproducibility, however. Apart from this, the calculation of the albumin concentration depends on the determination of the total protein, which leads to further variations according to the composition of the mixture. Nonetheless, electrophoresis has the advantage of yielding more diagnostic information than an isolated albumin determination. Immuno-electrophoretic methods with various antisera (if necessary, in a mixture) have the advantage of specificity at high resolution and of diagnostic polyvalency. If albumin is determined on its own, this can only be with the aim of discovering a lowering in the albumin values. An elevated value is found only at increased haem concentration. Here we have described the indicator-error method as an alternative to rocket immunoelectrophoresis. Extensive automation and rapid read-out are essential for precise and accurate results. The method described is based on the modified method of Doumas et al.[3] and that of Gustafsson.[4] If the values are below 2.5 g per 100 ml an electrophoretic, or better, an immunoelectrophoretic, determination should be carried out.

3.10.4.6.3. Principle

Albumin is bound with bromocresol green and this changes the extinction of the dye. The albumin concentration is proportional to the increase in extinction at 628 nm.

3.10.4.6.4. Reagents

Stock solutions. Succinate buffer, 0.1 *mol* l^{-1}, *pH* 4.0. Dissolve 11.8 g of succinic acid in *ca*. 800 ml of water. Adjust the pH to 4.0 with NaOH and make up to 1 litre with water. Keep at 4 °C.

Bromocresol green solution, 0.58 *mmol* l^{-1}. Dissolve 419 mg of bromocresol green in 10 ml of 0.1 mol l^{-1} NaOH and make up to 1 litre with water. Keep at 4 °C.

Working solution. Dilute 1 volume of bromocresol green with 3 volumes of succinate buffer, add 4.0 ml per litre of 30% Brij-35 solution and adjust the pH to 4.20 ± 0.05.

Standard human albumin solution (in sodium azide solution, 0.5 g l^{-1}; *poisonous!*), 2.0–6.0 g per 100 ml.

3.10.4.6.5. Measurement

Conditions. The final concentrations in the reaction mixture are succinate 75 mmol l^{-1}, bromocresol green 0.14 mmol l^{-1}, and Brij-35 1.19 g l^{-1}; the temperature is 25 °C.

3.10.4.6.6. Procedure

	Sample, ml	Reagent blank, ml	Standard, ml
Working solution	2.0	2.0	2.0
Plasma, serum	0.010	—	—
Standard	—	—	0.010
DM-water	—	0.010	—

Mix and read off the extinction against DM-water exactly 1 min after adding the plasma

3.10.4.6.7. Calculation

$$\text{Albumin concentration} = \frac{E(S) - E(RB)}{E(ST) - E(RB)} \cdot C(St) \text{ g per 100 ml.}$$

3.10.4.6.8. Specificity

By taking the reading within a short time and working at room temperature, other proteins, in particular globulins, are unable to bond to the bromocresol green owing to their lower affinity for the dye; Gustafsson[4] recommended the use of a recorder and extrapolation of the initial increase in the extinction to zero time.

Neither bilirubin (up to 80 mg per 100 ml) nor salicylate (up to 60 mg per 100 ml) interfere. Haemoglobin leads to an apparent increase of 0.1 g per 100 ml for 100 mg of Hb per 100 ml. With extremely lipaemic samples an analysis blank may be conducted using succinate buffer (74 mmol l^{-1}).

3.10.4.6.9. Reference values

4.65 ± 0.31 ($\bar{x} \pm s$, $n = 54$) g per 100 ml; range 3.8–5.1 mg per 100 ml.[3]

3.10.4.6.10. Comments on the method

1. The instrument reading must be taken after a precisely defined period—1 min at the latest. Incubation may not be carried out at 37 °C owing to the interference from proteins of lower affinity.

2. Our own comparisons with rocket immunoelectrophoresis have shown that with the bromocresol green reaction albumin concentrations of less than 2.5 g per 100 ml give extinctions which are systematically too high even if the reading is taken after 10 s (25 °C).

3. For the reference values cited above, Doumas et al.[3] used human serum albumin with a prescribed moisture content.

3.10.4.6.11. Diagnostic significance

A lowering of the albumin concentration is found in cases of reduced production (infections, trauma, liver damage) or loss (kidneys, gut, body cavities).

References

1. Laurell, C.-B., Electroimmunoassay, *Scand. J. clin. Lab. Invest.*, **29**, suppl. 124, 21 (1972).
2. Laurell, C.-B., Quantitative estimation of proteins by electrophoresis in agarose gel containing antibodies, *Analyt. Biochem.*, **15**, 45 (1966).
3. Doumas, B. T., Watson, W. A., and Biggs, H. G., Albumin standards and the measurement of serum albumin with bromcresol green, *Clin. chim. Acta*, **31**, 87 (1971).
4. Gustafsson, J. E. C., Improved specificity of serum albumin determination and estimation of 'acute phase reactants' by use of the bromcresol green reaction, *Clin. Chem.*, **22**, 616 (1976).

3.11. LIPID METABOLISM

3.11.1. SURVEY

The terms lipids covers those substances which are not soluble in water but which may be extracted with non-polar, organic solvents (ether, chloroform, hydrocarbons). The most preponderant fractions in human plasma are classified according to their chemical composition in Fig. 106.

The main fraction consists of fatty acids (about 50%, saturated and unsaturated), triglycerides, cholesterol (cholesterol esters and free cholesterol), and phospholipids (phosphatides). The minor fraction includes non-esterified fatty acids, acetalphosphatides (plasmalogens), and the sphingomyelins.

The analysis of the individual serum lipids has gained in significance in recent years, particularly in the field of preventive medicine. This applies to the determination of the cholesterol, the triglycerides, and, to a certain extent, the phospholipids.

Special attention has been devoted to the transport system of lipids in the blood, the plasmalipoproteins. Plasma lipids are water-insoluble and must therefore be transported as plasmalipoproteins formed by bonding to proteins—carrier proteins. The main task of the plasmalipoproteins is to shunt the exogenous and endogenous lipids between the liver, fatty tissues and other organs. The plasmalipoproteins (Fig. 107) may be classified according to their physicochemical and chemical properties.[1,2] These include the mobility in an electrical field (lipoprotein electrophoresis) and the rate of sedimentation in the ultracentrifuge. The commonest classification is based on electrophoresis:

---

1. *Chylomicrons*
These are produced in the mucous of the intestinal mucous membrane, mainly from exogenous triglycerides.

Density: 0.9–0.94 g ml^{-1}

Composition	%
Triglycerides	85–90
Cholesterol	6
Proteins	1
Phospholipids	5

Size: 75–1000 nm
Electrophoretic mobility: chylomicrons remain stationary at the point of application

419

420

2. Pre-β-lipoproteins

Pre-β-lipoproteins (very low-density lipoproteins, VLDL) are synthesized predominantly in the liver and consist largely of endogenous triglycerides.

Density: 0.94–1.006 g ml^{-1}

Composition	%
Triglycerides	50–60
Cholesterol	15–19
Proteins	8–10
Phospholipids	15–18

Size: 30–70 nm
Electrophoretic mobility: they are also termed α_2-lipoproteins and may be recognized as the second fraction from the starting point

3. β-lipoproteins

These low-density lipoproteins (LDL), formed by the organism from the very low-density lipoproteins (VLDL), transport the main part of the cholesterol.[3]

Density: 1.006–1.063 g ml^{-1}

Composition	%
Triglycerides	10
Cholesterol	42–45
Proteins	20–26
Phospholipids	23

Size: 15–25 nm
Electrophoretic mobility: they are characterized as the first fraction from the start

4. α-Lipoproteins

Where the high-density lipoproteins (HDL) are produced is not exactly known (probably the liver). They consist predominantly of protein and phospholipids.

Density: 1.063–1.21 g ml^{-1}

Composition	%
Triglycerides	2–5
Cholesterol	18–20
Proteins	45–50
Phospholipids	30

Size: 7.5–10 nm
Electrophoretic mobility: these migrate the furthest, often separated into two fractions corresponding to the α-globulins, whence the term α-lipoproteins

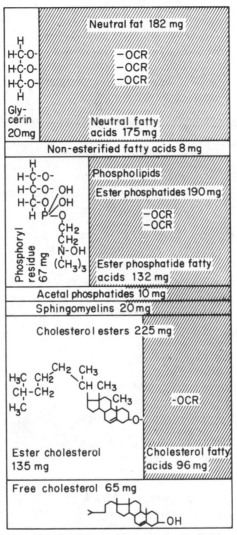

Fig. 106. Division of human serum lipids.

   The heterogeneity of the plasmalipoproteins is expressed in the different particle sizes and densities, the sugar composition of the carbohydrate part, and the amino acid composition and configuration of the protein moiety.[2,4] The classification into various *apolipoproteins* is based on the carrier proteins. The compositional relationship of the apolipoproteins to the plasmalipoproteins is as follows:

Chylomicrons	
Pre-β-lipoproteins	Apo A, B, C
β-Lipoproteins	Apo B (A)
α-Lipoproteins	Apo C (A, B)

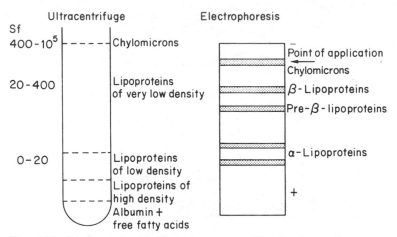

Fig. 107. Lipoprotein fractions in serum. Sf = Svedberg flotation constant, 10^{-13} cm s^{-1} dyn^{-1} g^{-1}.

A further classification of the plasmalipoproteins may be made on the basis of this subdivision. This has a special significance particularly with genetic disturbances of the fat metabolism (e.g. abetalipoproteinaemia).[5]

For an unambiguous *interpretation of lipid investigations* in serum the following points demand attention:[6]

1. Use of fasting serum (12 h).
2. Discontinuation of medication and special diets (for at least 2 weeks).
3. The age, weight, and size of the patients should be known.
4. A second investigation is advisable after a period of 2 weeks.

The following methods have been described for the detection and differentiation of hyperlipoproteinaemia in fat-metabolism disturbances:

1. Consideration of the serum (after allowing to stand for 12 h at 4 °C).
2. Precipitation methods (heparin–MgCl$_2$ precipitation).
3. Analysis for lipids (cholesterol, triglycerides, phospholipids).
4. Lipoprotein electrophoresis.
5. Immunological methods (differentiation of the apolipoproteins).[7]
6. Ultracentrifuge (differentiation: VLDL, LDL, HDL).

The *classification of genetic (familial) hyperlipoproteinaemias* into five different forms has been described by Fredrickson and co-workers.[8,9] The basis of the classification is lipoprotein electrophoresis. Table 73 provides a survey of the different hyperlipoproteinaemias.

The differentiation between hyperlipoproteinaemias of types IIb and IV is simplified by means of the nomogram designed by Fredrickson[10] (Fig. 108).

With elevated cholesterol and triglyceride values ranging between 150 and 400 mg per 100 ml, differentiation of types IIb and IV may be made by estimating the cholesterol bound to β-lipoproteins.

Table 73. Classification of hyperlipoproteinaemias.

Division of types according to Fredrickson and co-workers[8,9]

	Type I	Type IIa	Type IIb	Type III	Type IV	Type V
Appearance of the serum after 12 h at 4 °C	Upper layer creamy, lower phase clear	Clear	Slightly turbid	Slightly turbid, possibly thin upper layer	Turbid	Upper layer creamy, lower phase turbid
Electrophoresis (serum lipoproteins with elevated fraction)	Chylomicrons	β-Lipoproteins	Pre-β- and β-lipoprotein	Overlapping of the pre-β- and β-lipoprotein fractions	Pre-β-lipoprotein	Chylomicrons, pre-β-lipoprotein
Triglycerides	Greatly elevated	Normal	Elevated	Elevated	Elevated	Greatly elevated
Cholesterol	Normal to elevated	Slightly elevated	Elevated	Elevated	Normal to elevated	Slightly elevated
Ultracentrifugation	Chylomicrons present	LDL	LDL + VLDL	VLDL + LDL 'floating lipoproteins'[a]	VLDL	Chylomicrons + VLDL

For abbreviations see text.
[a] Abnormal lipoprotein with β-electrophoretic mobility.

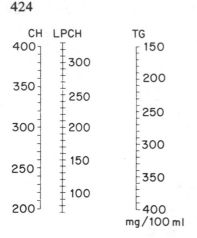

Fig. 108. Nomogram for differentiating between types IIb and IV. The known values for cholesterol and triglycerides on the scales CH (cholesterol) and TG (triglycerides) are joined by a straight line. The intersection of this line with the scale LPCH gives the concentration of cholesterol bound to β-lipoproteins. Evaluation: LPCH > 180 mg per 100 ml = type IIb; LPCH < 180 mg per 100 ml = type IV.

3.11.1.1. Diagnostic significance

For disturbances of the lipoprotein metabolism, distinction is made between primary (genetically conditioned) and secondary (actuated by another root illness) hyperlipoproteinaemias.

Patients with *primary hyperlipoproteinaemias* often manifest early arteriosclerosis, pancreatitis, or a prevalence of heart infarctions in the familial anamnesis.[5,11] The actuation of secondary hyperlipoproteinaemias by particular root diseases may be frequent, rare, or occasional.[12-14]

Frequent	Diabetes mellitus
	Nephrotic syndrome
	Hypothyroidism
	Cholestatic hepatosis
Rare	Pancreatitis
	Alcoholic hepatitis
	Kidney insufficiency
	Hyperammonaemia
	Ovulation inhibitors
	Pregnancy
Occasional	Analbuminaemia
	Malignant diseases
	Porphyria
	Idiopathic hypercalcaemia
	Essential hypertonus
	Stress
	Infections

References

1. Seidel, D., Plasmalipoproteine: biochemische und klinische Aspekte, *Münch. med. Wschr.*, **15**, 613 (1973).
2. Seidel, D., in Schettler, *Fettstoffwechselstörungen, ihre Erkennung und Behandlung*, Thieme, Stuttgart, 1971.
3. Levy, I., Bilheimer, D. W., and Eisenberg, S., The structure and metabolism of chylomicrons and VLDL plasma lipoproteins, *Biochemical Society Symposia*, No. 33, Academic Press, New York, 1971.
4. Ewing, A. M., Freeman, N. K., and Lindgren, F. T., The analysis of human serum lipoprotein distribution, *Adv. Lipid Res.*, **3**, 25 (1965).
5. Fredrickson, D. S. and Levy, R. I., Familial hyperlipoproteinemia, in Standbury, Wyngaarden, and Fredrickson, *The metabolic basis of inherited disease*, 3rd ed., McGraw-Hill, New York, 1972, p. 531.
6. Kattermann, R., Laboratoriumsdiagnostik der Fettstoffwechselstörungen, *Münch. med. Wschr.*, **15**, 633 (1973).
7. Seidel, D., Biochemische und pathophysiologische Grundlagen von Fettstoffwechselstörungen, *Ärztl. Lab.*, **21**, 275 (1975).
8. Fredrickson, D. S. and Lees, R. S., A system for phenotyping hyperlipoproteinemia, *Circulation*, **31**, 321 (1965).
9. Fredrickson, D. S., Levy, R. I., and Lees, R. S., Fat transport in lipoproteins on integrated approach to mechanism and disorders, *New Engl. J. Med.*, **276**, 34, 94, 148, 215, 273 (1967).
10. Fredrickson, D. S., Modern concepts of cardiovascular disease, **16**, 31 (1972).
11. Greten, H., Die primären Hyperlipoproteinämien, *Münch. med. Wschr.*, **15**, 639 (1973).
12. Schwandt, P., Die sekundären Hyperlipoproteinämien, *Münch. med. Wschr.*, **15**, 644 (1973).
13. Hartmann, G. and Wyss, F., *Die Hyperlipidämien in Klinik und Praxis*, Huber, Berne, 1970.
14. Lewis, B., *The hyperlipidaemias*, Blackwell, London, 1976.

3.11.2. CHOLESTEROL

3.11.2.1. Introduction

The determination of cholesterol is one of the most frequently performed analyses for the diagnosis of disturbances in lipid metabolism. Cholesterol, an alcohol, is esterified by fatty acids and owing to its poor solubility in water it is mainly bound to β-lipoproteins (Fig. 109).

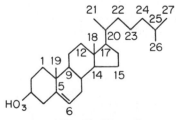

Fig. 109. Cholesterol.

426

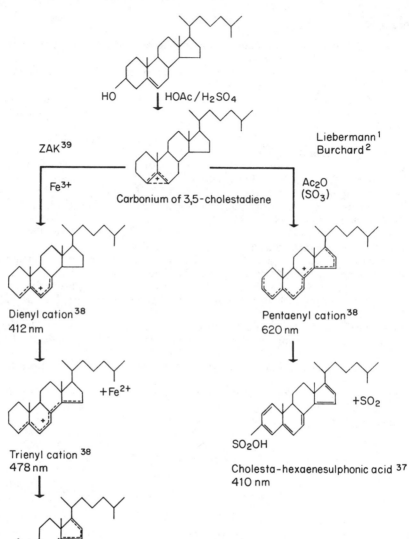

HO — HOAc / H_2SO_4

ZAK[39]

Liebermann[1]
Burchard[2]

Fe^{3+}

Ac_2O
(SO_3)

Carbonium of 3,5-cholestadiene

Dienyl cation[38]
412 nm

Pentaenyl cation[38]
620 nm

$+Fe^{2+}$

$+SO_2$

Trienyl cation[38]
478 nm

SO_2OH

Cholesta-hexaenesulphonic acid[37]
410 nm

Tetraenyl cation[38]
563 nm

Fig. 110. Reaction mechanism of the determination of cholesterol according to Liebermann,[1] Burchard,[2] and Zak.[39]

The determination of greatest diagnostic importance is that of the concentration of total cholesterol (free cholesterol and cholesterol ester) in serum.

3.11.2.2. Choice of method

A large number of methods are available and these are based on one of the long-known reactions of Liebermann[1] and Burchard[7]. Cholesterol reacts with acetic anydride and concentrated sulphuric acid with the production, under anhydrous conditions, of polymeric unsaturated hydrocarbons.[1] This reaction mechanism was reinvestigated by Burke et al.,[3] who showed that an enyl-carbonium ion was produced by cleavage of water and protonation of the cholesterol, and this ion subsequently undergoes multiple dehydration. The dehydration process results in increasingly conjugated unsaturated systems. Their absorption maxima lie at 412, 478, 563, and 620 nm and are characteristic of the dienyl, trienyl, tetraenyl, and pentaenyl cations, respectively (Fig. 110).

In the Liebermann and Burchard methods a distinction is generally made between one-stage and multi-stage methods (Table 74). The latter involve extraction and isolation of the cholesterol followed by a colour reaction. In the one-stage methods the serum is subjected to the colour reaction directly. These do not achieve the same degree of specificity as the extraction methods but combine a high degree of precision with simplicity of procedure. Richmond[10,11] and Flegg[12] were the first to describe an enzymatic method: following preliminary saponification of the cholesterol esters the cholesterol is enzymatically oxidized to hydrogen peroxide and Δ-4-cholestenone by cholesterol oxidase, or to Δ-4-cholestenone alone by the dehydrogenase. The proneness to interference with these methods was attributable predominantly to the saponification. Since about 75% of the cholesterol in plasma is present in the esterified form, Röschlau et al.,[13] preceded the oxidation reaction by a cholesterol esterase-catalysed cleavage of the cholesterol esters. This cholesterol esterase was originally isolated from the pancreatic juices of pigs and rats;[14] today it is obtained from micro-organisms.[15] This opened up better possibilities for a fully enzymatic determination of the total cholesterol. A survey of the development of cholesterol analysis is provided in Table 74.

3.11.2.3. Cholesterol: fully enzymatic determination

3.11.2.3.1. Principle

By the action of cholesterol esterase the cholesterol esters present in the serum are converted to free cholesterol and fatty acids. Together with the free cholesterol in the serum, the liberated cholesterol is converted to

Table 74. Methods for determining cholesterol.

Method	Disadvantages, possible sources of error	References
Liebermann–Burchard principle		
Multistage method		
Extraction, saponification	Caustic reagents	4
Digitonin precipitation, colour reaction	Troublesome determination of desmosterol,	5
Saponification, extraction, colour reaction	cholesterol	6
Direct determination		
Glacial acetic, FeCl$_3$, sulphuric acid or	Moisture, solvents	7
Glacial acetic, acetic anhydride, sulphuric acid, colour reaction	Light, temperature, bilirubin, haemoglobin	8,9
Enzymatic determination	—cholesterol oxidase	
Saponification—cholesterol oxidase H$_2$O$_2$ colour reaction		10,11
Saponification—measurement of the Δ-4-cholestenone with cholesterol dehydrogenase		12
Cholesterol esterase, oxidase Catalase colour reaction		13
Cholesterol esterase, oxidase Peroxidase, 4-aminophenazone Colour reaction		16 17
	No interference from drugs, haemoglobin, bilirubin	
Cholesterol esterase, oxidase Peroxidase, o-dianisidine Colour reaction		19
Cholesterol esterase, oxidase Peroxidase, fluorimetric measurement with homovanilic acid		20
Ion-selective electrodes: cholesterol esterase, oxidase peroxidase, measurement of the iodoxidation or measurement of the O$_2$ consumption		18 41
Gas-chromatographic determination		
Extraction, saponification, removal of the ether, drying, gas chromatography	Specific	21

Δ-4-cholestenone and hydrogen peroxide by the action of cholesterol oxidase in the presence of oxygen (Fig. 111).

The reaction may be measured in many different ways:

1. Oxygen consumption (measurement with a pO$_2$ electrode).
2. Measurement of the formed Δ-4-cholestenone at 240 nm.

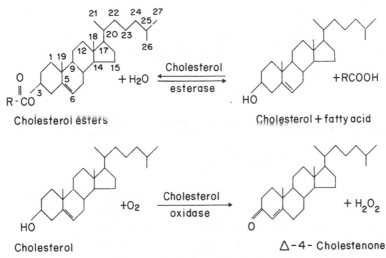

Fig. 111. Reaction mechanism for the enzymatic determination of cholesterol.

3. Measurement of the enzymatically produced H_2O_2 by means of:
   3.1. catalse (Hantzsch reaction);[13]
   3.2. peroxidase reaction;[16,17]
   3.3. ion-selective electrodes:[18] $H_2O_2 + 2\,I^- + 2H^+ \xrightarrow{\text{Mo(VI)}} I_2 + H_2O$

Note to 3.1. In the presence of catalase the hydrogen peroxide produced oxidizes methanol to formaldehyde and in the presence of ammonium ions the latter reacts with acetylacetone to form a yellow pigment, 3,5-diacetyl-1,4-dihydrolutidine; the intensity of the colour is proportional to the cholesterol concentration and is measured at 405–415 nm (Fig. 112). This system has also been used for the enzymatic determination of uric acid[22] and is distinguished by its high specificity and relative freedom from interference.

By omitting the cholesterol esterase and performing the assay with cholesterol oxidase only, it is possible to determine the serum concentration of the free cholesterol. The cholesterol esters are then given by the difference between the total and free cholesterol.

Note to 3.2. Using the Trinder[23] peroxidase reaction, the reaction

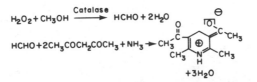

3,5–Diacetyl–1,4–dihydrolutidine

Fig. 112. Mechanism of the catalase reaction.

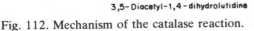

430

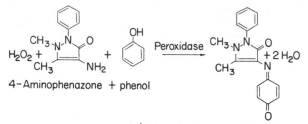

4-(ρ-Benzoquinonemonoimino) phenazone

Fig. 113. Mechanism of the peroxidase reaction.

proceeds as follows (as in the determination of glucose in serum). Under the catalytic action of the peroxidase, hydrogen peroxide oxidizes 4-aminophenazone (4-aminoantipyrine, PAP) and 2,4-dichlorophenol to a 4-(ρ-benzoquinonemonoimino)phenazone. The intensity of this red chromogen is directly proportional to the cholesterol concentration and can be measured between 500 and 550 nm (Fig. 113).

3.11.2.3.2. Measurement

Conditions. In the methods for the enzymatic determination of cholesterol described by Röschlau et al.[13] and Stähler et al.[17] the reagents have the final concentrations in the reaction mixture shown in Table 75.

3.11.2.3.3. Specificity

The enzymatic methods for cholesterol are not absolutely specific for this compound. Steroids present in the serum also react but since their serum concentration is very much smaller than that of cholesterol the interference is minimal.[16,24]

Table 75. Final concentrations of the reagents in the reaction mixture.

Reagents	Catalase reaction[13]	Peroxidase reaction[17]
Ammonium phosphate buffer, pH 7.0	0.544 mol l^{-1}	—
Potassium phosphate buffer, pH 7.7	—	0.394 mmol l^{-1}
Methanol	1.7 mol l^{-1}	1.82 mmol l^{-1}
Acetylacetone	0.019 mol l^{-1}	—
4-Aminophenazone	—	0.985 mmol l^{-1}
Phenol	—	9.85 mmol l^{-1}
Hydroxypolyethoxydodecane	0.099%	0.19%
Catalase	$661 \cdot 10^3$ U l^{-1}	—
Peroxidase	—	39.4 U l^{-1}
Cholesterol esterase	18.8 U l^{-1}	197 U l^{-1}
Cholesterol oxidase	20 U l^{-1}	59 U l^{-1}

3.11.2.3.4. Comments on the method

1. The catalase method[13,25] exhibits linearity up to 1000 mg per 100 ml (25.86 mmol 1^{-1}), whereas with the peroxidase reaction linearity is not maintained above 500 mg per 100 ml (12.93 mmol 1^{-1}).[17] At higher concentrations the sample must be diluted. With these determinations the calculation of the cholesterol concentration is made on the basis of aqueous primary cholesterol standards.

2. Either serum or plasma (EDTA, heparin) may be used.

3. The accuracy of both of these fully enzymatic methods has been established by the determination of cholesterol in standard solutions.[25] The cumulative studies on human sera[13,17] and the prescribed value derived from them gave excellent results.

4. The reaction of the catalase method requires incubation for 60 min at 37 °C, which poses a disadvantage as regards adaptation for automatic analysers with final-value determination. The period of incubation for the peroxidase method is given[17] as about 15 min at 37 °C, the reaction having come to a halt after 10 min. The latter method is thus the more practical for automatic operation.

5. Interference due to metabolites. (a) A bilirubin concentration of up to 40 mg per 100 ml (680 μmol 1^{-1}) produces no interference.[25,26] (b) Haemoglobin at concentrations above 200 mg per 100 ml has been observed to interfere in the peroxidase reaction.[25] This interference is not observable with the catalase reaction.[25,26] (c) Lipaemic sera have no effect on the reaction, even when the triglyceride concentration is high.[25,26] The turbid sera are clarified during the incubation by the action of the esterase.[13] This applies to both the catalase and peroxidase methods. (d) Comparative studies on cholesterol determinations,[40] as analysed by the methods of Abell et al.[6] and Huang et al.,[8] and the enzymatic methods, the catalase,[13] and peroxidase[16,17] reactions, showed that the catalase reaction is subject to greater methodic variation and haemoglobin interference than described in the literature.[13,24,25] On the other hand, the peroxidase reaction evinced a high precision, was relatively free from interference, and was suited to adaptation for automation.[44]

6. Interference from drugs. More than 45 preparations from different fields of use were tested in vitro at concentrations corresponding to the maximum daily dosage. No significant effects were established.[25,26] In vitro, novaminesulphonic acid and α-methyl-dopa interfered in the fully enzymatic test.[17,25] According to Stähler et al.,[25] clofibrate, ascorbic acid, and nicotinamide have no effect on the catalase reaction, but it has been established that ascorbic acid interferes in the peroxidase reaction.[16,27] Robinson et al.[28] observed a significant difference detectable at ascorbic acid concentrations as low as 5 mg per 100 ml. Neither of the fully enzymatic methods is affected by the anticoagulants sodium citrate, oxalate, sodium and ammonium heparinate, or fluoride .[17,25,26] Our own

researches[29] have shown that fluoride interferes in the peroxidase reaction when present in blood at a final concentration above 1 mg ml^{-1} (dose-dependent decrease in the cholesterol concentration).

7. Pesce and Bodourian[43] have detailed the interferences in five different p-aminophenazone methods. The following interferences have been established for the Stähler et al.[17] modification: (a) uric acid, up to 20 mg per 100 ml no interference; (b) haemoglobin, up to 100 mg per 100 ml no interference; (c) ascorbic acid, lower results above 2.5 mg per 100 ml; (d) bilirubin, differences of about 10% are to be expected with concentrations of 5–20 mg per 100 ml; the cholesterol results obtained are lower with a serum blank and higher without one; (e) no interferences were found with the drugs clofibrate (1 g l^{-1}), phenobarbital (650 mg l^{-1}) and the oral contraceptive Ovral-28 (250 μg l^{-1}) in vitro.

3.11.2.4. Total cholesterol: Liebermann–Burchard reaction

3.11.2.4.1. Choice of method

Hitherto, cholesterol determinations have most commonly been based on the Liebermann–Burchard reaction. On treating cholesterol with acetic anhydride and concentrated suphuric acid in anhydrous media an intense blue–green colour is produced due to the formation of polymeric unsaturated hydrocarbons. All these methods suffer from the following disadvantages: the intensity of the colour is dependent on the amount of water in the reagents, the concentration of the sulphuric acid, the illumination, and the period between adding the reagents and the measurement of the intensity. The acetic anhydride extracts the cholesterol, precipitates the proteins, and guarantees an anhydrous medium. The addition of p-toluenesulphonic acid disperses the proteins. Since heat is evolved by the dehydration of the proteins in this first step, the heating produced by the addition of suphuric acid is diminished accordingly. This reduces an important source of error—irregular heating. Although the intensity of the colour is greatest at 620–650 nm, it is measured at 560 nm since the colour is much more stable at this wavelength. p-Toluenesulphonic acid is now usually replaced with 2,5-dimethylbenzenesulphonate as impurities in the former occasioned mild explosions. The development of heat is more readily controlled by carrying out the reaction in a water-bath.

In 1961, Huang et al.[8] showed that glacial acetic acid, acetic anhydride and sulphuric acid could be mixed and stabilized with anhydrous sodium sulphate. In this way the number of reagents could be reduced even further. A modification of the Huang et al reagent[9] has proved very successful.

3.11.2.4.2. Reagents

Cholesterol reagent. The liquid reagents are first cooled to 4 °C and then added in the following order in an ice-bath and with thorough mixing (magnetic stirrer): 300 ml of glacial acetic acid, 600 ml of acetic anhydride, and 100 ml of concentrated sulphuric acid. To this mixture are added 20 g of anhydrous sodium sulphate and 10 g of 2,5-dimethylbenzenesulphonic acid. In a dark bottle at 4 °C the reagent is stable for at least 2 months. A slight yellowing does not interfere with the reaction.

Cholesterol standard, 250 mg per 100 ml, 6.5 mmol l^{-1}. Dissolve 250 mg of cholesterol in glacial acetic acid and make up to 100 ml with further acetic acid. The solution keeps indefinitely at room temperature.

3.11.2.4.3. Procedure

(Water-bath 37 °C)	Sample, ml	Reagent blank, ml	Standard, ml
Plasma, serum	0.02	—	—
DM-water	—	0.02	—
Standard solution	—	—	0.02
Cholesterol reagent	1.00	1.00	1.00

Mix briefly and allow to stand in the water-bath at 37 °C for exactly 10 min. Then mix again and read the extinction at 560–580 nm against the cholesterol reagent within 5 min

3.11.2.4.4. Calculation

$$\text{Concentration} = \frac{E(S) - E(RB)}{E(ST) - E(RB)} \cdot 250 \text{ mg per 100 ml}$$

$$= \frac{E(S) - E(RB)}{E(ST) - E(RB)} \cdot 6.5 \text{ mmol } l^{-1}$$

3.11.2.4.5. Reference range

Results on the reference range of the serum cholesterol are very difficult to compare since they differ in each population as a function of diet, age, sex, and bodily activity (Table 76). Reference ranges for enzymatic methods have been reported by Weisshaar[30] and Kaffarnik et al.[31]

1. The serum cholesterol concentration was studied in a population of 183 women and 306 men[30] by means of the catalase reaction, paying attention to sex, age, and weight and checking for GPT, γ-GT,

434

Table 76. Reference values for cholesterol in serum.

Age, years	Men		Women		Range, %	Reference
	mg per 100 ml	mmol l^{-1}	mg per 100 ml	mmol l^{-1}		
Up to 20	100–230	2.59–5.95	100–200	2.59–5.69		30
21–30	130–250	3.36–6.47	130–240	3.36–6.21	5–95	
31–45	150–270	3.88–6.98	130–250	3.36–6.47		
46–60	160–280	4.13–7.24	150–280	3.88–7.24		
≥60	170–320	4.39–8.28	170–330	4.39–8.53		
Up to 30	156–220	4.03–5.69	157–239	4.06–6.18		31
31–40	162–238	4.19–6.15	161–237	4.16–6.13	5–95	
41–50	168–240	4.34–6.21	173–257	4.47–6.64		
51–60	175–237	4.52–6.13	194–270	5.02–6.98		
0–13	145–250	3.75–6.47	150–250	3.88–6.47		34
14–19	140–260	3.62–6.72	155–260	4.01–6.72	5–95	
20–29	170–310	4.39–8.02	175–310	4.53–8.02		
30–39	175–325	4.53–8.40	180–280	4.65–7.24		
40–49	180–320	4.65–8.28	190–325	4.91–8.40		
50–59	185–320	4.78–8.28	190–355	4.91–9.18		
60–69	190–330	4.91–8.53	220–375	5.69–9.70		
≥70	180–320	4.65–8.28	210–360	5.43–9.31		

0–13	118–234	3.05–6.05		122–234	3.15–6.05	35
14–19	129–245	3.33–6.33		134–242	3.46–6.26	
20–29	143–263	3.69–6.80	±2s	139–255	3.59–6.59	
30–39	147–271	3.80–7.01		149–281	3.85–7.26	
40–49	165–277	4.27–7.16		167–283	4.32–7.32	
20–29	154–256	3.98–6.62		144–290	3.72–7.50	46
30–39	174–284	4.49–7.34		160–290	4.13–7.50	
40–49	194–292	5.02–7.55	2.5–97.5	172–296	4.44–7.65	
50–59	174–280	4.49–7.24		196–299	5.07–7.73	
≥60	158–297	4.09–7.68		184–299	4.76–7.73	
Children						
1st–10th day	70–125	1.81–3.23				33
10 day–5th year	85–190	2.20–4.91				
4–6 years	124–239	3.21–6.18				32
7–9	124–244	3.21–6.31	1–99			
10–12	127–233	3.28–6.03				
13–14	127–221	3.28–5.72				
11–19	124–236	3.2 –6.39				42
5–16	118–255	3.05–5.81				45

haemoglobin, uric acid, bilirubin, and blood sedimentation rate (Table 76). The following observations were made: (a) cholesterol shows an age dependence for both sexes. (b) There is a relationship between increasing body weight and the cholesterol concentration of the plasma. (c) Up to the 60th year of life, the values for men are higher than those for women. Above this age women have higher values.

2. Kaffarnik et al.[31] investigated the cholesterol content of the serum from 265 men and 274 women. The screening of the subjects was made on the basis of prior checking of clinical and chemical laboratory data. The upper limit of the reference range for women above 40 years of age was higher than that for men; this is attributable to hormonal action during the menstrual cycle.

3. The data of Häberlin[32] and Hering[33] cover the period of childhood. Both authors determined the cholesterol by non-enzymatic methods.

4. Around 300 ambulant patients were investigated by Werner et al.[34] using the automated method according to Huang et al.[8] The ranges between 5 and 95% are presented in Table 76. There is a significant age dependence and sex difference.

5. Brown and Daudiss[35] used blood donors as subjects. The cholesterol was determined by the method of Block et al.[36] The ranges were established as $2s$ limits. No significant sex difference was observed but there was a markedly significant age dependence.

6. Reference ranges for the population of Vienna were set up by Bayer et al.[46] using the cholesterol–PAP method and a sample of 622 women and 117 men.

The marked differences in the reference ranges for the cholesterol concentration in serum are attributable to the analytical methods (enzymatic, non-enzymatic), as is clear from Table 76.

3.11.2.4.6. Comments on the method

1. The given method can also be performed on a micro-scale. In this case all volumes should be increased five-fold. The results of the micro-scale method are not so reliable as those of the ultramicro-scale method since more heat is evolved.

2. Use only well cleaned, dry glassware, since even very small amounts of moisture affect the reaction.

3. Haemolytic sera (plasma haemoglobin greater than 100 mg per 100 ml) should not be used.

4. With icteric sera substantial discrepancies only arise for a bilirubin content greater than 10 mg per 100 ml. Bilirubin at a concentration of 1 mg per 100 ml produces an increase in extinction corresponding to about 2.5 mg of cholesterol per 100 ml.

5. The Liebermann–Burchard reaction is a relatively non-specific reaction of steroids. This is irrelevant, however, for the determination of

cholesterol in the plasma since apart from cholesterol the only steroids present in the plasma are 1–3% of dihydrocholesterol and 0.5–1.4% of Δ-7-cholesterol. Following treatment with cholesterol depressants the plasma contains the ultimate precursor, desmosterol. All of these substances show a positive Liebermann–Burchard reaction.

References

1. Liebermann, C., Über das Oxychinoterpen, *Ber. dt. chem. Ges.*, **18**, 1803 (1885).
2. Burchard, H., Beiträge zur Kenntnis des Cholesterins, *Chem. Zentbl.* **61**, 25 (1890).
3. Burke, W., Diamondstone, B. I., Velapoldi, R. A., and Menis, O., Mechanisms of the Liebermann–Burchard and Zak color reactions for cholesterol, *Clin. Chem.*, **20**, 794 (1974).
4. Schonheimer, R. and Sperry, W., A micromethod for the determination of free and combined cholesterol, *J. biol. Chem.*, **106**, 745 (1934).
5. Sperry, W. M. and Webb, M., A revision of the Schonheimer–Sperry method for cholesterol determination, *J. biol. Chem.*, **187**, 97 (1950).
6. Abell, L. L., Levy, B. B., Brodie, B. M., and Kendall, F. F., A simplified method for estimation of total cholesterol in serum and demonstration of its specificity, *J. biol. Chem.*, **195**, 357 (1952).
7. Zlatkis, A. and Zak, B., Study of new cholesterol reagent, *Analyt. Biochem.*, **29**, 143 (1969).
8. Huang, T. C., Chen, C. P., Wefler, V., and Raftery, A., A stable reagent for the Liebermann–Burchard reaction—application to rapid serum cholesterol determination, *Analyt. Chem.*, **33**, 1405 (1961).
9. Richterich, R. and Lauber, K., Bestimmung des Gesamt-Cholesterins im Serum. VIII. Mitteilung über Ultramikromethoden im klinischen Laboratorium, *Klin. Wschr.*, **40**, 1252 (1962).
10. Richmond, W., The development of an enzymic technique for the assay of cholesterol in biological fluids, *Scand. J. clin. Lab. Invest.*, **29**, suppl., 126 (1972).
11. Richmond, W., Preparation and properties of a cholesterol oxidase from *Nocardia* sp. and its application to the enzymatic assay of total cholesterol in serum, *Clin. Chem.*, **19**, 1350 (1973).
12. Flegg, H. M., An investigation of the determination of serum cholesterol by an enzymatic method, *Ann. clin. Biochem.*, **10**, 79 (1973).
13. Röschlau, P., Bernt, E., and Gruber, W., Enzymatische Bestimmung des Gesamtcholesterins im Serum, *Z. klin. Chem. klin. Biochem.*, **12**, 403 (1974).
14. Hernandez, H. H. and Chaikoff, I. L., Purification and properties of pancreatic cholesterol esterase, *J. biol. Chem.*, **228**, 447 (1957).
15. Beaucamp, K., Lang, G., and Möllering, H., (Boehringer, Mannheim), unpublished observations.
16. Allain, Ch. C., Poon, L. S., Chan, C. S. G., Richmond, W., and Fu, P. C., Enzymatic determination of total serum cholesterol, *Clin. Chem.*, **20**, 470 (1974).
17. Stähler, F., Gruber, W., Stinshoff, K., and Röschlau, P., Eine praxisgerechte enzymatische Cholesterin-Bestimmung, *Med. Lab., Stuttg.*, **30**, 29 (1977).
18. Papastathopoulos, D. S. and Rechnitz, G. A., Enzymatic cholesterol determination using ion-selective membrane electrodes, *Analyt. Chem.*, **47**, 1792 (1975).

438

19. Tarbutton, P. N. and Gunter, C. R., Enzymatic determination of total cholesterol in serum, *Clin. Chem.*, **20**, 724 (1974).
20. Huang, H.-S., Kuan, J.-C. W., and Guilbault, G. G., Fluorometric enzymatic determination of total cholesterol in serum, *Clin. Chem.*, **21**, 1605 (1975).
21. Curtius, H. Ch. and Bürgi, W., Gaschromatographische Bestimmung des Serumcholesterins, *Z. klin. Chem. klin. Biochem.*, **4**, 38 (1966).
22. Kageyama, N., A direct colorimetric determination of uric acid in serum and urine with uricase-catalase system, *Clin. chim. Acta.*, **31**, 421 (1971).
23. Trinder, P. A., Determination of glucose in blood using glucose oxidase with an alternative oxygen acceptor, *Ann. clin. Biochem.*, **6**, 24 (1969).
24. Knob, M. and Rosenmund, H., Enzymatische Bestimmung des Gesamtcholesterins im Serum mit Zentrifugalanalyzern, *Z. klin. Chem. klin. Biochem.*, **13**, 493 (1975).
25. Stähler, F., Munz, E., and Kattermann, R., Enzymatische Bestimmung von Gesamtcholesterin im Serum—Richtigkeit und Methodenvergleich, *Dt. med. Wschr.*, **100**, 876 (1975).
26. Ziegenhorn, J., Enzymatische Bestimmung des Gesamt-Cholesterins im Serum mit Analysen-automaten, *Z. klin. Chem. klin. Biochem.*, **13**, 109 (1975).
27. Rodriquez-Castellon, J. A., Robinson, C. A., Smith, M. S., and Frye, J. H., Evaluation of an automated glucose oxidase procedure, *Clin. Chem.*, **21**, 1513 (1975).
28. Robinson, C. A., Jr., Hall, L. B., and Vasiliades, J., Evaluation of an enzymatic cholesterol method, *Clin. Chem.*, **22**, 1542 (1976).
29. Bachmann, C., Peheim, E., Beck, E., and Perritaz, R., Interferenzen bei der Cholesterin PAP Methode, *Unpublished* (1977).
30. Weisshaar, D., Enzymatische Cholesterinbestimmung. Norm(Referenz)-bereiche, *Medsche Welt, Stuttg.*, **26**, 940 (1975).
31. Kaffarnik, H., Schneider, J., Zoefel, P., Mühlfellner, G., Mühlfellner, O., Hausmann, L., Schubotz, R., and Meyer-Bertenrath, J. G., Vergleich der enzymatischen Cholesterin-Bestimmung mit der Methode nach Liebermann–Burchard bei Normalperonen, *Med. Lab., Stuttg.*, **30**, 38 (1977).
32. Häberlin, H. R., Normalwerte des Serumcholesterins im Kindesalter, *Diss. Universität Zürich*, 1960.
33. Hering, S. E., Das Verhalten der Lipidfraktionen und Fettsäuren im Serum im Kindesalter. Gaschromatographische Untersuchungen, *Helv. paediat. Acta.*, **21**, 423 (1966).
34. Werner, M., Tolls, R. E., Hultin, J. V., and Mellecker, J., Influence of sex and age of the normal range of eleven serum constituents, *Z. klin. Chem. klin. Biochem.*, **8**, 105 (1970).
35. Brown, D. F. and Daudis, K., Hyperlipoproteinemia. Prevalence in a free-living population in Albany, New York, *Circulation*, **47**, 558 (1973).
36. Block, W. D., Jarrel, J. K., and Levine, J. B., Use of single color reagent to improve the automated determination of serum total cholesterol, in Skeggs, *Automation in analytical chemistry*, Medical Inc., New York, 1965.
37. Fieser, L. F. and Fieser, M., *Steroids*, Reinhold, New York, 1959.
38. Sørensen, T. S., The preparation and reactions of a homologous series of aliphatic polyenylic cations, *J. Am. chem. Soc.*, **87**, 5075 (1965).
39. Zak, B., Simple rapid microtechnic for serum total cholesterol, *Am. J. clin. Path.*, **27**, 583 (1957).
40. Gent, C. M. van, Voort, H. A. van der, Bruyn, A. M., and Klein, F., Cholesterol determinations. A comparative study of methods with special reference to enzymatic procedures, *Clin. chim. Acta*, **75**, 243 (1977).
41. Noma, A. and Nakayama, K., Polarographic method of microdetermination of cholesterol with cholesterol esterase and cholesterol oxidase, *Clin. Chem.*, **22**, 336 (1976).

42. Court, J. M. and Dunlop, M., Plasma lipid values and lipoprotein patterns during adolescence in boys, *J. Pediat.*, **3**, 453 (1975).
43. Pesce, M. A. and Bodourian, S. H., Interference with enzymic measurement of cholesterol in serum by use of five reagent kits, *Clin. Chem.*, **4**, 757 (1977).
44. Paula, R., Sonntag, A., Martin, G., and Kaiser, E., Serumcholesterin. Vergleichende Untersuchungen zur Zuverlässigkeit verschiedener Bestimmungsmethoden, *Med. Lab., Stuttg.*, **30**, 250 (1977).
45. Srinivasan, R. S., Frerichs, R. R., and Berenson, G. S., Serum lipid and lipoprotein profile in school children from a rural community, *Clin. Chim. Acta*, **60**, 293 (1975).
46. Bayer, P. M., Dorda, W., Gabl, F., Gergely, Th., Junker, E., Schnack, H., and Zyman, H., Cholesterin und Triglyceride im Serum. Referenzbereiche für eine städtische Bevölkerung, *Med. Lab., Stuttg.*, **30**, 270 (1977).

3.11.3. TRIGLYCERIDES

3.11.3.1. Introduction

Neutral fat (*exogenous triglycerides*) taken in with the diet is split up in the intestine by the pancreatic and intestinal lipases. The triglycerides are broken down to free acids, glycerine, and mono- and diglycerides, and absorbed by the cells of the intestinal mucous membrane.

In the mucosa cells of the intestine the triglycerides are resynthesized from the reabsorbed free fatty acids and α-glycerophosphate (glucose metabolism). As a component of the chylomicrons the triglycerides so formed pass into the blood stream via the ductus thoracicus. The chylomicrons are particles with a very low proportion of proteins (2–6%) and act as a means for transporting the reabsorbed triglycerides. After a meal rich in fats this leads to turbidity of the serum.

The degradation of the chylomicrons takes place in the fat tissues and in the muscle in the endothelium of the capillaries by the action of lipoprotein lipase which mainly derives from extrahepatic tissue, and of triglyceride lipase of hepatic origin;[1-4] the products are free fatty acids and glycerine.

The *endogenous triglycerides* are synthesized in the fat tissue and in the liver, from the reabsorbed free fatty acids, and from α-glycerophosphate produced in the carbohydrate metabolism. The endogenous triglycerides are brought into the blood-stream bound to the very low-density lipoproteins (pre-β-lipoproteins).[5]

Hydrolysis is effected by the lipoprotein lipase and triglyceride lipase, predominantly in the fatty tissue, and once again the products are free fatty acids and glycerine. The unesterified or free fatty acids which are shunted between the liver and the fatty tissue are transported in the plasma bound to albumin.

Up to two thirds of the free fatty acids formed from the exogenous and endogenous triglycerides are reabsorbed in the cells of the fat tissue. The other third, bound to albumin, reaches the liver via the blood-stream.

The determination of triglycerides is of great practical importance to the clinician and, together with the cholesterol determination, it forms the basis for differentiating between the various forms of hyperlipoproteinaemia.

3.11.3.2. Choice of method

The indirect methods for the determination of triglycerides mentioned below should be rejected on principle since the results obtained are never very exact, even when the analyses are carried out with great care.

1. *Determination of the total fat, the ester cholesterol, the free cholesterol, and the lipid phosphorus.* Neutral fat = total lipids − (free cholesterol + 1.68 ester cholesterol + 25 lipid phosphorus). The factor 1.68 serves to convert the esterified cholesterol to cholesterol ester. Since the phosphorus content of the phosphatides is about 4%, the lipid phosphorus must be multiplied by 25.

2. *Determination of the total fatty acids, the ester cholesterol, and the lipid phosphorus.* Neutral fat = 1.04 (total fatty acids) − [(0.72 ester cholesterol + 17.9 lipid phosphorus)]. The factor 1.04 serves to convert oleic acid to triolein, the factor 0.72 calculates the fatty acids bound to cholesterol from the ester cholesterol, and the factor 17.9 gives the fatty acids from the lipid phosphorus figure.

3. *Determination of the ester fatty acids and of lipid fatty acids.* In this case the neutral fats correspond to the difference between the total fatty acid esters and the fatty acids bound in the phosphatides and cholesterol esters. Neutral-fats = ester fatty acids − (0.764 ester cholesterol + 14.2 lipid phosphorus). The factor 0.764 is for calculating the fatty acids bound to cholesterol and the factor 14.2 for conversion of the fatty acids bound to phosphatides.

In recent years methods have been developed which allow the estimation of triglycerides by determining the glycerine. The commonest direct methods for triglyceride in current use are listed in Table 77.

The determination of triglyceride is usually carried out by determining the glycerine. The latter occurs in the plasma as: 1, free glycerine; 2, a component of mono-, di-, and triglycerides; and 3, as a component of the phosphatides.

From the practical point of view, these methods can be divided into two groups: 1, colorimetric and fluorimetric methods; and 2, enzymatic methods. In the methods of the first group the glycerine is oxidized to formaldehyde and subsequently subjected to a suitable colour reaction. In the enzymatic determination of triglyceride, developed by Eggstein[11] and Eggstein and Kreutz,[17] the triglycerides are cleaved to glycerine and fatty acids, either by alkaline hydrolysis or by enzymatic hydrolysis[14,15] with

Table 77. Methods for the determination of triglycerides.

Method	References
1. *Colorimetric determination*	
1.1 Extraction, adsorption, saponification, oxidation of glycerine to formaldehyde and chromotropic acid	6, 7
1.2 Extraction, transesterification, oxidation of glycerine to formaldehyde, reaction with acetylactone + NH_3 (Hantzsch reaction)	8, 9
2. *Fluorimetric determination*	
As for 1.1 up to production of the formaldehyde, then:	
2.1 +o-aminophenol + sulphuric acid	33
2.2 acetylacetone + NH_3	10
3. *Enzymatic determination*	
3.1 Hydrolysis with ethanolic KOH, optical test, $NADH_2$ reaction	11–13, 17
3.2 Fully enzymatic hydrolysis with lipase and esterases, then as for 3.1	14–16, 26, 27
3.3 Fully enzymatic hydrolysis with lipase and α-chymotrypsin, glycerokinase, α-glycerinephosphatc dehydrogenase + $NAD^+ \rightarrow$ dihydroxyacetone phosphate (DHAP) + $NADH + H^+$ DHAP + hydrazine $\rightarrow$ DHAP hydrazone + H_2O	31, 32

vegetable lipase and esterase, and the hydrolysis products measured with the optical test. By reasons of its simplicity, specificity, and sensitivity, these methods are to be preferred.[35]

3.11.3.3. Triglycerides: enzymatic determination as glycerine following alkaline hydrolysis

3.11.3.3.1. Principle

The triglycerides are determined as total glycerine following alkaline hydrolysis. The total glycerine is compounded of the glyceride glycerine and the free glycerine already present in the serum. The latter can be measured directly without any preliminaries in a second serum assay and subtracted from the total glycerine. The difference corresponds to the glyceride glycerine or the triglyceride or neutral fat concentration. The method is based on the principles of Eggstein and Kreutz[11,17] as modified according to Da Fonseca Wollheim[13] and Dauwalder and Colombo.[18]

The triglyceride determination may be regarded as a coupled, enzymatic optical test and is performed in two stages, as follows.

1.1. Saponification of the triglycerides with ethanolic KOH:

Triglycerides $\xrightarrow[70\ °C]{\text{ethanolic KOH}}$ Glycerine + 3 fatty acids

1.2. Measurement of the glycerine in the optical test using glycerokinase; pyruvate kinase acts as auxiliary enzyme and lactate hydrogenase is the indicator enzyme:

$$\text{Glycerine + ATP} \xrightarrow{\text{Glycerokinase}} \text{Glycerine-3-phosphate + ADP}$$

$$\text{ADP + Phosphoenol pyruvate} \xrightarrow{\text{Pyruvate kinase}} \text{ATP + Pyruvate}$$

$$\text{Pyruvate + NADH + H}^+ \xrightarrow{\text{Lactate dehydrogenase}} \text{Lactate + NAD}^+$$

The work is considerably simplified by running a sample blank in parallel with each sample. This cuts down the time required for each measurement since it is then unnecessary to record the progress of the reaction (see Procedure, Section 3.11.3.3.4).

3.11.3.3.2. Reagents

Ethanolic potassium hydroxide solution, $0.5\,mol\ l^{-1}$. Dilute 11.2 ml of 25% potassium hydroxide solution with 95% ethanol and make up to 100 ml with further ethanol. The solution keeps indefinitely at room temperature.

Standard glycerine solution, $2\,mmol\ l^{-1}$. Dilute 92.09 mg of glycerine to 500 ml with doubly distilled water. The solution keeps for about 1 month at 4 °C.

Standard triglyceride solution, $2\,mmol\ l^{-1}$. Dissolve 88.5 mg of triolein in 50 ml of ethanol. The solution keeps for about 1 week at 4.°C.

Magnesium sulphate solution, $200\,mmol\ l^{-1}$. Dissolve 4.94 g of $MgSO_4.7H_2O$ in doubly distilled water and make up to 100 ml. The solution keeps indefinitely at 4 °C.

Triethanolamine–EDTA buffer, $200\,mmol\ l^{-1}$, $pH\ 7.5$. Dissolve 37.13 g of triethanolammonium chloride and 7.44 g of EDTA in about 800 ml of doubly distilled water. Adjust the pH to 7.5 with 1 N NaOH and make up to 1 litre with doubly distilled water. The solution keeps for about 2 months at 4 °C.

Potassium chloride solution, $2\,mol\ l^{-1}$. Dissolve 14.9 g of KCl in doubly distilled water and make up to 100 ml. The solution keeps indefinitely at 4 °C.

$NADH_2$ solution, $3\,mmol\ l^{-1}$. Dissolve 21.3 mg of nicotinamide adenine dinucleotide (reduced), disodium salt, in 10 ml of 1% $NaHCO_3$ solution. The solution keeps for about 1 week at 4°C.

Adenosine triphosphate solution, $160\,mmol\ l^{-1}$. Dissolve 9.68 g of ATP-$N_2H_2.3H_2O$ in doubly distilled water and make up to 100 mol. Store frozen in separate small portions.

Phosphoenol pyruvate solution, $30\,mmol\ l^{-1}$. Dissolve 1.39 g of

PEP-(CHA)$_3$ in doubly distilled water and make up to 100 ml. Store frozen in separate small portions.

Pyruvate kinase, about 100 U ml^{-1}. Dilute 1.0 ml of stock solution (specific activity about 200 U ml^{-1}) with 2.1 mmol l^{-1} ammonium sulphate solution and make up to 20 ml with further ammonium sulphate solution. The solution keeps for about 3 months at 4 °C.

Lactate dehydrogenase, 825 U ml^{-1}. Dilute 5 ml of stock suspension (specific activity about 5500 U ml^{-1}) with 2.1 mmol l^{-1} ammonium sulphate solution and make up to 10 ml with further ammonium sulphate solution. The solution keeps for about 3 months at 4 °C.

Glycerokinase, 85 U ml^{-1}. Dilute 1 part glycerokinase suspension (specific activity 425 U ml^{-1}) with 4 parts of ammonium sulphate solution. The solution keeps for about 3 months at 4 °C.

Magnesium sulphate solution, 150 mmol l^{-1}. Dissolve 3.7 g of MgSO$_4$.7H$_2$O in doubly distilled water and make up to 100 ml.

Working solutions (prepare fresh each day). Triethanolamine–EDTA buffer 38 parts, MgSO$_4$ solution 4 parts, KCl solution 3 parts, NADH$_2$ solution 8 parts, PK solution 2 parts, ATP solution 1 part, PEP solution 3 parts, LDH solution 1 part. Adjust the pH of the solution to 7.5. This solution is stable for about 24 h at 4 °C.

3.11.3.3.3. Measurement

Conditions. Temperature 25 °C; wavelength 365 nm; final concentration of the solutions (mmol l^{-1}), buffer (pH 7.5) 94.5, EDTA 9.41, MgSO$_4$ 9.9, KCl 74.6, PEP 1.12, NADH$_2$ 0.3, ATP 1.99, PK 12.4 μg ml^{-1}, 2.48 U ml^{-1}, LDH 18.6 μg ml^{-1}, 10.26 U ml^{-1}, GK 10 μg ml^{-1}, 0.25 U ml^{-1}.

3.11.3.3.4. Procedure

Hydrolysis of the triglycerides.[18] The assay mixtures are prepared in sealable test-tubes according to the following scheme:

	Sample, ml	Standard, ml	Reagent blank, ml
Serum	0.2	—	—
Triolein standard	—	0.2	—
Doubly distilled water	—	—	0.2
Ethanolic KOH	0.5	0.5	0.5

Mix and hydrolyse in the sealed tubes for 30 min in a heating block at 70 °C

444

After cooling to room temperature, add:

MgSO$_4$ (150 mmol l^{-1})	1.0	1.0	1.0

Mix, allow to stand for a short period, and centrifuge strongly. The supernatant is used for the determination of the total glycerine. The pH should be 7.5

Total glycerine determination. After the hydrolysis, the supernatant is mixed with the working solution and the further procedure is as follows:

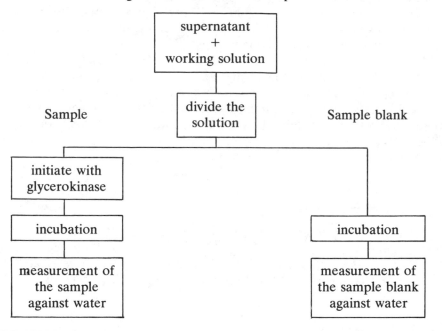

The assay is performed in test-tubes

Supernatant, ml	0.5
Working solution, ml	1.5

Mix well and divide:

	Sample, ml	Sample blank, ml
Reaction mixture	1.0	Remainder, about 1.0
Glycerokinase	0.1	—

Mix and incubate for 15 min at 25 °C. Measure the extinction of sample and sample blank at 365 nm against doubly distilled water

Free glycerine. For the determination of the free glycerine, the serum is used directly for the optical test:

	Sample, ml	Standard, ml	Reagent blank, ml
Serum	0.1	—	—
Glycerine standard	—	0.1	—
Doubly distilled water	—	—	0.1
Working solution	1.5	1.5	1.5

Mix and divide

	Sample, ml	Sample blank, ml
Reaction mixture	0.90	Remainder, about 0.7
Glycerokinase	0.01	—

Mix and incubate at 25 °C for 15 min. Measure the extinction of the sample and sample blank at 365 nm against doubly distilled water

3.11.3.3.5. Calculation

Total glycerine. Using the molar extinction coefficient for NADH $\varepsilon_{365} = 3400 \ \text{l mol}^{-1} \ \text{cm}^{-1}$:[21]

$$E(\text{SB}) - E(\text{S}) - E(\text{RB}) \cdot \frac{1}{3400} \cdot \frac{1.01}{1.0} \cdot \frac{2.0}{0.5} \cdot \frac{1.7}{0.2} \cdot 10^3 \ \text{mmol l}^{-1}$$

$$E(\text{SB}) - E(\text{S}) - E(\text{RB}) \cdot 10.1 \ \text{mmol l}^{-1}$$

Free glycerine.

$$E(\text{SB}) - E(\text{S}) - E(\text{RB}) \cdot \frac{1}{3400} \cdot \frac{0.91}{0.9} \cdot \frac{1.6}{0.1} \cdot 10^3 \ \text{mmol l}^{-1}$$

$$E(\text{SB}) - E(\text{S}) - E(\text{RB}) \cdot 4.76 \ \text{mmol l}^{-1}$$

Triglyceride = glyceride glycerine, mmol l⁻¹.

Glyceride glycerine = total glycerine − free glycerine

Triglyceride, mg per 100 ml.

Glyceride glycerine mmol $l^{-1} \cdot 87.5$

The calculation assumes an average molecular mass of 875 for the triglycerides.[28] If the calculation was referred to triolein, the relative molecular mass would be 885.

3.11.3.3.6. Specificity

Since the glycerine contained in phosphatides is not cleaved off in the alkaline hydrolysis, the method may justifiably be considered specific for triglycerides. Only traces of mono- and diglycerides are present in serum.

3.11.3.4. Triglycerides: fully enzymatic determination

3.11.3.4.1. Principle

Triglycerides undergo a stepwise enzymatic cleavage in the presence of the enzymes lipase and esterase:[15,16]

$$\text{Triglycerides} \xrightarrow[\text{(dodecylsulphate)}]{\text{Lipase}} \text{1,2-Diglycerides + Fatty acids}$$

$$\text{1,2-Diglycerides} \xrightarrow{\text{Lipase, esterase}} \text{2-Monoglycerides + Fatty acids}$$

$$\text{2-Monoglycerides} \xrightarrow{\text{Esterase}} \text{Glycerine + Fatty acids}$$

The glycerine produced is subjected to the glycerokinase-catalysed reaction, exactly as in the method using hydrolysis with ethanolic KOH. It is not possible to determine the free glycerine by this method. To take the latter into account, following recommendations in the literature,[15,36] 10 mg per 100 ml (0.11 mmol l^{-1}) should be subtracted from the calculated triglyceride value. For triglyceride concentrations of 150–250 mg per 100 ml (1.71–2.85 mmol l^{-1}) the possible limits of error for these methods have been stated as +3.5 to −7 mg per 100 ml (+0.04 to −0.08 mmol l^{-1}).[34] From our own experiences a deduction of 7% should be made from the calculated triglyceride value to account for the free glycerine.

3.11.3.4.2. Measurement

Conditions. Modification according to Wahlefeld:[15] temperature 20–25 °C; wavelengths 366, 340, 334 nm; FV = 1.025 ml, SV = 0.02 ml. The reaction mixture buffer–cosubstrate–enzyme is prepared in the proportions 50:1:1.

Final concentrations of the solutions (mmol l^{-1}). *Buffer*: triethanolamine–EDTA buffer (pH 7.0) 46.8, $MgSO_4$ 3.75, Na dodecylsuphate 0.328. *Cosubstrate*: $NADH_2$ 0.188, ATP 0.413, PEP 0.34. *Enzyme*: LDH 5.6 U ml^{-1}, PK 0.94 U ml^{-1}, lipase 75 U ml^{-1}, esterase 0.56 U ml^{-1}, GK 0.73 U ml^{-1}.

3.11.3.4.3. Procedure

1. *Enzymatic hydrolysis.* Mix the serum, cosubstrate, buffer, and enzymes (apart from GK) and allow to stand for 15 min at 20–25 °C; then read the extinction, E_1.

447

2. *Glycerine determination.* Glycerokinase is added to the hydrolysate and the extinction, E_2, is read after 10 min at 20–25 °C.

3.11.3.4.4. Calculation

$$E_1 - E_2 = \Delta E \cdot \frac{1}{\varepsilon d} \cdot \frac{FV}{SV} \cdot 10^3 \text{ mmol l}^{-1}$$

3.11.3.4.5. Specificity

The method is specific for triglycerides; no other glycerides are hydrolysed under these conditions.

3.11.3.4.6. Reference values

The evaluation of the reference values (Table 78) is complicated by the fact that the values are not normally, and in all probability not

Table 78. Reference values for triglycerides in serum.

Sex	Number	Age, years	mmol/l $\bar{x} \ll 2s$	mg/100 ml $\bar{x} \pm 2s$	Reference
Women	25	20–29	1.53 ± 1.13	134 ± 100	17
Women	8	30–39	1.54 ± 2.27	135 ± 199	
Men	17	20–29	1.15 ± 0.70	101 ± 61	
Men	6	30–39	1.2 ± 0.81	105 ± 71	
Men and women	447	5–16	0.55 ± 0.63	48 ± 55	20
Men and women	99	6–15	0.25 ± 1.55	22 ± 136	19
Men	153	6–20	0.28 ± 1.89	25 ± 165	
Women		6–20	0.29 ± 1.99	25 ± 174	
Men	214	11–19	0.52 ± 1.74	46 ± 154	30
Men	117	Interval: 2.5–97.5%			
		20–29	0.62–2.01	54–176	
		30–39	0.64–1.94	56–170	
		40–49	0.75–2.1	66–184	
		50–59	0.86–2.03	76–178	
		≥60	0.66–2.23	58–195	37
Women	622				
		20–29	0.64–1.81	56–159	
		30–39	0.51–1.82	45–160	
		40–49	0.50–1.99	44–174	
		50–59	0.65–1.99	57–174	
		≤60	0.62–2.06	54–180	
	Glycerine				
	57		0.12 ± 0.065	10.5 ± 5.7	17

log-normally, distributed.[19,34] Extensive results for values for school-age children may be found in ref. 20.

3.11.3.4.7. Comments on the method

1. With extinction differences above 0.400 (365 nm) in the not fully enzymatic method, 1 part of serum or hydrolysate should be diluted with 1 part of buffer and the result multiplied by 2. For the fully enzymatic method, if the extinction difference is greater than 0.400 (365 nm) or 0.800 (340 and 334 nm), 0.1 ml of serum should be mixed with 0.9 ml of 0.9% NaCl solution and the result multiplied by 10.

2. The standard is only analysed to monitor the reagents.

3. Stability of triglycerides in serum: Henry[29] reported a fall of up to 10% in the triglyceride concentration on keeping at room temperature for 6 days. Frings et al.[22] established that there was no significant change after 3 days at room temperature or +4 °C. Martin[24] investigated sera which had been kept for 4 weeks at +4 °C and, using the method as modified by Soloni,[8] showed that there was no change in the values. Allowing serum to stand may result in the liberation of glycerine: Chu and Turkington[23] found an increase in free glycerine of:

	Up to 3 days, %	Up to 7 days, %
At room temperature	35–80	50–180
At +4 °C	10–42	33–100

However, Stinshoff et al.[34] were unable to discover any increase in the free glycerine after up to 96 h at +4 °C. In spite of this result, it is advisable to freeze sera for triglyceride determination if they are not to be analysed immediately. Frozen sera are not suitable for electrophoretic separation.

4. Interferences. Haemolysis of up to 2.3%, albumin 10 g per 100 ml, glucose 1000 mg per 100 ml, and lecithin[24] appeared to have no effect on the determination of triglyceride by the method as modified by Soloni.[8] Haemoglobin at above 200 mg per 100 ml and high concentrations of protein interfere in the fully enzymatic reaction.[15] The lactate concentration (up to 5 mmol l^{-1}) is without effect on the determination of glycerine by the fully enzymatic method.[32]

5. By adding trioctanoin (tricaprylin) to human or bovine serum[25] a clear control and reference material may be prepared. In the lyophilized state this triglyceride control is stable for 1 year, and in the dissolved form for 5 days at 5 °C.

References

1. Greten, H., Walter, B., and Brown, W. V., Purification of a human post heparin plasma triglyceride lipase, *FEBS Lett.*, **27**, 306 (1972).

2. Greten, H., Sniderman, A., Chandler, J. C., Steinberg, D., and Brown, W. V., Evidence for the hepatic origin of a canine post heparin plasma triglyceride lipase, *FEBS Lett.*, **42**, 157 (1974).
3. Seidel, D., Müller, P., Felchin, R., Lambrecht, J., Agostini, B., Wieland, H., and Rost, W., Hyperglyceridemia secondary to liver disease, *Eur. J. clin. Invest.*, **4**, 419 (1974).
4. Augustin, J., Middelhoff, G., and Brown, W. V., Metabolismus der Lipoproteine, in *Handbuch der inneren Medizin, Band 7, Fettstoffwechsel*, Springer, Berlin, 1976.
5. Havel, R. J. and Kane, J. P., Quantification of triglyceride transport in blood plasma: a critical analysis, *Fed. Proc.*, **13**, 2250 (1975).
6. Carlson, L. A. and Wadström, L. B., Determination of glycerides in blood serum, *Clin. chim. Acta*, **4**, 197 (1959).
7. Handel, E. van and Zilversmit, D, B., Micromethod for the direct determination of serum triglycerides, *J. Lab. clin. Med.*, **50**, 152 (1957).
8. Soloni, F. G., Simplified manual micromethod for determination of serumtriglycerides, *Clin. Chem.*, **17**, 529 (1971).
9. Giegel, J. L., Ham, A. B., and Clema, V., Manual and semi-automated procedures for measurement of triglycerides in serum, *Clin. Chem.*, **11**, 1575 (1975).
10. Kessler, G. and Lederer, H., Fluorometric measurement of triglycerides, in Skeggs *et al.*, *Automation in analytical chemistry*, Technicon Symposia 1965, Eds. Medical, New York, 1966, p. 361.
11. Eggstein, M., in Zöllner and Eberhagen, *Untersuchung und Bestimmung der Lipoide im Blut*, Springer, Berlin, 1965, p. 289.
12. Schmidt, F. H. and Dahl, K. von, Zur Methode der enzymatischen Neutralfett-Bestimmung in biologischem Material, *Z. klin. Chem. klin. Biochem.*, **6**, 156 (1968).
13. Da Fonseca Wollheim, F., Vereinfachte enzymatische Triglyceridbestimmung, *Ärztl. Lab.*, **19**, 65 (1973).
14. Bucolo, G. and David, H., Quantitative determination of serum triglycerides by the use of enzymes, *Clin. Chem.*, **19**, 476 (1973).
15. Wahlefeld, A., Methods of enzymatic analysis, in Bergmeyer, *Triglycerides determination after enzymatic hydrolysis*, 3rd ed., Verlag Chemie, Weinheim, 1974, p. 1831.
16. Wahlefeld, A. W., Möllering, H., Gruber, W., Bernt, E., and Röschlau, B., *Deutsche Patentmeldung*, P 22 29 849.3 (1972).
17. Eggstein, M. and Kreutz, F. H., Eine neue Bestimmung der Neutralfette im Blutserum und Gewebe, *Klin. Wschr.*, **44**, 262 (1966).
18. Dauwalder, H. and Colombo, J. P., Enzymatische Triglycerid- und Glycerinbestimmung, *Interne Mitteilung*, No. 23, Chemisches Zentrallabor Inselspital Berne, 1974.
19. Dyerber, J. and Hjörne, N., Plasma lipid and lipoprotein levels in childhood and adolescence, *Scand. J. clin. Lab. Invest.*, **31**, 473 (1973).
20. Srinivasan, S. R., Frerichs, R. R., and Berenson, G. S., Serum lipid and lipoprotein profile in school children from a rural community, *Clin. chim. Acta*, **60**, 293 (1975).
21. Bergmeyer, H. U., Neue Werte für die molaren Extinktions-Koeffizienten von NADH und NADPH zum Gebrauch im Routine-Laboratorium, *Z. klin. Chem. klin. Biochem.*, **13**, 507 (1975).
22. Frings, Ch. S., Neri, B. P., Freeman, K., and Fendley, T. W., Stability of triglycerides in serum, *Clin. Chem.*, **20**, 87 (1974).
23. Chu, S. Y. and Turkington, V. E., Stability studies of triglycerides in serum, *Clin. Biochem.*, **8**, 145 (1975).

24. Martin, P. J., A quick inexpensive method for serum triglycerides, *Clin. chim. Acta*, **62**, 79 (1975).
25. Bonderman, D. P., Proksch, G. J., and Bonderman, P., Addition of triglyceride to serum for use in quality control and reference, *Clin. Chem.*, **22**, 1299 (1976).
26. Bucolo, G., McCroskey, R., and Whittaker, N., Lipase-triggered kinetic assay of serum triglycerides, *Clin. Chem.*, **21**, 424 (1975).
27. Chin, H. P. and Abdel-Megud, S. S., Evaluation of an enzymatic method for determination of serum and plasma triglycerides, *Biochem. Med.*, **7**, 460 (1973).
28. Lipert, H., Sl-Einheiten in der Medizin, Urban & Schwarzenberg, Munich, 1976.
29. Henry, R. J., *Clinical chemistry, principles and techniques*, 2nd ed., Hoeber, New York, 1974, p. 1460.
30. Court, J. M. and Dunlop, M., Plasma lipid values and lipoprotein patterns during adolescence in boys, *J. Pediat.*, **3**, 453 (1975).
31. Bublitz, C. and Kennedy, E. P., Synthesis of phosphatides in isolated mitochondria. III. The enzymatic phosphorylation of glycerol, *J. biol. Chem.*, **211**, 951 (1954).
32. Wakayama, J. E. and Swanson, R. J., Ultraviolet spectrometry of serum triglycerides by a totally enzymic method adapted to a centrifugal analyzer, *Clin. Chem.*, **23**, 223 (1977).
33. Mendelson, D. and Antonis, A., A fluorimetric glycerol method and its application to the determination of serum triglycerides, *J. Lipid Res.*, **2**, 45 (1961).
34. Stinshoff, K., Weisshaar, D., Staehler, F., Hesse, D., Gruber, W., and Steier, E., Relation between concentrations of free glycerid and triglycerides in human sera, *Clin. Chem.*, **23**, 1029 (1977).
35. Keller, H., Triglycerid-Bestimmungsmethoden, *Med. Lab., Stuttg.*, **30**, 239 (1977).
36. Müller, P. H., Schmülling, R. M., Liebich, H. M., and Eggstein, M., Vollenzymatische Triglyceridbestimmung: Präzision, Richtigkeit, Methodenvergleich, *Z. klin. Chem. klin. Biochem.*, **15**, 457 (1977).
37. Bayer, P. M., Dorda, W., Gabl, F., Gergely, Th., Junker, E., Schnack, H., and Zyman, H., Cholesterin und Triglyceride im Serum. Referenzbereiche für eine städtische Bevölkerung, *Med. Lab., Stuttg.*, **30**, 270 (1977).

3.12. NUCLEIC ACID METABOLISM

3.12.1. URIC ACID

3.12.1.1. Introduction

In the degradation of nucleotides, purine bases liberated by the hydrolysis of nucleotides are, in part, reincorporated in the metabolism, but a fraction is oxidized to uric acid via xanthine. In most mammals the poorly water-soluble uric acid is degraded further to the very soluble allantoin and rapidly eliminated through the kidneys.[1,2,23,24] However, in humans and in the anthropoid apes, the necessary enzyme, uricase (urate:oxygen oxidoreductase, E.C. 1.7.3.3), is lacking. Consequently, the uric acid must be excreted as the end-product of the purine metabolism. The maximum solubility of uric acid in plasma is stated as 8.5–8.8 mg per 100 ml. At pH 7.4 about 98% of the uric acid is present as the monopotassium salt. Higher stable concentrations have been observed, however.[3] This is understandable since the solubility of uric acid in the plasma depends *inter alia* on the interaction of uric acid with the macromolecules. As *in vitro* studies have shown, these include albumin and the uric acid-binding α-1,2-globulin, a glycoprotein. Normally about 25% of the plasma uric acid is involved in interaction with the macromolecules. Certain forms of primary gout appear to be accompanied by changes in the urate-binding globulin.

Uric acid is also taken up by the erythrocytes. In equilibrium, the concentration in the erythrocytes is about half that in the plasma.[4] Owing to its poor solubility in body fluids, uric acid is potentially dangerous since it can lead to the formation of crystals.[5]

3.12.1.2. Choice of method

As early as 1894 it was observed that uric acid could reduce an alkaline solution of phosphotungstic acid to a blue compound of unknown constitution.[6] Various methods for the determination of uric acid were developed based on this finding. Such methods have serious disadvantages, however: because the coloured substance is not a well defined compound, it is always necessary to run a blank; proteins interfere with the reaction; the Lambert–Beer law is not always obeyed. But the decisive factor is that

452

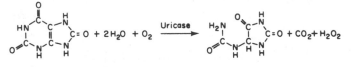

Fig. 114. Uricase reaction.

none of these methods is specific for uric acid. Phosphotungstic acid is reduced to a blue substance by numerous other compounds present in the organism, e.g. glutathione, cystein, and glucose in high concentrations. Drugs such as salicylate, caffeine, theophylline, and theobromine likewise give a positive reaction.

The first real progress was made with the development of the enzymatic determination of uric acid. The enzyme uricase—obtained from the liver of warm-blooded animals—specifically oxidizes uric acid and its salts to allantoin (Fig. 114).

Under the appropriate conditions, this highly specific reaction proceeds quantitatively from left to right, offering the following analytical possibilities:

Determination of the uric acid degraded by uricase. (a) Direct photometric determination of the uric acid in ultraviolet light.[7] Uric acid (but not allantoin, H_2O_2, or CO_2) has an absorption maximum at 293 nm, so the ΔE_{293} before and after the reaction with uricase represents the concentration of uric acid. However, the direct measurement of the decrease in absorption at 293 nm is complicated by the high background absorption of serum at this wavelength, and this must be compensated for. This places relatively high demands on the quality of the instrument. Apart from this, the total extinction of a serum diluted with alkaline buffer does not remain constant, since ascorbic acid present in the serum is (spontaneously) oxidized during the period of incubation with uricase and this alters the background absorption. This necessitates running a serum blank which is treated in the same way as the sample but without the addition of uricase. This method is taken as the *reference method* (operation, see p. 000).

(b) Indirect determination of the uric acid. The phosphotungstic acid reaction is carried out before and after the action of the uricase, the difference corresponding to the enzymatically degraded uric acid. The residual blue colour represents the non-specific fraction. This method does not require an expensive ultraviolet photometer.[8]

Determination of the H_2O_2 formed in the uricase reaction, and calculation of the uric acid. Various methods have been developed on these lines, as follows.

(a) Visualization of the produced H_2O_2 by oxidation of a suitable leuco-dye, e.g. *o*-dianisidine in the presence of peroxidase (POD)[9] (Fig. 115), or, more recently, 2,4-dichlorophenol and 4-aminophenazone (PAP) by the Trinder reaction (see pp. 367 and 429).

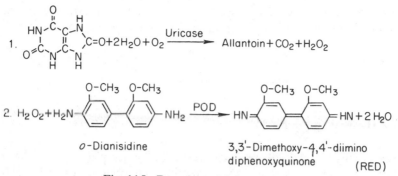

Fig. 115. Reaction with leuco-dye.

Methods of this kind, differing in the nature of the leuco-dye used, generally require a preliminary deproteination since the indicator reaction is affected by various components of the biological material. Uric acid may be lost during the deproteination, however.

More recently, fluorimetric methods for the determination of uric acid have also been described. The H_2O_2 oxidizes non-fluorescent substrates to fluorescent compounds or their precursors, the latter being converted to fluorescent compounds by simple reactions.[10,11]

(b) Determination of the H_2O_2 by enzymatic oxidation (catalase) of methanol to formaldehyde (Kagayema reaction)[12,13] and subsequent non-enzymatic determination of the produced formaldehyde by a Hantzch condensation (Fig. 116). The yellow reaction product has an absorption maximum at 410 nm. At 405 nm, its molar extinction coefficient is $\varepsilon = 7.46 \times 10^3$ l mol^{-1} cm^{-1}.[14]

The reaction is slow even at 37 °C, so that automation is difficult. Although the molar extinction coefficient of the yellow dye is known, it is advisable to run a standard in parallel. In contrast with the peroxidase methods, no deproteination is necessary.

(c) Determination of the H_2O_2 by enzymatic oxidation (catalase) of

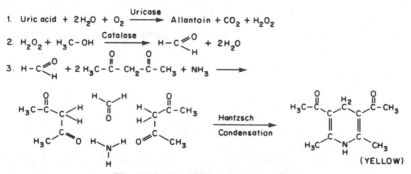

Fig. 116. The Kageyama reaction.

ethanol to acetaldehyde followed by enzymatic determination of the acetaldehyde (aldehyde dehydrogenase) by means of the optical test:[15]

1. Uric acid + 2 H_2O + O_2 $\xrightarrow{\text{Uricase}}$ Allantoin + CO_2 + H_2O_2

2. H_2O_2 + $H_3C\!-\!CH_2\!-\!OH$ $\xrightarrow{\text{Catalase}}$ $H_3C\!-\!C{\overset{\textstyle O}{\underset{\textstyle H}{<}}}$ + 2 H_2O

3. $H_3C\!-\!C{\overset{\textstyle O}{\underset{\textstyle H}{<}}}$ + NAD^+ $\xrightarrow{\text{Aldehyde-DH}}$ $H_3C\!-\!COOH$ + NADH + H^+

The aldehyde dehydrogenase reaction is rapid even at room temperature. Since the molar extinction coefficient of the $NADH_2$ produced is known very accurately, it is not necessary to run a standard. It is of particular significance that up to now no physiological components of serum, or any drugs, have been observed to interfere.

(d) Microcalorimetric measurement of the heat of reaction for the decomposition of H_2O_2 by catalase.[25]

Determination of the oxygen consumed in the uricase reaction. In this method, carrier-bound uricase is packed into a column through which buffer flows. The oxygen pressure of the buffer solution is measured at the end of the column. If a solution containing uric acid is added to the top of the column, the oxygen pressure of the medium falls as a result of the oxidation process, and the difference in the oxygen tensions represents the quantity of the oxidized substrate, i.e. uric acid.

3.12.1.3. Uric acid: ultraviolet spectrophotometry with uricase

3.12.1.3.1. Principle

Uric acid has an extinction maximum at 293 nm with a molar extinction coefficient ε of 12.6×10^3 1 mol^{-1} cm^{-1}. In the presence of uricase it undergoes the following reaction:

Uric acid + O_2 + $2H_2O$ $\longrightarrow$ Allantoin + H_2O_2 + CO_2

Since allantoin, hydrogen peroxide, and carbon dioxide do not absorb at 293 nm, the concentration of uric acid may be concluded directly from the reduction in the extinction. For uricase the optimum pH is 9.4–9.5.

3.12.1.3.2. Reagents

Ammediol buffer, 50 *mmol* l^{-1}, *pH* 9.4, *EDTA* 5 *mmol* l^{-1}. See Appendix 1. Add 2 g of EDTA to 1 litre.

Uricase solution (100 *U* l^{-1}). 18 U ml^{-1} (specific activity 9 U mg^{-1}) in 50% glycerine solution.

Standard solution, 5 mg per 100 ml. Mix 100 mg of uric acid, 80 mg of lithium carbonate and 15 ml of water, heat to 60 °C and make up to

Table 79. Determination of uric acid.

	Sample, ml	Sample blank, ml	Reagent blank, ml	Standard, ml	Standard blank, ml
Plasma buffer	1.0	1.0	—	—	—
Buffer	—	—	1.0	—	—
Standard buffer	—	—	—	1.0	1.0
Uricase solution	0.01	—	0.01	0.01	—
DM-water	—	0.01	—	—	0.01

100 ml with DM-water. Dilute this solution to give a standard of 5 mg per 100 ml.

3.12.1.3.3. Measurement

Conditions. Plasma 24.4 μl, ammediol buffer (pH 9.4) 50 mmol l^{-1}, EDTA 5 mmol l^{-1}, uricase 180 U l^{-1}.

3.12.1.3.4. Procedure for plasma and serum

(a) *Dilution.* A 100-μl volume of plasma (serum) or standard solution is pipetted into 4.0 ml of ammediol buffer. The sample and sample blank must be pipetted from the same dilution mixture.

(b) *Determination.* The details are given in Table 79. Mix well and allow to stand at room temperature for 45–60 min. Read off the extinction differences between sample blank and sample at 293 nm in quartz cuvettes (if necessary set the sample at zero). Take separate readings of the extinctions of the reagent blank, standard, and standard blank, all against buffer.

3.12.1.3.5. Calculation

1. *By means of the standard solution.*

$$\text{Concentration of uric acid} = \frac{E(\text{SB}) - E(\text{S}) + E(\text{RB})}{E(\text{STB}) - E(\text{ST}) + E(\text{RB})}$$

$$\cdot \text{ 5 mg per 100 ml}$$

2. *Using the percentage or molar extinction coefficient ($a = 750$, $\varepsilon = 12.6 \times 10^3$).*

$$\text{Concentration of uric acid} = \frac{[E(\text{SB}) - E(\text{S})] + E(\text{RB})}{ad} \cdot \frac{FV}{SV}$$

$$10^3 \text{ mg per 100 ml}$$

$$= \frac{[E(\text{SB}) - E(\text{S})] + E(\text{RB})}{750 \cdot 1} \cdot \frac{4.1}{0.1}$$
$$\cdot 10^3 \text{ mg per 100 ml}$$

Concentration of uric acid $= [E(\text{SB}) - E(\text{S})] + E(\text{RB})$
$$\cdot 54.7 \text{ mg per 100 ml}$$
$$= [E(\text{SB}) - E(\text{S})] + E(\text{RB})$$
$$\cdot 3254 \ \mu\text{mol l}^{-1}$$

Urine. First 1 part of urine is diluted and thoroughly mixed with 9 parts of ammediol buffer. Thereafter the determination is carried out exactly as for plasma. The result is finally multiplied by 10.

3.12.1.3.6. Reference values

These are given in Table 80.

3.12.1.3.7. Physiological variability

The reference range of the uric acid concentration in serum is likely to be subject to considerable genetic and environmentally conditioned variation from one individual to the next. Large-scale studies in various parts of the world have shown sizeable differences even with collectives of different social and ethnic backgrounds. In addition to genetic factors, both age and sex play a determining role.[16] New-born infants show values higher than those of adults, but these revert to the adult norm by the end of the first year of life. On average, men show higher uric acid concentrations than women, but this sex difference is age dependent. For men the uric acid in the plasma tails off slowly with increasing age, whereas with women it increases. After the menopause the values for women are practically identical with those for men.[5,17]

For a single individual, day-to-day variations of about 0.5 mg per 100 ml are observed. A diurnal rhythm is also detectable; the values are lower at night than in the day-time.

That the uric acid values are markedly affected by diet has been known for many years. Thus, a purine-free diet can lead to a fall in the uric acid

Table 80. Reference values valid for Central Europe.

	Units	Reference range	Limit range	Hyperuricaemia above:
Women	μmol l^{-1}	149–363	369–387	387
	mg per 100 ml	2.5–6.1	6.2–6.5	6.5
Men	μmol l^{-1}	208–387	387–422	422
	mg per 100 ml	3.5–6.5	6.5–7.1	7.1

concentration in the blood to 0.8 mg per 100 ml.[18] Conversely, with purine-rich foods more than 1000 mg of uric acid per day is excreted in the urine. Bodily activity has a pronounced effect on the concentration of uric acid in the plasma.[18] Hard work can result in an elevation of 2.5 ml per 100 ml. The excretion of uric acid in the urine increases in parallel with this. Similar results have been observed in cases of excessive psychic and mental stress.[18]

The *uric acid excretion in the urine* is dependent on the purine content of the diet. Uric acid in the glomerular filtrate is reabsorbed in the tubules but it is also secreted in the tubules and, in part, reabsorbed again.[19] For healthy persons receiving a normal diet it may be taken as a general rule that not more than 250–700 mg is excreted in 24 h. An elevated excretion of uric acid in the urine is also found in Lesch–Nyhan syndrome. This is characterized by chorioathetosis, mental retardation, self-inflicted injuries, and hyperuricaemia conditioned by a deficiency in enzymic activity of hypoxanthine guanine phosphoribosyl transferase (HGPRT).[20] A diagnostic test of proved effectiveness for patients with Lesch–Nyhan syndrome is the determination of the *uric acid/creatinine ratio* in freshly excreted morning urine.[21] Normal adult individuals show a ratio of 0.34 ± 0.10 $(\bar{x} \pm s)$, new-born infants 1.55, and 10-year olds 0.61; patients suffering from Lesch–Nyhan syndrome with complete enzyme defect have a ratio of 3.19 ± 1.00, or 1.06 ± 0.43 if the enzyme defect is only partial. The ratio is age dependent. Sucklings show high values which decline during the course of childhood.[20] Drugs which affect the excretion of uric acid may falsify the quotient.

3.12.1.3.8. Diagnostic significance

1. *Pathophysiology of uric acid.* The total amount of uric acid circulating in the plasma is dependent on many factors; these are represented in a simplified diagram in Fig. 117.[1]

The three most important factors which regulate the uric acid concentration in the plasma are:[23] 1, the endogenous synthesis and degradation of purine derivatives; 2, the exogenous supply of purines; and

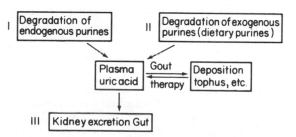

Fig. 117. Factors regulating the concentration of uric acid in the plasma.

3, the urate clearance of the kidneys. On the other hand, in terms of the amount involved, the excretion through the gut is small and constant. These three factors, presumably often in combination, play a decisive role in the primary forms of gout.

2. *Hyperuricaemia. Primary gout* is principally a disease which afflicts young men following puberty. It occurs in women after the menopause, but even then it is rarer than in men. It is almost never observed in women of reproductive age.

The risk of a manifest gout with one or several attacks of gout increases with increasing concentration of uric acid in the plasma. The experiences of large-scale field studies show that for patients with uric acid values between 6 and 7 mg per 100 ml, the risk of a gout attack is only 1.8%. This risk rises to 11.8% for values between 7 and 7.9 mg per 100 ml, to 36% for values between 8 and 8.9 mg per 100 ml and to near 100% for values above 9 mg per 100 ml.

Secondary hyperuricaemia has no typical age and sex distribution and results from a basic disorder not causally connected with gout: 1, increased purine metabolism due to (a) myeloproliferative diseases such as polycythaemia vera, myeloid leukaemia, and myelosclerosis, (b) other forms of leukaemia, (c) other neoplasms, or (d) cytostatic treatment; 2, diminished excretion of uric acid due to (a) primary kidney damage, (b) drug-conditioned retention of uric acid, e.g. by diuretics of the thiacid group or tuberculostatics of the pyrazinamide type, or (c) diseases which lead to retention of uric acid, e.g. hyperparathyroidism, hyperthyroidism, or glycogen storage disease.

With these secondary forms of gout priority is given to treatment of the basic disorder.

Hyperuricaemia is also found with Lesh–Nyhan syndrome.[20] Elevated serum uric acid values have also been reported in hypertriglyceridaemia, complete fasting, respiratory acidosis, alcoholism, Down syndrome, and pregnancy toxicosis.

3. *Hypouricaemia.* Various drugs (aspirin, allopurinol) and X-ray contrast agents can lead to a reduced concentration of uric acid in the serum (<2 mg per 100 ml). Low values have also been observed in various neoplasias,[22] in rare, familial isolated defects of the renal tubules,[26] and in two congenital metabolic disorders—xanthineoxidase deficiency (xanthinuria)[27] and purine nucleoside phosphorylase deficiency.[28]

References

1. Keller, H., Harnsäure und Gicht, *Med. Lab., Stuttg.,* **28**, 73 (1975).
2. Müller, M. M. and Kaiser, E., Die Regulation des Purinstoffwechsels und seine medikamentöse Beeinflussung, *Dt. med. Wschr.,* **100**, 198 (1975).
3. Alvsaker, J. O., Urate-plasma protein interactions, *Scand. J. clin. Lab. Invest.,* **30**, 345 (1972).
4. Lang, F., Greger, R., Silbernagel, H., Günther, R., and Deetjen, P., Aufnahme

von 2-C14 Harnsäure in die Erythrocyten von Patienten mit Hyperurikämie und Gicht, *Klin. Wschr.*, **53**, 261 (1975).

5. Lentner, C. and Eggstein, M., Leitsymptom Laborwert, *Documenta Geigy*, Geigy, Basle, 1976.
6. Offer, T. R., *Zentbl. Physiol.*, **8**, 801 (1894).
7. Praetorius, E. and Poulsen, H., Enzymatic determination of uric acid with detailed directions, *Scand. J. clin. Lab. Invest.*, **5**, 273 (1953).
8. Eisenwiener, H. G. and Ferrat, R., Kombinationsverfahren zur spezifischen Bestimmung der Harnsäure, *Bull. schweiz. Ges. klin. Chem.*, **1**, 24 (1973).
9. Lorentz, K. and Berndt, W., Enzymic determination of uric acid by a colorimetric method, *Analyt. Biochem.*, **18**, 58 (1967).
10. Kuan, J.-C. W., Kuan, S. S., and Guilbault, G. G., An alternative method for the determination of uric acid in serum, *Clin. chim. Acta*, **64**, 19 (1975).
11. Kamoun, P., Lafourcade, G., and Jerome, H., Ultramicromethod for determination of plasma uric acid, *Clin. Chem.*, **22**, 964 (1976).
12. Kageyama, N., A direct colorimetric determination of uric acid in serum and urine with uricase–catalase system, *Clin. chim. Acta*, **31**, 421 (1971).
13. Kageyama, N., Bestimmung der Harnsäure mit dem Reagenziensatz "Uricolor 400", *Med. Lab., Stuttg.*, **26**, 264 (1973).
14. Forstmeyer, H., Enzymatischer Farbtest zur Bestimmung der Harnsäure, *Ärztl. Lab.*, **20**, 125 (1974).
15. Haeckel, R. and Heinz, F., Die Bestimmung der Harnsäurekonzentration mittels Uricase, Katalase und NADH-abhängiger Aldehyddehydrogenase, *Z. klin. Chem. klin. Biochem.*, **13**, 244 (1975).
16. Thefeld, W., Hoffmeister, H., Busch, E. W., Koller, U. P., and Vollmar, J., Normalwerte der Serumharnsäure in Abhängigkeit von Alter und Geschlecht mit einem neuen enzymatischen Harnsäurefarbtest, *Dt. med. Wschr.*, **98**, 380 (1973).
17. Bentgsson, C. and Tibblin, E., Serum uric acid levels in women, *Acta med. scand.*, **196**, 93 (1974).
18. Wyngaarden, J. B. and Kelley, W. N., Gout, in Stanbury, Wyngaarden and Fredrickson, *The metabolic basis of inherited disease*, 3rd ed., McGraw-Hill, New York, 1972.
19. Steetc, T. H. and Rieselbach, R. E., Renal urate excretion in normal man, *Nephron*, **14**, 21 (1975).
20. Thorpe, W. P., The Lesch–Nyhan syndrome, *Enzyme*, **12**, 129 (1971).
21. Kaufman, J. M., Greene, M. L., and Seegmiller, J. E., Urine uric acid to creatinine ratio—a screening test for inherited disorders of purine metabolism, *J. Pediat.*, **73**, 583 (1968).
22. Ramsdell, M. C. and Kelley, W. N., The clinical significance of hypouricemia, *Ann. intern. Med.*, **78**, 239 (1973).
23. Balis, M. E., Uric acid metabolism in man, *Adv. clin. Chem.*, **18**, 213 (1976).
24. Wyngaarden, J. B., Metabolic defects of primary hyperuricemia and gout, *Am. J. Med.*, **56**, 651 (1974).
25. Rehak, N. N., Janes, G., and Young, D. S., Calorimetric enzymic measurement of uric acid in serum, *Clin. Chem.*, **23**, 195 (1977).
26. Benjamin, D., Sperling, O., Weinberger, A., Pinkhas, J., and De Vries, A., Familial hypouricemia due to isolated renal tubular defect, *Nephron*, **18**, 220 (1977).
27. Watts, R. W. E., Disorders of purine metabolism, in Neale, *8th Symposium on Advanced Medicine, London*, 1972.
28. Cohen, A., Doyle, D., Martin, D. W., and Ammann, A. J., Abnormal purine metabolism and purine overproduction in a patient deficient in purine nucleoside phosphorylase, *New Engl. J. Med.*, **295**, 1449 (1976).

3.13. ENZYME METABOLISM AND GENERAL ENZYME DIAGNOSTICS

3.13.1. ENZYME METABOLISM

3.13.1.1. Physiology

In the living organism the reactions of metabolic processes are catalysed by enzymes. On their own, these reactions would proceed far too slowly and they are accelerated by the action of the enzymes. It is only through their agency that an adequate conversion of substances is at all possible.

Enzymes have all the typical properties of proteins. They are sensitive to outside influences (they should therefore be determined as quickly as possible), are macromolecular (consequently present in urine only in very small amounts), and highly specific in respect of their antigen nature and their catalytic action. This specificity applies both to the molecule and to the action of the enzyme. Thus the enzymes belong to the group of organ-specific proteins and there may be subunits, the isoenzymes, even within the same enzyme. As a rule the enzymes are also extremely specific in respect of their action insofar as the majority catalyse only a single chemical reaction. This catalytic action is phenomenal: 1 molecule of acetylcholine esterase cleaves about 1 million acetylcholine molecules per minute into choline and acetic acid. One practical consequence of this catalytic action is that enzymes are readily determined. On the basis of an activity determination it is possible to detect as little as 1 pg, i.e. $1/1000$ μg, of enzyme without great difficulty. The concentration of plasmatic enzymes is less than 100 μg l^{-1}. In terms of weight this is a very small amount, but it is relatively easy to determine by measurement of the activity.[1]

3.13.1.2. Intracellular topography

The sphere of action of the enzymes, is the cell. They are partly to be found in different cell compartments and they are not exchangeable. They often form integrating components of the cell structures. Most of our knowledge on the location of the intracellular enzymes has been gained from cell fractionations. In some cases, enzymes which function cooperatively are even to be found in common cell structures. The outer membrane of the *mitochondria* contain enzymes for the activation of fatty acids and the monoamino oxidases, while the inner membrane is the domain of the principal enzyme for respiration and the oxidative phosphorylation. The matrix contains the enzymes of the citric acid cycle, glutamate dehydrogenase, certain transaminases, and enzymes for the degradation of fatty acids. The *lysosomes* are rich in degradative acid hydrolases which play a part in the breakdown of the cell and the removal of foreign substances. The *cytoplasm* contains the glycolysis and hydrogen transfer enzymes.[2]

3.13.1.3. Enzymes in the extracellular space

Enzymic activities are measured in the intravasal space. This, together with the interstitial space, is a component of the extracellular space.

The enzymic activity in the plasma is the resultant of the following processes:[3,4]

1. Issue of the enzymes from the cells into the extracellular space.
2. Dilution of the enzymes in the extracellular space.
3. The speed with which the enzyme passes into the intravasal space.
4. The elimination of the enzyme activity from the intravasal space.

3.13.1.3.1. Issue of the enzymes from the cells

Very little is known about the mechanism of this process as it occurs in the normal organism. In pathological situations membrane defects must arise. The issue of the enzymes is probably also dependent on the energy or the adenosine triphosphate content of the cell. Low intracellular concentrations of adenosine triphosphate appear to increase the membrane permeability.

The concentration gradient of enzymes between the cell and the extracellular space is enormous. For example, the intracellular concentration of lactate dehydrogenase is 10 times and that of creatine kinase about 10^6 times as great as that in the plasma. Obviously, the liberation of enzymes from a relatively small amount of tissue is sufficient to lead to measurable changes in the plasma. Thus, 30 000–50 000 U of creatine kinase can be liberated from 1 g of skeletal muscle. Even after distribution in the extracellular space (about 10 litres) this affords 3000–5000 U l^{-1}. Since the biological half-life of this enzyme is very short (about 18 h), about 30–50 U l^{-1} is the concentration normally detected.

3.13.1.3.2. Actual elimination

Since the concentration of enzymes in the plasma remains constant it may be concluded that the issue from the cell, the entry into the intravasal compartment, and the elimination from the plasma are relatively constant processes. In man the biological half-life of the individual enzymes is relatively short, e.g. creatine kinase 15 h, GOT 17 h, GPT 47 h, LDH isoenzyme-5 10 h, and LDH isoenzyme-1 113 h. The following points have been established: (a) the elimination is independent of the absolute concentration of enzyme; (b) it is independent of the molecular weight of the enzyme; and (c) enzymes are eliminated as native, active enzyme molecules.

Using animals it was possible to show that, following administration of ^{14}C-labelled LDH isoenzyme-5, the radioactivity and the enzyme activity disappeared from the plasma at the same time.[6] A direct elimination from the plasma by filtration in the kidneys can only be assumed for enzymes with a molecular weight less than 60 000 (α-amylase). Elimination from the intravasal compartment can proceed in one or in two phases. In the one-phase process the elimination is limited by the permeability of the capillary membrane; with two-phase kinetics the rate of the elimination of the enzyme from the extracellular space is also involved.[5,6]

Following their emergence or after liberation by cell breakdown, under physiological conditions the cell enzymes generally reach the interstitium first and are then transported into the intravasal compartment via the lymphatic system. Only enzymes with low molecular weight can also reach the intravasal space directly via the capillary membrane and this condition is not fulfilled for the majority of the diagnostically important enzymes.[5]

3.13.1.4. Classification of the plasma enzymes

3.13.1.4.1. Plasma-specific enzymes

The following enzymes are fed to the plasma and exert their action there: lipoprotein lipase, prothrombin complex, cholinesterase, Factors V, VII, and X, lecithin, and cholesterol acyltransferase.

3.13.1.4.2. Secreted enzymes

These include the enzymes of the glandular secretions: pancreatic and parotic α-amylase, prostatic phosphatase, and pepsinogen.

3.13.1.4.3. Organ non-specific enzymes of the energy metabolism

The production of energy in living things occurs by way of three metabolic paths: glycolysis, oxidation of glucose, and cell respiration. The processes which yield energy proceed in fundamentally the same manner in all types of cell. Thus, the enzyme pattern for both glycolysis and cell respiration is qualitatively identical in skeletal muscle, heart muscle, liver, and the kidneys. The catalysts which participate in these processes are thus justifiably termed organ non-specific enzymes. In certain organs these also include the enzymes of the transamination reaction such as the glutamate–oxaloacetate and glutamate–pyruvate transaminases. The diagnostic significance of the individual enzymes is therefore small. The organ specificity of certain organ non-specific enzymes can be increased by measurement of their isoenzymes, e.g. in the case of lactate dehydrogenase and of creatine kinase.

3.13.1.4.4. Organ specific enzymes of the energy metabolism

This group includes the various isoenzymes of individual principal chain enzymes. These are enzymes which fulfil the same biological function but are chemically different.

3.13.1.4.5. Organ-specific enzymes of the cell function

Just as the morphology and physiology of each organ, each tissue and each cell exhibit peculiarities, so the enzyme pattern of these structures must have its own properties. For example, the synthesis of α-amylase and lipase is a specific function of the pancreas. The key enzyme of gluconeogenesis, glutamate–pyruvate transaminase, is to a large extent found in the liver.

The occurrence of such enzymes in the plasma allows the inference of lesions in the corresponding organ, e.g. bones: alkaline phosphatase; pancreas: α-amylase, lipase; muscle: creatine kinase; prostate gland: acid phosphatase; liver: these include enzymes for metabolic processes which take place predominantly in the liver, e.g. the synthesis of urea. Attempts have been made to measure such enzymes in the plasma, enzymes such as ornithine–carbamyl transferase, the argininosuccinate lyases and the arginases. Recently, xanthine oxidase has also been used. Whereas the quantification of these enzymes would be organ-specific, the method is too costly for serial determinations. Also, the activities are low as a result of the dilution in the extracellular space. For the liver there is simply no organ-specific enzyme directly detectable in the plasma. The nearest candidate is GPT and, under certain conditions, γ-glutamyl transpeptidase.

3.13.2. ENZYME DIAGNOSTICS

The *aim of enzyme diagnostics* is to identify cell damage by the quantification of the enzyme activity in the plasma and, if possible, to locate the morbidity in a particular organ and assess its subsequent evolution. The practice of enzyme diagnostics demands a certain knowledge of the pathophysiology of the enzymes in the plasma.

3.13.2.1. Elevated and reduced enzyme activity in the plasma

Elevated activity is found in the following situations:

1. Increased synthesis, e.g. increase in alkaline phosphatase in rickets.

464

2. Necrosis of the cells which produce the enzyme: increase in GOT, aldolase, and creatine kinase in myocardial infarction.
3. Increased cell permeability: increase of the creatine kinase in progressive muscular dystrophy, increase of the transaminases in viral hepatitis.
4. Reduced elimination, e.g. increase of α-amylase in secretion blockage of the pancreatic duct.

Reduced activity has been established in the following cases:

---

1. Reduction in the number of cells producing the enzyme, e.g. pepsinogen deficiency following gastrectomy; reduced activity of cholinesterase and caeruloplasmin in cirrhosis of the liver.
2. Selective deficiencies in synthesis, e.g. lack of caeruloplasmin in Wilson's disease, hypophosphatasia, genetically conditioned hypocholine-sterasaemia.
3. Novel elimination routes, e.g. lowering of the caeruloplasmin in nephrotic syndrome by loss into the urine.
4. Inhibition of the enzyme, e.g. inhibition of trypsin by antitrypsin.

Usually, enzyme diagnostics is concerned with an elevated activity of the enzyme in the plasma conditioned by increased egress from the cells. Various pathogenetic factors are responsible: hypoxia, anoxia, necrosis and autolysis, infectious agents, chemical and physical poisons, endocrinal action, metabolic defects, immunoreactions, stasis, abnormal diet, and malignant growth. Furthermore, the enzyme activity is dependent on sex and age differences. Changes in activity of up to about 15% may be attributed to the displacement of fluids between the intra and extravasal compartments,[5] as occurs e.g. during venous stasis, or owing to changes in the position of the body. This probably also accounts for the increase in enzymes following bodily exertion.[7]

References

1. Richterich, R., Enzymstoffwechscl, in Siegenthaler, *Klinische Pathophysiologie*, Thieme, Stuttgart, 1970, p. 159.
2. Rapoport, S. M., *Medizinische, Biochemie. Topochemie der Zelle*, SEB-Verlag, Berlin, 1975, p. 163.
3. Friedel, R., Mattenheimer, H., Trautschold, I., and Forster, G., Der vorgetäuschte Enzymaustritt. Verteilung und Transport von Zellenzymen im extrazellulären Raum. I, *Mitt. Z. klin. Chem. klin. Biochem.*, **14**, 109 (1976).
4. Friedel, R., Bode, R., Trautschold, I., and Mattenheimer, H.: Die Lymphe als Verteilungsraum für Zellenzyme. Verteilung und Transport von Zellenzymen im extrazellulären Raum. II. *Mitt. Z. klin. Chem. klin. Biochem.*, **14**, 119 (1976).
5. Friedel, R., Bode, R., and Trautschold, I., Verteilung heterologer, homologer und autologer Enzyme nach intravenöser Injektion. Verteilung und Transport von Zellenzymen im extrazellulären Raum. III, *Mitt. Z. klin. Chem. klin. Biochem.*, **14**, 129 (1976).

6. Bär, U., Friedel, R., Heine, H., Mayer, D., Ohlendorf, S., Schmidt, F. W., and Trautschold, I., Studies on enzyme elimination. III. Distribution, transport, and elimination of cell enzymes in the extracellular space, *Enzyme,* **14**, 133 (1972/73).
7. Siest, G. and Galteau, M. M., Variations of plasmatic enzymes during exercise, *Enzyme,* **17**, 179 (1974).

3.14. PHARMACOLOGY AND TOXICOLOGY

3.14.1. INTRODUCTION

In recent times more and more determinations of pharmaceuticals in body fluids (plasma and urine) have been introduced to the clinical chemistry laboratory. These are predominantly for monitoring the medication therapy and for the diagnosis of intoxication.[1-3] Since many drugs interfere with clinical–chemical reactions even at the *in vivo* stage, a knowledge of their concentration would be desirable even from this point of view. Nevertheless, relatively little has yet been done in this field since the methods for the determination of the majority of pharmaceuticals are, as a rule, very complicated, and no methods have yet been developed for many drugs and their metabolites.

In addition to classical photometric methods, the methods which are coming into use are those such as thin-layer and gas chromatography, the latter often combined with mass spectrometry. More recently, radio- or enzyme immunoassays have been applied to the determination of pharmaceuticals.[4,5] A chapter on the determination of pharmaceuticals would exceed the scope of this book and we shall therefore confine our attention to a few classical photometric methods for the determination of *p*-aminophenol derivatives, phenacetin, salicylates, and barbiturates since the reaction mechanism on which the methods are based are of general interest.

References

1. Geldmacher-von Mallinckradt, M., *Einfache Untersuchungen auf Gifte im klinisch—chemischen Laboratorium*, Thieme, Stuttgart, 1976.
2. Baselt, R. C., Wright, J. A., and Cravey, R. H., Therapeutic and toxic concentrations of more than 100 toxicologically significant drugs in blood, plasma, or serum: a tabulation, *Clin. Chem.*, **21**, 44 (1975).
3. Martin, E. N., *Hazards of medication*, Lippincott, Philadelphia, 1971.
4. Toxicology and drug assay, *Clin. Chem.*, **20**, 2 (pp. 111–316) (1974).
5. Cleeland, R., Christenson, J., Usategui-Gomez, M., Heveran, J., Davis, R., and Grunberg, E., Detection of drugs of abuse by radioimmunoassay: a summary of published data and some new information, *Clin. Chem.*, **22**, 712 (1976).

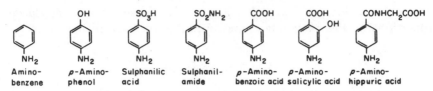

Fig. 118. A few medicinally important *p*-aminobenzene derivatives.

3.14.2. *p*-AMINOBENZENE DERIVATIVES

3.14.2.1. General

p-Aminobenzene derivatives are of great medicinal interest. A few characteristic representatives of this group are presented in Fig. 118. Derived from sulphanilamide we have the large group of sulphonamides with their bacteriostatic, tuberculostatic (*p*-aminosalicylic acid), antidiabetic, and diuretic actions. A common requirement is the monitoring of the course of the treatment by determination of the plasma concentration. Again, these compounds have occasionally been observed to produce intoxications and, in such cases, determination of the concentration in the plasma permits an assessment of its severity and course. Yet another *p*-aminobenzene derivative is *p*-aminohippuric acid (PAH), used as clearance substance for the measurement of kidney function.

3.14.2.2. Principle of the method

Marshall and Litchfield[1] and Bratton and Marshall[2] have worked out a sensitive and technically simple method for the determination of all of these derivatives. The principle of this Bratton–Marshall reaction is illustrated in Fig. 119. The *p*-aminobenzenes react with nitrite in acidic solution to form the corresponding diazonium salt. The excess nitrite is destroyed with ammonium sulphamate and the diazonium salt subsequently coupled with *N*-(1-naphthyl)ethylenediamine to give a stable azo dye. The maximum

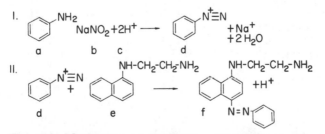

Fig. 119. Mechanism of the Bratton–Marshall reaction. a = Aminobenzene; b = nitrite; c = acid; d = benzene-diazonium cation; e = *N*-(1-naphthyl)ethylenediamine; f = azo dye.

intensity of colour is about 550 nm. The amount of dye produced is proportional to the *p*-aminobenzene concentration over a wide range.

3.14.2.3. Specificity

A positive Bratton–Marshall reaction is given by all *p*-aminobenzene derivatives with a free amino group. If the test is positive it is important to find out whether or not the patient has received other drugs in addition to the one being determined. In order to analyse for the amino-substituted derivatives, e.g. acetylated sulphonamides, the amino group must first be freed.

3.14.2.4. Comments on the method

1. The extinction coefficient of a given compound is constant, and in principle, therefore, running a standard is superfluous. On the other hand, the extinction coefficients of the individual compounds differ greatly. Consequently, it is always necessary to begin by preparing a calibration graph and determining the extinction coefficient for each compound to be determined (Table 81).

2. The reaction is highly sensitive to light and it is essential to exclude sunlight.

3. Normal human blood contains a small amount of substances which give a positive Bratton–Marshall reaction. Their concentration lies between 0.1

Table 81. Percentage and molar extinction coefficients of some *p*-aminobenzenes and related compounds (measurement with monochromatic light at 546 nm).

Compound	Molecular weight	Extinction coefficients	
		%	Molar
p-Aminobenzoic acid	137.13	3650	50 000
p-Aminohippuric acid	194.19	2550	49 500
p-Aminosalicylic acid	153.13	10	100
Azosulphamide	494.19	10	100
Phthalylsulphacetamide	362.36	1290	46 600
Phthalylsulphathiazole	403.43	1100	44 300
Succinylsulphathiazole	373.41	70	2 650
Sulphabenzamide	276.31	1720	47 600
Sulphacetamide	254.25	2200	56 000
Sulphasiazine	250.28	1950	49 400
Sulphaguanidine	214.25	2150	50 000
Sulphamerazine	264.30	1950	51 600
Sulphamethazine	278.32	1700	47 400
Sulphanilamide	172.21	2750	47 400
Sulphapyridine	249.29	1950	48 600
Sulphathiazole, sodium salt	304.33	1550	24 000

and 0.5 mg per 100 ml. Consequently, in determining medicines or p-aminohippurate (clearance) as reflected in their blood concentrations, the sample concentration before administration is taken as the sample blank and the value is subtracted from the result for the sample following administration of the substance. The concentration of these p-aminobenzene derivatives is relatively constant in each individual and is subject to only very slight daily variations.

4. p-Aminosalicylic acid (Fig. 118) is one p-aminobenzene derivative that cannot be quantified by the Bratton–Marshall reaction (Table 81). A sensitive and rapid method for its determination has been described by Rieder.[3]

References

1. Marshall, E. K. and Litchfield, J. T., The determination of sulfanilamide, *Science*, **88**, 85 (1938).
2. Bratton, A. C. and Marshall, E. K., A new coupling reagent for sulfanilamide determination, *J. biol. Chem.*, **128**, 537 (1939).
3. Rieder, H. P., Bestimmung freier p-Aminosalicylsäure im Fingerbeerenblut, *Klin. Wschr.*, **39**, 813 (1961).

3.14.2.5. Sulphonamides, Bratton–Marshall reaction

3.14.2.5.1. Principle of the method

Since the distribution ratio of the sulphonamides between the erythrocytes and the plasma differs from one preparation to the next, it is advisable to carry out the determination in whole blood and serum (plasma).[1] The sample is deproteinated with trichloroacetic acid prior to the Bratton–Marshall reaction. Deproteination is superfluous in the case of urine or cerebrospinal fluid.

3.14.2.5.2. Reagents

Trichloroacetic acid, 0.61 *mol l^{-1}* (10%). Dissolve 10 g of Trichloroacetic acid in DM-water and make up to 100 ml. The solution keeps indefinitely at room temperature.

Sodium nitrite, 14.5 *mmol l^{-1}* (0.1%). Dissolve 100 mg of sodium nitrite in DM-water and make up to 100 ml. The solution may be kept for 1 week if frozen.

N-(1-Naphthyl)ethylenediamine, 3.86 *mmol l^{-1}* (0.1%). Dissolve 100 mg of the dihydrochloride in DM-water and make up to 100 ml. The solution may be kept for 1 month if stored frozen in a dark bottle. Discard brown-coloured solutions.

Ammonium sulphamate solution, 4.38 *mmol l^{-1}*. Dissolve 500 mg of ammonium sulphamate in DM-water and make up to 100 ml.

Standard, 1 mg per 100 ml. Dissolve 10 mg of the sulphonamide (or the corresponding quantity of the sodium salt) in trichloroacetic acid (0.61 mol l^{-1}) and make up to 1000 ml. The solution keeps indefinitely.

3.14.2.5.3. Procedure

Preparation

(*a*) *Blood, plasma, serum.* Add 20 μl of blood, serum, or plasma to 0.2 ml of trichloroacetic acid. Mix well, allow to stand for 5 min, and centrifuge powerfully.

(*b*) *Urine, cerebrospinal fluid.* Dilute 0.1 ml of centrifuged urine with 1 ml of trichloroacetic acid.

	Sample, ml	Reagent blank, ml	Standard, ml
Sodium nitrite	0.2	0.2	0.2
Supernatant	0.1	—	—
Trichloroacetic acid	—	0.1	—
Standard	—	—	0.1
Ammonium sulphamate	0.2	0.2	0.2
N-(1-Naphthyl)ethylenediamine	0.5	0.5	0.5

After adding each reagent, mix well and wait 2 min. The development of the colour is complete after 10 min and the intensity of the colour remains stable for at least 1 h. Read the extinction against water at 546 nm.

3.14.2.5.4. Calculation

$$\text{Concentration} = \frac{E(S) - E(RB)}{E(ST) - E(RB)} \cdot C(ST) \cdot \frac{FV(\text{deprot.})}{SV(\text{deprot.})}$$

$$= \frac{E(S) - E(RB)}{E(ST) - E(RB)} \cdot 1 \cdot 11 \text{ mg per 100 ml}$$

$$\text{Concentration} = \frac{E(S) - E(RB)}{E(ST) - E(RB)} \cdot 11 \text{ mg per 100 ml}$$

3.14.2.5.5. Specificity

(See p. 468.)

3.14.2.5.6. Comments on the method

1. Normal human blood contains a small amount of substances which give a positive Bratton–Marshall reaction (0.1–0.5 mg per 100 ml). In

monitoring by determining the blood concentration, the concentration at time zero (prior to administration) is taken as the sample blank and the value is subtracted from the results of analysis following the administration of the drug. The concentration of these p-aminobenzene derivatives is relatively constant in each individual and shows only slight daily variations.

2. For the quantification of amino-substituted derivatives, e.g. acetylated sulphonamides, the amino group must first be freed. With the sulphonamides this is accomplished by first heating 0.1 ml of supernatant with 0.1 ml of 5 mol l^{-1} hydrochloric acid (dilute concentrated hydrochloric acid 1:1 with distilled water) at 100 °C for 30 min. A 0.1-ml volume of this solution is added to 0.2 ml of sodium nitrite solution and the determination continued as described above. The result is finally multiplied by 2. The period of hydrolysis and the concentration of the hydrochloric acid are critical stages prior to the Bratton–Marshall reaction.[2]

3. A standard is analysed only as a control on the method. The calculation of the results can be made with the aid of the experimentally determined extinction coefficients.

4. If no colour develops the nitrite solution should be prepared afresh. If a colour develops but fades rapidly, fresh ammonium sulphamate solution should be prepared.

References

1. Annino, J. S., Sulfonamides, *Stand. Meth. clin. Chem.*, **3**, 200 (1961).
2. Brown, N. D., Lofberg, R. T., and Gibson, T. P., A study of the Bratton and Marshall hydrolysis procedure utilizing high performance liquid chromatography, *Clin. chim. Acta*, **70**, 239 (1976).

3.14.2.6. p-Aminohippurate: Bratton–Marshall reaction

3.14.2.6.1. Principle

Blood must be deproteinated before the Bratton–Marshall reaction. This is not necessary for urine (clearance technique, see p. 673).

3.14.2.6.2. Reagents (see p. 469)

p-Aminohippurate standard, 1 mg per 100 ml. Dissolve 0.90 mg of free p-aminohippuric acid (do not use the sodium salt as it is very hygroscopic) in trichloroacetic acid $(0.612 \text{ mol } l^{-1})$ and make up to 100 ml with further trichloroacetic acid (corresponds to the administration of 1 mg of sodium salt in the clearance technique). The solution keeps indefinitely at room temperature.

3.14.2.6.3. Deproteination

(*a*) *Plasma, serum.* Add exactly 0.1 ml to 0.5 ml of trichloroacetic acid. Mix well, allow to stand for 5 min, and centrifuge strongly.

(b) *Urine*. Dilute 0.02 ml of urine with exactly 1 ml of trichloroacetic acid.

3.14.2.6.4. Procedure

	Sample, ml	Reagent blank, ml	Standard, ml
Sodium nitrite	0.2	0.2	0.2
Supernatant, urine	0.1	—	—
Trichloroacetic acid	—	0.1	—
PAH standard	—	—	0.1
Ammonium sulphamate	0.2	0.2	0.2
N-(1-Naphthyl)ethylenediamine	0.5	0.5	0.5

After the addition of each reagent, mix well and wait 2 min. The development of the colour is complete after 10 min and its intensity remains stable for at least 1 h. Read the extinction against water between 540 and 560 (546) nm.

(a) *Plasma, serum*.

$$\text{Concentration} = \frac{E(S) - E(RB)}{E(ST) - E(RB)} \cdot C(ST) \cdot \frac{FV(\text{deprot.})}{SV(\text{deprot.})}$$

$$= \frac{E(S) - E(RB)}{E(ST) - E(RB)} \cdot 1 \cdot \frac{0.6}{0.1} \text{ mg per 100 ml}$$

$$\text{Concentration} = \frac{E(S) - E(RB)}{E(ST) - E(RL)} \cdot 6 \text{ mg per 100 ml}$$

(b) *Urine*.

$$\text{Concentration} = \frac{E(S) - E(RB)}{E(ST) - E(RB)} \cdot C(ST) \cdot \frac{FV(V)}{SV(V)}$$

$$= \frac{E(S) - E(RB)}{E(ST) - E(RB)} \cdot 1 \cdot \frac{1.02}{0.02} \text{ mg per 100 ml}$$

$$\text{Concentration} = \frac{E(S) - E(RB)}{E(ST) - E(RB)} \cdot 51 \text{ mg per 100 ml}$$

3.14.2.6.5. Comments on the method

1. The blood and urine samples (S_0 and U_0) collected at time zero serve as analysis blanks. These values are subtracted from the analysis results.

2. Since *p*-aminohippurate penetrates the erythrocytes only poorly, for

clearance studies the concentration should be determined in the plasma or serum, not in whole blood.

3. A standard is analysed only as a control on the method. The results may be calculated by means of a table or using factors.

4. If 'odd results' are obtained it is important to find out if the patient has received some other drugs which give a positive Bratton–Marshall reaction.

5. If no colour develops, the nitrite solution should be prepared afresh. If a colour develops but fades rapidly, fresh ammonium sulphamate solution should be prepared.

3.14.3. SALICYLATE: TRINDER'S METHOD

3.14.3.1. Choice of method

Two reactions are available of the determination of salicylate in biological fluids:

1. Salicylates give a blue colour with Folin reagent in strongly alkaline solution.
2. In weakly acidic solution salicylates react with iron(II) salts to produce a purple–red substance which is attributable to the formation of a chelate between the iron(III) ions and the phenol residue.

The disadvantage of the first group of methods is that the measurements include other substances present in the body fluids of healthy persons (i.e. without intake of salicylates), which also give a positive reaction. This 'sample blank' can amount to as much as 8 mg per 100 ml. This source of error is much less pronounced with the second reaction.

3.14.3.2. Principle

If the reagent described by Trinder[1] is used, the deproteination and colour reaction are carried out simultaneously. The modifications proposed by Hanok[2] allow the determination to be made on 50 μl of blood or serum.

3.14.3.3. Reagents

Modified Tinder reagent (highly poisonous!). Dissolve 40 g of mercury(II) chloride (HgCl$_2$, corrosive sublimate, poisonous) in 850 ml of hot DM-water. Add exactly 120 ml of hydrochloric acid (1 mol l^{-1}) and 40 g of iron(III) nitrate monohydrate [Fe(NO$_3$)$_3$.H$_2$O] and make up to 1000 ml. The solution keeps indefinitely at room temperature.

Salicylate standard, 25 mg per 100 ml. Dissolve 29 mg of sodium salicylate in DM-water and make up to 100 ml. Add 0.5 ml of chloroform.

3.14.3.4. Procedure

	Sample, ml	Reagent blank, ml	Standard, ml
Trinder reagent	1.0	1.0	1.0
Blood, serum, plasma, urine	0.05	—	—
DM-water	—	0.05	—
Standard	—	—	0.05

Mix well, allow to stand for 5 min, then centrifuge strongly. The development of the colour is complete after 20 min. Read the extinction against water at 546 nm. The intensity of the colour remains constant for at least 60 min.

3.14.3.5. Calculation

$$\text{Concentration} = \frac{E(S) - E(RB)}{E(ST) - E(RB)} \cdot 25 \text{ mg per 100 ml}$$

3.14.3.6. Specificity

The reaction is relatively specific. The nature of the endogenous chromogen has not yet been elucidated. Up to now, the only metabolite known to give a positive reaction is acetoacetate.

3.14.3.7. Reference values

With 'normal' blood, serum, plasma, cerebrospinal fluid, and urine, apparent salicylate concentrations of up to 2 mg per 100 ml are obtained.

3.14.3.8. Comments on the method

1. For scientific investigations it is advisable to extract the salicylate with ethylene dichloride.[3] After the extraction the sample blank is zero.

2. The therapeutic concentrations lie between 20 and 50 mg per 100 ml (depending on the indication). In cases of poisoning values of up to 120 mg per 100 ml can be observed.[4]

3. In the organism, acetylsalicylate (aspirin) is rapidly cleaved to salicylate and acetic acid by the action of esterases.

References

1. Trinder, P., Rapid determination of salicylate in biological fluids, *Biochem. J.*, **57**, 301 (1954).

2. Hanok, A., The ultramicro determination of salicylates in biologic fluids, *Clin. Chem.*, **8**, 400 (1962).
3. Routh, J. I. and Dryer, R. L., Salicylate, *Stand. Meth. clin. Chem.*, **3**, 194 (1961).
4. Keller, W. J., A rapid method for the determination of salicylates in serum or plasma, *Am. J. clin. Path.*, **17**, 415 (1947).

3.14.4. PHENACETIN

3.14.4.1. Introduction

Phenacetin is used as an analgesic and antipyretic. Paracetamol (Fig. 120) belongs to the same group.

Following ingestion, phenacetin is rapidly deacetylated to paracetamol, which is then conjointly excreted in the urine.

3.14.4.2. Choice of Method

Of the two most commonly used methods for the determination of phenacetin derivatives in urine, that of Brodie and Axelrod[1] is based on extraction of the derivative and hydrolysis to p-aminophenol, which is then diazotized and coupled with α-naphthol to produce a red–violet dye. The method described below works without extraction. Following hydrolysis, the p-aminophenol is converted to an indophenol dye.[2]

3.14.4.3. Principle

Phenacetin, paracetamol, and their corresponding conjugates are hydrolysed to p-aminophenol by hydrochloric acid. The product reacts with hypobromite and phenol to give a blue indophenol dye.

The extinction is measured at 620 nm.[2-4] A similar reaction is used in the determination of urea.

3.14.4.4. Reagents

Hydrochloric acid, 4 mol l^{-1}. Dilute 33.2 ml of concentrated HCl to 100 ml with DM-water.

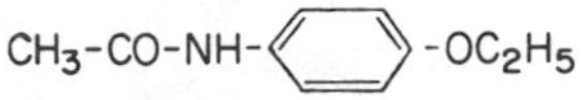

CH_3-CO-NH-⟨⟩-OC_2H_5

Phenacetin
(p -ethoxyacetanilide)

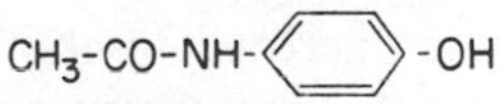

CH_3-CO-NH-⟨⟩-OH

Paracetamol
(p -hydroxyacetanilide)

Fig. 120. Phenacentin and paracetamol.

Sodium hydroxide solution, $0.2\,mol\,l^{-1}$. Dissolve 8 g of NaOH in DM-water and make up to 1000 ml.

Saturated bromine solution. Sufficient bromine to provide a marked excess is shaken for a long period with 100 ml of DM-water and allowed to stand for 24 h. The saturated supernatant is used as the reagent. The solution keeps indefinitely in a refrigerator.

Sodium carbonate–sodium bromide solution. Dissolve 10.6 g of anhydrous Na_2CO_3 (p.a. grade, Merck) in DM-water and make up to 100 ml. Add 15 ml of saturated bromine solution to 100 ml of the sodium carbonate solution. Prepare freshly each day.

Phenol solution, 1%. Dilute 1 ml of phenolum liquidum (88%, p.a.) with DM-water to make 88 ml. Prepare freshly each day.

Reaction mixture. Mix 80 ml of $0.2\,mol\,l^{-1}$ NaOH with 10 ml of 1% phenol solution and 10 ml of sodium carbonate–sodium bromide solution. Prepare freshly for each assay.

Standard solutions. Paracetamol solutions of e.g. 20, 40, 60, and 80 mg in 100 ml of DM-water are prepared for making a calibration graph.

3.14.4.5. Measurement, Procedure

A 1-ml volume of urine or DM-water is pipetted into a graduated test-tube, 4 ml of HCl ($4\,mol\,l^{-1}$) and 1 ml of DM-water are added and the solutions mixed. The tubes are covered (e.g. with glass sleeves), the contents hydrolysed for 60 min in a boiling water-bath, cooled, made up to 10 ml with DM-water and mixed well. In each case, 1 ml of the diluted hydrolysate or blank assay is treated with 10 ml of reaction mixture, mixed, and allowed to stand for 1 h at room temperature. In the presence of *p*-aminophenol a blue colour develops. The extinction is read against the blank at 620 nm.

3.14.4.6. Calculation

Two standards of, e.g. 20 and 40 mg per 100 ml, may be run parallel with each assay. The calculation is made by means of a calibration graph prepared in the laboratory, paying due regard to values for the standards.

3.14.4.7. Specificity

As a derivative of aniline and nitrobenzene, *p*-aminophenol can give a positive reaction.[3]

3.14.4.8. Reference values

About 75% of the administered dose of phenacetin is excreted in the urine as conjugated paracetamol in the first 24 h.[2,4] Normal urine shows a non-specific extinction of up to 0.056,[4] which corresponds to roughly 5 mg per 100 ml of *p*-aminophenol. Extinctions above this value are considered as indicative of phenacetin intake.

3.14.4.9. Comments on the method

1. The reagents will keep for at least 4 weeks at 4 °C.
2. Allowing the urine to stand at room temperature for 24 h or keeping it frozen for 4 weeks has no effect on the results.[4]
3. The method, preferably following extraction, can also be used for the determination in serum.[5]

References

1. Brodie, B. B. and Axelrod, J., The estimation of acetanilide and its metabolic products, aniline, *N*-acetyl-*p*-aminophenol and *p*-aminophenol (free and total conjugated) in biological fluids and tissues, *J. Pharmac. exp. Ther.*, **94**, 22 (1948).
2. Welch, R. M. and Conney, A. H., A simple method for the quantitative determination of *N*-acetyl-*p*-aminophenol (APAP) in urine, *Clin. Chem.*, **11**, 1064 (1965).
3. Geldmacher-von Mallinckrodt, M., *Einfache Untersuchungen auf Gifte im klinisch–chemischen Laboratorium*, Thieme, Stuttgart, 1976.
4. Dubach, U. C., *p*-Aminophenol-Bestimmung im Urin als Routinemethode zur Erfassung der Phenacetineinnahme, *Dt. med. Wscher.*, **92**, 211 (1967).
5. Tompsett, S. L., The detection and determination of phenacetin and *N*-acetyl *p*-aminophenol (paracetamol) in blood serum and urine, *Ann. clin. Biochem.*, **6**, 81 (1969).

3.14.5. BARBITURATES: ULTRAVIOLET SPECTROPHOTOMETRY

3.14.5.1. Choice of method

In the clinical chemistry laboratory barbiturates were formerly most often quantified by the method of Zwikker in one of its modifications. The barbiturates were first extracted by the Stas–Otto procedure and then subjected to a colour reaction utilizing the formation of a blue complex between cobalt salts and the malonylurea ring in isopropylamine solution. However, this method has numerous disadvantages: it is non-specific, time consuming and gives poorly reproducible results. The ring structure of the 5,5-substituted barbiturates leads to an intensive absorption in the ultraviolet. Since these compounds are present as lactams in acidic solution and as lactims under alkaline conditions (Fig. 121), the ultraviolet spectrum is pH dependent. Walker *et al.* were the first to suggest using the difference in the extinction at pH 10 and pH 2 as a measure for the barbiturate concentration. Since the typical barbituric acid spectrum with the peak at 240 nm disappears at pH 2 but the non-specific plasma absorption remains, this technique eliminates a series of interfering substances. That this method

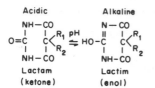

Acidic   Alkaline

Lactam   Lactim
(ketone)   (enol)

Fig. 121. The lactim and lactam forms of barbiturate derivatives.

is not restricted to pure barbiturate solutions but may also be applied to body fluids was first demonstrated by Lous. Dybing improved the extraction procedure and Zak suggested recording the spectrum directly with the double-beam spectrophotometer. The method described below has been modified for the analysis of small quantities of serum (see ref. 1 for the literature on this subject).

3.14.5.2. Principle

The body fluids are extracted with chloroform at pH 7.4. The barbiturates from the chloroform extract are taken up in sodium hydroxide solution and the absorption spectrum is recorded. The pH is subsequently reduced to 2 by the addition of hydrochloric acid and the spectrum plotted again. The barbituric acid concentration is calculated by means of the extinction coefficients but a standard may be employed if necessary. The differences between the extinction coefficients of individual barbiturates are significant (Table 82). Owing to their position in the middle of the range, cyclohexenylethyl- and allylisobutylbarbituric acids are especially suitable as reference barbiturates.

3.14.5.3. Reagents

Chloroform, p.a. grade.
Borate buffer, pH 8.0. Dissolve 6.2 g of boric acid in 0.1 mol l^{-1}

Table 82. Maxima of the extinction differences and percentage extinction coefficients of some common barbituric acid derivatives.

Systemic name	Common name	Maximum, nm	Percentage extinction coefficient (240 nm)
Diethylbarbituric acid	Barbital	339	466
Allylisopropylbarbituric acid	Aprobarbital	240.5	416
Diallylbarbituric acid	Allobarbital	240.5	400
Allylisobutylbarbituric acid	Allylbarbital	240	393
Cyclohexenylethylbarbituric acid	Cyclobarbital	240	390
Cycloheptenylethylbarbituric acid	Heptobarbital	239.5	383
Phenylethylbarbituric acid	Phenobarbital	241	360
Ethylmethylbutylbarbituric acid	Pentobarbital	240	400
Cyclohexenyldimethylbarbituric acid	Hexobarbital	245	270 (245 nm)

potassium chloride solution, make up to 1000 ml with further potassium chloride solution, and add 100 ml 0.1 of mol l^{-1} NaOH (adjust the pH to 8.0). The solution keeps indefinitely if frozen.

Phosphate buffer, pH 7.4. See appendix 1.

Sodium hydroxide solution, 0.1 *mol* l^{-1}.

Hydrochloric acid, 10 *mol* l^{-1}.

Barbiturate standard, 10 *mg per* 100 *ml.* Dissolve 10 mg of allylbarbital or cyclobarbital in DM-water and make up to 100 ml. If frozen, the solution may be kept for several weeks.

3.14.5.4. Extraction

1. Aliquots of 0.2 ml of plasma, serum, urine, or gastric juices and standard solution are treated with 0.3 ml of phosphate buffer (pH 7.4).

2. The solutions are extracted twice in a separating funnel using 30 ml of chloroform each time and shaking for 3 min. The chloroform extracts are combined.

3. The chloroform extracts are extracted for 3 min with 4 ml of 0.1 mol^{-1} sodium hydroxide solution in a separating funnel.

4. The sodium hydroxide solution is centrifuged strongly.

5. A 3-ml volume of the sodium hydroxide solution is treated with 7 ml of borate buffer. The pH must be 10.

3.14.5.5. Measurement

Single-beam photometer. About 3 ml of the buffered solution are transferred to a quartz cuvette and the extinction is measured at 225, 230, 235, 240, 245, 250, 255, and 260 nm. Then 1 drop of 10 mol l^{-1} of hydrochloric acid is added, which should bring the pH to 2. The extinctions are measured again at the same wavelengths used for pH 10. The difference in extinction between the two curves is plotted as a function of wavelength.

Double-beam spectrophotometer. About 3 ml of the pH 10 solution are placed in the quartz cuvette illuminated by the reference beam and a further 3 ml plus 1 drop of 10 mol l^{-1} hydrochloric acid are placed in the quartz cuvette to be illuminated by the measurement beam. The differential spectrum is recorded directly or plotted manually.

3.14.5.6. Interpretation

First the curve is analysed. The difference spectrum for barbiturates shows a peak at 240 nm and, at alkaline pH, a peak between 250 and 260 nm. Hexobarbital is an exception, its difference maximum being at 245 nm. Sulphonamides constitute the interference of greatest practical importance, but most of these show a second maximum at 260 nm in addition to the typical barbiturate maximum. The main conclusion to be derived from the presence of multiple peaks is the simultaneous presence of different substances.

3.14.5.7. Calculation

1. The percentage extinction coefficient of allylisobutylbarbituric acid ($a = 393$) is taken as average for the barbiturates. A few other extinction coefficients are listed in Table 82.

2. Concentration $= \dfrac{[E_{240}(\text{pH } 10) - E_{240}(\text{pH } 2)]}{ad} \cdot \dfrac{FV}{SV} \cdot \dfrac{FV'}{SV'}$

$$\cdot \, 10^3 \text{ mg per 100 ml}$$

$$= \dfrac{[E_{240}(\text{pH})10) - E_{240}(\text{pH } 2)]}{393 \cdot 1} \cdot \dfrac{10}{0.2} \cdot \dfrac{4}{3}$$

$$\cdot \, 10^3 \text{ mg per 100 ml}$$

Concentration $= [E_{240}(\text{pH } 10) - E_{240}(\text{pH } 2)] \cdot 169 \text{ mg per 100 ml}$

3.14.5.8. Specificity

The method is not specific for barbiturates since other pharmaceuticals can show a similar absorption spectrum. In cases of doubt, attempts are made to identify the compounds present by means of other spectral properties (ultraviolet and infrared).[2] In practice, the sulphonamides are the chief source of interference but these may easily be determined, quantitatively and selectively, by means of the Bratton–Marshall reaction. No interference is found with morphine derivatives, digitalis derivatives, a barbiturate-free soporific ($\alpha\alpha$-phenylethylglutaric acid imide), and antibiotics (penicillin, streptomycin, oxytetracyclin).[1]

3.14.5.9. Reference values

Normally, the 'barbiturates' level in plasma is 0.1 mg per 100 ml at the highest, but in cases of poisoning the values lie between 5 and 40 mg per 100 ml.

3.14.5.10. Comments on the method

1. The spectrum of a pure barbiturate solution is not identical with that of an extract from biological fluids. Consequently, it is hardly possible to identify the barbiturate from the spectral curve as has been suggested by various researchers. For identification, other methods, particularly thin-layer chromatography, must be employed.

2. In order to extract at least 95% of the barbiturates the ratio of the aqueous to the chloroform phase needs to be 1:50.

3. Using a very similar method, Schumann et al.[3] found a coefficient of variation in the series of 5.4% and a sensitivity in patients' sera of 0.33 mg per 100 ml. The method was linear at concentrations up to 10.5 mg per 100 ml. In any case, it is advisable to dilute samples with concentrations greater than 6 mg per 100 ml. Such plasma concentrations are generally toxic and lead to a comatose condition. Agreement with gas-chromatographic methods and the EMIT system was good. Glutethimide and meprobamate produced no interference.

References

1. Richterich, R., Die quantitative Bestimmung von Barbitursäure-Derivaten in Körperflüssigkeiten durch Ultraviolett-Spektrophotometrie, *Clini. chim. Acta,* **3**, 183 (1958).
2. Sunshine, I., Barbiturates, *Stand. Meth. clin. Chem.,* **3**, 46 (1961).
3. Schumann, G. B., Lauenstein, K., Le Fever, D., and Henry, J. B., Ultraviolet spectrophotometric analysis of barbiturates, *Am. J. clin. Path.,* **66**, 823 (1976).

4. ORGAN-SPECIFIC INVESTIGATIONS

4.1. INTRODUCTION

Many of the investigations carried out in the clinical chemistry laboratory are directed towards:

1. *locating disease processes* in particular organs (organ diagnostics) or cells (cell diagnostics); and
2. *testing the performance* of a particular organ, the so-called function testing.

All of these organ-specific investigations have one thing in common: the result as it stands says nothing about the aetiology or the pathological substrate of a morbid process, but merely establishes that particular biochemical symptoms or syndromes are present in particular organs. The following three criteria are of importance in assessing the practical value of such tests:

1. The *organ specificity*. The more specific a test for diseases of an individual organ or cell type, the greater is its diagnostic value. An example of a relatively organ-specific test is provided by the determination of the plasma creatine kinase which shows elevated plasma concentrations in particular diseases of the cardiac and skeletal muscle.
2. The *sensitivity*. An ideal test always gives normal results for healthy subjects but a pathological value in 100% of the cases when particular diseases are present. One example is glutamate–pyruvate transaminase, which is always present in abnormally high concentrations in anicteric, icteric, or cholangiolitic hepatitis, while positive results are a rarity for healthy subjects or for patients with non-hepatic diseases.
3. The *selectivity*. Where possible, a test should also allow the detection of particular pathogenetic syndromes or provide decisions in certain problems of differential diagnosis, as, for example, in the differentiation between obstructive and parenchymal icterus in diseases of the liver. The more selectively a test responds to one only of the two groups of diseases, the greater is its practical diagnostic importance.

Where there is a choice, preference is given to those methods of investigation for which as many as possible of these criteria are fulfilled.

485

Table 83. Organ specificity of 'indicator enzymes'.

Organ	Enzyme	Specificity
Pancreas	Lipase	+++
	Amylase	++
	Glutamate–oxalacetate transaminase	+
	Leucine aminopeptidase	+
Salivary glands	Amylase	++
Bones	Alkaline phosphatase	++
Prostate gland	Acid phosphatase	+++
Cardiac muscle	2-Hydroxybutyrate dehydrogenase	++
	Creatine kinase	++
	Total lactate dehydrogenase	+
	Glutamate–oxaloacetate transaminase	+
Skeletal muscle	Creatine kinase	++
	Aldolase	+
	Total lactate dehydrogenase	+
	Glutamate–oxaloacetate transaminase	+
	Glutamate–pyruvate transaminase	+
Liver	Ornithine–carbamyl transferase	+++
	Cholinesterase	+++
	γ-Glutamyltranspeptidase	+++
	Glutamate–pyruvate transaminase	++
	Glutamate–oxaloacetate transaminase	+
	Total lactate dehydrogenase	+
	Aldolase	+
Bile duct	Leucine aminopeptidase	+++
	γ-Glutamyltranspeptidase	+++
	5'-Nucleotidase	+++
	Alkaline phosphatase	++
	Caeruloplasmin	++
Stomach	Pepsinogen	+++
Erythrocytes	Total acid phosphatase	++
	Total lactate dehydrogenase	+
	Glutamate–oxaloacetate transaminase	+

In recent years the detection of organ-specific lesions has relied more and more frequently upon enzymes which pass from the cells into the plasma in the course of the disease. The most important of these 'indicator enzymes' are collected in Table 83, which also shows their specificity in respect of organ diagnostics.

4.2. BONES

4.2.1. ALKALINE PHOSPHATASE

4.2.1.1. Introduction

The total activity of alkaline phosphatase in serum (orthophosphoric monoester phosphohydrolase, alkaline optimum, E.C. 3.1.3.1) includes various isoenzymes of low substrate specificity which hydrolyse various phosphate esters at alkaline pH. In man, alkaline phosphatase has been detected in the bones (osteoblasts), in the intestinal and renal-tubule epithelium, in the bile-duct epithelium of the liver, in the placenta, and in the white blood cells.[1] The alkaline serum phosphatase can be separated into its component isoenzymes by physical, chemical, immunological, and electrophoretic methods.[2] The methods most frequently used are thermal inactivation, inhibition by L-phenylalanine, and electrophoresis with various support media, more recently polyacrylamide gel.[3-5] The serum of normal individuals generally contains isoenzymes from the liver, bones, intestines, and placenta. The placenta phosphatase appears in the mother's circulation during the first 3 months of pregnancy and is still detectable up to the 12th week following birth.[6] It shows pronounced genetic polymorphism and electrophoresis reveals various zones differing in their mobility.[7] The electrophoretic separation of normal sera shows that the greater part of the measurable activity is attributable to the isoenzymes of the liver. In children the isoenzyme of the bones predominates;[8] the liver isoenzyme begins to develop after the first trimenon and is concluded by the end of the first year of life.[12] In individuals with blood groups O and A the intestinal isoenzyme can also frequently be detected. The activity of this isoenzyme can be elevated by assimilation of fat.[9] A further isoenzyme of the serum alkaline phosphatase, the so-called Regan enzyme,[10] which shows a biochemical and immunological identity with the placenta phosphatase, can occur in cases of various malignant tumours. It is regarded as belonging to the group of tumour-associated antigens like the carcinoembrionic antigen and the α-fetoprotein.[11] Although there are many isoenzymes of alkaline phosphatase their physiological function in the organism has not been elucidated. Their natural substrate is also unknown. Their localization at the cell interfaces concerned with active transport processes leads one to guess that they may be connected with this function.

Table 84. Customary methods for the determination of alkaline serum phosphatase; the last column gives the calculated conversion factors.

Method	Incubation conditions					Definition of units	Factors for converting to international units $(U\ l^{-1})$ $\mu mol\ min^{-1}\ l^{-1}$
	Amount of serum, ml	Buffer	pH	Duration, min	Temperature, °C		
Incubation with α-glycerophosphate: determination of the liberated phosphate							
Method of Bodansky[15] and Aebi[16]	0.5	Sodium diethyl-barbiturate	9.3	60	37	mg phosphorus per 60 min per 100 ml of serum	1 Bodansky unit = 5.35 U l^{-1}
Incubation with phenylphosphate: determination of the liberated phenol							
Method of King and Armstrong,[17] Kirchberger and Martini[18]	0.2	Sodium diethyl-barbiturate	9.0	15	37	mg phenol per 15 min per 100 ml of serum	1 King–Armstrong unit = 7.1 U l^{-1}

Incubation with phenolphthalein phosphate: determination of the liberated phenolphthalein

Method		Buffer					
Method of Huggins and Talalay,[19] Linhardt and Walter[20]	0.5	Sodium diethyl-barbiturate	9.7	120	37	0.1 mg of phenolphthalein per 60 min per 100 ml of serum	1 Huggins–Talalay unit = 4.8 U l^{-1}

Incubation with p-nitrophenylphosphate: determination of the liberated p-nitrophenol

Method		Buffer					
Method of Bessey et al.,[14] Sommer[21]	0.2	Glycocoll	10.5	30	37	mmol h^{-1} l^{-1} serum	1 Sommer unit = 16.7 U l^{-1}
Richterich and Gautier[22]	0.02	Ammediol	10.0	30	37	μmol min^{-1} l^{-1} serum	= U l^{-1}
Deutsche Gesellschaft für klinische Chemie[23]	0.05	Diethanol-amine	9.8	<3	25	μmol min^{-1} l^{-1}	
Scandinavian Society for Clinical Chemistry and Clinical Physiology[24]	0.01	Diethanol-amine	9.8	<3	37	μmol min^{-1} l^{-1}	

490

4.2.1.2. Choice of method

A large number of methods are available for the determination of alkaline serum phosphatase. All of these techniques provide results of comparable diagnostic value, but the work involved varies enormously.

The experimental conditions for the most common methods are listed in Table 84, which clearly shows the superiority of those methods which use *p*-nitrophenylphosphate as substrate.

4.2.1.3. Principle

p-Nitrophenylphosphate was first used for the determination of phosphatases by Ohmori,[13] but the dissemination of the method was hindered by the lack of sufficiently pure preparations. In 1946 Bessey *et al.*[14] used *p*-nitrophenylphosphate as substrate for the determination of the alkaline serum phosphatase in 5 μl of serum. Their method has since been subjected to various modifications. Although *p*-nitrophenylphosphate absorbs in the ultraviolet region it is colourless in the visible range. Under the action of the enzyme the substrate is cleaved to phosphate and *p*-nitrophenol, taking up 1 molecule of water in the process (Fig. 122). The reaction product, *p*-nitrophenol, is an indicator and has an intense yellow colour in alkaline solution. The kinetic measurement at 405 nm has proved a particularly rapid and accurate technique.

4.2.1.4. Measurement

Conditions. These are given in Table 85.

4.2.1.5. Calculation

Calculation of the volume activity in μmol min^{-1} l^{-1} = U l^{-1} (p. 234). Molar extinction coefficient for *p*-nitrophenol: $\varepsilon = 18\,600$ l mol^{-1} cm^{-1}, pH 10, 25 °C.[24]

4.2.1.6. Reference values

The level of alkaline phosphatase is known to be strongly age dependent. In childhood the values are higher than those for adults by a

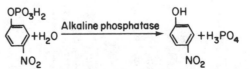

Fig. 122. Scission of *p*-nitrophenyl phosphate by alkaline phosphatase.

Table 85. Determination of alkaline phosphatase; final concentrations in the reaction mixture.

Deutsche Gesellschaft für klinische Chemie[23]	Scandinavian Society for Clinical Chemistry and Clinical Physiology[24]
Diethanolamine buffer (pH 9.8)	1 mol l^{-1}
$MgCl_2$	0.5 mmol l^{-1}
p-Nitrophenylphosphate	10 mmol l^{-1}
Volume fraction of the serum 1:61	1:111
Temperature of measurement 25 °C	37 °C
Wavelength	405 nm, $d = 1$ cm
Duration of measurement ΔE is recorded for 1–3 min	ΔE is recorded for at least 30 s

factor of about 3.[22,25,26] The activity increases in the 2nd–3rd months and falls only very slightly up to puberty. After puberty, once the enchondral bone growth has ceased, adult levels are attained; as a rule these are lower (Tables 86 and 87).

Table 86. Reference values for alkaline phosphatase in serum, measured according to the recommendations of the Deutschen Gesellschaft für klinische Chemie[23] (U l^{-1}, 25 °C).

Age	Sex	Number	Reference interval, 2.5–97.5%	References
3–14 years	f	26	149–473	25
3–15 years	m	37	152–468	
Adults	m		70–180	27
	f		50–140	
Adults	m	708	Up to 180	28
	f	668	Up to 170	
Adults	m	332	50–190	29
	f	363	50–190	
Adults	m	4790	70–175	30
Up to 50 years	f		55–147	
Over 50 years			60–170	
Children				
Umbilical cord blood		15	69–308	22[a]
2nd and 3rd month		17	126–490	
4th–6th month		14	107–509	
7th–12th month		15	107–440	
2nd–15th year		142	119–433	
Adults				
20–50 year old		100	41–141	
Pregnant women				
(final trimester)		16	88–364	

[a]Corrected with a temperature factor.

Table 87. Reference values for alkaline phosphatase in serum; determination according to the recommendations of the Committee on Enzymes of the Scandinavian Society of Clinical Chemistry and Clinical Physiology[24] $(U \ l^{-1}, 37 \ °C)$.

Age	Sex	Reference interval, 2.5–97.5%	References
Newly born		150–600	31
6th–24th month		250–1000	
2–5 years		250–850	
6–7 years		250–1000	
8–9 years		250–750	
10–11 years	m	250–750	
	f	250–950	
12–13 years	m	275–875	
	f	200–730	
14–15 years	m	170–970	
	f	170–460	
16–18 years	m	125–720	
	f	75–270	
18 years	m	60–250	
	f	50–200	

4.2.1.7. Comments on the method

1. Since human serum contains various isoenzymes of the alkaline phosphatase, each with their individual reaction conditions, it is not possible to establish a set of optimal conditions valid for every serum. Experience has shown that the exact stipulation of the measurement conditions is more important than their optimal position.

2. Alkaline phosphatase is relatively stable. The serum may be stored for up to 4 h at room temperature. Deep-freezing of the serum is not recommended since the activity increases continually after thawing out.[32]

3. Running a serum blank is necessary only with markedly icteric sera. Haemolytic sera should only be used with reservations since phosphomonoesterases may be liberated from the erythrocytes.

4. Heparin can be used as an anticoagulant but EDTA inhibits the reaction. Serum and plasma give the same results.

5. The value of the extinction of the substrate reagent before the reaction begins may be used as a quality control on the substrate. If the extinction, measured against water, is higher than 0.100, fresh p-nitrophenylphosphate solution should be used.

6. Since inorganic phosphate inhibits alkaline phosphatase the substrate p-nitrophenylphosphate should contain less than $1 \ mol \ l^{-1}$ of phosphate.[32]

7. The alkaline buffer takes up CO_2 from the air. Bottles should be kept well sealed.

8. The enzyme activity changes according to the buffer used. The

highest activity is measured with 2-ethylaminoethanol and diethanolamine;[33] this is attributable to a transphosphorylation in which the buffer functions as the phosphate acceptor. The diethanolamine must be free of ethanolamine since the latter inhibits the reaction.[24]

9. The molar extinction coefficient for p-nitrophenol in diethanolamine buffer (1 mol^{-1}, pH 10) at 25 °C is 18 600 l mol^{-1} cm^{-1};[24] Walter and Schütt[34] give the value as 18 500 l mol^{-1} cm^{-1}.

10. The intra-individual stability of alkaline phosphatase is remarkable. Measurements for the same individual spanning 5 years gave activities in the range of 42–51 U l^{-1}.[32]

4.2.1.8. Diagnostic significance

The level of alkaline phosphatase is of significance for the diagnosis of diseases of the bones and the liver.[35,37]

Hepatopathia. With the majority of patients suffering from obstructive jaundice the enzyme concentration in the plasma is elevated, whereas no such increase is observed with simple parenchymal jaundice. Particularly striking elevations, often accompanied by a normal bilirubin concentration, may be observed in cases of intrahepatic, and frequently local, obstructions. If there is jaundice, a distinction between parenchymal lesions and obstruction can usually be made by the simultaneous determination of alkaline phosphatase and transaminases. High transaminase levels, accompanied by normal or slightly elevated phosphatase values, indicate the presence of a parenchymal jaundice, and obstructive jaundice is indicated by a highly abnormal phosphatase value at slightly elevated transaminase levels. Difficulties may be encountered if the hepatitis is cholangiolitic, when both the transaminase values and those of alkaline phosphatase are abnormally high. In haemolytic jaundice the alkaline phosphatase is normal. If there is both a bone affection (e.g. osteomalacia, osteoblastic metastases) and a liver affection (e.g. metastasis), then additional liver-specific enzymes such as γ-glutamyl transpeptidase or leucine aminopeptidase and 5'-nucleotidase must be determined in order to differentiate between the conditions. This largely dispenses with the necessity to determine the alkaline phosphatase isoenzymes. With children, it should be rememberd that there is always a physiological high enzyme activity during puberty.[36]

If the biliary tract is blocked, the elevated level of alkaline phosphatase in the serum is not due to a retention of the serum phosphatases excreted through the bile, as is often assumed, but is caused by the efflux of liver phosphatase into the extracellular space.[1] It is conjectured that the stoppage of the biliary tract leads to an increased activity of the alkaline phosphatase due to an increased synthesis of the enzyme within the liver. If this synthesis is inhibited, no alkaline phosphatase is found in the serum, despite the stoppage of the biliary tract.

494

Table 88. Behaviour of alkaline phosphatase in diseases of the bones.

Elevated or normal phosphatase levels	
Hypo- and hypervitaminosis	
Vitamin D deficiency or resistance	+−+++
Osteomalacia, rickets	
Steatorrhoea	
Renal tubular syndromes	
Franconi syndrome	
Renal acidosis	
Endocrinopathia	
Primary or secondary hyperparathyroidism	n−+++
Hyperthyrosis	n−+
Acromegaly	n−+
Bone tumours	
Osteogenic sarcoma	n−++
Osteolytic metastasis syndrome	n−+
Osteoblastic metastasis syndrome	+−+++
Sundry diseases of the bones	
Osteoporosis	n
Paget's disease	+−+++
Ostitis fibrosa generalisata	n−+
Calcinosis universalis	n
Ectopic ossification	n−+
Tuberculosis of the bones	n−+
Sarcoidosis	n−+
Healing fractures	n−+
Hodgkin's disease	n−+
Myeloma	n−+
Lowered phosphatase	
Familial hypophosphatasia	
Hypothyrosis	
C-hypovitaminosis	
Undernourishment	

n = normal; + = slightly elevated; ++ = elevated to a moderate degree; +++ = strongly elevated.

Osteopathia. The behaviour of alkaline phosphatase in diseases of the bones is presented in Table 88.

References

1. Kaplan, M. M., Alkaline phosphatase, *Gastroenterology*, **62**, 452 (1972).
2. Fishman, W. H., Perspectives on alkaline phosphatase isoenzymes, *Am. J. Med.*, **56**, 617 (1974).
3. Tschanz, C., Dauwalder, H., and Colombo, J. P., Plasma alkaline phosphatase. Reference values for total activity, L-phenylalanine inhibited fraction, dependency on age, sex and blood grouping, *Clin. Biochem.*, **7**, 68 (1974).
4. Dingjan, P. G., Postma, T., and Stroes, J. A. P., Quantitative differentiation of human serum alkaline phosphatase isoenzymes with polyacrylamide disc gel electrophoresis, *Z. klin. Biochem.*, **11**, 167 (1973).

495

5. Balant, L., Fabre, J., Jung, A., and Rosenbusch, C. A., Les isoenzymes de la phosphatase alcaline, *Schweiz. med. Wschr.*, **105**, 601 (1975).
6. Sussmann, H. H., Bowman, M., and Lewis, J. L., Jr., Placental alkaline phosphatase in maternal serum during normal and abnormal pregnancy, *Nature, Lond.*, **218**, 359 (1968).
7. Bottini, E., Lucarelli, P., Pigram, P., Palmarino, R., Spennati, G. F., and Orzalesi, M., Interaction between placental alkaline phosphatase and ABO system polymorphisms during intrauterine life, *Am. J. hum. Genet.*, **24**, 495 (1972).
8. Afonja, A. O. and Baron, D. N., Plasma alkaline phosphatase isoenzymes in hepatobiliary disease, *J. clin. Path.*, **27**, 916 (1974).
9. Warnock, M. L., Intestinal alkaline phosphatase and fat absorption, *Proc. Soc. exp. Biol. Med.*, **129**, 768 (1968).
10. Lehmann, F. G., Regan-Isoenzym der alkalischen Phosphatase im Serum bei malignen Tumoren, *Klin. Wschr.*, **53**, 585 (1975).
11. Belliveau, R. E., Yamamoto, L. A., Wassell, A. R., and Wiernik, P. H., Regan Isoenzyme in patients with hematopoietic tumors, *Am. J. clin. Path.*, **62**, 329 (1974).
12. Hosenfeld, D. and Paulsen, H., Die Isoenzyme der alkalischen Serumphosphatase im Neugeborenen- und Säuglingsalter, *Klin. Pädiat.*, **188**, 55 (1976).
13. Ohmori, Y., Über die Phosphomonoesterase, *Enzymologia*, **4**, 217 (1937).
14. Bessey, O. A., Lowry, O. H., and Brock, M. A., A method for a rapid determination of alkaline phosphatase with five cubic millimeters of serum, *J. biol. Chem.*, **164**, 321 (1946).
15. Bodansky, A., Phosphatase studies. II. Determination of serum phosphatase. Factors influencing the accuracy of the determination, *J. biol. Chem.*, **101**, 93 (1933).
16. Aebi, H., Die Bestimmung der alkalischen Serumphosphatase in Theorie und Praxis, *Schweiz. med. Wschr.*, **82**, 135 (1952).
17. King, E. J. and Armstrong, A. R., A convenient method for determining serum and bile phosphatase activity, *Can. med. Ass. J.*, **31**, 376 (1934).
18. Kirchberger, E. and Martini, G. A., Bestimmungsmethode und klinische Auswertung der Phosphataseaktivität im Blut, *Dt. Arch. klin. Med.*, **197**, 268 (1950).
19. Huggins, C. and Talalay, P., Sodium phenolphthalein phosphate as a substrate for phosphatase tests, *J. biol. Chem.*, **159**, 399 (1945).
20. Linhardt, K. and Walter, K., Vergleichende Versuche und Betrachtungen über die Bestimmung der Serumphosphatase, *Med. Wschr.*, **5**, 22 (1951).
21. Sommer, A. J., The determination of acid and alkaline phosphatases using p-nitrophenyl phosphate as substrate, *Med. Bull. St. Louis Univ.*, **4**, 165 (1952).
22. Richterich, R. and Gautier, E., Ultramikromethoden im klinischen Laboratorium. IV. Bestimmung der alkalischen Serumphosphatase, *Schweiz. med. Wschr.*, **92**, 781 (1962).
23. Empfehlungen der Deutschen Gesellschaft für klinische Chemie. Standardisierung von Methoden zur Bestimmung von Enzymaktivitäten in biologischen Flüssigkeiten, *Z. klin. Chem. klin. Biochem.*, **10**, 182 (1972).
24. Recommended methods for the determination of four enzymes in blood, *Scand. J. clin. Invest.*, **33**, 291 (1974).
25. Szasz, G. and Rautenburg, H. W., Der Normbereich diagnostisch bedeutsamer Serumenzyme im Kindesalter, *Z. Kinderheilk.*, **111**, 233 (1971).
26. Sitzmann, F. C. and Wendler, H., Normalwerte der alkalischen Phosphatase im Serum bei Kindern, *Päd. Prax.*, **13**, 105 (1972).

496

27. Kübler, W., *Symposium über Normwerte*, Mainz, 1973.
28. Thefeld, W., Hoffmeister, H., Busch, E.-W., Koller, P. U., and Vollmar, J., Referenzwerte für die Bestimmungen der Transaminasen GOT und GPT sowie der alkalischen Phosphatase im Serum mit optimierten Standardmethoden, *Dt. med. Wschr.*, **99**, 343 (1974).
29. Schlebusch, H., Rick, W., Lang, H., and Knedel, N., Normbereiche der Aktivitäten klinisch wichtiger Enzyme, *Dt. med. Wschr.*, **99**, 765 (1974).
30. Weisshaar, D., Gossrau, E., and Faderl, B., Normbereiche von α-HBDH, LDH, AP und LAP bei Messung mit substratoptimierten Testansätzen, *Medsche Welt, Stuttg.*, **9**, 387 (1975).
31. Penttilä, I. M., Jokela, H. A., Viitala, A. J., Heikkinen, E., Nummi, S., Pystynen, P., and Saastamoinen, J., Activities of aspartate and alanine aminotransferases and alkaline phosphatase in sera of healthy subjects, *Scand. J. clin. Lab. Invest.*, **35**, 275 (1975).
32. Bowers, G. N., Jr., and McComb, R. B., Measurement of total alkaline phosphatase activity in human serum, *Clin. Chem.*, **21**, 1988 (1977).
33. McComb, R. B. and Bowers, G. N., Jr., Study of optimum buffer conditions for measuring alkaline phosphatase activity in human serum, *Clin. Chem.*, **18**, 97 (1972).
34. Walter, K. and Schütt, Ch., Acid and alkaline phosphatase in serum, in Bergmeyer, *Methods in enzymatic analysis*, Vol. 2, Academic Press, New York, 1974, p. 857.
35. Dubach, U. C., Klinik der Alkalischen Phosphatase, *Praxis*, **51**, 1106 (1962).
36. Salz, J. L., Daum, F., and Cohen, M. I., Serum alkaline phosphatase activity during adolescence, *J. Pediat.*, **82**, 537 (1973).
37. Greuner-Sigusch, P., Alkalische Phosphatase, *Ärztl. Lab.*, **23**, 44 (1977).

4.2.2. CALCIUM

4.2.2.1. Introduction

Despite the irregular supply of calcium to the body, its concentration in the blood is kept within narrow limits. The calcium homeostasis, which is closely linked with that of phosphate, is ensured by the parathyroid hormone secreted by the parathyroid gland and calcitonin produced in the parafollicular cells of the thyroid gland. The main part of the body calcium (about 98%) is deposited in the bones, and consequently only a small fraction is present in the plasma. The total calcium concentration in the serum amounts to about 10 mg per 100 ml (2.5 mmol l^{-1}). Of this, about 3.5 mg per 100 ml (0.9 mmol l^{-1}) is bound to protein and 6.5 mg per 100 ml (1.6 mmol l^{-1}) is retained on ultrafiltration. Of the latter fraction, 5.3 mg per 100 ml (1.3 mmol l^{-1}) is present in the ionized form, and 1.2 mg per 100 ml (0.3 mmol l^{-1}) is complexed by phosphate, bicarbonate, and citrate. The ionized calcium is the biologically active fraction and it is active at membranes, predominantly those of the nerve and muscle cells. Nowadays this can be measured with ion-selective electrodes. At low protein content the calcium concentration falls, and at high protein concentrations it increases.

4.2.2.2. Choice of method

The multiplicity of methods for the determination of calcium shows that the problem has not yet been solved to complete satisfaction. The most important methods are: complexometric titration with an indicator; colorimetric determination, usually in alkaline medium with various indicators which form a coloured complex with calcium;[1] flame photometry; and atomic-absorption spectrophotometry.[2] If a suitable flame photometer is available, the emission spectrophotometric determination is still one of the most suitable methods for the routine laboratory. The method usually employs the CaO band at 622 nm. Interference from sodium is almost completely eliminated by choosing an acetylene flame to produce a high temperature (2400 °C). Atomic-absorption spectrophotometry is the recognized reference method.

4.2.2.3. Reference values

These are given in Table 89. Calcium values in serum are normally distributed and show no significant sex differences in adults.[2]

4.2.2.4. Comments on the method

1. Measurements in heparinized plasma give lower results with flame photometry than with atomic-absorption spectrophotometry.[2]
2. In the measurement of calcium by flame photometry and atomic-absorption spectrophotometry the procedure recommended by the manufacturer should be adhered to.

4.2.2.5. Calcium in urine

The excretion of calcium in the urine is strongly dependent on the quantity and type of food and on the kidney function. In order to obtain results which may be evaluated it is necessary to analyse the diet for its calcium content or to feed the patient with a calcium-deficient diet for at least 3 days. With a calcium feed of 800–1000 mg per 24 h the calcium excretion is 150–200 mg per 24 h in women and 150–300 mg per 24 h in men. If, with a normal kidney filtration, the plasma concentrations are less than 8 mg per 100 ml, then most of the calcium is reabsorbed. Under such circumstances the excretion in urine falls below 30 mg per 24 h.[5] In children less than 1 year old, the elimination in the urine for a supply of 70–162 mg kg^{-1} per 24 h is 1.1–7.3 mg kg^{-1} per 24 h, and in larger children (supply 21–162 mg kg^{-1} per 24 h) it is 1.1–7.4 mg kg^{-1} per 24 h.[6]

Table 89. Calcium in serum in children and adults.

Age	Number	$\bar{x} \pm s$, mg per 100 ml	Reference range		Method	References
			mg per 100 ml	mmol l^{-1}		
Newly born, at term 12–36 h	60	8.5 ± 0.5	7.5–9.5	1.87–2.38	Flame photometry	7
Children up to 3 months	184	9.9 ± 0.5	9.0–10.9	2.25–2.72	Atomic absorption	3
3 months–1 year		10.0 ± 0.3	9.5–10.8	2.38–2.70		
1–2 years		9.9 ± 0.6	9.0–11.0	2.25–2.75		
2–3 years		10.0 ± 0.5	9.3–10.9	2.33–2.72		
3–4 years		9.9 ± 0.4	8.9–10.6	2.23–2.65		
4–5 years		9.9 ± 0.4	9.1–10.7	2.28–2.68		
5–6 years		9.8 ± 0.4	9.1–10.6	2.28–2.65		
6–7 years		9.7 ± 0.2	9.2–10.2	2.30–2.55		
7–8 years		9.6 ± 0.6	8.9–10.7	2.23–2.68		
8–14 years		9.4 ± 0.6	8.8–10.5	2.20–2.63		
Adults	50	9.9 ± 0.4	9.9–10.9	2.48–2.72	Atomic absorption	4
Adults (18–45 years)	84	9.4 ± 0.4	9.1–10.2	2.28–2.55	Atomic absorption	2
Adults (20–60 years)			8.6–10.2	2.15–2.55	Flame photometry	Personal investigations
Adults	312	2.5 ± 97.5%	9.2–11.2	2.29–2.79		

4.2.2.6. Diagnostic significance

1. *Hypercalcaemia*
   Primary bone tumours
   Non-metastasing malignant tumours (mammillary and bronchial carcinoma, sarcoma, reticuloses)
   Metastasing malignant tumours (thyroid gland, gastric, prostate, mammillary, and bronchial carcinoma, hypernephroma, plasmacytoma, leucoses)
   Oestrogen and androgen therapy for malignant tumours
   Primary hyperparathyroidism and other endocrine disturbances (hyperthyrosis, adrenal insufficiency, e.g. following adrenalectomy, Addison's disease in childhood, acromegaly)
   Vitamin D_3 intoxication and overdosing with dihydrotachysterine
   Boeck's sarcoidosis
   Immobilization (particularly in childhood and with Paget's disease)
   Milk alkali syndrome (Burnett syndrome)
   Idiopathic hypercalcaemia in childhood
   Therapy with thiazides
2. *Hypocalcaemia*
   Hypoparathyroidism
   Pseudohypoparathyroidism
   Vitamin D deficiency, steatorrhoea, nephrotic syndrome
   If the total calcium concentration falls below 6–7 mg per 100 ml (1.5–1.75 mmol l^{-1}) a tetany may develop; tetanies occurring at normal calcium concentrations can be due to hypomagnesaemia.

References

1. Barnett, R. N., Skodon, S. B., and Goldberg, M. H., Performance of 'kits' used for clinical chemical analysis of calcium in serum, *Am. J. clin. Path.*, **59**, 836 (1973).
2. Paschen, K., *Die Bestimmung des Calciums und seiner Fraktionen im Serum*, Thieme, Stuttgart, 1975.
3. Liappis, N., Brodehl, J., Dotchev, D., and Jäkel, A., Die Calcium-Konzentration im Serum, Urin und Stuhl von Säuglingen und Kindern, *Ärztl. Lab.*, **18**, 80 (1972).
4. Hurst, R. E., High precision high-speed analysis for calcium and magnesium in serum and urine, *Clin. chim. Acta*, **45**, 105 (1973).
5. Heidbreder, E., Röckel, A., and Heidland, A., Niere und Calciummetabolismus, *Dt. med. Wschr.*, **99**, 537 (1974).
6. Paunier, L., Borgeaud, M., and Wyss, M., Urinary excretion of magnesium and calcium in normal children, *Helv. paediat. Acta*, **25**, 577 (1970).
7. Rösil, A. and Fanconi, A., Neonatal hypocalcemia, *Helv. paediat. Acta*, **28**, 443 (1973)

4.2.3. INORGANIC PHOSPHORUS

4.2.3.1. Introduction

Of the approximately 800 g (25.8 mol) of phosphorus in the mature organism, 70–80% is located in the bones and teeth and 20–30% is present in the intracellular space, mostly in the form of organically bound phosphorus. Only about 1% is present in the body fluids as inorganic phosphorus. It must be assumed that inorganic phosphorus constitutes the most important transport form of phosphorus in the body and is responsible for the equalization of the phosphorus levels between individual compartments and places where it is needed. At the same time, serum phosphate acts as a buffer substance (cf. p. 280). As with calcium, the reabsorption of phosphate in the intestines is subject to endocrine regulation. The excretion of phosphate in the urine is influenced predominantly by the parathyroid hormone which inhibits reverse reabsorption in the proximal tubule and thus promotes the tubular excretion. In the normal way about 70% of the ingested phosphate is excreted in the urine in quantities of 0.3–1.0 g per 24 h. About one third

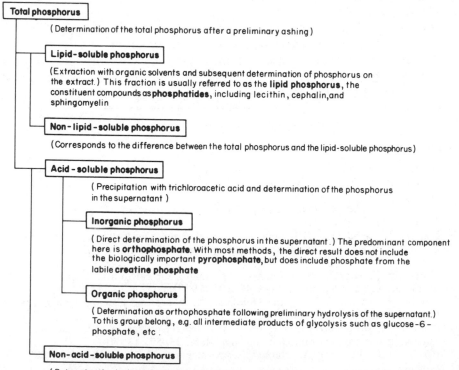

Fig. 123 Phosphorus fractions.

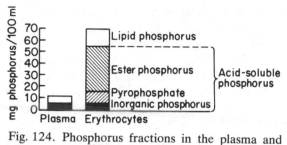

Fig. 124. Phosphorus fractions in the plasma and
in the erythrocytes.

of this consists of amorphous phosphates (calcium and magnesium salts), the rest being made up of sodium, potassium, and other phosphates.[1,17]

From the chemical point of view, the phosphorus in biological materials may be classified into the fractions shown in Fig. 123.

As shown in Fig. 124, illustrating a typical distribution in man, organic phosphorus compounds, especially phosphate esters, make up the preponderant intracellular fraction, whereas phospholipids and inorganic phosphorus are detectable in the plasma. The term 'inorganic phosphorus' as used in the clinic refers exclusively to orthophosphate present in the plasma, of which only the phosphorus is quantified in the determination.

The differential determination of the more than 50 known compounds belonging to the group of organic—or, better, organically bound—phosphorus was previously accomplished either by utilizing the differential behaviour of the barium salts, or by the method introduced by Lohmann based on the different lengths of time required for hydrolysis of the phosphate esters on boiling with hydrochloric acid.[2] Today, these relatively unspecific methods have been largely replaced by chromatographic techniques.

The isolation and determination of individual phosphorus compounds in cellular material can be extremely costly. In plasma the situation is simpler insofar as practically the only non-lipid-soluble, acid-soluble compounds present are orthophosphate and pyrophosphate, given the correct treatment—rapid separation of the plasma and cells—of the investigational material. However, the customary techniques do not register the pyrophosphate, so they may be considered as relatively specific for orthophosphate. In most of the methods of determination described in the literature the plasma is subjected to a preliminary deproteination. This is not necessary, however, and in the two methods described below deproteination is superfluous.

4.2.3.2. Choice of method

For the determination of phosphorus, whether directly or following a preliminary hydrolysis, there are in the main, three methods in current use.

1. Determination as ammonium molybdophosphate: if orthophosphate is treated with ammonium molybdate the initial product is a yellow heteropolyacid which is converted to the blue molybdenum blue on adding a reducing agent. The first step in this reaction, first described by Sonnenschein[3] in 1851, is preferably conducted in an acidic medium. The reaction is extremely complex so that it is scarcely worth giving equations. Many substances have been used as reducing agents, e.g. tin(II) chloride, phenylhydrazine, hydroquinone, hexacyanoferrate(II), ascorbic acid, aminoaphthosulphonic acid, p-methylaminophenyl sulphate, N-phenyl-p-phenylenediamine, and o-phenylenediamine (for literature, see ref. 4). The fact that several new modifications of the ammonium molybdate method are published every year suggests that all of these methods suffer from certain defects. The chief disadvantage is undoubtedly the low sensitivity, one consequence of which is the relatively frequent incidence of turbidities.

2. The formation of a coloured complex between ammonium molybdate and certain basic dyes in the presence of phosphorus: Itaya and Ui[5] were the first to describe (in 1966) the formation of an intesively coloured complex between phosphoammonium molybdate and malachite green. The method has since undergone several modifications;[6-8] in place of malachite green, Brilliant Green OO and Rhodamine B have also been recommended.[9,10] The sensitivity of this method is about 10 times higher than that of the ammonium molybdate methods.

3. Enzymatic determination of phosphate. The enzymatic methods must be mentioned as probably the most specific of the methods for the determination of inorganic phosphorus. They involve the principle of the optical test and the measurements are photometric or fluorimetric or are made by 'enzymatic cycling'. The determinations are based on the following series of reactions:

(a) $\text{Glycogen} + \text{P.}_4^{2-} \xrightarrow{\text{Phosphorylase A}} \text{Glucose-1-phosphate}$

$\text{Glucose-1-phosphate} \xrightarrow{\text{Phosphoglucomutase}} \text{Glucose-6-phosphate}$

$\text{Glucose-6-phosphate} + \text{NADP}^+ \xrightarrow{\text{Glucose-6-phosphate dehydrogenase}}$
$\qquad\qquad\qquad \text{6-Phosphogluconate} + \text{NADPH} + \text{H}^+ \qquad \text{(ref. 11)}$

(b) $\text{P.}_4^{2-} + \text{Glyceraldehyde-3-phosphate} + \text{NAD}^+ \xrightarrow[\text{Phosphate dehydrogenase}]{\text{Glyceraldehyde}}$
$\qquad\qquad\qquad \text{1,3-Diphosphoglycerate} + \text{NADH} + \text{H}^+$

$\text{1,3-Diphosphoglycerate} + \text{ADP} \xrightarrow{\text{Phosphoglycerate kinase}}$
$\qquad\qquad\qquad \text{3-Glycerophosphate} + \text{ATP}$

$\text{ATP} + \text{Glucose} \xrightarrow{\text{Hexokinase}} \text{Glucose-6-phosphate} + \text{ADP}$

$$\text{Glucose-6-phosphate} + \text{NADP}^+ \xrightarrow{\text{Glucose-6-phosphate dehydrogenase}}$$

$$\text{6-Phosphogluconate} + \text{NADPH} + \text{H}^+ \quad \text{(ref. 12)}$$

Certain authors used the same enzyme system but starting from the scission of fructose-1, 6-diphosphate with aldolase.[13]

The reactions are specific and sensitive, but they are not yet suitable for routine use owing to the difficulties attending the preparation of the reagents (contamination by phosphorus).

4.2.3.3. Molybdenum blue method

4.2.3.3.1. Principle

The basic reaction for this method of phosphorus determination is the reduction of molybdophosphate to molybdenum blue. Ascorbic acid and hydroquinone are used as reducing agents.

The inorganic phosphate is treated with molybdic acid without deproteination and converted to a phosphorus–molybdic acid complex, which is then reduced to molybdenum blue by ascorbate and measured photometrically. Borax is added to stabilize the colour and sodium pyrosulphite inhibits the non-specific formation of a complex salt. A carbonate/sulphite solution keeps the proteins in solution (pH 8.7). The Lambert–Beer law is valid over a range of 1–20 mg of inorganic phosphorus per 100 ml.

4.2.3.3.2. Reagents

Pyrosulphite–borate solution, borax $52\,mmol\,l^{-1}$, *pyrosulphite* $95\,mmol\,l^{-1}$. Dissolve 20 g of sodium tetraborate ($Na_2B_4O_7 \cdot 10\ H_2O$) and 18 g of sodium pyrosulphite in DM-water and up to 1000 ml. The solution keeps for a few months at 4 °C.

Molybdic acid solution, ammonium molybdate $40.5\ mmol\,l^{-1}$, H_2SO_4 $0.5\,mol\,l^{-1}$. Dissolve 5 g of finely powdered ammonium molybdate in $0.5\ mol\,l^{-1}\ H_2SO_4$ and make up to 100 ml with further H_2SO_4. The solution keeps for a few months at 4 °C.

Hydroquinone–ascorbate solution, hydroquinone $90\,mmol\,l^{-1}$, *ascorbate* $5\,mmol\,l^{-1}$. Dissolve 1 g of hydroquinone and 1 ampoule of Redoxon Roche (100 mg of sodium ascorbate) in DM-water and make up to 100 ml. Prepare freshly daily.

Carbonate–sulphite solution, carbonate $396\,mmol\,l^{-1}$, *sulphite* $56\,mmol\,l^{-1}$. Dissolve 7 g of sodium sulphite (anhydrous) and 42 g of sodium carbonate (anhydrous) in DM-water and make up to 1000 ml. The solution is filtered to remove any turbidity; it keeps for a few months at 4 °C.

Phosphorus standard, 5 *mg per* 100 *ml*, 1.61 *mmol l⁻¹*. Dissolve 22.0 mg of potassium dihydrogen phosphate in about 70 ml of distilled water, add 5 ml of 0.5 mol l⁻¹ sulphuric acid, and dilute to 100 ml with DM-water. The solution keeps indefinitely.

4.2.3.3.3. Measurement

Procedure.

(In test-tubes)	Sample, ml	Reagent blank, ml	Standard, ml
Pyrosulphite–borate solution	0.2	0.2	0.2
Plasma	0.02	—	—
DM-water	—	0.2	—
Standard solution	—	—	0.02
Molybdate solution	0.05	0.05	0.05
Hydroquinone–ascorbate solution	0.05	0.05	0.05

Allow to stand at room temperature for 15 min

| Sulphite–carbonate solution | 0.5 | 0.5 | 0.5 |

Take readings of the extinction against water at 578 nm between 5 and 30 min

4.2.3.3.4. Calculation

$$\text{Concentration} = \frac{E(S) - E(RB)}{E(ST) - E(RB)} \cdot 5 \text{ mg of inorganic phosphorus per } 100 \text{ ml}$$

$$= \frac{E(S) - E(RB)}{E(ST) - E(RB)} \cdot 1.61 \text{ mmol l}^{-1}$$

4.2.3.3.5. Specificity

The phosphate fractions present in the plasma are shown in Fig. 123. The method is specific for inorganic orthophosphate, i.e. phosphate present in an ionized, unbound form. The result also includes creatine phosphate, which is rapidly cleaved to creatine and orthophosphate by the molybdate reagent. However, since no measureable quantities of creatine phosphate are present in the serum, plasma, and urine, this does not restrict the specificity of the method. In order to measure the actual inorganic phosphorus, i.e. without simultaneous determination of the creatine phosphate, the orthophosphate must first be precipitated with a solution of calcium. The creatine phosphate remains in solution.

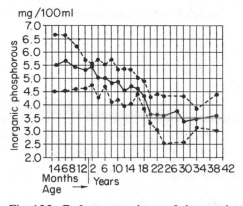

Fig. 125. Reference values of inorganic serum phosphorus as a function of age (from ref. 14).

4.2.3.3.6. Reference values

The reference values of the inorganic plasma phosphorus are illustrated in Fig. 125. The values are markedly higher during the growth period than after formation of the bones is complete. In adults values between 3 and 4.5 mg per 100 ml or 0.96–1.45 mmol^{-1} may be considered as normal. In the first 24 h of life the values fluctuate between 6.7 and 7.1 mg per 100 ml or 2.16 and 2.29 mmol l^{-1}.[15]

4.2.3.3.7. Comments on the Method

1. Values in plasma are lower than those in serum.[16]

2. The serum or plasma must be separated from the erythrocytes within 30 min, otherwise cleavage of phosphate esters leads to liberation of inorganic phosphorus.

3. A slow increase in inorganic phosphorus is also observed if serum or plasma is allowed to stand for longer periods, probably owing to the action of phospholipases and phosphatases. The determination should therefore be carried out within 6 h of collecting the blood.

4. The method is suitable for the determination of inorganic phosphorus in cerebrospinal fluids and in urine. The amount excreted in urine is 0.34–1.0 g per 24 h or 110–329 mmol l^{-1}. The concentration in cerebrospinal fluid amounts to 0.9–2.0 mg per 100 ml or 0.29–0.64 mmol l^{-1}.[1]

5. The method can also be conducted as a macro determination in which case all volumes are multiplied by 5 (0.1 ml of serum, etc.).

6. By reading the extinctions at higher wavelengths (600–650 nm) the sensitivity of the method is increased by a factor of up to 2.

7. Turbidities constitute the main source of error; they are not uncommon and a sample blank does not offer adequate correction.

4.2.3.4. Malachite green method

4.2.3.4.1. Principle

One principle for measuring the concentration of inorganic phosphorus has been described[5] which suffers from neither of the main disadvantages of the molybdenum blue method—low sensitivity and the occurrence of turbidities which may not be corrected for. Certain basic dyes such as malachite green react with phosphomolybdate to form coloured complexes with spectra differing from that of the starting materials:

1. Inorganic phosphorus + Ammonium heptamolybdate → molybdophosphate.
2. Molybdophosphate + Dye → Complex.

Malachite green has a shallow extinction maximum at 640 nm. Addition of phosphorus considerably intensifies the extinction and leads to a slight bathochromic shift of the extinction maximum. The measurement can be made at 578, 623, or 640 nm. The precise mechanism of the reaction is not known.

Although most authors recommend that the determination be carried out on deproteinated material, it is possible to circumvent the deproteination by modifying the reagent and by preliminary dissolution of the plasma or serum in 6 mol l^{-1} urea solution.

4.2.3.4.2. Reagents

Urea solution, 6 *mol l^{-1}*. Dissolve 360 g of urea in DM-water and make up to 1000 ml. The solution keeps indefinitely at room temperature.

Malachite green reagent. Dissolve 680 mg of malachite green in about 1000 ml of DM-water. To this solution, add 500 ml of a solution of 21 g of ammonium heptamolybdate in 500 ml of 5 mol l^{-1} hydrochloric acid (1:2 dilution of 37% HCl). Add 1 ml of undiluted Sterox and make up to 2000 ml with DM-water. Stir for about 30 min. The reagent is ready for use after it has been allowed to stand for 2 h; it keeps indefinitely in a glass bottle at room temperature.

Standard phosphorus solution, 5 *mg per* 100 *ml* (1.61 *mmol l^{-1}*). See p. 504.

4.2.3.4.3. Measurement

Procedure.

	Sample, ml	Reagent blank, ml	Standard, ml
Plasma, serum	0.02	—	—
DM-water	—	0.02	—

	Sample, ml	Reagent blank, ml	Standard, ml
Standard solution	—	—	0.02
Urea solution	1.0	1.0	1.0
Mix well			
Malachite green reagent	5.0	5.0	5.0

Mix well and allow to stand at room temperature for at least 20 min. Read the extinction against water at 578 nm within 1 h.

4.2.3.4.4. Calculation

$$\text{Concentration} = \frac{E(S) - E(RB)}{E(ST) - E(RB)} \cdot 5 \text{ mg per 100 ml}$$

$$= \frac{E(S) - E(RB)}{E(ST) - E(RB)} \cdot 1.61 \text{ mmol } l^{-1}$$

4.2.3.4.5. Specificity

Investigations have shown that no phosphorus compounds other than orthophosphate and creatine phosphate give a positive reaction.[9]

4.2.3.4.6. Reference values

The values obtained with this method are systematically 0.5–1 mg per 100 ml lower than with the molybdenum blue method described above (Table 90). However, it is our belief that the results of the malachite green method are more accurate; the faint turbidities, frequently only recognizable from the Tyndall effect, are probably responsible for the higher results with the molybdenum blue method. The results show a Gaussian normal distribution.[6]

4.2.3.4.7. Comments on the method

1. Not all batches of malachite green are suitable for the phosphorus determination. Faulty preparations may be recognized by deviations from the Lambert–Beer law, which is otherwise valid over a wide range of concentrations (0.5–20 mg per 100 ml); in addition, the calibration graph fails to pass through the origin. Consequently, each batch of the dye must be tested for suitability by first preparing a calibration graph.

Table 90. Reference values of inorganic phosphorus in serum (Malachite Green method).

Age	Sex	Number	$\bar{x} \pm s$		Reference range		References
			mg per 100 ml	mmol l^{-1}	mg per 100 ml	mmol l^{-1}	
Sucklings up to 12 months	m, f	41	3.39	1.27	1.82–5.76	0.59–1.86	6
Children 2–14 years	m, f	43	3.59	1.16	1.79–4.98	0.58–1.61	6
Adults	m, f	613	2.93 ± 0.6	0.95 ± 0.20	1.72–4.14	0.55–1.34	6
Adults	m, f	313	2.5–97.5%		1.74–4.0	0.65–1.29	Personal investigations

2. One disadvantage of the dye is that it readily precipitates on glass and plastic surfaces. The cuvettes must therefore be cleaned with acetone after each series of analyses.

3. Since the method is very sensitive, it is best to use only disposable tubes where possible. Detergents usually contain phosphorus and can therefore lead to wrong results.

4. Measurement of the day-to-day precision showed a coefficient of variation of 5.62%,[6] alternatively 4.3%,[7] or, in the series, 4.09%.[6]

4.2.3.4.8. Diagnostic significance

Hyperphosphataemia. This is less common and occurs e.g. in hypervitaminosis D, healing of fractures, kidney insufficiency, hypoparathyroidism, pseudo-hypoparathyroidism, diabetic ketoacidosis, and Paget's disease.

Hypophosphataemia is found in hyperparathyroidism, osteomalacia, and rickets, steatorrhoea (diminished resorption of vitamin D in the intestines), Franconi syndrome, ane renal acidosis of the Albright type.

References

1. Vanderlinde, R. E. and Kuwalski, P., The clinical biochemistry of phosphorus, *Clin. Biochem.*, **4**, 76 (1971).
2. Umbreit, W. W., Burris, R. H., and Stauffer, J. F., *Manometric techniques*, 2nd ed., Burgess, Minneapolis, 1959.
3. Sonnenschein, F. L., Über einige molybdänsaure Salze und die Anwendung der Molybdänsäure zur Bestimmung der Phosphorsäure, *J. prakt. Chemie*, **53**, 339 (1851).
4. Morin, L. G. and Prox, J., New and rapid procedure for serum phosphorus using o-phenylenediamine as reductant, *Clin. chim. Acta*, **46**, 113 (1973).
5. Itaya, K. and Ui, M., A new micromethod for the colorimetric determination of inorganic phosphate, *Clin. chim. Acta*, **14**, 361 (1966).
6. Hohenwallner, W. and Wimmer, E., The malachite green micromethod for the determination of inorganic phosphate, *Clin. chim. Acta*, **45**, 169 (1973).
7. Kallner, A., Determination of phosphate in serum and urine by a single step malachite-green method, *Clin. chim. Acta,* **59**, 35 (1975).
8. Anner, B. and Moosmayer, M., Rapid determination of inorganic phosphate in biological systems by a highly sensitive photometric method, *Analyt. Biochem.*, **65**, 305 (1975).
9. Van Belle, H., New and sensitive reaction for automatic determination of inorganic phosphate and its application to serum, *Analyt. Biochem.*, **33**, 132 (1970).
10. Garcic, A. and Kratochvila, J., Bestimmung von anorganischem Phosphor in biologischem Material mit einem einstufigen Reagens, enthaltend Rhodamin B, *Clin. chim. Acta*, **62**, 29 (1975).
11. Schulz, D. W., Passonneau, J. V., and Lowry, O. H., An enzymic method for the measurement of inorganic phosphate, *Analyt. Biochem.*, **19**, 300 (1967).
12. Fawaz, E. N. and Tejirian, A., A new enzymatic method for the estimation of inorganic phosphate in native sera, *Z. klin. Chem. klin. Biochem.*, **10**, 215 (1972).

13. Scopes, R. K., A new enzymatic method for inorganic phosphate determination, *Analyt. Biochem.*, **49**, 88 (1972).
14. Bullock, J.K., Physiologic variations in the inorganic blood phosphorus content at different age periods, *Am. J. Dis. Child.*, **40**, 725 (1930).
15. Jakarainen, E., Plasma magnesium levels during the first five days of life, *Acta paediat. scand.*, suppl., 222 (1971).
16. Carothers, J. E., Kurtz, N. M., and Lemann, J., Jr., Error introduced by specimen handling before determination of inorganic phosphate concentrations in plasma and serum, *Clin. Chem.*, **22**, 1909 (1976).
17. DeLuca, H. F., Vitamine D endocrine system, *Adv. clin. Chem.*, **19**, 125 (1977).

4.2.4. MAGNESIUM

4.2.4.1. Introduction

Of the intracellular cations of highest concentration, magnesium ranks next to potassium, which heads the list; 98%, i.e. about 1000 mmol, of the total body magnesium is found intracellularly. Of this, about half is deposited in the bones and the bulk of the remainder is in the muscles and the liver. Of the magnesium ingested (10–20 mmol per day), about one third is reabsorbed and the remainder excreted in the stool. The largest part of the reabsorbed magnesium is excreted through the kidneys, which regulate the magnesium concentration. The biological activity of magnesium is manifold. It serves as activator for many enzyme reactions, namely, those of the phosphate metabolism, also those of the intramitochondrial oxidative phosphorylation. It also plays a part in the regulation of the neuromuscular excitation process.[1,2] Only 1% of the total concentration of magnesium is present in the extracellular space, about one quarter of this being in the plasma; 55% of the total magnesium in the plasma is in the form of free ions, 32% is bound to protein (predominantly albumin), and the remaining 13% is complexed to phosphate, citrate, and other substances.[3]

4.2.4.2. Choice of method

Until recently, the methods for the determination of magnesium in biological fluids were unsatisfactory, largely because the alkaline earth metals, calcium and magnesium, behave very similarly in several respects.

Today, atomic-absorption spectrophotometry is the method of choice. Serum is aspirated into an acetylene flame and vaporized, and the absorption of the emission band of the magnesium hollow-cathode lamp is recorded at 285 nm. If no atomic-absorption spectrophotometer is available, then the method of Mann and Yoe[4] has proved workable. In 1956 these authors described a colour reagent which can be used for the specific determination of micro-quantities of magnesium. Bohuon[5] used this method with serum and urine without prior deproteination. However, our own studies[6] showed that the method was more reliable if a preliminary deproteination with perchloric acid was included and higher results were obtained in this way than with the direct determination.

4.2.4.3. Principle

The dye used by Mann and Yoe,[4] sodium 1-azo-2-hydroxy-3-(2,4-dimethyl-carboxanilido)naphthalene-1'-(2-hydroxybenzene)-5-sulphonate, has a blue colour. In the presence of magnesium in an aqueous alcoholic medium at pH 9–10 it produces a red dye.

4.2.4.4. Reagents

Mann and Yoe reagent, 8 mg per 100 ml. Dissolve 8 mg of the dye in 75 ml of absolute ethanol (under reflux, if necessary). When all the dye has dissoved, make up to 100 ml with absolute ethanol. The solution keeps indefinitely at room temperature.

Sodium tetraborate, 52 mmol l^{-1}. Dissolve 20 g of sodium tetraborate ($Na_2B_4O_7 \cdot 10H_2O$) in 500 ml of hot DM-water. Make up to 1000 ml with DM-water. The solution keeps indefinitely at room temperature.

Magnesium standard, 2 mg per 100 ml (0.834 mmol l^{-1}). Dissolve 202.8 mg of magnesium sulphate ($MgSO_4 \cdot 7H_2O$) in DM-water and make up to 1000 ml. Add 1 ml of chloroform. The solution keeps indefinitely at room temperature.

Perchloric acid, 0.33 mmol l^{-1} (3.3 g per 100 ml). Dilute 2.85 ml of 70% perchloric acid with DM-water to 100 ml. The solution keeps indefinitely.

4.2.4.5. Procedure

Deproteination. Add 20 µl of serum and standard to separate 0.2 ml portions of cold perchloric acid. Centrifuge briskly.

	Sample, ml	Reagent blank, ml	Standard, ml
Sodium tetraborate	0.5	0.5	0.5
Sample supernatant	0.1	—	—
Perchloric acid	—	0.1	—
Standard supernatant	—	—	0.1
Colour reagent	1.0	1.0	1.0

Always shake well after the addition of the reagents. The development of the colour is complete after 10 min. Read the extinctions against DM-water between 480 (492) and 520 nm. The colour is stable for 24 h.

Table 91. Concentration of magnesium in serum.

Age	Sex	Number	$\bar{x} \pm s$ mg per 100 ml	mmol l^{-1}	References
Newly born					
Umbilical cord blood	m, f	22[a]	1.93 ± 0.35	0.81 ± 0.15	2
Capillary blood	m, f	22[a]	2.20 ± 0.32	0.91 ± 0.13	2
Up to 128 h	m, f	30	2.05 ± 0.28	0.85 ± 0.11	2
1–3 months	f	39–40	2.04 ± 0.24	0.85 ± 0.1	7
	m	40–47	2.04 ± 0.24	0.85 ± 0.1	7
4th month	f	41	2.04 ± 0.24	0.85 ± 0.1	7
	m	50	2.16 ± 0.24	0.90 ± 0.1	7
Adults	m, f	12	2.17 ± 0.22	0.90 ± 0.09	8
Adults	m, f	42	2.16 ± 0.07	0.90 ± 0.03	9
			Range		
Adults	m, f		1.92–2.76	0.8–1.15	1
Adults	m, f		1.56–2.52	0.65–1.05	

[a]Same population.

4.2.4.6. Calculation

$$\text{Concentration} = \frac{E(S) - E(RB)}{E(ST) - E(RB)} \cdot 2 \text{ mg per 100 ml}$$

$$= \frac{E(S) - E(RB)}{E(ST) - E(RB)} \cdot 0.83 \text{ mmol l}^{-1}$$

4.2.4.7. Specificity

The method is specific for magnesium. At the concentrations present in biological fluids calcium does not affect the reaction.

4.2.4.8. Reference values

Values in serum are given in Table 91.

The value in cerebrospinal fluid is $1.19 \pm 0.12 \text{ mol l}^{-1}$.[8] The amount excreted into the urine is dependent on diet and kidney function. Standard values are, $0.68–7.0 \text{ mmol l}^{-1}$ per 24 h ($0.72–12 \text{ mmol l}^{-1}$).

In the erythrocytes the concentration of magnesium is roughly one third to one quarter of that in other cells and amounts to $2.65 \pm 0.26 \text{ mmol l}^{-1}$ or $2.21 \pm 0.39 \text{ mmol l}^{-1}$.[10]

4.2.4.9. Comments on the method

1. Running a sample blank is superfluous since there are no interfering chromogens present.

2. If the determination is conducted without deproteination the scatter is liable to be greater. Since the ratio of serum to alcohol is critical, the directions given by Bohuon[5] should be observed in this case. Deproteination is mandatory for lipaemic and icteric sera in any case.

3. For an exact standardization it is advisable to weigh out magnesium metal instead of magnesium sulphate.

4. The determination can also be conducted on the macro-scale, in which case all volumes must be multiplied by 5.

5. The method gives results which compare well with those obtained by atomic-absorption spectrophotometry.[1]

4.2.4.10. Diagnostic significance

Hypomagnesaemia.[11]

Deficient supply or absorption
    Malabsorption syndrome (primary or secondary)
    Protein–calorific malnutrition
    Pancreatitis

Renal loss
    Diuretics
    Osmotic diuresis
    Hyperaldosteronism
    Bartter syndrome
    Hypercalciuria and hypercalcaemia
    Administration of certain antibiotics and antimycotics

Hyperparathyroidism, hypoparathyroidism

Other causes
    Hyperthyroidism, chronic alcoholism, idiopathic hypomagnesaemia

Hypomagnesaemia in newly born children
    Hypomagnesaemia of the mother
    Hyperparathyroidism of the mother
    Maternal diabetes
    Hypoparathyroidism of the child
    Hyperphosphataemia of the newly born
    Exchange transfusion with citrate blood
    Specific Mg malabsorption

Hypermagnesaemia
    Rare; the commonest cause is chronic kidney insufficiency, parenteral Mg administration

---

514

References

1. Batsakis, J. G., Magnesium, *Am. Soc. clin. Pathologists check sample program*, 1972.
2. Jukarainen, E., Plasma magnesium levels during the first five days of life, *Acta paediat. scand.*, suppl., 222 (1971).
3. Walser, M., Ion association. VI. Interactions between calcium, magnesium, inorganic phosphate, citrate and protein in normal human plasma, *J. Univ. Invest.*, **40**, 723 (1961).
4. Mann, C. R. and Yoe, H. J., Spectrophotometric determination of magnesium with sodium 1-azo-hydroxy-3-(2,4-dimethyl-carboxanilido)-naphthalene-1'-(2-hydroxybenzene-5-sulfonate), *Analyt. Chem.*, **28**, 202 (1956).
5. Bohuon, C., Micro-determination of magnesium in various biological media, *Clin. chim. Acta*, **7**, 811 (1962).
6. Lauber, K., Unpublished observations.
7. Fomon, S. J., Filer, R. J., Jr., Thomas, L. N., and Roders, R. R., Growth and serum chemical values of normal breast-fed infants. *Acta paediat. scand.*, suppl., 202 (1969).
8. Iida, C. K., Fuwa, K., and Wacker, W., A general method for magnesium analysis in biological materials by atomic absorption spectroscopy, *Analyt. Biochem.*, **18**, 18 (1967).
9. Hunt, B. J., The estimation of magnesium in plasma, muscle and bone, by atomic absorption spectrophotometry, *Clin. Chem.*, **15**, 979 (1969).
10. Feenders, O., Dominick, H. Chr., and Bachmann, K. D., Die Magnesiumkonzentration der Erythrozyten und des Plasmas im Kindesalter, *Dt. med. Wschr.*, **102**, 1065 (1977).
11. Paunier, L., Troubles du transport du magnésium chez l'enfant, *Bull. schweiz. Ges. klin. Chem.*, 1, 2 (1977).

4.3 CARDIAC AND SKELETAL MUSCLE

4.3.1. CREATINE KINASE: OPTICAL TEST

4.3.1.1. Introduction

Creatine kinase (E.C. 2.7.3.2) catalyses the synthesis and cleavage of creatine phosphate (phosphagen):

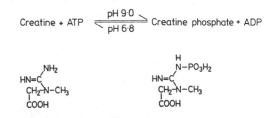

At pH 7, the equilibrium lies to the left-hand side; phosphorylation requires a pH of 9. Either direction can be used for the measurement of the enzyme activity but the reverse reaction (right to left) proceeds about 10 times faster than the forward reaction.

Creatine kinase is present in easily measurable concentrations, principally in skeletal muscle, cardiac muscle, and the brain.[1] The ATP produced in the reaction is mostly cleaved in the contraction of the muscles and must be continually renewed from the creatine phosphate reservoir. The creatine phosphate is reconstituted during the relaxation phase of the muscles. Intracellularly, creatine kinase is present in the cytoplasm but it is also said to be present in the mitochondria.[2] In the body fluids it is to be found in the plasma and also in cerebrospinal fluid.[3]

Three isoenzymes of creatine kinase are known. Each is a dimer composed from the subunit M or B. Consequently, it exists in three forms: MM, BB, and MB. On electrophoresis the MM enzyme migrates the slowest, together with the γ-globulins, the BB enzyme migrates the fastest, and the MB enzyme is intermediate (Fig. 125a). The MM enzyme occurs principally in skeletal and cardiac muscle, the MB enzyme in cardiac muscle, and the BB enzyme in the brain. The three isoenzymes differ in respect of their kinetic properties.[4] Electrophoretic, ion-exchange, kinetic, immunological, and radioimmunological methods have been developed for separating the isoenzymes.[5,60] The last methods appear to be the most

515

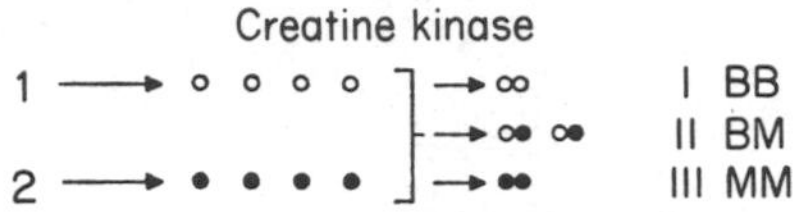

Fig. 125a. Isoenzymes of creatine
kinase.

sensitive and the most selective.[6] In the plasma, both the MM and the MB enzyme are normally measurable, but the BB enzyme is generally not measurable.[7]

During the ontogenesis of the muscle fibres in man, the BB isoenzyme appears at the beginning of gestation and is subsequently superseded by the MB and, principally, the MM isoenzyme.[8]

The total activity of the creatine kinase, as measured in the plasma for diagnostic purposes, is a combination of MM and MB activities, but the weak MB activity can be masked by that of the MM enzyme.

4.3.1.2. Choice of method

The known methods for the measurement of creatine kinase are listed in Table 92. The original methods used were those nowadays referred to as forward-reaction methods. Today, precedence is given to the reverse reaction, which is even more sensitive. Measurement is made by means of the optical test via the coupled hexokinase and glucose-6-phosphate dehydrogenase reaction. The method has been standardized.[17] In its active centre, creatine kinase contains one thiol group (cysteine) for each subunit.[18] The enzyme is rapidly inactivated by oxidation of these

Table 92. Methods for measuring creatine kinase.

Forward reaction

Creatine + ATP $\longrightarrow$ Creatine phosphate + ADP

1. Determination of the creatine phosphate
   Photometric method of Okinaka *et al.*[9]
2. Determination of the ATP
   Titration–pH method of Cho *et al.*[10]
   Determination by means of the optical test according to Tanzer and Gilvarg[11]

Reverse reaction

Creatine phosphate + ADP $\longrightarrow$ Creatine + ATP

1. Measurement of the creatine produced
   1.1. Photometry: method of Dreyfus *et al.*;[12] method of Hughes[13]
   1.2. Fluorimetry of the creatine, method of Conn and Anido[14]
2. Determination of the formed ATP by means of the optical test
   Oliver[15]
   Nielsen and Ludvigsen[16]

sulphydryl groups but the process can be reversed by the addition of thiol compounds to the incubation mixture. This reactivation lasts for at least 24 h at 20 °C, for up to 7 days at 4 °C, and for up to 14 days at −20 °C.[19,20] For practical purposes, the loss in activity at 4 °C and −20 °C within the stated time may be neglected.[20] Initially cysteine, and later glutathione, were proposed as thiol activators.[17,21] Today, the recommended activator is N-acetylcysteine, since it is stable, readily soluble, and likewise guarantees maximum activation.[22] Glutathione should be rejected as an activator since it interferes with the creatine kinase determination. Glutathione reductase, which is present in serum,[23] oxidizes the reduced glutathione so that there is insufficient for the activation; it also consumes $NADPH_2$ produced in the creatine kinase reaction. Thus, the measured creatine kinase values are too low.[24] Further, creatine kinase requires Mg^+ ions for the reaction; these form a complex with ATP.

4.3.1.3. Principle

1. Primary reaction: creatine kinase

$$\text{Creatine phosphate + ADP} \xrightleftharpoons{\text{Creatine kinase}} \text{Creatine + ATP}$$

2. Auxiliary reaction: hexokinase

$$\text{ATP + Glucose} \xrightleftharpoons{\text{Hexokinase}} \text{ADP + Glucose-6-phosphate}$$

3. Indicator reaction: glucose-6-phosphate dehydrogenase

$$\text{Glucose-6-phosphate + NADP}^+ \xrightleftharpoons{\text{Glucose-6-phosphate dehydrogenase}}$$
$$\text{6-Phosphogluconate + NADPH + H}^+$$

4.3.1.4. Measurement

Conditions. These are given in Table 93.

4.3.1.5. Calculation

Volume activity in μmol min^{-1} l^{-1} = U l^{-1} (for calculation, see p. 234).

4.3.1.6. Specificity

The above method is specific for creatine kinase.

4.3.1.7. Reference values

A compilation is given in Table 94. The values are neither normally nor log-normally distributed.[20,21]

Table 93. Determination of creatine kinase, final concentrations in the reaction mixture.

	Szasz et al.;[22] Deutsche Gesellschaft für klinische Chemie[59]	Scandinavian Society for Clinical Chemistry and Clinical Physiology[56]
Imidazole acetate buffer, pH 6.8 (25 °C)	100 mmol l⁻¹	pH 6.5 (37 °C) 100 mmol l⁻¹
Creatine phosphate	30 mmol l⁻¹	30 mmol l⁻¹
ADP	2 mmol l⁻¹	2 mmol l⁻¹
Magnesium acetate	10 mmol l⁻¹	10 mmol l⁻¹
D-Glucose	20 mmol l⁻¹	20 mmol l⁻¹
NADP	2 mmol l⁻¹	2 mmol l⁻¹
N-Acetylcysteine	20 mmol l⁻¹	20 mmol l⁻¹
AMP	5 mmol l⁻¹	5 mmol l⁻¹
Diadenosine pentaphosphate	10 μmol l⁻¹	10 μmol l⁻¹
Hexokinase	2500 U l⁻¹ (25 °C)	3500 U l⁻¹ (37 °C)
Glucose-6-phosphate dehydrogenase	1500 U l⁻¹ (25 °C)	2000 U l⁻¹ (37 °C)
Volume fraction of the serum (plasma)	1:26	1:23
Temperature of measurement	25 °C	37 °C
Wavelength	340 (334;365) nm, $d = 1$ cm	340 nm, $d = 1$ cm
Pre-incubation	3–5 min	5 min
Duration of measurement	1–5 min, ΔE is recorded throughout this period	ΔE is recorded after a 90-s lag phase

Table 94. Reference values for creatine kinase (U l⁻¹, 25 °C).

Age	Sex	Number	Reference interval, 2.5–97.5%	References
Prematurely born infants			60–83	25
Full-term newly born			87–119	
Sucklings				
Up to the 3rd month			45–66	
3rd–9th month			31–40	
9th–12th month			27–37	
Infants			28–34	
School children				
6–9 years			28–35	
9–15 years			26–43	
Adults		1429	0–50	26
Ambulant patients	m	204	10–70	20
Stationary patients	f	184	10–65	
Ambulant patients	f	206	10–60	
Stationary patients	f	204	7–55	
Adults	m	129	4–75	27
	f	163	2–37	

The reference values for adult women are markedly lower than those for men. The same applies to boys over 10 years old. The sex difference is possibly genetically conditioned and is also attributable to the different bodily activity and muscular mass of the two sexes. Scarcely any sex difference is detectable in younger children. Individuals with exceptionally heavy muscles may show higher values.[28] During pregnancy, especially during the last 3 months, the enzyme activity increases, the highest values being observed during birth. The values quickly return to normal during puerperium.[29,61,62]

Values measured in the venous plasma of new-born children within the first 24 h are higher than corresponding values for umbilical cord blood.[21,30] On the other hand, the values for umbilical cord blood are higher than those for adults. They reach adult values after about 10 days.[58]

In umbilical cord blood, both the total activity of the creatine kinase and that of the MB isoenzyme fraction are greater than in maternal serum.[58] In both samples, MB and BB fractions can be found in addition to the MM isoenzyme. This is probably connected with the distribution of the isoenzymes in the placenta and uterus (uterus 2% MM, 1% MB, and 97% BB isoenzyme; placenta 15% MM, 22% MB, and 63% BB isoenzyme).[58,61]

Children born naturally have a higher creatine kinase activity than children born by Caesarian section; this is attributable to the birth trauma.[30] During the 1st year of life the values fall[21,31] (Table 94). There is a slight increase prior to puberty. There also appear to be race differences. Investigations at a psychiatric hospital showed that male negroes showed higher values than negresses or whites of either sex.[32] Environmental factors such as alcohol consumption and the use of oral contraceptives likewise appear to influence the creatine kinase activity, if only slightly.[33] It has even been established that creatine kinase values are higher at the beginning of the week.[33]

4.3.1.8. Comments on the method

1. Serum and plasma give the same results. In collecting the blood, muscle-tissue juices can cause higher creatine kinase values.[21]

2. The reaction is linear up to 1000 U l^{-1}. Haemolysis can lead to an apparent increase in the creatine kinase activity since the adenylate kinase liberated from the erythrocytes produces additional ATP: $2ADP \rightarrow ATP + AMP$. Furthermore, 6-phosphogluconate dehydrogenase liberated from the erythrocytes can lead to the production of $NADPH_2$.

It has been claimed[34] that alkaline phosphatase might lower the activity of the creatine kinase by cleavage of glucose-6-phosphate, but this view has been contradicted.[35] Besides, thiols inhibit alkaline phosphatase. Various authors have found that dilution of the plasma leads to an increase in the creatine kinase activity;[36] this has been explained as due to the 'diluting out' of an inhibitor. However, other investigations cast doubt on this.[20] Uric acid in the serum has been found to act as an inhibitor for creatine kinase. This inhibition is reversible by thiol compounds.[37] The adenylate kinase present in the serum (from the erythrocytes, muscles, and liver) is inhibited by adenosine-5'-monophosphate and diadenosine pentaphosphate. Creatine kinase values in the serum are influenced by the position of the body during collection of the blood.[38] It appears that the creatine kinase in the serum can be activated by chelating agents such as EDTA.[63,64]

4.3.1.9. Diagnostic significance

Since creatine kinase is an enzyme of the cross-striped musculature (cardiac and skeletal muscle), various harmful conditions which affect these tissues can lead to an increase in the creatine kinase activity in the plasma. Only the most important will be mentioned here.

Strenuous physical work of long duration (e.g. long-distance running) leads, in contrast to a very short spell of work, to a noticeable elevation of the enzyme in the plasma.[39,40]

The increase of enzyme in *cardiac infarction* is of diagnostic importance. The creatine kinase activity is only rarely elevated in the first 3 h after the occurrence, but 6 h later it is elevated virtually without exception. It continues to increase up to 24 h and then falls to the reference-interval value between the 3rd and 5th days.[41] In addition to the MM enzyme, the plasma contains increased amounts of the MB enzyme, which is typical—but not specific—for cardiac infarction. As a rule there should be no increase in cases of angina pectoris attacks, though the values may show continuous variability. With pulmonary embolism the rise in creatine kinase is slight but it is true that in rare cases it may attain the level found with cardiac infarction. The gradient of the drop in creatine kinase, as determined by repeated measurement, permits conclusions to be drawn on the extent of the infarction.[42] Elevated creatine kinase values may also be found in cardioversion.[43]

Of the *diseases of the skeletal musculature* for which measurements of creatine kinase have acquired a diagnostic significance, progressive muscular dystrophy of the Duchenne type heads the list. This is an X-chromosomal recessive, inherited disease, carried by the mother who passes it on to her sons. The illness is progressive and is accompanied by increasing atrophy of the muscles of the extremities and of the torso. Without exception, these patients show a very highly elevated activity of the creatine kinase in the plasma with an increase of the MM and also of the MB enzyme. Measurement of the creatine kinase allows the illness to be diagnosed at the preclinical, asymptomatic stage.[44] Thus, new-born babies later to suffer from progressive muscular dystrophy already show elevated values. However, it is best to confirm this finding by repeating the measurement at the end of the child's 1st year.[30] The enzyme activity decreases in the course of evolution of the disease. The heterozygotic (female) carriers of the pathological gene can, to a large extent, be identified with the aid of measurements of the creatine kinase activity. In about 70% of the cases, adult carriers show values higher than the norm. The quota identified amongst young school-children is probably higher.[45] In this context the activated and optimized method for creatine kinase determination has proved to afford better discrimination than the older method.[46] Inflammatory changes in the musculature, such as polymyositis, dermatomyositis, and Coxsackie virus myocarditis, also lead to elevated creatine kinase activities.[47] An increase in the enzyme is only observed in neurogenic muscular atrophias if a myopathia is present concurrently.[48]

There is liable to be an increase in the enzyme if muscle tissue perishes (Crush syndrome, burns, surgical trauma). A transient increase in the activity of the creatine kinase in the plasma and in cerebrospinal fluid has been found during psychotic conditions.[3] Attacks of cramp produce an elevation not of the BB enzyme but of the MM enzyme. The BB enzyme can occur in the plasma following neurosurgical intervention and anoxic brain damage.[49] Studies with precipitating antibodies for B isoenzyme subunits were unable to confirm this finding.[55]

It is now known that even a single *intramuscular injection* of certain pharmaceuticals leads, after 12–24 h, to an increase in the creatine kinase activity (MM enzyme) to values 2–6 times the norm. The drugs concerned include narcotics, analgesics, barbiturates, phenothiazines, and antibiotics (e.g. lidocaine, procaine, secobarbital, phenobarbital, diazepam, chlorpromazine, pentazocine, diamorphine, penicilline G, and digoxin).[50,51] The cause is not clear but it is probably conditioned by a local reaction of the muscle fibre to the drug, leading to a necrosis.[52] It is important to know that such pharmaceuticals are frequently administered in the treatment of myocardial infarction. A differential diagnosis may be made by measuring other enzymes such as GOT, LDH, 2-hydroxybutyrate dehydrogenase, and creatine kinase isoenzyme (MB enzyme), which do not increase following the injection.

Elevated values of the creatine kinase in plasma, which return to normal on treatment, are also found in hypothyrosis;[53] also in strokes, malignant hypothermia, CO poisoning, exposure to cold, and other *illnesses with unspecified elevations*.[54] Although the creatine kinase activity can exceed the reference range in various diseases, the main diagnostic significance is undoubtedly in myocardial infarction and in diseases of the skeletal musculature.[57]

References

1. Colombo, J. P., Richterich, R., and Rossi, E., Serum-Kreatin-Phosphokinase: Bestimmung und diagnostische Bedeutung, *Klin. Wschr.*, **40**, 37 (1962).
2. Jacobus, W. E. and Lehinger, A. L., Creatine kinase of rat heart mitochondria, *J. biol. Chem.*, **248**, 4803 (1973).
3. Wale, S., Espejel, A., Calcaneo, F., Ocampo, J., and Diaz-de-Leon, J., Creatine phosphokinase, *Archs Neurol., Chicago*, **30**, 103 (1974).
4. Witteveen, S. A. G. J., Sobel, B. E., and DeLuca, M., Kinetic properties of the isoenzymes of human creatine phosphokinase, *Proc. natn. Acad. Sci. USA*, **71**, 1384 (1974).
5. Wong, R. and Swallen, T. O., Cellulose acetate electrophoresis of creatine phosphokinase isoenzymes in the diagnosis of myocardial infarction, *Am. J. clin. Path.*, **64**, 209 (1975).
6. Jockers-Wretou, E., Grabert, K., and Pfleiderer, G., Quantitative immunologische Bestimmungen der Isoenzyme der Kreatinkinase im Serum, *Z. klin. Chem. klin. Biochem.*, **13**, 85 (1975).
7. Nealon, D. A. and Henderson, A. R., Separation of creatine kinase isoenzymes in serum by ion-exchange column chromatography, *Clin. Chem.*, **21**, 392 (1975).
8. Wiesmann, U., Moser, H., and Mumenthaler, M., Ontogenese der Laktatdehydrogenase und der Kreatinkinase im foetalen menschlichen Muskelgewebe, *Pädiat. Pädol.*, **4**, 114 (1968).
9. Okinaka, S., Sugita, H., Momoi, H., Toyokura, Y., Watanabe, T., Ebashi, F., and Ebashi, S., Cysteine stimulated serum creatine kinase in health and disease, *J. Lab. clin. Med.*, **64**, 299 (1964).
10. Cho, A. K., Haslett, W. L., and Jenden, D. J., A titrimetric method for the determination of creatine phosphokinase, *Biochem. J.*, **75**, 115 (1960).
11. Tanzer, M. L. and Gilvarg, C., Creatine and creatine kinase measurement, *J. biol. Chem.*, **234**, 3201 (1959).
12. Dreyfus, J. C., Schapira, G., Resnais, J., and Scebat, L., La créatine kinase sérique dans le diagnostic de l'infarctus myocardique, *Revue fr. Etud. clin. biol.*, **5**, 386 (1960).
13. Hughes, B. T., A method for the estimation of serum creatine kinase and its use in comparing creatine kinase and aldolase activity in normal and pathological sera, *Clin. chim. Acta*, **7**, 597 (1962).
14. Conn, R. B., Jr. and Anido, V., Creatine phosphokinase determination by the fluorescent ninhydrin reaction, *Am. J. clin. Path.*, **46**, 177 (1966).

15. Oliver, J. T., A spectrometric method for the determination of creatine phosphokinase and myokinase, *Biochem. J.*, **61**, 116 (1955).
16. Nielsen, L. and Ludvigsen, B., Improved method for determination of creatine kinase, *J. Lab. clin. Med.*, **62**, 159 (1963).
17. Empfehlungen der Deutschen Gesellschaft für klinische Chemie. Standardisierung von Methoden zur Bestimmung von Enzymaktivitäten in biologischen Flüssigkeiten, *Z. klin. Chem. klin. Biochem.*, **8**, 658 (1970).
18. Watts, D. C., Creatine kinase, in Boyer, *The enzymes*, Academic Press, New York, 1973, p. 442.
19. Rotthauwe, H. W. and Kowalewski, S., Aktivierung und Alterung der Serum-Kreatin-Phosphokinase, *Klin. Wschr.*, **45**, 387 (1967).
20. Szasz, G., Laboratory measurement of creatine kinase activity, *Proc. 2nd Int. Symp. clin. Chem., Chicago, 1975*, American Association of Clinical Chemistry, Washington, 1976.
21. Wiesmann, U., Colombo, J. P., Adam, A., and Richterich, R., Determination of cysteine activated creatine kinase in serum, *Enzymol. biol. clin.*, **7**, 266 (1966).
22. Szasz, G., Gruber, W., and Bernt, E., Creatine kinase in serum. 1. Determination of optimum reaction conditions, *Clin. Chem.*, **22**, 650 (1976).
23. Weidemann, G., Optimierte Bestimmung und Eigenschaften der NADP-abhängigen Glutathionreductase im Serum, *Z. klin. Chem. klin. Biochem.*, **13**, 123 (1975).
24. Rosalki, S. B., Seeking the way: Richterich memorial lecture, in Anido, Rosalki, van Campen, and Rubin, *Quality control in clinical chemistry*, de Gruyter, Berlin, 1975, p. 175.
25. Sitzmann, F. C., *Das Labor in der Praxis*, Marseille-Verlag, Munich, 1974, p. 74.
26. Szasz, G., Busch, E. W., and Farohs, H. B., Serum-Kreatinkinase. I. Methodische Erfahrungen und Normalwerte mit einem neuen handelsüblichen Test, *Dt. med. Wschr.*, **95**, 829 (1970).
27. McCormick, D., The normal range for creatine phosphokinase, *Ir. J. med. Sci.*, **145**, 86 (1976).
28. Garcia, W., Elevated creatine phosphokinase levels associated with large muscle mass, *J. Am. med. Ass.*, **228**, 1395 (1974).
29. Kastinen, A. and Pyörälä, T., Serum enzyme activity in late pregnancy, at delivery and during puerperium, *Scand. J. clin. Lab. Invest.*, **15**, 429 (1963).
30. Bodensteiner, J. B. and Zellweger, H., Creatine phosphokinase in normal neonates and young infants, *J. Lab. clin. Med.*, **77**, 853 (1971).
31. Sitzmann, F. C. and Djayaputra, M., Untersuchung der "aktivierten" Creatinphosphokinase im Serum von Kindern, *Klin. Pädiat.*, **184**, 59 (1972).
32. Meltzer, H. Y. and Holy, P. A., Black–white differences in serum creatine phosphokinase (CPK) activity, *Clin. chim. Acta*, **54**, 215 (1974).
33. Paterson, Y. and Lawrence, E. F., Factors affecting serum creatine phosphokinase levels in normal adult females, *Clin. chim. Acta*, **42**, 131 (1972).
34. Mueller, R. G., Neville, K., Emerson, D. D., and Lang, G. E., Depressed apparent creatine kinase activity in sera with abnormally high alkaline phosphatase activity, *Clin. Chem.*, **21**, 268 (1975).
35. Wearne, J., Moore, R. W., and Caplan, B., Does alkaline phosphatase affect serum creatine kinase values?, *Clin. Chem.*, **21**, 1343 (1975).
36. Dobosz, I., The influence of the dilution effect on serum creatine phosphokinase activity in neuromuscular diseases, *Clin. chim. Acta*, **50**, 301 (1974).
37. Warren, W. A., Identification of a creatine kinase inhibitor in human serum, *Clin. Biochem.*, **8**, 247 (1975).

524

38. Röcker, L., Schmidt, H. M., Junge, B., and Hoffmeister, H., Orthostasebedingte Fehler bei Laboratoriumsbefunden, *Med. Lab., Stuttg.,* **28**, 267 (1975).
39. Griffiths, P. D., Serum levels of ATP: creatine phosphotransferase (creatine kinase). The normal range and effect of muscular activity, *Clin. chim. Acta,* **13**, 413 (1966).
40. Siest, G. and Galteau, M. M., Variations of plasmatic enzymes during exercise, *Enzyme.,* **17**, 179 (1974).
41. Wickert, P., Harm, K., Runge, M., and Monn, H., Enzymdiagnostik des Herzinfarktes unter besonderer Berücksichtigung differentialdiagnostischer Überlegungen bei erhöther Serum-CK, *Intensivmedizin,* **11**, 258 (1974).
42. Sobel, B. E., Serum creatine phosphokinase and myocardial infarction, *J. Am. med. Ass.,* **229**, 201 (1974).
43. Forssell, G., Nordlander, R., Nyquist, O., Orinius, E., Mandechi, T., and Gicl Kargul, W., Serum enzyme activities after cardioversion, *Br. Heart. J.,* **32**, 600 (1970).
44. Aebi, U., Richterich, R., Stillart, H., Colombo, J. P., and Rossi, E., Progressive Muskeldystrophie. III. Serumenzyme bei der Muskeldystrophie im Kindesalter, *Helv. paediat. Acta,* **16**, 543 (1961).
45. Moser, H. and Vogt, J., Follow-up study of serum creatinekinase in carriers of Duchenne muscular dystrophy, *Lancet,* **ii**, 661 (1974).
46. Moser, H., personal communication.
47. Brownlow, K. and Elevitch, F. R., Serum creatine phosphokinase isoenzyme (CPK) in myositis, *J. Am. med. Ass.,* **230**, 1141 (1974).
48. Achari, A. N. and Anderson, M. S., Serum creatine phosphokinase in amyotrophic lateral sclerosis, *Neurology, Minneap.,* **24**, 834 (1974).
49. Itano, M., The detection of CPK_1 (BB) in serum, *Am. J. clin. Path.,* **65**, 351 (1976).
50. Vorburger, C., Fässler, B., and Köhl, P., Serum-Kreatinphosphokinase und intramuskuläre Injektion, *Schweiz. med. Wschr.,* **103**, 927 (1973).
51. Sidell, F. R., Culver, D. L., and Kaminskis, A., Serum creatine phosphokinase activity after intramuscular injection, *J. Am. med. Ass.,* **229**, 1894 (1974).
52. Gloor, H. O., Serumkreatinphosphokinase (SCK)-Aktivität nach intramuskulärer Injektion von Benzoctamin, *Diss.,* Berne (1976).
53. Graig, F. A. and Smith, J. C., Serum creatine phosphokinase activity in altered thyroid states, *J. clin. Endocr. Metab.,* **25**, 723 (1965).
54. Nevins, M. A., Saran, M., Bright, M., and Lyon, L. K., Pitfalls in interpreting serum creatine phosphokinase activity, *J. Am. med. Ass.,* **224**, 1382 (1973).
55. Prellwitz, W., Neumeier, D., Knedel, M., Lang, H., Würzburg, U., Schönborn, H., and Schuster, H. P., Isoenzyme der Kreatinkinase bei extrakardialen Erkrankungen und nach diagnostischen und therapeutischen Eingriffen, *Dt. med. Wschr.,* **101**, 983 (1976).
56. Recommended method for the determination of creatine kinase in blood, *Scand. J. clin. Lab. Invest.,* **36**, 711 (1976).
57. Kämmerer, V., Creatinkinase, *Ärztl. Lab.,* **23**, 37 (1977).
58. Bayer, P. M., Gabl, F., Gergely, Th., Zazgornik, J., Widhalm, K., and Skalak, O., Isoenzyme der Creatinkinase in der Perinatalperiode, *Z. klin. Chem. klin. Biochem.,* **15**, 349 (1977).
59. Standard method for the determination of creatine kinase activity, *Z. klin. Chem. klin. Biochem.,* **15**, 255 (1977).
60. Roberts, R., Parker, C. W., and Sobel, B. E., Detection of acute myocardial infarction by radioimmunoassay for creatine kinase MB, *Lancet,* **ii**, 319 (1977).
61. Laboda, H. M. and Britton, V. J., Creatine kinase isoenzyme activity in human placenta and in serum of women in labor, *Clin. Chem.,* **23**, 1329 (1977).

525

62. McNeely, M. D. D., Berris, B., Papsin, F. R., Lyons, E., and Schipper, H., Creatine kinase and its isoenzymes in the serum of women during pregnancy and the peripartum period, *Clin. Chem.*, **23**, 1878 (1977).
63. Rollo, J. L., Ladenson, J. H., and McDonald, J. M., Stabilization and activation of creatine kinase (CK) in human serum by catio chelators, *Clin. Chem.*, **23**, 1119 (1977).
64. Sandifort, C. R. J., Effects of ethylenediaminetetraacetate on "CK-NAC" reagent stability and measured creatine kinase activities, *Clin. Chem.*, **23**, 2169 (1977).
65. MB Creatine kinase, *Lancet*, i, 313 (1978).

4.3.2. GLUTAMATE–OXALOACETATE TRANSAMINASE

4.3.2.1. Introduction

Glutamate–oxaloacetate transaminase (GOT, aspartate aminotransferase, E.C. 2.6.1.1; L-aspartate–2-oxoglutarate aminotransferase) is present at high concentrations in all organs but particularly in cardiac muscle, liver, and skeletal muscle. It catalyses the following reaction:

L-aspartate + 2-Oxoglutarate ⇆ Oxaloacetate + L-Glutamate.

GOT exists in two molecular forms, one deriving from the mitochondria and the other from the soluble cell fraction. The two forms differ in respect of their electrophoretic mobility, pH optimum, kinetic and immunological properties, and amino acid composition. There are two different enzyme proteins which probably also have separate structural gene loci.[1] Mature erthrocytes contain soluble GOT while reticulocytes and leucocytes contain both types of enzyme. Both are also present in the plasma.

4.3.2.2. Choice of method

In most of the methods of determination L-aspartate and 2-oxoglutarate are incubated at a suitable pH and the oxaloacetate produced is determined. Two methods are available for quantifying the oxaloacetate: colorimetric determination as a 2,4-dinitrophenylhydrazone and kinetic determination using the optical test. The latter is preferred nowadays. In our account of the determination we shall give consideration to the recommendations of the Deutsche Gesellschaft für klinische Chemie,[2] the Committee on Enzymes of the Scandinavian Society for Clinical Chemistry and Clinical Physiology,[3] and the International Federation of Clinical Chemistry.[4]

526

4.3.2.3. Principle

Primary reaction: GOT.

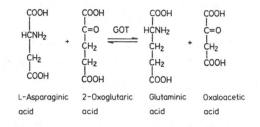

Indicator reaction: malate dehydrogenase (MDH).

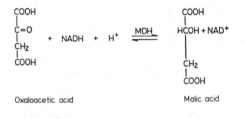

Conditions. The conditions of measurement for the three procedures are given in Table 95.

4.3.2.4. Calculation

Calculation of the volume activity in μmol min^{-1} l^{-1} = U l^{-1} (see p. 234).

4.3.2.5. Reference values

The reference values for the determination of GOT by the optical test, as quoted by the individual authors, are collected in Tables 96 and 97. It should be pointed out that the distribution is log-normal, not normal.[5,6] As a rule, men show higher values than women. The enzyme activity is weakly dependent on body weight. In new-born babies and young sucklings the GOT values are higher than in older children, but the difference is statistically not proved.[5] No sex difference could be established for GOT in children. The De Ritis GOT/GPT ratio is 0.77 for men and 0.90 for women.[6]

4.3.2.6. Comments on the method

1. Experience has shown a sample blank to be superfluous since, given sufficient time, the process always comes to a halt.

2. Haemolytic sera should not be used since transaminases from the erythrocytes pass into the plasma.

3. If the determination cannot be carried out within 4 h of collecting the blood the plasma or serum should be frozen.

Table 95. Determination of GOT, final concentrations in the reaction mixture.

	Deutsche Gesellschaft für klinische Chemie[2]	Scandinavian Society for Clinical Chemistry and Clinical Physiology[3]	International Federation for Clinical Chemistry[4]
Phosphate buffer	80 mmol l⁻¹		
Tris buffer		20 mmol l⁻¹	80 mmol l⁻¹
pH	7.4	7.7	7.8
EDTA		5 mmol l⁻¹	
Pyridoxal phosphate			0.10 mmol l⁻¹
L-Aspartate	200 mmol l⁻¹	200 mmol l⁻¹	200 mmol l⁻¹
NADH$_2$	0.18 mmol l⁻¹	0.15 mmol l⁻¹	0.18 mmol l⁻¹
2-Oxoglutarate	12 mmol l⁻¹	12 mmol l⁻¹	12 mmol l⁻¹
Malate dehydrogenase	600 U l⁻¹	600 U l⁻¹	10.0 μmol s⁻¹ l⁻¹
Lactate dehydrogenase	1200 U l⁻¹	200 U l⁻¹	10.0 μmol s⁻¹ l⁻¹
Volume fraction of the serum (plasma)	1:7.4	1:8.3	1:12
Temperature of measurement	25 °C	37 °C	30 °C
Wavelength	365 nm, $d = 1$ cm	340 nm, $d = 1$ cm	340 nm (Hg 334 nm), $d = 1$ cm
Pre-incubation	2–5 min		10 min
Duration of measurement recording ΔE for	3 min		300 s

Table 96. Reference values for the glutamate–oxaloacetate transaminase; determination according to the recommendations of the Deutsche Gesellschaft für klinische Chemie (U l^{-1}, 25 °C).

Age	Sex	Number	Reference interval, 5–95%	References
1 day–3 weeks	m, f	41	11–35	5
3 weeks–6 months	m, f	51	8–28	
6–12 months	m, f	40	10–28	
1–6 years	m, f	60	7–23	
6–14 years	m, f	60	6–20	
Adults	m	2384	6–21	6
	f	1595	5–18	
Adults	m	708	up to 19	7
	f	668	up to 15	
Adults 20–30 years	m	452	13–29 (2.5–97.5%)	8
Adults	m	442	5–17	18
	f	469	5–15 (2.5–97.5%)	

4. Identical results are obtained with serum and plasma.

5. The transaminases (GOT, GPT) have pyridoxal-5-phosphate as a common prosthetic group. In the transamination, the aldehyde group of the pyridoxal phosphate acts as acceptor for the amino group of the amino acid. This amino group is then transferred to the keto acids from the pyridoxamine phosphate produced. It has therefore been recommended

Table 97. Reference values for glutamate–oxaloacetate transaminase; determination according to the recommendations of the Committee on Enzymes of the Scandinavian Society of Clinical Chemistry and Clinical Physiology (U l^{-1}, 37 °C).

Age	Sex	Number	Reference interval, 2.5–97.5%	References
Pregnant women (39th–41st week)		70	<40	9
New-born babies	m, f	70	<40	
6–24 months	m, f	95	<50	
2–5 years 6–7 years 8–9 years	m, f	345	<45	
10–11 years 12–13 years 14–15 years 16–18 years	m, f	384	<35	
18 years	m		<35	
	f		<40	

that the GOT be activated by pre-incubation with pyridoxal-5-phosphate.[10,11] The activation can amount to as much as 30% with normal sera.[11] It is dependent on the temperature and on the nature of the buffer used, the activation being less in phosphate buffer than in tris buffer.[11,12]

6. The auxiliary enzyme and reagents should be free from ammonium sulphate so that the glutamate dehydrogenase reaction does not interfere. This can produce an interference in measurements on tissue homogenates with high concentrations of glutamate dehydrogenase (liver, kidneys, brain).

4.3.2.7. Diagnostic significance

The elevated activity in the plasma following cell damage is to a large extent due to the cytoplasmatic isoenzyme.

1. *Cardiac muscle.* A marked increase in GOT in the plasma is observed in 85–95% of myocardial infarctions. The activity increases within 4 h after the beginning of the infarction, reaches its highest value after 12–72 h, and returns to normal within 6–12 days.[13] The typical shape of the enzyme curve with a steep increase and a shallow fall is more important than the absolute value of the transaminase elevation. If a fresh cardial infarction is suspected, enzyme determinations should be made according to a fixed time schedule. The following procedure is recommended: first determination as soon as possible after the onset of the infarction, two further determinations after periods of 5–6 h, a fourth determination 12 h later, i.e. about 24 h after the beginning of the occurrence, and a further determination each day during the next 4 days. If there is no increase in GOT, the investigation may then be concluded. An infarction may be excluded with high probability. By artificial closure of coronary arteries in an animal, a relationship can be established between the degree of elevation of the transaminase and the extent of the infarcted region. In man too there is a sure correlation between the extent of myocardial infarction and the transaminase increase. Values above 170–200 U l^{-1} generally indicate extended necroses with a correspondingly less favourable prognosis. Above 250 U l^{-1} the prognosis is generally considered bad. There is a connection between the level of the transaminase value and the seriousness of the general picture insofar as massive infarction with high transaminasaemia is usually also accompanied by a correspondingly serious clinical condition. However, there are many exceptions to this rule, depending most of all on the localization of the infarction so that caution is advised in making prognoses on the basis of the transaminasaemia.[13]

Coronary conditioned angina pectoris without myocardial infarction is not, as a rule, accompanied by an increase in the enzyme. The GOT increase associated with cardiac insufficiency ought rather to be attributed to a liver constriction. Acute rheumatic and viral cardites, heart operations, cardiac catheterism, and heart massage can lead to an increase in the values. GOT can often be pathological in pulmonary embolisms which not uncommonly present a difficult problem in differential diagnosis. In doubtful cases determination of the creatine kinase enzyme may be of assistance.

2. *Liver*. For the diagnosis of hepatopathia the GOT should always be assessed in connection with the GPT.

Viral hepatitis. The plasma concentration rises even in the pre-clinical phase of the disease and reaches a peak in conjunction with the occurrence of the greatest subjective discomfort and hepatomegaly. Usually the icterus is then extremely pronounced also. As a rule the enzyme activity falls back within the reference interval after 4–7 weeks. The values of GOT measured in this disease are generally the highest. Lower values, ranged in decreasing order, are found in chronic aggressive hepatitis, chronic persistent hepatitis, and cirrhosis of the liver, in which case necrotic advances can lead to intermittent increases.[14,15] Increases can occur with infiltrative diseases of the liver such as carcinomas, leukaemias, lymphomas, and other metastases. Further elevations are observed in acute alcoholic intoxication, in acute mononucleosis (maximum after 2 weeks, falling about the 5th week), and in pancreatitis.

3. *Skeletal muscle, progressive muscular dystrophy*. The enzyme derives from the affected muscle fibres. The GOT activity is highest in infants and falls to lower values in the late stage, following the progress of the disease. The GOT cannot be used for detecting the conductors. An elevated GOT can be present in dermatomyosites and myoglobinurias.

4. *Pharmaceutical influences*. Elevated values of GOT have been observed after administration of salicylates, chlorpromazine, diproniazid, chlorpropamide, and codeine and morphine derivatives. The mechanism of the enzyme increase is unknown. The activity of GOT is not influenced by the intake of contraceptives.[9]

5. *Cerebrospinal fluid*. GOT is detectable in the cerebrospinal fluid. The enzyme activity may be elevated in encephalitis and progressive brain injuries.[16]

6. *Other diseases*. Diminished GOT enzyme activity has been observed in the serum of uraemic patients. In this case it has been established that for this group of patients the enzyme cannot rise above the reference range in cardiac complications.[17]

References

1. Chen, Shi-Han and Giblett, E. R., Genetic variation of soluble glutamic–oxaloacetic transaminase in man, *Am. J. hum. Genet.*, **23**, 419 (1971).
2. Empfehlungen der Deutschen Gesellschaft für klinische Chemie. Standardisierung von Methoden zur Bestimmung von Enzymaktivitäten in biologischen Flüssigkeiten, *Z. klin. Chem. klin. Biochem.*, **10**, 182 (1972).
3. Recommended methods for the determination of four enzymes in blood, *Scand. J. clin. Lab. Invest.*, **33**, 291 (1974).

4. IFCC method for aspartate aminotransferase (L-aspartate: 2-oxoglutarate aminotransferase, EC 2.6.1.1), *Clin. chim. Acta*, **70**, F19 (1976).
5. Sitzmann, F. C., Loew, C., Kalut, H., and Prestele, H., Normalwerte der Serumtransaminasen GOT und GPT mit neuen substratoptimierten Standardmethoden, *Klin. Pädiat.*, **186**, 346 (1974).
6. Laudahn, G., Hartmann, E., Rosenfeld, E. M., Weier, H., and Muth, H. W., Normalwerte der Serum-transaminasen bei Verwendung substratoptimierter Testansätze zur Aktivitätsmessung, *Klin. Wschr.*, **38**, 838 (1970).
7. Thefeld, W., Hoffmeister, H., Busch, E.-W., Koller, P. U., and Vollmar, J., Referenzwerte für die Bestimmungen der Transaminasen GOT und GPT sowie der alkalischen Phosphatase im Serum mit optimierten Standardmethoden, *Dt. med. Wschr.*, **99**, 343 (1974).
8. Siest, G., Schiele, F., Galteau, M., Panek, E., Steinmetz, J., Fagnani, F., and Gueguen, R., Aspartate aminotransferase and alanineaminotransferase activities in plasma: statistical distributions, individual variations and reference values, *Clin. Chem.*, **21**, 1077 (1975).
9. Penttilä, I. M., Jokela, H. A., Viitala, A. J., Heikkinen, E., Nummi, S., Pystynen, P., and Saastamoinen, J., Activities of aspartate and alanine aminotransferases and alkaline phosphatase in sera of healthy subjects, *Scand. J. clin. Lab. Invest.*, **35**, 275 (1975).
10. Rosalki, S. B. and Bayoumi, R. A., Activation by pyridoxal-5-phosphatase of aspartate transaminase in serum of patients with heart and liver disease, *Clin. chim. Acta*, **59**, 357 (1975).
11. Rej, R. and Vanderlinde, E., Effects of buffers on aspartate aminotransferase activity and association of the enzyme with pyridoxal phosphate, *Clin. Chem.*, **21**, 1585 (1975).
12. Jung, K., Lüdtke, B., and Egger, E., Der Einfluss von Pyridoxal-5'-phosphat auf das Temperaturverhalten der Aspartataminotransferase-Isoenzyme, *Z. klin. Chem. klin. Biochem.*, **13**, 179 (1975).
13. Müller, J. P., Die klinische Bedeutung der Serumtransaminase, *Diss.*, Berne, 1965.
14. Richterich, R., Enzymstoffwechsel, in Siegenthaler, *Klinische Pathophysiologie*, Thieme, Stuttgart, 1973, p. 169.
15. Richterich, R., Enzymdiagnostik der Leberkrankheiten, *Pädiat. Fortbildk. Praxis*, Vol. 15, Karger, Basle, 1965, p. 17.
16. Verrey, S., Die diagnostische Bedeutung der Glutamat–Oxalacetat-Transaminase, der Glutamat–Pyruvat-Transaminase, der Lactat-Dehydrogenase und ser Sorbit-Dehydrogenase des Liquors im Kindesalter, *Enzymol. biol. clin.*, **2**, 233 (1962/63).
17. Warnock, L. G., Stone, W. J., and Wagner, C., Decreased aspartate aminotransferase ('SGOT') activity in serum of uremic patients, *Clin. Chem.*, **20**, 1213 (1974).
18. Schlebusch, H., Rick, W., Lang, H., and Knedel, M., Normbereiche der Aktivitäten klinisch wichtiger Enzyme, *Dt. med. Wschr.*, **99**, 765 (1974).

4.3.3. LACTATE DEHYDROGENASE

4.3.3.1. Introduction

Lactate dehydrogenase (LDH; L-lactate; NAD oxidoreductase, E.C. 1.1.1.27), 'Warburg's reducing fermentation', catalyses the final step of glycolysis:

$$Pyruvate + NADH + H^+ \rightleftharpoons Lactate + NAD^+$$

532

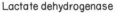

Lactate dehydrogenase

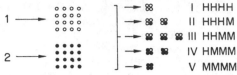

→ ⊠	I	HHHH
→ ⊠ ⊠	II	HHHM
→ ⊠ ⊠ ⊠	III	HHMM
→ ⊠ ⊠	IV	HMMM
→ ⊠	V	MMMM

Fig. 126. Heterogeneity of lactate dehydrogenase. Possible combinations for the tetrameric enzyme.[1]

The reaction is reversible, the equilibrium lying in the direction of lactate formation.

Lactate dehydrogenase is a tetramer made up of four single polypeptide chains (monomers). The tetramer can be composed of two kinds of monomer, the H-monomer (predominantly in cardiac muscle) and the M-monomer (predominantly in skeletal muscle and the liver); Fig. 126 shows the possible combinations of the monomers.

Accordingly, there are five possible tetramers, corresponding to the lactate dehydrogenase isoenzymes LDH_1 to LDH_5. In the electrophoretic separation, the LDH_1 isoenzyme migrates towards the anode the fastest and LDH_5 the slowest (Fig. 127). The lactate dehydrogenase is localized in the cell cytosome and consequently, in the event of cell damage, it rapidly passes into the extracellular space and thus into the plasma. Owing to its important role in glycolysis lactate dehydrogenase is present at high concentrations in all animal cells. In human organs it is found in the following, arranged in order of decreasing concentration of the enzyme: kidneys, heart, skeletal muscle, pancreas, spleen, liver, and lungs. It occurs in the serum, liquor cerebrospinalis, urine, and blood cells.

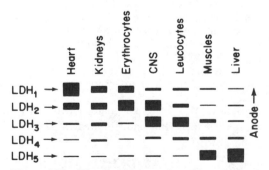

Fig 127. Distribution of the lactate dehydrogenase isoenzymes in extracts (cytosome) of various organs. The electrophoretic mobility of the isoenzymes as anions at pH 8 gradually decreases in the order LDH_1 (=HHHH) to LDH_5 (=MMMM).[1]

4.3.3.2. Choice of method

Nowadays the enzyme is determined almost exclusively by the optical test. In our experience the determination of lactate dehydrogenase in serum is the most unreliable of enzyme analyses and the most difficult to reproduce. The problem is that human serum contains five isoenzymes, all with different optimal conditions for reaction.[2] Depending on the choice of the $NADH_2$ and pyruvate concentrations, certain of these isoenzymes are inhibited. Since the isoenzyme composition in pathological sera is very different it is not possible to speak of 'optimal reaction conditions for serum lactate dehydrogenase'. This explains why the reference values show such great differences from one author to the next. In recording the enzyme activity, it transpires that the decrease in extinction in the serum lactate dehydrogenase does not proceed linearly but is composed of three or more linear parts with different slopes.[2] This is also probably attributable to the presence of isoenzymes. Every determination of the total lactate dehydrogenase in the serum consequently expresses a resultant of the activities of the individual enzymes. To determine the total lactate dehydrogenase, a compromise has to be made. The reaction conditions given in the described method are approximately optimal for the isoenzymes from the heart and the liver.[3]

4.3.3.3. Principle

$$
\begin{array}{l}
CH_3 \\
| \\
C{=}O \\
| \\
COO^-
\end{array}
+ NADH + H^+ \xrightarrow[\text{pH 7,5}]{\text{LDH}}
\begin{array}{l}
CH_3 \\
| \\
HCOH \\
| \\
COO^-
\end{array}
+ NAD^+
$$

Pyruvate   Lactate

4.3.3.4. Measurement

Conditions. Determination of lactate dehydrogenase, final concentrations in the reaction mixture:

	Deutsche Gesellschaft für klinische Chemie[4]	Scandinavian Society for Clinical Chemistry and Clinical Physiology[5]
Tris buffer		50 mmol l^{-1}
Phosphate buffer	50 mmol l^{-1}	
pH value	7.5	7.4
EDTA		5 mmol l^{-1}
Pyruvate	0.6 mmol l^{-1}	1.2 mmol l^{-1}
NADH	0.18 mmol l^{-1}	0.15 mmol l^{-1}
Volume fraction of the serum (plasma)	1:31.5	1:45

	Deutsche Gesellschaft für klinische Chemie[4]	Scandinavian Society for Clinical Chemistry and Clinical Physiology[5]
Temperature of measurement	25 °C	37 °C
Wavelength	340 nm, $d = 1$ cm (Hg 334 nm, Hg 365 nm)	340 nm, $d = 1$ cm
Duration of measurement	5–10 min, recording of ΔE during this period	

4.3.3.5. Calculation

Calculation of the volume activity in μmol min^{-1} l^{-1} = U l^{-1} (see p. 237).

4.3.3.6. Reference values

These are given in Table 98.

4.3.3.7. Comments on the method

1. Serum and plasma give identical results. There is no difference in the values obtained with capillary and venous plasma.[9]

2. The concentration of lactate dehydrogenase in the erythrocytes is about 100 times greater than that in the serum. Even a haemolysis which is scarcely detectable by eye (about 50 mg of haemoglobin per 100 ml of serum) leads to an elevation of the enzyme activity of 15–20%.

Table 98. Reference values of lactate dehydrogenase in serum.

Age	Sex	Number	Reference interval	References
Umbilical cord blood		29	70–542	6
1 month		6	118–494	
2 and 3 months		13	133–309	
4–6 months	m, f	12	135–299	
6–12 months		14	109–285	
2 years		12	116–344	
2–16 years		24	46–235	
Adults		49	60–230	
Blood donors		100	50–230	
Full-term new-born babies	m, f		270–915	12
1–30 days			190–755	
1–6 months			160–410	
7–12 months			170–350	
13–24 months			105–295	
Infants			150–280	7
School children			140–270	
Adults	m, f	4790	120–240	8

Consequently, no haemolyses may be tolerated. In strongly uraemic sera the lactate dehydrogenase activity may be diminished owing to the presence of inhibitors.

3. The anticoagulants heparin and EDTA do not interfere in the determination of lactate dehydrogenase. High concentrations of pyruvate can inhibit the lactate dehydrogenase, the effect being due to the inhibition of the cardiac isoenzyme.[10]

4. The lactate dehydrogenase reaction is not entirely specific—other keto acids may also be reduced. The reaction rate is inversely proportional to the chain length of the keto acid.

5. Compounds which inhibit lactate dehydrogenase may be produced in solutions of $NADH_2$ by the use of certain buffers, by high concentrations of hydrogen ions, by the action of ultraviolet light, and by freezing under poor conditions.[5]

6. Information on the stability of the enzyme in serum may be found on p. 106.

7. Lactate dehydrogenase can be inhibited by salicylate.[13]

4.3.3.8. Diagnostic significance

Since lactate dehydrogenase occurs in many types of cell the diagnostic evidence it supplies is non-specific, in contrast to that of the isoenzymes. Consequently the measurement of the total LDH activity is of relatively little value and is only rarely applied in the diagnosis of malignancies, certain anaemias, and cardiac infarction.

Malignant neoplasias can lead to an elevation of the lactate dehydrogenase in serum.[2,11] The increased activity is attributable to the necrosis of proliferating neoplastic cells which contain this enzyme. There is a relationship between the growth rate of the tumour and the elevation of the lactate dehydrogenase. Serous fluids in direct contact with malignant neoplasias, e.g. pleural effusions or ascites, also show an elevated enzyme concentration as a rule. Myeloid leukaemias show elevated lactate dehydrogenase activities, especially during an acute outbreak. In the remission phase this can be normal. In contrast, a normal activity by no means rules out a malignant process.

Megaloblastic anaemias of the pernicious type exhibit elevated serum lactate dehydrogenase levels. These diseases can produce the highest lactate dehydrogenase values measured. The enzyme is produced and released by the megaloblasts. Treatment with vitamin B_{12} leads to a rapid fall in the serum lactate dehydrogenase. Haemolytic anaemias, particularly with intravascular haemolysis, likewise lead to an elevation—though not so great—of the lactate dehydrogenase, since the enzyme occurs in high concentrations in the erythrocytes. A similar condition is observed in erythroblastosis fetalis.

During the first 12 h after cardiac infaction the enzyme increases from 2-

536

to 10-fold and reaches a maximum after 24–48 h. It returns to normal later than CK and GOT, between the 7th and 18th day.

References

1. Aebi, H., *Einführung in die praktische Biochemie*, Karger, Basle, 1971.
2. Richterich, R., Schafroth, P., and Aebi, H., A study of lactic dehydrogenase isoenzyme pattern of human tissues by adsorption-elution on Sephadex-DEAE, *Clin. chim. Acta*, **8**, 178 (1963).
3. Bergmeyer, H. U. and Berndt, E., Lactate dehydrogenase, in Bergmeyer, *Methods of enzymatic analysis*, Vol. 2, Academic Press, New York, 1974, p. 574.
4. Empfehlungen der Deutschen Gesellschaft für klinische Chemie. Standardisierung von Methoden zur Bestimmung von Enzymaktivitäten in biologischen Flüssigkeiten, *Z. klin. Chem. klin. Biochem.*, **10**, 182 (1972).
5. Recommended methods for the determination of four enzymes in blood, *Scand. J. clin. Lab. Invest.*, **33**, 291 (1974).
6. Gautier, E., Gautier, T., and Richterich, R., Valeur diagnostique d'anomalies d'activités enzymatiques du sérum en pédiatrie. I. Valeurs normales et influence des corticostéroïdes, *Helv. paediat. Acta*, **17**, 415 (1962).
7. Sitzmann, F. C., *Das Labor in der Praxis*, Marseille-Verlag, Munich, 1974, p. 77.
8. Weisshaar, D., Gossrau, E., and Faderl, B., Normbereiche von α-HBDH, LDH, AP und LAP bei Messung mit substratoptimierten Testansätzen, *Medsche Welt, Stuttg.*, **26**, 387 (1975).
9. Christoffersen, J., Marner, T., and Raabo, E., LDH-isoenzymes in children, *Acta paediat. scand.*, **64**, 822 (1975).
10. Everse, J., Reich, R. M., Kaplan, N. O., and Finn, W. D., New instrument for rapid determination of activities of lactate dehydrogenase isoenzymes, *Clin. Chem.*, **21**, 1277 (1975).
11. Ratliff, C. R., Culp, T. W., and Hall, F. F., Serum lactic dehydrogenase and other enzymes in malignant disease, *Am. J. Gastroent., N.Y.*, **56**, 199 (1971).
12. Veit, S., Sitzmann, F. C., and Prestele, H., Normalwerte für Lactat- und Glutamatdehydrogenase, sowie Leucinarylamidase, erstellt mit optimierten Standardansätzen, *Klin. Pädiat.*, **187**, 244 (1975).
13. Cheshire, M. R. and Park, M. V., The inhibition of lactate dehydrogenase by salicylate, *Int. J. Biochem.*, **8**, 637 (1977).

4.3.4. 2-HYDROXYBUTYRATE DEHYDROGENASE

4.3.4.1. Introduction

The so-called 2-hydroxybutyrate dehydrogenase or α-hydroxybutyrate dehydrogenase (HBDH; 2-hydroxybutyrate–NAD^+ oxidoreductase) is not an enzyme *sui generis*. It is, rather, a simple method for quantifying LDH_1 and LDH_2, the lactate dehydrogenase isoenzymes which migrate rapidly under electrophoresis. These isoenzymes, which occur predominantly in cardiac muscle, the brain and the erythrocytes, cleave 2-hydroxybutyrate, whereas the slower migrating isoenzymes (LDH_4 and LDH_5) characteristic of skeletal muscle and the liver, react much less readily with this substrate. Thus the lactate dehydrogenase to 2-hydroxybutyrate dehydrogenase ratio corresponds in practice to the ratio of the sum of all lactate dehydrogenase isoenzymes to cardiac muscle-specific isoenzymes. The smaller the ratio, the greater the fraction of isoenzymes deriving from cardiac muscle. The enzyme catalyses the following reaction:

$$2\text{-Oxobutyrate} + NADH + H^+ \xrightarrow{\text{pH 7.5}} 2\text{-Hydroxybuturate} + NAD^+$$

4.3.4.2. Choice of method

2-Hydroxybutyrate dehydrogenase is determined exclusively by the optical test.

4.3.4.3. Principle

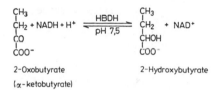

4.3.4.4. Measurement

Determination of 2-hydroxybutyrate dehydrogenase, final concentrations in the reaction mixture:

	Deutsche Gesellschaft für klinische Chemie[2]
Phosphate buffer	50 mmol l^{-1}
pH	7.5
2-Oxobutyrate	3.0 mmol l^{-1}
$NADH_2$	0.18 mmol l^{-1}
Volume fraction of the serum (plasma)	1:31.5
Temperature of measurement	25 °C
Wavelength	340 nm, $d = 1$ cm (Hg 334 nm, Hg 365 nm)

4.3.4.5. Calculation

Calculation of the volume activity in μmol min^{-1} l^{-1} = U l^{-1} (see p. 234).

538

Table 99. Reference values for 2-hydroxybutyrate dehydrogenase; determination according to the recommendations of the Deutsche Gesellschaft für klinische Chemie[2] (U l^{-1}, 25 °C).

Age	Sex	Number	Reference interval	References
Umbilical-cord blood			130–400	1
New-born babies			158–530	4
1–30 days			108–460	
1–6 months			100–275	
7–12 months			98–250	
Infants			90–200	1
School children			70–160	
Adults			50–140	
Adults	m, f	4790	68–135	3

4.3.4.6. Reference values

These are given in Table 99.

4.3.4.7. Comments on the method

1. Compounds which inhibit 2-HBDH may be present in NADH$_2$ preparations, as described for NAD in the case of lactate dehydrogenase.

2. Experience shows an analysis blank to be superfluous since the process always comes to a halt.

3. Haemolytic sera may not be used since the erythrocytes contain isoenzymes, predominantly those which migrate rapidly under electrophoresis.

4. The isoenzymes which migrate rapidly under electrophoresis are relatively stable. The serum may therefore be stored in a frozen condition; however, since the low-mobility isoenzymes are sensitive the ratio should not be determined after freezing the sample.

4.3.4.8. Diagnostic significance

Since a large part of this isoenzyme derives from cardiac muscle, 2-HBDH is principally concerned in the diagnosis of lesions of cardiac muscle, cardiac infarction being the most important. After the infarction the enzyme activity in the plasma generally increases within the first 12 h and reaches a maximum after 48–72 h. The activity then falls off slowly and, if there are no complications, returns to normal within 10–20 days. Elevated 2-HBDH values have also been observed in progressive muscular dystrophy.[1] Otherwise, the diagnostic criteria for 2-HBDH are similar to those for lactate dehydrogenase (see p. 535).

References

1. Sitzmann, S. C., Das Labor in der Praxis, Marseille-Verlag, Munich, 1974, p. 77.
2. Empfehlungen der Deutschen Gesellschaft für klinischè Chemie. Standardisierung von Methoden zur Bestimmung von Enzymaktivitäten in biologischen Flüssigkeiten, *Z. klin. Chem. klin. Biochem.*, **10**, 182 (1972).
3. Weisshaar, D., Gossrau, E., and Faderl, B., Normbereiche von α-HBDH, LDH, AP und LAP bei Messung mit substratoptimierten Testansätzen., *Medsche Welt, Stuttg.*, **9**, 387 (1975).
4. Veit, S., Sitzmann, F. C., and Prestele, H., Normalwerte für Lactat- und Glutamatdehydrogenase, sowie Leucinarylamidase, erstellt mit optimierten Standardansätzen, *Klin. Pädiat.*, **187**, 244 (1975).

4.4. HAEMATOPOIETIC SYSTEM

4.4.1. HAEMOGLOBIN AND ITS DERIVATIVES

4.4.1.1. General

The identification of haemoglobin derivatives is based on the characteristic absorption spectra. These are comprehensively illustrated in Fig. 128.

Haemoglobin consists of four protein chains with four haem groups. The adult haemoglobin HbA consists of two α- and two β-chains (molecular weight of $\alpha = 15\,128$ and of $\beta = 15\,869$). Besides this, normal adult blood contains 2.5% HbA$_2$ (two α- and two δ-chains) and 0.5% HbF (two α- and two δ-chains). One haemoglobin molecule is capable of binding four O$_2$ molecules. The degree of binding depends on the pH, the temperature, the pO$_2$ (see p. 314), and the concentration of 2,3-diphosphoglycerate in the erythrocytes.

In desoxy-, oxy-, and carbomonoxyhaemoglobin, the iron is present in the haem in its divalent form. If it is trivalent the term methaemoglobin is used. The characteristic absorption maxima are as follows:

	Soret bands, nm	Others, nm
Oxyhaemoglobin	412	540, 577
Desoxyhaemoglobin	430	555
Carbomonoxyhaemoglobin	418	569
Acid methaemoglobin	405	630
Methaemoglobin cyanide	419	540

4.4.1.2. Haematocrit

4.4.1.2.1. Definition of the haematocrit

The haematocrit represents the fraction of the corpuscular elements (erythrocytes and leucocytes) in whole blood, expressed as a percentage. It is determined by centrifuging whole blood until the cells have packed to a minimum volume ('packed cells'). The height of the corpuscle column is

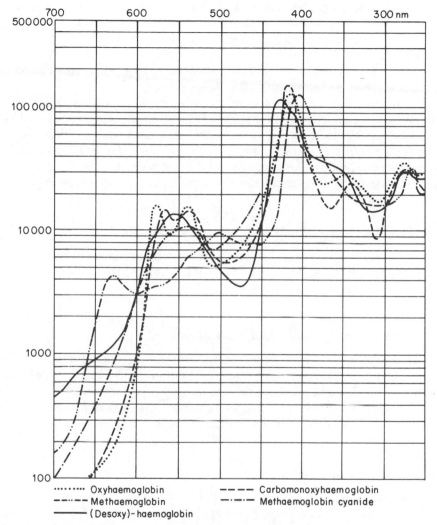

Fig. 128. Aborption spectra of some haemoglobin derivatives. Abscissa, wavelength in nm; ordinate, molar extinction (ε). The high extinction between 400 and 440 nm (Soret band) is common to all of the derivatives and is due to the presence of the porphyrin ring. Oxyhaemoglobin has two further maxima at 540 and 578 nm. The long-wave maxima of carbomonoxyhaemoglobin lie at 540 and 570 nm, i.e. compared with oxyhaemoglobin they are shifted slightly towards the short-wave region. Methaemoglobin has one peak at 500 nm and a second in the long-wave region at 630 nm. Haemoglobin has only a single peak, at 555 nm. Finally, the spectrum of methaemoglobin cyanide is also shown; this too has only a single peak, at 540 nm.

then compared with the height of the total column and the fraction of cells expressed as a percentage.

If, on centrifuging, the relative centrifugal force (RCF) is increased by enlarging the radius and the number of revolutions per minute (rpm), then the cells pack closer and closer together until a final volume is reached at infinitely great RCF. Wintrobe[1] proposes centrifugation at 3000 rpm and a radius of 15 cm, corresponding to an RCF of 1370 g. With this technique the sample must be centrifuged for at least 15 min.

The older methods, in most of which graduated glass tubes were employed for centrifuging, suffer from several disadvantages. A relatively large amount of blood was required, filling the tubes was troublesome, the centrifuging lasted 15–30 min, and the reading was relatively inaccurate. It is therefore advisable to conduct the haematocrit determination as an ultramicro method using capillaries and a high-performance centrifuge.

4.4.1.2.2. Micro determination of the haematocrit

The prime requirement for the micro determination is an efficient centrifuge. The RCF must be at least 10 000 g. The duration of the centrifuging can then be reduced to a few minutes.

4.4.1.2.3. Capillaries and reading instruments

The capillaries most frequently used conform to one of the following types: (a) 32 mm long, external diameter 0.8 mm; (b) 75 mm long, external diameter 1.2–1.4 mm. Both types are available commercially pretreated with ammonium heparinate as coagulation inhibitor. A reading instrument is essential for making exact readings of the capillary heights without strain.

4.4.1.2.4. Apparatus and instruments

Microcentrifuge with a haematocrit head.
Heparinized Guest–Weichselbaum capillaries, 75 × 1.2–1.4 mm diameter.
Plastic caps for the Guest–Weichselbaum capillaries, or plasticene.
Microhaematocrit reader.

4.4.1.2.5. Procedure

Collect the blood directly into 2–3 capillaries under gravity and capillary action (no suction). Mix the blood well with the anticoagulant by tilting the capillaries. Then close the tubes with plasticene or plastic caps and centrifuge for 3 min. Read off the haematocrit using the reader.

4.4.1.2.6. Reference values

women: 42 ± 5%
men: 47 ± 7%

4.4.1.2.7. Comments on the method

1. The duration of centrifuging is roughly inversely proportional to the RCF. If the RCF is doubled the duration of centrifuging may be halved. The quality of a centrifuge is judged on its RCF and not on the number of revolutions per minute.

2. The haematocrit capillaries may also be closed with sealing-wax or by melting the tip (match).

Reference

1. Wintrobe, M. M., *Clinical hematology*, 6th ed., Lea & Febiger, Philadelphia, 1967.

4.4.1.3. Haemoglobin as methaemoglobin cyanide

4.4.1.3.1. Choice of method

Of the various methods available for the determination of haemoglobin, the quantification as methaemoglobin cyanide (haemoglobin cyanide) is distinguished by the following advantages:

1. Haemoglobin cyanide is the only known stable haemoglobin derivative. This permits the preparation and posting of stable standard solution for control purposes.
2. The extinction maximum of haemoglobin cyanide at 540 nm is shallow (Fig. 128), so the determination does not require the isolation of monochromatic light and reliable results can be obtained with inexpensive photometers.
3. The Lambert–Beer law is valid over a wide range of values.
4. All haemoglobin derivatives are quantitatively converted to haemoglobin cyanide (oxyhaemoglobin, carboxyhaemoglobin, haemiglobin, and, to a large extent, even sulphaemoglobin).
5. Plasma proteins do not interfere, and by using the modified Drabkin solution the result may be read off after only 3–5 min.

Thus the method fulfils all the requirements of an ideal clinical chemical method.

Critique of the remaining methods of determination
1. *Hydrochloric acid haematin* (after Sahli). Irregular development of

poorly defined dyes, dependent on serum factors. Impossible to make an exact calibration of the instrument.

2. *Oxyhaemoglobin*. The colour is constant for only a short period; haemoglobin is gradually formed. The oxyhaemoglobin peak at 540 nm is so narrow that small deviations from the wavelength of measurement lead to considerable errors.

3. *Carbomonoxyhaemoglobin*. Stable derivative. However, meth-aemoglobin is excepted. Thus use of CO, or town gas, in the laboratory is unpleasant. In respect of the filter wavelengths, the same limitations apply as in the case of oxyhaemoglobin.

4. *Methaemoglobin*. This covers all derivatives and the colour is stable. However, the specific extinction of the colorant is very strongly dependent upon pH.

4.4.1.3.2. Principle

Haemoglobin is oxidized to haemiglobin with potassium hexacyanoferrate(III), and this is converted to haemiglobin cyanide by treatment with potassium cyanide. The following account adheres closely to the procedure of Zijlstra and van Kampen.[2,3]

4.4.1.3.3. Reagents

Modified Drabkin solution. Dissolve 50 mg of potassium cyanide, 200 mg of potassium hexacyanoferrate(III), 140 mg of monopotassium dihydrogen phosphate and 0.5 ml of Sterox in distilled water and dilute to 1000 ml. The solution keeps indefinitely. It is poisonous (cyanide). Do not freeze.

Haemiglobin cyanide standard ($\varepsilon = 11 \times 10^3$, monomer). Use only standards tested by the bureau. Its durability is at least 1 year in the dark. A photometer cuvette with a ground-glass stopper is filled with the standard solution and hermetically sealed. Store in a refrigerator; do not freeze. The standard may be used so long as its extinction remains stable (keep a control card). Prepare a fresh cuvette when the value falls.

4.4.1.3.4. Procedure

Place 5 ml of modified Drabkin solution in a clean test-tube and blow in 0.02 ml of blood and rinsings. After at least 3–5 min read off the extinction at 540–546 nm against Drabkin solution or water ($d = 1$ cm). Take a reading of the standard cuvette against water.

4.4.1.3.5. Calculation

Using a standard.

$$\text{Haemoglobin concentration} = \frac{E(\text{S})}{E(\text{ST})} \cdot C(\text{ST}) \cdot \frac{5.02}{0.02} \cdot \frac{1}{1000} \text{ g per 100 ml}$$

$$\text{Haemoglobin concentration} = \frac{E(S)}{E(ST)} \cdot C(ST) \cdot 0.251 \text{ g per 100 ml}$$

Using the extinction coefficient.

Molecular weight $= 64\ 458$; $\varepsilon = 44\ 000\ \text{l mol}^{-1}\text{cm}^{-1}$.

$$\text{Haemoglobin concentration} = \frac{E(S) \cdot 64458}{44\ 000} \cdot 10^{-1} \cdot \frac{5.02}{0.02} \text{ g per 100 ml}$$

$$\text{Haemoglobin concentration} = E(S) \cdot 36.8 \text{ g per 100 ml}$$

4.4.1.3.6. Reference values

See p. 548.

4.4.1.3.7. Comments on the method

1. Methods for preparing the standard were described by Zijlstra and van Kampen.[2,3] It is scarcely worthwhile preparing the standard, since cheap and accurately standardized solutions are commercially available. However, the only standards which should be used are those which have been tested by a Bureau. The purity of the standard can be checked by measuring the extinctions at 540 and 504 nm. Their ratio (540/504 nm) ought to lie between 1.58 and 1.62.[1]

2. For standardization, the iron determination is used almost without exception today; an iron content of 0.347% is assumed. Measurement of the oxygen-binding capacity is technically much more difficult. It gives lower results since 2–3% of the haemoglobin is present in an inactive condition, i.e. it will not bind oxygen. The results obtained with the manometric determination of the CO binding capacity are also about 2% lower than those from the iron determination.[3]

3. The original Drabkin solution has the following composition: 50 mg of potassium cyanide, 1 g of sodium bicarbonate and 200 mg of potassium hexacyanoferrate(III) in 1000 ml of DM-water. The disadvantage of using this and similar solutions is that it is necessary to wait 20 min before taking the reading. Also, it is said to give rise to turbidities on occasions.

4. Turbidities are practically excluded by buffering the solution and the addition of Sterox, following the instructions of Zijlstra and van Kampen.[2] This solution has the added advantage that the development of the colour is complete after only 3 min, whereas it is necessary to wait 20 min with Drabkin's solution.

5. Since the reagent contains cyanide, the work is performed with automatic pipettes exclusively. The use of automatic flushing cuvettes (filter-pump) is also advisable.

6. The method can also be used for the determination of free

haemo-globin in serum (2 ml of modified Drabkin solution + 0.5 ml of serum): values of up to 40 mg per 100 ml are normal.

Slight bodily exertions lead to a 3- to 5-fold increase and extreme physical stress to a 10- to 30-fold increase in the serum haemoglobin. More involved methods are necessary for quantifying concentrations below 20 mg per 100 ml.

7. For the quality control, sterile solutions of haemoglobin and haemoglobin cyanide are to be preferred, since then the dilution step can also be controlled, a stage which can obviously lead to considerable fluctuation. Photometric errors are already detected by control with haemoglobin cyanide solutions.[4]

References

1. International Committee for Standardization in Haemotology of the European Society of Haematology: Recommendations and requirements for haemoglobinometry in human blood, *J. clin. Path.,* **18**, 353 (1965).
2. Zijlstra, W. G. and Kampen, E. J. van, Standardization of hemoglobinometry. I. The extinction coefficient of hemiglobincyanide, *Clin. chim. Acta,* **5**, 719 (1960).
3. Zijlstra, W. G. and Kampen, E. J. van, Standardization of hemoglobinometry. III. Preparation and use of a stable hemiglobincyanide standard, *Clin. chim. Acta,* **7**, 96 (1962).
4. Assendelft, O. W. van, Buursma, A., Holtz, A. H., Kampen, E. J. van, and Zijlstra, W. G., Quality control in haemoglobinometry with special reference to the stability of haemiglobincyanide reference solutions, *Clin. chim. Acta,* **70**, 161 (1976).

4.4.1.4. Number and haemoglobin content of the erythrocytes

Wintrobe[1] invented a simple method for the precise description of the number and the haemoglobin content of the erythrocytes. The necessary parameters are haematocrit (%), haemoglobin concentration (g per 100 ml), and erythrocyte count [10^6 per mm^3 (μl)]. Of these three results the erythrocyte count—unless measured by automated methods—is notoriously unreliable. Formerly this served for calculating the colour index used in the assessment of anaemias. However, in the Wintrobe system this can be replaced by the *mean corpuscular haemoglobin concentration* (MCHC), or the mean haemoglobin concentration of the erythrocytes. To evaluate this quantity, only the haemoglobin concentration and the haematocrit value are necessary, two values which can be estimated very reliably today:

$$\text{MCHC} = \frac{\text{haemoglobin (g per 100 ml)} \cdot 100}{\text{haematocrit (\%)}} \text{ g per 100 ml erythrocytes}$$

The result gives the haemoglobin concentration of the erythrocytes in g per 100 ml. Instead of using the equation, the result may be calculated by using the nomogram in Fig. 129. This value should now be determined routinely in place of the colour index.

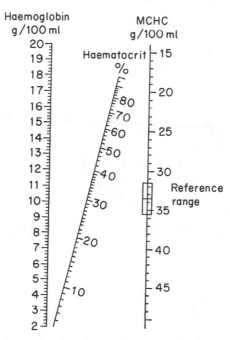

Fig. 129. Nomogram for evaluating the mean corpuscular haemoglobin concentration (MCHC) from the haemoblogin concentration and the value of the haematocrit. The haemoglobin concentration (left-hand scale) is joined by a straight line to the value of the haematocrit (central scale). The point of intersection of this line with the right-hand scale gives the mean corpuscular haemoglobin concentration.

The following quantities serve for further characterization:

Mean corpuscular volume (MCV). This is calculated by means of the equation

$$MCV\ (\mu m^3) = \frac{\text{haematocrit (\%)} \cdot 10}{\text{erythrocyte count } (10^6/mm^3)}$$

Mean corpuscular haemoglobin (MCH), or the mean haemoglobin content of a single erythrocyte. This is calculated by the equation

$$MCH\ (pg) = \frac{\text{haemoglobin (g per 100 ml)} \cdot 10}{\text{erythrocyte count } (10^6/mm^3)}$$

Mean cell diameter (MCD). This is evaluated by making a Price–Jones curve. The quantity is stated in micrometres (μm).

Mean cellular thickness (MCT). This is calculated from the equation

$$MCT\ (\mu m) = \frac{MCV}{\left(\dfrac{MCD}{2}\right)^2 \pi}$$

The normal values lie between 1.7 and 2.3 μm.

Table 100. Reference values.

Age	Erythrocyte count, ×10⁶ per mm³	Haemoglobin, g per 100 ml	Haemato-crit, %	MCV, μm³	MCH, pg	MCHC, g per 100 ml	MCD, μm
1st day	5.1 ± 1.0	19.5 ± 5.0	54.0 ± 10.0	106	38	36	8.6
2nd/3rd day	5.1	19.0	53.4	105	37	35	
4th–8th day	5.1	18.3 ± 4.0	52.5	103	36	35	
9th–13th day	5.0	16.5	49.0	98	33	34	
14th–60th day	4.7 ± 0.9	14.0 ± 3.3	42.0 ± 7.0	90	30	33	8.1
3rd–5th month	4.5 ± 0.7	12.2 ± 2.3	36.0	80	27	34	7.7
6th–12th month	4.6	11.8	35.5 ± 5.0	77	26	33	7.4
1st year	4.5	11.2	35.0	78	25	32	7.3
2nd year	4.6	11.5	35.5	77	25	32	
3rd year	6.5	12.5	36.0	80	27	35	7.4
4th year	4.6 ± 0.6	12.6	37.0	80	27	34	
5th year	4.6	12.6	37.0	80	27	34	
6th–10th year	4.7	12.9	37.5	80	27	34	7.4
11th–15th year	4.8	13.4	39.0	82	28	34	
Women	4.8 ± 0.6	14.0 ± 2.0	42.0 ± 5.0	87 ± 5	29 ± 2	34 ± 2	7.5 ± 0.3
Men	5.4 ± 0.8	16.0 ± 2.0	47.0 ± 7.0	87 ± 2	29 ± 2	32 ± 2	7.5 ± 0.3

The reference values for these quantities, taken from the results of Wintrobe,[1] are listed as a function of age in Table 100.

Reference

1. Wintrobe, M. M., *Clinical hematology*, 6th ed., Lea & Febiger, Philadelphia, 1967.

4.4.1.5. Methaemoglobin (haemiglobin): spectrophotometry

4.4.1.5.1. Choice of method

The spectrophotometric method of Evelyn and Malloy is distinguished by its simplicity and the reliability of its results.

4.4.1.5.2. Principle

Haemoglobin has a characteristic peak at 630 nm (red filter) (Fig. 128). This peak disappears on conversion of the haemiglobin to haemiglobin cyanide (treatment with potassium cyanide). In assay A, all of the haemoglobin is converted to haemiglobin by treatment with potassium hexacyanoferrate(III). After the addition of cyanide, the total (haemoglobin + haemiglobin) is then present as haemiglobin cyanide. In assay B, only the haemiglobin already present is converted to haemiglobin cyanide. The difference in extinction at 630 nm is directly proportional to the haemiglobin concentration.

4.4.1.5.3. Reagents

Potassium hexacyanoferrate(III) solution, $0.152 \, mol \, l^{-1}$. Dissolve 5 g in DM-water and make up to 100 ml. Keep in a dark bottle.
Potassium cyanide solution, $77 \, mmol \, l^{-1}$ *(highly poisonous!)*. Dissolve 500 mg of potassium cyanide in distilled water and make up to 10 ml. Do not pipette orally!
Collidine buffer, $50 \, mmol \, l^{-1}$, pH 7.0. See appendix 1.

4.4.1.5.4. Procedure

Preparation of the haemolysate. Haemolyse 0.2 ml of blood directly in 5 ml of DM-water. Allow to stand for 10 min, then add 5 ml of collidine buffer. Centrifuge briskly.

Assays.

	Cuvette	
	A	B
Supernatant, ml	3.0	3.0

Potassium hexacyanoferrate(III)	1 drop	—
DM-water	—	1 drop

Read the extinction at 630 nm: $E_1(A)$ and $E_1(B)$

Potassium cyanide solution	1 drop	1 drop

Wait for at least 5 min, then read the extinction at 630 nm again: $E_2(A)$ and $E_2(B)$

4.4.1.5.5. Calculation

$$\text{Haemoglobin (\%)} = \frac{[E_1(B) - E_2(B)] \cdot 100}{E_1(A) - E_2(A)}$$

4.4.1.5.6. Specificity

The method is specific form methaemoglobin. It fails only in the presence of certain HbM types.[2,3] In such cases the haemiglobin content has to be evaluated approximately from the determination of the oxygen-binding capacity.

4.4.1.5.7. Reference values

0.2–1.0%.

4.4.1.5.8. Comments on the method

A more expensive, but possibly more accurate, method has been described by Pilz et al.[1]

References

1. Pilz, W., Johann, I., and Boo. A. T., Die Bestimmung von Met-Hämoglobin im menschlichen Blut, *Int. J. environ. analyt. Chem.*, **2**, 179 (1973).
2. Wintrobe, M. M., Clinical hematology, 6th ed., Lea & Febiger, Philadelphia, 1967.
3. Tönz, O., The congenital methemoglobinemias, *Bibltheca haemat.*, No. 28, Karger, Basle, 1968.

4.4.1.6. Carbomonoxyhaemoglobin: spectrophotometry

4.4.1.6.1. Choice of method

Chemical, gas analytical, and spectrophotometric methods are available for the determination of carbomonoxyhaemoglobin. For laboratories possessing a spectrophotometer with a narrow, i.e. monochromatic, bandwidth, the spectrophotometric analyses are distinguished by their simplicity and reliability.

4.4.1.6.2. Principle

If two substances having different absorption spectra are present in the same solution, then the concentrations of both components may be calculated from the ratio of the extinctions at two different, appropriate wavelengths. In the present case we are concerned with evaluating the percentage of carbomonoxyhaemoglobin in the total haemoglobin. To this end, the haemoglobin (but not the carbomonoxyhaemoglobin) is converted to oxyhaemoglobin by reaction with ammonia. The carbomonoxyhaemoglobin fraction can now be evaluated by measurement of the extinction at two different wavelengths, as the absorption spectra of the two components are different. The method described conforms closely to that of Bruckner and Desmond.[1] The addition of saponin to the ammonia solution inhibits the occurrence of turbidities.

4.4.1.6.3. Reagents

Saponin solution. Dissolve 1.5 g of saponin in 100 ml of 0.1 ml l^{-1} ammonia solution. Always shake well before use. The solution keeps indefinitely.

4.4.1.6.4. Procedure

Add 0.02 ml of capillary blood to 5 ml of saponin solution in a centrifuge tube. Mix carefully (do not shake, since otherwise the carbomonoxyhaemoglobin dissociates). Centrifuge briskly for a brief period. Transfer the supernatant to a cuvette with a 1-cm path length and take readings of the extinction against saponin–ammonia solution at 541, 561, 573, 577, and 597 nm.

4.4.1.6.5. Calculation

1. Calculate the ratio $E_{573\,nm}/E_{597\,nm}$. The calculation may proceed further provided this value lies between 8.17 and 8.29.
2. The ratios $E_{541\,nm}/E_{561\,nm}$ and $E_{577\,nm}/E_{561\,nm}$ are calculated. The concentration of carbomonoxyhaemoglobin (in per cent) may be read off directly from the nomogram (Fig. 130).

4.4.1.6.6. Reference values

Physiologically, there should be less than 2% of carbomonoxyhaemoglobin in the blood. For non-smokers the values are usually less than 0.5%, but for smokers they may reach 6.5%.

552

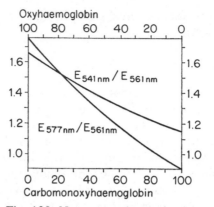

Oxyhaemoglobin

Fig. 130. Nomogram for evaluating the carbomonoxhaemoglobin concentration from the ratios and $E_{541\,nm}/E_{561\,nm}$ and $E_{577\,nm}/E_{561\,nm}$ (from ref. 1).

4.4.1.6.7. Comments on the method

1. The ratio of the isobestic points for oxyhaemoglobin and carbomonoxyhaemoglobin is the highest for all known haemoglobin derivatives. For comparison the quotients for a few other compounds and wavelengths are listed below:

Pigment	$E_{562\,nm}/E_{540\,nm}^{2}$	$E_{573\,nm}/E_{597\,nm}$
Oxyhaemoglobin	0.594	8.23
Carbomonoxyhaemoglobin	0.881	8.22
Haemoglobin	1.213	2.51
Haemoglobin (alkaline)	0.611	1.26
Sulphaemoglobin	1.026	1.15
Haemialbumin		1.14

2. The nomogram is calculated by means of the following equation:

$$\text{Intermediate ratio} = \frac{(E_{577\,nm}^{HbCO} \cdot c\%) + (E_{577\,nm}^{HbO_2} \cdot (100-c)\%)}{(E_{561\,nm}^{HbCO} \cdot c\%) + (E_{561\,nm}^{HbO_2} \cdot (100-c)\%)}$$

3. If the ratio $E_{573\,nm}/E_{597\,nm}$ is less than 8.17, then either there are other haemoglobin derivatives present or the solution is turbid. If necessary, the turbidity can be cleared by longer centrifugation, but in all such cases it is best to search for other haemoglobin derivatives at the same time.

4. Since the extinctions are temperature dependent, work must be carried out at constant temperature.

5. Blood may be kept several days before the determination, provided that it is well sealed.

6. If haemiglobin and carbomonoxyhaemoglobin are present simultaneously, the latter can still be determined by using a correction factor.

7. The above method yields reliable results only if a sophisticated spectrophotometer is used.

8. A reference method in which the HbCO is converted to CO_2 and the latter titrated has been described by Dijkhuizen et al.[2] Comparison has shown that the spectrophotometric method yields acceptable results.

References

1. Bruckner, J. and Desmond, F. B., A spectrophotometrical method for the estimation of carbon monoxide haemoglobin in blood, *Clin. chim. Acta,* **3**, 173 (1958).
2. Dijkhuizen, P., Buursma, A., Gerding, A. M., Kampen, E. J. van, and Zijlstra, W. G., Carboxyhaemoglobin, spectrophotometric determination tested and calibrated using a new reference method for measuring carbon monoxide in blood, *Clin. chem. Acta,* **80**, 95 (1977).

4.4.1.7. Carbomonoxyhaemoglobin: ratio method

4.4.1.7.1. Principle

If a line spectrophotometer is available, the carbomonoxyhaemoglobin concentration can also be calculated from the ratios of the extinctions at 546 and 578 nm.

4.4.1.7.2. Reagents

Sodium carbonate solution, 9.44 mmol l⁻¹. Dissolve 1.5 g of saponin and 100 mg of sodium carbonate (anhydrous) in DM-water and make up to 100 ml. The solution keeps indefinitely at room temperature.

4.4.1.7.3. Procedure

Add 0.02 ml of whole blood to 1.5 ml of sodium carbonate solution. Mix with care (do not shake since otherwise the carbomonoxyhaemoglobin dissociates). Centrifuge briskly for a short period. Read off the extinction at 546 nm against DM-water. Change the filter and re-set, then read the extinction at 578 nm against DM-water.

4.4.1.7.4. Calculation

Calculate $Q = E_{546\,nm}/F_{578\,nm}$ and read off the percentage of carbomonoxyhaemoglobin from Table 101.

Table 101. Conversion table.

$\dfrac{E_{546\ nm}}{E_{578\ nm}}$	Carbomonoxy haemoglobin, %	$\dfrac{E_{546\ nm}}{E_{578\ nm}}$	Carbomonoxy-haemoglobin, %	$\dfrac{E_{546\ nm}}{E_{578\ nm}}$	Carbomonoxy-haemoglobin, %
0.915	0	0.998	31	1.113	61
0.917	1	1.001	32	1.117	62
0.919	2	1.004	33	1.122	63
0.920	3	1.007	34	1.127	64
0.923	4	1.010	35	1.132	65
0.925	5				
0.928	6	1.014	36	1.136	66
0.931	7	1.017	37	1.141	67
0.934	8	1.021	38	1.146	68
0.936	9	1.024	39	1.151	69
0.939	10	1.028	40	1.156	70
0.941	11	1.031	41	1.991	71
0.944	12	1.035	42	1.166	72
0.946	13	1.038	43	1.172	73
0.949	14	1.042	44	1.177	74
0.952	15	1.046	45	1.182	75
0.955	16	1.050	46	1.188	76
0.957	17	1.054	47	1.193	77
0.960	18	1.058	48	1.199	78
0.963	19	1.062	49	1.205	79
0.966	20	1.066	50	1.211	80
0.968	21	1.070	51	1.217	81
0.971	22	1.074	52	1.223	82
0.974	23	1.078	53	1.229	83
0.977	24	1.082	54	1.235	84
0.980	25	1.086	55	1.241	85
0.983	25	1.091	56	1.248	86
0.986	27	1.095	57	1.255	87
0.989	28	1.100	58	1.262	88
0.992	29	1.104	59	1.269	89
0.995	30	1.109	60	1.276	90

4.4.1.7.5. Reference values

See p. 551.

4.4.1.7.6. Comments on the method

The extinction values are usually greater than 1.0, so the stepwise compensation must be used.

4.4.2. HAEMOGLOBIN SYNTHESIS AND PORPHYRIN METABOLISM

4.4.2.1. General

Porphyrins are precursors of haem. The physiologically important haem-containing proteins include haemoglobin, myoglobin, peroxidases, catalase, and cytochrome (especially cytochrome P_{450}, which is essential for the detoxication of many pharmaceuticals). Figure 131 shows the pathway for the synthesis of haem from glycine and succinyl-CoA. The synthesis of the δ-aminolaevulinic acid (ALA) and haem, the ultimate product, occurs in the mitochondrion, but that of the intermediate stages takes place in the cytosome so that the synthesis also involves additional transport steps. If there is an inherited defect or an acquired disturbance of an enzyme, there is an accumulation of the substrate preceding the block as well as a deficiency

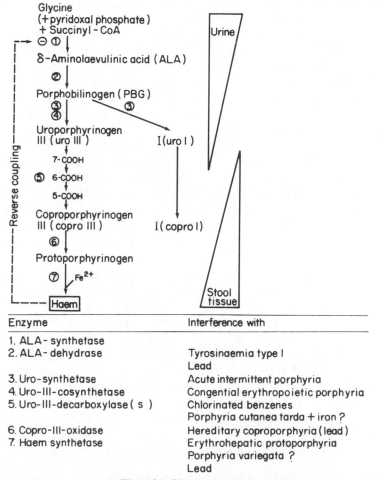

Enzyme	Interference with
1. ALA-synthetase	
2. ALA-dehydrase	Tyrosinaemia type I
	Lead
3. Uro-synthetase	Acute intermittent porphyria
4. Uro-III-cosynthetase	Congential erythropoietic porphyria
5. Uro-III-decarboxylase(s)	Chlorinated benzenes
	Porphyria cutanea tarda + iron ?
6. Copro-III-oxidase	Hereditary coproporphyria (lead)
7. Haem synthetase	Erythrohepatic protoporphyria
	Porphyria variegata ?
	Lead

Fig. 131. Haem synthesis.

in products. Accumulated metabolites are, if necessary, degraded via secondary pathways—uroporphyrinogen I (uro I), coproporphyrinogen I (copro I). For an appreciation of the findings, heed should be paid to the reverse coupling through haem on the ALA synthetase. If, owing to an enzyme block (and/or additional over-consumption of haem-containing proteins?), there should be a deficiency in haem, this would result in a de-inhibition of the synthesis of ALA and thus to an over-production of ALA and its metabolites up to the block.

The porphyrin precursors, ALA and porphobilinogen (PBG), as well as the polar porphyrins are excreted predominantly in the urine (uroporphyrins), whereas the lipophilic (decarboxylated) porphyrins are excreted predominantly in the stool (coproporphyrins) and tissue (protoporphyrin in the erythrocytes), as shown diagrammatically in Fig. 131. The porphyrinogens are irreversibly oxidized to porphyrins. The sensitivity to light and the instability of the various metabolites render laboratory diagnosis more difficult and necessitate the immediate processing of the material; in this connection it is advantageous to acidify the urine (e.g. with tartaric acid) for ALA determination but to make it alkaline (with 1% Na_2CO_3) for the determination of other metabolites. Elevations of the porphyrin concentration are encountered primarily in inherited defects of the enzyme participating in the haem synthesis, secondarily in disturbances of this enzyme due to other causes (secondary porphyrinurias). Of the inherited defects the hepatic forms are of greatest clinical importance. In acute cases these lead to symptoms which resemble those of many other diseases: psychic and neurological disturbances, acute abdominal pain, hypertonia, acidosis, and inadequate secretion of antidiuretic hormone. The exclusion of a hepatic porphyria by the Hoesch test is therefore essential to differential diagnosis. A series of pharmaceuticals can actuate such acute attacks. The observations of porphyria described by Stokvis in 1889[1] were made on patients who had taken the soporific sulphonal. The commonest of the *hepatic porphyria* is *acute intermittent porphyria*, the other forms being *porphyria variegata* and the very rare *coproporphyria*. In porphyria cutanea tarda the most prominent symptoms are in the skin. In addition to the hepatic porphyria there are *porphyria which predominantly affect the erythrocytes* (*congenital erythropoietic porphyria, erythrohepatic protoporphyria, erythropoietic coproporphyria*). The enzyme defects responsible for these hepatic and erythropoietic porphyrias are listed in the legend to Fig. 131. A secondary increase in porphyrin or its precursors in the urine, stool, or tissue occurs with tumours (hepatoma!), poisoning by heavy metals or chlorinated benzenes (tetrachlorodibenzodioxin, TCDD), and in infectious or cholestatic diseases of the liver.[2–5]

4.4.2.2. Recognition of porphyrias[2] (cf. Table 102)

During acute attacks, the *hepatic porphyrias* lead to an increased excretion of porphobilinogen and—to a smaller extent—of ALA. The

Table 102. Recognition of porphyrias.

	Acute intermittent porphyria	Congenital erythro-poietic porphyria	Porphyria cutanea tarda	Hereditary copro-porphyria	Erythrohepatic protoporphyria Porphyria variegata
ALA	↑↑		(↑)	(↑)	(↑)
PBG	↑↑			(↑)	(↑)
Uro III			↑	↑	
Copro III			(↑)	↑↑	↑
Proto					↑↑
Uro I		↑↑	↑		
Copro I		↑↑	↑		↑

↑↑ = Highly elevated; ↑ = elevated; (↑) = contingent elevation, especially during attacks.

detection of porphobilinogen is therefore suited to the diagnosis of these porphyrias.[5] Porphyria cutanea tarda forms an exception (liver disease).

If the preliminary test is positive, selective tests should be made for ALA, porphobilinogen, and the porphyrins in the urine, or in the stool and the erythrocytes, and, where necessary, these substances should be quantified. The determination of the porphyrins is necessary for the detection of the rarer hepatic porphyrias in the interval when no elevation of porphobilinogen is found and, particularly, in porphyria cutanea tarda. The enzyme assay[6–9] allows an unequivocal diagnosis.

Lead poisoning is suitably diagnosed by quantifying the ALA in the urine, or better by determination of the protoporphyrin zinc complexes in the erythrocytes.[10,11] Further causes of secondary increase of protoporphyrin in the erythrocytes include iron-deficiency anaemia and haemolytic anaemia.

Erythropoietic porphyrias. In the rare congenital erythropoietic porphyria, the wine-red colour of the urine should institute a directed investigation of the porphyrins. In the even rarer erythrohepatic protoporphyria, where the colour of the urine is not peculiar, the detection (fluorescence microscopy) of protoporphyrins in the erythrocytes is necessary.

4.4.2.2.1. Exploratory test for the detection of porphobilinogen

Since porphobilinogen (and δ-aminolaevulinic acid) is elevated during attacks of hepatic porphyrias, the detection of these metabolites plays an important role.

4.4.2.2.2. Choice of method

The Hoesch test is superior to the Watson–Schwarz test on account of its higher specificity and, consequently, the smaller likelihood of false positive results.[5,12,13]

4.4.2.2.3. Principle

In acidic media, porphobilinogen reacts with *p*-dimethyl-aminobenzaldehyde to give a cherry-red condensation product (reverse urobilinogen reaction).

4.4.2.2.4. Reagents

p-Dimethylaminobenzaldehyde, 2g in 100 ml of 6 mol l^{-1} HCl (modified Ehrlich-reagent). The solution may be kept for at least 9 months in a glass container.

4.4.2.2.5. Procedure

One or two drops of fresh urine are added to 1 ml of the modified Ehrlich reagent. The test is positive (porphobilinogen > 5 mg l^{-1}) if a brilliant pink to cherry-red coloration appears at the surface immediately on mixing the contents of the whole tube. More than 2 drops of urine should not be added as otherwise non-specific reactions interfere. A yellow or orange coloration is not due to an increase in porphobilinogen.

4.4.2.2.6. Specificity

Urobilinogen does not interfere (at least up to 20 mg per 100 ml).

4.4.2.2.7. Comments on the method

Since porphobilinogen is not stable, the test must be conducted on fresh urine.

4.4.2.2.8. Diagnostic significance

In the acute condition, all hepatic porphyrias lead to elevation of porphobilinogen. In the intervening remission other methods must be used for testing for uro-, copro- and protoporhyrin.[14–17] Pollack[18] recommends the Rimington test as an exploratory test for an elevation of uro- and coproporphyrin: 2 ml of urine are acidified with 5 drops of glacial acetic acid. The porphyrins are extracted into amyl alcohol (11 drops) and this upper, alcoholic layer is observed under ultraviolet light (366 nm). Porphyrins give a pink or red fluorescence. Porphobilinogen is not elevated in erythropoietic porphyria. In congenital erythropoietic porphyria, uroporphyrin (urine) and coprophorphyrin (faeces) are elevated; in erythropoietic protoporphyria, protoporphyrin must be detected in the erythrocytes or in the stool.

If the exploratory test is positive the porphyrin precursors and the various porphyrins in the urine and faeces or in the erythrocytes should be selectively detected and quantified. The semi-quantitative separations by means of thin-layer chromatography,[3,19] ion-exchange chromatography,[20] and the expensive extraction methods coupled with spectrophotometry are now being superseded by the separation by high-performance liquid chromatography of the native porphyrins or their methyl esters, the detection being accomplished spectrophotometrically or fluorimetrically.[21]

4.4.2.3. Isolation of δ-aminolaevulinic acid and porphobilinogen

4.4.2.3.1. Choice of method

Owing to interference, the direct quantification of δ-aminolaevulinic acid and porphobilinogen is inexact. For the isolation of these porphyrin precursors the ion-exchange chromatographic methods of Mauzerall and Granick[22] and Schaller et al.[23] have proved valuable. Ready-made columns are commercially available.

4.4.2.3.2. Separation of δ-aminolaevulinic acid and porphobilinogen

Porphobilinogen is adsorbed on to an anion-exchange column (AG2-X8, 100–200 mesh, acetate form) at pH 5–6. Urea, ALA, aminoacetone, amino acids, and chromogenes present in other than anionic forms are eluted with water. This eluate is transferred to a cation-exchange column for the determination of ALA. Porphobilinogen is eluted from the anion-exchange column with glacial acetic acid and determined colorimetrically. The ALA from the aqueous eluate of the anion-exchange column is adsorbed on the cation-exchange column (AG50-X8, 100–200 mesh, H^+ form). Urea is removed by washing the column with water. The ALA (together with aminoacetone) is eluted with sodium acetate buffer and condensed with acetylacetone to form pyrrole, which is determined by colorimetry.

4.4.2.4. δ-Aminolaevulinic acid: determination with Ehrlich reagent

4.4.2.4.1. Choice of method

δ-Aminolaevulnic acid can be determined with alkaline picric acid solution on the principle of the Jaffé reaction, but the more customary, and technically safer, method is the conversion of the δ-aminolaevulinic acid to a pyrrole which, on treatment with Ehrlich reagent, produces a dye which can be measured photometrically.

4.4.2.4.2. Principle

The δ-aminolaevulinic acid-containing eluate (see above) is treated with acetylacetone and heated. Under these conditions the δ-aminolaevulinic acid condenses with the acetylacetone and forms the monopyrrole, 2-methyl-3-acetyl-4-(3-propionic acid) pyrrole. This pyrrole gives a positive reaction with Ehrlich reagent. The result is best calculated by means of the molar extinction coefficient. If a photometer providing monochromatic radiation is available a δ-aminolaevulinic acid standard may also be used for comparison.

4.4.2.4.3. Reagents

Acetylacetone, p.a. grade.

δ-Aminolaevulinic acid, standard solution, 100 $\mu mol\ l^{-1}$. Dissolve 13.1 mg of δ-aminolaevulinic acid in 1 mol l^{-1} sodium acetate buffer (pH 4.6) and make up to 1000 ml with further buffer. Store in a frozen condition.

Acetate buffer, 1 *mol l^{-1}, pH* 4.6. Dilute 57 ml of glacial acetic plus 136 g of sodium acetate trihydrate to 1000 ml with DM-water. The solution keeps for a few weeks, if frozen.

Ehrlich reagent, modified according to Rimington et al.[24] Dissolve 60 g of p-dimethylaminobenzaldehyde in 260 ml of concentrated hydrochloric acid and dilute to 1000 ml with glacial acetic acid. The solution keeps for a few months at room temperature.

4.4.2.4.4. Procedure

Treat the eluted mixture (p. 559) with 0.2 ml of actylacetone and make up to 10 ml with acetate buffer. Close the measuring flask and place it in a boiling water-bath for 10 min. Allow to cool to room temperature. To 2 ml of the solution add 2 ml of modified Ehrlich reagent. After 15 min read the extinction at 555 nm against a blank consisting of 1 part of Ehrlich reagent and 1 part of acetate buffer. The colour is stable for about 15 min.

4.4.2.4.5. Calculation

Molecular weight = 131.15; $\varepsilon = 6.2 \times 10^4\ l\ mol^{-1}\ cm^{-1}$.

Concentration of δ-aminolaevulinic acid

$$= \frac{E(S) - E(B)}{\varepsilon d} \cdot \frac{FV}{SV} \cdot \frac{FV(\text{eluate})}{SV(\text{urine})} \cdot 10^6\ \mu mol\ l^{-1}$$

4.4.2.4.6. Specificity

A positive reaction is obtained not only with δ-aminolaevulinic acid, but also with aminoketones, glucosamine, ammonia, and glycine.

4.4.2.4.7. Reference values

$n = 87$	0.1–6 mg l^{-1}	0.76–46 $\mu mol\ l^{-1}$	Ref. 25

Doss[25] found an approximately log-normal distribution for the δ-aminolaevulinic acid excretion.

4.4.2.4.8. Stability

δ-Aminolaevulinic acid is very stable provided the pH of the urine is below 7. In this case the urine may even be kept at room temperature for a few days, but if the urine is alkaline the δ-aminolaevulinic acid remains stable for only a few hours at most. For storage (in the dark) the addition of tannic acid is recommended.[26]

4.4.2.4.9. Comments on the method

1. The δ-aminolaevulinic acid standard is used as follows: 1 ml of standard solution is treated with 0.2 ml of acetylacetone and made up to 10 ml with acetate buffer. The further procedure is as described in Section 4.4.2.4.4. The equation for calculating the concentration is

$$\delta\text{-Aminolaevulinic acid concentration} = \frac{E(S) - E(B)}{E(ST) - E(B)} \cdot \frac{10}{1} \cdot 100 \, \mu\text{mol}\,l^{-1}$$

Commercial δ-aminolaevulinic acid is seldom pure and it is advisable to estimate its purity by means of the molar extinction coefficient. The same standard solution can also be used for calibrating the porphobilinogen method (p. 563).

2. The Ehrlich reagent as modified by Rimington et al.[24] is more stable than the modification according to Mauzerall and Granick[22] and is only slightly less sensitive.

4.4.2.5. Porphobilinogen

4.4.2.5.1. Introduction

Porphobilinogen is detected by the Ehrlich reaction. The exact mechanism of this reaction is uncertain. The reaction of porphobilinogen with p-dimethylaminobenzaldehyde probably leads to the formation of a red condensation product, the so-called 'Ehrlich's porphobilinogen aldehyde' (Fig. 132). However, this dye is not stable but undergoes further reaction and colourless reaction products are formed in rapid succession. Consequently, when using the Ehrlich reaction for photometry, the time between the addition of the reagent and reading the extinction must be accurately observed.

The intensity of colour in the Ehrlich reaction is strongly dependent on the experimental conditions, particularly on the pH. This must be borne in mind when using the modified reagent. Some of the frequently used modifications are as follows:

Ehrlich's original reagent. Dissolve 20 g of p-dimethylaminobenzaldehyde

562

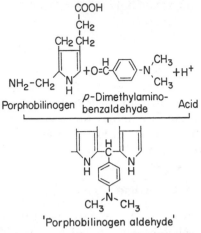

'Porphobilinogen aldehyde'

Fig. 132. 'Ehrlich's porphobilinogen aldehyde'.

in 6 mol l^{-1} hydrochloric acid and make up to 1000 ml with further hydrochloric acid. The solution keeps for a few months at room temperature.

Reagent as modified by Rimington et al.[24] Dissolve 60 g of p-dimethylamino-benzaldehyde in 260 ml of concentrated hydrochloric acid and dilute to 1000 ml with glacial acetic acid. The solution keeps for a few months at room temperature.

Reagent as modified by Mauzerall and Granick.[22] Dissolve 20 g of p-dimethylaminobenzaldehyde in 600 ml of glacial acetic acid, add 160 ml of 70% perchloric acid and dilute the solution to 1000 ml with glacial acetic acid. The reagent is not stable and must be prepared freshly each day.

When subjected to the Ehrlich reaction, porphobilinogen produces a characteristic red colour with an extinction maximum at 555 nm and a shoulder at 525 nm. For pure solutions, the ratio $E_{525\ nm}/E_{555\ nm}$ is 0.73. Most of the other Ehrlich-positive substances show a maximum in the region of 525 nm and therefore give different ratios. In doubtful cases it is advisable to measure the extinctions at 555 and 525 nm and thus verify the presence of porphobilinogen. The molecular extinction coefficients of the dyes are strongly dependent on the composition of the reagent. A few examples are given in Table 103.

4.4.2.5.2. Stability of porphobilinogen

Porphobilinogen is among the least stable of the compounds of diagnostic significance. At room temperature up to 70% of the prophobilinogen disappears in 10 h. The instability is due to the spontaneous conversion of porphobilinogen to the red ether-soluble pigment porphobilin on the one

Table 103. Molar extinction coefficients of the reaction products of porphobilinogen and δ-aminolaevulinic acid pyrrole (p. 559) with Ehrlich reagents (molecular weights: porphobilinogen 226.26, δ-aminolaevulinic acid 131.15).

Reagent	Molar extinction coefficient	
	Porphobilinogen	δ-Aminolaevulinic acid
Ehrlich	$3.5 \cdot 10^4$	?
Mauzerall and Granick[22]	$6.2 \cdot 10^4$	$7.2 \cdot 10^4$
Rimington et al.[24]	$5.4 \cdot 10^4$	$6.2 \cdot 10^4$

hand and to porphyrins, probably uroporphyrinogen and uroporphyrin, on the other. Porphyrin is most stable at a pH less than 1, but under these conditions δ-aminolaevulinic acid and the porphyrins are not stable. In practice, therefore, either the analyses must be conducted immediately, i.e. within 4 h, or the urine must be collected in a dark bottle and immediately frozen; in this case 5 g of sodium carbonate may be added per 24-h collection.

4.4.2.6. Porphobilinogen: photometric determination after isolation

4.4.2.6.1 Principle

Porphobilinogen is first adsorbed on to an anion-exchange column and subsequently eluted. The collected eluates are treated with Ehrlich reagent. Here, too, it is advisable to use the reagent as modified by Rimington et al.[24] The concentration is calculated by means of the molar extinction coefficient. If the monochromaticity of the available photometer is inadequate, a standard may be analysed in parallel (p. 561).

4.4.2.6.2. Reagents

Ehrlich reagent as modified by Rimington et al.[24] The preparation is described on p. 560.

4.4.2.6.3 Procedure

The eluates from the AG2 anion-exchange column are shaken together and diluted to 10 ml with DM-water. A 2-ml volume of the diluted solution is mixed with 2 ml of Ehrlich reagent. After exactly 5 min, the extinctions are read at 555 nm against DM-water.

4.4.2.6.4. Calculation

Molecular weight = 226.26; $\varepsilon = 5.4 \times 10^4 \, \mathrm{l \, mol^{-1} \, cm^{-1}}$.

$$\text{Porphobilinogen concentration} = \frac{E(S)}{\varepsilon d} \cdot \frac{FV}{SV} \cdot \frac{FV(\text{eluate})}{SV(\text{urine})} \cdot 10^6 \ \mu\text{mol l}^{-1}$$

4.4.2.6.5. Reference values

$n = 87$	0–1.6 mg l^{-1}	0–7.1 μmol l^{-1}	Ref. 25

4.4.2.6.6. Physiological variability

Little is known about the physiological scatter of values for porphobilinogen excretion. In the course of childhood the daily excretion gradually increases, attaining adult values at puberty. Sex differences are unknown.

4.4.2.6.7. Comments on the method

1. After the isolation of the porphobilinogen, the Lambert–Beer law is valid for measured values of the extinction up to 0.7. This is probably because inhibitors present in the urine are removed in the chromatography.
2. The method permits the detection of 3 μmol l^{-1}.
3. Instead of isolating the porphobilinogen by chromatography, a batch method may be employed. The results are identical.
4. If no absolute-measuring photometer providing monochromatic light is available, a standard is analysed in a parallel assay. Since porphobilinogen is unstable it is best to use 2-methyl-3-acetyl-4-(3-propionic acid)pyrrole, which is easily prepared from δ-aminolaevulnic acid and has photometric properties similar to those of porphobilinogen (for preparation of the standard, see p. 560). However, the extinction of porphobilinogen at 555 nm is higher than that of δ-aminolaevulnic acid; in fact, 1 mg of δ-aminolaevulnic acid corresponds to 0.73 mg of porphobilinogen.
5. A non-specific green coloration produced by reaction with phenothiazine metabolites should not be confused with the red coloration obtained with porphobilinogen.[27,28]

References

1. With, T. K., Acute porphyria, toxic and genuine in the light of history, *Dan. med. Bull.*, **18**, 112 (1971).
2. Lamon, J. M., Clinical aspects of porphyrin measurement, other than lead poisoning, *Clin. Chem.*, **23**, 260 (1977).
3. Elder, G. H., Gray, C. H., and Nicholson, D. C., The porphyrias: a review, *J. clin. Path.*, **25**, 1013 (1972).
4. Levere, R. D. and Kappas, A., Biochemical and clinical aspects of the porphyrias, *Adv. clin. Chem.*, **11**, 133 (1968).
5. Colombo, J. P. and Richterich, R., *Die einfache Urinuntersuchung*, Huber, Berne, 1977.
6. Meyer, U. A., Intermittierend akute Porphyrie. Klinische Bedeutung der

verminderten Aktivität der Uroporphyrinogen-I-Synthetase, *Schweiz. med. Wschr.,* **104**, 1874 (1974).

7. Watson, C. J., Hematin and porphyria, *New Engl. J. Med.,* **293**, 605 (1975).
8. Elder, G. H., Thomas, N., and Evans, J. O., The primary enzyme defect in hereditary coproporphyria, *Lancet,* **ii**, 1217 (1976).
9. Bloomer, J. R., Bonkowsky, H. L., Ebert, P. S., and Mahoney, M. J., Inheritance in protoporphyria comparison of haem synthetase activity in skin fibroblasts with clinical features, *Lancet,* **ii**, 226 (1976).
10. Hanna, T. L., Dietzler, D. N., Smith, C. H., Gupta, S., and Zarkowsky, H. S., Erythrocyte porphyrin analysis in the detection of lead poisoning in children: evaluation of four micromethods, *Clin. Chem.,* **22**, 161 (1976).
11. Lamola, A. A., Joselow, M., and Yamane, T., Zinc protoporphyrin (ZPP): a simple sensitive, fluorometric screening test for lead poisoning, *Clin. Chem.,* **21**, 93 (1975).
12. Hoesch, K., Über die Auswertung der Urobilinogenurie und die "umgekehrte" Urobilinogenreaktion, *Dt. med. Wschr.,* **72**, 704 (1947).
13. Lamon, J., With, T. K., and Redeker, A. G., The Hoesch test: bedside screening for urinary porphobilinogen in patients with suspected porphyria, *Clin. Chem.,* **20**, 1438 (1974).
14. Rimington, C., Quantitative determination of porphobilinogen and porphyrins in urine and porphyrins in faeces and erythrocytes, *Broadsheet Ass. clin. Path.,* **70**, 1 (1971).
15. Grinstein, M., Simplified method for the determination of porphyrins in body fluids, *Analyt. Biochem.,* **77**, 577 (1977).
16. Elder, G. H., Differentiation of porphyria cutanea tarda symptomatica from other types of porphyria by measurement of isocoproporphyrin in faeces, *J. clin. Path.,* **28**, 601 (1975).
17. Doss, M. and Schmidt, A., Kurzmitteilung, *Z. klin. Chem. klin. Biochem.,* **10**, 230 (1972).
18. Pollack, A., Diagnosing porphyria: why aren't labs interested? *Med Labs Mgmt int.,* **1/4**, 27 (1976).
19. Doss, M., Look, D., Henning, H., Lüders, C. J., Dölle, W., and Strohmeyer, G., Chronische hepatische Porphyrien, *Z. klin. Chem. klin. Biochem.,* **9**, 471 (1971).
20. Sobel, C., Cano, C., and Thiers, R. E., Separation and quantitation of coproporphyrin and uroporphyrin in urine, *Clin. Chem.,* **20**, 1397 (1974).
21. Adams, R. F., Slavin, W., and Williams, A. R., Porphyrin experiments in urine using high pressure liquid chromatography and fluorescence detection, *Chromatogr. Newsl.,* **4**, 24 (1976).
22. Mauzerall, D. and Granick, S., The occurrence and determination of δ-aminolevulinic acid and porphobilinogen in urine, *J. biol. Chem.,* **219**, 435 (1956).
23. Schaller, K.-H., Lehnert, G., and Szadkowski, D., Praxisgerechte δ-Aminolävulinsäurebestimmungen im Harn, *Arbeitsmed. Sozialmed. Arbeitshyg.,* **4**, 329 (1969).
24. Rimington, C., Krol, S., and Tooth, B., Detection and determination of porphobilinogen in urine, *Scand. J. clin. Lab. Invest.,* **8**, 251 (1956).
25. Doss, M. O., Porphyrins and porphyrin precursors, in Curtius and Roth, *Clinical biochemistry,* de Gruyter, Berlin, 1974, p. 1323.
26. Vincent, W. F. and Ullmann, W. W., The preservation of urine specimens for δ-aminolevulinic acid determination, *Clin. Chem.,* **16**, 612 (1970).
27. McEwen, J. and Paterson, C., Drugs and false-positive screening tests for porphyria, *Br. med. J.,* **i**, 421 (1972).
28. Reio, L. and Wetterberg, L., False porphobilinogen reactions in the urine of mental patients, *J. Am. med. Ass.,* **207**, 148 (1969).

4.5. GASTRO-INTESTINAL TRACT

4.5.1. INVESTIGATION OF GASTRIC SECRETION

4.5.1.1. Introduction[1-7]

The gastric juice is a mixture of at least three different glandular products: a solution of hydrochloric acid, a buffer solution, and a neutral diluent. There is no doubt that the formation of hydrochloric acid takes place in the cells of the lining (parietal cells). In human beings about 50 million of these cells produce 1 mmol of hydrochloric acid per hour, under maximal histamine stimulation. The production of the enzyme-containing buffer solution occurs in the principal cells of the stomach. The enzymes produced in the stomach include pepsin and gastricsin. The former has an optimal pH less than 1 and this is certainly independent of the substrate. Gastricsin has an optimal pH of 3. Both enzymes exist in several molecular forms (isoenzymes); particular observations indicate that these isoenzymes are produced in different sections of the stomach.

There is much uncertainty about the origin of the non-parietal secretion and its composition. Possible sites of production are the glandular cells of the cardiac sphincter and the pylorus, the principal cells of the fundus glands, the mucous cells of the glandular throats, and the cells of the surface epithelium. The expressions 'mucous secretion', 'diluent secretion', 'buffer secretion', 'alkaline component', and 'non-parietal component' probably all refer to the same fluid. However, the composition of this solution is heterogeneous and thus the possibility that it is a mixture of products from different glands cannot be ruled out. The most important components of this mixture are mucus, acid mucopolysaccharides, neutral glycoproteins, acid glycoproteins, blood-group substances, plasma proteins, and the 'intrinsic factors'. Nowadays the only measurement of diagnostic importance is that of the function of the acid-producing parietal cells. In this context it should be noted that in investigations of the efficiency of secretion of the gastric glands it is not the concentration of a particular component in the secretion which is of significance, but rather the rate of secretion in unit time. The efficiency of secretion of the gastric glands should therefore be referred to minutes or hours and should not be expressed as a concentration.

References

1. Babkin, B. P., *Secretory mechanism of the digestive glands*, 2nd ed., Hoeber, New York, 1950.
2. Ivy, A. C., Grossman, M. I., and Bachrach, W. H., *Peptic ulcer*, Blakiston, Philadelphia, 1950.
3. Wolf, S. and Wolf, H. G., *Human gastric function*, Oxford University Press, New York, 1943.
4. Conway, E. J., *The biochemistry of gastric secretion*, Thomas, Springfield, 1952.
5. Gregory, R. A., *Secretory mechanisms of the gastrointestinal tract*, Arnold, London, 1962.
6. Bockus, H. C., *Gastroenterology*, Saunders, Philadelphia, 1963.
7. Sun, D. C. H., Chemistry and therapy of peptic ulcer, Thomas, Springfield, 1966.

4.5.1.2. Notions of acid concentration

The presence of hydrochloric acid in the gastric juices was demonstrated by Prout[1] in 1824. After years of controversy over whether the gastric acid might not more probably be an organic acid, his finding was finally fully confirmed.[2] Prout[1] also introduced the concepts 'free acid', 'bound acid', and 'total acid'. The origin of these terms is uncertain; possibly they stem from one of the older theories of acids. Nevertheless, they have been retained, without precise definition, to the present day.

In biological fluids (mixtures of ions), hydrogen ions are present in three degrees of activity, as follows.

4.5.1.2.1. Bound hydrogen ions

These are hydrogen ions which are bound to anions (H^+X^-) and thus appear to be electrically neutral. A fraction of these hydrogen ions can be measured by titration. As the pH rises, more and more hydrogen ions dissociate and are neutralized by the titration with alkali. It is not clear where the precise end-point must lie. As the titration curve shows, any choice is arbitrary. We believe that pH 7.4 is correct, since a pH of 7.4 must be a valid starting point in matters concerning acids and bases.

4.5.1.2.2. Free hydrogen ions

These are hydrogen ions existing free in the solution. Their concentration is termed the hydrogen-ion concentration, C_{H^+}. It cannot be measured directly but can only be calculated. These free hydrogen ions possess a positive electric charge of 1.0, etc., only in infinitely dilute solutions. In concentrated solutions, for the same hydrogen-ion concentration, the degree of ionization is suppressed by inter-ionic interactions: the ionic activity is less than 1. Consequently, there is no longer any agreement between the ionic concentration and the ionic activity. The concentration of the free ions cannot therefore be determined by means of a pH measurement.

There is a widely disemminated misconception which ought to be corrected here. Time and again authors write that the pH electrode measures ionic concentrations, i.e.

$$pH = -\log C_{H^+}$$

This is incorrect. What is measured with the glass electrode is not the prevailing hydrogen-ion concentration but the hydrogen-ion activity (a_{H^+}), i.e.

$$pH = -\log a_{H^+}$$

Thus, the pH meter does not measure the concentration of all free hydrogen ions, but only their activity. The activity is only equal to the concentration in the case of pure and infinitely dilute solutions, and under these conditions the measurement of pH gives the concentration of all free hydrogen ions directly.

The relationship between the ionic activity and the ionic concentration is

$$a_{H^+} + C_{H^+}\gamma$$

where γ is the activity coefficient. The size of this activity coefficient is dependent on the nature of the ions present in the solution and on the ionic strength (p. 92). Figure 133 shows a few activity coefficients plotted as a function of the ionic concentration. The more concentrated a solution, the smaller is the activity coefficient and the smaller is the ionic activity. With extremely dilute solutions the activity coefficient approximates to 1.0, i.e. the ions are present in fully active form and the concentration of the free ions can be measured directly with the pH meter. This is the case in practice, e.g. with blood.

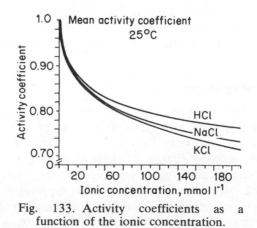

Fig. 133. Activity coefficients as a function of the ionic concentration.

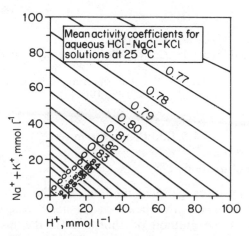

Fig. 134. Determination of the concentration of free hydrogen ions.

If, for gastric juice, the total ionic concentration $[Na^+] + [H^+] + [K^+] + [Cl^-]$ and the ionic activity (pH measurement) are known, then the concentration of free hydrogen ions can be calculated with the aid of empirical activity coefficients. Equations and tables for the rapid evaluation of these data may be found in Moore and Scarlata[3] and in Fig. 134. Consequently, in connection with gastric juice, the expression 'concentration of free hydrogen ions' no longer refers to an arbitrary value determined by titration but to a value calculated on the basis of the ionic activity and the activity coefficients. It was formerly assumed that this value could be determined by titration to pH 2.8–3.6 (phenolphthalein) or 3.5 (bromocresol blue). This is wrong, however, as has often been demonstrated. There is no theoretical or practical possibility of determining the concentration of free hydrogen ions by titration. The clinical concept of 'free acid' is an arbitrarily fixed quantity without any physiological significance for the secretion of hydrogen ions and ought therefore to be abandoned. If gastric juice is titrated, then the result is a measure of all the hydrogen ions present, including the ion-active, the non-ion-active, and the bound ions. Titration to pH 8–10 (phenolphthalein), pH 7 (phenol red) or, better, pH 7.4 (pH meter) is indubitably a measure of the concentration of acid. If such titrations are automatically recorded, then the curve obtained is virtually that for the titration of a pure hydrochloric acid solution. Deviations from this type of curve are attributable to the remainder of the basal secretions, food residues, or (previously) to the buffer action of the Ewald test meal.

To sum up, the only concentration of interest today is the total hydrogen-ion concentration. Whether the concentration of free hydrogen ions, as determined from the ionic activity and the activity coefficient, has a clinical significance is still unknown.

References

1. Prout, W., *Philos. Trans.*, part I, 45 (1824).
2. Bidder, F. H. and Schmidt, C., *Die Verdauungssäfte und der Stoffwechsel*, Leipzig, 1852.
3. Moore, E. W. and Scarlata, R. W., The determination of gastric acidity by the glass electrode, *Gastroenterology,* **49**, 178 (1965).

4.5.1.3. Pentagastrin test

4.5.1.3.1. Choice of method

Among the many investigational methods for the measurement of acid secretion in the stomach, the pentagastrin test[1] has, in recent years, established itself as the method of choice. It is the task of this test to provide a maximal acid secretion, thereby reducing the physiological scatter of this function to a minimum. The test provides insight into the efficiency of the mucous membrane of the stomach; this is of interest in certain diseases of the gastro-intestinal tract. The secretion of gastric acid obtained under maximal stimulation is proportionally related to the mass of the functionally competent gastric parietal cells.

4.5.1.3.2. Principle

The substance gastrin, a polypeptide formed from 17 amino acids, is produced in the G-cells of the antrum of the stomach. Passing into the circulation, it is capable of stimulating the cells of the lining of the stomach. Not all of the molecule is active in producing secretion. The biological active part proves to be a C-terminal peptide with the amino acid sequence tryptophan–methionine–asparaginic acid–phenylalanine. Accordingly, the substance currently in use for stimulating the gastric secretion of acid is a fully synthetic polypeptide, pentagastrin (*N-tert*-butyloxycarbonyl-β-alanyl-L-tryptophanyl-L-methionyl-L-aspartyl-L-phenylalaninamide). The pentagastrin test should be carried out by a doctor; it is very laborious. If there is no possibility of performing the test carefully, it should be dispensed with. Experience shows that the physiological scatter of the results is intrinsically so great that useful results can only be obtained if the method is carried out with precision.

4.5.1.3.3. Instruments

1, X-ray-dense gastric probe with several openings near the tip, size Ch. 14; 2, a 20-ml aspirator syringe; 3, an aspirator pump with maximum suction of 70 cm water (facultative); 4, a 100-ml measuring cylinder; 5, six test-tubes labelled 1a, 1b, and 2a–d; 6, a 2-ml injection syringe; 7, a 100-ml burette with 0.1-ml graduations, if possible an automatic titrator; 8, graduated 5-ml pipettes; 9, six Erlenmeyer flasks (20 ml) labelled 1a, 1b, and 2a–d.

4.5.1.3.4. Reagents

Phenol red indicator solution, 0.1 g *per* 100 *ml DM-water*. Dissolve by adding alcohol.
NaOH solution, 0.05 *mol* l^{-1}.

4.5.1.3.5. Operation

1. The patient may not receive either solid or liquid food after 20.00 hours on the day previous to the test. All medication (especially of antacids and spasmolytics) should be discontinued.
2. On the morning of the day of the test, the patient's nasal cavity should be anaesthesized and a Levine-type probe inserted and temporarily affixed to the nose with plaster.
3. The position of the probe is checked on the screen. The tip of the probe must rest against the side of greatest curvature at the lowest point of the gastric fundus. When the correct positition has been obtained, the probe is securely fixed.
4. The fasting secretion is aspirated with a syringe with the patient lying on his right and on his left; this material is discarded.
5. Determination of the basal secretion: continual aspiration of the gastric juice (preferably with a pump, or manually, with a syringe) for twice 30 min. Record the volumes after the 30-min periods and transfer 5–10 ml to test-tubes 1a and 1b.
6. Injection (s.c. or i.m.) of 6 μg of pentagastrin per kilogram of body-weight (1 ml of Peptavlon = 250 μg of pentagastrin).
7. Determination of the acid-secretion capacity: continual aspiration of the gastric juice for four times 15 min. Record the individual volumes and transfer *ca*. 5–10 ml of each liquid to test-tubes 2a–d.

4.5.1.3.6. Measurement

Titration. Pipette 2.5 ml from each of the six samples into the appropriate Erlenmeyer flask. Add 1–2 drops of phenol red solution and titrate the shaken solution with 0.05 mol l^{-1} NaOH to the red colour change or, with an automatic titrator, to pH 7.4.

4.5.1.3.7. Calculation

Acid concentration of the individual fractions:

$$H^+ (mmol\ l^{-1}) = \frac{ml\ 0.05\ mol\ l^{-1}\ NaOH \cdot 1000 \cdot 0.05}{aliquot\ in\ ml}$$

$$= \frac{\text{ml}\,0.05\,\text{mol}\,\text{l}^{-1}\,\text{NaOH} \cdot 50}{2.5} = \text{ml}\,0.05\,\text{mol}\,\text{l}^{-1}\,\text{NaOH} \cdot 20$$

Production of acid in unit time (per separate fraction):

$$\text{H}^+ \,(\text{mmol l}^{-1}) = \frac{\text{volume in ml} \cdot \text{value of the concentration}}{1000}$$

The values of acid production per unit time for the separate fractions are summed to give the acid production per hour.

4.5.1.3.8. Acid secretion (H^+/h)

Various collecting periods and modes of expression have been proposed for assessing the maximal acid production. That which is most customary today is to sum the secretion during the four pentagastrin periods and to take the output in 1 h as a meausre for the acid output.[2,3] The acid secreted during 1 h of fasting secretion is also called the basal acid output (BAO), and that during 1 h following pentagastrin is called the maximal acid output (MAO).

4.5.1.3.9. Reference values

These are given in Table 104.

4.5.1.3.10. Physiological variability

The variability of this gastric function test is very high, as Table 104 clearly shows. There is no doubt that the acid secretion in women is lower than in men. It decreases with increasing age.

4.5.1.3.11. Diagnostic significance

To discuss the diagnostic possibilities and the limits of the pentagastrin test would be to exceed the scope of this book. The reader is referred to the specialist literature.[1,2,4]

Table 104. Reference values for secretion of gastric acid[5] (acid production per hour $\pm\,1\,s$).

	Men	Women
Basal secretion (BAO)	2.8 ± 4.4	1.0 ± 2.5
Acid secretion capacity (MAO)	18.0 ± 10.5	12.0 ± 7.6

References

1. Halter, F., Der Pentagastrin-Test, in Englhardt and Lommel, *Malabsorption, Maldigestion*, Verlag Chemie, Weinheim, 1974.
2. Hämmerli, U. P., Bircher, J., Gassmann, R., Keel, H. J., Hefti, M. I., Hafter, E. and Meier, M. S., Notwendiges und Überflüssiges in der Magendiagnostik, *Schweiz. med. Wschr.*, **98**, 1032 (1968).
3. Halter, F. and Funk, H. U., Der Pentagastrintest als Routinetest zur Bestimmung der Magensäuresekretion, *Schwiez. med. Wschr.*, **98**, 1149 (1968).
4. Multicentre pilot study—pentagastrin as a stimulant of maximal gastric acid response in man, *Lancet*, **i**, 291 (1967).
5. Halter, F., personal communication.

4.5.2. REABSORPTION TESTS

4.5.2.1. D-Xylose tolerance

4.5.2.1.1. Introduction

The D-xylose tolerance may be used for testing the intestinal reabsorption function. Two methods have been adopted for testing for a malabsorption. Following oral administration of D-xylose the fraction absorbed is either concluded from the amount of this pentose excreted in the urine or is determined by measuring the concentration *in the blood* after a prescribed time.

Xylose is a pentose with a molecular weight of 150. It is mainly absorbed in the duodenum and jejunum and is taken into the cells by active transport. Following administration of D-xylose, the concentration in the blood shows a maximum increase after 1–2 h, falling to its initial value after 5 h. Of 25 g of xylose administered, about 16 g (60–65%) are absorbed and 6–7 g are excreted in the urine (about 25% of the dose given). With a normal kidney function, this fraction excreted is relatively constant. A small quantity of the D-xylose is obviously reabsorbed in the renal tubulus. The remaining 40% of the amount administered undergoes metabolic conversion being partly excreted as CO_2 and partly channeled into the pentosephosphate cycle[1-3] and excreted in the urine as D-threitol (C_4-polyalcohol).[11]

4.5.2.2. Xylose tolerance with determination of xylose in the urine

D-Xylose is administered orally. The urine is collected over a 5-h period and the total excretion of this sugar in the urine in this time is determined. It is best to carry out the test on fasting patients in the morning. The bladder should be as empty as possible at the start of the experiment. For adults, we give a xylose dose of 25 g, for children a dose corresponding to 15 g m^{-2} body surface area (for determination of the body surface area, see Appendix 3), dissolved in half the necessary quantity of water (volume of water used during the test: 1000 ml or 600 ml m^{-2} body surface area). The patient lies quietly in bed throughout the duration of the experiment and takes no food. At 1 and 2 h after the start of the experiment we allow the patient to drink each time half the water still remaining. The urine is collected throughout the 5 h of the test, the volume is measured, and it is stored in a refrigerator for further processing.

Example. Child with body surface area 0.5 m^2. At the beginning of the test 7.5 g of D-xylose dissolved in 150 ml of water are administered, followed by 75 ml of water after 1 h and again after 2 h.

With this dosage slight diarrhoea occurs only exceptionally. The principal likely source of error is loss of urine during the collecting period. The test is unreliable if the kidney function is disturbed, if there is insufficient diuresis and in cases of delayed emptying of the stomach.

4.5.2.2.1. Determination of D-xylose in urine: principle

Urine is heated at 70 °C for 10 min with a reagent containing glacial acetic acid, thiourea, and *p*-bromoaniline. This treatment converts the pentose, by removal of water, to a furfurol, which reacts with the *p*-bromoaniline to form a dye. The thioruea functions as an antioxidant.[4]

4.5.2.2.2. Reagents

Glacial acetic acid/thiourea solution. A cold, saturated solution of thiourea in glacial acetic acid (about 4 g per 100 ml). The solution keeps indefinitely.

p-Bromoaniline reagent. Dissolve 2 g of *p*-bromoaniline in glacial acetic acid/thiourea solution and make up to 100 ml with this solution. Keep in a dark bottle. The solution keeps for several weeks if frozen.

Xylose standard, 0.5 g *per* 100 ml. Dissolve 500 mg of D-xylose in cold, saturated benzoic acid solution and make up to 100 ml with further benzoic acid solution.

4.5.2.2.3. Measurement

Clear urine is used direct. Turbid urine is first filtered.

	Sample, ml	Reagent blank, ml	Standard, ml
Reagent	5.0	5.0	5.0
Urine	0.02	—	—
Distilled water	—	0.02	—
Xylose standard	—	—	0.02

Heat on a water-bath at 70 °C for 10 min. Allow to stand in the dark for 60–120 min. Within 3 h of heating, read the extinction at 546 nm against water.

4.5.2.2.4. Calculation

Concentration.

$$\text{Concentration} = \frac{E(S) - E(RB)}{E(ST) - E(RB)} \cdot C(ST) \text{ g of pentose per 100 ml}$$

$$\text{Concentration} = \frac{E(S) - E(RB)}{E(ST) - E(RB)} \cdot 0.5 \text{ g of pentose per 100 ml}$$

Absolute xylose excretion.

$$\text{Amount} = \frac{E(S) - E(RB)}{E(ST) - E(RB)} \cdot 5 \cdot \text{volume of urine in ml (mg pentose)}$$

Excretion as a percentage of the amount administered.

$$\% \text{ excreted} = \frac{\text{g xylose excreted} \cdot 100}{\text{dose administered (g)}} \% \text{ xylose}$$

4.5.2.2.5. Reference values

These are given in Table 105.

4.5.2.3. Xylose tolerance by determination in the plasma or serum

4.5.2.3.1. Procedure

For the confirmation of coeliac disease in childhood, the 1 hour blood-xylose test has been adopted.[5-8] After a period of fasting of at least 6 h, usually at 8 a.m., 5 g of D-xylose in 100 ml water are administered orally. A venous blood sample is collected exactly 1 h later.

4.5.2.3.2. Principle

See p. 573.

4.5.2.3.3. Reagents

See p. 574.

Table 105. Xylose excretion in the urine.

Population	Dose, g	Excretion		References
		g	%	
Adults	25	5.6–8.2	22–23	1
Children	15 g m ^{2}		15–37	2

Xylose standard, 20 *mg per* 100 *ml*. Dissolve 20 mg of D-xylose in cold, saturated benzoic acid solution and make up to 100 ml with further benzoic acid solution.

Trichloroacetic acid, 0.61 *mol* l^{-1}. Dissolve 10 g of trichloroacetic acid in DM-water and make up to 100 ml.

4.5.2.3.4. Measurement

Procedure. Add 0.1 ml of plasma or serum to 0.2 ml of trichloroacetic acid, mix well, allow to stand for 5 min and centrifuge sharply.

	Sample, ml	Reagent blank, ml	Standard, ml
p-Bromoaniline reagent	1.0	1.0	1.0
Supernatant	0.1	—	—
Trichloroacetic acid	—	0.1	—
Xylose standard, 20 mg per 100 ml	—	—	0.1

Heat in a water-bath at 70 °C for 10 min. Allow to stand in the dark at room temperature for 60 min, then read the extinctions against water at 515 nm (Hg 546 nm).

4.5.2.3.5. Calculation

$$\text{Concentration} = \frac{E(S) - E(RB)}{E(ST) - E(RB)} \cdot 20 \cdot \frac{0.3}{0.1} \text{ mg per 100 ml}$$

$$= \frac{E(S) - E(RB)}{E(ST) - E(RB)} \cdot 60 \text{ mg per 100 ml}$$

4.5.2.3.6. Reference values

After administration of D-xylose, the concentration after 1 h should amount to 20 mg per 100 ml or more. These values are valid for childhood (up to about 8 years) and for patients with a body-weight of up to 30 kg. For patients with a body-weight exceeding 30 kg it is probably necessary to increase the dose of xylose administered orally and to restandardize the reference values.[8] However, following administration of 25 g of D-xylose, values in adults of over 30 mg per 100 ml after 2 h have been reported.[3]

4.5.2.3.7. Comments on the method

1. The blood may also be collected from capillary blood using a heparinized Caraway capillary which is closed with a special stopper, and centrifuged in an ordinary centrifuge tube. The plasma is used for analysis.

2. After deproteination of the plasma the supernatant may be stored deep-frozen for up to 4 days.

3. The absorption maximum of the dye lies at 515 nm. By taking the reading at wavelength 546 nm (Hg) in a spectral-line photometer there is no appreciable loss in absorption.[10]

4. Linearity of the reaction is guaranteed up to a xylose concentration of 100 mg per 100 ml. This is sufficient, since in normal individuals measured plasma concentrations after oral loading with xylose are in the range 40–50 mg ml^{-1}.

5. The day-to-day precision over 6 months is good, with a coefficient of variation of 1.5%.

6. In testing for accuracy, no statistically significant difference was found between the measured and expected values.

7. Comparative determinations of xylose in whole blood and plasma showed no significant difference. It must be assumed that the xylose is evenly distributed between the plasma and erythrocyte water.

8. The method is of limited specificity: other sugars can interfere. Accordingly, interference with glucose was tested. For a blood-sugar level of 109 mg per 100 ml the average recovery of xylose was 101%, i.e. for this range of glucose concentrations there is no interference in the xylose determination.[10] For the determination of xylose, o-toluidine has also been recommended.[9]

4.5.2.3.8. Diagnostic significance

A reduced concentration of D-xylose in the plasma and a diminished excretion in the urine following oral administration is found with disturbed absorption function of the upper small intestine.

The test is mainly conducted for clarification of a malabsorption in gluten-sensitive enteropathias (coeliac disease, sprue).

Pathological results of absorption tests are also found with other, less common diseases of the upper small intestine.[2]

Diseases of the kidneys associated with decreased glomerular filtration lead to a diminished excretion of xylose coupled with elevated plasma concentrations and simulate a pathological result.

A normal outcome of the xylose test is generally found in digestive insufficiency conditioned by chronic pancreatitis and by disturbances in the flow of bile acid, and also in cirrhosis of the liver, regional enteritis, malnutrition, enterocolitis, and further affections of the lower small intestine.

References

1. Roe, J. H. and Rice, E. W., A photometric method for the determination of free pentoses in animal tissues, *J. biol. Chem.*, **173**, 507 (1948).

578

2. Oestreicher, R., Richterich, R., and Rossi, E., Der diagnostishce Wert der Xylose-Belastung bei Kindern, *Dt. med. Wschr.,* **89**, 1111 (1964).
3. Miller, B., Der D-Xylose-Test, in Englhard and Lommel, *Malabsorption, Maldigestion,* Verlag Chemie, Weinheim, 1974.
4. Hindmarsh, J. T., Xylose absorption and its clinical significance, *Clin. Biochem.,* **9**, 141 (1976).
5. Rolles, C. J., Kendall, M., Nutter, S., and Anderson, Ch. M., One-hour blood-xylose screening test for coeliac disease in infants and young children, *Lancet,* **ii**, 1043 (1973).
6. Rolles, C. J., Anderson, Ch. M., and McNeish, A. S., Confirming persistence of gluten intolerance in children diagnosed as having coeliac disease in infancy, *Archs Dis. Childh.,* **50**, 259 (1975).
7. Schaad, U., Gaze, H., Hadorn, B., Pedrinis, E., Cottier, H., Lorenz, E., and Colombo, J. P., Der Blutxylosetest im Kindesalter: Korrelation des 1-Studen-Blutxylosewertes mit der Anzahl intraepithelial gelegener Lymphozyten in der Dünndarmschleimhaut bei der Zöliakie, *Helv. paediat. Acta,* **30**, 331 (1975).
8. Christiansen, P. A., Kirsner, J. B., and Ablaza, J., D-Xylose and its use in the diagnosis of malabsorptive states, *Am. J. Med.,* **27**, 443 (1959).
9. Buttery, J. E., Kus, S. L., and De Witt, G. F., The *ortho*-toluidine method for blood and urine xylose, *Clin. chim. Acta.,* **64**, 325 (1975).
10. Lorenz, E. and Colombo, J. P., Die Bestimmung der D-Xylose im Plasma, *Med. Labor. Stuttg.,* **28**, 198 (1975).
11. Pitkänen, E., The conversion of D-xylose into D-threitol in patients without liver disease and patients with portal liver cirrhosis, *Clin. chim. Acta,* **80**, 49 (1977).

4.6 PANCREAS AND SALIVARY GLANDS

4.6.1. GENERAL

The pancreas functions both as an endocrine and an exocrine gland. It releases insulin and glucagon into the blood stream (endocrine) and it excretes pancreatic juices into the duodenum via the pancreatic duct (exocrine function). The pancreatic juices contain electrolytes (Na^+, K^+, Ca^{2+}, Zn^{2+}, HCO_3^-, Cl^-, HPO_4^-, SO_4^{2-}) and various enzymes and enzyme precursors. In addition to amylase, these enzymes include the lipases and the nucleases. The enzyme precursors include trypsinogen and chymotrypsinogen A and B, and the phospholipases and proelastases. In the duodenum enterokinase, an enzyme from the duodenal mucous membrane, catalyses the conversion of trypsinogen to trypsin in the presence of calcium ions. This in turn activates the formation of chymotrypsin A and B, and of the carboxypeptidases, from their precursors.

The formation of the above enzymes in the cells of the pancreas is stimulated via the vagus nerve, but also hormonally by gastrin (formed in the stomach) and cholecystokinin/pancreozymin (duodenum). The latter two hormones act simultaneously to effect the discharge of secretion from the glandular cells. A further hormone which influences the composition of the pancreatic juices is secretin. It is discharged into the bloodstream from the duodenum when the contents of the latter are acidic; in the pancreas it activates the release of water and electrolytes into the pancreatic duct.[1] In pancreatitis the cell membranes of the pancreatic-gland cells become damaged. The various enzymes listed above are converted to an active form and, together with kinins, they attack the walls of the cells and vessel. This causes pancreatic oedema and, later, haemorrhages in the pancreas. From the interstitium the pancreatic enzymes reach the lymph system and thence, via the ductus thoracicus, enter the bloodstream. The extent to which they also pass directly into the bloodstream is uncertain. The detection of elevations of lipase and amylase in the plasma (and urine) is accordingly used for the diagnosis of pancreatitis.

Reference
1. Kowlessar, O. D., Pathogenesis of pancreatitis, in Clearfield and Dinoso, *Gastrointestinal emergencies*, Grune & Stratton, New York, 1976, p. 223.

4.6.2. α-AMYLASE

4.6.2.1. Introduction

The α-amylases (α-1,4-glucan-4-glucanohydrolases, E.C. 3.2.1.1) are secretoral enzymes which are formed principally in the salivary glands and the pancreas. Their molecular weight is around 55 000, and their size is around 2.9 nm. The salivary and pancreatic amylases are genetically transmitted from two loci on chromosome 1.[1] These gene products are further modified in the cells; many variants of the enzymes are produced by glycosation, deglycosation, and deamidation. In 90% of the population at least two salivary and two pancreatic isoenzymes may be separated using acrylamide electrophoresis. More variants are present in a smaller section of the population, and this can make the interpretation of isoenzyme electrophoreses more difficult.[2,3] The cleavage of the 1,4-glucosidic linkages within polysaccharide molecules, as effected by the α-amylase, is disordered, i.e. not from the end of the chain on, as is the case with vegetable β-amylases. This gives rise to fragments of various sizes, dextrins, maltotetroses and maltotrioses, maltose, and dextrose. The optimum pH for the reaction lies between 6.7 and 7.5. Halogens and other anions activate the reaction ($Cl > Br > NO_3 > I$) and divalent cations stabilize the enzyme ($Ca > Mg > Zn$). Consequently, neither oxalate- nor citrate-blood can be used for the determination. For starch the K_m value has been reported as 0.6 g l^{-1}, and for amylose 8 g l^{-1}. The activation energy is about 13 000 cal mol^{-1}.[4-6] The half-life of amylase in the blood is very short (about 2 h). Part of the amylase is excreted through the kidneys, the pancreatic isoenzymes having a higher clearance than the salivary isoenzymes. Accordingly, the ratio of the isoenzymes in the urine and in serum is different (serum, 60% salivary; urine, 65% pancreatic). It has been shown experimentally that part of the amylase is degraded in the reticuloendothelial system.[7] The activity of the individual isoenzymes differs from one to the next according to the chosen substrate.[8-10]

4.6.2.2. Choice of method

The methods of determination are based either on measurement of the consumption of substrate—*amyloclastic methods*—or on the determination of the cleavage products—saccharogenic and chromogenic methods. The former group include the detection of starch by the starch–iodine reaction,[11,12] as well as viscosimetric, turbidimetric, and nephelometric determinations of the decrease in starch.[13,14]

For manual operation, viscosimetric, turbidimetric, and nephelometric methods should be rejected on the grounds of their low reproducibility.

The determination of starch by the starch–iodine reaction is based on the behaviour of helical glucose polymers with more than 18 units: these combine with iodine, by complex formation and adsorption, and are thus measurable by photometry. The wavelength of the absorption maximum changes with the chain-length and decreases rapidly for fragments with less than 72 units.[4,15] For this reason, and because optimal substrate concentrations cannot be selected in many methods,[4] amyloclastic methods are linear over only a very limited range. Interferences are known to be caused by iodine-bonding proteins, lipoproteins, and X-ray contrast agents.

The second group of methods include the saccharogenic reactions—measurement of the formation of reducing groups,[16,17] maltose, or glucose determinations[18,19]—and chromogenic methods which involve measurement of water-insoluble dye–saccharide complexes.

Saccharogenic methods are less sensitive than the amyloclastic and, if reducing groups are measured, are less accurate than chromogenic methods. Because α-amylase does not attack the polysaccharide from the end of the chain but first produces larger cleavage fragments, enzymatic methods of detection for glucose or maltose should also be rejected.

In all these methods difficulties are also presented by the substrate. As a mixture of amylose and amylopectin, starch is chemically ill-defined. Amylose cannot be dissolved to provide concentrations permitting a V_{max} for the reaction. Amylopectin is better suited for this purpose but even here the composition changes according to the preparation and there is variation from one batch to the next.

Nowadays, the preferred methods are *chromogenic reactions* using synthetic dye–polysaccharide complexes.[20-25] The amylases act upon water-insoluble polysaccharides covalently bonded to dye groups to cleave off water-soluble fragments with their attached dye groups; these can be determined photometrically in the supernatant when the incubation is complete. These chromogenic methods exhibit a higher precision than amyloclastic and saccharogenic methods.[25] A more restricted choice of tests is commercially available in the form of prepared packs, but it should be borne in mind that here too we are dealing with dyes coupled to starch and the quality varies from one batch to the next. In our opinion it is better to give preference to those methods in which the activity is not read from the standard curves supplied but may be determined by measurement of a dye standard. For the rest, the choice of chromogenic method is determined by considerations of practicability. Chromogenic methods have one disadvantage in operation: the water-insoluble part of the substrate has to be separated from the water-soluble reaction products. Up to now, experiments with fully synthetic substrates have failed owing to the poor substrate affinity of the amylase.[27] For carrying out the calculation and reporting the reference range, the reader is referred to the pack-literature as supplied by the manufacturer. For those readers who do not have access to chromogenic methods, a procedure for an amyloclastic method is outlined below.

4.6.2.3. α-Amylase by an amyloclastic method[12]

4.6.2.3.1. Principle

Starch is degraded by α-amylase (diastase) to dextrins and sugars (maltotriose, maltose, glucose). On addition of iodine, starch (but not its degradation products) gives a blue colour (starch–iodine addition compound); the intensity of its colour is directly proportional to the prevailing starch concentration. The enzyme activity is determined by measuring the decrease in the amount of starch during incubation of the enzyme (serum, urine, cerebrospinal fluid, etc.). The optimal pH is 7.5. Sodium chloride is added as an activator. The substrate solution is stabilized by addition of sorbic acid.[28]

4.6.2.3.2. Reagents

Hydrochloric acid, 0.5 mol l^{-1}

Buffer solution: collidine HCl, 50 mmol l^{-1}; *sodium chloride*, 50 mmol l^{-1}, *pH* 7.5. Mix 250 ml of DM-water, 250 ml of collidine buffer stock solution (200 mmol l^{-1}), 250 ml of hydrochloric acid (0.1 mol l^{-1}) and 2.92 g of sodium chloride in a 1000-ml volumetric flask. Adjust the pH to 7.5 and dilute to 1000 ml with DM-water. The solution keeps for several months at room temperature.

Iodine reagent: potassium iodine, 108 mmol l^{-1}; *iodine*, 14 mmol l^{-1}. Dissolve 18 g of potassium iodide and 1.8 g of iodine in DM-water and make up to 1000 ml. The solution keeps for several months at room temperature in a dark, well sealed bottle.

Substrate solution: starch, 6.2 mmol l^{-1}; *sorbic acid*, 18 mmol l^{-1}. Prepare about 1000 ml of cold, saturated solution of sorbic acid (about 2%) and allow it to stand for 1 week. Filter, and suspend 1 g of starch in about 50 ml of this solution in a 1000-ml volumetric flask. Add three times 300 ml of boiling sorbic acid solution and mix well. After cooling to room temperature, make up to 1000 ml with sorbic acid solution. The solution keeps for several months at room temperature.

4.6.2.3.3. Measurement

Final concentrations in the reaction mixture. The final concentrations are enzyme 20 μl, collidine.HCl buffer 50 mmol l^{-1} (pH 7.5), chloride 25 mmol l^{-1}, starch 3.1 mmol l^{-1}, and sorbic acid 9 mmol l^{-1}.

4.6.2.3.4. Procedure

Working solution. Mix 1 part of buffer and 1 part of substrate solution. Always prepare freshly before use.

Iodine solution. Dilute 1 ml of iodine reagent with DM-water to 200 ml. Prepare freshly each day.

(Test-tubes)	Sample, ml	Sample blank, ml
Working solution	1.0	1.0
Hydrochloric acid, 0.5 mol l^{-1}	—	0.5
Incubate at 37 °C for 5 min.		
Serum, urine, etc.	0.02	0.02
Incubate at 37 °C for exactly 30 min (stop-watch)		
Hydrochloric acid, 0.5 mol l^{-1}	0.5	—
Iodine solution	10.0	10.0

Between 10 and 120 min after the addition of the iodine reagent, take readings of the extinction of the sample blank and sample against DM-water at a wavelength between 560 and 620 nm (red filter) (room temperature).

4.6.2.3.5. Calculation

Definition of the unit. An International Unit (U) corresponds to the number of micromoles of the substrate which are converted per minute at 37 °C. As applied to the special case of the α-amylases, this corresponds to the number of glucosidic linkages cleaved per minute.

$$U = \frac{E(SB) - E(S)}{E(SB)} \cdot \frac{1000}{0.02} \cdot 3.1 \cdot \frac{1}{30} \ \mu\text{mol min}^{-1} \text{l}^{-1}$$

$$U = \frac{E(SB) - E(S)}{E(SB)} \cdot 5166 \ \mu\text{mol min}^{-1} \text{l}^{-1}$$

Note: For the purpose of calculation, the starch is assumed to be degraded to glucose monomer (molecular weight 162).

4.6.2.3.6. Reference values

The distribution of α-amylase activity in the plasma and urine is not normal. Amylase activity is independent of sex or age except in the neonatal period, when the amylase values are low.[29,30] During pregnancy there is a physiological increase in serum amylase.[31]

Reference values are given in Table 106.

Table 106. Reference values obtained with the amyloclastic method[12].

Serum
   $n = 312$; 2.5–97.5%; 435–1904 U l^{-1} (author's laboratory)

Urine
   $n = 152$; 2-hour collecting period: 35–450 U h^{-1} (ref. 32)
   $n = 140$; 24-hour collecting period: 30–550 U h^{-1} (ref. 32)

4.6.2.3.7. Comments on the method

1. The method described[12] is a modification of the technique of Smith and Roe.[11] The substrate is stabilized according to the direction of Rice.[28]

2. The method may be conducted as an ultramicro or macro method by proportionally scaling up or scaling down all of the volumes.

3. For high activities, i.e. if the ratio $[E(\text{SB}) - E(\text{S})]/E(\text{SB})$ is more than 0.5, the determination should be repeated with a 5-min incubation period and the final result should be multiplied by 6.

4. Since abnormally high results are occasionally caused by incidental addition of saliva (pipetting, speaking, sneezing), it is advisable to perform the determination in duplicate.

5. The determination in serum, urine, transudates, exudates, and human milk is carried out analogously. A 20-μl sample may be used in each case. The activity in cerebrospinal fluid is smaller, so a 0.1-ml sample should be used. The amylase concentration in saliva and duodenal juices varies greatly (between 10 and 10 000 times the serum concentration). In this case the convenient activity concentration (dilute with physiological saline solution) must always be predetermined by a preliminary experiment.

6. The intensity of the blue colour is dependent on the temperature.

7. The serum amylase may be kept for at least 1 week, either frozen or at 4 °C. In contrast, the stability of the amylase in urine appears to be less, even at -20 °C.

8. Factors for interconversion of Somogyi, Caraway, Smith–Roe, dye and International Units are variously reported in the literature, some of theoretical derivation, others determined experimentally. These may be of considerable help in comparing ranges of values in terms of orders of magnitude, but they should not be used for making exact conversions, e.g. of patient data. The various isoenzymes of α-amylase each have different activities according to the substrate used.[33] Consequently, the ratio of isoenzymes influences the total activity measured. It is therefore not surprising that factors which have been determined for comparing two methods with normal urines and sera should not be applicable to comparison studies on material from pancreatitis patients and other patients with different proportions of isoenzymes.[34]

4.6.2.3.8. Should the amylase be determined in serum or urine?

The amylase should be determined both in the serum (or heparin plasma) and the urine. In pancreatitis the α-amylase remains elevated in the urine for longer than in the plasma. False-negative, i.e. normal, values are found with pancreatitis more frequently in the plasma than in the urine.[35] However, the determination of the rate of excretion of amylase, both in 24-hour and 1-hour urine collectives, also gives false-negative values occasionally.[35,36] The determination in short-term urine portions is more reliable than that in 24-hour urine, besides which it is often not feasible to perform the determination on 24-hour urine because the therapeutic decisions cannot wait that long. The determination in urine may be expressed as concentration, as rate of excretion (this presupposes that the time of last micturition is known), or it may be referred to creatinine.

Levitt et al.[37] have shown that (in 2-hour urine portions) the amylase activity based on creatinine allows a better detection of amylase elevation than if the 2-hour excretion rate or the concentration of amylase in the serum is measured. This accords with our own experience. If the limiting value for the chromogenic amylase determination is fixed at 1.4 U mg^{-1} of creatinine a pancreatitis can be detected when it might be missed by determination of the plasma amylase of by urine determination based on volume of urine. In spite of this, the number of elevated values of the amylase/creatinine ratio in urine in non-pancreatitic diseases is smaller than that for the determination of amylase per unit volume or unit time in the plasma or urine. Patients with a severe kidney insufficiency constitute an exception. However the amylase activity in the urine is expressed, the finding is not specific for pancreatitis. This is also clear from the summary of the diagnostic significance. Nor can pancreatitis be concluded from the level of the amylase activity. Patients with cholecystitis, etc., can easily show higher activities in the plasma or urine than patients with acute pancreatitis. The improved discrimination afforded by the urinary amylase/creatinine ratio as compared with amylase per unit volume of urine or plasma results from the fact that not only is the elevation measured more or less organ specific, but also relies on a secondary action of pancreatitis. This concerns the absorption of small proteins in the renal tubulus which is disturbed in pancreatitis.

In recent years, various investigations[37–41] have thus indicated a new way of identifying acute pancreatitis. It was possible to show that in acute pancreatitis the amylase clearance, referred to the creatinine clearance, gave higher values than with other syndromes. Further investigations showed that glomerular filtration of amylase is not affected in pancreatitis. Nevertheless, the reabsorption in the renal tubulus is diminished. Consequently, referring the amylase clearance to the creatinine clearance results in a relative elevation of the former. This may be done by conducting a creatinine determination in the plasma and associated urine at the same time as the determination of the amylase in these fluids. In that case it is no longer necessary to measure the excretion period and the quantity of urine since these are eliminated in the calculation.

$$\frac{\dfrac{\text{Amylase activity (urine)}}{\text{Amylase activity (plasma)}} \cdot \dfrac{\text{Urine volume}}{\text{Period}}}{\dfrac{\text{Creatinine (urine)}}{\text{Creatinine (plasma)}} \cdot \dfrac{\text{Urine volume}}{\text{Period}}} = \frac{\text{Amylase clearance}}{\text{Creatinine clearance}}$$

$$= \frac{\text{Amylase (urine)}}{\text{Amylase (plasma)}} \cdot \frac{\text{Creatinine (plasma)}}{\text{Creatinine (urine)}} \cdot 100$$

= percentage of the amylase clearance based on the creatinine clearance

When the amylase clearance is expressed as a percentage of the creatinine clearance, we are no longer attempting to detect an elevation of a more or less organ-specific enzyme but to identify the diminished reabsorption of a small protein in the renal tubulus. Thus, e.g. β_2-microglobulin would serve the purpose just as well. The cause of this diminished tubular reabsorption of small proteins is unknown. Reference values for the amylase clearance as a percentage of the creatinine clearance are given only occasionally; 4.7–5.5% is reported as the upper limit with saccharogenic methods.[37,38] Salt and Schenker[42] have pointed out that the dye/starch complex method underestimates the urinary amylase activity by up to 40% in comparison with the saccharogenic or amyloclastic methods. Our own studies using chromogenic determinations[20] gave a limiting value of 4%. We found slightly elevated values (4 –8%) in patients with kidney insufficiency (creatinine clearance less than 40 ml min^{-1} per 1.73 m^2) but no pancreatitis. The determination of the percentage amylase clearance referred to the creatinine clearance allows the virtually specific identification of acute pancreatitis, especially during the first 4 days. This also applies to pancreatitis patients suffering from kidney insufficiency (creatinine clearance less than 25 ml min^{-1} per 1.73 m^2).[43] However, elevations have been observed in isolated cases with diabetic ketoacidosis and burns.[42] The determination of the amylase activity and the creatinine concentration in one urine portion and in a plasma sample collected at the same time (allowing the amylase clearance referred to the creatinine clearance to be calculated) appears currently to be the most useful clinical–chemical parameter for the diagnosis of acute pancreatitis. Next in line (less specific) comes the determination in urine, referred to its creatinine concentration.

4.6.2.3.9. Diagnostic significance

1. *Elevations of amylase* are found in a large number of diseases:

1. Pancreas: acute pancreatitis
   chronic recidivistic pancreatitis
   chronic pancreatitis with pseudocysts or abscess
   certain cancers of the pancreas
   trauma (including endoscopy)
2. Salivary glands: inflammation (mumps)
   injuries
3. Intra-abdominal processes: cholelithiasis, perforated ulcer, intestinal infarction, obstructive ileus, appendicitis, peritonitis, ruptured ectopic pregnancy
4. Drugs, especially opiates, leading to a closure of the sphincter of Oddi
5. Ectopic formation of amylases in neoplasms
6. Disturbed excretion of amylase through the kidneys: kidney insufficiency, macroamylasaemia
7. Miscellaneous: burns, diabetic ketoacidosis, injuries to the skull.

Macroamylasaemia involves a bonding of the amylase to plasma proteins (IgA, IgG) of the patient.[44] This has no pathological consequences for the patient, but can lead to wrong diagnoses since the serum amylase is elevated but owing to the bonding to plasma proteins the amylase excreted in the urine is minimal. The amylase clearance referred to the creatinine clearance is less than 1% for such a condition. A similar mechanism is possibly present in bisalbuminaemia and with sialic acid-containing isoamylases.[45]

2. A *diminution* of pancreatic amylase is found in the exocrine pancreas insufficiency of mucoviscidosis patients. This is most elegantly detected by immuno-antibodies following directed *in vitro* inactivation of the salivary isoenzymes.[46]

3. A survey of the clinical significance of the α-amylase determination has been compiled by Salt and Schenker.[42] In certain cases the determination of isoenzymes may be used for differential diagnosis.[47,48] It is interesting that elevations of the salivary isoenzymes have been found in pancreatic carcinomas and chronic hepatitis.[49,50]

References

1. Merritt, A. D., Lovrien, E. W., Rivas, M. L., and Conneally, P. M., Human amylase loci: genetic linkage with the Duffy blood group locus and assignment to linkage group I, *Am. J. hum. Genet.*, **25**, 523 (1973).
2. Lehrner, L. M., Ward, J. C., Karn, R. C., Ehrlich, C. E., and Merritt, A. D., Isozyme differentiation in patients with hyperamylasemia, *Am. J. clin. Path.*, **66**, 576 (1976).
3. Karn, R. C., Rosenblum, B. B., Ward, J. C., and Merritt, A. D., Genetic and post-translational mechanisms determining human amylase isozyme heterogeneity, in Markert, *Isoenzymes IV. Genetics and evolution*, Academic Press, New York, 1975, p. 745.

588

4. Lorentz, K., Zander, A., and Adlung, J., Untersuchung zur amyloklastischen α-Amylase-Bestimmung, *Z. klin. Chem. klin. Biochem.*, **7**, 241 (1969).
5. Gryszkiewicz, A., Isolation and properties of human serum amylase, *Acta biochim. polon.*, **9**, 301 (1962).
6. Kovacs, L. and Tuba, J., A note on the energy of activation of the amylase in various human body fluids, *Can. J. Biochem. Physiol.*, **34**, 6 (1956).
7. Hiatt, N. and Bonorris, G., Removal of serum amylase in dogs and the influence of reticuloendothelial blockade, *Am. J. Physiol.*, **210**, 133 (1966).
8. Meites, S. and Rogols, S., Serum amylases, isoenzymes, and pancreatitis. I. Effect of substrate variation, *Clin. Chem.*, **14**, 1176 (1968).
9. Hall, F. F., Ratliff, C. R., Hayakawa, T., Culp, T. W., and Hightower, N. C., Substrate differentiation of human pancreatic and salivary alpha-amylase, *Am. J. dig. Dis.*, **15**, 1031 (1970).
10. Geddes, R., Rates of attack of some α-amylases upon various substrates, *Carbohyd. Res.*, **7**, 493 (1968).
11. Smith, B. W. and Roe, J. H., A micromodification of the Smith and Roe method for the determination of amylase in body fluids. *J. biol. Chem.*, **227**, 357 (1957).
12. Richterich, R. and Colombo, J. P., Ultramikro-Methoden im Klinischen Laboratorium. I. α-Amylase (Diastase), *Ärztl. Lab.*, **8**, 33 (1962).
13. Zinterhofer, L., Wardlaw, S., Jatlow, P., and Seligson, D., Nephelometric determination of pancreatic enzymes. I. Amylase, *Clin. chim. Acta*, **43**, 5 (1973).
14. Shipe, J. and Savory, J., Kinetic nephelometric procedure for measurement of amylase activity in serum, *Clin. Chem.*, **18**, 1323 (1972).
15. Bailey, J. M. and Whelan, W. J., Physical properties of starch. I. Relationship between iodine stain and chain length, *J. biol. Chem.*, **236**, 969 (1961).
16. Henry, R. J. and Chiamori, N., Study of the saccharogenic method for the determination of serum and urine amylase, *Clin. Chem.*, **6**, 434 (1960).
17. Searcy, R. L., Wilding, P., and Berk, J. E., An appraisal of methods for serum amylase determination, *Clin. chim. Acta*, **15**, 189 (1967).
18. Guilbault, G. G. and Rietz, E. B., Enzymatic, fluorometric assay of α-amylase in serum, *Clin. Chem.*, **22**, 1702 (1976).
19. Proelss, H. F. and Wright, B. W., New, simple maltogenic assay for mechanized determination of alpha-amylase activity in serum and urine, *Clin. Chem.*, **21**, 694 (1975).
20. Klein, B., Foreman, J. A., and Searcy, R. L., New chromogenic substrate for determination of serum amylase activity, *Clin. Chem.*, **16**, 32 (1970).
21. Babson, A. L., Tenney, S. A., and Megraw, R. E., New amylase substrate and assay procedure, *Clin. Chem.*, **16**, 39 (1970).
22. Sax, S. M., Bridgewater, A. B., and Moore, J. J., Determination of serum and urine amylase with use of Procion Brillinat Red M-2BS amylopectin, *Clin. Chem.*, **17**, 311 (1971).
23. Rinderknecht, H., Wilding, P., and Haverback, B. J., A new method for the determination of α-amylase, *Experientia*, **23**, 805 (1967).
24. Ceska, M., Hultman, E., and Ingelman, B. G. A., A new method for determination of α-amylase, *Experientia*, **25**, 555 (1969).
25. Ceska, M., Birath, K., and Brown, B., A new and rapid method for the clinical determination of α-amylase activities in human serum and urine. Optimal conditions, *Clin. chim. Acta*, **26**, 473 (1969).
26. Gergely, T., Bayer, P. M., and Gabl, F., Vergleichende Amylasebestimmungen mittels handelsüblicher Testpackungen, *Med. Lab., Stuttg.*, **29**, 47 (1976).
27. Jansen, A. P. and Wydeveld, P. G. A. B., α-(p-Nitrophenyl)maltoside as a substrate for the assay of amylase, *Nature, Lond.*, **182**, 525 (1958).

28. Rice, E. W., Improved spectrophotometric determination of amylase with a new stable starch substrate solution, *Clin. Chem.*, **5**, 592 (1959).
29. Laha, P. N. and Pandya, K. V., Urinary diastase in newborn infants, *Indian J. Pediat.*, **27**, 153 (1960).
30. Searcy, R. L., Berk, J. E., Hayashi, S., Ackermann, B. D., Serum amylase activity in the newborn, *Pediatrics.*, **39**, 294 (1967).
31. Kaiser, R., Berk, J. E., Fridhandler, L., Montgomery, K., and Wong, D., Serum amylase changes during pregnancy, *Am. J. Obstet. Gynec.*, **122**, 283 (1975).
32. Eberhagen, D., Issmer, A., and Brandmaier, B., Über die α-Amylase-Ausscheidung im Urin bei gesunden Personen, *Z. klin. Chem. klin. Biochem.*, **8**, 284 (1970).
33. Take, S., Berk, J. E., and Fridhandler, L., Observations on a new simplified dye method for assesssing amylase activity, *Clin. chim. Acta*, **26**, 533 (1969).
34. Ceska, M., Brown, B., and Birath, K., Ranges of α-amylase activities in human serum and urine and correlations with some other α-amylase methods, *Clin. chim. Acta*, **26**, 445 (1969).
35. Saxon, E. I., Hinkley, W. C., Vogel, W. C., and Zieve, L., Comparative value of serum and urinary amylase in the diagnosis of acute pancreatitis, *Archs intern. Med.*, **99**, 607 (1957).
36. Calkins, W. G., A study of urinary excretion in patients with acute pancreatitis, *Am. J. Gastroent.*, **49**, 415 (1968).
37. Levitt, M. D., Rapoport, M., and Cooperbrand, S. R., The renal clearance of amylase in renal insufficiency, acute pancreatitis and macroamylasemia, *Ann. intern. Med.*, **71**, 919 (1969).
38. Lesser, P. B. and Warshaw, A. L., Differentiation of pancreatitis from common bild duct obstruction with hyperamylasemia, *Gastroenterology*, **68**, 636 (1975).
39. Warshaw, A. L. and Fuller, A. F., Specificity of increased renal clearance of amylase in diagnosis of acute pancreatitis, *New Engl. J. med.*, **292**, 325 (1975).
40. Warshaw, A. L. and Lee, K.-H., The mechanism of increased renal clearance of amylase in acute pancreatitis, *Gastroenterology*, **71**, 388 (1976).
41. Johnson, S. G., Ellis, C. J., and Levitt, M. D., Mechanism of increased renal clearance of amylase/creatinine in acute pancreatitis, *New Engl. J. Med.*, **295**, 1214 (1976).
42. Salt, W. B. and Schenker, S., Amylase—its clinical significance: a review of the literature, *Medicine, Baltimore,* **55**, 269 (1976).
43. Pedersen, E. B., Brock, A., and Kornerup, H. J., Serum amylase activity and renal amylase activity clearance in patients with severely impaired renal function and in patients treated with renal allotransplantation, *Scand. J. clin. Lab. Invest.*, **36**, 137 (1976).
44. Harada, K., Nakayama, T., Kitamura, M., and Sugimoto, T., Immunological and electrophoretical approaches to macroamylase analysis, *Clin. chim. Acta*, **59**, 291 (1975).
45. Sudo, K. and Kanno, T., Sialic acid containing abnormal amylases in human sera, *Clin. chim. Acta*, **64**, 303 (1975).
46. Kenny, D., Cooke, A. M., O'Donnell, M. D., Tempany, E., and Mc'Geeney, K. F., Use of an inhibitor to estimate pancreatic and salivary α-amylase in serum from cystic fibrosis patients, *Abstr. 2nd Eur. Congr. clin. Chem., Prague,* 1976.
47. Legaz, M. E. and Kenny, M. A., Electrophoretic amylase fractionation as an aid in diagnosis of pancreatic disease, *Clin. Chem.*, **22**, 57 (1976).
48. Benjamin, D. R. and Kenny, M. A., Clinical value of amylase isoenzyme determination, *Am. J. clin. Path.*, **62**, 752 (1974).

590

49. Shimamura, J., Fridhandler, L., and Berk, J. E., Nonpancreatic-type hyperamylasemia associated with pancreatic cancer, *Am. J. dig. Dis.,* **21**, 340 (1976).
50. MacGregor, I. L. and Zakim, D., A cause of hyperamylasemia associated with chronic liver disease, *Gastroenterology,* **72**, 519 (1977).

4.7. LIVER

4.7.1. INTRODUCTION

The liver is the central chemical laboratory of the human body. There is scarcely a chemical reaction which cannot be performed by the parenchymal cells of the liver. Consequently, it is hardly surprising that up to the present several hundred tests of function have been described for this organ. Even so, the laboratory diagnosis of diseases of the liver is still not completely satisfactory since one occasionally encounters chronic hepatophathia for which all tests show normal values. Here too it must be stressed that it is only in exceptional cases that any correlation can be detected between particular tests and particular aetiologies or pathological anatomical changes. Usually, these three aspects—test result, aetiology, and pathology—are only loosely related to one another.

The tests of liver function may be classified roughly according to the following pathomechanisms:

1. Tests which identify disturbances in the 'membrane' permeability of the liver parenchymal cells: increase in glutamate–pyruvate transaminase, glutamate–oxaloacetate transaminase, fructose-1-phosphate aldolase, sorbitol dehydrogenase, leucine aminopeptidase and ornithine–carbamoyl transferase in the plasma.
2. Tests which indicate a necrosis of the liver cells: the appearance of glutamate dehydrogenase and of M-glutamate–oxaloacetate transaminase in the plasma.
3. Tests which point to an insufficiency of the protein-producing mechanisms in the endoplasmic reticulum of the parenchymal cells: the decline in cholinesterase, albumin, individual clotting factors, and fibrinogens in the plasma.
4. Tests which indicate an obstruction in the region of the intra- or extrahepatic bile duct: increase of alkaline phosphatase, caeruloplasmin, leucine aminopeptidase, and certain bile acids in the plasma.
5. Tests which indicate a proliferation of the mesenchymal fraction in the liver: increase of the α_2- and γ-globulins on the electropherogram, abnormal result of lability tests.
6. Complex function tests in which different pathogenetic factors can be responsible for a pathological result: the determination of the plasma

591

Table 107. Pathogenic syndromes of the liver (after ref. 1).

Type	Enzyme	Syndrome a	b	c	d	e	f	a + e
↗	Glutamate–oxaloacetate transaminase	+++	+	+	++	+	+	+++
	Glutamate–pyruvate transaminase	+++	0	+	+	0	0	+++
	Leucine aminopeptidase	++	++	++	++	+++	++	+++
	Ornithine–carbamoyl transferase	+++	++	+	++	++	+	+++
	γ-Glutamyl transpeptidase	++	++	+	+	+++	+++	++
↗	Glutamate dehydrogenase	+	0	+++	+++	++	+	+
↗	Alkaline phosphatase	0	0	+	++	+++	++	+++
	Caeruloplasmin	0	0	+	+	+++	++	+++
	Cholesterol	0	0	0	0	+++	0	+++
↙	Cholinesterase	0	++	+++	0	0	0	0
	Albumin	0	++	+++	0	0	0	0
	Clotting factors	0	++	+++	0	0	0	0
	Bilirubin	++	+	+	+	+++	0	+++
	Bromsulphalein elimination	++	+++	+++	+		+	

a = Acute diffuse membrane damage (type: Hepatitis acuta); b = chronic diffuse parenchymal damage (type: compensated cirrhosis); c = disseminated necroses (type: necrotizing advance in cirrhosis); d = dystrophy of the parenchymal cells (type: subacute dystrophy); e = obstruction of the biliary tract (type: extrahepatic closure by calculus); f = local intra-hepatic obstruction (type: liver metastasis); a + 3 = type: cholangiolitic form of acute hepatitis.

bilirubin, the bromsulphalein elimination, and the rate of galactose elimination.

A few typical results for the most common diseases of the liver are collected in Table 107. It should be noted that it is the biochemical syndromes which are identified, the clinical picture (e.g. hepatitis epidemica) only rarely.

Reference

1. Richterich, R., Enzymdiagnostik der Leberkrankheiten, *Schweiz. med. Wschr.*, **93**, 1363 (1963).

4.7.2. BILIRUBIN AND ITS DERIVATIVES

4.7.2.1. General

Up to about 300 mg of bilirubin per day are produced as a degradation product of haem through the degradation of mature, aged erythrocytes in the reticulo-endothelial system (spleen, liver, lymph glands) (Fig. 135). A small part of the daily bilirubin production is also formed from the haem

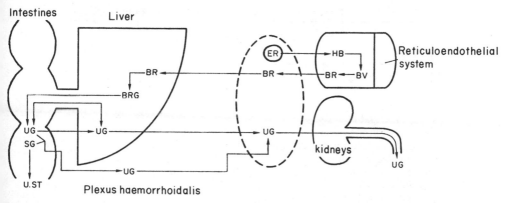

Fig. 135. Diagram of the bilirubin metabolism. ER = erythrocytes; HB = haemoglobin; BV = biliverdin; BR = bilirubin; BRG = bilirubin diglucuronide; UG = urobilinogen; SG = stercobilinogen; U = urobilin; ST = stercobilin.

enzymes of the liver (cytochrome, catalase) and from the bone marrow by the degradation of immature erythrocytes. Scisson of the tetraphyrrole ring of haemoglobin produces biliverdin, the precursor of bilirubin. This reaction is accelerated by haemoxygenase which requires oxygen and NADPH$_2$ for this purpose.[1] The biliverdin is reduced to bilirubin by the action of a second enzyme, biliverdin reductase.[2]

Owing to its low solubility in water, almost all of the bilirubin in plasma is bound to albumin. For the same reason, this bilirubin cannot be excreted through the kidneys or through the bile duct and, in cases of excessive production (haemolysis), it becomes concentrated in the lipid-rich structures (brain, subcutis). In new-born babies irreversible cell damage (nuclear jaundice) can be produced by passage of the blood–liquor barrier. The intestines and the placenta can also be passed.

By a specific transport system the bilirubin bound to albumin is taken up from the blood stream into the parenchymal cells of the liver, is subsequently bound to macromolecular carrier proteins (not precisely identified), and is actively transported further.[3]

In the cells of the liver the bilirubin is coupled to two molecules of glucuronic acid. Bilirubin is conjugated to bilirubin diglucuronide by the action of uridyl–glucuronyl transferase in the microsomes (endoplasmic reticulum). This bilirubin is water-soluble, non-toxic, and is excreted through the liver cells in the bile capillaries and, in pathological cases, through the kidneys. Following excretion through the bile into the intestines glucuronide is cleaved by bacterial action and reduced. The 'urobiligens' so formed are partly reabsorbed (enterohepatic cycle) and partly reduced further, to urobilin or stercobilin. Normally urobilinogen enters the great cycle, for the most part via the Plexus haemorrhoidalis, and is excreted through the kidneys.

594

Table 108. Differences between the two types of bilirubin in plasma.

Properties	Unesterified bilirubin	Bilirubin ester
Van den Bergh reaction	Indirect	Direct
Solubility in water	0	+++
Solubility in lipids	+++	0
Sensitivity to light	+++	+
Bonding to plasma albumin	+++	+
Bonding to denatured protein	0	+
Affinity to nerve cells	+++	0
Occurrence in urine	0	+++
Occurrence in bile	0	+++
Occurrence in plasma		
Normal	+	0
Haemolysis	+++	0
Obstructive jaundice	+	+++
Parenchymal jaundice	+	+++

0 = None; + = slight; +++ = high.

The terminology of the bilirubin derivatives has been subjected to various criticisms in recent years. The proposals of Fog and Bakken[4] and of the IFCC Expert Panel on Bilirubin[5] are regarded as acceptable and should be used as far as possible.

1. Total bilirubin
2. Bilirubin (formerly unconjugated, unesterified, 'indirect reacting' bilirubin)
2.1. Bilirubin bound to protein
2.2. Bilirubin–phospholipid–albumin bonding
2.3. 'Free' bilirubin (not bound to protein)
3. Bilirubin ester (formerly conjugated, 'direct reacting' bilirubin)
3.1. Bilirubin mono- or diester
3.2. Bilirubin ester-phospholipid–albumin bonding

The most important physical and biological properties of bilirubin and bilirubin ester are collected in Table 108.

References

1. Tenhunen, R., Marvek, H. S., and Schmid, R., Microsomal heme oxygenase; characterization of the enzyme, *J. biol. Chem.*, **244**, 6388 (1969).
2. Fleischner, G. J. and Arias, M., Recent advances in bilirubin formation, transport, metabolism and excretion, *Am. J. Med.*, **49**, 576 (1967).
3. Scherlock, S., *Diseases of the liver and biliary system*, 4th ed., Blackwell, London, 1968.

595

4. Fog, J. and Bakken, F. A., Definition of human bilirubin and bilirubin compounds, *Z. klin. Chem. klin. Biochem.*, **12**, 562 (1972).
5. *Internationl Federation of Clinical Chemistry (IFCC) Newsletter*, No. 13, March 1976.

4.7.2.2. Bilirubin standards and standard solutions

4.7.2.2.1. Properties of bilirubin

Bilirubin is readily isolated from the bile or gall-stones of animals and man. Pure bilirubin is completely soluble in chloroform and dimethyl sulphoxide (DMSO). On storage, crystalline bilirubin undergoes changes which are manifest, on the one hand, by a brownish discoloration of the crystalline powder and, on the other, by the presence of a residue on dissolving the crystals in chloroform or DMSO.[1-4] Chloroform or DMSO solutions should be prepared with exclusion of light, the DMSO solution being the more stable.[4]

4.7.2.2.2. Optical properties of bilirubin dissolved in chloroform

The purity of commercial bilirubin preparations has improved substantially in recent years. Thanks to the efforts of the International Federation of Clinical Chemistry (IFCC), the Expert Panel on Bilirubin, the American College of Clinical Pathologists, and the National Bureau of Standards (NBS), bilirubin preparations are now commercially available which conform to the necessary requirements. The optical properties of bilirubin solutions in chloroform or DMSO are used as a criterion of purity.

According to the American Association of Clinical Chemists (AACC) and American College of Clinical Pathologists,[5-7] the molar extinction coefficient (ε) of such solutions at a temperature of 25 °C and a wavelength of 453 nm is $60\,700 \pm 800\,\mathrm{l\,mol^{-1}\,cm^{-1}}$, and this should characterize pure bilirubin. These results have subsequently been confirmed by various authors.[4,8-10] The molar extinction coefficient of the azo dye from this chloroform solution of bilirubin is $70\,380 \pm 280\,\mathrm{l\,mol^{-1}\,cm^{-1}}$.[11]

The equation for evaluating the degree of purity of bilirubin is

$$\varepsilon = \frac{E}{Cm \cdot d} \cdot \frac{FV}{SV} = \frac{E}{d} \cdot \frac{MW}{C\% \cdot 10} \cdot \frac{FV}{SV}$$

If the observed value (ε_{obs}) is related to the prescribed value (ε_{pres}), then the percentage purity is given by the equation

$$\text{Purity (\%)} = \frac{\varepsilon_{obs} \cdot 100}{\varepsilon_{pres}} = \frac{60\,700 \cdot 100}{\varepsilon_{pres}} \cdot \frac{\varepsilon_{obs} \cdot 100}{60\,700}$$

For the preparation of bilirubin standards to be used for optimization of methods, it is still advisable to use only those preparations (standard reference materials) supplied by the National Bureau of Standards.

4.7.2.2.3. Optical properties of bilirubin in plasma

The absorption spectrum of bilirubin in plasma differs from that in chloroform. These optical properties are attributable to the bonding of bilirubin to albumin and phospholipids. The experiments of Wennberg and Cowger[12] showed that the bonding of bilirubin to albumin is strongly dependent on the pH of the solution. At pH 8.5, 1 mol of bilirubin is bound per mole of albumin.[12] The absorption maximum of bilirubin at pH 8.5 lies at 472 nm and the molar extinction coefficient is $\varepsilon_{472\,nm} = 67\,000 \pm 2000\,l$ $mol^{-1}\,cm^{-1}$. A fall in pH is likely to be attended by a shift of the maximum from 472 to 454 nm.[12]

4.7.2.2.4. Preparation of a standard solution of bilirubin (20 *mg per* 100 *ml*)

The best solvents to use are solutions of albumin, or bilirubin-free plasma or sera.[16] The most accurate results are probably obtained by dissolving the bilirubin in a solution of human albumin (5 g per 100 ml); bovine albumin can also be used.[17] It is imperative that the work be carried out in a dark-room under red light. Pure bilirubin (40 mg) is dissolved in 0.1 mol l^{-1} sodium carbonate solution to make exactly 20 ml; the mixture should be shaken as little as possible during solution (to avoid foaming). A 1-ml volume of this solution is added to 8 ml of solvent and the volume made up to 10 ml with 0.1 mol l^{-1} acetic acid. Bilirubin determinations on the solvent and the standard solution are carried out at once, the extinction of the solvent being subtracted as a blank. This procedure can be used for the calibration of azo methods.[13]

4.7.2.2.5. Stability of bilirubin solutions

Bilirubin is always unstable in the dissolved form. This applies to standard solutions in chloroform or plasma and to biological fluids containing bilirubin. Bilirubin is exceptionally sensitive to the action of light, whether this be daylight, ultraviolet light, or artificial light.[1,14] This decomposition is retarded at higher concentrations of albumin.[15] Several processes seem to be involved: shifts between free and conjugated bilirubin, oxidation to biliverdin, and decomposition to dipyrroles. For this reason, bilirubin should always be determined as quickly as possible or at least be kept in the dark until the determination is carried out.

References

1. Gambino, S. R. and Di Re, J., *Bilirubin assay*, American Society of Clinical Pathologists, Chicago, 1968.

2. Fog, J. and Bugge Asperheim, B., Stability of bilirubin, *Nature, Lond.*, **203**, 756 (1964).
3. Fog, J., Bilirubin—purification—purity, *Scand. J. clin. Lab. Invest.*, **16**, 49 (1964).
4. Heine, W. and Tittelbach-Helmrich, W., Zur Darstellung stabiler Bilirubin-standardlösungen mit Dimethylsulfoxid, *Z. med. Labortech.*, **14**, 332 (1973).
5. American Association of Clinical Chemists and College of American Pathologists, Recommendation on a uniform bilirubin standard, *Pediatrics, Springfield*, **31**, 878 (1963).
6. Recommendation on a uniform bilirubin standard, *Clin. Chem.*, **8**, 405 (1962).
7. Recommendation on a uniform bilirubin standard, *Stand. Meth. clin. Chem.*, **5**, 75 (1965).
8. Billing, B., Haslam, R., and Wald, N., Bilirubin standards and the determination of bilirubin by manual and Technicon AutoAnalyzer methods. *Ann. clin. Biochem.*, **8**, 21 (1971).
9. Colombo, J. P., Peheim, E., Kyburz, S., and Hoffmann, J. P., The determination of total bilirubin in plasma with 2,4-dichloraniline, *Chem. Rdsch.*, **27**, 23 (1974).
10. Doumas, B. T., Perry, B. W., Sasse, E. A., and Straumfjord, J. V., Jr., Standardization in bilirubin assays: evaluation of selected methods and stability of bilirubin solutions, *Clin. Chem.*, **9**, 984 (1973).
11. Meites, S. and Traubert, J. W., Use of bilirubin standards, *Clin. Chem.*, **11**, 691 (1965).
12. Wennberg, R. P. and Cowger, M. L., Spectral characteristics of bilirubin–bovine albumin complexes, *Clin. chim. Acta*, **43**, 55 (1973).
13. Richterich, R., Direkte spektrophotometrische Bestimmung des Gesamt-Bilirubins im Plasma von Neugeborenen, *Klin. Wschr.*, **41**, 778 (1963).
14 With, T. K., *Bile pigments, chemical, biological and clinical aspects*, Academic Press, New York, 1968.
15. Ente, G., Lanning, E. W., Cukor, P., and Klein, R. M., Chemical variables and new lamps in phototherapy, *Pediat. Res.*, **6**, 246 (1972).
16. Franzini, C. and Panseri, C., Diluted human serum as a suitable diluent in the preparation of bilirubin reference solutions, *Clin. chim. Acta*, **69**, 73 (1976).
17. Zebelmann, A. M., Kenny, M. A., and Sunday, C. G., Bilirubin standard solutions, *Clin. Chem.*, **22**, 934 (1976).

4.7.2.3. Determination of bilirubin ester and of total bilirubin as azobilirubin

4.7.2.3.1. Introduction

Bilirubin occurs in the plasma in various molecular forms. Part is present as bilirubin ester (mono- and diglucuronide), part as unesterified bilirubin. The glucuronides are water-soluble; unesterified bilirubin is insoluble in water and, in addition, is relatively strongly bound to plasma proteins, particularly albumin.

4.7.2.3.2. Choice of method

Ehrlich[1] in 1883 was the first to point out that urinary bilirubin, i.e. bilirubin glucuronide, would couple with diazotized sulphanilic acid and that the azo dyes formed possessed indicator properties. They are red at neutral pH and blue at strongly acidic and strongly alkaline pH (double indicator). Since bilirubin in urine is present almost exclusively as the water-soluble bilirubin glucuronies, the latter react with the diazotized sulphanilic acid directly, without the addition of other chemicals.

Hijmans-van den Bergh and Snapper[2] were the first (1916) to apply this principle to the détermination of the plasma bilirubin. In the course of their experiments they observed that in addition to a fraction which reacted directly there was a variable amount of bilirubin which reacted only after pretreatment with alcohol. They called this fraction 'indirect reacting bilirubin'. It consists of unesterified bilirubin (protein-bound bilirubin).

Today there are a large number of modifications of the determination of bilirubin as azo dyes. In the first methods to be described the determination was subsequent to deproteination, but since the precipitate always entrains variable quantities of bilirubin glucuronides which have a high affinity for denatured proteins, the yields with these methods are unsatisfactory.

Of the methods of determination not involving a deproteination, a distinction may be made between those methods which use methanol to liberate the bilirubin bound to protein[3] and those in which bilirubin is released not by methanol denaturation of the proteins, but by displacement by anions, using such substances as sodium acetate, caffeine, diphylline, and sodium benzoate.[4] A simple method for the determination of the total bilirubins, distinguished by high stability of the reagents, good linearity, and rapid reaction, has been described by Rand and Di Pasqua.[5] 2,4-Dichloraniline is used as dye and Brij 35 as accelerator.[6,7]

Coupling of the bilirubin to form azobilirubin can also be done using the diazonium salt of 2,5-dichlorobenzene.[8] In this method the preparation of the reagent is particularly simple.

Among the methods in which the unesterified bilirubin is released by displacement, those techniques in which the measurement is made at alkaline pH have an advantage in that the extinction coefficient of the azo dye at alkaline pH is about twice as high as that at neutral pH. The described method is based on the principle of the technique reported by Jendrassik and Grof[4] as modified by Nosslin[9] and incorporating the alterations recommended by Michaelson.[10]

4.7.2.3.3. Principle of the method using sulphanilic acid

Bilirubin forms a water-soluble azo dye with diazotized sulphanilic acid. The reaction mechanism of this process has now largely been elucidated.

Formation of diazonium chloride. Sulphanilic acid reacts with sodium nitrite in concentrated hydrochloric acid solution to form the diazonium chloride (diazotization). This substance is unstable and must therefore always be prepared freshly (Fig. 136).

Formation of the azo dye (coupling). The mechanism for the formation of the dye differs slightly, depending on which bilirubin derivative is

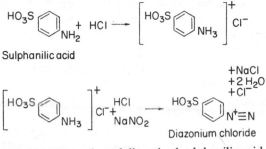

Fig. 136. Formation of diazotized sulphanilic acid.

concerned:

(a) Bilirubin ester (diglucuronide) (pigment II). Once the methylene bridge has been opened, one dipyrrole reacts directly with diazonium chloride to form a molecule of pigment B. The second, isomeric dipyrrole rearranges to form a hydroxypyrromethenecarbinol. In a second stage, this molecule is also converted to an (isomeric) pigment B by reaction with a second molecule of diazonium chloride (Fig. 137).

(b) Unesterified bilirubin. This water-insoluble pigment is first liberated from the protein bonding by the addition of accelerators. Again there is a two-stage production of isomeric azo dyes known as pigment A.

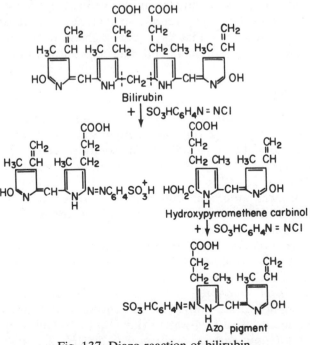

Fig. 137 Diazo reaction of bilirubin.

(c) Bilirubin ester (monoglucuronide) (pigment I). The aglucuronide half of the molecule produces one molecule of pigment A and the glucuronide half produces one molecule of pigment B.

The total bilirubin comprises both the unesterified bilirubin and the bilirubin ester (monoglucuronide) (pigment I) and (digluruconide) (pigment II). A direct reaction gives the bilirubin diglucuronide as well as part of the bilirubin monoglucuronide.

One benefit conferred by the use of diphylline/sodium acetate as accelerator is that turbidities occur very rarely, owing to the high solubility of diphylline (in contrast to caffeine). The purpose of adding ascorbic acid is two-fold. Firstly, it is intended to destroy any remaining diazonium chloride before the mixture is made alkaline. Secondly, ascorbic acid blocks the inhibiting action of haemoglobin derivatives on the formation of dye. It is known that the intensity of the azo dye is much lower in the presence of haemoglobin.

The changes in colour which accompany the reaction require a few words of explanation.

(a) Coupling of bilirubin with diazonium chloride produces the typical red azo dye.

(b) On the addition of sodium ascorbate this reacts with the diazonium chloride to form an intensely yellow dye. This brings the development of colour to an end. The visible colour is a mixture of the colours of the azo dye and the yellow pigment.

(c) On rendering the solution alkaline the red azo dye turns blue whereas the yellow by-product remains yellow, so that the mixture appears green. However, the spectra show that the yellow pigment does not absorb at the wavelength used for measurement and thus does not interfere.

4.7.2.3.4. Reagents

Diphylline acetate solution: diphylline, 0.35 *mol* l^{-1}; sodium acetate, 1.83 *mol* l^{-1}. Dissolve 9 g of diphylline and 25 g of sodium acetate (with heating) in DM-water and make up to 100 ml. The solution keeps indefinitely. Some precipitation of the diphylline does not interfere with the reaction. If necessary, store at 25–37 °C.

Sodium ascorbate solution, 260 *mmol* l^{-1}. Dissolve 50 mg of sodium ascorbate in 1 ml of water, or use one ampoule of Redoxone Roche (100 mg per 2 ml). Prepare freshly each day.

Sodium nitrite solution, 3.68 *mmol* l^{-1}. Dissolve 25.4 mg of sodium nitrite in DM-water and make up to 100 ml. The solution keeps for a few weeks in a tightly closed dark bottle.

Sulphanilic acid solution, 103 *mmol* l^{-1}. Dissolve 2.2 g of sodium sulphanilate monohydrate and 15 ml of concentrated hydrochloric acid by heating with about 80 ml DM-water. Make up to 100 ml with DM-water.

Fehling II solution: tartrate, 1.49 *mol* $^{-1}$; *NaOH,* 2.5 *mol* l^{-1}. Dissolve

420 g of potassium sodium tartrate and 100 g of sodium hydroxide in DM-water and make up to 1000 ml. The solution keeps indefinitely.

4.7.2.3.5. Measurement

Conditions. Incubation at room temperature, wavelength 578 nm. The final concentrations in the reaction mixture (mmol l^{-1}) are as follows:

Diphylline	130.5
Sodium acetate	683
Sodium ascorbate	3.88
Sodium nitrite	0.137
Sulphanilic acid solution	3.84
Tartrate	556
NaOH	933

4.7.2.3.6. Procedure

Diazo reagent. Mix equal parts of the sodium nitrite and sulphanilic acid solutions. Always prepare freshly.

	Total bilirubin, ml	Bilirubin ester, ml	Blank, ml	Standard[a], ml
Distilled water	0.20	0.20	0.20	0.20
Serum	0.03	0.02	—	St. 0.02
Diphylline reagent	0.50	—	0.50	0.50
Diazo reagent	0.10	0.10	0.10	0.10

Wait 10–15 min

	Total bilirubin, ml	Bilirubin ester, ml	Blank, ml	Standard,[a] ml
Sodium ascorbate	0.02	0.02	0.02	0.02
Serum	—	—	0.02	0.02
Diphylline solution	—	0.50	—	—
Fehling II solution	0.50	0.50	0.50	0.50

Read the extinction against DM-water at 578 nm

[a]Only necessary if an absolute-measuring photometer providing a wavelength of 578 nm is not available.

4.7.2.3.7. Calculation

Using the molar extinction coefficient $\varepsilon_{578\,nm} = 60\,550\,l\,mol^{-1}\,cm^{-1}$ (ref. 12).

Total bilirubin.

$$\text{Concentration} = \frac{E(TB) - E(B) \cdot 584.68}{60\,550 \cdot 1 \cdot 10} \cdot \frac{1.34}{0.02} \cdot 10^3 \text{ mg per 100 ml}$$

Concentration = $E(TB) - E(B) \cdot 64.7$ mg per 100 ml

$$\text{Concentration} = \frac{E(TB) - E(B)}{60\,550} \cdot \frac{1.34}{0.02} \cdot 10^6 \ \mu mol\ l^{-1}$$

Concentration = $E(TB) - E(B) \cdot 1106.5 \ \mu mol\ l^{-1}$

Bilirubin ester ('direct reacting bilirubin').

Concentration = $E(BE) - E(B) \cdot 64.7$ mg per 100 ml

$$\text{or} \cdot 1106.5 \ \mu mol\ l^{-1}$$

Bilirubin ('indirect reacting bilirubin').

Total bilirubin − bilirubin ester.

Using a standard solution.

$$\text{Concentration} = \frac{E(TB) - E(B)}{E(ST)} \cdot c(ST) \text{ mg per 100 ml or } \mu mol\ l^{-1}$$

4.7.2.3.8. Reference values

Total bilirubin, adults. The distribution of the reference values for total bilirubin is of the log-normal type. The deviation from the straight line in Fig. 138 is probably attributable to the technical unreliability of the method

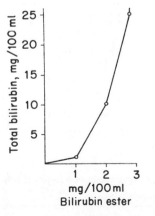

Fig. 138. Nomogram for ascertaining whether bilirubin ester ('direct reacting bilirubin') is present (disturbance in the elimination of bilirubin diglucuronide). If the value for the 'direct reacting bilirubin' falls to the right of the line-plot, bilirubin ester is probably present.[9]

Table 109. Total bilirubin in new-born babies ($\bar{x} \pm 2s$).

Age	Full-term birth		Premature birth	
	mg per 100 ml	μmol l^{-1}	mg per 100 ml	μmol l^{-1}
0–24 h	Up to 5.0	Up to 85.5	Up to 8.0	Up to 137
24–48 h	Up to 9.0	Up to 154	Up to 12.0	Up to 205
3–5 days	Up to 12.0	Up to 205	Up to 24.0	Up to 410
Older than 1 month	Up to 1.5	Up to 25.7	Up to 1.5	Up to 25.7

in determining very low values. The majority of the slightly elevated values are not—as is repeatedly and incorrectly assumed—attributable to diseases of the liver, for example, but are proabably due to constitutional anomalies of the bilirubin and haemoglobin metabolism.

Total bilirubin, new-born babies. Values are given in Table 109.

Bilirubin ester ('direct reacting bilirubin'), new-born babies. If the values lie to the right of the line drawn in Fig. 138, this is indicative of the presence of conjugated bilirubin.

4.7.2.3.9. Comments on the method

1. The molar extinction coefficients at 600 and 578 nm, as calculated from the results of various authors, are collected in Table 110.

The differences, though slight, would seem to be due to the use of different bilirubin preparations for standardization.

2. Minimal elevations of the total bilirubin (up to about 2 mg per 100 ml) are much more rarely of clinical significance than is commonly assumed. In contrast, the precise measurement of concentration between 15 and 25 mg per 100 ml is of great practical importance, both in children—as indicative of exchange transfusion—and in adults—for following the progress of the disease. The present method has therefore been so designed—that results at the upper limit of the normal range, although usable, are not very exact. However, it does provide very accurate values in the range between 3 and 25 mg per 100 ml. Such a compromise was unavoidable since no method yields reliable results over a wide range of measurements (0.5–25 mg per 100 ml).

Table 110. Molar extinction coefficients.

Author	$\varepsilon_{600 \, nm}$	$\varepsilon_{578 \, nm}$
Fog[11]	65 500	57 500
Nosslin[9]	69 400	62 460
Michaelson[10]	67 500	60 750
Richterich et al.[12]	67 300	60 550

3. The terms 'direct reacting' and 'indirect reacting' bilirubin have no absolute validity. Markedly different values are obtained depending on the method used. Accordingly, we prefer to follow the suggestion of Nosslin[9] and confine the statement to whether 'direct reacting bilirubin' (i.e. bilirubin ester) is actually present. This is the case when the result lies to the right of the line drawn in Fig. 138.

4. Sera for bilirubin determination should not be exposed to light, particularly sunlight, since the results may be low.

5. We would advise against the use of standard solutions of dyes[14] since their spectra do not agree with that of bilirubin.

6. Lipaemic sera with a triglyceride content above 100 mg per 100 ml produce considerable interference in the diazotization.[13]

7. Haemolysis does not adversely affect the determination.

4.7.2.3.10 Principle of the 2,4-dichloroaniline method

2,4-Dichloroaniline reacts with nitrite and hydrochloric acid to yield a stable diazo reagent. The total bilirubin is coupled with a diazo compound to give the corresponding azo dye. Brij 35 is used as accelerator, allowing the determination of water-soluble bilirubin as well as the bilirubin bound to protein. The sample blank and partial reagent blank determinations are carried out by omitting nitrite from the blank reagent.

4.7.2.3.11. Reagents

Stock solutions

Sodium nitrite solution, $217\,mmol\,l^{-1}$. Dissolve 150 mg of sodium nitrite in DM-water and make up to 10 ml. Prepare freshly each week.

Dichloroaniline solution: 2,4-dichloroaniline, $24.7\,mmol\,l^{-1}$; *hydrochloric acid,* $500\,mmol\,l^{-1}$. Dissolve 400 mg of 2,4-dichloroaniline in 0.5 mol l^{-1} hydrochloric acid and make up to 100 ml with further hydrochloric acid. Prepare freshly each week.

Brij 35 solution, $50\,g\,l^{-1}$, *hydrochloric acid* $250\,mmol\,l^{-1}$. Dissolve 50 g of Brij 35 in 800 ml of hot (*ca.* 50 °C) DM-water, add 25 ml of concentrated hydrochloric acid and make up to 1000 ml with DM-water. The solution is stable for at least 1 year.

Working solutions

Diazo reagent: Brij 35, $44.5\,g\,l^{-1}$; *2,4-dichloroaniline,* $2.47\,mmol\,l^{-1}$; *sodium nitrite,* $2.17\,mmol\,l^{-1}$; *hydrochloric acid,* $272\,mmol\,l^{-1}$. Mix 10 ml of dichloroaniline solution with 1 ml of sodium nitrite solution. Allow to stand in the ice-bath for 15 min and make up to 100 ml with Brij 35 solution. The solution is stable for at least 1 week at 4 °C.

Blank reagent: Brij 35, $44.5\,g\,l^{-1}$; *2,4-dichloroaniline,* $2.47\,mmol\,l^{-1}$; *hydrochloric acid,* $275\,mmol\,l^{-1}$. Dilute 10 ml of dichloroaniline solution to 100 ml with Brij 35 solution. The solution is stable for at least 1 week at 4 °C.

4.7.2.3.12. Measurement

Conditions. Incubation at room temperature, wavelength 546 nm; the colour is stable for about 2 h. The final concentrations are Brij 35 42.8 g l⁻¹, 2,4-dichloroaniline 2.37 mmol l⁻¹, sodium nitrite 2.08 mmol l⁻¹, and hydrochloric acid 261.5 mmol l⁻¹. The final volume is 0.52 ml and the sample volume is 0.02 ml.

4.7.2.3.13. Procedure

	Sample, ml	Sample blank, ml
Plasma, serum	0.02	0.02
Diazo reagent	0.5	—
Blank reagent	—	0.5

Mix and after 10 min measure the extinction against DM-water at 546 nm

4.7.2.3.14. Calculation

Using the molar extinction coefficient $\varepsilon_{546\,nm}$ = 54 167 l mol⁻¹ cm⁻¹ (ref. 6).

$$\text{Concentration} = \frac{E(S) - E(SB) \cdot 584.68}{54\ 167 \cdot 1 \cdot 10} \cdot \frac{0.52}{0.02} \cdot 10^3 \text{ mg per 100 ml}$$

Concentration = $E(S) - E(SB) \cdot 28.1$ mg per 100 ml

Concentration = $E(S) - E(SB) \cdot 480$ μmol l⁻¹

4.7.2.3.15. Specificity

The determination is specific for the total bilirubin in plasma or serum.

4.7.2.3.16. Reference values

Adults: 0.2–1.5 mg per 100 ml; 3.4–25.7 μmol l⁻¹ (author's laboratory).

4.7.2.3.17. Comments on the method

1. Haemoglobin concentrations above 400 mg per 100 ml interfere with the reaction.[6]

2. The method is linear up to bilirubin concentrations of 40 mg per 100 ml (684 μmol l⁻¹).

3. Determination of the bilirubin ester is not possible by the 2,4-dichloroaniline method. The Jendrassik–Grof method[4] is designed for this purpose (cf. p. 598).

References

1. Ehrlich, P., Sulfodiazobenzol, ein Reagens auf Bilirubin, *Centbl. klin. Med.*, **4**, 721 (1883).
2. Hijmans-van den Bergh, A. A. H. and Snapper, J., Die Farbstoffe des Blutserums. I. Eine quantitative Bestimmung des Bilirubins im Blutserum, *Arch. klin. Med.*, **110**, 540 (1916).
3. Malloy, H. T. and Evelyn, K. A., The determination of bilirubin with the photoelectric colorimeter, *J. biol. Chem.*, **119**, 481 (1937).
4. Jendrassik, L. and Grof, P., Vereinfachte photometrische Methode zur Bestimmung des Blutbilirubins, *Biochem. Z.*, **297**, 81 (1938).
5. Rand, R. N. and Di Pasqua, A., A new diazo method for the determination of bilirubin, *Clin. Chem.*, **8**, 570 (1962).
6. Colombo, J. P., Peheim, E., Kyburz, S., and Hoffmann, J. P., The determination of the bilirubin in plasma with 2,4-dichloraniline, *Chem. Rdsch.*, **27**, 23 (1974).
7. Colombo, J. P., Peheim, E., Kyburz, S., and Hoffmann, J. P., A new accelerator for the determination of total bilirubin in plasma, *Clin. chim. Acta*, **51**, 217 (1974).
8. Wahlefeld, A. W., Herz, G., and Bernt, E., Modification of the Malloy–Evelyn method for a simple determination of total bilirubin in serum, *Scand. J. clin. Lab. Invest.*, **29**, suppl., 126 (1972).
9. Nosslin, B., The direct diazo reaction of bile pigments in serum: experimental and clinical studies, *Scand. J. clin. Lab. Invest.*, **12**, suppl., 49, p. 1 (1960).
10. Michaelson, M., Bilirubin determination in serum and urine. Studies on diazo methods and a new copper-azo pigment method, *Scand. J. clin. Lab. Invest.*, **13**, suppl., 56, p. 1 (1961).
11. Fog, J., Icterus index determined with the spectrophotometer corrected for turbidity and hemoglobin, *Scand. J. clin. Lab. Invest.*, **10**, 246 (1958).
12. Soini, R., Dauwalder, H., and Richterich, R., Ein Vergleich physikalischer und chemischer Methoden zur Bestimmung des Plasmabilirubins, *Schweiz. med. Wschr.*, **99**, 1784 (1969).
13. Chan, G., Merrills, K., and Schiff, D., Bilirubin quantitation with lipemic plasma, *Clin. Biochem.*, **9**, 96 (1976).
14. Bilissis, P. K. and Speek, R. J., A stable bilirubin standard, *Clin. Chem.*, **9**, 552 (1963).

4.7.3. GLUTAMATE–PYRUVATE TRANSAMINASE

4.7.3.1. Introduction

Glutamate–pyruvate transaminase (GPT, alanine aminotransferase, L-alanine : 2-oxoglutarate aminotransferase, E.C. 2.6.1.2) occurs predominantly in the liver but is also found, in order of decreasing abundance, in the kidneys, heart, skeletal muscle, and other organs. It catalyses the conversion of alanine to pyruvate via transamination, also the reverse reaction, and it occupies an important key position in gluconeogenesis:

$$\text{L-Alanine} + \text{2-Oxoglutarate} \xrightleftharpoons{\text{GPT}} \text{L-Glutamate} + \text{Pyruvate}$$

Intracellularly, the enzyme occurs in the mitochondria and predominantly (50–85%) in the cytosome. The two enzymes differ in respect of their kinetic, physical, and electrochemical properties.[1] Glutamate–pyruvate transaminase is found in the plasma, in liquor cerebrospinalis, and in the urine. It is present in very small concentrations in the erythrocytes.

4.7.3.2. Choice of method

The quickest and most reliable method for determining glutamate–pyruvate transaminase is the optical test using lactate dehydrogenase as auxiliary enzyme.

4.7.3.3. Principle

Primary reaction: GPT.

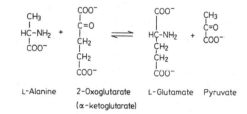

L-Alanine   2-Oxoglutarate   L-Glutamate   Pyruvate

(α-ketoglutarate)

Indicator reaction: lactate dehydrogenase (LDH).

$$\underset{\text{Pyruvate}}{\begin{array}{c}CH_3\\|\\C=O\\|\\COO^-\end{array}} + NADH + H^+ \underset{}{\overset{LDH}{\rightleftharpoons}} \underset{\text{Lactate}}{\begin{array}{c}CH_3\\|\\HCOH\\|\\COO^-\end{array}} + NAD^+$$

4.7.3.4. Measurement

Conditions. Preference is given here to the determinations according to the recommendations of the Deutsche Gesellschaft für klinische Chemie[2] and of the Scandinavian Society of Clinical Chemistry and Clinical Physiology[3] (Table 111).

Table 111. Determination of glutamate–pyruvate transaminase; final concentrations in the reaction mixture.

	Deutsche Gesellschaft für klinische Chemie[2]	Scandinavian Society for Clinical Chemistry and Clinical Physiology[3]
Phosphate buffer	80 mmol l^{-1}	
Tris buffer		20 mmol l^{-1}
pH	7.4	7.4
EDTA		5 mmol l^{-1}
L-Alanine	800 mmol l^{-1}	400 mmol l^{-1}
NADH$_2$	0.18 mmol l^{-1}	0.15 mmol l^{-1}
2-Oxoglutarate	18 mmol l^{-1}	12 mmol l^{-1}
LDH	1200 U l^{-1}	2000 U l^{-1}
Volume fraction of the serum (plasma)	1:7.4	1:8.3
Temperature of measurement	25 °C	37 °C
Wavelength	365 nm, $d = 1$ cm	340 nm, $d = 1$ cm

Table 112. Reference values for glutamate–pyruvate transaminase; determination according to the recommendations of the Deutsche Gesellschaft für klinische Chemie[2] (U l^{-1}, 25 °C).

Age	Sex	Number	Reference interval 5–95%	References
1 day–3 weeks	m, f	41	5–24	4
3 weeks–6 months	m, f	47	5–28	
6–12 months	m, f	41	5–20	
1–6 years	m, f	60	4–20	
6–14 years	m, f	60	4–20	
Adults	m	2384	6–33	5
	f	1595	5–21	
Adults	m	708	Up to 22	6
	f	668	Up to 18	
Adults 20–30 years	m	452	13–37 (2.5–97.5%)	7
Adults	m	446	5–23	11
	f	470	5–19 (2.5–97.5%)	

4.7.3.5. Calculation

Calculation of the volume activity in μmol min^{-1} l^{-1} = U (p. 234).

4.7.3.6. Reference values

These are given in Tables 112 and 113.

Laudahn et al.[5] found a log-normal distribution of the values, but other authors were unable to confirm this finding.[6,7] The values for women are lower than those for men. In men the glutamate–pyruvate transaminase activity is markedly dependent on weight, overweight being accompanied by an enzyme increase; this is less pronounced in women.[6,7] Women who take contraceptives show higher values than those who do not.[8]

4.7.3.7. Comments on the method

1. Serum and plasma give identical results.

2. Erythrocytes are low in glutamate–pyruvate transaminase. Consequently, slight haemolysis does not interfere with the determination of this enzyme. Bodily exercise does not lead to an increase in the enzyme activity in the plasma.

3. The specificity of the method is high. In the measurement of the enzyme activities in plasma, no side-reactions occur provided that the preliminary treatment is sufficiently long.

Table 113. Reference values for glutamate–pyruvate transaminase; determination following the recommendations of the Committee on Enzymes of the Scandinavian Society of Clinical Chemistry and Clinical Physiology[3] (U l^{-1}, 37 °C).

Age	Sex	Number	Reference interval 2.5–97.5%	Reference
Pregnant women (39th–41st week)		70		
New-born babies	m, f	70		
6–24 months	m, f	95		
2–5 years 6–7 years 8–9 years	m, f	345	<35	8
10–11 years 12–13 years 14–15 years 16–18 years	m, f	384		
18 years	m		<35	
	f		<40	

4. The auxiliary enzyme and reagents should be free from ammonium sulphate in order that there be no interference in the glutamate–dehydrogenase reaction.

5. Interference can occur in the measurement of the enzyme in tissue as a consequence of the production of ammonia in the presence of higher glutamate dehydrogenase activities (liver, kidneys, brain):

$$2\text{-Oxoglutarate} + NADH + H^+ + NH_4^+ \rightleftharpoons \text{L-Glutamate} + NAD^+ + H_2O$$

In such cases it is advisable to run a blank with 2-oxoglutarate (but without alanine) and to use the difference in extinction between sample and blank for calculating the units.

6. It is necessary to specify the organ of origin of the lactate dehydrogenase used as auxiliary enzyme since the individual isoenzymes obviously influence the kinetics of the reaction to different extents.[9]

7. The method is suitable for screening blood donors to exclude hepatitis in the pre-icteric, anicteric, or convalescent stage.[10]

4.7.3.8. Diagnositic significance

Glutamate–pyruvate transaminase (GPT), in association with glutamate–oxaloacetate transaminase (GOT), is of diagnostic significance in the field of *diseases of the liver*. The highest GPT activities are observed in acute viral hepatitis after the onset of jaundice. Its increase exceeds that of GOT. The GOT/GPT ratios falls below 0.7 (norm: 0.77 for men, 0.90 for women).[5] There is usually an increase in this enzyme even in the preclinical,

610

asymptomatic stage of hepatitis. This may also take an anicteric course without any massive increase of enzyme. Normalization of the process of the disease can only be assumed to have begun following a marked and persistent diminution of the transaminases. In the cholestatic course of a hepatitis, the icterus and the elevation of the transaminases persist for a longer period. Chronic forms of hepatitis are accompanied by lesser elevations of the transaminases, GPT predominating in the chronic persistent forms. Cirrhoses of the liver show only slightly elevated GPT values, which can increase with the GOT if there is a necrotic advance. With chronic progress the increases in activity are not constant but fluctuate. Even normal enzyme values may be observed at times. In alcoholic hepatitis the increase in GOT is generally greater than that of GPT. A survey of the enzyme constellations for the various pathogenetic syndromes of the liver is given in Table 107.

References

1. De Rosa, G. and Swick, W., Metabolic implications of the distribution of the alanine aminotransferase isoenzymes, *J. biol. Chem.*, **250**, 7961 (1975).
2. Empfehlungen der Deutschen Gesellschaft für klinische Chemie. Standardisierung von Methoden zur Bestimmung von Enzymaktivitäten in biologischen Flüssigkeiten, *Z. klin. Chem. klin. Biochem.*, **8**, 658 (1970).
3. Recommended methods for the determination of four enzymes in blood, *Scand. J. clin. Lab. Invest.*, **33**, 291 (1974).
4. Sitzmann, F. C., Loew, C., Kalut, H., and Prestele, H., Normalwerte der Serumtransaminasen GOT und GPT mit neuen substratoptimierten Standardmethoden, *Klin. Pädiat.*, **186**, 346 (1974).
5. Laudahn, G., Hartmann, E., Rosenfeld, E. M., Weyer, H., and Muth, H. W., Normalwerte der Serumtransaminasen bei Verwendung substratoptimierter Testansätze zur Aktivitätsmessung, *Klin. Wschr.*, **48**, 838 (1970).
6. Thefeld, W., Hoffmeister, H., Busch. E.-W., Koller, P. U., and Vollmar, J., Referenzwerte für die Bestimmungen der Transaminasen GOT und GPT sowie der alkalischen Phosphatase im Serum mit optimierten Standardmethoden, *Dt. med. Wschr.*, **99**, 343 (1974).
7. Siest, G., Schiele, F., Galteau, M., Panek, E., Steinmetz, J., Fagnani, F., and Gueguen, R., Aspartate aminotransferase and alanine aminotransferase activities in plasma: statistical distributions, individual variations and reference values, *Clin. Chem.*, **21**, 1077 (1975).
8. Pennttilä, I. M., Jokela, H. A., Viitala, A. J., Heikkinen, E., Nummi, S., Pystynen, P., and Saastamoinen, J., Activities of aspartate and alanine aminotransferases and alkaline phosphatase in sera of healthy subjects, *Scand. J. clin. Lab. Invest.*, **35**, 275 (1975).
9. Chang, M. M. and Chung, R. W., Effect of lactate dehydrogenase isoenzymes in the coupled enzymatic assay for alanine aminotransferase activity, *Clin. Chem.*, **21**, 330 (1975).
10. Stampfli, K., Neiger, A., Messerli, H., Halle, G., and Richterich, R., Serumenzyme bei Blutspenden. III. Resultat der Bestimmung der Glutamat-Pyruvat-Transaminase, *Schweiz. med. Wschr.*, **92**, 781 (1962).
11. Schlebusch, H., Rick, W., Lang, H., and Knedel, M., Normbereiche der Aktivitäten klinisch wichtiger Enzyme, *Dt. med. Wschr.*, **99**, 765 (1974).

4.7.4. GLUTAMATE DEHYDROGENASE

4.7.4.1. Introduction

Glutamate dehydrogenase (GLDH; L-glutamate : NAD(P) oxidoreductase, deaminating, E.C. 1.4.1.3) occurs predominantly in the liver and is localized in the mitochondrial matrix of the cells. It catalyses the reaction

2-Oxoglutarate + NH_4^+ + NADH + H^+ $\leftrightarrows$ L-Glutamate + NAD^+ + H_2O

The equilibrium favours the formation of L-glutamate.

4.7.4.2. Choice of method

Measurement by means of the optical test, as originally described by Olson and Anfinsen,[1] is currently the method most generally used.

4.7.4.3. Principle

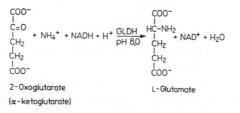

2-Oxoglutarate
(α-ketoglutarate)

L-Glutamate

4.7.4.4. Measurement

Conditions. These are given in Table 114.

Table 114. Determination of glutamate dehydrogenase, final concentrations in the reaction mixture.[2]

	mmol l^{-1}
Triethanolamine buffer, pH 8.0	50
2-Oxoglutarate	7
Ammonium acetate	100
EDTA	2.5
ADP	1
NADH	0.2
Lactate dehydrogenase, U l^{-1}	2000

Volume fraction of the serum (plasma) 1:6.2
Temperature of measurement 25 °C
Wavelengths 340 nm (Hg 334 nm, Hg 365 nm), $d = 1$ cm
Pre-incubation 5 min.
Duration of measurement 5–10 min, recording ΔE throughout this time
Result μmol min^{-1} l^{-1} = U l^{-1}

Table 115. Reference values for glutamate dehydrogenase (U l^{-1}, 25 °C).

Age	Sex	Reference interval	References
Adults	m	Up to 4	3
	f	Up to 3	
Adults		Up to 6	4
Adults	m	Up to 4	7
	f	Up to 3	

4.7.4.5. Calculation

Calculation of the volume activity in μmol min^{-1} l^{-1} = U l^{-1} (p. 234).

4.7.4.6. Reference values

These are given in Table 115.

4.7.4.7. Comments on the method

1. Pyruvate interferes and is reduced by the lactate dehydrogenase added in the preliminary treatment.
2. Adenosine diphosphate activates the enzyme. Since the erythrocytes do not contain any glutamate dehydrogenase, slight haemolysis does not interfere with the determination. The enzyme activity is not affected by bodily exertion.

4.7.4.8. Diagnositic significance

On account of its localization in the organ, glutamate dehydrogenase finds application in the *diagnosis of liver conditions*. Since it occurs within the mitochondria, its presence in the serum is an indication of considerable destruction of the cells, usually accompanied by necrosis. If these necroses are disseminated the function of the organ need not be severely impaired.

In fresh extrahepatic closure of the bile duct by tumours or calculus, especially with intermittent closure, the glutamate dehydrogenase in the serum can increase at relatively low transaminase values.[5,6] Even small increases in pressure in the biliary ducts with corresponding fluctuations can be recognized by a rapid rise and fall in serum activity. With intrahepatic cholestates the increase comes later.

With a fresh hepatitis, elevated values may, albeit rarely, be found even earlier than in acute advances of chronic hepatitis and of active cirrhoses. In liver metastases and in acute cholangitis an elevation of the glutamate dehydrogenase in the serum can be observed to a varying degree owing to the combined action of both mechanisms, namely injury to the liver cells

and cholestasis. It is frequently elevated in the serum in alcoholic fatty
liver. In this case the cause is much more likely to consist of a functional
disturbance of the mitochondria rather than in extensive cell damage.
Extremely high serum activities are observed in poisoning accompanied by
acute cell necroses (dystrophy). Persistent elevations of the glutamate
dehydrogenase are generally regarded as prognostically unfavourable
signs. The behaviour of the enzyme in the various pathogenetic liver
syndromes is outlined in Table 107.

References

1. Olson, J. A. and Anfinsen, C. B., The crystallization and characterization of
   L-glutamic acid dehydrogenase, *J. biol. Chem.*, **197**, 67 (1952).
2. Empfehlungen der Deutschen Gesellschaft für klinische Chemie.
   Standardisierung von Enzymaktivitäten in biologischen Flüssigkeiten, *Z. klin.
   Chem. klin. Biochem.*, **10**, 182 (1972).
3. Schmidt, E., Glutamate dehydrogenase, in Bergmeyer, *Methods of enzymatic
   analysis*, Vol. 2, Academic Press, New York, 1974, p. 65.
4. Kurzanleitungen, *Roche Diagnostica*, 1976.
5. Filippa, G., Die diagnostische Bedeutung der Glutamat-Dehydrogenase-
   Aktivität im Serum, *Enzymol. biol. clin.*, **3**, 97 (1963).
6. Schmidt, E. and Schmidt, F. W., Enzym-Bestimmungen im Serum bei
   Leber-Erkrankungen, *Enzymol. biol. clin.*, **3**, 1 (1963).
7. Schlebusch, H., Rick, W., Lang, H., and Knedel, M., Normbereiche der
   Aktivitäten klinisch wichtiger Enzyme, *Dt. med. Wschr.*, **99**, 765 (1974).

4.7.5. γ-GLUTAMYL TRANSPEPTIDASE

4.7.5.1. Introduction

γ-Glutamyl transpeptidase [GGTP; γ-glutamyltransferase, (γ-glutamyl)-
peptide : amino acid γ-glutamyl transferase, E.C. 2.3.2.2] transfers the
glutamyl radical from peptides to L-amino acids or other peptides. The
best known peptide is glutathione, which is also probably the natural
substrate for this enzyme: reaction of γ-glutamyl transpeptidase:

Glutathione + Amino acid → γ-Glutamylamino acid + Cysteinylglycine

The enzyme contains SH groups and is a glycoprotein. It occurs in the
following human organs, in order of decreasing concentration: kidneys,
prostate, pancreas, liver, caecum, and brain. The enzyme has also been
detected in seminal fluid, prostate secretion, bile, cerebrospinal fluid, and
urine, as well as in the erythrocytes, leucocytes, and plasma. Isoenzymes
have been distinguished in plasma.[1,2]
The physiological importance of the enzyme is not yet completely clear.
Meister[3] has postulated a cyclic process in which γ-glutamyl transpeptidase
acts as a membrane-bound enzyme responsible for the admission of amino
acids to the cells. This transport mechanism is of greatest importance in the

renal tubulus, at the blood–brain boundary, and possibly also in other types of cells such as the hepatocytes.

4.7.5.2. Choice of method

Various methods of measurement are possible:

(a) release of naphthylamide from γ-glutamyl-α-naphthylamide and measurement of the product following diazotization;[4]

(b) release of aniline from γ-glutamylanilide and determination of the aniline following diazotization;[5]

(c) incubation with γ-glutamyl-p-nitroanilide. The enzyme is incubated with γ-glutamyl-p-nitroanilide, the colourless substrate yielding a γ-glutamyl residue and a p-nitroaniline.[6] The extinction of p-nitroaniline is measured in an alkaline medium (Fig. 139).

4.7.5.3. Principle

The last method has been investigated more closely by Szasz[7] using the kinetic test. Glycylglycine serves as γ-glutamyl acceptor. Transpeptidation occurs even in the absence of an acceptor but leads to the formation of polyglutamyl polymers. Apart from the transpeptidation there is hydrolysis to glutamate and p-nitroaniline.

The low solubility of the substrate γ-glutamyl-p-nitroanilide (glupa) is a drawback in this method. It is often necessary to use heat or to dissolve it in acid. The addition of detergent has even been proposed.[8] However, it has been found possible to increase the solubility by substituting the substrate with sulphone or carboxyl groups. One such substrate still in current use is L-γ-glutamyl-3-carboxy-4-nitroanilide, which exhibits an absorption maximum at 317 nm (in tris buffer, 200 mmol l^{-1}, pH 8.2). At 405 nm the absorption still amounts to only 74% of the maximum for the unsubstituted substrate, but this is without any practical effect on the enzyme determination.[9] The 5-amino-2-nitrobenzoate formed following the scission has an absorption maximum at 380 nm.[28] At a substrate concentration of 4 mmol l^{-1} the enzyme activity was identical with glupa and glupa carboxylate.[30] Both methods are in current use, that with the

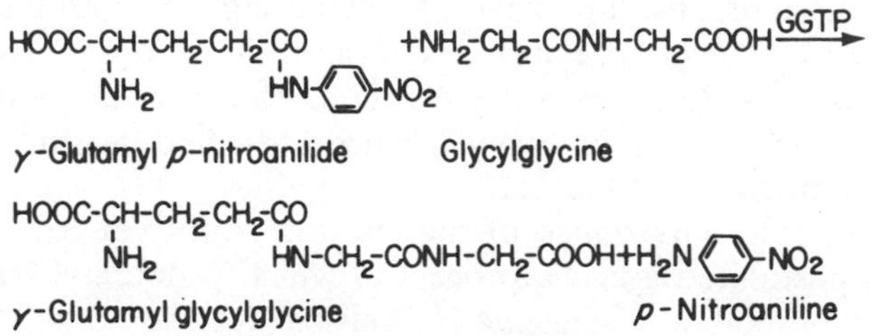

Fig. 139. Reaction of γ-glutamyl transpeptidase.

Table 116. Determination of γ-glutamyl transpeptidase; final concentrations in the reaction mixture.

2-Amino-2-methylpropane-1,3-diol	pH 7.4, 30 °C pH 8.0, 25 °C		
Tris–HCl buffer	pH 8.25 100 mmol l^{-1}	pH 7.6, 100 mmol l^{-1}	pH 8.2 100 mmol l^{-1}
Glycylglycine	100 mmol l^{-1}	75 mmol l^{-1}	150 mmol l^{-1}
L-γ-Glutamyl-3-carboxy-4-nitroaniline	4 mmol l^{-1}		6 mmol l^{-1}
L-γ-Glutamyl-p-nitroaniline		4 mmol l^{-1}	
MgCl$_2$		10 mmol l^{-1}	
Volume fraction of the serum (plasma)	1:11	1:11	1:16
Temperature of measurement	25 °C	37 °C	25 °C
Wavelength	400 nm (Hg 405 nm)	405–410 nm	405 nm
Path length	$d = 1$ cm	1 cm	1 cm
Duration of measurement: recording of ΔE during 1–5 min			Up to 10 min

carboxylated substrate being generally preferred.[28] However, certain authors give preference to the unsubstituted substrate, since at higher concentrations the carboxylated substrate has an inhibiting effect on the γ-glutamyl transpeptidase.[27]

4.7.5.4. Measurement

Conditions. These are given in Table 116.

4.7.5.5. Calculation

Calculation of the volume activity in μmol min^{-1} l^{-1} = U l^{-1} (p. 234).

Molar extinction coefficient for p-nitrobenzoate: $\varepsilon = 9.51 \times 10^3$ l mol^{-1} cm^{-1} (ref. 10).

Molar extinction coefficient for p-nitroaniline: $\varepsilon = 9.9 \times 10^3$ l mol^{-1} cm^{-1} (405 nm)[31] and 10.05×10^3 l mol^{-1} cm^{-1} (409 nm).[32]

4.7.5.6. Reference values

According to certain authors, the reference values obviously have a log-normal distribution.[11] The values for γ-glutamyl transpeptidase in plasma show significant sex differences[11] (Table 117).

As a rule, the values in men are higher than those in women. Higher activities are found in the plasma in early childhood as well as in new-born babies and sucklings.[14,15] The values appear to be especially high for premature births. One should therefore take care not to diagnose a cholestasis (bile-duct atresia). An excess of oestrogens lowers the enzyme activity, gestagens increase it. Accordingly, the effect of contraceptives on the enzyme activity depends on their composition.[16]

4.7.5.7. Comments on the method

1. Both γ-glutamyl-p-nitroaniline and γ-glutamyl-3-carboxy-4-nitro-anilide and glycylglycine can act as inhibitors unless optimal concentrations are used.

Table 117. Reference values for γ-glutamyl transpeptidase (U l^{-1}, 25 °C).

Age	Sex	Reference interval	References
New-born		33–100	15
Sucklings		6–35	
Infants		5–13	
Children		6–13	
Juveniles		6–19	
Adults	m	6–28	17
	f	4–18	

2. It has been maintained that glutamate activates the γ-glutamyl transpeptidase and must, consequently, be added to the incubation mixture.[19] However, it has been shown that at various temperatures and concentrations of the γ-glutamyl transpeptidase this is not the case.[20,28]

3. With the Persijn method[28] the coefficient of variation for the duplicate determination was 4.2% at 20–40 U l^{-1} and 1% at 100–140 U l^{-1}.

4. Haemolysis: concentrations of serum haemoglobin greater than 300 mg per 100 ml lead to a reduced enzyme activity.[28]

4.7.5.8. Diagnostic significance

Although the highest γ-glutamyltranspeptidase (GGTP) activity is found in the kidneys, the determination of the enzyme in serum has contributed nothing to the diagnosis of diseases of the kidneys. The chief role of such measurements is in the diagnosis of diseases of the liver since in such conditions an increase is almost always (i.e. in more than 90% of the cases) observed. However, a distinction must be made between diseases in which the parenchyma is predominantly involved and those in which discharge is obstructed in the region of the bile duct.

In acute viral hepatits (type A or B) the transaminases have still greater diagnostic significance. In our experience they are a more sensitive indication. GGTP increases to only a tenth or a fifth of the transaminase activity. The first indications that a hepatitis is receding are a fall in the transaminases and the bilirubin. GGTP falls more slowly and generally returns to normal values later than all other parameters. If a cholestatic phase occurs during the course of a hepatitis there is a noticeable increase in GGTP. A hepatitis may be considered cured only when both the transaminase and GGTP values have returned to normal.

Among the chronic inflammations of the liver the alcohol-conditioned fatty-liver hepatitis is above all distinguished by particularly high values.

If the chronic inflammation has resulted in a cirrhosis, then the more frequent alcoholic cirrhosis differs from the post-hepatitic form in showing substantially higher GGTP values. The transaminases are slightly elevated in both cases. Necrotic advances are recognized by a strong increase in GGTP. Alcohols with no pronounced liver damage show only sporadic increases in GGTP. On the other hand, if there is alcoholic liver damage, the activity of GGTP is frequently higher than that in patients with chronic hepatitis and cirrhoses from other causes. Certain authors are of the opinion that every chronic hepatitis or cirrhosis of the liver accompanied by highly elevated GGTP activities in the plasma should be considered as probably conditioned by alcohol, provided there is no cholestasis or carcinoma of the liver.[21–23] The enzyme is also valuable for assessing the success of an alcohol-deprivation cure. The fall of the values in the course of such a treatment and the prompt rise following a relapse are indications of the high sensitivity.[24]

GGTP values in the plasma are always elevated in both primary carcinoma of the liver and liver metastases. There is a good, positive correlation of GGTP with the findings of hepatic scintigraphy. Often, however, an elevated GGTP value on its own is sufficient to excite suspicion of a malignant tumour of the liver or in the region of the gastro-intestinal tract, above all the pancreas.

A high concentration of GGTP in the bile is indicative of an elimination via this path. It is therefore not surprising that initially, the principal use of GGTP in clinical diagnosis was as an enzyme indicating cholestasis. A pronounced increase in GGTP activity may be detected in all cases for patients with intra- and extrahepatic cholestasis. However, a differentiation between intra- and extrahepatic closure is not possible. In 1968 we compared the serum concentrations of three typical liver excretion enzymes (alkaline phosphatase, leucine aminopeptidase, and GGTP) in respect of their value in diagnosis of diseases of the bile duct. We investigated 25 patients with extrahepatic jaundice, 13 patients with liver metastases, and 12 patients with intrahepatic cholestasis. GGTP was, on average, 6 times more sensitive than alkaline phosphatase and 9 times more sensitive than leucine aminopeptidase. GGTP was the only enzyme to give pathological results with all patients, while with alkaline phosphatase 14% and with leucine aminopeptidase 18% were within the normal range.[25] This finding has since been confirmed many times.

The high sensitivity of GGTP in diseases of the liver has shown that the measurement of this enzyme in blood donors represents a more sensitive exploratory test for hepatopathias than the glutamate–pyruvate transaminase.

In both acute and chronic pancreatitis there is a short-term elevation of GGTP in the plasma. This is insufficient for making an organ diagnosis, however, and again it is necessary to rely on the α-amylase and lipase values. Nevertheless, carcinoma of the head of the pancreas, attended by no symptoms, can often only be discovered by an elevation of GGTP.

In contrast to alkaline phosphatase, GGTP may be used for the differential diagnosis of diseases of the liver and the bones. In the latter, in contrast to alkaline phosphatase, there is no change in the GGTP activity in the plasma. Pharmaceuticals such as barbiturates, anti-epileptic drugs, and aminopyrine have an activating influence on GGTP and are able to induce this enzyme.[26,33] The consumption of drugs and alcohol should be carefully controlled in patients who show elevated GGTP activity yet do not exhibit any conspicuous, clinical symptoms of a disease of the liver.

References

1. Jones, D. D., Williams, G., and Prochazka, B., Multiple molecular forms of γ-glutamyl transpeptidase during human pregnancy, *Enzyme,* **17**, 139 (1974).
2. Degenaar, C. P., Thijssen, C., Wal, G. van der, and Berends, G. T.,

Electrophoresis of γ-glutamyltranspeptidase on cellogel. The appearance of the a_2-β band in positive LP-X sera, *Clin. chim. Acta*, **67**, 79 (1976).

3. Meister, A., On the enzymology of amino acid transport, *Science*, **180**, 33 (1973).
4. Orlowski, M. and Szesczuk, A., Colorimetric determination of γ-glutamyltranspeptidase activity in human serum and tissues with synthetic substrates, *Acta biochem. polon.*, **8**, 189 (1961).
5. Goldbary, J. A., Friedman, O. R., Pineda, E. P., Smith, E. E., Chatterji, R., Stein, E. H., and Rutenberg, A. M., The colorimetric determination of γ-glutamyl transpeptidase with a synthetic substrate, *Archs Biochem. Biophys.*, **95**, 271 (1961).
6. Orlowksi, M. and Meister, A., γ-glutamyl-*p*-nitroanilide: a new convenient substrate for determination and study of L- and D-γ-glutamyltranspeptidase activity, *J. biol. Chem.*, **240**, 338 (1965).
7. Szasz, G., A kinetic photometric method for serum γ-glutamyltranspeptidase, *Clin. Chem.*, **15**, 124 (1969).
8. Tamaoki, H., Minato, S., Takei, S., and Fujisawa, K., A clinical method for the determination of serum gamma-glutamyl transpeptidase, *Clin. chim. Acta*, **65**, 21 (1975).
9. Szasz, G., Weimann, G., Stähler, F., Wahlefeld, A.-W., and Persijn, J.-P., New substrates for measuring γ-glutamyl transpeptidase activity, *Z. klin. Chem. klin, Biochem.*, **12.**, 228 (1974).
10. Adolph, L., Neuer Test zur Bestimmung der γ-Glutamyl-Transferase, *Arztl. Lab.*, **22**, 121 (1976).
11. Kley, S. and Kley, R., Serum-γ-Glutamyl-Transpeptidase, Serum-Glutamat-Pyruvat-Transaminase und Australia(SH)-Antigen-Antikörper-Bestimmung bei der Auswahl von Blutspendern, *Dt. med. Wschr.*, **98**, 1946 (1973).
12. Adjarow, D. and Iwanow, E. D., Neue Aspekte der klinischen Bedeutung der γ-Glutamyl-Transpeptidasebestimmung im Serum, *Acta hepato-gastroenterol.*, **20**, 315 (1973).
13. Mayr, K., Die Bedeutung der γ-Glutamy-Transpeptidase-Aktivität in der klinischen Diagnostik, *Med. Lab., Stuttg.*, **26**, 125 (1973).
14. Richterich, R. and Cantz, B., Normal values of plasma γ-glutamyl transpeptidase in children, *Enzyme*, **13**, 257 (1972).
15. Sitzmann, F. C. and Bierschenk, M., Der diagnostische Wert der Bestimmung der Gamma-Glutamyl-Transpeptidase im Serum von Kindern, *Pädiat. pädol.*, **7**, 145 (1972).
16. Feldmann, H. U., Pfeiffer, R., and Hirche, H., Zur Hormonabhängigkeit der γ-Glutamyl-Transpeptidase, *Dt. med. Wschr.*, **99**, 1171 (1974).
17. Szasz, G., γ-Glutamyltranspeptidase, in Bergmeyer, *Methods of enzymatic analysis,* Vol. 2, Academic Press, New York, 1974.
18. Recommended method for the determination of γ-glutamyltransferase in blood, *Scand. J. clin. Lab. Invest.*, **36**, 119 (1976).
19. Bondar, R. J. L. and Moss, G. A., Enhancing effect of glutamate on apparent serum γ-glutamyltranspeptidase activity, *Clin. Chem.*, **20**, 317 (1974).
20. Gerhardt, W., Persijn, P., and Rosalki, S. B., Lack of significant effect of L(+)-glutamate on serum γ-glutamyltransferase activity determined in the presence of glycylglycine, *Clin. Chem.*, **21**, 1515 (1975).
21. Colombo, J. P., Gamma-Glutamyltranspeptidase, ein altes Enzym neu in der Leberdiagnostik, *Praxis*, **63**, 3 (1975).
22. Colombo, J. P., γ-Glutamyltranspeptidase, Pathophysiologie und Diagnostik, *Chem. Rdsch.*, **27**, No. 26 (1974).
23. Rosalki, S. B., γ-Glutamyltranspeptidase, *Adv. clin. Chem.*, **17**, 53 (1975).

620

24. Wietholtz, H. and Colombo, J. P., Das Verhalten der Gamma-Glutamyltranspeptidase und anderer Leberenzyme im Plasma während der Alkohol-Entziehungskur, *Schweiz. med. Wschr.*, **106**, 981 (1976).
25. Lukasik, S., Richterich, R., and Colombo, J. P., Der diagnostiche Wert der alkalischen Phosphatase, der Leucinaminopeptidase und der γ-Glutamyl-Transpeptidase bei Erkrankungen der Gallenwege, *Schweiz. med. Wschr.*, **98** 81 (1968).
26. Bartels, H., Evert, W., Hauck, W., Petersen, C., Putzki, H., and Schulze, W., Significance of increased serum gamma-glutamyltransferase activity during long-term anticonvulsive treatment. Clinical and experimental studies, *Neuropädiatrie*, **6**, 77 (1975).
27. Theodorsen, L. and Strömme, J. H., γ-Glutamyl-3-carboxy-4-nitroanilide: the substrate of choice for routine determinations of γ-glutamyltransferase activity in serum, *Clin. chim. Acta*, **72**, 205 (1976).
28. Persijn, J. P. and Van der Slik, W., A new method for the determination of γ-glutamyltransferase in serum, *Z. klin. Chem. klin. Biochem.*, **14**, 421 (1976).
29. Zinser, W., Gamma-Glutamyltransferase (γ-GT), *Arztl. Lab.*, **23**, 31 (1977).
30. Shaw, L. M., London, J. W., Fetterolf, D., and Garfinkel, D., γ-Glutamyltransferase: kinetic properties and assay conditions when γ-glutamyl-4-nitroanilide and its 3-carboxy derivative are used as donor substrates, *Clin. Chem.*, **23**, 79 (1977).
31. Rosalki, S. B. and Tarlow, D., Optimized determination of γ-glutamyltransferase by reaction-rate analysis, *Clin. Chem.*, **20**, 1121 (1974).
32. Huseby, N. E. and Strömme, J. H., Practical points regarding routine determination of γ-glutamyl transferase (γ-GT) in serum with a kinetic method at 37°C, *Scand. J. clin. Lab. Invest.*, **34**, 357 (1974).
33. Ohnhaus, E. E., Martin, J., Kinser, J., and Colombo, J. P., Enzyme induction of renal function in man, *Br. J. clin. Pharmacol.*, **4**, 33 (1977).

4.7.6. CHOLINESTERASE

4.7.6.1. Introduction

The term cholinesterase embraces a poorly defined group of enzymes which are present in all organs and tissues.[16] According to Pilz,[1] at least eleven different cholinesterases and two acetylcholinesterases occur in the plasma. They differ in respect of substrates, pH optima, electrophoretic mobilities, and kinetics.[2] Cholinesterases (acylcholine acylhydrolase, E.C. 3.1.1.8) principally cleave organic esters of short-chain fatty acids, such as

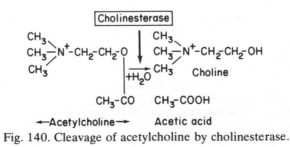

Fig. 140. Cleavage of acetylcholine by cholinesterase.

acetylcholine, benzoylcholine, tributyrin, acetylsalicylate, procaine and succinylcholine (Fig. 140).

Because of the lack of a specific substrate, this group of enzymes have been called by various names, such as pseudo-cholinesterase, s-type cholinesterase, non-specific esterase, tributyrinase, procainesterase, and aspirinesterase. However, in each case the enzyme is the same but its activity has been measured using different substrates.

The enzymes described as cholinesterases should not be confused with acetylcholinesterase, which cleaves acetylcholine exclusively and is found in the nervous system at the pre- and post-ganglionic synapses, at the motoric end-plates in the muscles, and at the parasympathetic, post-ganglionic synapses. It cleaves the acetylcholine necessary for depolarization of the synapse.

The functional task of cholinesterase in the metabolism is not known. It certainly plays a part in the detoxication of certain drugs such as procaine, succinylcholine, and acetylsalicylic acid. One way, among others, in which this detoxication takes place is by hydrolytic cleavage of the esters. It is also possible that the enzyme has a role in the metabolism of lipids, but this still requires precise confirmation. The principal sites of synthesis are the hepatocytes and Kupffer's cells of the liver.

4.7.6.2. Choice of method

The commonest methods are measurement in ultraviolet light using benzoylcholine as substrate and measurement of the cleavage of sulphur-containing choline ester in the visible range using the Ellman reaction as indicator reaction. The choline released by cleavage of benzoylcholine can also be oxidized to betaine by choline oxidase. The H_2O_2 formed at the same time can be measured with peroxidase by the Trinder reaction (p. 367).[17]

4.7.6.3. Cholinesterase and dibucaine number by ultraviolet spectrophotometry

4.7.6.3.1. Principle

An elegant method for the determination of cholinesterase and of the fraction of atypical enzyme has been developed by Kalow and Genest.[3] The selected substrate is benzoylcholine, which exhibits a much higher extinction in the ultraviolet region than the cleavage product benzoic acid. The fall in the extinction at 240 nm following incubation of plasma with benzoylcholine can therefore be taken as a direct measure of the enzyme activity. The typical and atypical cholinesterases differ, above all, in the extent to which they are inhibited by dibucaine. While normal cholinesterase at a final concentration of 10^{-5} mol l^{-1} is very strongly

Table 118. Determination of cholinesterase; final concentrations in the reaction mixture.

Benzoylcholine chloride	$5 \cdot 10^{-5}$ mol l^{-1}
Phosphate buffer, pH 7.4	66.7 mmol l^{-1}
±-Dibucaine	$1 \cdot 10^{-5}$ mol l^{-1}

Volume fraction of the serum (plasma) 1:151
Temperature of measurement 25 °C
Wavelength 240 nm, $d = 1$ cm
Duration of measurement 30 s to 2 min (depending on the activity)

Procedure

	Sample, ml	Inhibitor solution, ml	Blank, ml
Substrate solution	3.0	—	—
Inhibitor solution	—	3.0	3.0
Serum, plasma	0.02	0.02	—

Set the photometer to zero using the blank, then record (manually or automatically) the extinction of the sample and the inhibitor solution at 240 nm at 1-min intervals

inhibited by dibucaine, the atypical enzyme is scarcely affected. The degree of inactivation is expressed by the so-called dibucaine number: the value is around 20 for atypical cholinesterase and around 80 for the normal enzyme.

4.7.6.3.2. Reagents

Benzoylcholine chloride solution, 1×10^{-3} mol l^{-1}. Dissolve 24.4 mg of benzoylcholine chloride in Sørensen phosphate buffer (66.7 mmol l^{-1}) and make up to 100 ml with further buffer. Store frozen.

Dibucaine solution, 4×10^{-4} mol l^{-1}. Dissolve 15.2 mg of Nupercaine (Ciba) in Sørensen phosphate buffer (66.7 mmol l^{-1}) and make up to 100 ml with further buffer. Store frozen.

Substrate solution. Mix 2 parts of benzoylcholine chloride with 38 parts of buffer. Prepare freshly each day.

Inhibitor solution. Mix 2 parts of benzoylcholine chloride solution, 1 part of dibucaine solution and 37 parts of buffer. Prepare freshly each day.

4.7.6.3.3. Measurement

Conditions. These are given in Table 118.

4.7.6.3.4. Calculation of the volume activity

Using the extinction coefficient ($\varepsilon = 6600 \ l \ mol^{-1} \ cm^{-1}$).

$$\text{Concentration} = \frac{\Delta E/\text{min}}{6600 \cdot 1} \cdot 10^6 \cdot \frac{3.02}{0.02} \, \mu\text{mol min}^{-1} \, l^{-1}$$

$$\text{Concentration} = \Delta E/\text{min} \cdot 22\,880 \text{ U } l^{-1}$$

By the graphical method (p. 237, $c = 1144$).

$$\text{Concentration} = \tan \alpha \cdot 1144 \, \mu\text{mol min}^{-1} \, l^{-1}$$

Calculation of the dibucaine number (DN):

$$\text{DN} = \left(1 - \frac{\text{inhibited cholinesterase}}{\text{uninhibited cholinesterase}}\right) \cdot 100\%$$

4.7.6.3.5. Specificity

The specificity of the serum cholinesterase and its substrate preference have not yet been sufficiently elucidated.

4.7.6.3.6. Reference values

These are given in Table 119.
The reference values in children using this technique are not known. However, at ages above 1 year the values do not deviate appreciably from the adult values.

4.7.6.3.4. Comments on the method

1. Our reference values are markedly lower than those of Kalow and Genest.[3] This is partly explicable in terms of the different temperature used (25 instead of 37 °C), but there must be other reasons for the difference. $Q_{10} = 1.92$.[2]

2. The value $\varepsilon = 6600 \text{ l mol}^{-1} \text{ cm}^{-1}$, here referred to as the molar extinction coefficient, is not the coefficient for the substrate itself but corresponds to the difference between the extinctions of benzoylcholine and benzoic acid at 240 nm.

Table 119. Reference values for cholinesterase.

	Collective	Number	Reference interval	
			$\bar{x} \pm s$	$\bar{x} \pm 2s$
Cholinesterase	Normal	60	1026 ± 233	559–1493
	Heterozygotic			140–386
	Homozygotic			<100
Dibucaine number	Normal		76 ± 7	62–90
	Heterozygotic			30–65
	Homozygotic			<30

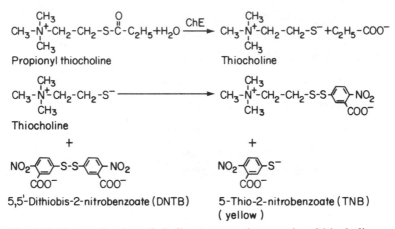

Fig. 141. Determination of cholinesterase using propionylthiocholine.

4.7.6.4. Cholinesterase and dibucaine number, determination with propionylthiocholine in the visible range

4.7.6.4.1. Principle

Propionylthiocholine is hydrolysed by cholinesterases. During the hydrolysis of the substrate, the thiocholine produced reacts with 5,5′-dithiobis-2-nitrobenzoate (DTNB) to form a yellow anion, 5-thio-2-nitrobenzoate (TNB). The extinction of this yellow anion may be read at 405 nm and is proportional to the amount of substrate hydrolysed (Ellman reaction)[4] (Fig. 141).

4.7.6.4.2. Reagents

Tris buffer, 50 mmol l^{-1}, *pH* 7.4. Dissolve 6.06 g of tris in about 900 ml of DM-water. Adjust the pH to 7.4 with 1 N HCl and make up to 1000 ml with DM-water. The solution keeps indefinitely at 4 °C.

Tris–DNTB buffer, 0.25 *mmol l^{-1} pH* 7.4. Dissolve 100 mg of DTNB in 1000 ml of tris buffer, stirring continuously. The solution keeps for several weeks at 4 °C.

Dibucaine–tris–DTNB buffer, 0.02 *mmol l^{-1}, pH* 7.4. Dissolve 38 mg of dibucaine in 100 ml of water. Dilute 10 ml of this solution to 500 ml with tris–DNTB buffer.

Substrate solution: propionylthiocholine iodide, 2.25 *mmol l^{-1}*. Dissolve 682.2 mg of propionylthiocholine iodide in 100 ml of DM-water. Transfer to tubes (10-ml portions). The solution keeps for several weeks at 4 °C.

4.7.6.4.3. Measurement

Conditions. These are given in Table 120.

Table 120. Determination of cholinesterase; final concentrations in the reaction mixture.

	mmol l^{-1}
Tris buffer, pH 7.4	45
DTNB	0.227
Dibucaine	0.018
Propionylthiocholine iodide	2.03

Procedure

	Sample, ml	Inhibitor solution, ml
Tris–DTNB buffer	2.0	—
Dibucaine–tris–DTNB buffer	—	2.0
Serum	0.02	0.02
Incubate for at least 2 min		
Substrate solution	0.2	0.2

Set the photometer at zero using water and record, either manually or with an automatic recorder, the extinction of the sample and inhibitor solution at 405 nm at 1–2–min intervals

4.7.6.4.4. Calculation

Volume activity, general equation. If v = paper speed and b = paper width (p. 234) then

$$\mu\text{mol min}^{-1}\,l^{-1} = \tan\alpha \cdot \frac{10^6}{13.6 \cdot 10^3} \cdot \frac{FV}{SV} \cdot \frac{v}{b}$$

$$\mu\text{mol min}^{-1}\,l^{-1} = \tan\alpha \cdot \frac{10^6}{13.6 \cdot 10^3} \cdot \frac{2.22}{0.02} \cdot \frac{5}{20}$$

$$\mu\text{mol min}^{-1}\,l^{-1} = \tan\alpha \cdot 2040$$

Dibucaine number.

$$\left(1 - \frac{\text{inhibited cholinesterase}}{\text{uninhibited cholinesterase}}\right) \cdot 100\%$$

The molar extinction coefficient of 5-thio-2-nitrobenzoate is $\varepsilon = 13.6 \times 10^3\,l\,mol^{-1}\,cm^{-1}$.[5]

4.7.6.4.5. Reference values

These are given in Table 121.

Table 121. Reference values for cholinesterase.

	Collective	Number	Reference interval $\bar{x} \pm s$	Reference interval $\bar{x} \pm 2s$
Cholinesterase	Normal	60	4100 ± 932	2200–6000
	Heterozygotic			560–1500
	Homozygotic			<400
Dibucaine number	Normal		76 ± 7	62–90
	Heterozygotic			30–65
	Homozygotic			<30

4.7.6.4.6. Comments on the method

1. Sodium fluoride (glycolysis inhibitor) inhibits the enzyme. The molar exinction coefficient of reduced DTNB as calculated from synthetic TNB has recently been checked and has been reported to be 13.6×10^3 l mol^{-1} cm^{-1}.[5]

2. Although the Ellmann reaction should theoretically respond to all SH-containing compounds in the plasma, the latter are present in such small concentrations that in practice this is irrelevant.

4.7.6.4.7. Diagnostic significance

Hypocholinesterasaemia. Hypocholinesterasaemia is diagnostically more important than an increase in the enzyme.

1. Physiologically during *pregnancy:* a lowering of cholinesterase is observed from the 2nd trimester until about 6 weeks postpartum.[6]

2. *Hepatopathia.* Acute hepatitis: cholinesterase may be lowered during the early stages of the disease, but in general it does not change very much. A fall in the enzyme is a bad sign indicating a necrotic advance.

Chronic hepatitis and cirrhosis of the liver: here progressive observations are important. The cholinesterase can be lowered and an early fall can indicate a dystrophic advance or a liver coma. In chronic hepatitis the values are in the lower reference interval, and in cirrhosis of the liver frequently below this. The forms of cirrhosis cannot be differentiated on the basis of the cholinesterase. In alcoholic fatty liver the cholinesterase is at the upper limit of the reference interval or even slightly elevated. In contrast, in alcoholic hepatitis and cirrhosis lowered values are the rule. Low values are often observable with malignant tumours of the liver.[7-10]

3. *Negative nitrogen balance.* Undernourishment and/or incorrect diet (protein deficiency) leads to a decline in the cholinesterase. This has regularly been observed in patients with kwashiorkor and marasmus.[11] The measurement of the cholinesterase is useful in the diagnosis and prognosis of these diseases. A steady increase in the enzyme activity is to be observed on

healing. The prognosis is unfavourable for patients with very low initial values.

4. *Atypical cholinesterases*. Cholinesterase cleaves muscle relaxants of the succinylcholine type (e.g. succamethonium) to succinic acid and choline, neither of which acts as a relaxant. However, there are individuals who show an atypical cholinestesa in the plasma and are unable to degrade these drugs. This leads to periods of narcosis, frequently with prolonged apnoea. Normal individuals take 8 min to degrade a dose of succinylcholine, for which individuals with an atypical cholinesterase would require 2 h. There are three genetically determined mutants (three mutant alleles) of normal cholinesterase known which are incapable of converting succinylcholine when present in pharmacological doses. The combination of the four alleles which are produced at the same gene location as the normal enzyme, gives ten phenotypes as the resultant of the corresponding cholinesterase enzyme proteins.[12,13] This peculiarity is inherited recessively so that three groups of individuals can be distinguished: healthy with normal cholinesterase, heterozygotic with one half normal and one half atypical cholinesterase, and homozygotic with exclusively atypical cholinesterase. In heterozygotic individuals the total cholinesterase activity is at the lower limit of the reference interval but is markedly lower in homozygotic individuals. If the cholinesterase values are low, it is necessary to check whether this is actually a consequence of an affection of the liver or is due to the presence of an atypical cholinesterase and the liver is sound.

Carriers of an atypical cholinesterase may be recognized from the fact that, in contrast to normal individuals, their cholinesterase is resistant to the inhibitor dibucaine. The normal enzyme is up to about 80% inhibited by dibucaine, as found in about 96% of all individuals. The dibucaine number shows how much of the initial cholinesterase activity can be inhibited by dibucaine. If a homozygotic disease is discovered the family should be investigated and the patients should be made aware of their defect. As a rule, heterozygotic individuals are capable of degrading succinylcholine.

5. *Poisoning by cholinesterase inhibitors*. Drugs: parasympathicomimetics such as physostigmine, prostigmin, eserine, myatin, and mintacol can reversibly inhibit the cholinesterase.

Organic phosphoric acid esters: these are used as insecticides (e.g. parathion, phosphamidon) and can lead to cases of poisoning in individuals (factory and farm workers) receiving inordinate exposures and insufficient protection.

Nerve poisons: these are of the trilon type (e.g. soman, sarin, tabun) and are designed for use in chemical warfare. They are extremely toxic, their action depending on the inhibition of the intrinsic acetylcholinesterase at the synapses and end-plates for which the cholinesterase in the plasma represents an 'indicator'. The inhibition of the acetylcholinesterase leads to a rapid increase in the concentration of acetylcholine and, consequently, to a high stimulation of the parasympathetic system and a disturbance of the

628

neuromuscular transference of stimulation. Death can occur in a very short time—within minutes—as a result of respiratory paralysis. Stimulation of the parasympathetic system leads to contracted pupils (diagnosis!).

6. *Hypocholinesterasaemia due to other causes.* Lowered values are observed in cardiac infarction, dermatomyositis, hypothyroidisim, and severe anaemias. Elevated, normal, and low activities have been observed in myasthenia gravis.[13] Contraceptives lead to a slight decrease in the activity.[14]

7. *Hypercholinesterasaemia. Nephrotic syndrome:* in nephrotic syndrome of most varied aetiology, the proteinuria leads to a compensatory increase in the synthesis of albumin in the liver, accompanied by a parallel increase in the production of cholinesterase. Since, in contrast to albumin, the cholinesterase is not excreted in the urine, a hyperenzymia occurs. It has also been maintained that the elevated cholinesterase serves to stabilize the lipoproteins.[15]

Exudative enteropathia. The elevations of serum cholinesterase occasionally observed with this disease are probably attributable to a pathomechanism similar to that in nephrosis (loss of protein in the gut). The linking of albumin synthesis to the synthesis of cholinesterase is purely phenomenological and experimental proof is still wanting.

In interpreting the findings it should be remembered that the values in plasma are found to be be lower in women than in men.[9]

References

1. Pilz, W., Specific acetylcholinesterasis of human serum, *6th Int. Congr. clin. Chem., Munich, 1966*; Vol. 2, *Clinical enzymology*, Karger, Basle, 1968, p. 121
2. La Motta, U. R. V. and Woronick, C. L., Molecular heterogeneity of human serum cholinesterase, *Clin. Chem.,* **17**, 135 (1971).
3. Kalow, W. and Genest, K., A method for the detection of atypial forms of human serum cholinesterase. Determination of dibucaine numbers, *Can. J. Biochem. Physiol.,* **35**, 339 (1957).
4. Ellman, G. L., Courtney, K. D., Andres, V., Jr., and Featherstone, R. M., A new and rapid colorimetric determination of acetylcholinesterase activity, *Biochem. Pharmac.,* **7**, 88 (1961).
5. Silverstein, R. M., The determination of the molar extinction coefficient of reduced DTNB, *Analyt. Biochem.,* **63**, 281 (1975).
6. Richterich, R., Enzym-Diagnostik für den praktischen Arzt. I. Serum-Cholinesterase, *Praxis, 50,* 624 (1961).
7. Siders, D. B., Batsakis, J. G., and Stiles, D. E., Serum cholinesterase activity. A colorimetric microassay and some clinical correlations, *Am. J. clin. Path.,* **50**, 344 (1968).
8. Finterlmann, V. and Lindner, H., Diagnostische Bedeutung der Serum-Cholinesterase bei Lebererkrankugen, *Dt. med. Wschr.,* **95**, 469 (1970).
9. Prellwitz, W., Kapp, S., and Muller, D., Vergleich von Methoden zur Aktivitätsbestimmung der Serumcholinesterasen (Acylcholin-acylhydrolase E.C. 3.1.1.8) und deren diagnostische Wertigkeit, *Z. klin. Chem. klin. Biochem.,* **14**, 93 (1976).

10. Tamarelle, C., Quinton, A., Bancons, J., and Dubarry, J.-J., La cholinesterérase sérique, test d'insuffisance cellulaire hépatique, *Sem. Hôp. Paris,* **49**, 859 (1973).
11. Barclay, G. P. T. and Path, M. R. C., Pseudocholinesterase activity as a guide to prognosis in malnutrition, *Am. J. clin. Path.,* **59**, 712 (1973).
12. Goedde, H. W., Therapie und Genetik, *Materia med. Nordmark,* **27**, 237 (1975).
13. Goedde, H. W., Doenicke, A., and Altland, K., *Pseudocholinesterasen,* Springer, Berlin, 1967.
14. Sidell, F. R. and Kaminskis, A., Influence of age, sex, and oral contraceptives on human blood cholinesterase activity, *Clin. Chem.,* **21**, 1393 (1975).
15. Way, R. C., Hutton, C. J., and Kutty, K. M., Relationship between serum cholinesterase and low density lipoproteins in children with nephrotic syndrome, *Clin. Biochem.,* **8**, 103 (1975).
16. Schuh, F. T., Über Cholinesterasen—eine anästhesiologisch wichtige Enzymgruppe, *Anästhesist,* **25**, 501 (1976).
17. Okabe, H., Sagessaka,, K., Nakajima, N., and Noma, A., New enzymatic assay of cholinesterase activity, *Clin. chim. Acta,* **80**, 87 (1977).

4.7.7. COMPLEX TESTS OF FUNCTION

4.7.7.1 General

The defining criterion of complex function tests is that instead of measuring a partial function of the liver they are dependent on a whole series of physiological, morphological, and biochemical factors. Consequently, these tests are generally very sensitive but less specific and less selective. The most commonly used complex tests are the galactose tolerance test, the *p*-aminohippuric acid test and the bromsulphlein elimination test but we now restrict our own investigation to the latter. The complexity of the processes participating in the excretion of bromsulphalein from the plasma into the bile may be seen from Fig. 142. The result is dependent upon the following:

1, the flow of blood through the liver;
2, the mass of the hepatic parenchymal cells;
3, a mechanism for the uptake of the dye by the parenchymal cells;
4, a storage organ for the dye in the parenchymal cells;
5, the conjugation of the dye to amino acids by specific enzymes;
6, the ejection of the dye into the bile; and
7, an unconstricted bile duct.

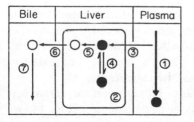

Fig. 142. Factors affecting the elimination of bromsulphalein. 1 = Flow of blood through the liver; 2 = liver mass; 3 = active uptake into the liver parenchymal cells; 4 = storage in the liver; 5 = conjugation of the bromsulphalein; 6 = ejection of the conjugate into the bile duct; 7 = elimination of the conjugate by the bile.

Similar processes must be involved in the other dye tests (and for bilirubin also).

4.7.7.2. Bromsulphalein elimination

4.7.7.2.1. Introduction

The bromsulphalein (BSP) test was introduced to clinical diagnostics in 1925 by Rosenthal and White[1] as a method of testing partial functions of the liver. Following intravenous injection, the dye is initially bound to the plasma albumin but there are no structural changes in the latter;[2] it is then actively taken up into the liver cells through the liver-cell membranes and, after storage and conjugation in the liver cells, it is ultimately excreted into the bile.

Clinical experience has shown that the determination of the elimination of this dye from the plasma is of proven value and it may be considered as a sensitive, organ-specific, and simple test for the identification of certain hepatopathia.

4.7.7.2.2. Choice of method

Numerous variant procedures have been described for performing the BSP test and almost all of them use a dose of 5 mg of bromsulphalein per kilogram body-weight. The most commonly used variants are:

1. Percentual retention of the BSP dye after 45 min: results as percentages of the intitial concentration, often referred to a value determined 3 min after injection (>5% pathological). But there are also methods which dispense with a first blood sampling for calculation of the initial concentration.

2. Measurement of the concentration of BSP in plasma (mg per 100 ml) 45 and 60 min after injection: in the 45-min test[3,4] BSP values above 0.5 mg per 100 ml and in the 60-min test[5,6] values above 0.2 mg per 100 ml are pathological.

3. Zimmer's two-dye test:[7] 10 min after the injection of two dyes (bromsulphalein and trypan red) the bromsulphalein concentration is measured in relation to the trypan red concentration and the ratio is evaluated by means of a nomograph to give the elimination capacity as a percentage (<80% pathological).

The 45-min and 60-min tests have generally been favoured in practice. A particular merit of these tests is their simplicity, since in healthy persons the elimination of BSP is generally complete after 60 min. Methods of determination utilizing plasma elimination curves are of interest for scientific problems but are less suited to routine investigations.

The diagram constructed by Hennecke and König[8] can be used for comparing the results of the different methods of determination (Fig. 143).

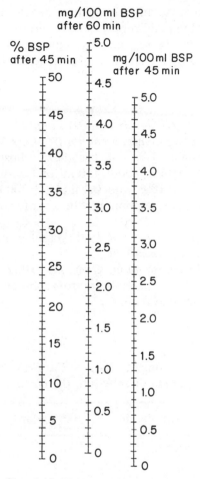

mg/100 ml BSP
after 60 min

% BSP
after 45 min

mg/100 ml BSP
after 45 min

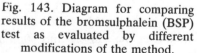

Fig. 143. Diagram for comparing
results of the bromsulphalein (BSP)
test as evaluated by different
modifications of the method.

For the determination of the BSP elimination the method of Seligson *et al.*[3,4] is greatly preferred. In this modification there is no need to collect blood at time zero. It takes into consideration the change in the absorption spectrum of BSP produced by bonding to the plasma proteins, predominantly to α_1-lipoprotein and albumin.

4.7.7.2.3. Principle

The residual dye concentration in the plasma is determined exactly 45 min after the intravenous injection of 5 mg of bromsulphalein per kilogram body-weight. The bromsulphalein turns an intense red (indicator) when the plasma is made alkaline. Acidification causes decoloration and permits the plasma blank (haemoglobin, bilirubin) to be determined. The formation of a bromsulphalein–albumin bond, and a consequent reduction in the extinction, is prevented by the addition of p-toluenesulphonate.

4.7.7.2.4. Reagents

Alkaline buffer, $Na_2HPO_4.2H_2O$, 91 $mmol\ l^{-1}$; $Na_3PO_4.12\ H_2O$, 9.3 $mmol\ l^{-1}$; p-*toluenesulphonate*, 33 $mmol\ l^{-1}$. Dissolve 8.1 g of disodium monohydrogen phosphate, 1.7 g of trisodium phosphate, and 3.2 g of sodium p-toluenesulphonate in about 450 ml of DM-water. Adjust the pH of the solution to exactly 10.6 with 0.1 mol l^{-1} NaOH amd make up to 500 ml with DM-water. The solution may be kept for more than a month at room temperature.

Acidic buffer, $NaH_2PO_4.H_2O$, 2 $mol\ l^{-1}$. Dissolve 69 g of $NaH_2PO_4.H_2O$ in DM-water and make up to 250 ml.

Standard bromsulphalein solution, 5 mg per 100 ml. Dissolve 5 mg of bromsulphalein (sodium phenoltetrabromophthaleindisulphonate) in DM-water and make up to 100 ml.

4.7.7.2.5. Measurement

Collect blood from the finger, using a Caraway capillary (p.101), and centrifuge. Pipette the plasma directly into the cuvette.

	Sample, ml	Standard, ml
Alkaline buffer	0.7	0.7
Plasma	0.1	—
Standard	—	0.1

Mix and read the extinction of the sample and standard at 578 nm against DM-water as reference

Acidic buffer	0.02	—

Mix and read the extinction blank against DM-water

4.7.7.2.6. Calculation

Using the standard solution.

$$\text{Bromsulphalein concentration} = \frac{E(S) - E(B)}{E(ST)} \cdot 5 \text{ mg per 100 ml}$$

Using the molar extinction coefficient, $\varepsilon = 65\,080\;l\,mol^{-1}\,cm^{-1}$.

$$\text{Bromsulphalein concentration} = \frac{[E(S) - E(B)] \cdot 838.1}{65\,080 \cdot 10} \cdot 10^3 \cdot \frac{0.8}{0.1}$$

mg per 100 ml

$$= E(S) - E(B) \cdot 10.3 \text{ mg per 100 ml}$$

Calculation of the so-called 'percentage retention'. The result in mg per 100 ml is multiplied by 10.

4.7.7.2.7. Reference values

The residual bromsulphalein after 45 min should be at most 0.2–0.5 mg per 100 ml of plasma, corresponding to a retention of <5%.

4.7.7.2.8. Comments on the method

1. The concept of 'percentage retention' is based on the following considerations: it is assumed that the plasma volume in which the bromsulphalein is distributed makes up 5% of the body-weight, i.e. 50 ml per kilogram body-weight. After an injection of 5 mg kg^{-1} body-weight, 50 ml ought to contain 5 mg of bromsulphalein (assuming that the dye is distributed only in the plasma volume), i.e. 10 mg per 100 ml plasma. Thus, 1 mg per 100 ml of plasma corresponds to a retention of 10% of the dose administered, and 0.1 mg per 100 ml to a retention of 1%. However, the concept 'percentage retention' is not recommended since it depends on several assumptions not proved in individual cases.

2. In new-born babies the method can lead to false negative results (higher plasma volume). It is advisable to determine the BSP concentration several times during 60 min and to express the 45-min value as a percentage of the value extrapolated for zero time.[9]

3. The BSP elimination is known to be influenced by various substances and drugs.[10,11] (Table 122).

4. Bromsulphalein tests and oral cholecystography with triiodated contrast substances should not be performed on the same day since the contrast agents obviously interfere with the elimination of the bromsulphalein.[11]

4.7.7.2.9. Diagnostic significance

1. Symptoms of irritation have been described as side-effects of the administration of bromsulphalein by paravenous injection (thrombophlebitis). There have also been cases of anaphylactic reactions, partly recurrent, with fatal outcome.[12] Patients with asthma or other allergies are particularly prone to such reactions.

2. The test should be performed with the patient lying down.

634

Table 122. Interfering factors to which the BSP test
is sensitive and which limit its applicability.

Haemolysis	+	
Lipaemia	+	
Dysproteinaemia	+	−
Incorrect blood-collection times	+	−
Contamination by dye residues	+	
Fever	+	
X-ray contrast agents	+	
Sulphonylurea		−
Anabolics	+	
Oral contraceptives	+	
Steroid hormones	+	
Antibiotics	+	
Morphine	+	
Chlorpromazine		−
Phenobarbital	+	
Prednisolone		−

+ = False positive result, delayed elimination.
− = False negative result, accelerated elimination.

3. The test should only be performed by a doctor, and then only if
preparations have been made to take quick remedial action in the event of
anaphylactic shock.

4. The bromsulphalein retention may be regarded as the 'rough estimate'
of the excretory function of the liver. It correlates well with galactose
elimination capacity.[13] Unequivocal positive results have been observed in
patients with *cirrhosis of the liver* (alcoholic, cryptogenic, primary, biliary),
hepatitis (chronic, aggressive), *Gilbert's syndrome, Crigler-Najjar syndrome*,
and *cholestasis*.[13]

References

1. Rosenthal, S. M. and White, E. C., Clinical application of the bromsulphalein test for hepatic function, *J. Am. med. Ass.*, **84**, 1112, (1925).
2. Deutschmann, G., Gratzl, M., and Ullrich, V., Binding of bromsulphthalein to serum albumin, *Biochem. biophys. Acta*, **371**, 470 (1974).
3. Seligson, D., Marino, J., and Dodson, E., The determination of sulfo-bromophthalein in serum, *Clin. Chem.*, **3**, 638 (1957).
4. Seligson, D. and Marino, J., Sulfobromophthalein (BSP) in serum, *Stand. Meth. clin. Chem.*, **2**, 186 (1958).
5. Pezold, F. A. and Kessel, M., Der Bromsulphaleintest in der Leberdiagnostik. Untersuchungen zur Methodik, Bewertung und klinischen Brauchbarkeit, *Dt. Arch. klin. Med.*, **204**, 518 (1957).
6. Südhof, H., Kuhlmann, H., and Kapinski, B., Zur praktischen Durchführung des Bromthaleintestes, *Medsche Welt, Stuttg.*, **10**, 533 (1963).
7. Zimmer, V., Der Zwei-Farbstoff-Test, *Ärztl. Lab.*, **2**, 285 (1956).
8. Hennecke, A. and König, K. J., Zur Vergleichbarkeit verschiedener Modifikationen des Bromsulfalein-Tests, *Dt. med. Wschr.*, **98**, 934 (1973).

9. Donath, A., Eduah, S., and Colombo, J. P., Der Bromsulphaleintest im Säuglingsalter, *Helv. paediat. Acta,* **25**, 634 (1970).
10. Wernze, H. and Speck, H. J., Veränderungen der Brom-sulphthalein(BSP)-Elimination durch Pharmaka, *Materia med. Nordmark*, **26**, 306 (1974).
11. Dubach, U. C., Increased bromsulphalein retention after cholecystography, *Schweiz. med. Wschr.,* **92** 393 (1962).
12. Wernze, H. and Speck, H. J., Funktionsdiagnostik der Leber mit Farbstoffen, *Dt. med. Wschr.,* **101**, 620 (1976).
13. Buesch, B., Häcki, W., and Bircher, J,, Was misst die BSP-Retention?, *Schweiz. med. Wschr.,* **103**, 397 (1973).

4.8. ENDOCRINE GLANDS

4.8.1. ENDOCRINE FUNCTION OF THE PANCREAS

The following particulars have been taken from a brochure prepared by the Medical Commission of the Swiss Diabetes Association and the Commission for Standardization of the Swiss Association for Clinical Chemistry.[1]

4.8.1.1. Introduction

Considering the frequency of diabetes, the procedure chosen for early diagnosis must be as economic as possible. In addition to the detection of glucose in the urine this includes the determination of the blood-glucose concentration as well as a simple oral glucose tolerance test capable of being performed in every practice and every hospital. Of the large number of glucose tolerance tests—according to a statement issued by the World Health Organization (WHO) 115 different exploratory methods have already been published in 1968—these requirements appear to have been fulfilled most adequately by the proposal of the European Diabetes Epidemiology Study Group. This tolerance test allows a rough division of individuals who, at the time of the investigations, are classifiable within normal, suspect, or pathological ranges of the carbohydrate metabolism. It also permits a selection of those individuals for whom further investigations are advisable. As with all tests of this kind, in addition to the laboratory results the clinical aspects are obviously crucial in determining the further procedure. This may involve a periodic control of the patients, a repetition of the tolerance test, or carrying out an investigation of the carbohydrate metabolism under standard conditions.

4.8.1.2. Definitions

The following is a survey of the definitions of the genetically conditioned forms of diabetes mellitus in accordance with the proposals of the American Diabetes Association.[2]

1. *Manifest diabetes*. Certain diabetes with fasting hyperglycaemia of the ketotic or keto-resistant type. Diabetic symptoms may be present (a glucose loading is superfluous).

2. *Latent (= chemical) diabetes*. This is an asymptomatic diabetes. The fasting blood-sugar is usually normal but it may be elevated and so may the postprandial value. An oral or intravenous glucose tolerance gives pathological values, excluding situations of stress.

3. *Suspected diabetes*. (including stress hyperglycaemia). Individuals showing a temporary carbohydrate intolerance under physiological or pathological situations are suspected of diabetes, particularly if there is a familial history of diabetes.

Important stress situations associated with pathological glucose tolerance:

(a) *Pregnancy*. A 'pregnancy diabetes' returns to normal after parturition. However, there is a very high risk of a diabetes developing later. A diabetes should also be suspected if the case history reveals 'heavy' children (over 4500 g), frequent abortions, still-born children, neonatal mortality, and hydramnios.

(b) *Overweight* with pathological glucose tolerance and normalization following reduction in weight.

(c) *'Stress'*. Infections, trauma, affections of the circulation, burns, dietary deficiency, psychological disturbances.

(d) *Drugs*, such as corticosteroids or thiazides, ovulation inhibitors.

(e) *Endocrinopathia*. Acromegaly, Cushing's disease, thyrotoxicosis, phaeochromocytoma.

(f) Diabetes must also be suspected in *elderly persons* who fulfil the diagnostic criteria for younger individuals.

4. *Prediabetes*. Prediabetes does *not* denote a latent diabetes but the period from the fertilization of the egg to the detection of a pathological glucose tolerance in an individual genetically predisposcd to develop diabetes. The glucose tolerance is normal during the prediabetic phase. It is still not possible to diagnose a prediabetes except in the case of identical twin children of a diabetic, and probably, in the offspring of two diabetic parents. It concerns the so-called *potential diabetic*.

4.8.1.3. Simple screening methods

4.8.1.3.1. Detection of glucose in urine

The detection of a glucosuria is the simplest screening test for the presence of a diabetes. Unfortunately, it has already been established that a positive result is not proof that diabetes is present, nor can the presencc of the disease be excluded if the result is negative.

The following methods are available for the detection of glucose in urine:

(a) *Reduction tests*. The reduction tests (e.g. Benedict, Trommer, Fehling) respond to other reducing sugars (all physiological mono- and disaccharides other than saccharose) bcsides glucose and a large number of

638

drugs are also liable to give false-positive results.[3] In spite of this disadvantage, these tests can be used for the semi-quantitative assessment of the course of the glucosuria in known diabetics.

(b) *Enzymatic with glucose oxidase/peroxidase*. This method is used almost exclusively in commercial test papers. The advantage of these test papers lies in their convenience in use (simple and rapid), their specificity for glucose, and their relatively high sensitivity. This lies below 30 mg per 100 ml (0.03%). They suffer from interference to the reaction by high concentrations of physiological substances and drugs.[3]

(c) *Polarimetry*. This is the customary method for the semi-quantitative determination of urine glucose. The sensitivity of this method lies between 100 and 300 mg per 100 ml (0.1–0.3%). The results obtained with this method are very imprecise since there are numerous optically active substances present in the urine, even physiologically. Many drugs, particularly antibiotics can lead to gross distortion of the results.

(d) *Enzymatic with hexokinase/glucose-6-phosphate dehydrogenase*. This technically exacting method yields the most exact results and is absolutely specific for glucose. It must be regarded as the reference method but for the present it belongs to the specialized laboratory (p. 370).

A glucosuria is mostly to be expected 1–2 h post-prandial and after loading with glucose. In either case, the presence of a glucosuria requires further elucidation.

4.8.1.3.2. Fasting blood-glucose concentration

It is still not possible to give any absolutely valid norms for the fasting blood-glucose concentration. The following values are therefore to be interpreted as *approximate values* incorporating results from all methods:

	mg per 100 ml	mmol l^{-1}
Normal	50–90	2.78–5.00
Suspect	90–130	5.00–7.22
Pathological	>130	>7.22

If the result is suspect or pathological, the determination must be repeated. If at least two fasting blood-glucose values are in the pathological range, diagnosis of a diabetes is probable. If several determinations lie in the suspect range a further investigation is necessary.

4.8.1.3.3. Post-prandial blood-glucose concentration

If the urinary glucose test is positive, it is best initially to perform a determination of the blood-glucose concentration 1–2 h after a carbohydrate-rich meal, making simultaneous tests for urinary glucose and acetone.

It is a more frequent practice to attempt to base the diabetes screening solely on a post-prandial blood-sugar value. Since it is not possible to define the post-prandial condition more closely (differing rates of degradation and absorption of the different carbohydrates, variable composition of the diet, and, consequently, of the secretion of intestinal factors which induce insulin secretion) the suspect range must, naturally, be very broad.

The following criteria are customary for assessing capillary whole-blood glucose concentrations 1–2 h after a carbohydrate-rich meal:

above 180 mg per 100 ml: diabetes very probable;

between 130 and 180 mg per 100 ml: diabetes suspected;

less than 130 mg per 100 ml: diabetes improbable but not excluded.

4.8.1.4. Comments on the method

1. For venous collection the stated values may be arbitrarily reduced by 20 mg per 100 ml.

2. If the post-prandial blood-sugar is 180 mg per 100 ml or more, a glucose tolerance test is contra-indicated.

3. The diagnosis 'diabetes mellitus' can never be made on the basis of a post-prandial blood-sugar value.

References

1. Ärztekommission der Schweizerischen Diabetesgesellschaft und Standardisierungskommission der Schweizerischen Vereinigung für klinische Chemie: Richtlinien zur Diagnostik des Diabetes mellitus, *Schweiz. med. Wschr.*, **101**, 345, 390 (1971).
2. Special report: classification of genetic diabetes mellitus, *Diabetes*, **16**, 540 (1967).
3. Colombo, J. P. and Richterich, R., *Die einfache Urinuntersuchung*, Huber, Berne, 1977.

4.8.1.5. Oral glucose tolerance test: screening test with 50 g of glucose

4.8.1.5.1. Principle

As the most customary method, the investigation of the glucose tolerance performs valuable service in the diagnosis of a diabetes in the early stages. Although it has been criticized on account of its poor reproducibility, the glucose tolerance, as recommended in numerous modifications, has remained very popular. Since no glucose tolerance test can conform to the requirements of all the specialists, practitioners and hospital doctors it seems sensible for the present to use a standard loading with 50 g of glucose during 2 h, according to the criteria of a Study Group of the European Diabetes Association.[1,2]

4.8.1.5.2. Indications

Suspect fasting and/or post-prandial blood-glucose value. Clinical suspicion when the fasting and/or post-prandial blood-glucose value is not elevated, e.g. in disturbances of the arterial circulation, adiposity, familial history of diabetes. Every glucosuria (even traces) at normal blood-glucose concentration, including pregnancy glucosuria.

4.8.1.5.3. Contra-indications

With unequivocally pathological fasting and/or post-prandial hyperglycaemia, a tolerance test is not indicated. It is frequently performed, however, for observing the course of the illness in spite of slightly pathological values, but in such cases a glucose loading can result in a diabetes decompensation.

Clinical contra-indications are: acute gastro-intestinal diseases, status febrilis, greatly reduced general condition and undernourishment, and complex pharmaceutical therapies.

4.8.1.5.4. Operation

Dietetic preparations. An adequate feed of carbohydrate prior to the test increases the strength of the testimony.

Time at which the test should be conducted. It is best to observe a nocturnal fasting period of 12 h without eating, drinking, or smoking. Bodily activity (route to hospital, scooter-riding, night-work) impair the strength of the testimony.

Test drink. A 50-g amount of glucose is dissolved in about 300 ml of water. The drink may be flavoured with a few drops of lemon juice and should be drunk within 5 min.

Position of the patient during the test. The investigation is best carried out with the patient sitting, in bed for those who are bed-ridden, but children may sit at a table and play.

4.8.1.5.5. Blood collection

See p. 376.

4.8.1.5.6. Time at which the blood should be collected

See Table 123.

Table 123. Glucose concentration.

	Fasting		60 min		120 min	
	mg per 100 ml	mmol l^{-1}	mg per 100 ml	mmol l^{-1}	mg per 100 ml	mmol l^{-1}
Normal	<90	<5.0	<160	<8.9	<120	<6.7
Suspect	90–130	5.0–7.2	160–220	8.9–12.2	120–150	6.7–8.3
Pathological	>130	>7.2	>220	>12.2	>150	>8.3

Assessment

Normal. All three values lies in the normal range (later diabetes not excluded!).

Suspect (border range). One to three values in the suspect range and/or one value in the pathological range. Such a test demands monitoring.

Pathological. Two or more values in the pathological range. This generally signifies diabetes.

4.8.1.5.7. Comments on the method

1. This *exploratory test* constitutes a glucose tolerance test with *relatively little standardization*. Both in hospitals and in general practice, it is frequently not possible, on economical grounds, to investigate the subjects under standard conditions. Consequently, in the assessment given here we have paid no attention to age, sex, body-weight, drugs, body temperature, inactivity, accompanying illnesses, and other factors. Equally little consideration has been given to the technique of the blood collection or the method of the blood-sugar determination, provided one of the given methods is used. Differences between whole blood, plasma, and serum do not need to be considered in this coarse classification.

2. With this test the clinical diagnosis 'Diabetes: yes or no?' should not be prejudiced in cases of suspect blood-sugar values. In practice, a normal glucose tolerance is usually compatible with 'no diabetes at the time of investigation', while a pathological glucose tolerance warrants the diagnosis 'diabetes mellitus' in most cases.

3. If all three values lie in the normal range but a glucosuria is detectable, the test is assessed as *temporarily suspect* since it is possible that the time of the glucose peak value was passed or there is some error in the determination. Subsequent repetition of the test according to clinical urgency.

4. If there is a marked glucosuria even in the fasting condition and all three blood-sugar values are unequivocally in the reference range, a *renal glucosuria* may be indicated. In this case it is recommended that the question be resolved by a more thoroughly standardized glucose tolerance test.

5. For a venous blood collection, 20 mg per 100 ml can be subtracted from the listed value for the 1-hour or peak-value and 10 mg per 100 ml from the fasting or 2-hour value.

4.8.1.5.8. Diagnostic significance (from ref. 3)

Normal
1. Fasting concentration. For the determination of true glucose, this amounts to less than 100 mg per 100 ml (5.5 mmol l^{-1}).
2. After 60 min, <160 mg per 100 ml; after 120 min, <120 mg per 100 ml.
3. After 2 h the blood-glucose concentration falls back to the fasting concentration.
4. It sometimes happens that the glucose concentrations are transiently lower than the fasting concentration. The difference amounts to 10–15 mg per 100 ml. This decrease is proportional to the height and the steepness of the increase of the peak of the curve. It is directly dependent upon the amount of glucose administered and the speed with which it is absorbed.

Pathology
1. Reduced glucose tolerance
   1.1 Fasting hyperglycaemia (see blood-glucose)
   1.2. Abnormally high maximum:
      1.2.1. Increased glucose absorption in the intestines:
         1.2.1.1. 'Alimentary'. Abnormally high glucose feed.
         1.2.1.2. Hyperthyroidism, Earlier, higher peak followed by a more rapid return to the norm.
         1.2.1.3. Following gastrectomy, gastroenterostomy, or vagotomy. More rapid, earlier peak and subsequent more rapid decline to subnormal concentrations. The hyperglycaemic phase can last for a prolonged period and is probably indicative of an abnormal production of insulin in answer to the very rapid absorption of glucose.
         1.2.1.4. 'Lag' curve. As 1.2.1.3, but no gastrectomy or gastroenterostomy.
      1.2.2. Increased glycogenolysis and gluconeogenesis:
         1.2.2.1. Hyperthyrosis.
         1.2.2.2. Hyperadrenalism as a consequence of excitement or a phaeochromocytoma.
         1.2.2.3. Toxaemia with infections
         1.2.2.4. Gravidity. Above all, a slight decrease in glucose

tolerance is frequently observed in the 3rd trimester.

1.2.3. Inability to form glycogen from the glucose administered:

    1.2.3.1. Severe hepatopathia

    1.2.3.2. Glycogenoses.

    In both diseases the glucose fasting concentration is low and there is a hypoglycaemia after the test.

1.2.4. Inability of the tissues to utilize the glucose

    1.2.4.1. Diabetes mellitus.

    1.2.4.2. Haemochromatosis with simultaneous diabetes mellitus.

    1.2.4.3. Steroid diabetes caused by the presence of a Cushing syndrome, an administration of ACTH or of adrenal-cortex corticoids, e.g. hydro-cortisone.

    1.2.4.4. Head injuries and intracranial lesions can lead to damage of, or a pressure on, the hypothalamic region. With this group of diseases the blood-glucose concentration rises to an abnormally high peak, remains high for longer, and falls more slowly to the fasting value.

1.3. Return to the fasting blood-glucose concentrations. With the diseases in 1.2.4.1 and 1.2.4.2 the fasting blood-glucose concentration is achieved within 3 h. In diabetes mellitus and in steroid diabetes (1.2.4.4) the fasting glucose concentration is frequently abnormally high and is not attained for a few hours.

1.4. Simultaneous glucosuria. With a normal kidney function, the filtered glucose is almost completely reabsorbed. The reabsorption capacity is only exceeded at blood-glucose concentrations above about 160 mg per 100 ml. The critical concentration or kidney threshold depends, amongst other things, on the flow of urine.

    1.4.1. Alimentary glucosuria. Fasting urine contains no glucose but glucose is detectable at least during the final urine collection.

    1.4.2. Diabetic glucosuria. As 1.4.1, except that the fasting glucose concentration lies higher than the kidney threshold. In this case glucose is found in all samples.

    1.4.3. Renal glucosuria. Glucose is detectable in the urine even when the fasting blood-glucose concentration is lower than the kidney threshold.

2. Increased tolerance.

2.1. Fasting hypoglycaemia (see blood-glucose). For the identification of a true hypoglycaemia it is necessary to use arterial or

free-flowing capillary blood for the analysis and a specific method of glucose determination.
2.2. Flattened curve.
   2.2.1. Disturbed intestinal glucose absorption:
      2.2.1.1. Addison's disease.
      2.2.1.2. Hypopituitarism with a secondary insufficiency of the adrenal cortex.
      2.2.1.3. Intestinal diseases steatorrhoea (e.g. coeliac disease, idiopathic steatorrhoea, sprue, tuberculous enteritis, Whipple's disease).
      2.2.1.4. Hypothyroidism.
   2.2.2. Abnormal insulin secretion:
      2.2.2.1. Hyperplasia of the islet cells of the pancrease, e.g. after keeping to an excessive carbohydrate feed for long.
      2.2.2.2. Pancreas islet-cell adenoma.
      2.2.2.3. Carcinoma of the pancreas islet-cells.
   2.2.3. Exaggerated hypoglycaemic phase. After the blood-glucose concentration has returned to the fasting value it can fall by a further 20 mg per 100 ml or more and remain depressed for one or more hours; symptoms of hypoglycaemia appear. This is particularly to be observed in islet-cell hyperplasia, adenoma, and carcinoma, or in chronic pancreatitis. If one of these diseases is suspected, the glucose tolerance test should by extended to 8 h. This permits detection of the severe hypoglycaemic phase. A similar pattern is observed in hypopituitarism and hepatopathia.

References

1. European Diabetes Epidemiology Study Group, *Diabetologia,* **6**, 646 (1970).
2. Teuscher, A. and Richterich, R., Neue schweizerische Richtlinien zur Diagnose des Diabetes mellitus, *Schweiz. med. Wschr.,* **101**, 345 (1971).
3. Eastham, R. D., *Interpretation klinisch-chemischer Laborresultate*, translated and revised by R. Richterich and J. P. Colombo, Karger, Basle, 1970.

4.8.2. THYROID GLANDS

4.8.2.1. Introduction

In recent years the introduction of radioimmunoassays has led to a considerable transformation in thyroid-gland diagnostics. The determination of the thyroid hormones *thyroxine* (T_4) and 3,5,3'-L-triiodothyronine (T_3) (Fig. 144), as well as of the functional condition of the transport pro-

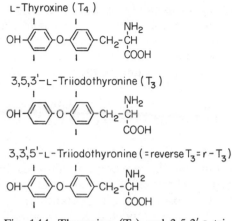

Fig. 144. Thyroxine (T_4) and 3,5,3'-L-tri-
iodothyronine (T_3).

teins, have, together with the identification of the thyroid-stimulating basic hormone (= thyrotrophin = TSH) and the thyrotrophin-releasing hormone (TRH), largely replaced the non-specific determinations such as butanol-extractable or protein-bound iodine, and rendered certain nuclear medicinal *in vivo* investigations unnecessary.

4.8.2.2. Thyroid hormones and their transport proteins in the blood

The thyroid hormones T_3 and T_4 are formed from iodine and the amino acid tyrosine in the thyroid glands. Bound to thyroglobulin—a protein specific to the cells of the thyroid glands—these hormones are stored extracellularly as a colloid in the thyroid glands. When needed they are taken up by the cells of the thyroid glands again, cleaved from thyroglobulin and secreted into the bloodstream.[1] In the blood the largest fraction of T_3 and T_4 is bound to transport proteins. Whether these play a part in the uptake of the hormones by the cells of the periphery is not clear. The thyroid-gland hormones are also stored in peripheral cells, in particular those of the liver. Receptors have been detected in the cytoplasm and in the cell nucleus. In the periphery the hormones are deiodinated with the production (from T_4) of T_3 (3,5,3'-L-triiodothyronine) or reverse-T_3 (3,3',5'-L-triiodothyronine = r-T_3), according to which ring is deiodinated. The contribution of the thyroid glands to T_3 formation is about 60%, the peripheral fraction 40%.[2] Whether the hormonally active T_3 or the inactive r-T_3 is produced from T_4 does not appear to be randomly determined. In conditions of hunger, severe illnesses or operations and also at the neonatal age, the degradation to r-T_3 is favoured.[3-6] About 10% of the thyroxine and 75% of the triiodothyronine is converted per day.[2] The ratio in the blood is of diagnostic importance. T_4 is present at higher concentration than T_3 (T_4, 100 nmol l^{-1}; T_3, 2 nmol l^{-1}). However, the biological activity of T_3 substantially exceeds that of T_4. It is assumed that the free part of the thyroid hormones is responsible for hormonal action, i.e. about 0.4⁰/oo of the total serum concentration for T_4 and about 3⁰/oo for T_3. The whole of the remainder is bound to transport proteins in the blood. Up to 60% of the T_4 is bound to *thyroxine-binding globulin* (TBG), up to about 30% on pre-albumin, and up to 10% on albumin. The T_3 is also bound to TBG and albumin.[2] The distribution between the bound and free parts obeys the Law of Mass Action and the bonding is dependent on pH and temperature. The binding constant for T_4 on TBG at physiological pH and 37 °C amounts to $9 \times 10^9 \, l \, mol^{-1}$. The bonding is effected at a single bonding site but for T_3 two bonding sites have been found. The bonding of the latter is about 20 times as weak as that for T_4.[7,8] The determination of the hormonally important free fraction of the thyroid-gland hormones is very involved and remains the province of the specialist laboratory (equilibrium dialysis). The measurement of the total serum concentration of thyroid hormones is only conditionally of diagnostic value. Thus, e.g. an elevation of the serum T_4 concentration is only indicative of hyperthyrosis if the concentration of the transport proteins is normal. But if the concentration of the transport proteins is elevated, e.g. as on intake of oestrogen-containing ovulation inhibitors, the elevated T_4 concentration is of no significance. Consequently, in addition to the determination of the total T_4, it is essential to determine the degree of saturation of the transport proteins by the so-called T_4 uptake test. In all the tests described care must be taken to ensure that the patient has not previously received radioactive iodine, e.g. in a nuclear-medicinal investigation.

4.8.2.3. Regulating hormones (Fig. 145)

The synthesis of the thyroid hormones and their release from the thyroid glands is regulated by the *thyroid-stimulating hormone thyrotrophin* (*TSH*), which is synthesized in the anterior pituitary. It is a glycoprotein (molecular weight 26 000) consisting of two polypeptide chains. The β-chain is the unit specific for TSH. The α-chain is identical with a polypeptide chain also found in luteinizing hormone (LH), in follicle-stimulating hormone (FSH), and in human chorionic gonadotrophin (HCG). Activated by adenyl cyclase, the TSH in the thyroid glands causes an increased uptake of iodine, and increased synthesis of iodothyronines, an increased uptake of colloid, and cleavage of the thyroid hormones from thyroglobulin followed by release of the hormone from the cells.[9,10]

The long-term effect of elevated TSH activity is to increase the division of the thyroid-gland cells (goitre). TSH is determined by radioimmunoassay. It serves principally for the identification and clarification of hypothyrosis. The synthesis of TSH and its release from the hypophysis is stimulated by the *thyrotrophin-releasing hormone* (*TRH*) and

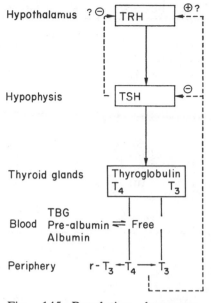

Fig. 145. Regulating hormones. TRH = thyrotrophin-releasing hormone; TSH = thyroid-stimulating hormone; TBG = thyroxine-binding globulin; $T_3 = 3,5,3'$-L-triiodothyronine; T_4 = thyroxine; r-T_3 = reverse T_3.

modulated by the thyroid hormones. High T_3 and T_4 inhibit the release of TSH from the pituitary (negative feedback). TRH is a tripeptide (pyroglutamylhistidylprolinamide). It is formed in the hypothalamus and transported to the hypophysis via a portal circulation. In addition to stimulating the synthesis and secretion of TSH it also effects the release of prolactin and growth hormone in the hypophysis. It possibly also has a direct effect on mood and behaviour. Giving oestrogen intensifies the action of TRH doses while the application of corticosteroids or medication with dopa inhibits it. In new-born babies and infants chilling causes a release of TRH. It is not clear to what extent TSH, T_3, and T_4 take part in the regulation of the TRH synthesis.[2,11] TRH-degrading enzymes have been found in many tissues but not in the pituitary body.[12] The half-life of TRH in the blood is about 4 min. A synthetic TRH is commercially obtainable and is used for performing the TRH test. After giving a peroral or intravenous dose of TRH the increase in TSH is measured by radioimmunoassay. In the peroral test it is sufficient to take blood samples at zero time and 3 h after application.[13-15] Table 124 gives information on the outcome of the TRH test for various pathological pictures. The introduction of the TRH test together with the determination of the total

Table 124. Thyroid gland diagnostics.

	T_4	T_3	T_3 uptake test[a]	TSH	TRH test
Euthyroid	n	n (↑)	n	↑ (n)	n (protracted when necessary)
Struma					
Hyperthyrosis	↑	↑	↑	↓	↓
T_3 thyrotoxicosis	n	↑	n	n (↓)	
Hypothyrosis					
thyrogenic	↓	↓ (n)	↓	↑	↑
hypophyseal	↓	↓ (n)	↓	↓	↓
hypothalamic	↓	↓ (n)	↓	↓	n
Hypalbuminaemia	↓	↓	↑	n	n
Gravidity					
Ovulation inhibitors	↑	(↑)	↓	n (↑)	n

n = normal.
↑ = elevated value.
↓ = lowered value/answer.
[a] Uptake of the ^{125}I-T_3 not bound to transport protein by the secondary binder (resin), referred to the total activity applied.

T_3 have rendered the so-called T_3 suppression test involving an *in vivo* application of radioisotopes superfluous.[2]

The determination of further biochemical parameters for the identification of rarer diseases of the thyroid gland or pathogenetic mechanisms remains the province of the specialist laboratory.

4.8.2.4. Measurement of thyroid hormones

4.8.2.4.1. Determination of the total triiodothyronine in serum: T_3 radioimmunoassay

The determination of the T_3 in the serum is effected by radioimmunoassay. For the measurement of the total T_3, the bound T_3 must first be liberated from the transport proteins. Detailed studies on different reagents for this purpose and their interference with the antigen–antibody reaction were described by Hüfner and Hesch.[16] Eastman et al.[17] gave the following clinical indications for a T_3 determination:

1. Thyrotoxicosis, especially T_3 thyrotoxicosis.
2. Assessment of acute changes in the secretion of thyroid hormone following thyroidectomy, change of thyrostatic therapy, or hypophysectomy.
3. Assessment of the thyroid-hormone substitution.
4. Investigation of the autonomy of the thyroid glands or reserve capacity, in the eventuality, following dosing with TRH or TSH.

Results for T_3 concentrations for selected clinical pictures are given in Table 124. Of the T_3 and T_4 determinations, the former appears better for discriminating between eu- and hyperthryosis while the latter is better for demarcating hypothyrosis.[18-20] The T_3 determination is suitable for the identification of compensatory T_3 overproduction in diminution of the T_4 reserve (example: endocrinal exophthalmopathy). The elevation of T_3 in heroin addicts should also be mentioned. The determination of the T_3 concentration in the serum should not be confused with the T_3 uptake test (T_3 resin uptake test, Hamolsky test).

Rigorous quality control by means of control sera is essential.[21,30]

4.8.2.5. Determination of the total thyroxine (T_4)

The non-specific determinations of the organic iodide as a measure of the T_4 concentration have been superseded. Thyroxine determinations are based on the principles of saturation analysis (cf. Section 2.8.3). The binding protein chosen depends on the method; the so-called competitive protein binding (CPB) method uses thyroxine-binding protein but anti-T_4 antibody is better (T_4 RIA).

4.8.2.5.1. Competitive protein binding method

The determination is conducted on serum since heparin can interfere with the uptake. The T_4 in the patient's serum is released from its bonding to transport proteins and separated, preferably by extraction with ethanol or methanol, resulting in a simultaneous denaturation of the proteins. Alternatively, small Sephadex columns may be used or the T_4 adsorbed on silica gel at acidic pH. The extracted T_4 (usually after evaporation of the ethanol) is added to a limited amount of thyroxine-binding protein and [^{125}I]thyroxine. During incubation the known quantity of tracer and the patient's T_4 compete for the binding sites of the thyroxine-binding protein. The free part of the T_4 is separated off. Ekins,[22] who introduced this method, accomplished this by electrophoresis, but the separation may be effected more practically with Sephadex or with ion-exchange resins (as granules, on sponges or strips). The radioactivity of either this fraction or that bound to thyroxine-binding protein is measured and compared with a standard graph. Standards and control sera should be included in every determination (including the extraction), since the position of the equilibrium is both temperature and time dependent. In particular, the equilibrium may also be affected by the agents used for making the separation.[31]

The chief disadvantage of this method is the need for extraction and the consequent uncertainty as to whether the standard sera and samples are extracted with the same efficiency. In addition, the method uses a greater quantity of serum than is required for the radioimmunoassay. Dosing with iodide does not interfere but there is certainly interference from free fatty acids, which also compete for bonding to the protein; they lead to falsely high T_4 values, while haemolyses yield low results.[23,24] Consequently, owing to lipolysis, sera to be determined by this method should not be stored at room temperature or sent through the post.

The determination of T_4 by radioimmunoassay may currently be considered the method of choice, since it does not suffer from the above-mentioned disadvantages. The partly hypothetical results from ring experiments show how important it is to safeguard the quality of the analyses by using control sera and external ring experiments.[21]

4.8.2.5.2. Determination of thyroxine by radioimmunoassay

The determination is carried out on serum. The principle has been thoroughly described in Section 2.8.3.2. For the separation of the T_4 from the TBG in the patient's serum, 8-anilinonaphthalene-1-sulphonic acid or ethylmercurithiosalicylate (merthiolate, thiomersal) is added to the incubation buffer. The barbital buffer most commonly used for the incubation blocks the binding to prealbumin. The high dilution of the serum and the affinity of the anti-T_4 antibody also prevent the albumin from interfering. Various methods are used for separating the free and bound fractions. The use of polyethylene glycol appears advantageous in comparison with active charcoal, ion-exchange resins, and solid-phase methods.[25-27]

4.8.2.6. Diagnostic significance

Table 124 provides a survey of the changes in T_4 with various diseases. Drugs which interefere *in vivo* with the binding to the transport proteins lead to depressed total T_4 results. Diphenylhydantoin (TBG) and high doses of salicylate (prealbumin) are worth mentioning. Since the total amount of T_4 is also dependent on the concentration of transport proteins, an elevation of TBG (inherited or after dosing with oestrogens during pregnancy or in the early phase of a hepatitis) can lead to an elevation of the total T_4. A depression of the transport proteins (inherited, mongolism, androgeny, hypoproteinaemia through loss or disturbed synthesis) is associated with low total T_4.[28]

4.8.2.7. T_3 uptake test

Instead of making the laborious determination of TBG, prealbumin and albumin the transport-protein function is measured by means of the T_3 uptake test. This is indispensible for the correct interpretation of the total T_4 since the largest part of the thyroid hormones is present in the bound condition.

The T_3 uptake test should not be confused with the determination of the total T_3 in the blood. The reagent employed here is triiodothyronine labelled with ^{125}I. Part of the transport protein in the serum is not fully saturated with thyroid hormones. If now an amount of T_3 exceeding the transport capacity of the transport proteins is added to the serum *in vitro*, then the free binding capacity of the transport proteins in the serum can be measured from the amount which does not become bound. This unbound fraction is taken up by a second binder. Previously erythrocytes were used for this purpose (Hamolsky test), but today the free T_3 is adsorbed on ion-exchange resins, Sephadex, or albumin-coated silicates; after separation, the radioactivity taken up by the second binder is measured. Labelled triiodothyronine is used because it has a smaller affinity for TBG than T_4, so that the latter is not extensively displaced from its binding to the transport proteins. It should be noted that we are dealing here with artificial, *in vitro* behaviour (non-physiological pH, room temperature, excess of T_3, second binder). Hence it would be wrong to attempt to calculate the absolute concentration of free thyroid hormones in the serum by the application of the Law of Mass Action to the free binding capacity of the transport proteins as determined above. Nevertheless, the results of the T_3 uptake test are an aid in estimating the relative fractions of free T_4 and T_4 bound to transport proteins. Thus, e.g. a lowered uptake of T_3 in the second binder, at normal total T_4 concentration, indicates hypothyrosis (elevated TBG concentration showing that the TBG in equilibrium with the free T_4 has not been fully saturated, i.e. the free T_4 is probably lowered).

The manner of calculating the results creates considerable confusion. Each of the different manufacturers of test packs express this in a different way.[29] The radioactivity taken up by the second binding reagent is referred to the total activity or this percentage is yet again referred to the percentage uptake in a reference pooled serum. Since the trend of the corresponding results may be in a contrary sense, the manner of calculation must be specified and it is best if, in addition to stating the reference interval, there is an indication of whether hypo- or hyperthyroid values lie above or below this reference interval. There have also been attempts to express the result of T_4 and T_3 uptake tests mathematically as an index. The so-called 'free thyroxine index' (FTI), which is by no means identical with the measured free thyroxine, is obtained by multiplication of the T_4 concentration by the percentage resin uptake referred to the total activity (also called T_7). Other FTI are produced in respect of a reference serum result. A more elegant method is an actual (procedural) combined determination of T_4 and T_3 uptake test (effective thyroxine ratio, ETR, or compensated T_4, CT_4). The first stage is as in the determination of T_4 by the competitive protein binding assay. The T_4 is extracted from the serum and a radio-labelled tracer is added together with a limited quantity of binding reagent (TBG). When an equilibrium has been established

652

between TBG-bound and free T_4, a small quantity of the patient's serum is added. The transport proteins in this serum compete for the remaining free T_4. The free radioactive T_4 still remaining is taken up by a secondary binder and counted. The addition of the small amount of patient's serum (order of magnitude 10 μl) thereby compensates the T_4 determination in accordance with its capacity for taking up thyroid hormone. The advantage of this combined test over the calculated compensation is that overall there are fewer pipetting errors to contribute to the sum of the variances. However, these tests have the same disadvantages as the competitive protein determination of T_4. They cannot replace the determination of T_4. In addition, information on the individual components is lost in the combination of T_4 and T_3 uptake tests.[2] It is also questionable whether the test will find favour in view of its cost.

References

1. Studer, H., Schilddrüse, in Siegenthaler, *Klinische Pathophysiologie*, 3. Aufl., Thieme, Stuttgart, 1976, p. 307.
2. Wellby, M. L., The laboratory diagnosis of thyroid disorders, *Adv. clin. Chem.*, **18**, 103 (1976).
3. Chopra, I. J., Chopra, U., Smith, S. R., Reza, M., and Solomon, D. H., Reciprocal changes in serum concentrations of 3,3',5'-triiodothyronine (reserve T_3) and 3,3',5'-triiodothyronine (T_3) in systemic illnesses, *J. clin. Endocr. Metab.*, **41**, 1043 (1975).
4. Nicod, P., Burger, A., Staeheli, V., and Vallotton, M. B., A radioimmunoassay for 3,3',5'-triiodo-L-thyronine in unextracted serum: method and clinical results, *J. clin. Endocr. Metab.*, **42**, 823 (1976).
5. Burger, A., Nicod, P., Vagenakis, A., and Vallotton, M. B., Reduced active thyroid hormones in acute infection, *Eur. J. clin. Invest.*, **6**, 326 (1976).
6. Burr, W. A., Black, E. G., Griffiths, R. S., and Hoffenberg, R., Serum triiodothyronine and reverse triiodothyronine concentrations after surgical operation, *Lancet*, **ii**, 1277 (1975).
7. Korcek, L. and Tabachnick, M., Thyroxine–protein interactions, *J. biol. Chem.*, **251**, 3558 (1976).
8. Refetoff, S., Fang, V. S., and Marshall, J. S., Studies on human thyrozine-binding globulin (TBG). IX. Some physical, chemical, and biological properties of radioiodinated TBG and partially desialylated TBG. *J. clin. Invest.*, **56**, 177 (1975).
9. Orgiazzi, J., Mécanisme d'action de la TSH, *Lyon méd.*, **235**, 217 (1976).
10. Broughton, A., Clinical usefulness of radioimmunoassay of thryotropin: a review, *Sth. med., Nashville*, **69**, 702 (1976).
11. Hall, R. and Gomez-Pan, A., The hypothalamic regulatory hormones and their clinical applications, *Adv. clin. Chem.*, **18**, 173 (1976).
12. Prasad, D. and Peterkofsky, A., Demonstration of pyroglutamylpeptidase and amidase activities toward thyrotropin-releasing hormone in hamster hypothalamus extracts, *J. biol. Chem.*, **251**, 3229 (1976).
13. Gemsenjäger, E., Untersuchungen der Schilddrüsenfunktion mittels TRH-Tests bei blander Struma vor und nach Strumektomie, *Schweiz. med. Wschr.*, **106**, 1084 (1976).
14. Staub, J. J., Girard, J., Gemsenjäger, E., and Müller, J., Oral TRH as a simple

diagnostic test for thyroid investigation with special regard to patients with low response, *Eur. J. clin. Invest.*, **6**, 317 (1976).

15. Wildmeister, W., Über Methoden und Bewertung von Radioimmunoassays bei Schilddrüsenerkrankungen, *Ärztl. Lab.*, **21**, 324 (1975).

16. Hüfner, M. and Hesch, R.-D., A comparison of different compounds for TBG-blocking used in radioimmunoassay for tri-iodothyronine, *Clin. chim. Acta*, **44**, 101 (1973).

17. Eastman, C. J., Corcoran, J. M., Ekins, R. P., Williams, E. S., and Nabarro, J. D. N., The radioimmunoassay of triiodothyronine and its clinical application, *J. clin. Path.*, **28**, 225 (1975).

18. Seth, J., Toft, A. D., and Irvine, W. J., Simple solid-phase radioimmunoassays for total tri-idodothyronine and thyroxine in serum, and their clinical evaluation, *Clin. chim. Acta*, **68**, 291 (1976).

19. Stafford, J. E. H., Lees, S., and Watson, D., Serum triiodothyronine determination in clinical use, *J. clin. Path.*, **29**, 642 (1976).

20. Brunelle, Ph., Laine, G., Nouel, J.-P., and Bohuon, C., Le dosage radio-immunologique de la triiodothyronine sérique et ses applications cliniques, *Annls. Biol. clin.*, **32**, 323 (1974).

21. Horn, K., Marschner, I., and Scriba, P., Erster Ringversuch zur Bestimmung der Konzentration von L-Trijodthyronin (T_3) und L-Thyroxin (T_4) im Serum: Bedeutung für die Erkennung methodischer Fehlerquellen, *Z. klin. Chem. klin. Biochem.*, **14**, 353 (1976).

22. Ekins, R. P., The estimation of thyroxine in human plasma by an electrophoretic technique, *Clin. chim. Acta.*, **5**, 453 (1960).

23. Nye, L., Yeo, T. H., Chan, V., Goldie, D., and Landon, J., Stability of thyroxine and triiodothyronine in biological fluids, *J. clin. Path.*, **28**, 915 (1975).

24. Liewendahl, K. and Helenius, T., Effect of fatty acids on thyroid function tests *in vitro* and *in vivo*, *Clin. chim. Acta*, **72**, 301 (1976).

25. Nye, L. Hassan, M., Willmott, E., and Landon, J., Introduction of a rapid, simple radioimmunoassay and quality control scheme for thyroxine, *J. clin. Path.*, **29**, 452 (1976).

26. Cheung, M. C. and Slaunwhite, W. R. Jr., Use of polyethylene glycol in separating bound from unbound ligand in radioimmunoassay of thyroxine, *Clin. Chem.*, **22**, 299 (1976).

27. Kruse, V., Production and evaluation of high-quality thyroxine antisera for use in radioimmunoassay, *Scand. J. clin. Lab. Invest.*, **36**, 95 (1976).

28. Kohler, H., Riek, M., and Studer, H., Die Beurteilung der Schilddrüsenfunktion anhand der Konzentration und der Transportverhältnisse der Hormone im Serum, *Ther. Umsch.*, **30**, 701 (1973).

29. Horn, D. B., Available assays for serum thyroxine and for serum uptake tests, *J. clin. Path.*, **28**, 218 (1975).

30. Staub, J. J. and Peyer, P., Moderne Labormethoden zur Beurteilung der Schilddrüsenfunktion, *Schweiz. Z. med.-tech. Lab. Pers.*, **4**, 543 (1977).

31. Krüskemper, H. L., Rudorff, K.-H., and Herrmann, J., Schilddrüssenhormone. Irrtumsmöglichkeiten bei der Bestimmung und Interpretation von Messergebnissen, *Dt. med. Wschr.*, **102**, 526 (1977).

4.9. NERVOUS SYSTEM

4.9.1. CEREBROSPINAL FLUID

4.9.1.1. Introduction

Cerebrospinal fluid (Table 125) is the inner liquor filling the ventricles of the brain; as external fluid in the arachnoid space it also surrounds the brain and the spinal cord. The fluid is secreted by the choroid plexus which, conversely, can take up substances from the fluid. Additional fluid is formed in the ependyma of the ventricles and in the subarachnoid space. The fluid space is connected with the intercellular space. The liquor flows into the blood via the arachnoid villi but part appears to reach the blood by another route. Cerebrospinal fluid is not an ultrafiltrate of the plasma.[1,2,6]

References

1. Thorn, L., Zur Bildung und Resorption des Liquor cerebrospinalis, *Dt. med. Wschr.*, **98**, 2253 (1973).
2. Cerebrospinal fluid: the lymph of brain?, *Lancet*, **ii**, 444 (1975).
3. Fremont-Smith, F. and Dailey, M. E., *The human cerebrospinal fluid*, Hoeber, New York, 1924.
4. Katzenelbogen, S., *The cerebrospinal fluid and its relation to the blood*, Johns Hopkins Press, Baltimore, 1935.
5. Documenta Geigy, *Liquor cerebrospinalis. Wissenschaftliche Tabellen*, 7. Aufl., Geigy, Basle, 1968, p. 631.
6. Katzman, R. and Pappius, H. M., *Brain electrolytes and fluid metabolism*, Williams & Wilkins, Baltimore, 1973.

4.9.1.2. Cerebrospinal-fluid pigments

As shown in Table 126, the determination (or at least the identification) of pigments in xanthochromic or blood-stained fluid has a differential diagnostic significance.

We use the following methods for the identification of the pigments:

Bilirubin. Determination with the dichloroaniline method (p. 604).

Oxyhaemoglobin. Direct measurement of the extinction at 578 or 546 nm and calculation by means of the following equations:

$$\text{Concentration} = E_{578\,nm} \cdot 104 \text{ mg haemoglobin per 100 ml}$$
$$\text{Concentration} = E_{546\,nm} \cdot 109 \text{ mg haemoglobin per 100 ml}$$

Table 125. Synopsis of the most important physical properties and the chemical composition of cerebrospinal fluid[3-6] (lumbar fluid) (the composition is slightly different according to the origin of the fluid—lumbar, sub-occipital, ventricular).

	Component	CSF $\bar{x} \pm 2\,s$	Plasma mean
Physical properties	Colour	Crystal clear	Yellowish
	Amount, ml	100–200	3000
	Daily production, ml	*ca.* 700	—
	pH	7.4–7.5	7.45–7.50
	Pressure:		
	children, cmH$_2$O	5–10	—
	adults, cmH$_2$O	7–20	—
Chemistry		mg per 100 ml	mg per 100 ml
	Total protein	20–40	6500
	Albumin	15–30	4000
	Globulin	4–9	2500
	Fibrinogen	0	200
	Amino acid nitrogen	1.6–2.7	4.4
	Urea nitrogen	7.5–15.0	25
	Uric acid	0.5–2.6	4.0
	Creatinine	1.0–1.5	1.0
	Glucose	45–80	70
	Cholesterol	0.06–0.5	200
	Bilirubin	0.2	0.8
	Calcium	4.1–5.9	10
	Inorganic phosphorus	1.3–1.5	3.5
		mmol l^{-1}	mmol l^{-1}
	Sodium	129–153	145
	Potassium	2.06–3.86	4
	Chloride	120–130	104

Table 126. Pigments in cerebrospinal fluid.

Disease	Pigments
Haemorrhages (subarachnoidal or ventricular)	Oxyhaemoglobin + bilirubin
Subdural haematoma	Haemiglobin
Block (spinal, subarachnoidal, ventricular)	Bilirubin
Jaundice in diseases of the liver	Bilirubin
Subdural discharge	Bilirubin
Cystic fluid	Bilirubin
Controls	Negative
Traumatic puncture (supernatant)	Negative

656

If necessary the fluid should be diluted with ammonia solution prior to measurement. The result must then be multiplied by the appropriate factor.

Haemiglobin. Quantitative determination by the method described on p. 543.

4.9.1.3. Comments on the method

Blood is found in the cerebrospinal fluid (a) when there is bleeding in the central nervous system and the blood reaches the subarachnoid space, or (b) iatrogenically, if the puncture has been performed badly. With fresh bleeding the two causes cannot be differentiated. After centrifugation of the blood, provided there is no haemolysis, the supernatant of the fluid is clear in the case of fresh bleeding. If the bleeding took place more than 2 h earlier, the supernatant after centrifuging may show a yellowish tinge (xanthochromic fluid).[2]

References

1. Barrows, L. J., Hunter, F. T., and Banker, B. Q., The nature of clinical significance of pigments in the cerebrospinal fluid, *Brain,* **78**, 59 (1955).
2. Van der Meulen, J. P., Cerebrospinal fluid xanthochromia: an objective index, *Neurology, Minneap.,* **16**, 170 (1966).

4.9.2. CEREBROSPINAL FLUID PROTEINS BY THE BIURET METHOD

4.9.2.1. Choice of method

The concentration of proteins in cerebrospinal fluid is lower than that in plasma by a factor of 200, placing special demands on the sensitivity of the method. The following techniques are currently in common use.

Kafka's method. The cerebrospinal fluid is placed in a special graduated tube, the protein is precipitated with a picrate–citrate mixture, and the height of the column of precipitate is read off from the scale.

Kjeldahl determination. Separate determinations of non-protein nitrogen and total nitrogen and calculation of the protein from the difference.

Folin–Ciocalteu method.[1] The high sensitivity of this method renders it particularly suitable for determination of the proteins in cerebrospinal fluid.

Biuret method. The sensitivity of this method is not great enough to permit its use in a direct determination, but if the protein is first precipitated and the biuret reaction performed on the precipitate it can give very good results.

Measurement of the extinction at 280 nm. This method is unsuitable since cerebrospinal fluid contains many substances which absorb strongly in the ultraviolet region.

Comparative studies have shown that Kafka's method is very unreliable and generally need not be considered. The Kjeldahl analysis suffers from two disadvantages: the large amount of cerebrospinal fluid required and the work involved. The Folin–Ciocalteu method[1] also has its drawbacks: there are a number of drugs, particularly salicylates (*p*-aminosalicylic acid, acetylsalicylic acid, sodium salicylate), which give a positive reaction with the phenol reagent and thus yield falsely high results; secondly, the method is technically difficult. For these reasons we prefer the determination with the biuret reagent.

4.9.2.2. Principle

The cerebrospinal fluid proteins are precipitated with phosphotungstic acid. The precipitate is completely dissolved in sodium hydroxide solution and then the biuret reagent is added.

4.9.2.3. Reagents

Precipitating reagent. Dissolve 2 g of phosphotungstic acid in 2 mol l^{-1} H_2SO_4. The turbidity can be removed by centrifuging or filtration.

Base reagent. Dissolve 36 g of urea in 2 mol l^{-1} NaOH and make up to 100 ml with further NaOH.

Biuret reagent: copper sulphate 34 *mmol* l^{-1}, *potassium iodide* 21.7 *mmol* l^{-1}, *sodium carbonate* 472 *mmol* l^{-1}; *sodium citrate* 349 *mmol* l^{-1}; *urea* 5.99 *mol* l^{-1}. Dissolve 8.5 g of $CuSO_4$ and 3.6 g of potassium iodide separately in about 50 ml of DM-water each. Dissolve 87 g of sodium citrate and 50 g of sodium carbonate in about 500 ml of DM-water and stir the solution while first the $CuSO_4$ solution and then the potassium iodide solution are added. Dissolve 360 g of urea in this solution with constant stirring and make up to 1000 ml with DM-water.

Protein standard. 50 *mg per* 100 *ml* (p. 411).

4.9.2.4. Measurement

Conditions. The final concentrations in the reaction mixture are as follows:

	mmol l^{-1}
$CuSO_4$	15.2
Potassium iodide	9.69
Na_2CO_3	211
Sodium citrate	156
Urea	5.348
NaOH	670

Procedure. The precipitation of the protein and the biuret reaction are carried out in the same vessel.

	Sample, ml	Standard, ml	Blank, ml
Phosphotungstic acid	0.5	0.5	—
Cerebrospinal fluid	0.5	—	—
Standard, 200 mg per 100 ml	—	0.5	—

Thoroughly mix on a vibrator for about 1 min. Allow to stand for 15 min (until the precipitate begins to settle). Centrifuge briskly for 5 min and discard the supernatant

Precipitate	+++	+++	—
DM-water	—	—	0.02
Base reagent	0.5	0.5	0.5

Agitate the precipitate with a vortex stirrer until the precipitate has completely dissolved (1–2 min)

Biuret reagent	0.5	0.5	0.5

Mix and allow to stand at room temperature for 30 min. Read the extinction against water at 546 nm

4.9.2.5. Calculation

By means of the standard.

$$\text{mg per 100 ml} = \frac{E(S) - E(B)}{E(ST) - E(B)} \cdot 50$$

By means of the extinction coefficient (a = 3.12).

$$\text{mg per 100 ml} = \frac{E}{ad} \cdot \frac{FV}{SV} = \frac{E}{3.12 \cdot 1} \cdot \frac{1.02}{0.5} \cdot 50$$

Concentration $= E(S) - E(B) \cdot 654$ mg per 100 ml

4.9.2.6. Reference values

There are few reliable results for the reference values for the concentration of cerebrospinal fluid proteins and their fractions. As emerges from Tables 127 and 128, values of up to 100 mg per 100 ml can

Table 127. Reference values for total protein (mg per 100 ml) in children of various ages (g = geometric mean, assuming a log-normal distribution).

Age	Sex	Number	g	$g \pm 2 \log s$	References
1st year					
1 month		10	48.4	23.6 ± 99.5	3
2 months		8	26.9	11.6 ± 62.4	
3 months		7	25.4	13.6 ± 47.4	
4 months		9	27.9	15.7 ± 49.3	
5–6 months	m + f	25	22.9	10.2 ± 51.6	
7–12 months		54	20.7	8.7 ± 49.1	
2nd–16th year					
Healthy		239	18.1	7.2 ± 45.3	
Retarded		52	16.1	5.5 ± 47.1	
With cerebral paralysis		41	18.0	10.1 ± 32.2	
Total		332	17.7	7.1 ± 44.3	
				$\bar{x} \pm 2 s$	
0–6 months		16		7.9–70.0	4
6 months–2 years		14		4.3–39.3	
2–4 years		17		8.7–29.1	
4–6 years	m + f	14		4.2–32.6	
6–8 years		18		6.5–37.8	
8–10 years		17		10.8–31.5	
10–13 years		13		11.4–34.4	
Adults	m + f	389		8.6–32	9

be taken as valid for the 1st month and concentrations between 7 and 44 mg per 100 ml at ages above this.[3-7]

4.9.2.7. Comments on the method

1. By precipitating with phosphotungstic acid, low-molecular-weight glycoproteins, which are increased during the acute phase, are included in the determination. Non-proteins which interfere in the biuret reaction are eliminated.

2. The described method is linear up to 600 mg per 100 ml. Before precipitating the proteins it is advisable to perform a qualitative test with a test paper. For +++-protein an appropriately smaller amount of cerebrospinal fluid (e.g. 0.2 ml) must be used for the determination. If only traces are present, then it is advantageous to perform the precipitation with 1 ml of phosphotungstic acid. The appropriate corrections must then be made in the calculation.

3. The percentual extinction coefficient $a = 3.12$ has been determined with bovine serum albumin (p. 410).

4. It is possible to increase the sensitivity by measuring the extinction in

660

Table 128. Reference values for specific protein fractions in cerebrospinal fluid (mg per 100 ml).

	Age	Sex	Number	$\bar{x} \pm 1s$	References
IgG	1–24 months	m		0.75 ± 1.07	5
	1–24 months	f		0.22 ± 0.30	
	2–14 years	m		0.80 ± 0.67	
	2–14 years	f		0.89 ± 0.61	
	0–6 months		16	2.56 ± 1.43	4
	6–24 months		14	1.18 ± 0.63	
	2–4 years		17	1.29 ± 0.59	
	4–6 years	m + f	14	1.18 ± 0.37	
	6–8 years		18	1.38 ± 0.61	
	8–10 years		17	1.62 ± 0.64	
	10–13 years		13	1.80 ± 0.57	
	Adults			Up to 2.80	6
γ-Globulin	0–6 months		16	0–5.31	4
	6–24 months		14	0.49–3.99	
	2–4 years		17	0.20–2.41	
	4–6 years	m + f	14	0–2.97	
	6–8 years		18	0.08–3.46	
	8–10 years		17	0–3.53	
	10–13 years		13	0.57–3.16	
	Adults			Up to 4.10	6
Albumin	1–24 months	m		13.59 ± 7.02	5
	1–24 months	f		9.25 ± 3.69	
	2–14 years	m		10.25 ± 2.44	
	2–14 years	f		10.70 ± 2.89	
	Adults			Up to 29.0	6
Prealbumin				2.50	
a_1				2.70	
a_2				3.10	
β				4.30	
τ				2.70	
IgA				0.34	
IgM				0	

the ultraviolet region at 340 nm, but it is then necessary to use an analysis blank.[2]

5. Maurer[8] has described a biuret method which does not include a deproteination.

References

1. Daughaday, W. H., Lowry, O. H., Rosebrough, N. J., and Fields, W. S., Determination of cerebrospinal fluid protein with the Folin phenol reagent, *J. Lab. clin. Med.*, **39**, 663 (1952).
2. Bürgi, W. and Kaufmann, H., Die Bestimmung von Gesamteiweiss im Liquor

661

cerebrospinalis mit der Biuretmethode im ultravioletten Bereich, *Schweiz. med. Wschr.*, **104**, 1720 (1974).

3. Ammon, J. and Richterich, R., Die Ermittlung, von Normalwerten der Konzentration von Glukose, Protein und Zellen im Liquor des Kindes, *Schweiz. med. Wschr.*, **100**, 1317 (1970).
4. Krause, H. D. and Wisser, H., Normalbereich des Gesamteiweisses und der Eiweissfraktionen des Liquor cerebrospinalis bei Kindern, *Z. klin. Chem. klin. Biochem.*, **13**, 137 (1975).
5. Liappis, N. and Jäkel, A., Normalbereich der mittels radialer Immundiffusion bestimmten Albumin- und IgG-Konzentration im Liquor cerebrospinalis von Kindern, *Klin. Pädiat.*, **188**, 267 (1976).
6. Glasner, H., Barrier impairment and immune reaction in the cerebrospinal fluid, *Eur. Neurol.*, **13**, 304 (1975).
7. Siemes, H., Siegert, M., and Rating, D., Das Liquorproteinprofil normaler Kinder und seine Abhängigkeit vom Lebensalter. Untersuchungen mittels CAF- und Agarosegel-Elektrophorese, *Neuropädiatrie*, **6**, 383 (1975).
8. Maurer, C., Schnellbestimmung von Gesamtprotein im Liquor cerebrospinalis ohne Fällung mit einer modifizierten Biuret-Methode, *Z. klin. Chem. klin. Biochem.*, **6**, 217 (1968).
9. Hohenwallner, W., Sommer, R., and Wimmer, E., Der normale Liquor: Referenzwerte von Gesamteiweiss, Eiwesselektrophorese, Glucose und Chlorid, *12. Seminar der Österreichischen Gesellschaft für klinische Chemie, Linz*, 1977.

4.9.3. CEREBROSPINAL FLUID GLUCOSE

4.9.3.1. Choice of method

In principle the concentration of glucose in cerebrospinal fluid may be determined by any of the methods described in Section 3.9, but the enzymatic techniques using glucose oxidase/peroxidase or hexokinase/glucose-6-phosphate dehydrogenase (p. 370) have proved the best. In both cases the determination is carried out directly on the sample, i.e. without prior deproteination. Cerebrospinal fluid (20 μl) is used in place of plasma (20 μl).

4.9.3.2. Reference values

Reliable, recent studies on the glucose concentration in cerebrospinal fluid are scarce. The lower limit of the norm is particularly significant in connection with the differential diagnosis of meningitis. Values of between 40 and 70 mg per 100 ml are to be found in the literature,[1] but these are doubtless too high. More recent investigations using the glucose oxidase/peroxidase method[1] and polarography[2] gave the results listed in Table 129. The glucose concentration is normally distributed and is practically independent of age. The lower limit of the norm was found to be 30 mg per 100 ml. The much higher values found in earlier studies are probably attributable to residual reduction, which has been little investigated in cerebrospinal fluid but is nonetheless obviously considerable.

Table 129. Reference values for the glucose concentration (mg per 100 ml) in children of various ages (it should be noted that the composition of the cerebrospinal fluid can vary slightly according to its origin—lumbar, sub-occipital, ventricular).

Age	Sex	Number	$\bar{x}$	$\bar{x} \pm 2s$	References
1st year					
1 month		9	58.8	29.8–87.8	1
2 months		8	65.5	30.7–100.3	
3 months		8	59.8	29.8–89.9	
4 months		9	51.3	32.5–70.1	
5–6 months		25	60.7	32.5–88.9	
7–12 months		54	64.0	29.2–98.8	
2nd–16th year					
Healthy		238	62.0	33.8–90.2	
Retarded		56	54.5	30.9–78.1	
Paralysed		41	54.6	31.2–78.0	
Total		335	57.0	32.0–82.0	
				$\bar{x} \pm 1s$	
1–12 months	m			57.0 ± 13.7	2
	f			57.0 ± 10.4	
1–14 years	m			63.0 ± 12.5	
	f			61.5 ± 12.2	

4.9.3.3. Diagnostic significance

Pathology
Elevation
1. Diabetic hyperglycaemia
2. Epidemic encephalitis
3. Syphilis of the central nervous system (slight increase)
Depression
1. Meningitis
  1.1. Bacterial
      1.1.1. purulent
      1.1.2. tuberculous
  1.2. Fungal
2. Other diseases with pleocytosis (e.g. syphilitic meningitis)
3. Metastases in the meninges, rare:
  3.1. Carcinoma
  3.2. Sarcoma
  3.3. Lymphoma
4. Hypoglycaemia (see blood-glucose)

Note. The glucose concentration of cerebrospinal fluid may be normal both in viral infections and with tuberculous meningitis.

References

1. Ammon, J. and Richterich, R., Die Ermittlung von Normalwerten der Konzentration von Glucose, Protein und Zellen im Liquor des Kindes, *Schweiz. med. Wschr.*, **100**, 1317 (1970).
2. Liappis, N., Polarographische Bestimmung der Glukose-Konzentration im Liquor des Kindes, *Klin. Pädiat.*, **188**, 51 (1976).

4.10. MALE GENITAL ORGANS

4.10.1 PROSTATIC ACID PHOSPHATASE: INHIBITION BY TARTRATE

4.10.1.1. Introduction

Acid phosphatase consists of a group of enzymes which cleave phosphate esters at acidic pH. In humans they have been detected in the following locations:[1] prostata, pancreas, spleen, testicles, kidneys, lungs, liver, bones, intestines, suprarenal capsules, skeletal muscle, ovaries, thymus, cardiac muscle, parathyroid glands, uterus, gall bladder, skin, leucocytes, thrombocytes, and bone marrow. The enzymes in the tissues are heterogeneous: 1–5 isoenzymes have been detected by electrophoresis on polyacrylamide gel.[1] At the sub-cellular level, except in the erythrocytes, the acid phosphatases are located together with other hydrolases in the lysosome membranes. The natural substrate is unknown. The acid hydrolases are involved in the degradation of cell enzymes and proteins.

Several acid phosphatases occur in human plasma; they originate from different types of cell but they are not always detectable:[2]

Thrombocyte phosphatase. The major part of the phosphatases derives from the thrombocytes. It is released into the plasma during coagulation of the blood. The preferred substrate for this enzyme is β-glycerophosphate but it also cleaves p-nitrophenylphosphate and α-naphthylphosphate.[3,4] The enzyme is inhibited by formaldehyde but not by L-tartrate.

Erythrocyte phosphatase. The enzyme activity in the erythrocytes is relatively small but more is present in the reticulocytes and erythroblasts. There are three known enzymes giving a genetic polymorphism with six phenotypes.[5] In non-haemolytic plasma the role of the erythrocytes as a source of enzyme is smaller than was formerly assumed. The erythrocyte enzyme is distinguished by its inefficiency in cleaving α-naphthylphosphate. It can be inhibited by formaldehyde but not by L-tartrate.

Prostate phosphatase. It is possible that this enzyme does not occur in the plasma of healthy persons but only appears if a prostate carcinoma is present. The prostate phosphatase can be selectively inhibited by L-tartrate.

Bone phosphatase. It is not uncommon for an increase in acid phosphatase to be observed in diseases of the bone, particularly in meta-

stasizing cancer of the breast, Paget's disease, and hyperparathyroidism. The enzyme possibly originates in the bones (osteoclasts?).
The preferred substrates for this enzyme are phenylphosphate and
α-naphthylphosphate. It is not inhibited by L-tartrate.

'*Gaucher*' *phosphatase*. This enzyme is present in the lysosomes of the
Gaucher cells. It is also found in the reticulocytes in this storage disease.
The origin of the acid phosphatase occurring in this lipidosis is not clear
(spleen?). The enzyme is not inhibited by L-tartrate.

4.10.1.2. Choice of method

Gutman and Gutman[6] introduced the determination of the acid
phosphatase in 1938 as an aid in the diagnosis of prostate carcinoma.
However, the determination of this total phosphatase allowed a sure
diagnosis in only about 50% of all cases. Consequently, the development of
a method for the specific determination of prostate phosphatase by
Fishman and Lerner[7] in 1953 was a considerable advance. Prostate
phosphatase is selectively inhibited by means of L(+)-tartrate. The amount
of the prostate phosphatase may be calculated from the difference
between the total and the tartrate-inhibited phosphatase. According to
these authors' results, about 90% of all patients with prostate carcinoma
are chemically identifiable using the improved method. Unfortunately, the
stated method is time consuming, complicated, and affords poor
reproducibility. Attempts to simplify the method have been unsuccessful.

In principle, all of the methods which have been described for the
determination of the alkaline serum phosphatase can also be applied to the
measurement of the total acid phosphatase. However, evidence that the
prostate phosphatase can be selectively determined with these methods,
using tartrate inhibition, has, as yet, only been furnished for the following
techniques:

Method of Fishman and Lerner.[7] Incubation of the serum with
phenylphosphate (±-tartrate). Precipitation of the proteins. Determination
of the liberated phenol with the Folin–Ciocalteu reagent.

Method of King and Jegatheesan.[8] Incubation of the serum with
phenylphosphate (±-tartrate). Determination of the liberated phenol with
4-aminoantipyrine.

Methods of Jacobsson[9] *and Richterich et al.*[10]

α-Glycerophosphate and phenolphthalein phosphate are not suitable as
substrates in conjunction with the tartrate inhibition.

α-Naphthylphosphate has also been recommended for the determination
of the prostatic serum phosphatase.[11] Since this substrate is cleaved
adequately by the prostate enzyme but hardly at all by the erythrocyte
phosphatase, the authors believe that this may be applied to the selective
determination of the prostate fraction without using tartrate inhibition.
However, they pay no attention to the fact that the major part of the

serum enzyme does not come from the erythrocytes but from the thrombocytes and that the thrombocyte enzymes cleave this substrate.

An obvious candidate for use as substrate for the determination of prostate phosphatase is p-nitrophenylphosphate, which proved excellent for the determination of alkaline phosphatase. However, the first methods to be developed with this substrate were inadequately researched. Further experiments have been discontinued, probably on account of Fishman and Lerner's advice[7] that p-nitrophenylphosphate is unsuitable for the determination of the L-tartrate-inhibited phosphatase. Jacobson[9] and Richterich et al.[10] have proved that this substrate is, in fact, very well suited for the determination of prostate phosphatase.

4.10.1.3. Principle

Where possible plasma should be used instead of serum, since this avoids liberation of the interfering thrombocyte phosphatase. For the determination, 20 μl of plasma or serum are incubated for 30 min at 37 °C with p-nitrophenylphosphate as substrate, at pH 4.9. Under the action of the acid phosphatase the substrate is cleaved to p-nitrophenol and phosphate. The reaction is interrupted by the addition of sodium hydroxide. As a result of rendering the solution alkaline, the p-nitrophenol, which is colourless at pH 4.9, turns deep yellow (indicator!). The intensity of the yellow colour, measured at 400 nm, is directly proportional to the enzyme activity. In a parallel assay the prostatic acid phosphatase is inhibited by the addition of L(+)-tartrate. The difference between the two analyses (total phosphatase and non-prostate phosphatase) corresponds to the activity of the enzyme deriving from the prostata.

4.10.1.4. Reagents

Sodium hydroxide solution, 0.1 *mol* l^{-1}.
Sodium hydroxide solution, 1 *mol* l^{-1}.
Standard p-*nitrophenol solution*, 0.2 *mmol* l^{-1}. Dissolve 27.8 mg of p-nitrophenol in DM-water and make up to 1000 ml. The solution keeps indefinitely at 4 °C or in a frozen condition.
p-*Nitrophenylphosphate solution*, 15.2 *mmol* l^{-1}. Dissolve 56.4 mg of 4-nitrophenylphosphate (disodium salt, hexahydrate) in DM-water and make up to 10 ml. The solution keeps indefinitely if frozen in a dark bottle, or for at least 1 week at 4 °C. A slight yellow colour does not interfere with the reaction (see 'Comments on the method').
Citrate buffer, 90 *mmol* l^{-1}, *pH* 4.9. Dissolve 19 g of citric acid in about 700 ml of DM-water, add 180 ml of 1 N sodium hydroxide solution, adjust the pH to 4.9, and make up to 1000 ml with DM-water. The solution keeps indefinitely if frozen, or for a few weeks at 4 °C.
Citrate–tartrate buffer: citrate 90 *mmol* l^{-1}; *tartrate* 40 *mmol* l^{-1}, *pH* 4.9.

Dissolve 19 g of citric acid and 6 g of L(+)-tartaric acid in about 700 ml of DM-water, add 180 ml of 1 M sodium hydroxide solution, adjust the pH to exactly 4.9, and make up to 1000 ml with DM-water. The solution keeps indefinitely if frozen or for a few weeks at 4 °C.

ACD solution. Contains 2.5 g of glucose, 0.47 g of citric acid and 1.6 g of sodium citrate per 100 ml (p. 103).

Acetic acid, $1.75 \, mol \, l^{-1}$. Dilute 10 ml of glacial acetic acid to 100 ml with DM-water.

Working solutions.

Substrate–citrate solution. Mix equal volumes of citrate buffer and *p*-nitrophenylphosphate solution. Prepare freshly each day.

Substrate–tartrate solution. Mix equal volumes of citrate–tartrate buffer and *p*-nitrophenylphosphate solution. Prepare freshly each day.

Stabilization of the enzyme. There are two possible ways of stabilizing the acid phosphatase:

1. *Using plasma*. The ACD tubes containing 1 ml of ACD solution are weighed (W_1). About 5 ml of blood are added to each tube and the contents mixed. The tubes are then weighed again (W_2). Each tube is now centrifuged and the plasma removed. The plasma can be stored for up to 3 days at 4 °C, or for 1 day at room temperature, without any change in the enzyme activity. In order to correct for the dilution the results must be multiplied by the following dilution factor:

$$\text{Dilution factor} = \frac{W_2 - W_1 + 1}{W_2 - W_1} \quad \text{(result stated in grams).}$$

2. *Using serum*. Exactly 20 μl of 10% acetic acid are added to 1 ml of serum. The pH falls to about 5.5 in the process, thus stabilizing the enzyme which can be stored for up to 3 days at 4 °C, for 1 day at room temperature and for several months at −20 °C. The enzyme can also be stabilized with sodium hydrogen citrate or by $NaHSO_4 \cdot H_2O$ (5 mg ml^{-1} serum).

4.10.1.5. Measurement

Conditions. The final concentrations in the reaction mixture are as follows:

Total phosphatase	
Plasma, serum	0.02 ml
p-Nitrophenylphosphate	7.6 mmol l^{-1}
Citrate buffer, pH 4.9	45 mmol l^{-1}

Duration of incubation: 30 min
Temperature: 37 °C

Non-prostate phosphatase

Plasma, serum	0.02 ml
p-Nitrophenylphosphate	7.6 mmol l^{-1}
Citrate buffer, pH 4.9	45 mmol l^{-1}
L(+)-Tartrate	20 mmol l^{-1}

Duration of incubation: 30 min
Temperature: 37 °C

Procedure.

(Test-tubes)	Total phosphatase, ml	Reagent blank I, ml	Non-prostate phosphatase, ml	Reagent blank II, ml	Sample blank, ml	Standard, ml
Substrate–citrate solution	0.1	0.1	—	—	—	—
Substrate–tartrate solution	—	—	0.1	0.1	—	—
DM-water	—	—	—	—	0.1	0.1
Pre-heat at 37 °C for 5 min						
Sample	0.02	—	0.02	—	0.02	—
DM-water	—	0.02	—	0.02	—	—
Standard solution	—	—	—	—	—	0.02
Incubate at 30 °C for exactly 30 min						
Sodium hydroxide solution, 0.1 mol l^{-1}	0.5	0.5	0.5	0.5	0.5	0.5

Read the extinction at 400 nm (Hg 405 nm) against water

4.10.1.6. Calculation

Using the standard.
Total phosphatase:

$$U \; l^{-1} = \frac{E(TP) - E(SB) - E(RBI)}{E(ST)} \cdot 0.0002 \cdot 10^6 \cdot \frac{1}{30}$$
$$\cdot \text{ dilution factor } \mu\text{mol min}^{-1} \, l^{-1}$$

$$U \; l^{-1} = \frac{E(TP) - E(SB) - E(RBI)}{E(ST)} \cdot 6.67 \cdot \text{ dilution factor } \mu\text{mol min}^{-1} \, l^{-1}$$

Non-prostate phosphatase:

$$U \; l^{-1} = \frac{E(NPP) - E(SB) - E(RBII)}{E(ST)}$$

$$\cdot 6.67 \cdot \text{ dilution factor } \mu\text{mol min}^{-1} \, l^{-1}$$

Prostatic phosphatase = total phosphatase − non-prostate phosphatase

Using the molar extinction coefficient: $\varepsilon_{405\,nm} = 186\,000\; l\; mol^{-1}\; cm^{-1}$.

Total phosphatase:

$$U\;l^{-1} = \frac{E(TP) - E(SB) - E(RBI)}{186\,000 \cdot 1} \cdot 10^6 \cdot \frac{0.62}{0.02} \cdot \frac{1}{30}$$

$$\cdot \text{dilution factor } \mu mol\; min^{-1}\; l^{-1}$$

$$U\;l^{-1} = [E(TP) - E(SB) - E(RBI)] \cdot 55.6 \cdot \text{dilution factor}$$
$$\mu mol\; min^{-1}\; l^{-1}$$

Non-prostate phosphatase:

$$U\;l^{-1} = [E(TP) - E(SB) - E(RBII)] \cdot 55.6 \cdot \text{dilution factor}$$
$$\mu mol\; min^{-1}\; l^{-1}$$

Prostatic phosphatase = total phosphatase − non-prostate phosphatase

4.10.1.7. Specificity

Despite improvements, the specificity of the method leaves much to be desired. Clearly it does not determine solely the prostate fraction in the serum but responds to small fractions of other acid phosphatases which are also measurable in female subjects.

4.10.1.8. Reference values

Blood donors.[10] Reference values for the acid serum and plasma phosphatases are given in Table 130. Each value given represents $\bar{x} \pm 2s$ for 50 duplicate determinations (U l^{-1}).

Urological patients.[12] Since prostatic hypertrophy and, possibly, increasing age, also lead to a slight increase in the enzyme, the upper limit of the reference values should be set higher in these patients (Table 131).

4.10.1.9. Comments on the method

1. In each series of analyses a standard and a reagent blank are included at the beginning and the end.

Table 130. Reference values for the acid phosphatase.

Collective	Sex	Total phosphatase	Non-prostate phosphatase	Prostate phosphatase
Serum	m	9.15 ± 4.40	7.40 ± 3.40	1.84 ± 1.80
	f	8.00 ± 2.94	6.16 ± 2.28	1.85 ± 1.58
Plasma	m	5.08 ± 3.54	4.76 ± 2.88	0.32 ± 0.64
	f	4.41 ± 1.60	4.10 ± 1.64	0.33 ± 0.60

Table 131. Reference values.

		Number	$\bar{x}$	$\bar{x} \pm 2s$
Total phosphatase				
(plasma)	Blood donors	45	4.2	1.5–8.6
	Prostatic hypertrophy	115	6.5	3.2–11.8
	Prostatic carcinoma			
	local	21	8.5	3.0–26.0
	metastasic	6	26.0	
Prostatic phosphatase				
(plasma)	Blood donors	45	0.2	0.3–1.0
	Prostatic hypertrophy	115	0.8	0.2–1.6
	Prostatic carcinoma			
	local	21	2.0	0.3–11.0
	metastasic	6	12.0	

2. It is advisable to make all the readings against water. If the extinction of the reagent blank is greater than 0.1 a fresh substrate solution should be prepared.

3. If the total phosphatase activity is more than $50\ U\ l^{-1}$, the determination should be repeated with a 10-min incubation and the result multiplied by 3.

4. If the determination of the total phosphatase is all that is required (in the confirmatory diagnosis of a prostatic carcinoma), the non-prostate phosphatase assay is omitted.

5. A sample blank is always necessary.

6. Haemolytic sera cannot be used.

7. Ammonium heparinate should not be used as anticoagulant since the pH becomes alkaline and the enzyme is inactivated. In addition, turbidity is observed on adding the plasma to the citrate buffer, although this clears on the addition of NaOH. Modder[3] maintains that heparin plasma acid phosphatase values which are too low are due to a partial precipitation of the enzyme by the heparin.

8. The K_m of acid phosphatase is 7.3×10^{-5}, as against the K_i for L-tartrate of 4×10^{-4}. The percentage inhibition of the acid phosphatase may be calculated from the following equation, which describes the competitive inhibition:

$$V_1 = \frac{V(S)K_i}{K_m K_i + K_m(I) + K_i(S)}$$

The substrate concentration $(S) = 6.3 \times 10^{-3}$. If the maximum rate, V, is set at 100%, then an actual rate, V_1, of 1.7% is calculated. Under the experimental conditions described above, the inhibition of the prostatic acid phosphatase consequently comes to more than 98%.

9. Large amounts of acid phosphatase are released from the

thrombocytes in the coagulation of the blood. This explains the significantly higher activities in serum in comparison with those in plasma.

10. In the instructions given above provision is made for reagent blanks for both total phosphatase and non-prostate phosphatase assays. If the extinctions of these blanks are identical, then one will suffice for the remaining analyses of the series. However, we have occasionally observed differences; these are obviously caused by the L(+)-tartrate used for preparing the solution.

4.10.1.10. Diagnostic significance

An elevation of the acid phosphatase activity in the plasma is found in the following cases.

Diseases of the prostate gland. Elevated values of the prostatic phosphatase are found in about 65–78% of all prostate carcinomas, in about 85–95% of all prostate carcinomas with skeletal metastasis, in 50–70% of all prostate carcinomas without radiographically detectable skeletal metastases, and in 5–10% of all prostate carcinomas where there is no clinical suspicion of metastasis.[12,13]

A normal acid phosphatase does not exclude a malignoma of the prostata. The effects of therapy on prostatic carcinoma can be reliably followed by determination of the acid phosphatase. On response to treatment the concentration of the enzyme in the plasma rapidly falls to its normal value and rises to pathological values again if there is a relapse. Anti-androgenic treatment likewise reduces the enzyme activity.

Following rectal investigation or prostatic massage the activity of the acid phosphatase increases but returns to its initial value after 24 h.

Osteopathia. Individual patients with diseases of the bones may not only show abnormally high alkaline phosphatase values but the acid phosphatase may also be elevated. The fact that this was already observed at an early stage, particularly in metastasizing cancer of the breast, excluded a prostatic origin from the outset. The enzyme derives from the bones, probably from the osteoclasts. In addition, elevated values have been observed in the following diseases of the bones: Piaget's disease, hyperparathyroidism, primary bone tumours, bone metastases, osteopetrosis, osteogenesis imperfecta, Albright's disease, and renal osteopathia.[13]

Gaucher's disease. In this sphyngolipidosis glucocerebrosides (ceramide glucose) are accumulated. An elevated phosphatase activity is found in the plasma.

Thrombocyteopathia. Thrombocytopenias can be divided into two large groups: those in which the formation of the thrombocytes is disturbed and those in which there is an increased destruction of thrombocytes. In the latter case there is a significant increase in the acid phosphatase in the plasma, and in the former the enzyme remains normal.

672

Thrombo-embolic diseases. In the viscous metamorphosis of the thrombocytes which initiates the coagulation of the blood, acid phosphatase is liberated. Consequently, thromboses and embolisms can give rise to a small increase, particularly with pulmonary embolism, in myocardial infarction and with peripheral venous and arterial thromboses. In these processes the enzyme partly derives from the erythrocytes.

Diseases of the blood. A slight elevation of the enzyme activity has been established in the following diseases:[13] chronic and acute myeloid leukaemia, myeloid metaplasia, myeloma, chronic lymphatic leukaemia, polycythaemia vera, and megaloblastic anaemia.

References

1. Lam, K. W., Li, O., Li, C. Y., and Yam, L. T., Biochemical properties of human prostatic acid phosphatase, *Clin. Chem.,* **19**, 483 (1973).
2. Richterich, R., *Enzympathologie. Enzyme in Klinik und Forschung*, Springer, Berlin, 1958.
3. Modder, C. P., Investigation on acid phosphatase activity in human plasma and serum. *Clin. chim. Acta*, **43**, 205 (1973).
4. Bodansky, O., Acid phosphatase, *Adv. clin. Chem.,* **15**, 43 (1972).
5. Pflugshaupt, R., Scherz, R., Trautwein, M., Richiger, U., and Bütler, R., Polymorphism of the red cell acid phosphatase in the Swiss population, *Hum. Genet.,* **8**, 354 (1970).
6. Gutman, A. B. and Gutman, E. B., An 'acid' phosphatase occurring in the serum of patients with metastasizing carcinoma of the prostate gland, *J. clin. Invest.,* **17**, 473 (1938).
7. Fishman, W. H. and Lerner, F., A method for estimating serum acid phosphatase of prostatic origin, *J. biol. Chem.,* **200**, 89 (1953).
8. King, E. J. and Jegatheesan, K. A., A method for the determination of tartrate-labile, prostatic acid phosphatase in serum, *J. clin. Path.,* **12**, 85 (1959).
9. Jacobsson, K., The determination of tartrate-inhibited phosphatase in serum, *Scand. J. clin. Lab. Invest.,* **12**, 367 (1960).
10. Richterich, R., Colombo, J. P., and Weber, H., Ultramikromethoden im klinischen Laboratorium, VII. Bestimmung der sauren Prostata-Phosphatase, *Schweiz. med. Wschr.,* **92**, 1496 (1962).
11. Babson, A. L. and Read, P. A., A new assay for prostatic acid phosphatase in serum, *Am. J. clin. Path.,* **32**, 88 (1959).
12. Richterich, R., Weber, H., Entner, B., and Niederhäusern, W. von., Die Früherfassung des Prostatacarcinoms durch Bestimmung der sauren Prostate-Phosphatase im Plasma, *Dt. med. Wschr.,* **88**, 1421 (1963).
13. Yam, L. T., Clinical significance of the human acid phosphatases, *Am. J. Med.,* **56**, 604 (1974).

4.11. KIDNEYS

4.11.1. INTRODUCTION

The primary tasks of the kidneys are the regulation of the body water and the electrolyte balance, the maintenance of the acid–base equilibrium, the elimination of metabolic wastes, and the excretion of alien substances. All of these functions involve three fundamental processes:

1. *Filtration*, i.e. the transfer of low-molecular-weight substances from the plasma into the primary urine.
2. *Reabsorption* of molecules and ions from the tubular urine into the plasma.
3. *Secretion* of molecules and ions from the plasma into the urine.

The processes are illustrated schematically in Fig. 146. Diseases of the kidney frequently occasion a reduction in the efficiency of the nephrons, a decrease in glomerular filtration and in tubular reabsorption and secretion. There may also be a restriction of the flow of blood through the kidneys. The development of the clearance concept has made it possible to make quantitative measurement of the glomerular filtration and the flow of blood through the kidneys.

4.11.2. GLOMERULAR FUNCTION: CLEARANCE

The expression 'clearance' was defined in 1928 by Van Slyke *et al.*[1] as that quantity of plasma which is 'cleared' of a particular substance by the kidneys in 1 min. It says nothing about the mechanism of the elimination of the substance which may take place by filtration and/or secretion. A few typical clearance quantities as functions of the plasma concentration are illustrated in Fig. 147. Up to concentrations of about 170 mg per 100 ml the glucose clearance is zero since glucose is reabsorbed after filtration. On the other hand, the clearance of creatinine, urea, inulin, and sodium thiosulphate is independent of the blood concentration: these substances are only filtered. The highest clearance values are observed with diodrast and *p*-aminohippurate (PAH). These substance are not only filtered but secreted. If:

U = concentration of a substance in the urine (mg ml^{-1});
V = urine volume per unit time (ml min^{-1});

674

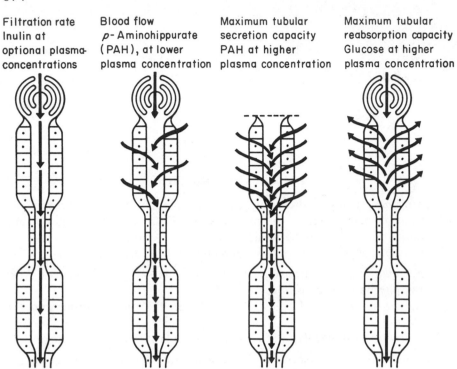

Filtration rate
Inulin at
optional plasma
concentrations

Blood flow
p-Aminohippurate
(PAH), at lower
plasma concentration

Maximum tubular
secretion capacity
PAH at higher
plasma concentration

Maximum tubular
reabsorption capacity
Glucose at higher
plasma concentration

Fig. 146. Schematic illustration of filtration, reabsorption, and secretion.

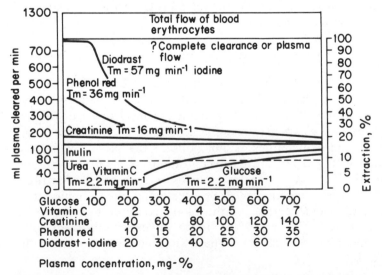

Fig. 147. Clearance of various substances as a function of the
plasma concentration (from ref. 2).

P = concentration of the substance in the plasma (mg ml^{-1});

C = clearance of this substance, i.e. that quality of plasma freed from the substance per minute;

then

$$UV = PC$$

Since U, V and P can readily be determined experimentally, the clearance of any substance can be calculated from the equation

$$C = \frac{UV}{P}$$

4.11.2.1. Glomerular filtration

A substance which is excreted exclusively by glomerular filtration must satisfy the following requirements:

1. It must be completely filterable in the glomerulus, i.e. its molecular weight must be so small that passage through the 'filter' is not hindered.
2. It should be neither secreted nor reabsorbed in the tubules.
3. It must be present in the plasma in true solution, i.e. it should not be bound to proteins which are not themselves filterable.
4. It must be pharmacologically inert, i.e. it should have no effect on the function of the kidneys.

In practice this means that the excretion of the substance must be constant and independent of the blood pressure, and that the urine volume per unit time should have no effect on the excretion.

After searching for a suitable substance for nearly 10 years, Smith,[2] in 1933, found in inulin a substance which ideally satisfied the above-mentioned conditions. Sodium thiosulphate proved to be equally safe.[3,4]

^{51}Cr-EDTA proved successful for the determination of the glomerular filtration both in animals[5] and in man.[6,7] ^{51}Cr is a γ-emitter and its compound with EDTA is relatively stable.

The determination by counting the radioactivity with a γ-counter is simple, is not time consuming, and may be carried out with high precision. The preparation is tolerated well *in vivo* and shows a good correlation with other standard clearance substances.[8,9]

4.11.2.2. Renal plasma flow (*p*-aminohippurate clearance)

The exact measurement of the flow of blood through the kidneys may be done using the Fick principle. However, this does entail the determination of the arterial and venous concentration of a test substance, i.e.

catheterization of a renal vein is necessary. It was therefore a considerable advance when Smith[2] showed that the PAH clearance was almost as large as the renal blood flow. Obviously, the kidneys are able to eliminate this substance completely, by filtration and secretion, in a single passage through the kidneys. Consequently, the disappearance of this substance is a function of the renal blood flow only. Since the PAH does not penetrate the erythrocytes, only the concentration in the plasma is determined.

$$\text{Renal plasma flow} = \frac{U_{PAH}V}{P_{PAH}}$$

Accordingly, the result corresponds not to the actual renal blood flow but to the effective renal plasma flow (ERPF), and is termed the PAH clearance.

4.11.2.3. Filtration fraction

If the PAH clearance corresponds to the glomerular blood flow, then from a simultaneous knowledge of the glomerular filtration (inulin, thiosulphate, or ^{51}Cr-EDTA clearance) it must be possible to calculate what quantity of liquid is filtered from the blood into the primary urine: this quantity, the filtration fraction (*FF*), corresponds to the ratio of the inulin and PAH clearances:[2]

$$FF = \frac{C_{inulin}}{C_{PAH}}$$

The determination of the filtration fraction is necessary for the differential diagnostic assessment of diseases of the kidney. A depression of the filtration (norm 0.2) is predominantly observed in cases of glomerular injury, whereas an elevation is predominantly indicative of vascular diseases.[3]

References

1. Möller, E., McIntosh, J. F., and Van Slyke, D. D., Studies of urea excretion; relationship between urine volume and rate of urea excretion by normal adults, *J. clin. Invest.*, **6**, 427 (1928).
2. Smith, H. N., *The kidney, structure and function in health and disease*, Oxford University Press, New York, 1951.
3. Reubi, F. C., *Clearance tests in clinical medicine*, Thomas, Springfield, 1963.
4. Gilman, A. F., Philips, S., and Koella, E. S., Metabolic reduction and nephrotoxic action of tetracyanate in relation to a possible interaction with sulfhydril compounds, *Am. J. Physiol.*, **147**, 115 (1946).
5. Stacy, B. D. and Thorburn, G. D., Chromium-51 ethylenediaminetetraacetate for estimation of glomerular filtration rate, *Science*, **152**, 1076 (1966).
6. Favre, H. R. and Wing, J., Simultaneous ^{51}Cr edetic acid, inulin and endogenous creatinin clearances in 20 patients with renal disease, *Br. med. J.*, **i**, 84 (1968).

7. Garnett, E. S., Parsons, V., and Veall, N., Measurement of glomerular filtration rate in man using a ^{51}Cr edetic acid complex, *Lancet*, **i**, 818 (1967).
8. Vorburger, C., Riedwyl, H., and Reubi, F., Vergleichende Studien zwischen den renalen Clearances von NaCr$_2$^{51}-Cr-EDTA, Inulin und Natriumthiosulfat beim Menschen, *Klin. Wschr.*, **47**, 415 (1969).
9. Stamp, T. C. B., Stacey, T. E., and Rose, G. A., Comparison of glomerular filtration rate measurement using inulin ^{51}CR-EDTA and phosphate infusion technique, *Clin. chim. Acta*, **30**, 351 (1970).

4.11.2.4. Simultaneous inulin and *p*-aminohippurate clearance

4.11.2.4.1. Choice of method

Reliable results are only obtained if the concentration of the clearance substance in the plasma is constant. If the inulin–PAH clearance cannot be performed, then a creatine clearance is recommended as next best, although this only yields conclusions on the glomerular efficiency.[1,2]

4.11.2.4.2. Principle

The subject's urine is collected over a certain test period and the urine volume per unit time (ml min^{-1}) is calculated. At the same time, the concentrations of the test substance in the plasma and the urine are determined. Two infusions are administered to maintain a constant blood level of inulin and PAH. The primary infusion (P) is for the rapid attainment of the desired blood level (inulin 40–80 mg per 100 ml, PAH 1–2 mg per 100 ml). The secondary infusion (S) serves to maintain this blood level.

4.11.2.4.3. Preconditions

The test is carried out on patients lying quietly, either in the morning, following a light breakfast, or in the afternoon at least 2 h after lunch. Sulphonamides, diuretics, and contrast agents complicate the determination and therefore they should not be administered in the last 3 days prior to the test (beware of slow-acting sulphonamides!).

4.11.2.4.4. Infusion solutions

P: primary infusion (priming). Dilute 40–50 ml of inulin (10%, pyrogen-free) and 3 ml of sodium *p*-aminohippurate (20%) to 100 ml with sterile physiological saline solution.
S: secondary infusion (sustaining). Dilute 30 ml of inulin (10%, pyrogen-free) and 5 ml of PAH (20%) to 200 ml with physiological saline solution.

4.11.2.4.5. Dosage

These dosages are reckoned for the average body surface area of adults (1.73 m²). For departures from the average, especially for children, the dosage should be raised or lowered in proportion to the body surface area (see Appendix 3). This applies to both infusions.

If there are indications of a kidney insufficiency accompanied by lowered clearance values, the dosage of the secondary infusion should be lowered in proportion to the expected reduction (only the secondary infusion is to be lowered!)

4.11.2.4.6. Clinical procedure

Catheterization of the bladder
Collection of 20 ml of bladder urine for determining the null-value of the clearance substance (U_0). After disinfecting the urethral aperture the catheter is inserted. This should lie with the tip just inside the bladder. The correct position is tested by swilling with sterile water and air; if the catheter is seated properly the air is discharged last. Clip the catheter shut and fix it with sticking plaster.

Collection of blood
Venous puncture and collection of blood (about 2 ml) for determining the null-value (P_0); add 1–2 drops of heparin to the syringe. Centrifuge immediately and take off the plasma.

Drip-feed infusion
Primary infusion. After the blood has been collected, begin the drip-feed primary infusion using the same needle. The whole volume should be fed within 10–15 min in order to achieve a rapid increase in the concentration.
Secondary infusion. Allow the first 30 ml to infuse at roughly the same rate as the primary infusion, then set the flow-rate at 1 drop s⁻¹ (4 ml min^{-1}). Wait about 20 min after beginning the infusion to allow the substance to reach an equilibrium distribution (plasma/extracellular fluid). If the kidney function is normal the concentration of the substance in the blood stays constant.

First collecting period
Empty the bladder completely and rinse several times with air and 20 ml of water. Clamp the catheter when it has been drained completely and start the stop-watch: beginning of the first urine collection period, which should last for 15 min. After 7 min from the beginning of this collection take a *blood sample* using a heparinized needle which is kept in place in the forearm: P_1. At the end of the collecting period (15 min from the beginning), empty the bladder and rinse twice with exactly 20 ml of sterile, luke-warm, distilled water and air. Collect the rinsings in the same vessel as the urine and record the volume, and the time of concluding the

emptying and rinsing of the bladder (stop-watch). Collected urine + rinsings = U_1.

Second collecting period

This should immediately follow the first period. The same procedure is used, giving U_2 and P_2.

The following data are required for the evaluation of the results: size and weight of the patient, duration of the collection periods in minutes, urine volumes per unit time in millilitres (without rinsings), and urine volumes including the rinsings in millilitres.

4.11.2.4.7. Side effects

Formerly, inulin frequently led to pyrogenic reactions but the commercial preparations now available no longer have this side-effect. In contrast, PAH can lead to severe complications in rare cases. Whereas normally the rapid infusion of PAH can engender a precordial feeling of warmth and, not uncommonly, a certain tightness similar to that following injection with calcium (which can, however, be relieved by temporarily slowing down the infusion), an over-sensitivity towards PAH can result in severe diarrhoea, skin reactions, and, occasionally, even anaphylactic shock. On the occurrence of such symptoms the infusion should be interrupted immediately and the infusion solution replaced by a saline infusion. Because of these complications the clearance should be directly supervised by a doctor.

4.11.2.4.8. Determinations and calculation

A detailed example of a clearance test is presented in Table 132. First the body surface area, BS (m^2), is evaluated from the height and weight of the patient, using the Dubois and Dubois nomogram[3] (Appendix 3) or the following equation:

$$BS \ (\text{m}^2) = \frac{(\text{weight in kg})^{0.425} \cdot (\text{height in cm})^{0.725} \cdot 71.84}{100}$$

The correction factor for the body surface area is calculated by means of the equation

$$BS \ \text{factor} = \frac{1.73}{BS}$$

in which 1.73 m^2 corresponds to the theoretical average body surface area of adults.

The actual volume of urine is obtained by subtracting the rinsing volume from the total urine volume. By dividing this figure (ml) by the length of

Table 132. Calculation of the results of an inulin and PAH clearance.

1. Patient

Name		Sex: m
Height, cm: 170	Weight, kg: 74	Age: 42
Body area, m²: 1.85		BS factor $= 1.73/BS$
		$= 0.935$

2. Urine volume per unit time $V = $ ml min^{-1}

	Period I	Period II
Length of period, min	10	11
Total urine volume, ml	63	66
Rinsing volume, ml	40	40
Urine volume, ml	23	26
Dilution factor, $VF = b/d$	2.74	2.54
Urine volume per unit time V, ml min^{-1}	2.30	2.36

3. Plasma concentrations P, mg per 100 ml

	Inulin		PAH	
	Period I	Period II	Period I	Period II
P_1, P_2	36.0	37.5	2.42	2.48
P_0	2.5	2.5	0.02	0.02
P	33.5	35.0	2.40	2.46

4. Urine concentrations U, mg per 100 ml

	Inulin		PAH	
	Period I	Period II	Period I	Period II
U_1, U_2	395	415	208	216
U_0	14	14	11	11
$U_{\text{uncorr.}}$	381	401	197	205
$U = VF \cdot U_{\text{uncorr.}}$	1044	1019	540	521

5. Clearance $C = UV/P$, ml min^{-1}

	Inulin		PAH	
	Period I	Period II	Period I	Period II
U	1044	1019	540	521
V	2.30	2.36	2.30	2.36
UV	2401	2405	1242	1230
P	33.5	35.0	2.40	2.46
$C_{\text{uncorr.}}$	71.7	68.7	517	500
$\bar{x}$ for $C_{\text{uncorr.}}$	70.2		509	
$C = BS$ factor $\cdot C_{\text{uncorr.}}$	65.6		476	

6. Filtration factor $= C_{\text{In}}/C_{\text{PAH}} = 0.138$

the collecting period (min), the urine volume per unit time (V ml min^{-1}) is obtained.

The inulin and PAH concentrations in the three urine samples, U_0, U_1, and U_2, and in the three plasma samples, P_0, P_1, and P_2, are determined in the laboratory by the methods described on pp. 385 and 471. The results are obtained in mg per 100 ml and are included in Table 132. Since the methods for the determination of inulin and PAH are not completely specific, the fasting values for urine (U_0) and plasma (P_0) are subtracted from the concentrations during the test period. For plasma this gives the true PAH concentration immediately, but for urine the dilution (with rinse liquid) still has to be taken into consideration, i.e. the results must be multiplied by the dilution factor.

The clearance is calculated by means of the equations

$$C_{In} = \frac{U_{In}V}{P_{In}}$$

and

$$C_{PAH} = \frac{U_{PAH}V}{P_{PAH}}$$

The clearance values are calculated separately for each test period, for both inulin and PAH, and the arithmetic mean of the individual periods is subsequently calculated. All that is necessary now is to correct the obtained results for the body surface area, by multiplying by the *BS* factor.

In the present example the inulin clearance, a measure of the glomerular filtration, amounts to 66 ml min^{-1}, i.e. it is significantly low. The same applies to the PAH clearance, 476 ml min^{-1}, which serves as a measure for the blood flow through the kidneys.

4.11.2.4.9. Reference values

The reference values for glomerular filtration, effective plasma flow (PAH), and filtration fraction (*FF*), are based on the investigations of Reubi.[4] In a study on 5000 persons, sodium thiosulphate was mostly used as test substance in place of inulin. With the same experimental conditions the results are identical (Table 133).[4]

Table 133. Reference values for tests of kidney function.

	Men and women $\bar{x}$; 2s range
1. Glomerular filtrate (C-sodium thiosulphate), ml min^{-1} per 1.73 m^2	126; 93–159
2. Plasma flow (C-PAH), ml min^{-1} per 1.73 m^2	646; 478–814
3. Filtration fraction sodium thiosulphate/C-PAH	0.195; 0.157–0.233

No consideration was given to age or sex. The clearance values in women are obviously 10–15% lower than in men.[2]

4.11.2.4.10. Comments on the method

1. The reduction in glomerular filtration has been stated as 1 ml per year of life.[5]

2. As a test substance, sodium thiosulphate has the advantages of being very cheap and easy to determine. The draw back is that there is a fair chance of incompatibility at high plasma concentrations. Differences between thiosulphate and inulin clearance have been observed at low plasma concentrations (<15 mg per 100 ml).[4] For this reason plasma concentrations of 20–40 mg per 100 ml are the most suitable.

3. Comparison of inulin and thiosulphate clearance with ^{51}Cr-EDTA clearance has shown that the results are practically identical. A correction factor of 1.073 has to be applied in order to calculate the actual glomerular filtration. No interference has been observed with drugs, glucose, or mannitol.

References

1. Reubi, F. C., *Clearance tests in clinical medicine*, Thomas, Springfield, 1963.
2. Goldring, W. and Chasis, H., *Hypertension and hypertensive disease*, Commonwealth Fund, New York, 1944.
3. Dubois, E. F. and Dubois, D., Clinical calorimetry. A formula to estimate the approximate surface area if height and weight be known, *Archs intern. Med.*, **17**, 863 (1921).
4. Reubi, F. C., *Nierenkrankheiten*, 2. Aufl., Huber, Berne, 1970.
5. Watkin, D. M. and Schock, N. W., Agewise standard value for C_{In}, C_{PAH}, and Tm_{PAH} in adult males, *J. clin. Invest.*, **34**, 969 (1955).
6. Vorburger, C., Riedwyl, H., and Reubi, F. C., Vergleichende Studien zwischen den renalen Clearances von NaCr$_2$^{51}-Cr-EDTA, Inulin und Natriumthiosulfat beim Menschen, *Klin. Wschr.*, **47**, 415 (1969).

4.11.3. CREATININE METABOLISM

Figure 148 illustrates the most important biochemical and physiological processes in the metabolism of creatine and creatinine. Premuscular, intramuscular, postmuscular, and renal stages may be distinguished. To simplify the picture the various processes may be described as follows:

1. Creatine is synthesized from arginine and glycine in various organs, including the pancreas, liver, and kidneys, via several intermediate steps. This creatine passes into the plasma and thence, in part, it reaches the kidneys.
2. In adults the creatine filtered by the kidneys is almost completely reabsorbed. One of the transport mechanisms involved in this process is the same as that in the reabsorbtion of amino acids.

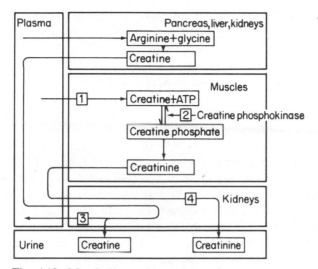

Fig. 148. Metabolism of creatinine (for description, see text).

3. The creatine is actively taken up from the plasma by muscle cells[1] and phosphorylated by the action of creatine kinase[2], producing creatine phosphate intramuscularly. The latter substance acts as an energy storer in a manner analogous to that of adenosine triphosphate (ATP). In muscular contraction the chemical energy of the creatine phosphate is converted into mechanical energy while creatine, phosphate, and adenosine diphosphate (ADP) are formed. If sufficient ATP is available the creatine can be phosphorylated again.

4. However, creatine phosphate also undergoes spontaneous decomposition to creatinine. As yet there is no enzyme known which catalyses the decomposition of creatine phosphate to creatinine and the metabolic decomposition obeys the same laws as the decomposition *in vitro*. About 1.5–2% of the body creatine is converted to creatinine daily.[1]

5. This creatinine, produced by spontaneous decomposition, reaches the plasma and is quantitatively excreted into the urine by filtration. There is probably even secretion of creatinine at high plasma concentrations; reverse reabsorbtion of creatinine is unknown.

The concentration of creatine in the plasma depends on its rate of formation, its uptake into the muscles, its storage in the muscles, and its reabsorption in the kidneys. Under physiologically normal conditions its concentration in the plasma is so low that, owing to technical limitations, it is scarcely possible to establish a lower limit. However, an elevation is observed when the muscular mass is decreased (e.g. in muscular atrophy, muscular dystrophy, or following amputations). Practically no creatinuria is

684

observed in adults but the condition can occur if the T_m, i.e. the reabsorbtion capacity of the renal tubules, is exceeded,[3] either through a disturbance in the reabsorbtion of the creatine or through an excessive demand for creatine. Such a demand is observed during the growth phase in children, in cases of reduced muscular mass (e.g. muscular dystrophy, muscular atrophy), or rapid muscular degradation (e.g. involution of the uterus during puberty). Hormonal inhibition of the reabsorption mechanism has been postulated in cases of creatinuria during pregnancy and in various endocrinepathia. Clinically there is only one indication for the measurement of creatinuria: as a measure for atrophia and regeneration of the muscular mass in myopathias. With all other conditions the measurement of the plasma and urinary creatine concentration and excretion is so unspecific that no clinically useful conclusions may be drawn from the results. The creatine concentration in the plasma is primarily a function of the glomerular filtration and the plasma concentration may serve as a rough measure for this parameter of the kidney function. The creatinine excretion in the urine may be reduced either if the kidney function is very highly curtailed or, alternatively, if the muscular mass is very much reduced, i.e. the decomposition of creatine phosphate into creatinine no longer occurs.[4] With reduced glomerular filtration and elevated concentration of the creatinine in the plasma, part of the creatinine is obviously excreted into the gut and metabolized there by bacterial flora.[2]

References

1. Crim, M. C., Calloway, D. H., and Margen, S., Creatine metabolism in men: urinary creatine and creatinine excretions with creatine feeding, *J. Nutr.*, **105**, 428 (1975).
2. Jones, J. D. and Burnett, P. C., Creatinine metabolism in humans with decreased renal function: creatinine deficit. *Clin. Chem.*, **20**, 1204 (1974).

4.11.3.1. Creatinine

4.11.3.1.1. Choice of method

Up to now the determination of creatinine in plasma and urine has been unsatisfactory. Many methods are so unspecific that the true creatinine cannot be unequivocally determined.

Creatinine

After Jaffé,[1] in 1886, had shown that creatinine reacts with picric acid in alkaline media to form an orange dye having an extinction maximum at

490 nm, Folin[2] carried out the first determinations of creatinine in urine. It has transpired that the Jaffé reaction is influenced by more than 50 different chromogenic substances. Many authors have concerned themselves with the mechanism of the Jaffé reaction. In 1974, Blass et al.[3] suggested the following structural formula for the compound of picric acid and creatinine formed in the Jaffé reaction:

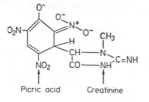

Vasiliades[31] showed that in the Jaffé reaction alkaline sodium picrate and creatinine form a 1:1 complex with an absorption maximum at 480 nm and that the kinetics of the reaction are 1st order.

The classical Jaffé reaction and its numerous modifications[2,4-11] respond not only to creatinine but also to other chromogens. The sum of the components determined in this way is therefore also termed the 'total creatinine' or, better, 'total chromogens'.

Since the various non-creatinine chromogens react with picric acid, their interference is greatest at low plasma creatinine concentrations (1 mg per 100 ml, 88.4 μmol l^{-1}) and is less important at high creatinine concentrations.

The most important non-creatinine chromogens in the plasma are acetoacetic acid, acetone, pyruvate, glucose, and ascorbic acid. Haemolysis may bring about a release of Jaffé-positive substances. The presence of these substances in the plasma causes the results to be too high. These interferences can be overcome in various ways:

1. By adsorption of the creatinine on Lloyd's reagent (Fuller's earth), an aluminosilicate, it is possible to separate off the greatest part of the chromogens.[12-15] This improves the specificity of the reaction.
2. The use of ion-exchange resins[16-19] permits a still more selective adsorption of the creatinine, eliminating the unspecific chromogens. These techniques yield results which are, for the most part, correct but they are technically relatively involved.
3. Dialysis (Technicon AutoAnalyser[7,8,20-23] affords creatinine results lower than those obtained by the simple methods, but it is worth questioning whether these low results are actually attributable to a higher specificity. Certainly the dialysis step has a certain selective action but it is not easy to see why the unspecific substances (predominantly of low molecular weight) should behave differently to creatinine on dialysis. In diabetic subjects it has been observed that the low-molecular-weight acetoacetic acid, above all, can simulate results

which are much too high, even with the AutoAnalyzer technique.[21] The lower results ought probably to be attributed to a relatively slight dialysis of plasma samples in comparison with the aqueous standard solutions customarily used. This has induced individual investigators to carry out a preliminary isolation using ion-exchange resins before submitting the samples to the AutoAnalyzer.[23]

4. The measurement may be made kinetically if consideration is given to the differential reaction rates of creatinine and the non-creatinine chromogens in the Jaffé reaction.[24–31] In such measurements the time of measurement must be observed precisely, and this means that they are better performed with the aid of automated analysis instruments than manually. This technique obviously reduces the effect of the chromogenic substances to a minimum.

5. Miller and Dubos[32,33] attempted to circumvent interference from non-specific chromogens by using enzymatic methods. They determined the plasma creatinine by the Jaffé reaction, before and after treatment with creatininase. After the preparation of a pure bacterial creatininase (creatine amidohydrolase) it was possible to develop a fully enzymatic test for the determination of the creatinine.[34–37]

The course of the reaction in the enzymatic measurement of creatinine is as follows:

$$\text{Creatinine} + H_2O \xrightarrow[\text{amidohydrolase}]{\text{Creatinine}} \text{Creatine}$$

$$\text{Creatine} + \text{ATP} \xrightleftharpoons[\text{kinase}]{\text{Creatine}} \text{Creatine phosphate} + \text{ADP}$$

$$\text{ADP} + \text{Phosphoenolpyruvate} \xrightleftharpoons[\text{kinase}]{\text{Pyruvate}} \text{ATP} + \text{Pyruvate}$$

$$\text{Pyruvate} + \text{NADH} + H^+ \xrightleftharpoons[\text{dehydrogenase}]{\text{Lactate}} \text{Lactate} + \text{NAD}^+$$

This reaction is specific for creatinine.

Heparin, citrate and ascorbic acid up to 100 mg per 100 ml and glucose up to 300 mg per 100 ml do not interfere with the enzymatic reaction. Oxalate and EDTA inhibit the reaction.[36]

6. Attempts are continually being made to replace the Jaffé principle by other reactions, for example the reaction of creatinine with 3,5-dinitrobenzoic acid[38] and the use of ion-selective electrodes.[39] The methods are technically involved and have not yet won recognition.

A summary of the methods for the determination of creatinine is presented in Table 134.

In view of the technical difficulties of all of the methods of isolation and

Table 134. Methods for the determination of creatinine.

	References
1. *Jaffé reaction*	
1.1. Colorimetric	2.4–8.20–23
End-point with deproteination or dialysis	
1.2. Colorimetric	9–11
End-point without deproteination	
1.3. Colorimetric	12–15
Adsorption on Fuller's earth	
1.4. Ion exchangers	16–19
Isolation of the creatinine from other	
Jaffé chromogens followed by colorimetric determination	
1.5. Kinetic	24–31
1.6. Colorimetric before and after enzymatic hydrolysis	32, 33, 40
2. *Enzymatic determination*	34–37
3. *Ion-selective electrodes*	
Creatinine $\xrightarrow{\text{Creatinase}}$ *N*-Methylhydantoin +	
Ammonia (measured)	39
4. *Reaction of creatinine with 3,5-dinitrobenzoyl chloride*	38

the complexity of the other techniques, the sole recourse is to improve the specificity of the Jaffé reaction by chemical means. This should be possible largely through choice of optimal reaction conditions. Particular attention should be paid to the concentration of the picric acid and the sodium hydroxide. The development of the colour is slower for the pseudo-creatinines than for creatinine, so that a precise observation of the time of measurement has a definite effect on the specificity of the reaction.[11,24–31]

4.11.3.2. Determination of creatinine in serum, plasma, and urine by a modified Jaffé method[11] without deproteination

4.11.3.2.1. Principle

This is a simple, well documented procedure for the kinetic determination of creatinine by means of alkaline picrate solution, without deproteination.

4.11.3.2.2. Reagents

Buffer solution: NaOH, 313 *mmol l^{-1}*; *Na$_2$HPO$_4$*, 12.5 *mmol l^{-1}*.
Picric acid, 8.73 *mmol l^{-1}*.
Standard creatinine solution, 88.4 *μmol l^{-1}* (1 *mg* per 100 *ml*), *HCl* 10 *mmol l^{-1}*.

4.11.3.2.3. Measurement

Conditions. The buffer solution and the picric acid must be brought to the correct temperature beforehand and the reaction temperature must be kept within the range 20–37 °C as accurately as possible. The reaction temperature for the analyses and the standards must be identical. The wavelength is 492 nm and the path-length 1 cm. The final concentrations in the reaction mixture are as follows: 1, buffer solution, NaOH 125.2 mmol l^{-1} and Na_2HPO_4 5 mmol l^{-1}; 2, picric acid 3.5 mmol l^{-1}.

Procedure.

	Sample, ml	Standard, ml
Plasma, or urine diluted 1:100	0.5	—
Standard	—	0.5
Picric acid	1.0	1.0

Mix and thermally equilibrate for about 5 min

Buffer solution	1.0	1.0

Mix and measure the extinction (E_1) within 1 min and the extinction (E_2) exactly 5 min after the first measurement. At temperatures above 30 °C, E_1 must be measured within 30 s.

The method is linear up to 5 mg per 100 ml (422 μmol l^{-1}). For concentrations above this, plasma and urine (1:100 dilution) are diluted 1 part to 6 with NaCl solution and, after repeating the determination, the result is multiplied by 6.

4.11.3.2.4. Calculation

$$\text{Creatinine concentration} = \frac{E_2(S) - E_1(S)}{E_2(ST) - E_1(ST)}$$

$$\cdot 88.4 \ \mu\text{mol } l^{-1} \quad \text{or} \quad 1 \text{ mg per 100 ml}$$

4.11.3.2.5. Specificity

This modification of the Jaffé reaction is distinguished by a higher specificity and has the advantage that, as a consequence of the low picric acid concentration, the measurement may be made at the extinction maximum and there is no interference from deproteination. Because of the rapidity of the reaction between creatinine and picric acid, side-reactions which come into play after the measurement are almost eliminated. The temperature and the time of measurement must be rigorously adhered to.

Table 135. Reference values for the plasma creatinine concentration.

Method	Number	Sex	mg/100 ml $\bar{x}$	mg/100 ml $\bar{x} \pm 2s$	µmol/l $\bar{x}$	µmol/l $\bar{x} \pm 2s$	References
Jaffé, manual	15	m	1.17	0.91–1.43	103.4	80.4–126.4	42
(total chromogens)	15	f	0.98	0.88–1.08	86.6	77.7–95.5	
	15	m	0.93	0.71–1.15	82.2	62.8–101.7	43
	15	f	0.77	0.51–1.03	68.1	45–91	
Jaffé, AutoAnalyzer	39	m	0.86	0.52–1.20	76	46–106	7
	39	f	0.71	0.55–0.87	69.8	48.6–76.9	
	65	m	0.99	0.78–1.2	87.5	68.9–106	20
	58	f	0.82	0.61–1.03	72.5	53.9–91	
	102	m		0.8–1.4		70.7–123.8	44
	123	f		0.7–1.2		61.9–106	
Jaffé, Fuller's earth	30	m	1.01	0.79–1.23	89.3	69.8–108.7	42
	30	f	0.82	0.64–1.0	72.5	56.6–88.4	
	63	m	0.82	0.64–1.05	72.5	56.6–92.8	14
	55	f	0.68	0.56–0.91	60	49.5–80.4	
Jaffé, ion exchanger	30	m	0.91	0.67–1.15	80.4	59.2–101.6	16
	22	f	0.69	0.53–0.85	61.0	46.8–75.0	
	83	m	1.07	0.79–1.41	94.4	70–125	19
	51	f	0.85	0.56–1.18	75.2	50–105	
Jaffé, kinetic	220	m	0.92	0.7–1.1	81.3	61.9–97.2	29
	191	f	0.77	0.6–0.9	68.1	53–79.6	
	64	m	0.99	0.67–1.31	87.5	59.2–115.8	41
	56	f	0.83	0.51–1.15	73.4	45.1–101.7	
	178	m		0.9–1.56[a]		80–138[a]	27
				0.78–1.32[b]		69–117[b]	
	301	f		0.7–1.37[a]		63–121[b]	37
				0.59–1.14[a]		53–101[b]	
Enzymatic	171	m		0.55–1.1		49–97	37
	173	f		0.47–0.9		42–80	

[a]Measurement of the reaction up to 1 min.
[b]Measurement of the reaction up to 2 min.

For this reason, adaptation of the method to automated analysis systems is particularly simple.

The values obtained with this modification are about 5–15% lower than those obtained with the AutoAnalyzer method.[11]

4.11.3.2.6. Reference values

Plasma. Many studies have been published on the reference values for the plasma creatinine concentration. As might be expected, the values differ greatly according to the method. The summary in Table 135 clearly shows that the reference values become lower and lower with increasing specificity of the technique employed.

In evaluating the reference values for creatinine, in addition to taking

690

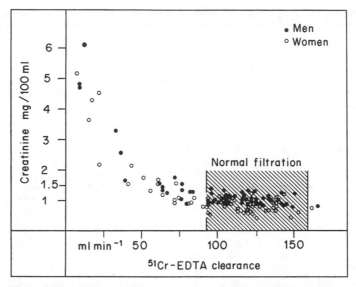

Fig. 149. Relationship between plasma creatinine and glomerular filtration. Determination of plasma creatinine with Greiner Selective Analyser GSA II on 112 patients.

sex differences into account, we believe it is necessary to make a simultaneous test of the renal glomerular filtration.[29,41] The reference values for the plasma creatinine have been determined with the Greiner Selective Analyzer GSA II (Jaffé, kinetic) in persons with a glomerular filtration (^{51}Cr-EDTA clearance) $\geqslant$ 93 ml min^{-1}.[41] The relationship between plasma creatinine and glomerular filtration is evident from Fig. 149.

Schwartz et al.[56] investigated the relationship between plasma creatinine reference values and age in 772 boys and 626 girls aged 1 to 20 years. They discovered the following empirical relationship:

Boys: creatinine (mg per 100 ml) = 0.35 + 0.025 · age (years)
Girls: creatinine (mg per 100 ml) = 0.37 + 0.018 · age (years)

Urine. Since the investigations of Folin,[45] the excretion of creatinine in the urine is reported as 1.36–1.77 g day^{-1}. In studies on the excretion of nitrogen-containing products with normal and protein-poor diets, Folin found a striking 'dichotomy' of the individual components. Although the excretion of urea fell sharply within a few days, the creatinine excretion remained largely constant despite the protein deficiency. The idea of 'constancy of creatininuria' derives from this period. Today, this relative constancy of the creatininuria is frequently used to check the completeness of the 24-hour urine collection; 24-hour urine collections are unreliable; individual investigators found up to 30% of such collections to be unusable.[53] The creatinine excretion is slightly dependent on the diet[46–48] and diuresis[50] (no fasting variations),[48] and can be slightly elevated by

Table 136. Creatinine coefficients (mg kg^{-1} body weight per 24 h).

Children	m	24–26	54
4–18	f	20–22	54
Adults	m	9–25	55
	f	7–21	55

intense physical exertion. Szadkowski *et al.*[51] have measured the daily creatinine excretion in 10 subjects over a period of 8 days and have shown that, on average, this was constant and so was the daily quantity of urine. A circadian rhythm could not be established for creatinine excretion. Nevertheless, it must now be considered clear that this 'constancy' is a relative concept, since day-to-day variations of 10–30% are observed.[49,50] Even so there is, as yet, no other substance known which is excreted as constantly as creatinine. Consequently, even today, it still has a great significance in practical clinical chemistry.

The problems of the 24-hour urine collection have led many laboratories to state their results not just in units per day but also in units per milligram of creatinine.

The determination of the *creatinine coefficient* is based on the observation that the creatinine excretion per kilogram body-weight is relatively constant in 24 h (Table 136). With a normal kidney function the synthesis of creatinine and the creatininuria is directly proportional to the muscular mass. The relationship between muscular mass and creatinine excretion is linear and uninfluenced by age or sex.[52]

Finally, we ought to discuss the '*creatinine equivalent*'. In the study of metabolites which occur in the urine in highly variable amounts it has become customary to make the determination not on a definite volume of urine but to adjust the volume to a definite quantity of creatinine, a 'creatinine equivalent'. This method is used, above all, in the chromatographic investigation of amino-acidurias. Although this practice has been criticized, it has the advantage that the diuresis factor, the volume variations, and the consequent variations in concentration can be at least partially eliminated.

References

1. Jaffé, M., Über den Niederschlag, welchen Pikrinsäure in normalem Harn erzeugt und über eine neue Reaktion des Kreatinins, *Z. physiol. Chem.*, **10**, 391 (1886).
2. Folin, O., On the determination of creatinine and creatine in blood, milk, and tissues, *J. biol. Chem.*, **17**, 475 (1914).
3. Blass, K. G., Thibert, R. J., and Lam, L. K., A study of the mechanism of the Jaffé reaction, *Z. klin. Chem. klin. Biochem.*, **12**, 336 (1974).
4. Popper, H., Mandel, E., and Mayer, H., Zur Kreatininbestimmung im Blute, *Biochem. Z.*, **291**, 354 (1937).

5. Bosnes, R. W. and Taussky, H. H., On the colorimetric determination of creatinine by the Jaffé reaction, *J. biol. Chem.*, **158**, 581 (1945).
6. Zank, W., Untersuchungen zur Qualitätsverbesserung der Serumkreatininbestimmung, *Z. med. Labortech.*, **15**, 136 (1974).
7. Zender, R., and Falbriard, A., Analyse automatique de la créatinine dans le serum et dans l'uriné Valeurs "normales" chez l'homme de la créatininémie et de la clearance, *Clin. chim. Acta*, **12**, 183 (1965).
8. Close-Moll, M. and Lines, J. G., Some observations and improvements on SMA 6/60 determinations of creatinine, *J. clin. Path.*, **25**, 603 (1972).
9. Heinegård, D. and Tiderström, G., Determination of serum creatinine by a direct colorimetric method, *Clin. chim. Acta*, **43**, 305 (1973).
10. Yatzidis, H., New method for direct determination of 'true' creatinine, *Clin. Chem.*, **20**, 1131 (1974).
11. Helger, R., Rindfrey, H., and Hilgenfeldt, J., Eine Methode zur direkten Bestimmung des Creatinins in Serum und Harn ohne Enteiweissung nach einer modifizierten Jaffé-Methode, *Z. klin. Chem. klin. Biochem.*, **12**, 344 (1974).
12. Owen, J. A., Iggo, B., Scandrett, F. J., and Stewart, C. P., The determination of creatinine in plasma or serum, and in urine, a critical examination, *Biochem. J.*, **58**, 426 (1954).
13. Løken, F., On the determination of creatinine in plasma by the Jaffé reaction, after adsorption to Lloyd's reagent, *Scand. J. clin. Lab. Invest.*, **6**, 325 (1954).
14. Knoll, E. and Stamm, D., Spezifische Creatininbestimmung im Serum, *Z. klin. Chem. klin. Biochem.*, **8**, 582 (1970).
15. Knoll, E. and Wisser, H., Kreatininbestimmung im Serum ohne Enteiweissung, *Z. klin. Chem. klin. Biochem.*, **11**, 411 (1973).
16. Teger-Nilsson, A. C., Serum creatinine determination using an ion exchange resin, *Scand. J. clin. Lab. Invest.*, **13**, 326 (1961).
17. Rockerbie, R. A. and Rasmussen, K. L., Rapid determination of serum creatinine by an ion-exchange technique, *Clin. chim. Acta*, **15**, 475 (1967).
18. Mitchell, R. J., Improved method for specific determination of creatinine in serum and urine, *Clin. Chem.*, **19**, 408 (1973).
19. Vedsö, S., Rud, C., and Place, J. F., Routine creatinine determination. An equilibrium technique involving reusable cation exchange membranes, *Scand. J. clin. Lab. Invest.*, **34**, 275 (1974).
20. Ditzel, J., Bang, H. O., and Thorsen, N., Low plasma creatinine in diabetic subjects free of renal disease, *Scand. J. clin. Lab. Invest.*, **20**, 360 (1967).
21. Watkins, P. J., The effect of ketone bodies on the determination of creatinine, *Clin. chim. Acta*, **18**, 191 (1967).
22. Chasson, A. L., Grady, H. J., and Stanley, M. A., Determination of creatinine by means of automatic chemical analysis, *Am. J. clin. Path.*, **35**, 83 (1961).
23. Polar, E. and Metcoff, J., True creatinine chromogen determination in serum und urine by semi-automated analysis, *Clin. Chem.*, **11**, 763 (1965).
24. Bartels, H. and Böhmer, M., Eine Mikromethode zur Kreatininbestimmung, *Clin. chim. Acta*, **32**, 81 (1971).
25. Bartels, H. and Böhmer, M., Eine kinetische Messmethode für Serumkreatinin, *Med. Lab., Stuttg.*, **26**, 209 (1973).
26. Bartels, H., Böhmer, M., and Heierli, C., Serum-Kreatininbestimmung ohne Enteiweissen, *Clin. chim. Acta*, **37**, 193 (1972).
27. Larsen, K., Creatinine assay by a reaction-kinetic principle, *Clin. chim. Acta*, **41**, 209 (1972).
28. Lustgarten, J. A. and Wenk, R. E., Simple, rapid, kinetic method for serum creatinine measurement, *Clin. Chem.*, **18**, 1419 (1972).
29. Ullmann, R. and Bonitz, K., Vollmechanisierte kinetische Messung von Kreatinin, *Med. Lab., Stuttg.*, **29**, 137 (1976).

30. Ward, P., Ewen, M., Pomeroy, J. A., and Leung, F. Y., Kinetic urine creatinine determination with the Gemsaec analyzer, *Clin. Biochem.*, **9**, 225 (1976).
31. Vasiliades, J., Reaction of alkaline sodium picrate with creatinine. I. Kinetics and mechanism of formation of the monocreatinine picric acid complex, *Clin. Chem.*, **22**, 1664 (1976).
32. Miller, B. F. and Dubos, R., Determination by a specific, enzymatic method of the creatinine content of blood and urine from normal and nephritic individuals, *J. biol. Chem.*, **121**, 457 (1937).
33. Dubos, R. and Miller, B. F., The production of bacterial enzymes capable of decomposing creatinine, *J. biol. Chem.*, **121**, 429 (1937).
34. Wahlefeld, A. W., Herz, G., and Bergmeyer, H. U., A completely enzymatic determination of creatinine in human sera or urine, *Scand. J. clin. Lab. Invest.*, suppl. 29, p. 126 (1972).
35. Meyer-Bertenrath, J. G. and Döbert, R., Enzymatische Schnellbestimmung von Kreatinin, *Ärztl. Lab.*, **20**, 41 (1974).
36. Moss, G. A., Bondar, R. J. L., and Buzzelli, D. M., Kinetic enzymatic method for determining serum creatinine, *Clin. Chem.*, **21**, 1422 (1975).
37. Szasz, G., Börner, U., Stähler, F., and Bablok, W. G., Reference values for creatinine in serum established by a highly specific enzymatic method, *Clin. Chem.*, **23**, 1172 (1977).
38. Parekh, A. C., Cook, S., Sims, C., and Jung, D. H., A new method for the determination of serum creatinine based on reaction with 3,5-dinitrobenzoyl chloride in an organic medium, *Clin. chim. Acta*, **73**, 221 (1976).
39. Thompson, H. and Rechnitz, G. A., Ion electrode based enzymatic analysis of creatinine, *Analyt. Chem.*, **46**, 246 (1974).
40. McLean, M. H., Gallwas, J., and Hendrixson, M., Evaluation of an automated creatininase creatinine procedure, *Clin. Chem.*, **19**, 623 (1973).
41. Peheim, E., Colombo, J. P., and Flury, W., Referenzwerte des Plasmacreatinins in Beziehung zur Cr^{51}-EDTA Clearance, *Schweiz, Ges. klin. Chem.*, 20. Jahresvers, Lausanne, 1976.
42. Doolan, P. D., Alpen, E. L., and Teil, G. B., A clinical appraisal of the plasma concentration and endogenous clearance of creatinine, *Am. J. Med.*, **32**, 65 (1962).
43. Schirmeister, J. J., Willmann, H., and Kiefer, H., Endogenes Kreatinin in Serum und Harn, *Klin. Wschr.*, **41**, 878 (1963).
44. Roberts, L. B., The normal ranges, with statistical analysis for seventeen blood constituents, *Clin. chim. Acta.*, **16**, 69 (1967).
45. Folin, O., Laws governing the chemical composition of urine, *Am. J. Physiol.*, **13**, 66 (1905).
46. Bleiler, R. A. and Schedl, H. P., Creatinine excretion: variability and relationship to diet and body size, *J. Lab. clin. Med.*, **59**, 945 (1962).
47. Ritchey, S. J., Derise, N. L., Abernathy, R. P., and Korslund, M. K., Variability of creatinine excretion in preadolescent girls consuming a wide range of dietary nitrogen, *Am. J. clin. Nutr.*, **26**, 690 (1973).
48. Pasternack, A. and Kuhlbäck, B., Diurnal variations of serum and urine creatine and creatinine. *Scand. J. clin. Lab. Invest.*, **27**, 1 (1971).
49. Zorab, P. A., Clark, S., and Harrison, A., Creatinine excretion, *Lancet*, ii, 1254 (1969).
50. Paterson, N., Relative constancy of 24 hour urine volume and 24 hour creatinine output, *Clin. chim. Acta*, **18**, 57 (1967).
51. Szadkovski, D., Jörgensen, A., Essing, H. G., and Schaller, K. H., Die Kreatinineliminationsrate als Bezugsgrösse für Analysen aus Harnproben, *Z. klin. Chem. klin. Biochem.*, **8**, 529 (1970).

694

52. Forbes, G. B. and Bruining, G. J., Urinary creatinine excretion and lean body mass, *Am. J. clin. Nutr.*, **29**, 1359 (1976).
53. Schirmeister, J., Willmann, H., Kiefer, H., and Hallauer, W., Für und wider die Brauchbarkeit der endogenen Kreatininclearance in der funktionellen Nierendiagnostik, *Dt. med. Wschr.*, **89**, 1640 (1964).
54. Clark, L. C., Thompson, H. L., Beck, E. I., and Jacobson, W. J., Excretion of creatine and creatinine by children., *Am. J. Dis. Child.*, **81**, 774 (1951).
55. Hopper, J., Creatinine clearance: simple way of measuring kidney function, *Bull. Univ. Calif. med. Center*, **2**, 315 (1951).
56. Schwartz, G. J., Haycock, G. B., Chir, B., and Spitzer, A., Plasma creatinine and urea concentration in children: normal values for age and sex, *J. Pediat.*, **88**, 828 (1976).

4.11.3.3. Creatinine clearance

Supposing that the creatinine was neither reabsorbed nor secreted in the renal tubuli, Rehberg[1] applied creatinine to the measurement of glomerular filtration. He made use of the so-called exogenous creatinine clearance, i.e. a clearance measurement following oral administration of creatinine. The advantage of this method lay in the simplicity of the clinical procedure as regards the patient and, in consequence of the relatively high plasma concentration (7–15 mg per 100 ml), the ease of the determination on the plasma. However, it soon proved that in man creatinine was not only filtered in the glomeruli but was also secreted in the tubules.[2] In evidence for this the following findings may be cited:

1. In man and the anthropoid apes the creatinine/inulin clearance ratio is almost always greater than 1.[1,4] In all other mammals investigated the ratio lies close to 1, so a secretion is improbable.[3]
2. The administration of phloridzin and PAH, a non-specific inhibitor of reabsorption and secretion processes, leads to a depression of the creatinine clearance in comparison with the inulin clearance.[3]

The observation that the creatinine clearance falls with increasing concentration of creatinine in the plasma led next to a study of the endogenous creatinine clearance, i.e. a clearance without special administration of creatinine.[5-7] With the endogenous creatinine clearance, as it is performed almost exclusively today, results were obtained which, in the reference range and at slight reduction of the glomerular filtration, actually correlated fairly well with the results of the inulin clearance. In contrast, at increased plasma creatinine concentrations, in uraemia for instance, the deficit in comparison with the inulin clearance becomes progressively less great (Fig. 150). This phenomenon is explained on the assumption of an increased tubular secretion dependent on the plasma creatinine concentration. This shows up as an ever-increasing creatinine/inulin clearance ratio. It may be taken as a rule of thumb that the creatinine clearance provides useful values in healthy persons and in patients showing less than 30% reduction in the glomerular filtration.

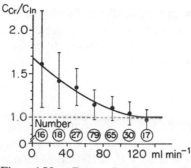

Fig. 150. Dependence of the creatinine/inulin clearance ratio on the inulin clearance ($\bar{x} \pm 1s$).[4]

Considerable differences from the inulin clearance are likely if the renal function is further reduced.

4.11.3.3.1. Procedure

The creatinine clearance may be carried out on patients at rest or ambulant patients. A blood sample is taken for determining the plasma creatinine and, at the same time, the urine is collected over a definite period. Periods of 1, 2, 5, 12 and 24 h have been recommended.[12] The results should be roughly the same independent of the collecting period. It is therefore advisable to use shorter periods, e.g. 4 h, for hospitalized patients but it is safer to use a 24 h collection with ambulant patients. Whatever type of creatinine clearance is employed, the main source of error lies in the incompleteness of the urine collection (it must be collected without losses). It is therefore vital to give the patient precise instructions.

4.11.3.3.2. Calculation

C_{CR} = creatinine clearance (ml min^{-1})

U_{CR} = urinary creatinine concentration (mg per 100 ml);

P_{CR} = plasma creatinine concentration (mg per 100 ml);

U_{vol} = urine volume (ml);

t = collecting period (min);

$V = U_{vol}/t$, i.e. urine volume per unit time (ml min^{-1});

BS = body surface area (m^2).

The simple clearance equation is

$$C_{CR} = \frac{U_{CR}V}{P_{CR}} \quad \text{or} \quad \frac{U_{CR}}{P_{CR}} \cdot \frac{U_{vol}}{t} \text{ ml min}^{-1}$$

696

Since clearance results are basically referred to a body surface area of 1.73 m², the result has to be multiplied by the *BS* factor, i.e. 1.73/*BS*. The body surface area is most simply evaluated from the weight and height by using a nomogram (Appendix 3).

4.11.3.3.3. Specificity

The application of the endogenous creatinine clearance is much favoured in the clinic on account of the simplicity of the method. Yet it is clear from the great number of publications that the creatinine clearance permits only a semiquantitative appraisal of the glomerular filtration.[8] Nevertheless, these methods can be used to obtain a useful parameter for coarse monitoring of the progress of diseases of the kidneys.

In healthy persons the creatinine/inulin clearance ratio should, theoretically, be 1. However, this numerical agreement is only apparent, since falsely high plasma values (non-creatinine chromogens) are counteracted by a certain amount of tubular secretion, roughly cancelling each other and simulating a pure glomerular filtration.

With clearance values less than 50 ml min^{-1} the C_{CR}/inulin or sodium thiosulphate clearance ratio can rise to 2, so that it is no longer possible to draw any quantitative conclusions about the kidney function with such results.

In a study of 83 kidney patients, Reubi[8] found a $C_{CR}/C_{Na\,thiosulphate}$ ratio of 1.47.

4.11.3.3.4. Reference values

Table 137 lists a few reference values taken from the literature. Despite the various methods used, the agreement of the results is relatively good. A value of about 120 ml min^{-1} may be taken as the mean and 85–160 ml min^{-1} as the reference interval. This is in relatively good agreement with

Table 137. Reference values for the creatinine clearance (ml min^{-1}), corrected to a body surface area of 1.73 m².

Number	Sex	$\bar{x}$	1s	$\bar{x} \pm 2s$	Method	References
20	m	123	24	75–171	Jaffé	4
12	f	119	13	93–145	Jaffé	4
11	m	135	14	107–163	o-Nitrobenzaldehyde	10
7	f	127	18	91–163	o-Nitrobenzaldehyde	10
30	m	123	19	85–161	Lloyd's reagent	11
30	m	114	16	82–146	Lloyd's reagent	11
127	m, f	119	18	83–155	AutoAnalyzer (Jaffé)	13
23	m	104	18	68–140	AutoAnalyzer (Jaffé)	9
24	f	107	18	71–143		

the values for the inulin clearance, for which 123 ml min^{-1} and about 80–160 ml min^{-1} are reported for the mean and the reference interval, respectively.[2] After they have corrected for the body surface area, most investigators find no sex difference. Since the body surface area for women is usually smaller than that for men, these results indirectly attest the existence of a sex difference in the plasma creatinine concentration.

4.11.3.3.5. Comments on the method

For the elucidation of kidney function a knowledge of the clearance of other substances (e.g. uric acid, phosphorus, electrolytes) is frequently also necessary. The determination is made by measurement of the substances concerned, in plasma and urine. The amount of filtration and reabsorption of the investigated substances may be calculated by using the creatinine clearance, measured simultaneously, as a measure of the glomerular filtrate.

Amount of filtration (mg min^{-1}): C_{CR} · plasma concentration of the substance (mg ml^{-1}).

Amount excreted (mg min^{-1}): urine concentration of the substance (mg ml^{-1}) · $\dfrac{\text{urine volume}}{t \ (\text{min})}$.

Amount reabsorbed (mg min^{-1}): amount filtered − amount excreted.

Amount reabsorbed (%): $100 \cdot \left(1 - \dfrac{\text{clearance of the substance}}{C_{CR}}\right)$

References

1. Rehberg, P. B., Sudies on kidney function. 1. The rate of filtration and reabsorption in the human kidney, *Biochem. J.*, **20**, 447 (1926).
2. Smith, H. W., *The kidney*, Oxford Press, New York, 1951
3. Smith, H. W., *Principles of renal physiology*, Oxford Press, New York, 1956.
4. Schirmeister, J., Willmann, H., Kiefer, H., and Hallauer, W., Für und wider die Brauchbarkeit der endogenen Kreatininclearance in der funktionellen Nierendiagnostik, *Dt. med. Wschr.*, **89**, 1640 (1964).
5. Findley, T., The excretion of endogenous 'creatinine' by the human kidney, *Am. J. Physiol.*, **123**, 260 (1938).
6. Popper, H. and Mandel, E., Filtrations- und Resorptionsleistung in der Nierenpathologie, *Ergebn. inn. Med. Kinderheilk.*, **53**, 685 (1937).
7. Steinitz, K. and Türkand, H., Determination of glomerular filtration by endogenous creatinine clearance, *J. clin. Invest.*, **19**, 285 (1940).
8. Reubi, F., *Nierenkrankheiten*, 2. Aufl., Huber, Berne, 1970.
9. Wibell, L. and Björsell-Östling, E. B., Endogenous creatinine clearance in apparently healthy individuals as determined by 24 hour ambulatory urine collection, *Uppsala J. med. Sci.*, **78**, 43 (1973).
10. Pilsum, J. F. van, and Seljeskok, E. L., Longterm endogenous creatinine clearance in men, *Proc. Soc. exp. Biol. Med.*, **97**, 270 (1958).
11. Doolan, P. D., Alpen, E. L., and Teil, G. B., A clinical appraisal of the plasma

698

concentration and endogenous clearance of creatinine, *Am. J. Med.*, **32**, 65 (1962).

12. Tobias, G. J., McLaughlin, R. F., and Hopper, J., Endogenous creatinine clearance. A. valuable clinical test of glomerular filtration and a prognostic guide in chronic renal disease, *New Engl. J. Med.*, **266**, 317 (1962).

13. Zender, R. and Falbriard, A., Valeurs "normales" chez l'homme de la créatininémie et de la clearance, *Clin. chim. Acta*, **12**, 183 (1965).

4.11.4. TUBULAR FUNCTION

4.11.4.1. Urinary acid–base regulation

4.11.4.1.1. Introduction

With a normal diet, 40–80 mmol of H^+ ions accrueing from the metabolism has to be excreted through the kidneys daily. The ions are removed by tubular secretion. This ability of the kidneys is termed acidogenesis. If all of the protons were to be excreted in the free form the resultant pH of the urine would be less than 2, but in fact, under normal conditions the urine pH is around 5–6. Consequently, the bulk of the protons must be buffered (Fig. 151).

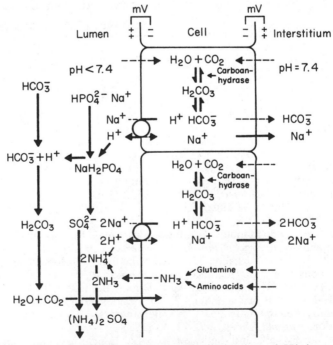

Fig. 151. Mechanism of the renal elimination of H^+ ions (for explanation, see text).

Hence the eliminated H^+ ions occur in various fractions in the final urine:[1,2]

1. *As free H^+ ions*. These are determined by measuring the pH of the urine with a pH meter.
2. *Bound* to phosphate, sulphate and organic proton acceptors. Monohydrogen phosphate (HPO_4^{2-}) is converted to primary dihydrogen phosphate ($H_2PO_4^-$) by the uptake of protons and at pH 4.5 the phosphate is almost exclusively present in the latter form (Fig. 151). Other proton acceptors, especially organic acids, are less likely candidates as buffer substances since their pH is too low. This fraction together with the free protons makes up the *titratable acidity* determined by back-titration with alkali.
3. *As ammonium* (NH_2^+) (Fig. 151). Ammonia is produced in the tubular cells by enzymatic reactions on substrates which are either brought there in the arterial blood or are synthesized in the cells themselves, principally glutamine. Some 43% of the urinary NH_4^+ derives from the amide-N of glutamine by the action of glutaminase I, and 18% from the glutamine amino group by the action of glutaminase II. The remainder comes from alanine, glycine, and glutamic acid.[3] The NH_3 formed in the tubulus cells diffuses into the lumen and reacts with protons to form NH_4^+, which can no longer pass through the cell membrane.

In cases of acid overload the ammonium buffer system is only fully effective after a few days while the phosphate system acts faster. Owing to the presence of these buffer systems it is possible to excrete H^+ ions which are normally completely reabsorbed, without having to draw on the bicarbonate (Fig. 151).

The total excretion of acid into the urine (total H^+) is obtained from the sum of the titratable acidity (TA) and the ammonium concentration:

Total H^+ = TA + ammonium ions

This assumes that all of the bicarbonate filtered by the glomeruli is reabsorbed, which is generally the case. If, however, the bicarbonate is excreted, it should be remembered that for each millimole of bicarbonate 1 mmol of H^+ ions is recovered for the organism. This must be taken into account in the excretion balance for H^+ ions. Consequently, the net H^+ ion excretion of the kidneys per day is given by

$[H^+] = TA + [NH_4^+] - [HCO_3^-]$ mmol per 24 h

A kidney functioning normally is in a position to compensate for an increased production of acid (which can, e.g. in diabetic coma, be many times the norm) by an increase in ammoniogenesis and titratable acidity. By the exchange of sodium for hydrogen ions the titratable acid increases. This conserves Na^+ ions since H^+ ions are excreted with phosphate and also with sulphate. The ability to increase the acidity of the urine can be tested by oral loading with ammonium chloride or arginine chloride. With chronic kidney insufficiency the ability to excrete H^+ ions decreases. This may lead to a renal acidosis which may occasion a positive H^+ ion balance in these patients. In such a case the ammoniogenesis and the titratable acidity may be defective independently of one another.

4.11.4.1.2. Three-stage titration for the quantitative determination of bicarbonate, titratable acid, and ammonia in urine

All of the components involved in the equation:

$$\text{Net } H^+ \text{ ions} = TA + [NH_4^+] - [HCO_3^-]$$

can be determined individually: the net H^+ ion excretion by the Jørgensen method,[4] the titratable acidity by titrimetry, the bicarbonate by gasometric determination of the total CO_2, giving consideration to the Hendersen–Hasselbalch formula and the ionic strength of the urine. Lin and Chan (5) describe a modified Jørgensen method whereby bicarbonate is determined titrimetrically. Following a suggestion by Chan,[6] ammonium can likewise be determined by titrimetry using a modified Jørgensen method. Lüthy et al.[7] have described a method which determines all the parameters (pH, NH_4^+, HCO_3^-) in a combined three-stage titration.

4.11.4.1.3. Principle

The urine pH is measured. Usually the pH value so found is smaller than the blood pH. In such cases the titratable acidity is determined in a 1st stage by titration with NaOH to the actual blood pH. Subsequently (stage 2), a known quantity of HCl is added and the specimen boiled. In this acidification stage, the H^+ ions from the HCl react with any HCO_3^- ions present to form carbonic acid, which then decomposes into gaseous CO_2 and H_2O. As a result, some of the HCl is used up and in the titration which now follows the HCl added can no longer be completely back-titrated. The difference corresponds to the concentration of the HCO_3^- ions. In the 3rd stage formaldehyde is added so as to render the H^+ present as NH_4^+ (and previously non-titratable) titratable at the blood pH: formalin titration by the method of Sørensen.[8] In the rare cases where the urine pH is greater than the actual blood pH, the 1st stage of the titration is superfluous and the 2nd stage is begun immediately. The amount of the titration signifies HCO_3^- and any titratables base (TB) which may be present. For this reason the HCO_3^- is additionally determined by gasometry in such situations. If the gasometrically determined bicarbonate concentration is smaller than that determined by titration the difference denotes TB.

4.11.4.1.4. Apparatus and other equipment

Automatic titrator with autoburette, pH meter, recorder, water-bath with titration stand (titration at 37 °C), plastic titration vessel, magnetic stirrer, pipettes (1 and 0.1 ml).

4.11.4.1.5. Reagents

All solutions are prepared with CO_2-free doubly distilled water.
NaOH, 0.1 *mol* l^{-1} *(Titrisol)*.
HCl, 1 *mol* l^{-1} *(Titrisol)*.
Formaldehyde, *8%*, *pH 7.4 (buffered with* Na_2HPO_4). A 500-ml volume of 8% formaldehyde is adjusted to pH 7.4 by adding about 10 drops of 10% Na_2HPO_4 solution. The pH should always be monitored before use and, if necessary, readjusted.
Standard solutions, 10, 20 *and* 30 *mmol* l^{-1}.
Sodium bicarbonate, 10 *mmol* l^{-1}. Dissolve 84.01 mg of $NaHCO_3$ (e.g. Merck, p.a.) in CO_2-free boiled-out doubly distilled water and make up to 100 ml.
Ammonium chloride, 10 *mmol* l^{-1}. Dissolve 53.5 mg of NH_4Cl (e.g. Merck, p.a.) in boiled-out CO_2-free doubly distilled water and make up to 100 ml.

4.11.4.1.6. Procedure

A magnetic stirrer is placed in each of the titration vessels and 1 ml of urine, 1 ml of CO_2-free water, or 1 ml of standard, is pipetted in.
First titration stage. Measurement of the pH and the *TA*. The urine sample, standard, and blank are titrated with 0.1 mol l^{-1} NaOH to pH 7.4. Even more accurate values may be achieved by titrating to the actual blood pH found during the collection of the urine.

Calculation.

$$\text{ml NaOH (sample)} - \text{ml NaOH (blank)} \cdot 100 = \mu\text{mol ml}^{-1}\ TA$$

$$\left(Factor\ 100: \frac{\text{NaOH } 0.1 \text{ mol } l^{-1} \cdot 1000}{1 \text{ ml sample under analysis}} = \text{mmol } l^{-1} = \mu\text{mol ml}^{-1} \right)$$

Note. If the urine pH is higher than the actual blood pH, this titration stage is omitted. Begin with the second stage straight away.
Second titration stage. Determination of bicarbonate. Pipette 100 μl of 1 mol l^{-1} HCl into the same vessel. This must bring the pH lower than 4, otherwise use twice the quantity of HCl (use 200 μl of HCl with the water blank too and insert this figure in the calculation). Place the samples in a boiling water-bath for 5–8 min, then allow to cool; titration with 0.1 mol l^{-1} NaOH as in the 1st stage.

Calculation.

ml NaOH (blank) − ml NaOH (sample) · 100 = μmol ml^{-1} HCO$_3^-$

Notes. 1. If the pipetting is exact the water blank will require 1 ml of 0.1 mol l^{-1} NaOH since this neutralizes 0.1 ml of 1 mol l^{-1} HCl. If the analysis has begun with the 2nd stage because the urine pH is higher than the blood pH, the result gives μmol ml^{-1} HCO$_3^-$ and titratable base. The bicarbonate concentration must be determined gasometrically on another sample so as to discover the amount of titratable base.

Calculation.

μmol ml^{-1} HCO$_3^-$ and titratable base (titrimetric) − μmol ml^{-1} HCO$_3^-$ (gasometric) = μmol ml^{-1} titratable base

Third titration stage. Determination of NH$_4^+$. After the titrimetric determination of HCO$_3^-$, 1 ml of 8% buffered formaldehyde is pipetted into the titration vessel; mix; titration with 0.1 mmol l^{-1} NaOH.

Calculation.

ml NaOH (sample) − ml NaOH (blank) · 100 = μmol ml^{-1} NH$_4^+$

4.11.4.1.7. Reference values

	mmol per 24 h	References
Urine pH 5.5–6.9	10–30	9
Titratable acidity	20–50	2
Ammonium ions	30–50	9

These values should be interpreted with caution since collecting periods of more than 24 h are too long (production of ammonia by bacteria, loss of uptake of CO$_2$). Consequently, the values are reliable only if the parameters are investigated immediately in the individual portions. Today preference is usually accorded to the timed urine specimen, i.e. a urine specimen accurately timed, collected over a short period (e.g. collect the urine for 2–5 h after emptying the bladder). The result is then expressed as an excretion rate in μmol min^{-1} per 1.73 m^2. The rate of excretion of the total acid into the urine should always be compared simultaneously with the acid–base status of the blood, as this is the only way to interpret the results of the H$^+$ ion excretion meaningfully. With a normal diet the number of equivalents of acid excreted into the urine in the normal plasma bicarbonate range should not exceed the number of equivalents of base. If the urine cannot be investigated immediately it may be kept in a deep-freeze until analysed.[6]

References

1. Davenport, H. W., *The ABC of acid base chemistry*, University of Chicago Press, Chicago, 1969.
2. Müller-Plathe, O., *Säure–Basen-Haushalt und Blutgase*, Thieme, Stuttgart, 1973.
3. Colombo, J. P., Congenital disorders of urea cycle and ammonia detoxication, *Monogr. Paediatr.*, Vol. 1, Karger, Basle, 1971.
4. Jørgensen, K., Titrimetric determination of the net excretion of acid/base in urine, *Scand. J. clin. Lab. Invest.*, **9**, 287 (1957).
5. Lin, S. L. and Chan, J. C. M., Urinary bicarbonate: a titrimetric method for determination, *Clin. Biochem.*, **6**, 207 (1973).
6. Chan, J. C. M., The rapid determination of urinary titratable acid and ammonium and evaluation of freezing as a method of preservation, *Clin. Biochem.*, **5**, 94 (1972).
7. Lüthy, Ch., Moser, C., and Oetliker, O., Dreistufige Säure–Basen-Titration im Urin, *Med. Lab., Stuttg.*, **30**, 174 (1977).
8. Sørensen, S. P. L., Enzymstudien. I. Über die quantitative Messung proteolytischer Spaltungen, *Biochem. Z.*, **7**, 45 (1908).
9. Krück, F., Renal bedingte Störungen des Säure-Basenhaushaltes, in *Handbuch der inneren Medizin*, Vol. 8, Teil I, Springer, Heidelberg, 1968.

4.11.4.2. Urinary protein: biuret method

4.11.4.2.1. Choice of method

The Esbach test, still frequently used, is not quantitative. It should be replaced by the much more reliable biuret method. Bacterial contamination leads to a degradation of protein, so the urine should be collected using thymol/isopropanol as additive. A 24 h urine specimen should be used where possible.

4.11.4.2.2. Principle

The urinary protein is precipitated with phosphotungstic acid and the precipitate dissolved in biuret reagent. The method may be performed using a standard or as an absolute method, using the extinction coefficient.

4.11.4.2.3. Reagents

See p. 657.
Precipitation reagent, base reagent, biuret reagent, protein standard.

4.11.4.2.4. Conditions

See p. 657.

4.11.4.2.5. Procedure

	Sample, ml	Standard, ml	Blank, ml
Phosphotungstic acid	0.5	0.5	—
Urine	0.5	—	—
Standard, 200 mg per 100 ml	—	0.5	—

Mix thoroughly on the vibrator for about 1 min. Allow to stand for 15 min (until the precipitate begins to settle). Centrifuge briskly for 5 min. Discard the supernatant

Precipitate	+++	+++	—
DM-water	—	—	0.02
Base reagent	0.5	0.5	0.5

Agitate the precipitate with the cortex-stirrer until it has completely dissolved (1–2 min)

Biuret reagent	0.5	0.5	0.5

Mix and allow to stand at room temperature for 30 min. Read the extinction against water at 546 nm

4.11.4.2.6. Calculation

Using the standard.

$$\text{mg per 100 ml} = \frac{E(S) - E(B)}{E(ST) - E(B)} \cdot 50$$

Using the extinction coefficient ($a = 3.12$).

$$\text{mg per 100 ml} = \frac{E}{ad} \cdot \frac{FV}{SV} = \frac{E}{3.12 \cdot 1} \cdot \frac{1.02}{0.5} \cdot 1\,000$$

Concentration $= E(S) - E(B) \cdot 6540$ mg l^{-1}

24-Hour excretion.

mg protein per 24 h $=$ mg protein l$^{-1} \cdot$ 24 hour urine volume in litres

4.11.4.2.7. Reference values

1. There have been very few results on the physiological proteinuria in recent times and these have been quite disparate. Heinemann *et al.*[1] set the

upper limit of the norm at an excretion of 150 mg per day and consider higher values to indicate proteinuria. Other authors give values of 48–150,[2] 141–247,[3] 43–127,[4] 82–207 (including mucoprotein),[5] even up to 300 mg per day.[6] In a later study, the last authors obtained values of 8–133 mg per day[7] and reported that no sex difference was found.[7,8] The nocturnal proteinuria (recumbent patients) varied between 4 and 100 (mean 24.5) μg min^{-1}, and in the day between 5 and 210 (mean 40.9) μg min^{-1}. Here it should be noted that the proteinuria is dependent on body position, being greater for standing than for recumbent patients.

2. Bodily exertion quickly leads to an increase in the proteinuria which can increase to 5 times the norm.

3. The higher day-time values of the proteinuria in comparison with nocturnal values is probably related to bodily activity.

4. In proteinurias, the amount excreted in the course of a day can vary very considerably. Accordingly, if a kidney disease is suspected it is always advisable to analyse a 24-hour urine specimen.

4.11.4.2.8. Comments on the method

Methods which precipitate the protein beforehand show the highest sensitivity.[14,15]

4.11.4.2.9. Diagnostic significance

The appearance of proteins in the urine is due to an increased glomerular permeability or to a disturbance of the tubular reabsorbtion or both.[9] In diseases associated with increased glomerular permeability, larger quantities of high-molecular-weight proteins (molecular weight > 60 000) pass into the primary urine and are either lost to the body as intact proteins or are partially reabsorbed by the tubulus cells and catabolized. The result is a decrease in the corresponding plasma protein.

If the proteinuria is due to a disturbance of the tubuli the amount of protein filtered by the glomeruli is normal. The filtered lower molecular weight proteins (molecular weight < 50 000) such as β_2-microglobulin, L-chains of immunoglobulins, and lysozyme, are merely inadequately taken up by the tubulus cells and they appear in the urine.

Glomerular proteinuria is found in various glomerular affections.[10] The most important are:

—acute, subacute and chronic glomerulonephritis of varied aetiology;
—inflammatory–degenerative glomerulopathia with nephrotic syndrome;
—degenerative–metabolic glomerulopathias (amyloidosis, diabetic glomerulosclerosis, pregnancy kidney);
—functional disturbances of the renal haemodynamic (kidney blockage in cardiac decompensation).

Tubular proteinuria is found in Fanconi syndrome, cystinosis, chronic cadmium poisoning, rejection reaction following kidney transplant, Wilson's disease, galactosaemia, and other, rarer tubular affections. It can also occur following an arginine infusion.[11]

Mixed proteinuria.[12] In pyelonephritis a purely tubular proteinuria is described,[13] but a glomerular loss of proteins has also been detected.[1]

In addition there are other special kinds of proteinuria. Functional proteinuria is temporary and hence does not indicate a nephropathia. It occurs most commonly following bodily exertion, fever, immersion in cold water, and following situations of stress.

The so-called orthostatic proteinuria should equally be classed among the benign proteinurias. It is most prominent in children. The frequency of orthostatic (benign) proteinuria in children varies between 5.8 and 60%. In children under 10 years, this type of proteinuria is rare. The frequency begins to increase in 12-year-olds, reaches its peak in 14- to 16-year-olds (up to 60% of all subjects investigated), subsequently falling to very low values again. In persons over 20 years, the incidence of orthostatic proteinuria should not exceed about 2%. Typically, benign proteinuria disappears when the patient lies down and attains its highest values in association with lordotic body posture (for literature, see ref. 9).

References

1. Heinemann, H. O., Maack, T. N., and Sherman, R. L., Proteinuria, *Am. J. Med.*, **56**, 71 (1974).
2. Savory, J., Sunderman, F. W., Jr., and Pu, P. H., Methods for measurement and fractionation of proteins in urine and serum, in Sunderman and Sunderman, *Laboratory diagnosis of kidney diseases*, Green, St. Louis, 1970.
3. Saifer, A. and Gerstenfeld, S., Photometric determination of urinary proteins., *Clin. Chem.*, **10**, 321 (1964).
4. Peterson, P. A., Evrin, P. E., and Berggaard, J. Differentiation of glomerular, tubular and normal proteinuria: determinations of urinary excretion of β_2-microglobulin, albumin and total protein, *J. clin. Invest.*, **48**, 1189 (1969).
5. Doetsch, K. and Gradsden, R. M., Determination of total urinary protein, combining Lowry sensitivity and biuret specificity, *Clin. Chem.*, **19**, 1170 (1973).
6. Balant, L., Mulli, J.-C., and Fabre, J., Urinary protein analysis with sodium dodecylsulfate polyacrylamide gel electrophoresis—a comparison with other analytical techniques, *Clin. chim. Acta*, 27 (1974).
7. Cao, A., Balant, L., Fabre, J., and Canavese, J.-C., La protéinurie physiologique., *Schweiz. med. Wschr.*, **105**, 421 (1975).
8. Hemmingsen, L. and Skaarup, P., The 24-hour excretion of plasma proteins in the urine of apparently healthy subjects, *Scand. J. clin. Lab. Invest.*, **35** 347 (1975).
9. Colombo, J. P. and Richterich, R., *Die einfache Urinuntersuchung*, Huber, Berne, 1977.
10. Reubi, F., *Nierenkrankheiten*, Huber, Berne, 1970, p. 140.
11. Mogensen, C. E., Vittunghus, E., and Sölling, K., Increased urinary excretion

of albumin, light chains and β_2-microglobulin after intravenous arginine administration in normal man, *Lancet*, **ii**, 581 (1975).

12. Manuel, Y. and Revillard, J. P., Study of urinary proteins by zone electrophoresis. Methods and principles of interpretation, in Manuel, Revillard and Beutal, *Proteins in normal and pathological urine*, Karger, Basle, 1970, p. 156.
13. Traeger, J., Revillard, J. P., and Manuel, Y., La protéinurie tubulaire au cours des pyélonéphrites chroniques, in Mertz and Kluthe, *Aktuelle Probleme der klinischen Nephrologie*, Thieme, Stuttgart, 1967, p. 39.
14. Yatzidis, H., New colorimetric method for quantitative determination of protein in urine, *Clin. Chem.*, **23**, 811 (1977).
15. Lizana, J., Brito, M., and Davis, M. R., Assessment of five quantitative methods for determination of total proteins in urine, *Clin. Biochem.*, **10**, 89 (1977).

4.11.4.3. Urinary ammonia: direct determination by Berthelot's method

4.11.4.3.1. Choice of method

Ammonia is present in the urine in such a large excess in comparison with substances liable to cause interference (amino acids, proteins) that the direct determination, without prior deproteination or isolation of the ammonia, is adequate for most investigations. Since the pK of the ammonium–ammonia system is around 9.4 and the pH of normal urine is less than 7, by far the largest part of the ammonia is present as ammonium ions. Today the determination is carried out almost exclusively by the Berthelot reaction (p. 392).

4.11.4.3.2. Collection of urine

The urine is collected and preserved, either by the addition of 5 ml of thymol–isopropanol solution or by deep-freezing individual portions (prevention of cleavage of urea by bacterial urease). Turbid urine is first filtered. Since the concentration of ammonia in urine can vary considerably it is advisable to perform assays on (a) an undiluted specimen, (b) a specimen diluted 1:10 (1 part urine + 9 parts DM-water).

4.11.4.3.3. Reagents

See p. 393.

Standard ammonium solution, $10 \, mmol \, l^{-1}$. Dissolve 0.6605 g of ammonium sulphate in DM-water and make up to 1000 ml. The solution keeps indefinitely if frozen.

4.11.4.3.4. Procedure

	Sample, ml	Reagent blank, ml	Standard, ml
DM-water	0.5	0.5	0.5
Urine	0.02	—	—
DM-water	—	0.02	—
Standard	—	—	0.02
Phenol solution	10.0	10.0	10.0
Hypochlorite–NaOH solution	1.0	1.0	1.0

After 10 min at room temperature, read the extinction against water between 540–546–580 nm. The colour is stable for 2 h

4.11.4.3.5. Calculation

$$\text{Concentration} = \frac{E(S) - E(RB)}{E(ST) - E(RB)} \cdot 10 \text{ mmol l}^{-1} \text{ (NH}_4^+)$$

4.11.4.3.6. Reference values

The excretion of ammonia in the urine is primarily dependent on the acid–base status. Thus, in order to obtain comparable results the determination must be made under standardized test conditions.

4.11.4.3.7. Specificity

The Berthelot reaction, as performed directly, is not absolutely specific for ammonia. A less intense colour reaction is obtained with uric acid, creatinine, amino acids, and sulphonamides.

4.11.4.3.8. Comments on the method

1. The DM-water used for the ammonia determination must be absolutely free of ammonia (if necessary, distil twice).

2. Urine specimens for ammonia determination may be kept in sterile vessels for 8 days at −20 °C.[1]

Reference

1. Hoge, J. H. C., Hazenberg, H. J. A., and Gips, C. H., Determination of ammonia in urine with an ammonium electrode and with a direct method, *Clin. chim. Acta*, **55**, 273 (1974).

4.11.4.4. Urinary glucose: enzymatic determination with hexokinase and glucose-6-phosphate dehydrogenase

An exact determination of the urinary glucose excretion is currently held to be necessary in the following situations:

1. For unequivocal diagnosis and monitoring the course of diabetes.
2. The complete absence of glucose in the urine is evidence for the presence of a bacteriuria.[1] From the experience gained so far, a reduction in the urinary glucose concentration is at least of equal value to the customary screening tests. For screening for bacteriuria it is necessary to investigate a concentrated, i.e. morning urine, specimen.
3. An exact method for the determination of urinary glucose is necessary for the faultless differentiation between the not very rare renal glucosuria and diabetes.
4. In investigations on the physiological glucosuria, only the highly sensitive methods come into consideration.

At present, the only method which fulfils all these requirements is the determination with hexokinase/glucose-6-phosphate dehydrogenase.

4.11.4.4.1. Reference values for the concentration of glucose in urine

The use of this method has shown that healthy persons always excrete a certain amount of glucose in the urine.[2,3] The urinary glucose concentration should be 7.1 ± 3.8[2] or 6.3 ± 4.2 mg per 100 ml $(\bar{x} \pm s)$.[3] In a further study,[4] a mean value of 6.5 mg per 100 ml was found, 85% of the values lying between 1 and 15 mg per 100 ml. Here the upper limit of the norm was established as 30 mg per 100 ml.

There are few results on the excretion of glucose above 24 h. The values lie between 50 and 70,[4] or is 67.11 ± 14.75 mg per 24 h.[5] From our own experience, the normal urinary glucose concentration in adults lies between 5 and 15 mg per 100 ml and the urinary glucose excretion between 25 and 100 mg per 24 h.

It may be mentioned that the glucose excretion in the urine is inversely proportional to the flow-rate of urine;[2,3] the concentration can fall from 20 to 5 mg per 100 ml as a consequence of a severe diuresis.

4.11.4.4.2. Diagnostic significance

1. Physiological glucosuria

2. Increased glomerular filtration of glucose in hyperglycaemia (Diabetes mellitus)

3. Tubulopathia of the proximal tubulus
   Primary tubular syndromes:
   (a) Disturbance of the glucose reabsorbtion
       Familial renal glucosuria = diabetes renalis
   (b) Disturbance of several specific tubular functions:
       Glucosuria combined with amino- and phosphaturia
       Fanconi syndrome
       Diabetes glucophosphaticus
       Fam. glucosuria with aminoaciduria
       Fam. gluco-glycinuria
   Probable secondary tubular syndromes:
       Disturbances of several specific tubular functions:
       Glucosuria in hereditary metabolic diseases:
       Cystinosis
       Galactosaemia
       Wilson's disease
       Lowe's oculo-cerebro-renal syndrome
       Glycogenosis
       Glucosuria in toxic kidney damage

4. Symptomatic glucosuria, possibly combined with the excretion of other
   sugars (attendant mellliturias):
       Neo-natal mellituria
       Hunger glucosuria
       Emotional glucosuria
       Glucosuria in infectious diseases
       Attendant mellliturias in diseases of the liver and kidneys
       Hormonally conditioned mellliturias
       Pregnancy mellliturias
       Alimentary glucosuria

References

1. Scherstén, B. and Fritz, H., Subnormal levels of glucose in urine. A sign of urinary tract infection, *J. Am. med. Ass.*, **201**, 949 (1967).
2. Renschler, H. E., Weicker, H., and Bayer, H., Die obere Normgrenze der Glucosekonzentration im Urin Gesunder, *Dt. med. Wschr.*, **90**, 2349 (1965).
3. Schubert, G. E., Schuster, H. P., and Baum, P., Physiologische Glucosurie bei verschiedenen Diuresezuständen, *Klin. Wschr.*, **42**, 619 (1964).
4. Hannsch, A. W. and Kaiser, G., Untersuchung zur Glucose-Ausscheidung im Urin bei Stoffwechselgesunden, *Medsche Welt, Stuttg.*, **21**, 1328 (1968).
5. Vitek, V. and Vitek, K., Chromatography of sugars in body fluids, *Biochem. Med.*, **4**, 282 (1970).

APPENDIX 1. BUFFER SOLUTIONS

The buffer solutions described are classified according to whether they are prepared in accordance with constant ionic strength (I) (for electrophoretic investigations) or constant molarity (M) (for other purposes).

For the amine buffers (introduced by Gomori,[1,2] which are particularly suited for biological work, see *Ann. N.Y. Acad. Sci.*, **92**, art. 2 (1961).

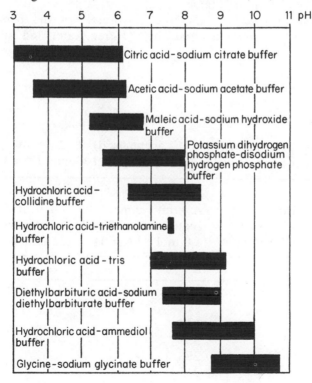

Citrate buffer[3,4]

23 °C; A(= ml 0.1 mol l^{-1} citric acid) + B(= ml 0.1 mol l^{-1} sodium citrate) are dissolved in DM-water and made up to 100 ml. M = constant 50 mmol l^{-1}.

pH	A	B	pH	A	B
3.0	46.5	3.5	4.6	25.5	24.5
3.2	43.7	6.3	4.8	23.0	27.0
3.4	40.0	10.0	5.0	20.5	29.5
3.6	37.0	13.0	5.2	18.0	32.0
3.8	35.0	15.0	5.4	16.0	34.0
4.0	33.0	17.0	5.6	13.7	36.3
4.2	31.5	18.5	5.8	11.8	38.2
4.4	28.0	22.0	6.0	9.5	41.5
			6.2	7.2	42.8

Solutions

Citric acid, 0.1 mol l⁻¹. Dissolve 21.02 g of citric acid monohydrate in DM-water and make up to 1000 ml.

Sodium citrate, 0.1 mol l⁻¹. Use the dihydrate, *not* the pentahydrate. Dissolve 29.41 g of trisodium citrate in CO_2-free water and make up to 1000 ml.

Acetate buffer[5]

25 °C; dissolve A(= ml 1.0 mol l⁻¹ acetic acid) + B(= ml 1.0 mol l⁻¹ sodium hydroxide) in DM-water and make up to 1000 ml. I = constant.

pH	$I = 0.05$ A = ml 1.0 mol l⁻¹ acetic acid + 50 ml 1.0 mol l⁻¹ NaOH	$I = 0.1$ A = ml 1.0 mol l⁻¹ acetic acid + 100 ml 1.0 mol l⁻¹ NaOH	$I = 0.2$ A = ml 1.0 mol l⁻¹ acetic acid + 200 ml 1.0 mol l⁻¹ NaOH
	A	A	A
3.6	650		
3.8	428	828	
4.0	288	559	
4.2	200	389	760
4.4	145	283	553
4.6	110	215	423
4.8	87.7	173	341
5.0	73.8	146	289
5.2	65.0	129	256
5.4	59.5	118	235
5.6	56.0	112	222
5.8	53.8	107	

Solutions
    Acetic acid, 1.0 mol l^{-1}. Dilute 57.5 ml of glacial acetic acid to 1000 ml with DM-water. Monitor by titration with alkali (indicator: phenolphthalein) and adjust if necessary.
    Sodium hydroxide, 1.0 mol l^{-1}. Preparation of a CO_2-free stock solution: prepare a concentrated solution of 100 g of NaOH in 100 ml of DM-water, allow to stand, and use the clear supernatant. Dilute 65 ml of the stock solution to 1000 ml with CO_2-free water. Monitor the concentration by titration with hydrochloric acid (indicator: methyl red) and adjust if necessary.

Maleate buffer[4,6]
23 °C; dissolve 50 ml of 0.2 mol l^{-1} monosodium maleate + B(= ml 0.2 mol l^{-1} sodium hydroxide) in DM-water and make up to 1000 ml.

pH	B	pH	B
5.2	7.2	6.0	26.9
5.4	10.5	6.2	33.0
5.6	15.3	6.4	38.0
5.8	20.8	6.6	41.6
		6.8	44.4

Solutions
    Sodium maleate, 0.2 mol l^{-1}. Dissolve either 23.22 g of maleic acid or 19.61 g of maleic anhydride in 0.2 mol l^{-1} NaOH and make up to 1000 ml.
    Sodium hydroxide, 0.2 mol/l^{-1}. Preparation of a CO_2-free solution, see acetic acid–sodium acetate buffer. Dilute 13 ml of the stock solution to 1000 ml with CO_2-free water. Check the concentration by titration with hydrochloric acid (indicator: methyl red) and adjust if necessary.

Potassium dihydrogen phosphate–disodium hydrogen phosphate buffer[5]
25 °C; dilute A(= ml 0.5 mol l^{-1} KH_2PO_4) + B(= ml 0.5 mol l^{-1} Na_2HPO_4) to 1000 ml with DM-water. I = constant.

pH	$I = 0.05$		$I = 0.1$		$I = 0.2$	
	A	B	A	B	A	B
5.6					333	22.4
5.8			159	13.8	303	32.4
6.0	74.2	8.58	142	19.5	265	44.8

ph	$I = 0.05$		$I = 0.1$		$I = 0.2$	
	A	B	A	B	A	B
6.2	64.6	11.8	121	26.4	222	59.4
6.4	53.4	15.5	98.2	34.0	176	74.6
6.6	42.0	19.3	75.6	41.4	133	89.2
6.8	31.4	22.8	55.4	48.2	95.2	102
7.0	22.4	25.8	39.0	53.6	65.8	111
7.2	15.4	28.2	26.4	57.8	44.2	119
7.4	10.3	30.0	17.6	60.8	29.2	124
7.6	6.74	31.0	11.5	62.8	18.9	127
7.8	4.36	31.8	7.38	64.2	12.1	129
8.0	2.80	32.4				

Solutions

Potassium dihydrogen phosphate, 0.5 mol l^{-1}. Dry for 1–2 h at 110 °C; dissolve 68.05 g in DM-water and make up to 1000 ml.

Disodium hydrogen phosphate, 0.5 mol l^{-1}. Dry the anhydrous salt for 1–2 h at 110 °C; dissolve 70.99 g in CO_2-free water and make up to 1000 ml.

Sörensen phosphate buffer[8,9], 66.7 mmol l^{-1} (isotonic)
18 °C; dilute A [ml 66.7 mmol l^{-1}(1/15 M) potassium dihydrogen phosphate] to 100 ml with 66.7 mmol l^{-1}(1/15 M) disodium hydrogen phosphate.

pH	A	pH	A
5.0	98.8	6.8	50.8
5.2	98.0	7.0	39.2
5.4	96.7	7.2	28.5
5.6	94.8	7.4	19.6
5.8	91.9	7.6	13.2
6.0	87.7	7.8	8.6
6.2	81.5	8.0	3.3
6.4	73.2	8.2	
6.6	62.7		

Solutions

Potassium dihydrogen phosphate, 66.7 mmol l^{-1}. Dry at 110 °C; dissolve 9.08 g of KH_2PO_4 in CO_2-free DM-water and make up to 1000 ml.

Disodium hydrogen phosphate, 66.7 mmol l^{-1}. Dry the anhydrous salt at 110 °C for 1–2 h; dissolve 9.47 g of Na_2HPO_4 in CO_2-free water and make up to 1000 ml.

Collidine buffer[1]
Dilute $A(=$ ml 0.1 mol l^{-1} hydrochloric acid) $+$ 25 ml of 0.2 mol l^{-1} collidine to 100 ml with DM-water. $M =$ constant 50 mmol l^{-1}.

A = ml HCl, 0.1 mol l^{-1}	Desired pH		A = ml HCl, 0.1 mol l^{-1}	Desired pH	
	23 °C	37 °C		23 °C	37 °C
5.0	8.35	8.25	27.5	7.30	7.23
7.5	8.18	8.10	30.0	7.22	7.14
10.0	8.00	7.94	32.5	7.13	7.05
12.5	7.88	7.80	35.0	7.03	6.95
15.0	7.77	7.70	37.5	6.92	6.84
17.5	7.67	7.60	40.0	6.80	6.72
20.0	7.57	7.50	42.5	6.62	6.54
22.5	7.50	7.40	45.0	6.45	6.37
25.0	7.40	7.32			

Solutions

Hydrochloric acid, 0.1 mol l^{-1}. Dilute about 9 ml of concentrated hydrochloric acid (36% HCl) to 1000 ml with DM-water. Check the concentration by titration with alkali (indicator; methyl red) and adjust if necessary.

Collidine, 0.2 mol l^{-1}. Dilute about 27 ml of collidine to 1000 ml with CO_2-free water. Check the concentration by titration with hydrochloric acid (indicator: methyl red) and adjust if necessary.

Triethanolamine–HCl–EDTA buffer[7], 50 mmol l^{-1}, EDTA 5 mmol l^{-1}, pH 7.6
Dissolve 7.6 g of triethanolamine (redistilled) $+$ 15 ml of 2.0 mol l^{-1} hydrochloric acid $+$ 2 g of ethylenediaminetetraacetate, disodium salt, dihydrate to 1000 ml with quartz-distilled water and make up to 1000 ml.

Tris buffer[5]
25 °C; dissolve $A(=$ ml 1.0 mol l^{-1} hydrochloric acid) $+$ $B[=$ ml 1.0 mol l^{-1} tris(hydroxymethyl)aminomethane] in DM-water and make up to 1000 ml. $I =$ constant.

716

pH	$I = 0.05$ $A = 50$ ml 1.0 mol l^{-1} HCl + B = ml 1.0 mol l^{-1} tris B	$I = 0.1$ $A = 100$ ml 1.0 mol l^{-1} HCl + B = ml 1.0 mol l^{-1} tris B	$I = 0.2$ $A = 200$ ml 1.0 mol l^{-1} HCl + B = ml 1.0 mol l^{-1} tris B
7.0	53.6	107	214
7.2	55.7	111	222
7.4	59.1	118	235
7.6	64.4	128	257
7.8	72.9	144	290
8.0	86.1	169	342
8.2	107	208	421
8.4	141	270	550
8.6	194	367	738
8.8	279	524	
9.0	414	761	
9.2	627		

Solutions

Tris(hydroxymethyl)aminomethane, 1.0 mol l^{-1}. Recrystallize the amine by dissolving 500 g by heating with 250 ml of DM-water and 600 ml of methanol, filter through charcoal, cool, and filter off. Dissolve 121.14 g in CO_2-free DM-water and make up to 1000 ml. Check the concentration by titration with hydrochloric acid (indicator: methyl orange) and adjust if necessary.

Hydrochloric acid, 1.0 mol l^{-1}. Dilute about 90 ml of concentrated hydrochloric acid (36% HCl) to 1000 ml with DM-water. Check the concentration by titration with alkali (indicator: methyl red) and adjust if necessary.

Tris buffer[1]

Dilute $A(=$ ml 0.1 mol l^{-1} hydrochloric acid) + 25 ml of 0.2 mol l^{-1} tris to 100 ml with DM-water. M = constant 50 mmol l^{-1}.

A = ml 0.1 mol l^{-1}	Desired pH 23 °C	37 °C	A = ml 0.1 mol l^{-1} HCl	Desired pH 23 °C	37 °C
5.0	9.10	8.95	27.5	8.05	7.90
7.5	8.92	8.78	30.0	7.96	7.82
10.0	8.74	8.60	32.5	7.87	7.73

A = ml 0.1 mol l^{-1} HCl	Desired pH		A = ml 0.1 mol l^{-1} HCl	Desired pH	
	23 °C	37 °C		23 °C	37 °C
12.5	8.62	8.48	35.0	7.77	7.63
15.0	8.50	8.37	37.5	7.66	7.52
17.5	8.40	8.27	40.0	7.54	7.40
20.0	8.32	8.18	42.5	7.36	7.22
22.5	8.23	8.10	45.0	7.20	7.05
25.0	8.14	8.00			

Solutions

Hydrochloric acid, 0.1 mol l^{-1}. See hydrochloric acid–collidine buffer.

Tris(hydroxymethyl)aminomethane, 0.2 mol l^{-1}. Recrystallization and preparation, see tris–HCl buffer. Dissolve 24.23 g in CO_2-free DM-water and make up to 1000 ml.

Diethylbarbituric acid—sodium diethylbarbiturate buffer[5]

25 °C; dilute A(= ml 0.025 mol l^{-1} diethylbarbituric acid) + B(= ml 0.5 mol l^{-1} sodium diethylbarbiturate) + C(= ml 0.5 mol l^{-1} sodium chloride) to 1000 ml with DM-water. I = constant.

pH	$I = 0.05$			$I = 0.1$			$I = 0.2$		
	A	B	C	A	B	C	A	B	C
7.4	648	10	90	639	10	190	275	5	395
7.6	409	10	90	403	10	190	348	10	390
7.8	645	25	75	636	25	175	219	10	390
8.0	814	50	50	401	25	175	346	25	375
8.2	514	50	50	506	50	150	218	25	275
8.4	648	100	—	639	100	100	275	50	350
8.6	409	100	—	403	100	100	348	100	300
8.8	258	100	—	509	200	—	438	200	200
9.0	163	100	—	321	200	—	277	200	200

Solutions

Diethylbarbituric acid, 25 mmol l^{-1}. Recrystallize the substance from ethanol and dry over P_2O_5 in vacuum. Dissolve 4.604 g in DM-water and make up to 1000 ml.

Sodium diethylbarbiturate, 0.5 ml l^{-1}. Recrystallization: dissolve 100 g of substance in 300 ml of hot water, filter, add 300 ml of ethanol, cool to

718

5 °C, and filter. Dry in vacuum over P_2O_5. Dissolve 103.09 g in CO_2-free water and make up to 1000 ml. The solution does not keep well (decomposition).

Ammediol buffer[1]

0.05 mol l^{-1}; dilute $A (=$ mol 0.1 mol l^{-1} hydrochloric acid$) + $ 25 ml of 0.2 mol l^{-1} ammediol to 100 ml with DM-water. $M = $ constant 50 mmol.

$A = $ ml 0.1 mol l^{-1} HCl	Desired pH 23 °C	37 °C	$A = $ ml 0.1 mol l^{-1} HCl	Desired pH 23 °C	37 °C
5.0	9.72	9.62	27.5	8.70	8.58
7.5	9.56	9.45	30.0	8.60	8.50
10.0	9.38	9.27	32.5	8.50	8.40
12.5	9.26	9.15	35.0	8.40	8.30
15.0	9.15	9.03	37.5	8.30	8.20
17.5	9.05	8.94	40.0	8.18	8.07
20.0	8.96	8.85	42.5	8.00	7.90
22.5	8.87	8.76	45.0	7.83	7.72
25.0	8.78	8.67			

Solutions

Hydrochloric acid, 0.1 mol l^{-1}. See Collidine buffer.

2-Amino-2-methylpropane-1,3-diol, 0.2 mol l^{-1}. Dissolve 21.03 g in DM-water and make up to 1000 ml. For recrystallization and preparation, see following section.

Ammediol buffer[5]

25 °C; dilute $A (=$ ml 1.0 mol l^{-1} hydrochloric acid$) + B(=$ ml 1.0 mol l^{-1} 2-amino-2-methylpropane-1,3-diol$)$ to 1000 ml with DM-water. $I = $ constant.

pH	$I = 0.05$ $A = 50$ ml 1.0 mol l^{-1} HCl $+ B =$ ml ammediol B	$I = 0.1$ $A = 100$ ml 1.0 mol l^{-1} HCl $+ B =$ ml ammediol B	$I = 0.2$ $A = 200$ ml 1.0 mol l^{-1} HCl $+ B =$ ml ammediol B
8.2	61.7	120.9	239
8.4	68.2	133.7	260
8.6	79.0	154	294
8.8	95.9	183	348
9.0	122	229	441
9.2	164	303	606
9.4	232	424	
9.6	338	626	
9.8	511		
10.0	790		

Solutions

Hydrochloric acid, 1.0 mol l^{-1}. Preparation, see Tris buffer.

2-Amino-2-methylpropane-1,3-diol, 1.0 mol l^{-1}. Recrystallize 100 g from absolute ethanol. Dry in a vacuum desiccator. Dissolve 105.14 g in CO_2-free water and make up to 1000 ml. Check the concentration by titration with hydrochloric acid (indicator: methyl orange) and adjust if necessary.

Glycine buffer[5]

Dilute 1.0 mol l^{-1} sodium hydroxide solution $+ A (=$ ml 2.0 mol l^{-1} glycine) to 1000 ml with DM-water. $I =$ constant.

pH	$I = 0.05$ $A =$ ml 2.0 mol l^{-1} glycine $+$ 50 ml 1.0 mol l^{-1} NaOH A	$I = 0.1$ $A =$ ml 2.0 mol l^{-1} glycine $+$ 100 ml 1.0 mol l^{-1} NaOH A	$I = 0.2$ $A =$ ml 2.0 mol l^{-1} glycine $+$ 200 ml 1.0 mol l^{-1} NaOH A
8.8	234	467	
9.0	157	313	633
9.2	108	216	437
9.4	77.4	155	312

pH	$I = 0.05$ A = ml 2.0 mol l^{-1} glycine + 50 ml 1.0 mol l^{-1} NaOH A	$I = 0.1$ A = ml 2.0 ml l^{-1} glycine + 100 ml 1.0 mol l^{-1} NaOH A	$I = 0.2$ A = ml 2.0 ml l^{-1} glycine + 200 ml 1.0 mol l^{-1} NaOH A
9.6	58.1	116	234
9.8	45.8	91.7	185
10.0	38.1	76.3	153
10.2	33.2	66.5	134
10.4	30.1	60.3	121
10.6	28.0	56.4	113
10.8	26.7	53.8	108

Solutions

Glycine, 2.0 mol l^{-1}. Dry the substance for 1–2 h at 110 °C. Dissolve 150.14 g in DM-water and make up to 1000 ml.

Sodium hydroxide solution, 1.0 mol l^{-1}. See Acetate buffer.

Glycine buffer[8,9]

X = ml 0.1 ml l^{-1} glycine in 0.1 mol l^{-1} sodium chloride solution + $(100 - X)$ ml 0.1 mol l^{-1} sodium hydroxide solution (18 °C). M = constant 100 mmol l^{-1}.

pH	X	pH	X
8.4	96.5	10.6	53.7
8.6	94.8	10.8	52.2
8.8	92.1	11.0	51.0
9.0	88.5	11.2	50.3
9.2	84.2	11.4	49.7
9.4	79.0	11.6	48.8
9.6	73.2	11.8	47.8
9.8	67.7	12.0	46.2
10.0	63.0	12.2	43.6
10.2	59.0	12.4	40.0
10.4	56.1	12.6	33.4
		12.8	24.2
		13.0	7.5

721

Solutions

Glycine–NaCl, 0.1 mol l⁻¹. Dry the glycine and NaCl for 1–2 h at 110 °C. Dissolve 7.5 g of glycine and 5.85 g of NaCl in water and make up to 1000 ml.

Sodium hydroxide solution, 0.1 ml l⁻¹. For the preparation of the CO_2-free stock solution, see p. 713. Dilute 6.5 ml of the stock solution to 1000 ml with CO_2-free water. Check the concentration by titration with hydrochloric acid (indicator: methyl red) and adjust if necessary.

References

1. Gomori, G., *Proc. Soc. exp. Biol. Med.*, **62**, 33 (1946).
2. Gomori, G., *Proc. Soc. exp. Biol. Med.*, **68**, 354 (1948).
3. Lillie, R. D., *Histopathologic technique*, Blakiston, Philadelphia, 1948.
4. Gomori, G., in *Methods in enzymology*, Vol. 1, Academic Press, New York, 1955, p. 138.
5. Datta, S. P. and Grzybowski, A. K., in Long, *Biochemist's handbook*, Van Nostrand, London, 1961, p. 19.
6. Temple, J. W., *J. Am. chem. Soc.*, **51**, 1754 (1929).
7. Beisenherz, G., Boltze, H. J., Bücher, T., Czok, R., Garbade, K. H., Meyer-Arendt, E., and Pfleiderer, G., *Z. Naturforsch*, **8b**, 555 (1953).
8. Sörensen, S. P. L., *Biochem. Z.*, **22**, 352 (1909).
9. Sörensen, S. P. L., *Ergebn. Physiol.*, **12**, 393 (1912).

APPENDIX 2. TRANSMISSION/EXTINCTION

Relationship between the transmission in % (%T) and the extinction (E) (absorbance; optical density).

%T	E	%T	E	%T	E	%T	E
100	0.000	75	0.125	50	0.301	25	0.602
99	0.004	74	0.131	49	0.310	24	0.620
98	0.009	73	0.137	48	0.319	23	0.638
97	0.013	72	0.143	47	0.328	22	0.658
96	0.018	71	0.149	46	0.337	21	0.678
95	0.022	70	0.155	45	0.347	20	0.699
94	0.027	69	0.161	44	0.357	19	0.721
93	0.032	68	0.168	43	0.367	18	0.745
92	0.036	67	0.174	42	0.377	17	0.770
91	0.041	66	0.181	41	0.387	16	0.796
90	0.046	65	0.187	40	0.398	15	0.824
89	0.051	64	0.194	39	0.409	14	0.854
88	0.056	63	0.201	38	0.420	13	0.886
87	0.061	62	0.208	37	0.432	12	0.921
86	0.066	61	0.215	36	0.444	11	0.959
85	0.071	60	0.222	35	0.456	10	1.000
84	0.076	59	0.229	34	0.469	9	1.046
83	0.081	58	0.237	33	0.482	8	1.097
82	0.086	57	0.244	32	0.495	7	1.155
81	0.092	56	0.252	31	0.509	6	1.222
80	0.097	55	0.260	30	0.523	5	1.301
79	0.102	54	0.268	29	0.538	4	1.398
78	0.108	53	0.276	28	0.552	3	1.523
77	0.114	52	0.284	27	0.569	2	1.699
76	0.119	51	0.292	26	0.585	1	2.000

APPENDIX 3. NOMOGRAM FOR CALCULATING THE BODY SURFACE AREA

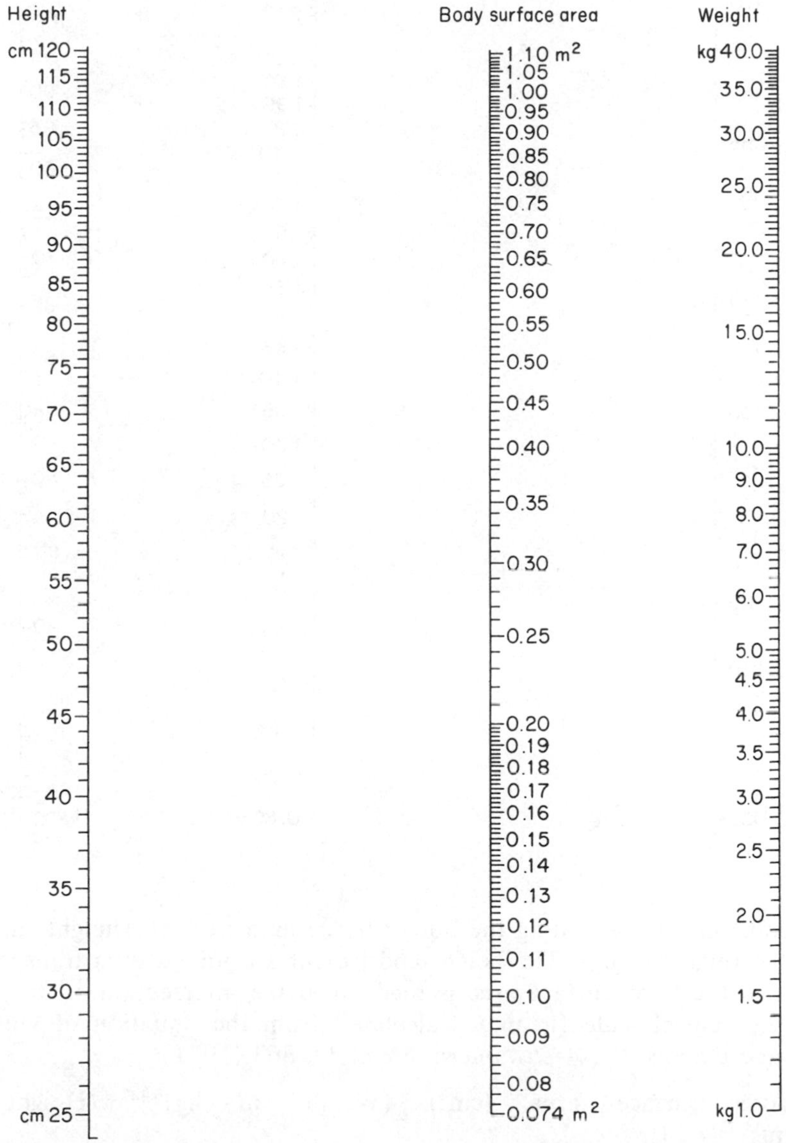

Height

cm 120
115
110
105
100
95
90
85
80
75
70
65
60
55
50
45
40
35
30
cm 25

Body surface area

1.10 m²
1.05
1.00
0.95
0.90
0.85
0.80
0.75
0.70
0.65
0.60
0.55
0.50
0.45
0.40
0.35
0.30
0.25
0.20
0.19
0.18
0.17
0.16
0.15
0.14
0.13
0.12
0.11
0.10
0.09
0.08
0.074 m²

Weight

kg 40.0
35.0
30.0
25.0
20.0
15.0
10.0
9.0
8.0
7.0
6.0
5.0
4.5
4.0
3.5
3.0
2.5
2.0
1.5
kg 1.0

Height	Body surface area	Weight
cm 200	2.80 m²	kg 150
195	2.70	145 / 140
190	2.60	135 / 130
185	2.50	125
180	2.40	120
175	2.30	115
170	2.20	110 / 105
165	2.10 / 2.00	100
160	1.95 / 1.90	95 / 90
155	1.85 / 1.80	85
150	1.75 / 1.70	80 / 75
145	1.65 / 1.60	70
140	1.55 / 1.50	65
135	1.45 / 1.40	60
130	1.35 / 1.30	55
125	1.25 / 1.20	50
120	1.15	45
115	1.10 / 1.05	40
110	1.00	
105	0.95	35
	0.90	
cm 100	0.86 m²	kg 30

Nomogram for evaluating the body surface area from the height (in cm) and the weight (in kg). The height and weight are joined by a transparent ruler and the body surface area is read off at the intersection of the ruler with the central scale (in m²). Calculated from the equation of Dubois, E. F. and Dubois, D., *Archs. intern. Med.*, **17**, 863 (1921):

Body surface area (cm^2) = (Weight in kg)$^{0.425}$ · (Height in cm)$^{0.725}$ · 71.84

APPENDIX 4. pCO_2 AND CO_2 NOMOGRAM

Nomogram for evaluating the pCO_2 from the barometric pressure (BP) and the volume per cent CO_2.

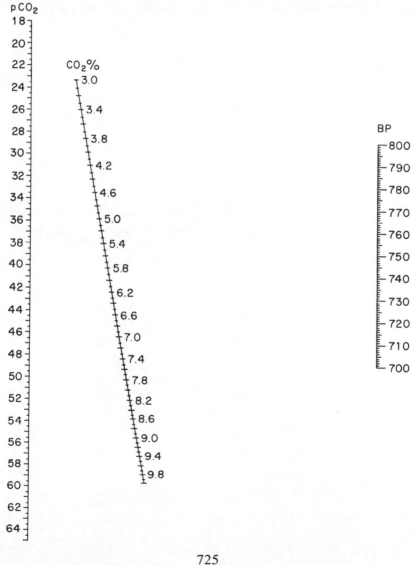

APPENDIX 5. ACID–BASE NOMOGRAM

Acid–base nomogram as constructed by Siggaard-Andersen, O. and Engel, K., *Scand. J. clin. Lab. Invest.*, **12**, 177 (1960), and Siggaard-Andersen, O., *Scand. J. clin. Lab. Invest.*, **14**, 598 (1962).

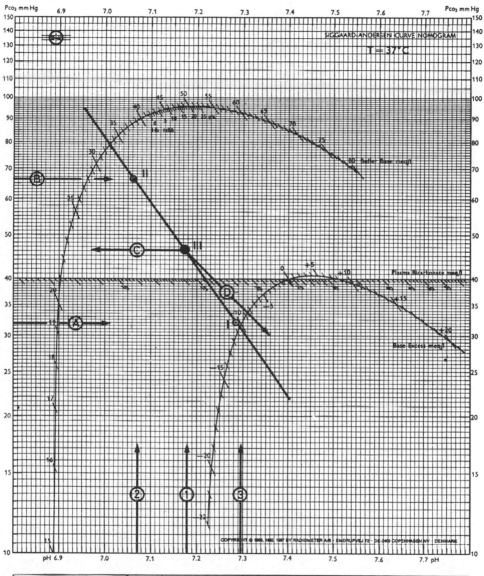

Patient's name:		Barometric pressure		766 mm Hg		READINGS			RESULTS		
		CO₂ percentage	Cylinder No 1: 4,4 %		Before equilibration	Actual pH: 7,18			Actual Pco₂	46	mm Hg
Dept:	Sample No.:		Cylinder No 2: 9,2 %						Base Excess	−11	mmol/l blood
Date:		CO₂ partial pressure	Cylinder No 1: 31,8 mm Hg		After equilibration	high Pco₂	pH: 7,07		Buffer Base	33,5	mmol/l blood
Hour of Sampling:			Cylinder No 2: 66,2 mm Hg			low Pco₂	pH: 7,29		Plasma Bicarb. at Pco₂=40 mm Hg	16	mmol/l plasma
Remarks:		Hemoglobin:		(8) g/100 ml		Readings made by:			Actual Bicarb.	15,5	mmol/l plasma
		Oxygen Saturation:		percent		Signature:			Total CO₂	17,9	mmol/l plasma

APPENDIX 6. SPECIFIC GRAVITY AND CONCENTRATION OF PERCHLORIC ACID AND TRICHLOROACETIC ACID

D = specific gravity at room temperature.
$C\%$ = concentration in g per 100 ml.
M = concentration in mol l^{-1}.
MW = molecular weight.

Perchloric acid $MW = 100.47$			Trichloroacetic acid $MW = 163.40$		
D	$C\%$	M	D	$C\%$	M
1.000	0		1.000	0	0
1.005	1.005	0.10	1.0025	0.50	0.031
1.011	2.022	0.20	1.0050	1.00	0.061
1.023	4.091	0.41	1.0076	1.50	0.092
1.035	6.209	0.62	1.0101	2.00	0.123
1.047	8.377	0.87	1.0126	2.50	0.155
1.060	10.60	1.12	1.0151	3.00	0.186
1.073	12.87	1.37	1.0177	3.50	0.218
1.086	15.20	1.64	1.0202	4.00	0.249
1.100	17.59	1.93	1.0227	4.50	0.281
1.114	20.04	2.22	1.0253	5.00	0.313
1.128	22.56	2.53	1.0278	5.50	0.345
1.143	25.14	2.86	1.0304	6.00	0.378
1.158	27.79	3.20	1.0329	6.50	0.410
1.174	30.52	3.57	1.0355	7.00	0.443
1.190	33.32	3.95	1.0380	7.50	0.476
1.207	36.20	4.35	1.0406	8.00	0.509
1.224	39.16	4.77	1.0431	8.50	0.542
1.242	42.22	5.22	1.0457	9.00	0.575
1.260	45.37	5.69	1.0483	9.50	0.608
1.279	48.62	6.19	1.0508	10.00	0.642

APPENDIX 7. CONVERSION FACTORS FOR SI UNITS

SI units for components of blood and urine, and for tests of function (conversion factors), after Stamm.[1]

Metabolite (relative molecular mass)	SI unit	Conversion factor From old to SI units	From SI to old units	Old unit
Components of blood				
Serum and plasma				
Albumins (69 000)	μmol l^{-1}	144.93	0.0069	g per 100 ml
α-Amino-nitrogen (14.0067)	mmol l^{-1}	0.7139	1.401	mg per 100 ml
Ammonia, NH_3 (17.03)	μmol l^{-1}	0.5872	1.703	μg per 100 ml
Bilirubin (584.65)	μmol l^{-1}	17.104	0.0585	mg per 100 ml
Lead (207.19)	μmol l^{-1}	0.0483	20.719	μg per 100 ml
Calcium (40.08)	mmol l^{-1}	0.2495	4.0080	mg per 100 ml
		0.5000	2.0000	mequiv l^{-1}
Chloride (35.453)	mmol l^{-1}	0.2821	3.5453	mg per 100 ml
		1.0000	1.0000	mequiv l^{-1}
Cholesterol (386.64)	mmol l^{-1}	0.0259	38.664	mg per 100 ml
Creatine (131.14)	μmol l^{-1}	76.254	0.0131	mg per 100 ml
Creatinine (113.12)	μmol l^{-1}	88.402	0.0113	mg per 100 ml
Iron (55.847)	μmol l^{-1}	0.1791	5.5847	μg per 100 ml
Iron-binding capacity (Fe: 55.847)	μmol l^{-1}	0.1791	5.5847	μg per 100 ml
Free fatty acids	μmol l^{-1}	1000.0	0.0010	mequiv l^{-1}
Fibrinogen	g l^{-1}	0.0100	100.00	mg per 100 ml
Fructose (180.16)	mmol l^{-1}	0.0555	18.016	mg per 100 ml
Galactose (180.16)	mmol l^{-1}	0.0555	18.016	mg per 100 ml
Glucose (180.16)	mmol l^{-1}	0.0555	18.016	mg per 100 ml
Haemoglobin	mg l^{-1}	0.1000	10.000	mg per 100 ml
Haemoglobin (Hb4: 16 144.5)	mmol l^{-1}	2.5907	0.3860	mg per 100 ml
Uric acid (168.11)	μmol l^{-1}	59.485	0.0168	mg per 100 ml
Urea (60.06)	mmol l^{-1}	0.1665	6.0060	mg per 100 ml
Urea-N (N: 14.0067)	mmol l^{-1}	0.3561	2.8080	mg per 100 ml
Haptoglobin	g l^{-1}	0.0100	100.00	mg per 100 ml
β-Hydroxybutyric acid (104.10)	μmol l^{-1}	96.062	0.0104	mg per 100 ml
Potassium (39.102)	mmol l^{-1}	0.2557	3.9102	mg per 100 ml
Copper (63.546)	μmol l^{-1}	0.1574	6.3546	μg per 100 ml
Lactate (90.08), lactic acid	mmol l^{-1}	0.1110	9.0080	mg per 100 ml
Lipids, total	g l^{-1}	0.0100	100.00	mg per 100 ml
Lipoproteins	g l^{-1}	0.0100	100.00	mg per 100 ml

Metabolite (relative molecular mass)	SI unit	Conversion factor		Old unit
		From old to SI units	From SI to old units	
Magnesium (24.312)	mmol l^{-1}	0.4113	2.4312	mg per 100 ml
		0.5000	2.0000	mequiv l^{-1}
Methaemoglobin (Hb/4: 16 114.5)	μmol l^{-1}	621.12	0.00161	g per 100 ml
Molality	mmol l^{-1}	1.0000	1.0000	mosm per kg
Myoglobin (17 100)	μmol l^{-1}	0.5848	1.7100	mg per 100 ml
Sodium (22.989)	mmol l^{-1}	0.4350	2.2989	mg per 100 ml
		1.0000	1.0000	mequiv l^{-1}
Oxyhaemoglobin	l	0.0100	100.00	%
Phospholipids (mean molecular mass: 774;	mmol l^{-1}	1.2920	0.7740	g per l
P: 30.9738)	mmol l^{-1}	0.3229	3.0974	mgP per 100m
Phosphorus (30.9738), inorganic (pH = 7.40)	mmol l^{-1}	0.3229	3.0974	mgP per 100 ml
		0.5556	1.8000	mequiv l^{-1}
Proteins	g l^{-1}	10.000	0.1000	g per 100 ml
C-reactive protein	mg l^{-1}	10.000	0.1000	mg per 100 ml
Pyruvate (88.06)	μmol l^{-1}	113.56	0.00881	mg per 100 ml
Sulphaemoglobin (Hb/4: 16 144.5)	μmol l^{-1}	621.12	0.00161	g per 100 ml
Transferrin	g l^{-1}	0.0100	100.00	mg per 100 ml
Triglycerides	mmol l^{-1}	0.0114	87.500	mg per 100 ml
(mean molecular mass: 875;				Triglyceride
glycerine: 92.09)		0.1086	9.2090	mg per 100 ml Glycerine
Zinc (65.37)	μmol l^{-1}	0.1530	6.5370	μg per 100 ml
Acid–base status				
pH	$-$log molc.		1.0000	$-$log molc.
Bicarbonate (61.02)	mmol l^{-1}	1.0000	1.0000	mequiv l^{-1}
pCO$_2$	kPa	0.1333	7.5020	mmHg
Base excess	mmol l^{-1}	1.0000	1.0000	mequiv l^{-1}
pO$_2$	kPa	0.1333	7.5020	mmHg
Hormones				
ACTH (4 541.1)	ρmol l^{-1}	0.2202	4.5410	ng l^{-1}
Cortisone (362.47)	μmol l^{-1}	0.0276	36.247	μg per 100 ml
Insulin (5 807.6)	ρmol l^{-1}	172.12	0.00581	ng per ml
PBI (iodine: 126.90)	nmol l^{-1}	78.802	0.0127	μg per 100 ml
Somatotropin (22 000)[a]	ρmol l^{-1}	45.454	0.0220	ng per ml
Growth hormone				
Thyroxine (776.93)	nmol l^{-1}	12.871	0.0777	μg per 100 ml

[a]amino acid sequence still insufficiently certain.

Vitamins				
Folic acid (441.41)	nmol l^{-1}	22.655	0.0441	μg per 100 ml
Carotine (536.89)	μmol l^{-1}	0.0186	53.689	μg per 100 ml
Vitamin A (286.46)	μmol l^{-1}	0.0349	28.646	μg per 100 ml
Vitamin B$_{12}$ (1 355.42)	μmol l^{-1}	0.7378	1.3554	ng per 100 ml
Vitamin C (176.13)	μmol l^{-1}	56.776	0.0176	mg per 100 ml

Metabolite (relative molecular mass)	SI unit	Conversion factor		Old unit
		From old to SI units	From SI to old units	
Haematology				
Erythrocytes	T l^{-1}	1.0000	1.0000	Mill. per mm^{-3}
(T = tera = 10^{12};				
M = mega = 10^6)		1.0000	1.0000	M per μl^{-1}
Erythrocytes (volume ratio)	l l^{-1}	0.0100	100.00	%
(haematocrit)				
Erythrocyte sedimentation rate	mm h^{-1}			
(BKS)				
Erythrocyte volume	fl	1.0000	1.0000	μ m^{-3}
(MCV = mean volume of the		1.0000	1.0000	μ^3
individual erythrocytes)				
Haemoglobin (Hb/4: 16 144.5)	g l^{-1}	0.1000	10.000	g per 100 ml
	mmol l^{-1}	0.0621	16.110	g per 100 ml
Ery-haemoglobin (Hb/4: 16 144.5)	fmol	0.0621	16.110	ρg
Hb$_E$, MCH		0.0621	16.110	$\gamma\gamma(\mu\mu$g)
Ery-haemoglobin (Hb/4: 16 114.5)	mmol l^{-1}	0.0621	16.110	g per 100 ml
molc.				
Leucocytes (G = giga = 10^9)	G l^{-1}		10^3	1 per mm^3
Erys-reticulocytes	1	0.0010	1000.0	$^0/_{00}$
Thrombocytes (G = giga = 10^9)	G l^{-1}	0.0010	1000.0	1 per mm^3
Pharmaceuticals and alcohol				
Ethanol (46.07)	mmol l^{-1}	0.0217	46.070	mg l^{-1}
Diethylbarbituric acid (184.20)	μmol l^{-1}	54.289	0.01842	mg per 100 ml
Lithium (6.939)	mmol l^{-1}	1.0000	1.0000	mequiv. l^{-1}
Phenacetin (179.22)	mmol l^{-1}	0.0558	17.922	mg per 100 ml
Salicylic acid (138.12)	mmol l^{-1}	0.0724	13.812	mg per 100 ml
Components of urine				
5-Aminolaevulinic acid (δ-amino-				
laevulinic acid (131.13)	μmol	7.6260	0.1311	mg
α-Amino-nitrogen (14.0067)	mmol	0.0714	14.007	mg
Ammonia (NH$_3$: 17.03)	mmol	0.0587	17.030	mg
Bence-Jones proteins	g l^{-1}	0.0100	100.00	mg ml^{-1}
(microparaprotein)				
Bicarbonate (61.02)	mmol	1.0000	1.0000	mequiv
Calcium (40.08)	mmol	0.5000	2.0000	mequiv
		0.0250	40.080	mg
Chloride (35.453)	mmol	1.0000	1.0000	mequiv
Relative density $\left(\dfrac{\text{urine, 20 °C}}{\text{H}_2\text{O, 20 °C}}\right)$		28.206	0.0355	g
	1	0.0010	1000.0	g l^{-1}
Creatine (131.14)	μmol	7.6254	0.1311	mg
Creatinine (113.12)	mmol	0.0088	113.12	mg
Glucose (180.16)	mmol	5.5506	0.1802	g
Gonadotrophin FSH, LH and HCG				IU
Uric acid (168.11)	mmol	0.0059	168.11	mg

Metabolite (relative molecular mass)	SI unit	Conversion factor		Old unit
		From old to SI units	From SI to old units	
Urea (60.06)	mmol	16.650	0.0601	g
Urea-nitrogen (N: 14.0067)	mmol	0.0357	28.020	mg
		35.698	0.02802	g
17-Hydroxycorticosteroids (cortisone: 362.47) total	μmol	2.7586	0.3625	mg
Hydroxyproline (131.13)	μmol	7.6260	0.1311	mg
Potassium (39.102)	mmol	1.0000	1.0000	mequiv
		25.5744	0.0391	g
Ketone body (acetone: 58.08)	μmol	17.212	0.0581	mg
Coproporphyrin (654.73)	nmol	1.5273	0.6547	μg
Magnesium (24.312)	mmol	0.5000	2.0000	mequiv.
		0.0411	24.312	mg
Myoglobin (17 100)	μmol	0.0585	17.100	mg
Sodium (22.99)	mmol	1.0000	1.0000	mequiv.
		43.497	0.0230	g
pH	$-$log molc.		1.0000	$-$log molc.
Phosphate (P: 30.97)	mmol	0.0323	30.970	mgP
Porphobilinogen (226.23)	μmol	4.4203	0.2262	mg
Urobilinogen (590.73)	μmol	1.6929	0.5907	mg
Uroporphyrins (830.77)	nmol	1.2037	0.8308	μg
Hormones				
Adrenalin (183.21)	nmol	5.4582	0.1832	μg
Aldosterone (360.45)	nmol	2.7743	0.3605	μg
5-Hydroxyindoleacetic acid (191.19)	μmol	5.2304	0.1912	mg
17-Ketosteroids, total (DHEP: 288.4)	μmol	3.4674	0.2884	mg
17-Ketogenic steroids, total (DHEP: 288.4)	μmol	3.4674	0.2884	mg
Noradrenalin (169.18)	nmol	5.9109	0.1692	μg
Oestradiol (272.39)	nmol	3.6712	0.2724	μg
Oestriol (288.39)	nmol	3.4675	0.2884	μg
Oestrone (270.37)	nmol	3.6986	0.2704	μg
Pregnanediol (320.5)	nmol	3.1201	0.3205	mg
Pregnanetriol (336.5)	μmol	2.9718	0.3365	mg
Testosterone (288.43)	nmol	3.4670	0.2884	μg
Vanillinmandelic acid (198.77) (VMA)	μmol	5.0309	0.1988	mg
Function tests				
Bromsulphalein test (BSP-Na$_2$)	l	0.0100	100.00	%
Congo Red test (696.68)	l	0.0100	100.00	%
Kidney clearance, urea clearance, diuresis >33 μl/s	ml s^{-1}	0.0167	60.000	ml min^{-1}
Kidney clearance (creatinine clearance) referred to a body surface area of 1.73 m^2	ml s^{-1}	0.0167	60.000	ml min^{-1}

Metabolite (relative molecular mass)	SI unit	Conversion factor		Old unit
		From old to SI units	From SI to old units	
Kidney plasma flow	ml s^{-1}	0.0116	86.400	1/a
(PAH clearance, ^{125}I-hippurate clearance referred to a body surface area of 1.73 m^2		0.0167	60.000	ml min^{-1}
Phenolsulphalein test (354.38) PSP, dosage 6 mg per 15 min	l	0.0100	100.00	%
Xylose reabsorption test (150.13), 25 g per 5 h	mmol	6.6667	0.1500	g
		1.6639	0.6010	%

References

1. Stamm, D., *Mitt. Dt. Ges. klin. Chem.*, Heft 1 (1975).
2. Lippert, H., *SI-Einheiten in der Medizin*, Urban & Schwarzenberg, Munich, 1976.

APPENDIX 8. CONVERSION SCALES FOR SI UNITS

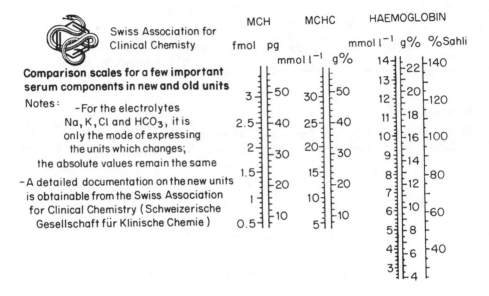

Swiss Association for Clinical Chemisty

Comparison scales for a few important serum components in new and old units

Notes:
- For the electrolytes Na, K, Cl and HCO$_3$, it is only the mode of expressing the units which changes; the absolute values remain the same
- A detailed documentation on the new units is obtainable from the Swiss Association for Clinical Chemistry (Schweizerische Gesellschaft für Klinische Chemie)

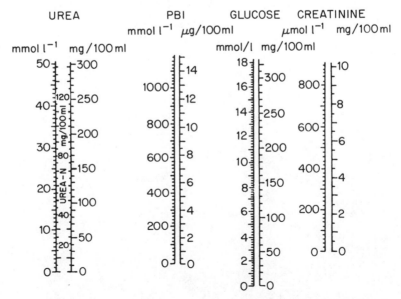

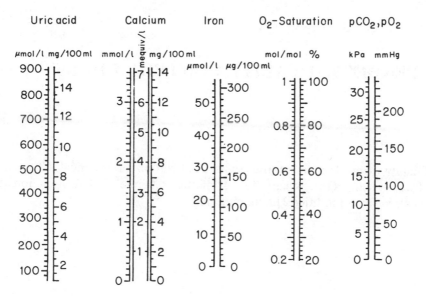

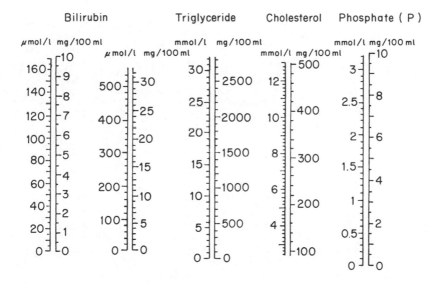

APPENDIX 9. QUALITY CONTROL CHART

Quality control chart of the Swiss Association for Clinical Chemistry (Schweizerische Gesellschaft für klinische Chemie). Determination of cholesterol by a modified Huang method.

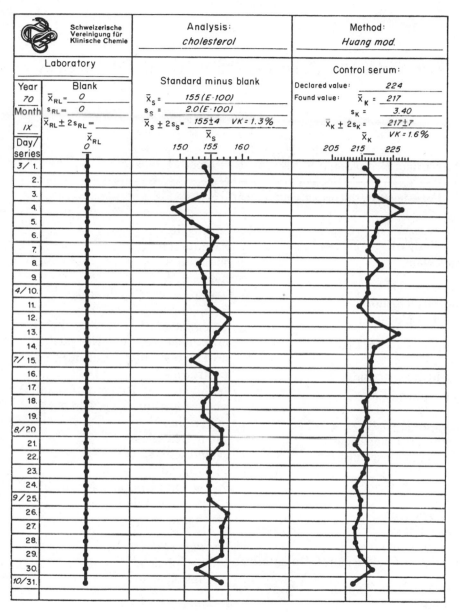

| | Schweizerische Vereinigung für Klinische Chemie | Analysis: *cholesterol* | Method: *Huang mod.* |

APPENDIX 10. INTERFERENCE FROM DRUGS

U = urine; P = plasma; S = serum; B = blood; F = faeces.

1. False-positive (incorrectly abnormal) results

Acetoacetic ester (U)
phenothiazine[3]
salicylate[3]

Acetone (U)
bromsulphalein[4]
phenolsulphalein[4]

Amino acids (P,S,U)
ACTH[3]
11-hydroxysteroids[3]
sulphonamides[3]

Ammonia (B,P,S)
ammonium heparinate[5]
oral cation exchangers[6]

α-Amylase (P,S,U)
codeine[7]
methylcholine[6]
morphine[8]
pancreozymin[6]
pethidine (e.g. Demerol, Dolantin)[9]

Blood, occult (F)
bromide[4]
ferriammonium citrate[9]
iodide[4]
copper[4]
permanganate[4]

Bromsulphalein (P,S)
amidon[12]
anabolic steroids[12]

androgens[12]
barbiturates[12]
choleretics[12]
halogens[12]
isocarboxacid (e.g. Marplan)[12]
morphine[6,14]
mycostatics[12]
norethandrolon (e.g. Nilevar)[14]
norethynodrel (e.g. Enavid)[10]
oestrogens[13]
pethidine (e.g. Demerol, Dolantin)[12]
phenazopyridine (e.g. Pyridazil, Pyridium)[2]
phenolphthalein[2]
phenolsulphalein[11]
probenecid[12]
X-ray contrast agents for the bile duct (e.g. Telepaque)[14]
serenium[2]

Creatine (P,S,U)
bromsulphalein[15–17]
phenolsulpalein[15–17]

Creatinine (P,S,U)
ascorbic acid[18]
bromsulphalein[15–17]
fructose infusions[19]
phenolsulphalein[15–17]

Chloride (P,S,U)
bromide[11,20]
chloride-containing infusions[5]

738

Cholesterol (P,S)
ACTH[3]
bromide[9,20,21]
cortisone[3]
vitamin A[22]

Catecholamines (U)
quinidine[23]
quinine[23]
erythromycin[24]
meprobromate (e.g. Miltown)[2]
methyldopa[24]
tetracyclin[2]

Diagnex Blue (U)
quinacrin · HCl (e.g. Antebrin)[5,20]
quinidine[5,20]
quinine[5,20]
methylene blue[5,20]
drugs containing Na, K, Al, Mg, Ca,
    Ba, and Fe[5,20]
drugs containing nicotine[5,20]

Iron (P,S)
iron–dextran complexes (e.g.
    Imferon)[2]

Iron binding capacity (P,S)
iron–dextran complexes (e.g.
    Imferon)[2]

Glucose (B,P,S)
ACTH[6]
glucose infusions[5]
physostygmin[25]

Glucose tolerance (B,P,S)
antihypertensives[26–29]
barbiturates[26–29]

*Glutamate–oxaloacetate
    transaminase (P,S)*
morphine[31]
salicylate[6,32]

Hamolsky test (P,S)[40–45]
ACTH
androgens
chlormadinone acetate (e.g.
    Gestafortin)
corticoids
dicoumarol
dilantin
heparin
contraceptive drugs
norethynodrel (e.g. Enovid)
orthonovum
phenylbutazone
provest
salicylate
triiodothyronine

Uric acid (P,S)
chlorothiazide[30,34]
cytostatics[6]
N-lost[6]
pyrazinamide[6]

Urea (P,S,U)
methyldopa[33]
triamterin[2]

5-Hydroxyindoleacetic acid (U)
acetanilide[2]
bananas[36]
chlorpromazine[2]
mephenesin[37]
methocarbamol[1]

17-Hydroxycorticoids
chlordiazepoxide (e.g. Librium)[49]
cortisone[1]
etryptamine[49]
hydroxyzine[49]
reserpine[49]

17-Ketogenic steroids
meprobamate (e.g. Miltown)[39]

Porphyrins (U)
ethanol[6]

antipyretics[6]
barbiturates[6]
chlorpromazine[25]
phenylhydrazine[6]
sedatives[3]
sulphonamides[3]

Porphbilinogen (U)
procaine[2]
sulphonamides[2]

PBI (protein-bound iodine)[40-45]
inorganic iodide (cough remedies,
  desquamatants, gargles,
  mouth-washes, anti-sunburn
  creams, etc.), 1–4 weeks after
  administration
barium sulphate
bromsulphalein
chlormadinone acetate (e.g.
  Gestafortin)
cholegrafin (gall bladder), 3–4
  weeks after administration
diodrast, 2 weeks after
  administration
iophendylate (e.g. Pantopaque,
  myelography), 3–12 months after
  administration
iopanoic acid (e.g. Telepaque, bile
  duct), 6–12 weeks after
  administration
iophenoxic acid (e.g. Teridax, bile
  duct), up to 30 years after
  administration
iodine-containing intestinal
  disinfection agents (e.g.
  Enterovioform)
iodine-containing vaginal
  suppositories, up to 4 weeks after
  administration
iodated oil (e.g. Lipiodol,
  bronchography), 1–2 years after
  administration
contraceptive drugs
cod-liver oil

sodium amidotrizoate (e.g.
  Urografin), 4–7 days after
  administration
norethindrone (e.g. Ortho-Novum)
norethynodrel (e.g. Enovid)
oestrogens, 2–4 weeks after
  administration
propyliodone (e.g. Dionosil,
  bronchography), 1–5 months
  after administration
provest
thyroid extract, 4–6 weeks after
  administration
thyroxine, 2–4 weeks after
  administration
triiodothyronine

Phenolsulphalein (U)
bromsulphalein[5]
phenazopyridine (e.g. Oyridazil,
  Pyridium)[2]
phenolphthalein[2]
serenium[2]

Protein (P,S)
bromsulphalein[1]

Protein-precipitation reaction (U)
carinamide[4]
iodine-containing contrast media[4,47]
penicillin (high doses)[3]
salicylate[3]
sulphisoxazole (e.g. Gantrisin)[2]
tolbutamide[48]

Prothrombin
barbiturates[2]

Phosphorus, inorganic (P,S)
pituitrin[48]
vitamin D[6]

Reducing substances (U), ('sugar')
ascorbic acid[3]
aureomycin[20]
carinamide[4]

chloral hydrate[4]
chloromycetin[20]
isoniacid[4]
p-aminosalicylic acid[4]
penicillin[4]
salicylate[4,20]
streptomycin[4,35]
sulphonamides[20]
tetracyclin[4]

Specific gravity (U)
X-ray contrast agents[46]

Urobilinogen (U)
antipyridine[4]
bromsulphalein[9]
chlorpromazine[25]
p-aminosalicylic acid[4]
phenazopyridine (e.g. Pyridium)[4]

2. False-negative (incorrectly normal) results

Alkaline phosphate (P,S)
corticoids[5]
parenteral albumin administration[2]

Amino acids (P,S)
adrenalin[3]
insulin[3,32]

Cholesterol (P,S)
ACTH[3,6]
androsterone[6]
cortisone[6]

Iron (P,S)
ACTH[6]
steroids[6]

Iron-binding capacity (P,S)
ACTH[6]
steroids[6]

Glucose tolerance (B,P,S)
contraceptive drugs[26–29]
meprobamate (e.g. Miltown)[26–29]
tranquilizers[26–29]

Glucose (U), enzymatic
ascorbic acid[2]

Uric acid (P,S)
dicoumarol[6]
piperazine[48]

17-Hydroxycorticoids (U)
chlorpromazine[49]
dexamethason[1]

5-Hydroxyindolacetic acid (U)
chlordiazepoxide[49] (c.g. Librium)
chlorpromazine[1]
phenothiazine[38]

[131]Iodine uptake (thyroid glands)[40–45]
ACTH, up to 1 week following administration
inorganic iodide, 3–4 weeks after administration
antihistamines, up to 1 week following administration
chlorpromazine
cholegrafin (gall bladder), up to 3 months following administration
corticosteroids, up to 1 week following administration
diodrast, up to 3 months after administration
iophendylate (e.g. Pantopaque, myelography), up to 1 year following administration
iopanoic acid (e.g. Telepaque, bile duct), up to 2 months following administration
iophenoxyc acid (e.g. Teridax, bile duct), up to 30 years following administration

742

isoniazid, up to 2 weeks following
administration
iodine-containing intestinal
disinfectants (e.g. Enterovioform)
iodine-containing vaginal
suppositories, up to 4 weeks after
administration
iodated oil (e.g. Lipiodol,
bronchography), up to 3 years
following administration
iodothiouracil, up to 1 week after
administration
sodium amidotrizoate (e.g.
Urografin), up to 2 weeks after
administration
p-aminosalicylic acid, up to 2 weeks
after administration
pentothal, up to 1 week after
administration
phenylbutazone
propyliodone (e.g. Dionosil,
bronchography), up to 5 months
after administration
propylthiouracil, up to 1 week after
administration
renografin, up to 2 weeks after
administration
salicylates
sulphonamides, up to 1 week after
administration
thiamazole (e.g. Tabazol)
thiocyanate, up to 1 week after
administration
thyroid extract, up to 2 weeks after
administration
thyroxine, up to 2 weeks after
administration
tolbutamide, up to 1 week after
administration
triiodothyronine

Hamolsky test (P,S)[40–45]
iodothiouracil
oestrogens
propylthiouracil

thiamazole (e.g. Tapazol)
thiocyanate
thyroid extract
thyroxine

PBI (P,S)[40–45]
androgens, 3 weeks after
administration
chlorate
corticoids, 2 weeks after
administration
dilantin, 10 days after
administration
gold salts, up to 4 weeks after
administration
Hg-diuretics, 3 days after
administration
p-aminobenzoic acid
p-aminosalicylic acid
phenylbutazone, up to 2 weeks after
administration
propylthiouracil, 5–7 days after
administration
salicylate
sulphonamide (?)
tapazole, 1 week after
administration
thiocyanate, 3 weeks after
administration
thiazide diuretics
triiodothyronine, 2–4 weeks after
administration

Phenolsulphalein (U)
carinamide[4]
penicillin[2]
salicylate[2]
sulphonamides[2]

Prothrombin (P,S)
antibiotics[48]
hydroxyzin (e.g. Vistaril)[2]
salicylate[6]
sulphonamides[48]

Phosphorus, inorganic (P,S)
adrenaline[49]

anaesthetics[48]
insulin[48]

Vanillin mandelic acid (U)
nialamide[49]

References

1. Caraway, W. T., *Am. J. clin. Path.*, **37**, 445 (1962).
2. Wirth, W. A. and Thompson, R. L., *Am. J. clin. Path.*, **35**, 579 (1956).
3. Miller, S. W., *A textbook of clinical pathology*, 6th ed., Williams & Wilkins, Baltimore, 1960.
4. Bray, W. E., *Clinical laboratory methods*, 5th ed., Mosby, St. Louis, 1957.
5. Richterich, R., *Klinische Chemie. Theorie und Praxis*, 2. Aufl., Karger, Basle, 1968.
6. Eastham, R. D., *Biochemical values in clinical medicine*, 2nd ed., Williams & Wilkins, Baltimore, 1963.
7. Gross, J. B., Comfort, M. W., Mathieson, D. R., and Power, M. H., *Proc. Staff Meet. Mayo Clin.*, **26**, 81 (1951).
8. Bogoch, A., Roth, J. L. A., and Bockus, H. L., *Gastroenterology*, **26**, 697 (1954).
9. Davidsohn, I. and Wells, B. B., *Clinical diagnosis by laboratory methods*, 13th ed., Saunders, Philadelphia, 1962.
10. Searle, *Circular on Enovid.*
11. Annino, J. S., *Clinical chemistry: principles and procedures*, 2nd ed., Little, Brown, Boston, 1960.
12. Hynson, Westcott & Dunning, Inc., *Circular on bromsulfalein solution ampoules.*
13. Mueller, M. N. and Kappas, A., *Med. Sci.*, February 1965, p. 6.
14. Schoenfeld, L. J. and Foulk, W. T., *J. clin. Invest.*, **43**, 1419 (1964).
15. Taussky, H. H., *Clin. chim. Acta*, **1**, 210 (1956).
16. Taussky, H. H., *Stand. Meth. clin. Chem.*, **3**, 99 (1961).
17. Tillson, E. K. and Schuckhardt, G. S., *J. Lab. clin. Med.*, **41**, 312 (1953).
18. Taussky, H. H., *J. biol. Chem.*, **208**, 853 (1954).
19. Paget, M., Gontier, M., and Liefooghe, J., *Annis Biol. clin.* **13**, 535 (1955).
20. Henry, R. J., *Clinical chemistry*, Hoeber, New York, 1964.
21. Rice, E. W. and Lukasiewicz, D. B., *Clin. Chem.*, **3**, 160 (1957).
22. Kinley, L. J. and Krause, R. F., *Proc. Soc. exp. Biol. Med.*, **99**, 244 (1958).
23. Sax, S. M., Waxman, H. E., Aarons, J. A., and Lynch, H. J., *Clin. Chem.*, **6**, 168 (1960).
24. Gifford, R. W. and Tweed, D. C., *J. Am. med. Ass.*, **182**, 493 (1962).
25. Levinson, S. A. and MacFate, R. P., *Clinical laboratory diagnosis*, 6th ed., Lea & Febiger, Philadelphia, 1961.
26. Hasselblatt, A. and Haun, G., *Klin. Wschr.*, **38**, 1108 (1960).
27. Hüdepohl, M. and Lederbogen, K., *Klin. Wschr.*, **41**, 245 (1963).
28. Opitz, K., *Klin. Wschr.*, **40**, 56 (1962).
29. Opitz, K. and Loeser, A., *Dt. med. Wschr.*, **87**, 105 (1962).
30. Gutmann, A. B. and Yü, T. F., *Am. J. Med.*, **13**, 744 (1952).
31. Foulk, W. T. and Fleisher, G. A., *Proc. Staff Meet. Mayo Clin.*, **32**, 405 (1957).
32. Boutwell, J. H., *Clinical chemistry*, Lea & Febiger, Philadelphia, 1961.
33. Merck, Sharp & Dohme, *Circular on Aldoril.*
34. Freeman, R. B. and Duncan, G. G., *Metabolism*, **9**, 1107 (1960).
35. Neuberg, W. II., *Am. J. clin. Path.*, **24**, 245 (1954).
36. Crout, J. R. and Sjoerdsma, A., *New Engl. J. Med.*, **262**, 1103 (1960).

744

37. Honet, J. C., Casey, T. V., and Runyan, J. W., *New Engl. J. Med.*, **261**, 188 (1959).
38. Ross, G., Neinstein, B., and Kabakow, B., *Clin. Chem.*, **4**, 66 (1958).
39. Salvesen, S. and Nissen-Meyer, R., *J. clin. Endocr. Metab.*, **17**, 914 (1957).
40. Davis, P. J., *Am. J. Med.*, **40**, 918 (1966).
41. Shultz, A. L., *Postgrad. Med.*, **31**, A-34 (1962).
42. Magliotti, M. F., Hummon, I. F., and Hierschbiel, E., *Am. J. Roentg*, **81**, 47 (1959).
43. Grayson, R. R., *Am. J. Med.*, **28**, 397 (1967).
44. Hamolsky, M. W., Golodetz, A., and Freesberg, A. S., *J. clin. Endocr. Metab.*, **19**, 103 (1959).
45. Ravel, R., *Bull. path.*, 172 (June 1967).
46. Hurt, R., *Am. J. med. Technol.*, **26**, 122 (1960).
47. Free, A. H. and Fancher, E. E., *Am. J. med. Technol.*, **24**, 64 (1958).
48. Frankel, S. and Reitman, S., in Gradwohl, *Clinical laboratory methods and diagnosis*, 6th ed., Mosby, St. Louis, 1963.
49. Borushek, S. and Gold, J. J., *Clin. Chem.*, **10**, 41 (1964).

APPENDIX 11. EQUATIONS FOR PHOTOMETRY

Abbreviations

E	Extinction	SV	sample volume (ml)
ΔE	change in extinction	FV	final volume (ml)
MW	molecular weight	C_m	molar concentration (mol l^{-1})
(S)	sample	$C\%$	percentual concentration
(B)	blank		(g per 100 ml)
(SB)	sample blank	d	path-length (cm)
(RB)	reagent blank	ε	molar extinction coefficient
(ST)	standard		($l\ mol^{-1}\ cm^{-1}$)
t	time (min)	a	percentual extinction
λ	wavelength (nm = nanometres)		coefficient (g cm^{-1})

Concentration

$$C_m = \frac{C\% \cdot 10}{MW} \qquad C\% = \frac{C_m \cdot MW}{10}$$

Molar extinction coefficient (ε)

$$\varepsilon = \frac{E}{C_m d} = \frac{E}{d} \cdot \frac{MW}{C\% \cdot 10}$$

Percentual extinction coefficient (a)

$$a = \frac{E}{C\% d} = \frac{E}{d} \cdot \frac{10}{C_m \cdot MW}$$

Conversion equations

$$\varepsilon = \frac{aMW}{10} \qquad a = \frac{\varepsilon \cdot 10}{MW}$$

Calculation of concentrations

By means of a standard (ST)

$$C_m(S) = \frac{E(S) - E(SB) - E(RB)}{E(ST) - E(RB)} \cdot C_m(ST)$$

$$C\%(S) = \frac{E(S) - E(SB) - E(RB)}{E(ST) - E(RB)} \cdot C\%(ST)$$

By means of the molar extinction (ε)

$$C_m(S) = \frac{E(S) - E(SB) - E(RB)}{\varepsilon d} \cdot \frac{FV}{SV} \qquad C\%(S) = C_m(S) \cdot \frac{MW}{10}$$

By means of the percentual extinction coefficient (a)

$$C\%(S) = \frac{E(S) - E(SB) - E(RB)}{ad} \cdot \frac{FV}{SV} \qquad C_m(RS) = C\%(S) \cdot \frac{10}{MW}$$

Calculation of enzyme activities

U = International Units

By means of a standard (ST)

$$U\, l^{-1} = \frac{E(S) - E(SB) - E(RB)}{E(ST) - E(RB)} \cdot C_m(ST) \cdot 10^6 \cdot \frac{1}{t}\ \mu\text{mol min}^{-1}\, l^{-1}$$

By means of the molar extinction coefficient (ε) in two-point determinations

$$U\, l^{-1} = \frac{E(S) - E(SB) - E(RB)}{\varepsilon d} \cdot \frac{FV}{SV} \cdot 10^6 \cdot \frac{1}{t}\ \text{mol min}^{-1}\, l^{-1}$$

By means of the molar extinction coefficient (ε) in kinetic determinations (optical test)

$$U\, l^{-1} = \frac{E(S)/\text{min} - E(B)/\text{min}}{\varepsilon d} \cdot \frac{FV}{SV} \cdot 10^6\ \mu\text{mol min}^{-1}\, l^{-1}$$

INDEX

748

750

754

756

758

760

ABBREVIATIONS

Units of measurement

Length		Weight		Volume		Moles	
		kg	kilogram				
m	metre	g	gram	l	litre	mol	mole
mm	millimetre	mg	milligram	ml	millilitre	mmol	millimole
μm	micrometre	μg	microgram	μl	microlitre	μmol	micromole
nm	nanometre	ng	nanogram	nl	nanolitre	nmol	nanomole
pm	picometre	pg	picogram	pl	picolitre	pmol	picomole

Chemical Abbreviations

ALT (GPT)	Alanine aminotransferase (glutamate–pyruvate transaminase)
AST (GOT)	Aspartate aminotransferase (glutamate–oxaloacetate transaminase)
CoA	Coenzyme A
CK	Creatine kinase
DNA	Desoxyribonucleic acid
GK	Glycerokinase
GLDH	Glutamate dehydrogenase
GGTP	γ-Glutamyl transpeptidase
GOD	Glucose oxidase
GOT (AST)	Glutamate–oxaloacetate transaminase (aspartate aminotransferase)
G-1-P	D-Glucose-1-phosphate
G-6-P	D-Glucose-6-phosphate
G-6-PDH	Glucose-6-phosphate dehydrogenase
GPT (ALT)	Glutamate–pyruvate transaminase (alanine aminotransferase)
GSH	Glutathione, reduced
3-HBDH	3-Hydroxybutyrate dehydrogenase
HK	Hexokinase
LDH	Lactate dehydrogenase, specific for L-(+)-lactate
MDH	Malate dehydrogenase

NAC N-Acetylcysteine
NAD   Nicotinamide adenine dinucleotide
$NADH_2$ Nicotinamide adenine dinucleotide, reduced
NADP   Nicotinamide adenine dinucleotide phosphate
$NADPH_2$ Nicotinamide adenine dinucleotide phosphate, reduced
2-OxoG   2-Oxoglutarate
PEP   Phosphoenol pyruvate
3-PG   3-Phosphoglycerate, D-Glycerate-3-phosphate
6-PG   6-Phosphogluconate, D-Gluconate-6-phosphate
P_i Inorganic phosphate
RNA   Ribonucleic acid

General Abbreviations

BS   Body surface area
BW   Body weight
DM-water   Demineralized water ($\approx$distilled water)
I Ionic strength
U   International enzyme unit, μmol min^{-1}
MW   Molecular weight, relative molecular mass
N   Normality
°C   Degrees Centrigrade

Statistical Abbreviations

$x\ (x_1, x_2, \ldots, x_n)$ Individual values (observations)
$\bar{x}$ Arithmetic mean
N Number of individual values
s Standard deviation
CV Coefficient of variation
$\Sigma \ldots$ Sum of ...
r Correlation coefficient
n Number of degrees of freedom in the t-test
p Probability of obtaining a correct result by chance
$p > a$ p is greater than a
$p < a$ p is less than a

ATOMIC WEIGHTS AND VALENCIES OF THE ELEMENTS

International Atomic Weights, 1961,[1,2] based on an exact value of 12 for the relative atomic mass of the pure ^{12}C isotope.

( )   Mass number of the most stable isotope
( )* Mass number of the best known isotope
a   The atomic weight varies owing to natural variation in the isotopic composition: B ± 0.003; C ± 0.00005; H ± 0.00001; O ± 0.0001; Si ± 0.001; S ± 0.003
b   The atomic weight values are assumed to be subject to the following experimental error: Br ± 0.0009; Cl ± 0.001; Cr ± 0.001; Fe ± 0.003; Ag ± 0.001

For all other atomic weights quoted, the last decimal figure is considered to be accurate to ±0.5.

References

1. International Union of Pure and Applied Chemistry, *Comptes Rendus de la 21^e Conference, Montreal 1961*, Butterworths, London, p. 281.
2. *Nat. Bur. Stand. Tech. New Bull.*, **49**, 74 (1965).

Element	Symbol	Atomic weight	Valency
Actinium	Ac	(227)	III
Aluminium	Al	26.9815	III
Americium	Am	(243)	III (IV) (V) (VI)
Antimony	Sb	121.75	III V
Argon	Ar	39.948	0
Arsenic	As	74.9216	III V
Astatine	At	(210)	(I III V VII)
Barium	Ba	137.34	II
Berkelium	Bk	(247)	III (IV)
Beryllium	Be	9.0122	II
Bismuth	Bi	208.980	III (V)
Boron	B	10.811^a	III
Bromine	Br	79.904^b	I (III) V (VII)
Cadmium	Cd	112.40	II

Element	Symbol	Atomic weight	Valency
Caesium	Cs	132.905	I
Calcium	Ca	40.08	II
Californium	Cf	(249)*	III
Carbon	C	12.01115[a]	(II) IV
Cerium	Ce	140.12	III IV
Chlorine	Cl	35.453[b]	I III V VII
Chromium	Cr	51.996[b]	II III VI
Cobalt	Co	58.9332	II III
Copper	Cu	63.546	I II
Curium	Cm	(247)	III
Dysprosium	Dy	162.50	III
Einsteinium	Es	(254)	III
Erbium	Er	167.26	III
Europium	Eu	151.96	(II) III
Fermium	Fm	(253)	III
Fluorine	F	18.9984	I
Francium	Fr	(223)	I
Gadolinium	Gd	157.25	III
Gallium	Ga	69.72	(II) III
Germanium	Ge	72.59	II IV
Gold	Au	196.967	I III
Hafnium	Hf	178.49	IV
Helium	He	4.0026	0
Holmium	Ho	164.930	III
Hydrogen	H	1.00797[a]	I
Indium	In	114.82	(I) (II) III
Iodine	I	126.9044	I (III) V VII
Iridium	Ir	192.2	III IV
Iron	Fe	55.847[b]	II III (IV) (VI)
Krypton	Kr	83.80	0
Lanthanum	La	138.91	III
Lead	Pb	207.19	II IV
Lithium	Li	6.939	I
Lutetium	Lu	174.97	III
Magnesium	Mg	24.312	II
Manganese	Mn	54.9380	(I) II (III) IV (VI) VII
Mendelevium	Md	(256)	III
Mercury	Hg	200.59	I II
Molybdenum	Mo	95.94	(II) III (IV) VI
Neodymium	Nd	144.24	III
Neon	Ne	20.183	0
Neptunium	Np	(237)	III (IV) (V) (VI)
Nickel	Ni	58.71	(0) (I) II (III)

Element	Symbol	Atomic weight	Valency
Niobium	Nb	92.906	(II) III (IV) V
Nitrogen	N	14.0067	III V
Nobelium	No	(253)	
Osmium	Os	190.2	(II) III IV VIII
Oxygen	O	15.9994[a]	II
Palladium	Pd	106.4	II (III) IV
Phosphorus	P	30.9738	III V
Platinum	Pt	195.09	(I) II (III) IV
Plutonium	Pu	(242)	III (IV) (V) (VI)
Polonium	Po	(210)*	(II) IV (VI)
Potassium	K	39.102	I
Praseodymium	Pr	140.907	III (IV)
Promethium	Pm	(147)*	III
Protactinium	Pa	(231)	(IV) V
Radium	Ra	(226)	II
Radon	Rn	(222)	0
Rhenium	Re	186.2	(I) II (III) IV (V) VI VII
Rhodium	Rh	102.905	(II) (III) IV (V)
Rubidium	Rb	85.47	I (II) (III) (IV)
Ruthenium	Ru	101.07	(0) (I) (II) III IV (V) (VI) (VII) VIII
Samarium	Sm	150.35	(II) III
Scandium	Sc	44.956	III
Selenium	Se	78.96	II IV VI
Silicon	Si	28.086[a]	IV
Silver	Ag	107.868[b]	I (II)
Sodium	Na	22.9898	I
Strontium	Sr	87.62	II
Sulphur	S	32.064[a]	II IV VI
Tantalum	Ta	180.948	(II) (III) (IV) V
Technetium	Tc	(99)*	(III) (IV) (VI) VII
Tellurium	Te	127.60	II IV VI
Terbium	Tb	158.924	III IV
Thallium	Tl	204.27	I III
Thorium	Th	232.038	IV
Thulium	Tm	168.934	(II) III
Tin	Sn	118.69	II IV
Titanium	Ti	47.90	II III IV
Tungsten	W	183.85	(II) (III) (IV) (V) VI
Uranium	U	238.03	III IV (V) VI
Vanadium	V	50.942	II III IV V
Xenon	Xe	131.30	0
Ytterbium	Yb	173.04	(II) III

Element	Symbol	Atomic weight	Valency
Yttrium	Y	88.905	III
Zinc	Zn	63.37	II
Zirconium	Zr	91.22	(II) IV